Removal of Emerging Contaminants Through Microbial Processes

Maulin P Shah
Editor

Removal of Emerging Contaminants Through Microbial Processes

Springer

Editor
Maulin P Shah
Division of Applied &
Environmental Microbiology
Environmental Microbiology Lab
Bharuch, India

ISBN 978-981-15-5900-6 ISBN 978-981-15-5901-3 (eBook)
https://doi.org/10.1007/978-981-15-5901-3

This Springer imprint is published by the registered company Springer Nature Singapore Pte Ltd.
The registered company address is: 152 Beach Road, #21-01/04 Gateway East, Singapore 189721, Singapore

Contents

About the Editor

Maulin P Shah is a senior scientist in the Environmental Microbiology Lab at Bharuch, Gujarat, India. He is an active editorial board member in more than 150 highly reputed journals in the field of environmental and biological sciences. He is on the Advisory Board of *CLEAN—Soil, Air, Water* (Wiley); editor of *Current Pollution Reports* (Springer Nature), *Environmental Technology & Innovation* (Elsevier), *Current Microbiology* (Springer Nature), *Journal of Biotechnology & Biotechnological Equipment* (Taylor & Francis), *Ecotoxicology (Microbial Ecotoxicology)* (Springer Nature), and *Current Microbiology* (Springer Nature); and associate editor of *GeoMicrobiology* (Taylor & Francis) and *Applied Water Science* (Springer Nature). He has edited 50 books in the areas of wastewater microbiology and industrial wastewater treatment with Elsevier, Springer, CRC Press, and Nova.

Aspects of Microbial Biofilms in Water Treatment

Riddha Dey and Richa Raghuwanshi

Abstract

Global water crisis created due to increasing population and its demands needs to ascertain proper water management practices. Biofilm, which is a conis the fundamental unit of atgregation of microbial cells that are irreversibly attached to the surface and confined within the matrix of polysaccharide material, plays a major role in water treatment and reuse technology. A biofilm structure has a specific architecture that consists of microbial cells and extracellular polymeric substance provided with an ideal environment for the exchange of genetic material between cells. The growth and attachment process depend on the growth medium, substratum, and the cell surface. Biofilms are the main component of membrane bioreactors. The microorganisms present in the biofilms participate actively in contaminant bioremediation and degrade the organic contaminants of the polluted water. The planktonic-biofilm transition is a highly regulated and complex process that depends on the phenotypic characteristics of the bacteria and environmental factors. The microorganisms that help in biofilm formation have specific regulatory genes and communicate through quorum sensing which in turn can initiate certain biofilm processes such as detachment. Recently, studies have identified the genes and regulatory circuits which are involved in initial cell surface interactions, biofilm maturation, detachment, and the conversion of biofilm microbial cells into planktonic mode of growth. This chapter explores the ways of formation and architecture of biofilms and deals with the ecology of the surface microbes, their growth control mechanisms, and their role in water reclamation system.

R. Dey
Department of Botany, Institute of Science, Banaras Hindu University, Varanasi, India

R. Raghuwanshi (✉)
Department of Botany, Mahila Mahavidyalaya, Banaras Hindu University, Varanasi, India

M. P. Shah (ed.), *Removal of Emerging Contaminants Through Microbial Processes*,
https://doi.org/10.1007/978-981-15-5901-3_1

Keywords

Biofilm architecture and ecology · Biofilters · Quorum sensing

1 Introduction

Modern civilizations have witnessed the process of rapid industrialization and urbanization which have not only affected the individual lifestyle and society but have also modulated the environment. Unlike the positive drive it has on the economical and social frames, the impact of both industrialization and urbanization has adverse effect on nature and its precious resources especially the air and water. Since water has guided the growth of various epical civilizations, its importance in structuring a stable society can never be overlooked. However, the discharge of toxic chemicals and untreated sewages from the rapidly growing industries and cities has contaminated a large portion of usable water, affecting the water quality and human heath at the same time. Contaminated water generally consists of industrial, agricultural, and domestic effluents which may be comprised of detergents, fats, oils, pesticides, and trace metals like lead, cadmium, mercury, nickel, zinc, etc., making it unfit for drinking. The lack of proper water treatment plants has also contributed to increased toxicity of water bodies to a large extent. According to the UNESCO (2017), over 80% of the effluents released worldwide in the water bodies are without sufficient treatment. Moreover, the WHO has quoted that 40% of the deaths and 30% of all the diseases have been resulted due to contaminated water (Kantawanichkul et al. 2009), causing it as a global concerned.

Deterioration of life and health quality due to polluted water and regular assessment of progressive contamination of water bodies by various researchers suggest the need of focusing toward a necessary cure for the existing situation. Treatment of contaminated water bodies in an efficient way to remove toxins has gained attention. The process of water purification starting with the filtration was appreciated due to its cost-effectiveness and practicability. However, the use of chemical filters or the contemporary physical methods had many drawbacks. The use of biological knowledge for wastewater treatment was introduced in the year 1893 in England as trickling filters (Metcalf and Eddy Inc 1991). The concept of utilizing microbial population for the metabolic degradation of organic wastes and chemicals and removal of trace metals from the effluents has gradually evolved and is being commercialized. The pioneering works in this area guided the development of biofilters and the successful application of biofilms for filtering both polluted air and water at industrial level.

Initially, the biofilters were developed on slag or rock as filter media; however, with the latest advancement, an array of filter media are purposefully used for obtaining requisite biomass preeminent toward metabolic degradation of pollutants. Progress in the knowledge of biofilms and their effective utilization using array of support media leads to the development of plastic-based media (bundled plastic transition to expanded polystyrene units) into various fixed film filtration units in the mid-twentieth century (Antonie and Aacken 1971; Antonie 1976). Subsequently, the advanced biofiltration unit was developed with high treatment efficacy by the purposeful implication of support media. Further, hybrid (suspended and fixed film system together) and upgraded fixed film systems were also introduced for better water treatment (Canler and Perret 1994). However, in biofiltration units, the optimum operational parameter, which includes pH, temperature, moisture, aeration, etc., should be properly maintained for better results. Govind (2009) percolated that biofiltration can effectively remove as high as 5000 ppmv concentration of contaminants if provided with structural and operational optima.

Utilization of biofilms for filtration has many advantages over the physical methods which include cost-effectiveness and filtration efficacy due to metabolic breakdown of biodegradable pollutants along with environmental benefits. In nature, microorganisms preferably form consortia for proper utilization of available nutrients and combat stress. This natural process gradually results in the form of biofilm which may consist of microbes belonging to different trophic levels. In a biofilm, complex organic contaminants can be utilized sequentially by various classes of microbes ultimately resulting into increased biomass and subsequent biodegradation of contaminants. Therefore, the utilization of biofilms in the membrane reactors and filtration units for the removal of pollutants has great scope. Today, modern biofilter technology is playing a paramount role over the quintessential physical methods for water reclamation and reuse processes. Nevertheless, the increasing demand of water due to rapid rise in the global population and following depletion of freshwater sources is urging for more efficient biofilters to increase the mass reuse of available water and prevention of water pollution.

2 Wastewater Treatment

Wastewater treatment is a process through which the contaminants present in the effluent are removed and converted into reusable form with minimum impact on the environment. Wastewater treatment can be broadly divided into three categories: physical treatment, chemical treatment, and biological treatment.

Physical Water Treatment: In this process, solid wastes are removed without involving any use of chemicals. Processes like screening, sedimentation, and skimming are used in which the insoluble heavy particles settle down at the bottom and the pure water is separated. Other effective methods are aeration (providing air through the water to provide oxygen) and filtration (wastewater is passed through the filters to separate the contaminants and insoluble particles). Sand filters are mostly used in this process. The major disadvantage is that not all the contaminants can be removed by using physical treatments. The particles having smaller size cannot be removed.

Chemical Water Treatment: This process involves the use of chemicals for water treatment. Chlorine is widely used for water treatment. Ozone is an excellent disinfectant and oxidizing agent and is used in water purification. Neutralization technique is also used where acid or bases are added to neutralize the pH of the water. Chemical treatments can check the growth of microorganisms, but its prolonged use can cause water hardness and make it turbid changing its taste and odor.

Biological Water Treatment: Biological treatments make use of microorganisms to break down organic wastes using normal cellular processes. Among the available treatment processes, application of biological processes is gradually gaining pace as it involves no chemical use and is ecofriendly, cost-effective, and efficient in lower level of contamination. It is mainly of two types: suspended and fixed film growth systems. Suspended growth systems have some definite disadvantages like washout and low biomass concentration (Metcalf and Eddy 2003). Fixed film systems have many advantages like:

1. They are closer to natural biofilm systems.
2. Prevent washout of biomass.
3. Increased mean cell retention time.
4. Enhanced biomass loading per unit reactor volume.
5. Easy solid and liquid separation.
6. Surface biodegradation facilitates providing resistance to shock loadings.
7. Higher biodegradation rates.
8. Higher active biomass.
9. Enhanced rates of genetic transfer resulting in stable gene pool.
10. Extensive microbial diversity.
11. Greater efficiency to degrade recalcitrant (Bisop et al. 1995).

The concept of biofilms was followed for the development of biofilters which are the most recent and promising technique possessing higher efficiency and better performance for biological wastewater treatment (Cohen 2001).

3 Biofiltration

Biofiltration system involves removal of pollutants through biological degradation unlike normal filtration technique where physical straining is used (Chaudhary et al. 2003). This breakdown of pollutants is carried out by microorganisms fixed to a porous medium. The microorganisms including (anaerobic, aerobic, and facultative) protozoa, bacteria, algae, and fungi suspended beside the medium particles gradually develop a slime layer known as biofilm over the surface of the filter media. To ensure large surface area for attachment and increase in nutrient supply, the filter bed medium consists of relatively inert substances. The potential of a biofilter depends on the properties and characteristics of the support medium including degree of compaction, porosity, ability to host microbial populations, and water retention capabilities. The parameters that determine the performance and successful operation of biofilters include the medium pH, microbial inoculation, moisture, temperature, and nutrient content (Devinny et al. 1999). Over the last few decades, fixed film systems like trickling filters and rotating biological contractors have been used for

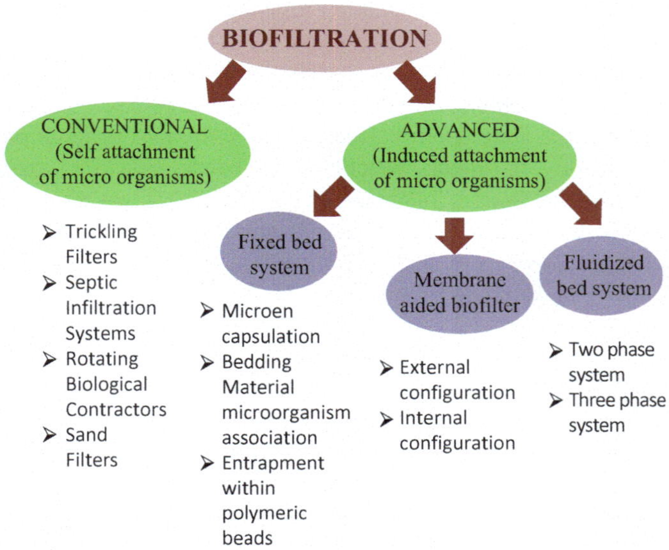

Fig. 1 Types of biofiltration processes

wastewater treatment (Antonie 1974; Pederson 1982; Hinton and Stensel 1994; Parker and Bratby 2001). The various abatement techniques related to biofiltration is illustrated in Fig. 1. As the performance of biofilter depends on the microbial activity, so availability of constant source of substrates is important to maintain and control a healthy biomass present on the surface of the filter.

4 Performance Parameters of Biofilters

The parameters that play a crucial role on the performance of a biofilter are briefly discussed below:

1. Physical factors
   (a) Filter media
   Filter media or packing is the fundamental unit of attached growth wastewater treating technology. It provides a surface on which the microbial growth takes place and forms a biofilm layer. The source and concentration of pollutants are the basis of selection of biofilter media which makes it crucial for efficient functioning of biofilters. Porosity, size, density, and resistance to erosion and chemicals are also important parameters for selection. The biofilter media should provide suitable and larger surface area for quick biomass growth and good quality surface texture to withstand shear and sloughing of biomass. The filter medium should be insoluble, durable, and resistant to chemicals (Christensson and Welander 2004). The hydrodynamic

conditions of the support material are affected by the geometry of the reactor, i.e., surface area and texture, affecting biofilm formation causing inappropriate wastewater treatment (Yu et al. 2008; Matos et al. 2011). A medium that has high void fraction, resistance to clogging, large passage diameter, high specific surface area, good mechanical strength, inert material of construction, flexibility in shape, wettability for better growth of biofilm, light attenuation, and ease of maintenance and is lightweight and cost-effective is considered as a good filter medium. Various researchers have reported the use of different synthetic and natural materials like polypropylene (Khatoon et al. 2014), polystyrene (Naz et al. 2013), and pebbles (Naz et al. 2015; Khan et al. 2015) as efficient media. For primary wastewater treatment, blast furnace slag, granite, and synthetic media can be used as filter media. The selection of filter media largely depends on the volume of wastewater. Moreover, for the tertiary wastewater treatment, granular activated carbon (GAC), filter coal, anthracite, and sand can be used. The use of GAC for the removal of organic substances from tertiary wastewater is better than using anthracite or sand as they are non-adsorptive media (Le Chevallier et al. 1992; Wang et al. 1995a, b). The biodegradable components are retained and adsorbed by GAC, leading to continuous bioregeneration additionally providing protection from shear loss of biomass. Certain spectroscopic techniques like X-ray photoelectron spectroscopy and energy-dispersive X-ray spectroscopy are utilized for the detection and quantification of the elemental composition of the filter medium.

(b) Filter backwash

Filter backwashing is a very crucial step for the maintenance of the biofilm, and the backwashing technique should be selected appropriately, or else it could damage the biomass attached to the surface (Ahmad et al. 1998; Bouwer and Crowe 1988; Bablon et al. 1988; Graese et al. 1987; Miltner et al. 1995). The hydrophobic biological particles are attached to filter media with greater force compared to that of nonbiological clay particle (Ahmad and Amirtharajah 1998). Backwashing in granular activated carbon biofilter showed no significant loss in vertical biomass profile (Servais et al. 1991).

(c) Empty bed contact time

The empty bed contact time (EBCT) is the main operating parameter of a biofilter. The concept dimensionless contact time incorporating EBCT was given by Zhang and Huck (1996). With the increase in contact time escalates the organic substance removal until it reaches its optimum value. The hydraulic loading and filter depth can be amended to enhance the EBCT. In a rapid filtration unit, organic removal is not affected by hydraulic loading for a given EBCT (Carlson and Amy 1995).

(d) Temperature

The effect of temperature on the bacterial activity is yet another parameter for biofiltration. The activities of bacterial community adapted at 10 °C and 20 °C but can increase with the rise in temperature by 10–30 °C.

2. Biological factors
   The formation of microbial biofilm is the fundamental principle of biofiltration technique. So it is required to understand the mechanism of the biological processes involved in the formation of biofilms. Microorganisms play a crucial role in removing anthropogenic contaminants from wastewater. The knowledge of composition of the microbial consortia, which form the biofilm, is preferably significant not only to construct highly efficient engineered biofilms but also to understand the basic mechanism of decomposition performed by them. Although microbial consortia are responsible for the biofilm formation, the methanogens are the key players in waste removal. The microbes recovered from anaerobic bioreactors mostly belong to the Euryarchaeota, Crenarchaeota, *Korarchaeota*, and Thaumarchaeota (Riviere et al. 2009; Leclerc et al. 2004; Collins et al. 2006; Dang et al. 2013). Along with Crenarchaeota, other potential ammonia-oxidizing organisms, including *Nitrosopumilus maritimus*, *Candidatus*, and *Nitrososphaera viennensis*, were also isolated from biofilms (Konneke et al. 2005; Tourna et al. 2011). Methanogenic archaea, dominantly consist of Methanobacteriaceae, Methanosarcinaceae, and Methanosaetaceae, get attached to the packing support materials carrying out methanogenesis efficiently (Zhang et al. 2011; Buzzini et al. 2006; Del Nery et al. 2008). Moreover, Leclerc et al. (2004) reported the consortia of *Methanobacterium* spp. and *Methanosaeta concilii* in the biofilm layer. However, in the stirred tanks and fixed film digesters, *Methanosarcina frisus* is the prevalent species, yet, in anaerobic sludge bed, *Methanosaeta* spp. is mostly found. Visser et al. (1991) revealed that even after a rise in temperature from 38 °C to 55 °C, several subpopulations including *Methanobrevibacter smithii*, *Methanospirillum hungatei*, *Methanobrevibacter arboriphilus*, *Methanobacterium thermoautotrophicum*, *Methanogenium cariaci*, and *Methanosarcina thermophila* continue to carry out their activity proficiently.

5 Microbial Biofilms

A biofilm is defined as an accumulation of microbial cells that are attached irreversibly with a surface enclosed in a matrix consisting primarily of polysaccharide. Biofilm matrix may also contain noncellular materials according to the environment in which the biofilm has developed. The water system biofilm contains clay material, corrosion products, filamentous bacteria, and freshwater diatoms making it more complex. There are multiple biofilm growth modes. They may grow in flat sheets, discontinuous patches, cluster shape, or columnar form. The mechanism of biofilm formation is illustrated in Fig. 2.

There are three main biological processes that can occur in a biofilter:

1. Attachment of microorganisms.
2. Growth of microorganisms.
3. Decay and detachment of microorganisms.

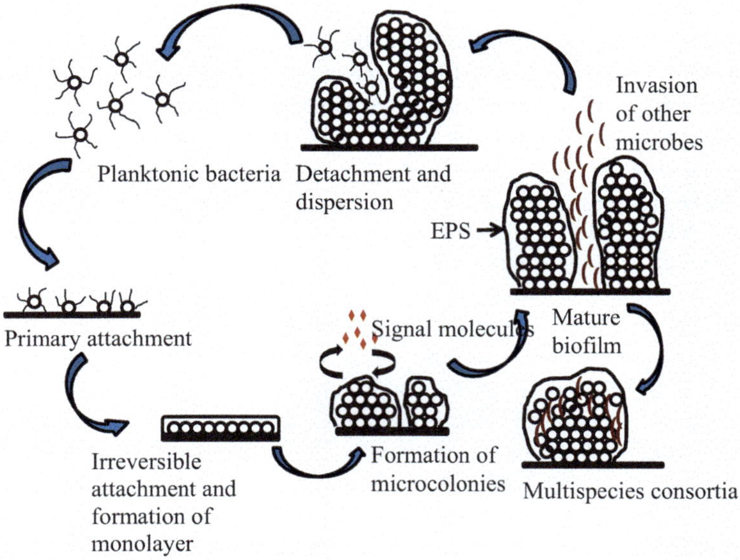

Fig. 2 Mechanism of biofilm formation

5.1 Attachment of Microorganisms

Attachment is a process in which the microorganisms attach and colonize over the filter media surface through transportation, initial adhesion, firm attachment, and colonization (Van Loosdrecht et al. 1990). The microorganisms then get transported toward the filter media surface further involving four processes: diffusion, convection, sedimentation due to gravity, and active mobility of the microorganisms. Depending upon the total interaction energy (sum of van der Waals forces and electrostatic force), the initial attachment can be reversible or irreversible. Irreversible adhesion of microorganisms occurs through production of extracellular polymeric substances. The polyhydroxyl groups in extracellular polymeric substances form hydrogen bonding through which bacterial colonization takes place (Kjelleberg et al. 2007). The process of adhesion of the microorganisms on the filter media surface stands by the DLVO (Derjaguin-Landau-Verwey-Overbeek) theory. The influent characteristics and surface properties of the filter media decide the processes of firm attachment and colonization. The method of surface attachment of microorganisms is greatly influenced by hydrophobicity of the microorganisms, steric effects, contact angle, and electrophoretic mobility values.

5.2 Growth of Microorganisms

Following the process of microbial attachment on the filter media surface, the formation of monolayer microcolonies takes place. The bulk and surface transport phenomenon governs the supply of organic substrate. Once the substrate is transported to the biofilms from the bulk liquid, it gradually diffuses into the biofilms initiating the process of metabolism. Transportation of mass substrate to the biofilm, rate of diffusion of the substrate into the biofilm, and kinetics of its utilization within the biofilm decide the rate of substrate utilization. A three-dimensional arrangement is formed by the attachment of the biofilms to debris from the adjacent environment. Growth yield of the substrate also influences the performance of a biofilm process. Invasion of new planktonic bacteria also takes place and forms multispecies consortia.

5.3 Detachment of Biomass

Detachment is a critical phenomenon where matrix-encased and sessile microbial cells are converted to free planktonic microorganisms that affect the maintenance of biomass. Some of the important detachment mechanisms are erosion of biomass (due to the fluid shear), abrasion of biomass (scraping the biocell off the surface by collision of external particle), sloughing (detachment of large patches of biomass), predation of protozoa (detachment of biomass on the outer surface of the biofilm), and filter backwashing. During backwashing, the biomass is affected by expansion of backwash bed, mode of backwash like filter effluent, chlorinated water, and air scour (Chaudhary et al. 2001). There is no loss of effective biomass during normal filter backwash (Ahmad and Amirtharajah 1998). However, it has been also reported that biomass loss occurs due to shear stress and is stimulated by cell-to-cell signaling mechanism through quorum sensing (Webb 2007).

6 Factors Promoting Biofilm Formation

The factors that promote the process of biofilm formation are as follows:

(a) Nutrients, pH, and temperature
    Nutrient condition is the major factor that plays a pivotal role in microbial growth in the biofilms. It ranges from enriched media to non-detectable one. The conversion of microbial cells from planktonic to biofilm state needs dense nutrient-rich environment, whereas the depletion of nutrients promotes biofilm cell detachment from surfaces. The microbial community obtains nutrients through various prominent ways like utilization of waste products from secondary colonizers, accumulating trace organic materials through extracellular polymeric substances, and biochemical resource pool (Sehar and Naz 2016).

    Alteration in pH affects the microbial growth and development. The bacteria present in the media modify the synthesis and activity of proteins associated with different cellular processes. Although the formation of polysaccharides and excretion of exopolymeric substances do not respond to pH variations, the optimum pH for the production of polysaccharide is around 7 (Oliveira et al. 1994).

Temperature is yet another factor which is very sensitive for microbial growth. A healthy growth of microbes needs optimum temperature, or else it will lead to reduction in bacterial enzyme reaction rates which will result in loss of bacterial growth efficiency. The optimum temperature for bacterial growth found in cooling water systems is about 40 °C (Ells and Hansen 2006).

(b) Surface topography

Bacterial adhesion to the surface is greatly governed by surface topography. The aqueous media with higher flow rate and rough surface reduce the shear force on microbial cells.

In the initial steps of attachment, a limited surface roughness enhances the adhesion of microbes by providing it increased surface area for attachment of cells. The surface exposed to the aqueous medium is coated with polymers resulting in chemical modifications that affect the rate of microbial attachment. Furthermore, the process of adhesion is influenced by factors like hydrophobicity, charge, and elasticity (Prakash et al. 2003)

(c) Hydrodynamics

The zone adjacent to the substratum which experiences negligible turbulent flow is known as the hydrodynamic boundary layer. The flow velocity in this zone is insufficient to remove the biofilm layer. The region outside this layer shows high level of turbulent flow and influences the attachment of the microbes on the surface. The microbial cells act as particles in a liquid, and the linear velocity of the liquid is the deciding factor for the rate of association of the cells with the submerged surface. Association of the cells largely depends on cell motility and cell size. The thickness of the boundary layer decreases with the increase in linear velocity. When the magnitude of the linear velocity is less, the cells travel through the hydrodynamic boundary layer, but as it increases, the boundary layer decreases. As a result, the microbial cells experience high turbulence level and mixing (Characklis 1990). Higher linear velocities can be considered the same as rapid association with the surface until the detachment of the cells occurs due to shear force caused by high velocities (Rijnaarts et al. 1993; Zheng et al. 1994). Simoes et al. (2007) showed that physical properties such as structure, thickness, and mass along with extra polymeric substance production and metabolic activities of biofilms are influenced by hydrodynamic conditions.

(d) Gene regulation

Initial attachment of the cells with the solid media involves upregulation and downregulation of a number of genes. *Pseudomonas aeruginosa* is an important bacterium which is involved in biofilm formation. Approximately 22% of its genes are upregulated, and 16% are downregulated during the process (Steyn et al. 2001). Bacterial alginate functions as extracellular matrix material which initiates differentiated biofilm formation. Genes like *algD*, *algU*, and *rpoS* which encode bacterial alginate formation and genes regulating polyphosphokinase synthesis are also upregulated during the process of biofilm formation by *Pseudomonas aeruginosa* (Prakash et al. 2003). In case of

Staphylococcus aureus, genes responsible for the synthesis of enzymes like phosphoglycerate mutase, triphosphate, and alcohol dehydrogenase used in glycolysis or fermentation are also upregulated (Becker et al. 2001)

(e) Extracellular polymeric substances
Extracellular polymeric substances are voluntarily secreted by microbial biofilms in nature which are, generally, composed of high-molecular-weight polymers, like polysaccharides, proteins, glycoproteins, DNA oligomers, phospholipids, and humic acids, comprising of 50% to 90% of the total organic carbon of the biofilms ((Flemming and Wingender 2010; Flemming et al. 2000). The uronic acids (D-galacturonic, D-glucuronic, and D-mannuronic acids) or ketal-linked pryruvates (Sutherland 2001) provide it the negative charge that facilitated their association with divalent cations causing cross-linking of the polymeric strands that provides greater binding force in the biofilm (Flemming et al. 2000). The extracellular polymeric substances are also highly hydrated by large amounts of water into their structure due to hydrogen bonding. The extensive hydrophobic interactions along with its bridging with multivalent cations aggregate the bacterial cell in the gel-like network. However, extracellular polymeric substances are not essentially required for the sustainability of microbial life, yet their provenance has significantly influenced the increased survival, plasticity, adaptability. and their metabolic efficacy. Although extracellular polymeric substances are considered as the secreted metabolic wastes, they crucially secure the biofilm attachment and facilitate organization of each microbial cell in the consortium. The extracellular polymeric substances, thus, not only enable the increase in the efficiency of biofilm for entrapping and gradual degradation of organics and nutrients from the surrounding but also increase cell-to-cell exchanges and signaling (Zhao et al. 2013; Decho and Gutierrez 2017). Moreover, it causes flocculation and granulation and further protects bacteria against environmental stresses (Zhao et al. 2013). The properties and primary characteristics of biofilm can be determined by the compositional and structural variations of the extracellular polymeric substances, since it is not uniform and rather alters on the basis of microbial composition along the spatial and temporal scales. Nevertheless, the extracellular polymeric substance production in microbial community is affected by nutrient status, where excess availability of carbon and limitation of nitrogen, potassium, or phosphate promote extracellular polymeric substance synthesis (Sutherland 2001). Retardation in bacterial growth also promotes extracellular polymeric substance production, likely, as a mechanism of tolerance. Further, it also contributes toward antimicrobial resistance of biofilms by slowing down the bulk transfer and exposure of antibiotics through the biofilm by binding directly to these agents (Donlan 2000). Notably, the increasing age of biofilms promotes the biosynthesis and secretion of extracellular polymeric substances significantly (O'Toole 2011).

(f) Extracellular DNA
Extracellular DNA serves very crucial role in various stages of biofilm formation including initial adhesion of bacterial cells followed by aggregation and

microcolony formation promoting wastewater treatment. It is an integral part in production of extracellular polymeric substances (Bockelmann et al. 2006). It also provides strength to biofilms along with protecting the biofilms from antibiotics, physical stress, and detergents (Das et al. 2013). Currently, it is also prominently used in engineered biofilms for environmental pollutant remediation and production of electricity or fuel in bioelectrochemical systems or bioreactors.

(g) Divalent cations

Divalent cations are abundant in aquatic environments and are involved in bacterial growth in biofilms. The extracellular DNA chelates divalent cations which results in amendment of surface properties of microbial cell providing resistance against detergents and antimicrobial agents (Mulcahy et al. 2008). Calcium which is a divalent cation contributes in initial attachment by chelating anionic sites on extracellular polymers between microbial aggregates of anaerobic sludge granules, activated sludge flocs, and biofilms (Kerchove and Elimelech 2008). Calcium also influences cellular and extracellular product formation, cell signaling, alginate regulation, and biofilm virulence (Sarkisova et al. 2005). Cations like sodium, calcium, lanthanum, and ferric ions cause reduction in the repulsive forces created between the negatively charged microbial cells and the surfaces of attachment (Fletcher 1988). Introduction of divalent cations enhances the thickness of the biofilms making it denser and mechanically more stable (Das et al. 2014).

7 Biofilm Ecology

The microcolonies are the fundamental unit of biofilm structure. The compaction and proximity of the microcolonies provide an ideal condition for exchange of genes, quorum sensing, and creation of nutrient gradients. The cycling of various nutrients like nitrogen, carbon, and sulfur occurs through redox reactions. Microbial interactions like predation and competition also take place in the biofilm.

(a) Gene transfer

Gene transfer among the bacterial communities occurs through horizontal gene transfer (HGT) events which generally includes the exchange of plasmid DNA. HGT events confer the transfer of novel properties like antibiotic resistance, hydrocarbon degradation, or stress tolerance among bacterial communities; however, the lower rate of such interactions due to spatial separation of planktonic cells is a prime barrier. In contrast, the biofilm provides an ideal niche for such genetic exchanges as cells are more or less spatially fixed in the space (Ehlers and Bouwer 1999; Roberts et al. 1999; Hausner and Wuertz 1999). The consensus established that the HGT in the biofilm mostly proceeds with the phenomenon of conjugation as the biofilm structure allows cell-to-cell contact requisite for it. Thus, biofilms are considered as HGT “hot spots” due to high frequency of plasmid exchanges (Van Elsas and Bailey 2002; Aminov 2011;

Fux et al. 2005; Madsen et al. 2012). Nevertheless, the several abiotic factors, nutrient status, and biofilm compositions influence the frequencies of conjugation. Conjugation takes place with conjugative pilus, encoded by the *tra* operon of the F plasmid that facilitates as adhesion factor between donor and recipient cells, forming three-dimensional biofilm (Ghigo 2001). Additionally, effective and enhanced conjugation may be obtained in the biofilm as a consequence of closer cell-to-cell contact facilitating minimal DNA shearing. The recipient organisms which lack plasmid only produce microcolonies without any further development, but plasmid-carrying donor strains convert them into biofilm forming organisms through plasmid transfer.

(b) Quorum sensing

Among all the chemical signals required for the formation and regulation of biofilms, quorum sensing is the most common one that facilitates biofilm homeostasis through regulated cell attachments and detachments. This mechanism of quorum sensing not only maintains the microfloral density but also modulates toxin productions, bioluminescence, secondary metabolite secretion, etc. (Harmsen et al. 2010). Microbiota densely packed in extracellular polymeric substances matrix release a density-dependent chemical signal that mediates the growth and development of biofilms on different surfaces via quorum sensing. Quorum sensing uses the transcriptional activator protein which acts with small autoinducer signaling molecules to stimulate expression of target genes, only after its significant accumulation, resulting in changes in physiochemical behavior (Xiong and Liu 2010). The cell-to-cell signaling systems in *Pseudomonas aeruginosa*, namely, *lasR-lasI* and *rhlR-rhlI*, are involved in biofilm formation (Davies et al. 1998). Biofilm differentiation is initiated when these signals reach their optimum concentrations required for the activation of genes. The microbes with double mutant although produce a biofilm lack typical biofilm architecture with much thinner cell layer and more densely packed cells. Moreover, the mutant biofilms are easily removed from surfaces. Quorum sensing mediates induction of genetic competence, i.e., enabling the uptake and incorporation of exogenous DNA by transformation in *S. mutans*, and increases it up to 10–600-fold than planktonic cells (Yung-Hua et al. 2001). The mechanism by which the bacterial communication is interrupted is known as quorum quenching. It enhances the bacterial dispersal across membrane during the early development. It suppresses the growth of specific microbial species (e.g., *Acinetobacter*). It also influences the community dynamics affecting correspondingly the ecological selection and dispersal processes (Jeong et al. 2020).

(c) Predation and competition

The microbial biofilms due to localized cell concentration are subjected to predation by protozoa, bacteriophage, *Bdellovibrio* sp., and polymorphonuclear leukocytes (PMNs). Murga et al. (2001) reported colonization of heterotrophic biofilms leading to predation by a free-living protozoon, *Hartmannella vermiformis*. McLaughlin-Borlace et al. (1998) demonstrated *Acanthamoeba* sp. as a predatory organism in contact lens storage case biofilms.

Competition is also a regular phenomenon observed within biofilms. James et al. (1995) noted that in spite of the notable count of *Hyphomicrobium* sp. in *Hyphomicrobium* biofilm, the invasion of *Pseudomonas putida* always results in the dominance of *Pseudomonas putida*. Biofilms containing *Klebsiella pneumoniae* and *Pseudomonas aeruginosa* show consortial growth in stable community, but when it comes to mixed culture biofilms, the growth of *Pseudomonas aeruginosa* is dominated by *Klebsiella pneumonia* (Stewart et al. 1997). *Pseudomonas aeruginosa* grow primarily as a base biofilm, but *Klebsiella pneumoniae* form localized microcolonies that are advantageous as they may have greater access to nutrients and oxygen. *Pseudomonas aeruginosa* compete by rapidly colonizing the surface, thereby establishing a long-term competitive advantage, but they are outcompeted by *Klebsiella pneumoniae* as they grow more rapidly by attaching themselves to the surface layer of *Pseudomonas aeruginosa* biofilm.

8 Use of Biofilms in Biofiltration

Biofilm system is an advanced technology that is used in biofiltration for the wastewater treatment. Here, solid media is provided to the suspended growth reactors which act as attachment surface for biofilms. The solid media increase microbial concentration which leads to enhancement in contaminant degradation. Besides from that, biofilm system also involves many processes like biodegradation, bioaccumulation, biosorption, and biomineralization (Pal et al. 2010). Along with the degradation of different pollutants, the microorganisms present in the biofilm break down the trapped pathogens present in the wastewater. The treated water after biofiltration is then reused for other recreational purposes. The mechanism of water treatment using microbial biofilm technology is schematically shown in Fig. 3.

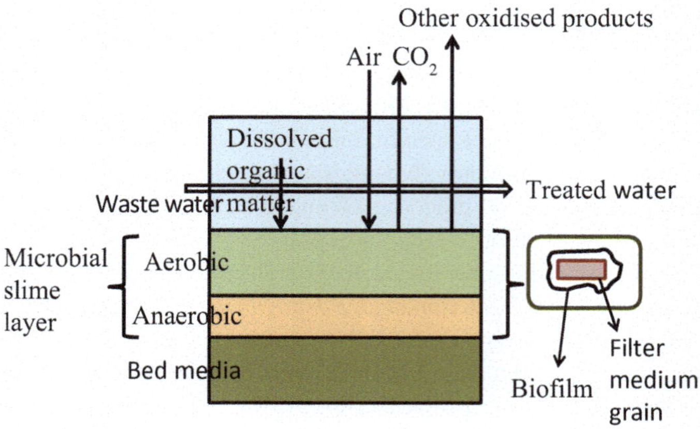

Fig. 3 The mechanism of water treatment using microbial biofilm technology

Membrane filtration technology is used in membrane bioreactors (MBR) in pilot plants. Its working principle is based on the combination of two basic processes, i.e., biological degradation and membrane separation. The biodegradation is done by the suspended solids and microorganisms, and the treated water is separated by a membrane filtration unit (Manem and Sanderson 1997). As a result of this, the slow-growing microorganisms and the particulate matters get accumulated, leading to the enhancement of recalcitrant organic treatment. Moreover, it causes retention of most pathogens which act like disinfectant. All biological solid wastes are trapped in the system, and the excess growth is removed as waste activated sludge reducing the turbidity as well as the suspended solids. Subsequently, the inorganic nutrients like nitrogen and phosphorous are also reduced (Di Giano et al. 2004).

Traditionally, membrane filtration is used in reverse osmosis process for water treatment. Reverse osmosis membrane filtration is more popular because it reduces the cost of desalination process using various improvements in the technology which decreases the energy consumption making the process more robust and cost-effective (Veerapaneni et al. 2007). Besides this, it can be used in water softening process also (Conlon et al. 1990). Instead of other traditional processes in water treatment like coagulation, sand and activated carbon filtration, or ion exchange, membrane filtration can be used without addition of chemicals to the water preventing the formation of harmful by-products. The main limitation of membrane applications is excessive biofouling which is prevailed by the use of reverse osmosis and nano-filtration (Vrouwenveldera et al. 2009). The advantages of using biofilm systems include low-space requirements, resilience to changes in the environment, operational flexibility, reduced hydraulic retention time, high active biomass concentration, increased biomass residence time, enhanced ability to degrade recalcitrant compounds, slower microbial growth rate, and lower sludge production. Over the year, biological membrane-based system has shown a great success in converting alternative water sources into potable water.

9 Research Status

Recent research works focus on exploiting the natural theory behind the microbial biofilm formation for wastewater treatment technology in pilot plants. Odegaard et al. (1994) reported the use of moving bed biofilm reactor for the first time in Norway during late 1980s and early 1990s. Later on, he also added that designing the biofilm carriers with increased specific surface area to escalate the future treatment capacity excludes the requirements of additional reactors (Odegaard 2000). Further, Maurer et al. (2001) used two types of biofilm carriers, sponge cubes and plastic tubes, while studying denitrification in a full-scale pilot plant. Laboratory models were also built with both attached growth biological reactor and suspended growth biological reactor to study the effect of combined reactors. It was deduced that combined reactors showed better oxygen transfer rate (Karamany Hesham 2001). Borghei and Hosseini (2004) pioneered the use of moving bed biofilm reactors in domestic and industrial wastewater treatment. The role of aeration system in rapid biodegradation was justified by Ahl et al. (2006). Biofilm reactors with freely moving carrier media were also operated on activated sludge treatment process (Odegaard 2006). Studies on organic phosphorus and nitrogen removal process have been examined in synthetic wastewater on a laboratory scale (Kermani et al. 2009). Sombatsompop et al. (2011) deduced the fact that when organic load is increased, the moving bed batch reactor showed more efficiency than conventional batch reactors. Yang et al. (2012) studied biosolids dynamics and explained biofilm growth and detachment, suggesting innovative design of reactors in membrane bioreactors which drew attention of researchers.

10 Future Research Prospects

Progressive application of microbial biofilm for enhancing not only water quality but also various fields has gained significant attention among the researches and industries. However, the lack of knowledge in understanding the molecular aspects and gene-dependent phenomenon of biofilm-associated organisms and their ecological interactions has somehow remained unappreciated. Microbiologist should focus on the dynamics of microbial interactions and also elucidate the genetic factor influencing such versatile phenotypes persisting among different biofilms. Recent researchers are escalating toward understanding the intrinsic and extrinsic factors that have driven versatility among the biofilm phenotypes. Nevertheless, it is crucial to explore the scope of utilizing biofilms in pathogenic resistance and treatment of chronic diseases as well. It is a matter of acknowledgment for the microbiologist as the pharmaceutical industries are embracing this novel biofilm-based approaches that may have potential benefits for the mankind.

References

Ahl RM, Leiknes T, Odegaard H (2006) Tracking particle size distributions in a moving bed biofilm membrane reactor for treatment of municipal wastewater. Water Sci Technol 53:33–42

Ahmad R, Amirtharajah A (1998) Detachment of particles during biofilter backwashing. J Am Water Works Assoc 90(12):74

Ahmad R, Amirtharajah A, Al-Shawwa A, Huck PM (1998) Effects of backwashing on biological filters. J Am Water Works Assoc 90(12):62

Aminov RI (2011) Horizontal gene exchange in environmental microbiota. Front Microbiol 2:158

Antonie RL (1974) Discussion-evaluation of a rotating disc wastewater treatment plant. J Water Pollut Control Fed 46:2792

Antonie RL (1976) Fixed biological surfaces-wastewater treatment. Chemical Rubber Company press, Cleveland

Antonie R, Aacken KV (1971) Rotating discs fulfill dual wastewater role. Water Waste Eng 8 (1):37–38

Bablon GP, Ventresque C, Ben Aim R (1988) Developing a sand- GAC filter to achieve high rate biological filtration. J Am Water Works Assoc 80(12):47

Becker P, Hufnagle W, Peters G, Herrmann M (2001) Detection of different gene expression in biofilm-forming versus planktonic populations of *Staphylococcus aureus* using microrepresentational difference analysis. Appl Environ Microbiol 67:2958–2965

Bisop PL, Zhang TC, Fu YC (1995) Effect of biofilm structure, microbial distribution and mass distribution on biodegradation processes. Water Sci Technol 31:197–203

Bockelmann U, Janke A, Kuhn R, Neu TR, Wecke J, Lawrence JR, Szewzyk U (2006) Bacterial extracellular DNA forming a defined network-like structure. FEMS Microbiol Lett 262:31–38

Borghei SM, Hosseini SH (2004) The treatment of phenolic wastewater using a moving bed biofilm reactor. Process Biochem 39(10):1177–1181

Bouwer EJ, Crowe PB (1988) Biological processes in drinking water treatment. J Am Water Works Assoc 80(9):82

Buzzini AP, Sakamoto IK, Varesche MB, Pires EC (2006) Evaluation of the microbial diversity in an UASB reactor treating wastewater from an unbleached pulp plant. Process Biochem 41:168–176

Canler JP, Perret JM (1994) Biological aerated filters: assessment of the process based on the sewage treatment plants. Water Sci Technol 29(1011):13–22

Carlson KH, Amy GL (1995) Relative importance of EBCT and HLR on the removal of BOM during biofiltration. In: Proceedings of the Journal of American Water Works Association. Water Quality Technology Conference, New Orleans

Characklis WG (1990) Microbial fouling. In: Characklis WG, Marshall KC (eds) Biofilms. Wiley, New York, pp 523–584

Chaudhary DS, Vigneswaran S, Ngo HH, Jegatheesan V (2001) Applicability of granular activated carbon (GAC) biofilter in wastewater treatment and reuse. In: Proceedings of Asian Water Quality IWA Asia-Pacific Regional Conference, Fukuoka, vol 2, p 685

Chaudhary DS, Vigneswaran S, Ngo HH, Shim WG, Moon H (2003) Biofilter in water and wastewater treatment. Korean J Chem Eng 20(6):1054–1065

Christensson M, Welander T (2004) Treatment of municipal wastewater in a hybrid process using a new suspended carrier with large surface area. Water Sci Technol 49(11–12):207–214

Cohen Y (2001) Biofiltration-the treatment of fluids by microorganisms immobilized into the filter bedding material: a review. Bioresour Technol 77:257–274

Collins G, Kavanagh S, McHugh S, Connaughton S, Kearney A, Rice O, Carrigg C, Scully C, Bhreathnach N, Mahony T (2006) Accessing the black box of microbial diversity and ecophysiology: recent advances through polyphasic experiments. J Environ Sci Health Part A 41:897–922

Conlon WJ, Hornburg CD, Watson BM, Kiefer CA (1990) Membrane softening: the concept and its application to municipal water supply. Desalination 78:157–175

Dang Y, Ye J, Mu Y, Qiu B, Sun D (2013) Effective anaerobic treatment of fresh leachate from MSW incineration plant and dynamic characteristics of microbial community in granular sludge. Appl Microbiol Biotechnol 97(24):10563–10574

Das T, Sehar S, Manefield M (2013) The roles of extracellular DNA in the structural integrity of extracellular polymeric substance and bacterial biofilm development. Environ Microbiol Rep 5 (6):778–786

Das T, Sehar S, Koop L, Wong YK, Ahmed S, Siddiqui KS, Manefield M (2014) Influence of calcium in extracellular DNA mediated bacterial aggregation and biofilm formation. PLoS One 9(3):e91935

Davies DG, Parsek MR, Pearson JP, Iglewski BH, Costerton JW, Greenberg EP (1998) The involvement of cell-to-cell signals in the development of a bacterial biofilm. Science 280:295–298

Decho AW, Gutierrez T (2017) Microbial extracellular polymeric substances (EPSs) in ocean systems. Front Microbiol 8:922

Del Nery V, Pozzi E, Damianovic MHRZ, Domingues MR, Zaiat M (2008) Granules characteristics in the vertical profile of a full-scale upflow anaerobic sludge blanket reactor treating poultry slaughterhouse wastewater. Bioresour Technol 99:2018–2024

Devinny JS, Deshusses MA, Webster TS (1999) Biofiltration for air pollution control. CRC Press, New York

Di Giano F, Andreottola G, Adham S, Buckley C, Cornel P, Daigger G, Fane A, Galil N, Pollice A, Rittmann B, Rozzi A, Stephenson T, Ujang Z (2004) Safe water for everyone: membrane bioreactor technology. Water Environ Technol 16(6):31–35

Donlan RM (2000) Role of biofilms in antimicrobial resistance. ASAIO J 46:S47–S52

Ehlers LJ, Bouwer EJ (1999) RP4 plasmid transfer among species of *Pseudomonas* in a biofilm reactor. Water Sci Technol 7:163–171

Ells TC, Hansen LT (2006) Strain and growth temperature influence *Listeria* spp. attachment to intact and cut cabbage. Int J Food Microbiol 111(1):34–42

Flemming HC, Wingender J (2010) The biofilm matrix. Nat Rev Microbiol 8:623–633

Flemming HC, Wingender J, Griegbe Mayer C (2000) Physico-chemical properties of biofilms. In: Evans LV (ed) Biofilms: recent advances in their study and control. Harwood Academic Publishers, Amsterdam, pp 19–34

Fletcher M (1988) The application of interference reflection microscopy to the study of bacterial adhesion to solid surfaces. In: Houghton DR, Smith RN, Eggins HOW (eds) Biodeterioration, vol 7. Springer, Dordrecht, pp 31–35

Fux CA, Costerton JW, Stewart PS, Stoodley P (2005) Survival strategies of infectious biofilms. Trends Microbiol 13:34–40

Ghigo JM (2001) Natural conjugative plasmids induce bacterial biofilm development. Nature 412 (6845):442–445

Govind R (2009) Biofiltration: an innovative technology for the future. University of Cincinnati, Cincinnati

Graese SL, Snoeyink VL, Lee RG (1987) Granular activated carbon filter-Adsorber systems. J Am Water Works Assoc 79(12):64

Harmsen M, Yang L, Pamp SJ (2010) An update on *Pseudomonas aeruginosa* biofilm formation, tolerance, and dispersal. FEMS Immunol Med Microbiol 59:253–268

Hausner M, Wuertz S (1999) High rates of conjugation in bacterial biofilms as determined by quantitative in situ analysis. Appl Environ Microbiol 65:3710–3713

Karamany Hesham (2001) Combined suspended/attached growth reactor: oxygen transfer rate. In: Sixth international water technology conference IWTC 2001, Alexandria

Hinton SW, Stensel HD (1994) Oxygen utilization of trickling filter biofilms. J Environ Eng 120:1284

James GA, Beaudette L, Costerton JW (1995) Interspecies bacterial interactions in biofilms. J Ind Microbiol 15(4):257–262

Jeong SY, Chung HL, Taewoo Y, Tae GK (2020) Effects of quorum quenching on biofilm Metacommunity in a membrane bioreactor. Microbial Ecol 79:84–97

Kantawanichkul S, Kladprasert S, Brix H (2009) Treatment of high-strength wastewater in tropical vertical flow constructed wetlands planted with *Typha angustifolia* and *Cyperus involucratus*. Ecol Eng 35(2):238–247

Kerchove AJ, Elimelech M (2008) Calcium and magnesium cations enhance the adhesion of motile and nonmotile *Pseudomonas aeruginosa* on alginate films. Langmuir 24:3392–3399

Kermani M, Bina B, Movahedian H, Amin MM, Nikaein M (2009) Biological phosphorus and nitrogen removal from wastewater using moving bed biofilm process. Iran J Biotechnol 7 (1):19–27

Khan ZU, Naz I, Rehman A, Rafiq M, Ali N, Ahmed S (2015) Performance efficiency of an integrated stone media fixed biofilm reactor and sand filter for sewage treatment. Desalin Water Treat 54(10):2638–2647

Khatoon N, Naz I, Ali MI, Ali N, Jamal A, Hameed A, Ahmed S (2014) Bacterial succession and degradative changes by biofilm on plastic medium for wastewater treatment. J Basic Microbiol 54(7):739–749

Kjelleberg S, Marshall KC, Givskov M (2007) The biofilm mode of life. In: Kjelleberg S, Givskov M (eds) The biofilm mode of life: mechanisms and adaptations. Horizon Bioscience, Norfolk, pp 5–21

Konneke M, Bernhard AE, De La Torre JR, Walker CB, Waterbury JB, Stahl DA (2005) Isolation of an autotrophic ammonia-oxidizing marine archaeon. Nature 437:543–546

Le Chevallier MW, Becker WC, Schorr P, Lee RG (1992) Evaluating the performance of biologically active rapid filters. J Am Water Works Assoc 84(4):136

Leclerc M, Delgenes JP, Godon JJ (2004) Diversity of the archaeal community in 44 anaerobic digesters as determined by single strand conformation polymorphism analysis and 16S rDNA sequencing. Environ Microbiol 6:809–819

Madsen JS, Burmolle M, Hansen LH, Sorensen SJ (2012) The interconnection between biofilm formation and horizontal gene transfer. FEMS Immunol Med Microbiol 65:183–195

Manem J, Sanderson R (1997) Membrane bioreactors. In: Mallevialle J, Odendall P, Wiesner M (eds) Water treatment membrane processes, vol 17. McGraw-Hill, New York, pp 1–31

Matos M, Alves C, Campos JL, Brito AG, Nogueira R (2011) Sequencing batch biofilm reactor: from support design to reactor operation. Environ Technol 32(10):1121–1129

Maurer M, Fux C, Graff M, Siegrist H (2001) Moving-bed biological treatment (MBBT) of municipal wastewater: denitrification. Water Sci Technol 43(11):337–344

McLaughlin-Borlace L, Stapleton F, Matheson M, Dart JKG (1998) Bacterial biofilm on contact lenses and lens storage cases in wearers with microbial keratitis. J Appl Microbiol 84:827–838

Metcalf and Eddy (2003) In: Tchobanoglous G, Burton FL, Stensel HD (eds) Wastewater engineering·treatment and reuse, 4th edn. McGraw·Hill, New York, pp 888–893

Metcalf and Eddy Inc (1991) Wastewater engineering: treatment, disposal and reuse (3rd edn) (revised by Tchobanoglous G, Burton F). McGraw-Hill, Singapore

Miltner RJ, Summers RS, Wang JZ (1995) Biofiltration performance: effect of backwashing. J Am Water Works Assoc 87(12):64

Mulcahy H, Charron-Mazenod L, Lewenza S (2008) Extracellular DNA chelates cations and induces antibiotic resistance in *Pseudomonas aeruginosa* biofilms. PLoS Pathog 4(11): e1000213

Murga R, Forster TS, Brown E, Pruckler JM, Fields BS, Donlan RM (2001) The role of biofilms in the survival of *Legionella pneumophila* in a model potable water system. Microbiology 147:3121–3126

Naz I, Batool SA, Ali N, Khatoon N, Atiq N, Hameed A, Ahmed S (2013) Monitoring of growth and physiological activities of biofilm during succession on polystyrene from activated sludge under aerobic and anaerobic conditions. Environ Monit Assess 185(8):6881–6892

Naz I, Sehar S, Perveen I, Saroj DP, Ahmed S (2015) Physiological activities associated with biofilm growth in attached and suspended growth bioreactors under aerobic and anaerobic conditions. Environ Technol 36(13):1657–1671

O'Toole GA (2011) Microtiter dish biofilm formation assay. J Vis Exp 47:1–2

Odegaard H (2000) Advanced compact wastewater treatment based on coagulation and moving bed biofilm process. Water Sci Technol 42(12):33–48

Odegaard H (2006) Innovations in wastewater treatment: the moving bed biofilm process. Water Sci Technol 53(9):17–33

Odegaard H, Rusten B, Westrum T (1994) A new moving bed biofilm reactor applications and results. Water Sci Technol 29:157–165

Oliveira R, Melo L, Oliveira A, Salgueiro R (1994) Polysaccharide production and biofilm formation by *Pseudomonas fluorescens*: effects of pH and surface material. Colloids Surf B Biointerfaces 2(1):41–46

Pal S, Sarkar U, Dasgupta D (2010) Dynamic simulation of secondary treatment processes using trickling filters in a sewage treatment works in Howrah, West Bengal, India. Desalination 253 (1):135–140

Parker DS, Bratby JR (2001) Review of two decades of experience with TF/SC process. J Environ Eng 127(5):380–387

Pederson K (1982) Method for studying microbial biofilm in flowing water system. Appl Environ Microbiol 43:6

Prakash B, Veeregowda BM, Krishnappa G (2003) Biofilms: a survival strategy of bacteria. Cur Sci 85(9):1299–1307

Rijnaarts HH, Norde W, Bouwer EJ, Lyklema J, Zehnder (1993) Bacterial adhesion under static and dynamic conditions. Appl Environ Microbiol 59:3255–3265

Riviere D, Desvignes V, Pelletier E, Chaussonnerie S, Guermazi S, Weissenbach J, Li T, Camacho P, Sghir A (2009) Towards the definition of a core of microorganisms involved in anaerobic digestion of sludge. ISME J 3:700–714

Roberts AP, Pratten J, Wilson M, Mullany P (1999) Transfer of a conjugative transposon, Tn5397 in a model oral biofilm. FEMS Microbiol Lett 177:636

Sarkisova S, Patrauchan MA, Berglund D, Nivens DE, Franklin MJ (2005) Calcium-induced virulence factors associated with the extracellular matrix of mucoid *Pseudomonas aeruginosa* biofilms. J Bacteriol 187:4327–4337

Sehar S, Naz I (2016) Role of the biofilms in wastewater treatment. In: Dhanasekaran D, Thajuddin N (eds) Microbial biofilms-importance and applications. Intech Open, Rijeka, pp 121–144

Servais P, Billen G, Ventresque C, Bablon GP (1991) Microbial activity in GAC filters at the Choisy-le-Roi treatment plant. J Am Water Works Assoc 83(2):62

Simoes M, Pereira MO, Silankorva S, Azeredo J, Vieira MJ (2007) The effect of the hydrodynamic conditions of the phenotype of *Pseudomonas fluorescens* biofilms. Biofouling 23(3/4):249–258

Sombatsompop K, Songpim A, Reabroi S, Inkong-ngam P (2011) A comparative study of sequencing batch reactor and moving bed sequencing batch reactor for piggery wastewater treatment. Maejo Int J Sci Technol 5(2):191–203

Stewart PS, Camper AK, Handran SD, Huang CT, Warnecke M (1997) Spatial distribution and coexistence of *Klebsiella pneumoniae* and *Pseudomonas aeruginosa* in biofilms. Microb Ecol 33:2–10

Steyn B, Oosthuizen MC, MacDonald R, Theron J, Brozel VS (2001) The use of glass wool as an attachment surface for studying phenotypic changes in *Pseudomonas aeruginosa* biofilms by two-dimensional gel electrophoresis. Proteomics 1(7):871–879

Sutherland IW (2001) Biofilm exopolysaccharides: a strong and sticky framework. Microbiology 147:3–9

Tourna M, Stieglmeier M, Spang A, Konneke M, Schintlmeister A, Urich T, Engel M, Schloter M, Wagner M, Richter A (2011) *Nitrososphaera viennensis*, an ammonia oxidizing archaeon from soil. Proc Natl Acad Sci U S A 108:8420–8425

UNESCO (2017) World water assessment programme. The United Nations World Water Development Report

Van Elsas JD, Bailey MJ (2002) The ecology of transfer of mobile genetic elements. FEMS Microbiol Ecol 42:187–197

Van Loosdrecht MC, Lyklema J, Norde W, Zehnder AJ (1990) Influence of interfaces on microbial activity. Microbiol Mol Biol Rev 54(1):75–87

Veerapaneni S, Long B, Freeman S, Bond R (2007) Reducing energy consumption for seawater desalination. J Am Water Works Assoc 99:95–106

Visser FA, Van Lier JB, Macario AJ, Conway de Macario E (1991) Diversity and population dynamics of methanogenic bacteria in a granular consortium. Appl Environ Microbiol 57:1728–1734

Vrouwenveldera JS, Graf von der Schulenburgc DA, Kruithofa JC, Johnsc ML, van Loosdrechtb MCM (2009) Biofouling of spiral-wound nanofiltration and reverse osmosis membranes: a feed spacer problem. Water Res 43:583–594

Wang JW, Summers RC, Miltner RJ (1995a) Biofiltration performance: relationship to biomass. J Am Water Works Assoc 87(12):55

Wang JW, Summers RC, Miltner RJ (1995b) Biofiltration performance: effect of backwashing. J Am Water Works Assoc 87(12):64

Webb JS (2007) Differentiation and dispersal in biofilms. In: Kjelleberg S, Givskov M (eds) The biofilm mode of life: mechanisms and adaptations, vol 7. Horizon Bioscience, Norfolk, pp 167–174

Xiong Y, Liu Y (2010) Biological control of microbial attachment: a promising alternative for mitigating membrane biofouling. Appl Environ Microbiol 86:825–837

Yang Q, Qiang H, Ibrahim HT (2012) Review on moving bed biofilm processes. Pak J Nutr 11 (9):706–713

Yu Y, Feng Y, Qiu L, Han W, Guan L (2008) Effect of grain-slag media for the treatment of wastewater in a biological aerated filter. Bioresour Technol 99(10):4120–4123

Yung-Hua L, Lau PCY, Lee JH, Ellen RP, Cvitkovitch DG (2001) Natural genetic transformation of *Streptococcus mutans* growing in biofilms. J Bacteriol 183:897–908

Zhang S, Huck PM (1996) Parameter estimation for biofilm processes in biological water treatment. Water Res 30(2):456

Zhang D, Li J, Guo P, Li P, Suo Y, Wang X, Cui Z (2011) Dynamic transition of microbial communities in response to acidification in fixed-bed anaerobic baffled reactors (FABR) of two different flow directions. Bioresour Technol 102:4703–4711

Zhao K, Tseng BS, Beckerman B, Jin F, Gibiansky ML, Harrison JJ, Luijten E, Parsek MR, Wong GCL (2013) Psl trails guide exploration and microcolony formation in *Pseudomonas aeruginosa* biofilms. Nature 497:388–391

Zheng D, Taylor GA, Gyananath G (1994) Influence of laminar flow velocity and nutrient concentration on attachment of marine bacterioplankton. Biofouling 8:107–120

New Bioremediation Technologies to Remove Heavy Metals and Radionuclides

Shubhalakshmi Sengupta, Uttariya Roy, Shamik Chowdhary, and Papita Das

Abstract

Environment of this planet is facing hazards from various pollutants, among which heavy metal and radionuclide pollution is of great importance. This pollution is a resultant of both geological and anthropogenic activities. Various industrial and municipal solid wastes have been a major source of heavy metal contamination in soil, water and also as atmospheric aerosols in air. Similarly, radioactive wastes from nuclear plants and places where radioactive materials are used (e.g., medical centers) are contributing to radionuclide pollution of the environment. These contaminants are harmful for living beings and cause various health hazards to them. Proper management of wastes from these sources is required along with environment-friendly remedial techniques. Phytoremediation has been used in this regard for many years. However, nowadays, novel biotechnological tools are used for achieving paths in bioremediation through microorganisms. Microbes possess the ability to biotransform, biosorb, and biomineralize these metals and radionuclides. Techniques are now being availed to identify the microorganisms and study their biological functions in order to use them in remediating these hazardous pollutants from the environment.

Keywords

Heavy metal · Radionuclide · Phytoremediation · Microbes · Biosorption

S. Sengupta
Department of Sciences and Humanities, Vignan's Foundation for Science Technology and Research (Deemed to be University), Guntur, Andhra Pradesh, India

U. Roy · S. Chowdhary · P. Das (✉)
Department of Chemical Engineering, Jadavpur University, Kolkata, West Bengal, India

M. P. Shah (ed.), *Removal of Emerging Contaminants Through Microbial Processes*,
https://doi.org/10.1007/978-981-15-5901-3_2

1 Introduction

Environmental pollution is a global concern of this century. One of the major areas of environmental pollution is the pollution caused by elements, which are now categorized by researchers as "emerging contaminants" (Yu et al. 2014). This pollution is mainly due to the presence of elements whose elimination results in harm and destruction of the natural environment. Elemental pollution is primarily due to anthropogenic activities involving rapid industrialization, advanced agricultural practices, and improper waste disposal methods. These practices have led to the increase in elemental concentration in the environment which can create immense toxicological and adverse effects on living organisms (Wijnhoven et al. 2007). Among the elemental pollution, heavy metals have emerged as a pollutant posing grave concern for the planet. These heavy metals have the capacity to enter the environment through high level single exposure and can also pose cumulative effect through low and high level exposures. However, when present in the environment, it can persist in its toxic form for a prolonged time period, causing ecological, nutritional, and environmental problems (Das and Osborne 2018). Heavy metal pollution comes from either natural or anthropogenic sources (Nagajyoti et al. 2010). But anthropogenic activities have resulted in heavy metal contamination in large parts of the world. In India, several researchers have conducted detailed studies on heavy metal contamination (Rajaganapathy et al. 2011; Paul 2017). Khan et al. (2005) reported heavy metal contamination in the surroundings of Neyveli Lignite Mines and associated industrial complex in Tamil Nadu. Similarly, Hejabi et al. (2011) reported pollution due to heavy metals discharged from industrial effluents in Kabini River of Karnataka. Similar to heavy metal pollution, another emerging pollutant of this world is radionuclide contamination (Pravalie 2014; Szarlowicz et al. 2019). Radionuclides are present in the environment from natural sources and in today's world majorly from anthropogenic sources. The advent of industrial revolution in the early nineteenth century and discoveries regarding nuclear weapons during World War II led to massive weapon production, experiments, and nuclear power generations. These things contributed to the increase of radionuclides in the environment (Hu et al. 2010). Both heavy metals and radionuclides pose detrimental threats to the living organisms (Das and Osborne 2018). Thus, heavy metal and radionuclide pollution is a worldwide phenomenon whose mitigation is of outmost importance.

Conventional methods employed in industries have been used for these purposes. Techniques like precipitation, oxidation, filtration, and adsorption have been used for removal of heavy metals and radionuclides from the environment. However, novel remedial technologies using organisms are being used. From conventional phytoremediation to the new age genetically modified microorganisms, various new approaches have been studied by the researchers in recent times. This chapter deals with such emerging technologies employed in remediation of heavy metals and radionuclides.

2 Heavy Metals and Its Sources

The term heavy metal is referred to as a metal element having atomic density more than 4 g/cm^3 imparting toxicity at low concentrations. These heavy metals (e.g., lead, cobalt, cadmium, iron, zinc, nickel, arsenic, manganese, chromium) naturally occur mostly in dispersed state in rocks. Industrialization and increasing urbanization of humans have resulted in their presence in the biosphere. Heavy metals are found dispersed mainly in soils and aquatic bodies and in lesser proportions as vapors or particulates in the atmosphere (Nagajyoti et al. 2010). Heavy metals are toxic to plants depending on parameters like metal concentration, pH, etc. but also are required by plants as essential elements. For example, metals like copper and zinc play the role of cofactors and activators of enzyme reactions. They also exert catalytic properties (e.g., prosthetic groups of metalloproteins). Besides these functions, they also play active role in redox reactions, nucleic acid metabolism, and electron transfer. However, metals like Cd, Hg, and As can also inhibit growth and death of organisms by targeting the metal-sensitive enzymes in organism's body (Nagajyoti et al. 2010).

In general, heavy metals are categorized as trace elements which are nonessential or class B and are highly toxic (e.g., nickel, mercury, lead). They accumulate and are not easily metabolized or eliminated from organism's body. These metals become a part of the food chain and remain in the ecosystem (Nagajyoti et al. 2010).

As mentioned earlier, there are different sources which contribute to heavy metal pollution in the environment. These can be broadly classified into:

(a) Natural sources.
(b) Anthropogenic sources.

2.1 Heavy Metals from Natural Sources

Heavy metals have their natural origin in the Earth's crust and are present naturally in the soil due to weathering. Thus, the geological parent rock is the primary source of heavy metals whose composition and concentration vary throughout the planet depending on the rock type and environmental conditions. In the soils, sedimentary rocks act as a lesser source of heavy metals in comparison to igneous rocks (Nagajyoti et al. 2010). Volcanoes, however, have considerable higher levels of heavy metals (Pb, Mn, Ni, Al, Zn, etc.). According to a very early research reported by Pacyna (1986), global emissions of various heavy metals are mainly from various sources which include windblown dust, volcanic particles, forest wild fires, vegetation, and sea salt.

2.2 Heavy Metals from Anthropogenic Sources

Various sources as a resultant of different anthropogenic activities contribute to the increase of heavy metals in the environment which are accessible to living organism at a level above their permissible limits. The various anthropogenic sources are illustrated in Fig. 1.

Agricultural sources contribute immensely to heavy metal pollution. Fertilizers both organic and inorganic serve as sources for heavy metal contamination. Fungicide contains heavy metals like Cd, Cr, Ni, Pb, and Zn in variable proportions (Nagajyoti et al. 2010). Pesticides like lead arsenate were used in Canada for six decades in the orchards which resulted in soil contamination with heavy metals. Again, Paul (2017) reported that the uncontrolled use of heavy metal-containing pesticides and fertilizers in agricultural fields have resulted in groundwater and surface water heavy metal contamination in India. It has even led to their presence in drinking water. Animal manure and sewage sludge used in agriculture add heavy metals like Mn, Zn, Cu, Cr, Pb, Ni, Cd, and Co in the soil. Liming process also contributes to this contamination (Nagajyoti et al. 2010). Various pedogenic and anthropogenic processes also lead to contamination of forest soils (Wuana and Okieimen 2011).

Industrial sources of heavy metal contamination occur due to processes like smelting, metal finishing and recycling, and transportation of ores. Mining processes also release metals in the environment. Coal mines act as sources of Cd, As, and Fe. Hg use in gold mines also results in Hg contamination (Lacerda 1997; Nagajyoti et al. 2010; Rajaganapathy et al. 2011). Heavy metal contamination of soil and aquatic bodies also occurs due to mine waste erosion, transportation of crude metals and metal leaching in water bodies. Thermal power plants and coal mining also contribute to these contamination processes.

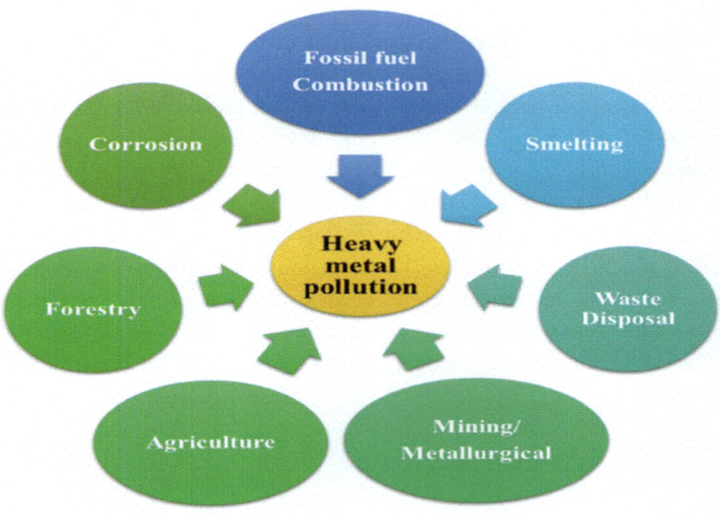

Fig. 1 Various anthropogenic sources of heavy metal contamination (Nagajyoti et al. 2010; Wuana and Okieimen 2011; Das and Osborne 2018)

Domestic effluents, solid wastes, and untreated wastewater from industries also contribute to this list. These liquid wastes primarily pollute the water bodies and also contaminate soils.

Again, the application of various biosolids like municipal sewage sludge, manures, and composts in soils contributes to heavy metal contamination. Airborne sources of metal contamination are also prevalent. In industries, metals such as As, Cd, and Pb volatilize during processing at very high temperature. These metals are emitted as aerosols in the air from the stacks (Wuana and Okieimen 2011).

3 Heavy Metal Toxicity

Heavy metals are often required by plants and animals as trace elements (ppb range to less than 10 ppm) depending upon multiple parameters. Some often play their role in biochemical and physiological functioning. Being constituents of enzymes, they take part in oxidation-reduction processes in the body. For example, copper takes part as a cofactor in enzymes like peroxidases, cytochrome c oxidase, etc. It is also present in enzymes required for hemoglobin formation. Paradoxically, its cycling between C (II) and C (I) state makes it toxic to living organisms leading to diseases like Wilson's disease in human beings. Metals have often been found to interact with cell components leading to the disruption of normal cellular activity. They can cause DNA damage, changes in cellular confirmation, and even carcinogenicity (Tchounwou et al. 2012). In Table 1, the toxicity imparted by heavy metals in living organisms specially in human beings is given.

Table 1 Heavy metals, its anthropogenic sources, and toxicity in humans (Tchounwou et al. 2012; Paul 2017; Sengupta et al. 2017)

Metals	Sources (anthropogenic)	Toxicity
Zinc	Refineries, metal plating	Anemia, skin problems
Nickel	Batteries, electroplating	Respiratory disorder, kidney problems, gastrointestinal distress, dermatitis, cancer
Mercury	Pesticide, mining, paper industry	Respiratory disorder, neurological disorder, kidney problems
Lead	Paint, pesticide, thermal power plants, mining	Anemia, neurological disorder, renal damage
Copper	Pesticide, mining, electroplating	Gastric and neurological disorders
Chromium	Mining, tannery, textile	Respiratory disorder, cancer
Arsenic	Pesticide, fungicide, smelting	Neurological disorder, cardiovascular disease, hematologic disorders, cancer
Cadmium	Pesticides, fertilizer, nuclear power plants, batteries	Oxidative stress, cancer
Manganese	Fuels, welding	Respiratory disease, neurological disorder, Parkinson's disease

4 Methods of Heavy Metal Removal

Several methodologies have been adopted by researchers to successfully mitigate heavy metals from soil and water. Among the conventional methods used in heavy metal remediation, chemical precipitation is a widely used method in industries. In this method, the chemicals (hydroxides and sulfides) formed precipitates, and the metal ions were subsequently removed from the system. Furthermore, chemical precipitation was combined with other methods (e.g., sulfide precipitation with nanofiltration) for this purpose. Another approach that was used was heavy metal chelation. Chelating agent like 1,3-benzenediamidoethanethiol ($BDET^2$) dianion was used to chelate out heavy metals. Then method like ion exchange was also used having high treatment capacity and efficiency. Synthetic resins were used for this purpose. Recent advancements led to technologies like membrane filtration. The different types of this method include ultrafiltration, reverse osmosis, nanofiltration, and electrodialysis. Again, adsorption of metals onto various adsorbents like activated charcoal, clay, graphene oxide, carbon nanotube, rice husk, cellulose, etc. is used effectively by researchers. However, using biological organisms for this purpose has emerged as a green and sustainable route for this purpose. Plants, animals like earthworms and microorganisms like bacteria, fungi, and algae are being used for remediation purposes. Biomass of microorganisms, living or dead, is being effectively used. Genetic engineering of these organisms is also being carried out (Sengupta et al. 2017; Das and Osborne 2018). Various approaches adopted for bioremediation of heavy metals are given in Fig. 2.

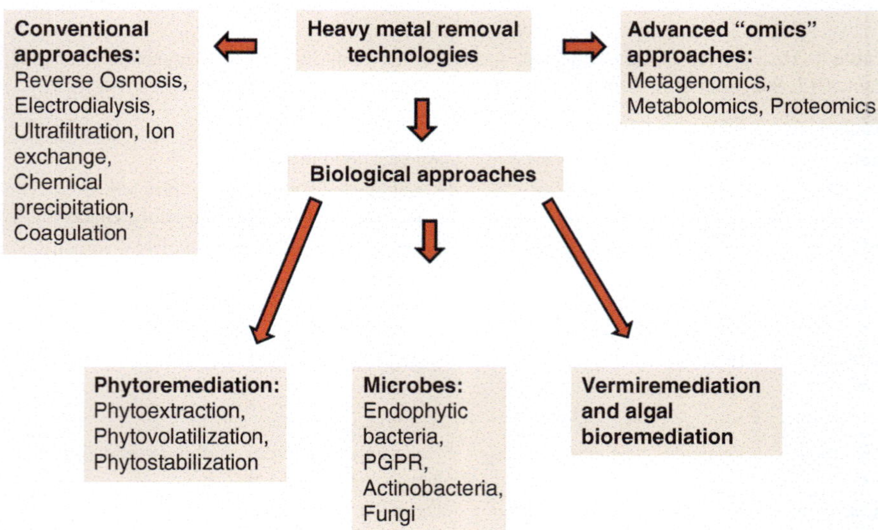

Fig. 2 Various heavy metal remediation approaches (Su et al. 2014; Das and Osborne 2018)

5 Bioremediation of Heavy Metals

Bioremediation of heavy metals is a widely sustainable, green, and sought-after method in today's age. This technique employs organisms to breakdown hazardous substances through biosorption, bioaccumulation, and biomineralization. The uses of biomasses are studied to be more effective than living organisms in case of heavy metal removal. Phytoremediation and use of earthworms are also some of the techniques which could be deployed individually or through integrated approach (Das and Osborne 2018).

5.1 Phytoremediation

Phytoremediation is one of the conventional bioremediation procedures, whereby plants and their physiological processes are used for metal remediation purposes. In nature, plants play the role of both accumulator and excluder of heavy metals. They possess the ability to biodegrade or biotransform the heavy metals in their tissues in inert forms. Whereas when plants exclude the heavy metals, they block the containment of the heavy metals in their tissues. The ion uptake mechanism through proton pumps (i.e., ATPases) plays an interesting role in the uptake of heavy metals also. This technique uses three processes, namely, phytoextraction, phytostabilization, and phytovolatilization (Tangahu et al. 2011).

(a) Phytoextraction is the uptake of heavy metals by plant roots and their storage in their biomass, e.g., *Brassica juncea* and Napier grass.
(b) Phytostabilization is another approach adopted by plants, whereby they immobilize metals in their rhizosphere, e.g., giant reed and silver grass.
(c) Phytovolatilization employs volatilization of heavy metals like Hg and Se, e.g., canola and Indian mustard (Das and Osborne 2018; Tangahu et al. 2011).
There are various factors which affect the phytoremediation process in plants like species of plant used, chemicals added as chelating agents in the soil, properties of the medium (pH, fertilizer added), bioavailability of the concerned heavy metal, root zone properties, and environmental conditions (Ginneken et al. 2007;Tangahu et al. 2011).

However, phytoremediation technologies are in general cheap, aesthetic, and eco-friendly methods for heavy metal containment, but as a method, it is time-consuming and dependent on various external and environmental parameters. Thus, researchers have been on the lookout of these remediation facilities by the use of microorganisms as they are more effective and employ less hassle-free technology.

5.2 Bioremediation Using Microorganisms

Bioremediation technologies using microorganisms like bacteria, fungi, and algae have been rapidly developed in recent times (Sengupta et al. 2017). In a recent review, Yin et al. (2019) highlighted various strategies adopted by microorganism for the effective remediation of heavy metals in the environment. They identified the anthropogenic and natural processes/activities as the primary origin of heavy metals in the environment emphasizing on the utility of bioremediation process to tackle the challenges related to heavy metal deposition. In that respect, both living and dead organisms have been used to get rid of the heavy metals. The straightforward and inexpensive protocols along with high adsorption capacity and abundant choice of the organisms have prompted the research to flourish at a faster rate than the other methods.

Among the living choices, bacteria, fungi, and algae are the primary candidates which are primarily considered for this remediation strategies. In case of bacteria, primarily two types of mechanisms for heavy metal adsorption have been identified: (1) through the functional groups (carboxyl, amino, phosphate, sulfate, etc.) on the polysaccharide slime layers of bacteria (Yin et al. 2016; Yue et al. 2015) and (2) via extracellular polymeric substances comprised of proteins, lipids, carbohydrates, nucleic acids, etc. (Fang et al. 2010). Though, the first case is more prevalent in the literature as exemplified by the recent reports of bioadsorption of Hg, Cr(VI)/Zn (II) by *Pseudomonas aeruginosa* (Yin et al. 2016) *and Staphylococcus epidermidis* (Quiton et al. 2018)., \respectively. The second type of adsorption have also started to emerge (Wang et al. 2014). After the bioadsorption, subsequent transport of the heavy metal into bacterial cells and through the corresponding enzyme-dependent metabolism pathways, the oxidation state of the metal is changed to a less toxic oxidation state. Commonly, the uptake capacity of bacteria for the heavy metals falls between 1 mg/g and 500 mg/g, depending on the types and concentration of the metal as well as the types of bacteria used for the experiment.

In case of fungi, due to its ability to survive in heavy metal concentrations, they have also been used in various cases for the successful bioadsorption of heavy metal. Both functional groups (amine, carboxyl, etc.) and the ionizable sites (glucuronic acid, chitin-chitosan complex, etc.) affect the adsorption capacity as well as the affinity toward a particular heavy metal. For example, *Termitomyces clypeatus* effectively absorbs Cr (VI) with the help of various functional groups like hydroxyl, imidazole, carboxyl, etc. on the surface (L. Ramrakhiani et al. 2011). *Saccharomyces cerevisiae*, the most common fungus, has been employed to effectively remove copper, zinc, and cadmium at high-salt environment (Li et al. 2013:46–52). Algae on the other hand adsorb heavy metal ions on different peptide which in turn protect algae from heavy metal toxicity. For example, the green microalga *Desmodesmus* sp. (Rungini et al. 2018) and *Fucus vesiculosus* (Demey et al. 2018) have been shown to effectively remove Cu (II)/Ni (II) and Pb (II), respectively, with high capacity.

The major part of nonliving microorganisms, which are routinely used for the removal of heavy metal, primarily originates from biomass produced by debris of microorganisms. Though sounds promising due to the high abundance of biomass, in order to achieve satisfactory remediation capability, they are often modified using different techniques. While acid treatment allows to incorporate additional adsorption sites (Mao et al. 2013), the base treatment increases the adsorption capability by increasing the surface negative charge for the electrostatic attraction between the surface and the positively charged heavy metals (Yan and Viraraghavan 2000). Apart from chemical treatments, thermal treatment (Mane et al. 2011) including heating in the presence of oxygen/air/argon/nitrogen, autoclave methods, etc. is also routinely used to alter the nature and the amount of surface functional groups which allow better interaction between the surface and the metal leading to higher adsorption efficiency.

Based on the available reports where the bioadsorption mechanisms and the strategies are clearly explained, it is well-understood that for detoxification, active bioorganisms are preferred because of their potential to convert the toxic metal ion to their corresponding nontoxic counterparts. However, in case the requirement is aimed toward removal of heavy metal, biomass-based systems with modified surface properties are more conducive. However, the protocols that are to be adopted for different heavy metal adsorptions are essentially purpose-based, and any generalization on this may lead to oversimplification of the complexity.

While uptake of heavy metal into microorganisms emerges as effective remedial method mainly due to the active participation in various metabolic and enzymatic processes, the overdose of the metal ion concentration can be detrimental for the microorganisms leading to enzyme inhibition, damage of DNA and cell membranes, etc. Also, in some cases, metals such as Hg and Pb can inherently be toxic for those microorganisms. The related studies of such toxicological effects of such heavy metals on the microorganisms have impacted significantly in microbial ecology where adoptability of microbial stains in the presence of various heavy metal is investigated (Huertas et al. 2014). In case of exposure, the microorganisms have their own detoxification strategies which are mainly enzyme detoxification, transportation of heavy metals, and sequestration of intacellular and extracellular enzymes, by which they try to circumvent heavy metal overdose (Fig. 3). In fact, thorough knowledge of the resistance mechanism of microorganisms in the presence of heavy metal can potentially lead to their evolutionary background and their future fate.

These informations and strategies shows the path for development of techniques which uses these microorganisms for remedial measures and are truly the emerging technologies of our times.

5.3 Emerging Microbial Bioremediation Technologies

In recent researches, heavy metal adsorption by microbial biosorbents is used. This uptake is mainly done through the electrostatic interactions of the cell wall components with the heavy metals. However, it is a non-metabolic process. Whereas, accumulation of the metals occurs by the functional groups of the

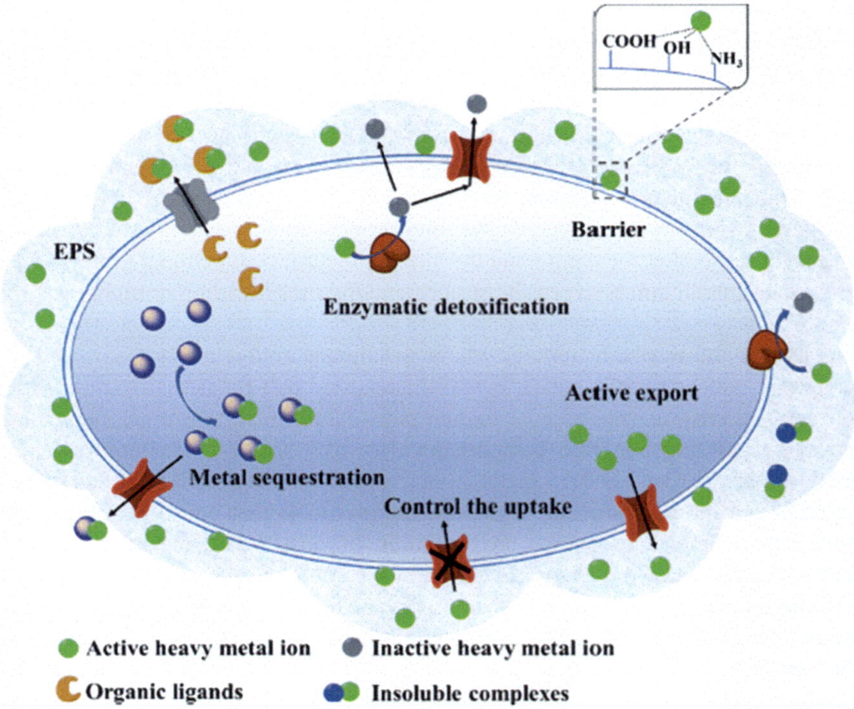

Fig. 3 Proposed detoxification pathways of microorganisms toward heavy metal ions. (Reproduced with permission from Yin et al. 2019)

polysaccharides/chitin/cellulose present in the cell walls (Yin et al. 2019). Thus, it is mainly an exchange process of the functional groups which makes these microorganisms effective absorbents. In this regard, genetic engineering plays an important role to improve and redesign the microorganisms having heavy metal uptake and sequestering properties. Genetic engineering can engineer the selectivity and also enhance the accumulative properties. Cell surface adsorption can be genetically engineered by embedding functional group proteins on the anchoring protein molecules on the cell surface. This cell surface engineering in bacterial cells has been achieved successfully. Again, cysteine-rich peptides like glutathione, phytochelatins, and metallothioneins have the ability to sequester metal ions. Genetically modified bacterial cells with overexpression of these proteins are able to absorb more metals. All these techniques can be used for better remediation of metal contamination (Sengupta et al. 2017).

With the advent of modern bioinformatics tools, information of microbial organisms and processes could be used more effectively. These information could be gathered through different approaches like metagenomics. By definition, metagenomics refers to the study of genetic material sampled directly from environmental sources. In the current context, metagenomics has been proven to be utilized to assess the correlation between the genetic materials for the microbial community and the potential of their remedial activity. This emergence of this field is majorly attributed to the fact that it provides direct information about the microbial communities irrespective of their culturability. Till date, two major classes of metagenomics techniques have been adopted for better understanding of interaction between the microorganisms and the foreign bodies/environment: (1) library-based targeted metagenomics and (2) direct sequencing. In the first case, isolated DNA from the environment is cloned inside a host (commonly *Escherichia coli*), followed by the selection and isolation of clone of interest depending on either their functions or their sequence homology to generate a metagenomic library. Despite the prevalence of *Escherichia coli* as a host, several other new strategies such as multiple hosts, etc. are also being adopted to overcome the difficulties related to gene expressions, toxicity, etc. In case of direct sequencing, the cloning step is omitted, and the genetic components of microorganisms are directly studied for the structural and functional information of the microbial communities as well their interaction with environment. For example, 16S rRNA genes and marker genes have been utilized for this approach. Recently, after the introduction of next-generation sequencing (NGS), the growth in the field of metagenomic has been quite sharp due to the possibility of parallel sequencing of genetic materials. All these approaches coupled with newer technological inventions are expected to help to identify the missing links of the complete picture of bioremediation process. This work has been carried out on various microbial species (e.g., *Lysinibacillus* sp. and *Rhodococcus* sp. for removal of Pb, Mn, and Cu). Similarly, metabolomics and proteomics approach designed to identify various chemical compounds, proteins, and metabolic pathways in remediation studies helps in investigating remedial studies. This information could be stored and used effectively for bioremediations of heavy metals in the future (Das and Osborne 2018).

6 Radioactive Elements and Its Sources

Radioactivity is a natural phenomenon present in the planet Earth. Radionuclides are found in the environment being present in air, water, soil, and living organisms. The natural radionuclides may be classified into three categories, based on their source:

(a) Primordial radioisotopes have presented since the Earth was originated, nearly 4.6×10^9 years ago.
(b) The second classes are produced from nuclear reactions, and its constituents are present in air or on the upper layer of this Earth.
(c) Third category includes the radioactive elements coming from different radioactive decay series, i.e., uranium-radium series (containing ^{226}Ra, ^{222}Rn, ^{218}Po, ^{214}Pb, ^{214}Bi, ^{214}Po, ^{210}Pb), uranium-actinium series (including ^{231}Th, ^{227}Ac, ^{223}Ra, ^{215}Po, ^{211}Bi, ^{207}Pb), and thorium series (such as ^{228}Ra, ^{228}Ac, ^{228}Th, ^{220}Rn, ^{212}Pb, 212Bi, ^{208}Tl) (Isaksson and Raaf 2016).

However, over the recent years, various anthropogenic activities have contributed largely to the increase of the amount of radionuclides in the environment and their exposure to living organisms. Different industrial processes like metal mining, mineral sands, activity associated with coal industry, etc. can give rise to drastically increased radiation exposures especially to naturally occurring radioactive material (NORM) (Szarlowicz et al. 2019). Another reason for exposure to radioactive elements is due to the commencement of the so-called atomic age which was a result of the experiments of mankind with nuclear weapons.

Various studies have been conducted regarding the presence of radioactive elements in the environment. There are reports on efforts being carried out to study the spatial boundary of highly radioactive sites of our planet and to note its exposure on human beings. (Pravalie 2014). Researchers have reported studies regarding the effects of radionuclides on living organisms especially on humans where the source of radioactivity is anthropogenic. Likewise, analyses have been carried out to correlate the incidence rate of the enhancement of thyroid cancer and the rate of the presence of radioactive isotope (^{131}I) at heavily contaminated nuclear testing sites of Nevada, USA. Radionuclide pollution was reported due to hydrogen bomb testing in Marshall Islands in the United States on the Bikini Atoll in 1954 (the Castle Bravo test) and also due to the radioactive experiments conducted in the Novaya Zemlya archipelago in the United States of Soviet Russia in 1961 (the Tsar test) (Goodby 2005). It was reported that most of the airborne fission products had been released through hydrogen bomb, vented and leaked over a great time span. Therefore, highly volatile fission outcomes like ^{129}Te, ^{131}I, ^{136}Cs, etc. were released into the atmosphere. These radioisotopes were also moved together with different air particles, and its subsequent wet and dry accumulation caused their clustering on the surface of the Earth. Radionuclides (iodine, caesium, and strontium), which yielded huge fission products, delivered a greater risk for inner radiation exposure through intake of contaminated farming products (Kinoshitaa et al. 2011).

Naturally occurring radionuclides are present either in their cosmogenic or terrestrial form in our surroundings. The major radioisotopes generated through the reaction of different gases with cosmic rays are basically ^{3}H, ^{7,10}Be, ^{14}C, ^{26}Al, and ^{39}Ar. The rocks, minerals, and also soil carry NORM which has been characterized by their greater half-life time spans (ATSDR 1999). The most potential terrestrial radionuclides are ^{238}U and ^{232}Th decline series, with ^{40}K. ^{226}Ra, ^{232}Th, and ^{40}K are mainly responsible for soil activity (UNSCEAR 2000). The term radioactive contamination defines the presence of unwanted or undesirable radioactive materials on the upper layer of soil or within solid particles, liquid substances, and different gaseous materials and also in several biotas (IAEA 2007). The source of any kind of NORM is associated with the generation of this planet. In other context, different anthropogenic activities regarding the designing of nuclear energy and its several usages have become significant source of radioactive pollution (Smičiklas and Šljivić-Ivanović 2016). Since the last century, radioactive isotopes have seemed through the discharge of anthropogenic radionuclides causing ionizing radiation one of the crucial environmental factors (Aleksakhin 2009). Each and every organism on the Earth is frequently exposed to natural ionizing radiation called background radiation. Origins of this type of radiation include cosmic rays coming from the Sun and stars, NORM found in rocks as well as in soil, radionuclides into animal tissues, and the products of radon, which are basically inhaled. Human beings are also exposed to background radiation from different anthropogenic activities, mainly through medical processes such as X-ray diagnostics. Therapy associated with radiation is normally focused only to the damaged tissues (Hazra 2018). Concern for radioactive contamination is enhanced by the invention of artificial radioactivity, deigning of nuclear weapons, and development of different nuclear reactors for producing electricity. Radioactive pollution coming from both common nuclear approaches and defense-associated nuclear activities poses an alarming problem for protecting and sustaining our environment for both present and future generations (Hazra 2018).

The radioactivity of soils relies not only on man-made activities. Soils and other naturally found objects carry NORM, and their presence is known as natural radioactivity. The natural radionuclides can be of different origins; its major sources are global fallouts of technogenic radionuclides due to greater radiation experiments and nuclear accidents with discharge of technogenic radionuclides into the environment. Special focus is imparted to the removal of technogenic radionuclides and their intake through soil-plant cover from the enterprises with nuclear fuel recycling plants. The soil-plant cover adjacent to nuclear contaminant disposal sites is subjected to radioactive pollution (Aleksakhin 2009). Some of the nuclear accidents resulted in the radioactive pollution of considerable areas (Aleksakhin et al. 2001). In 1986, after the most dangerous nuclear accident in Chernobyl, total area impacted by the radioactive contamination (with the degree of ^{137}Cs contamination above 1 Ci/km^2) was nearly 195,000 km^2 (Izrael et al. 1990, 1994). The Kyshtym accident in 1957 ensued in the generation of the East-Ural radioactive footprint (the level of the radioactivity with ^{90}Sr above 2 Ci/km^2) on the domain of approximately 23,000 km^2. Different mineral fertilizers and some agrochemical modifiers can become sources of several radioactive elements of the soil. Lower radioactive pollution can also happen when different phosphoric fertilizers and phosphogypsum containing ^{238}U, ^{232}Th, and their fallouts are used (Aleksakhin 2009).

7 Toxicological Effects of Radionuclide Pollution

Most of the radioactive elements did not exist naturally. Radioactive contamination has become a critical issue since nuclear bombs and reactors have been developed (Walker and Don 2013). Radionuclides have also been found in different types of seafood in India, variety of foods in the Balkans, and food as well as drinking water in Switzerland (Khan and Wesley 2011; Carvalho and Oliveira 2010; Brennwald and Dorp 2009). Risk assessments are carried out to ascertain that degree remains within permissible levels (Thompson and Darwish 2019). Moreover, several experimental approaches are undertaken to measure safety in ingestion pathways analyzing different food uptakes (Prohl et al. 2005). Radionuclide substances have many negative impacts on individual living creatures as well as whole ecosystems.

Radionuclide and their compounds contain double toxicity—chemical toxicity and toxicity induced by radioactive elements (known as radiotoxicity). Although the occurrence of radioactive pollution is relatively rare, it requires more attention because of utmost degrading effects of radioactive isotopes on different living cells (Smičiklas and Šljivić-Ivanović 2016). The degree of the detrimental effects of radioactive contamination depends on absorbed nuclear energy, permeable capability of radioactive ions, time period of exposure, as well as reproducibility of cells (ATSDR 1999). Exposure to radionuclides or ionizing radiation induces severe health effects including nausea, vomiting, headaches, etc. With more exposure, the victim may also suffer different physical abnormalities such as diarrhea, dizziness, disorientation, fatigue, fever, hair loss, weakness, low blood pressure, and eventually death. Fetuses are basically susceptible to the effects of radioactivity at the cellular level, which can ensue smaller brain formation, poorly developed eyes, retard growth, and mental problems (Bogutskaya et al. 2011; Al-Zoughool and Krewski 2009). Several contemporary literatures reported that long-term exposure to radioisotopes results in an elevated risk of leucopenia, leukemia, and genetic distortion which can cause lethal health issues that can be transferred into the next generation (Mohner et al. 2006). Radioactive isotopes present in aquatic system are components of uranium, thorium, and actinium radioactive series and also include radium and radon. These radioactive isotopes may cause many biological alterations (Bonavigo et al. 2009).

Uranium is an alpha-emitting, radioactive element that appears naturally in almost all rocks and soils. Uranium has its own carcinogenic property, and it also has the negative toxicological impact on kidneys. The main target organ of this radionuclide is the kidney. Most of the kidney damage reports have been found after inhaling or insertion of uranium compounds into the living body. On the other hand, kidney problems have not been consistently seen in militaries having radioactive metal fragments in their bodies for many years. Insertion of dissolved uranium compounds will mostly participate in kidney damage than following exposure to insoluble uranium. Persons who inhaled uranium hexafluoride have suffered with respiratory irritation as well as accumulation of fluid in the lungs. Adverse effects of uranium on the functioning of kidneys were previously reported including alterations in renal metabolism of xenobiotics (Souidi et al. 2005), homeostasis effect on vitamin D (Tissandié et al. 2007) and iron homeostasis (Berradi et al. 2008). More iron aggregation, apoptosis in the tubulointerstitial region, and oxidative stress influenced by uranium were also documented (Taulan et al. 2004). Renal tissue disruption was also reported mostly in the cortical part (Canu et al. 2011). Uranium increases blocking of osteoblastic activity, yielding decreased bone volume, and healing interference. In vitro studies revealed that uranium influences instability in genomes and neoplastic conversion in osteoblasts (Canu et al. 2011). This radioisotope alters oxidative metabolism and decreases bone structures (Tasat et al. 2007). According to Prat et al. (2010), uranium modifies the expression of the gene for a biomarker of bone resorption, osteopontin.

Ra stored in bones can cause different types of bone abnormalities and even cancers. Many contemporary researches investigated that radium has many detrimental effects on bone tissues (Canu et al. 2011). Both radioisotopes ^{226}Ra and ^{228}Ra rapidly enhance modifications in bone formation and hematopoiesis. Almost in all species bone sarcomas and leukemias were also reported after ^{224}Ra ingestion.

8 Conventional Methods for Removal of Radionuclide

Various conventional remedial technologies are available for removal of radioactive wastes (Fig. 4). Chemical-reducing agents can be introduced into radioactive contaminated soil or unconfined aquifers to form a subsurface blocker to trap or completely ruin target contaminants. Then water having associated by-products and other remaining reagents is pumped back out. The treatment barrier is a region of suitable redox potential. The aim is to efficiently convert dissolved metals and radionuclides to less soluble structures and to advance the demolition of organic compounds like chlorinated hydrocarbons. This advanced technology permits in situ treatment of groundwater pollutants and avoids disposal expenditure as radionuclides are immobilized in place. The techniques available for treating aqueous radioactive contaminants are generally chemical precipitation, evaporation, ion exchange, reverse osmosis, sorption, ultrafiltration, etc. Sedimentation, decantation, filtration, centrifugation, etc. are techniques applied mainly to remove the effluent wastes, insoluble particles, and other miscellaneous debris (Abdel et al. 2011; IAEA 2001).

Chemical precipitation are mainly applied for eliminating radionuclides from aqueous radioactive contaminants at fuel recycling facilities, different research centers, and power plants. Precipitation techniques are highly versatile and inexpensive, and it can cover a broad range of concentrations of liquid effluents. However, in order to improve the precipitation process, pretreatment steps like oxidation of organic contaminants, degradation of different composites, and alteration of pH and valence state should be carried out prior to the precipitation. Radioactive wastes can also be dispatched using coprecipitation or adsorption (Valdovinos et al. 2014; IAEA 2001).

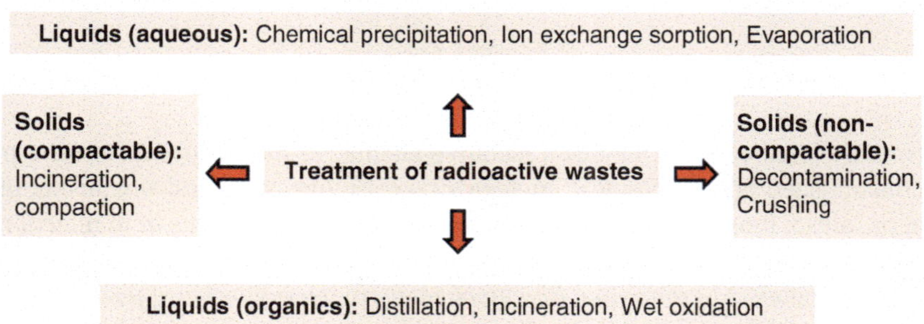

Fig. 4 Classification of different technologies applied to radioactive contaminants (Valdovinos et al. 2014)

Ion exchange process has broad use in order to dislodge soluble radioactive compounds fromliquid solutions generated from nuclear operations, radionuclide formation, and other research activities. It can significantly convert large proportions of radioactive substances into a trace amount of solid substance. Ion exchange technique exchanges positively or negatively charged ions in between solid matrices having ionizable groups and liquid system. Thus, ion exchangers may be reproduced and recovered with its high activity. If exchangers were completely utilized, they are mostly removed and treated as radioactive wastes (Valdovinos et al. 2014).

Varieties of substances are available for the ion exchange technology—(1) ion exchanges such as cellulose, collagen, clays, etc. and (2) synthetic materials like zeolites and oxides containing ionic groups—which has applicability in both continuous and batch systems. When high concentration of radionuclides are present, the water is pretreated before ion exchange process (IAEA 2001).

Evaporation is a potential approach for hazardous materials like radionuclide removal from different liquid wastes. This process is utilized for treatment of different levels of waste effluents. It may be conducted by the application of commercially available evaporation instrument (IAEA 2001).

Again, incineration is also used for removal of radioactive wastes (solid and liquid). This process burns contaminated substances at very high temperatures, producing gases like carbon dioxide, water vapor, sulfur dioxide, nitric oxide, and hydrogen chloride gases as end products of incineration procedure (Valdovinos et al. 2014).

Again, wet oxidation methods require insertion of chemicals which act as oxidizing sources to damage the contaminated compounds containing the radioactive wastes. In comparison to incineration, this method is economical as energy requirement is less (Chang 2001).

Distillation technology can reduce radionuclide volume of pretreating liquid solution and different solvent wastes in conventional equipment. This procedure is easy, well-known, and inexpensive as the valuable solvent is reused. The remaining part could be trapped and then destroyed by incineration (IAEA 2001).

Compaction is a process by which the volume of radioactive waste materials can be reduced and condensed and acts as a processing technique before further remedial measures can be used (Valdovinos et al. 2014; IAEA 2001).

9 Bioremediation Techniques for Removal of Radionuclide

Microbial bioremediation may be a lucrative alternative to excavate different types of contaminated soil. Biological strategies take facility of natural organisms like bacteria, algae, fungi, and plants to demolish or eliminate pollutants or to mineralize metallic pollutants and its associated radioactive elements, thus trapping them in a specific boundary. Microbes like *Rhodanobacter* sp., *Desulfuromusa ferrireducens*, etc. were reported to be capable to eliminate these contaminants (Green et al. 2012; Amachi et al. 2010). The interaction of bacteria induces solubility of modified radionuclides with addition or elimination of electrons, thus raising the motility of the radioactive pollutants and allowing it to be quickly irrigated from our surroundings (Amachi et al. 2010). Thus, microorganism-assisted biotransformation delivers the chances for bioremediation of radionuclides in this Earth, either to trap them in a particular place or to induce their removal. The bioremedial approaches for radioactive materials rely on the active metabolizing efficiencies of different types of microbial cells. Radionuclides may be solubilized by enzymatic oxidation-reduction reaction, alteration in pH and electronic activity, sorption by biomass, degradation of radionuclides using microbes, etc. (Hegazy and Emam 2011; Law et al. 2010; Holker et al. 2002). Activity of radionuclide-degrading microorganisms is highly induced by electron transfer mechanism, several nutrients, and different environmental factors.

Microbial degradation is basically employed to eliminate radioactive contamination from soil surface and also from groundwater where the other organic compounds are operating as chelating agents which influence contaminant movement (Gerber and Fayer 1994; CBCEC 1994). Demolition of the organic compounds assists to decrease rate of transportation which gives the time for radioactive contaminants to decay to harmless levels. Basically native or inoculated microbial species degrade organic pollutants in soil as well as groundwater under optimized either aerobic or anaerobic parameters. The in situ bioremediation of soil generally includes the percolation or insertion of uncontaminated water combined with nutrients and filled with dissolved oxygen. In different cases, microbial species and a source of oxygen are also injected. Many national initiative programs have begun for many years in order to study the bioremediation by microorganisms through noninvasive techniques to remove radioactive contamination. Bioremediation strategies require comparatively inexpensive, low-technology methods and produce very few or no residual waste, which normally have greater level of commercial acceptance (Lloyd and Renshaw 2005). Compared with different other more invading conventional approaches (IAEA 2004), biodegradation may offer a low-cost route for removing radionuclide from polluted soil and water. The conventional remediation approaches are expensive and cause ecological disturbances along with having other limitations. This has resulted in bioremediation becoming the most sought-after approach for radioactive waste removal. Microorganisms including *Deinococcus*, *Geobacter*, *Serratia*, *Kineococcus radiotolerans*, *Hymenobacter metalli* etc. are effective in removal of radioactive waste containing higher level of radionuclides (Roh et al. 2015).

The efficient bioremediation of radioactive contaminants depends on a complex interaction of biological, chemical, and physical procedures. Different types of mechanisms including oxidation, reduction, precipitation, sorption, etc. can influence the toxicity and transfer of radioisotopes in biogeochemical systems (IAEA 2004). Uranium (VI) and technetium (VII) have been reported to be prone to various microbial enzyme actions. The oxidized form of U and Tc is soluble in water, while the reduced ones are not. Thus, enzymatic reduction as mentioned in Fig. 5 is a possible approach for detection and removal of these from groundwater.

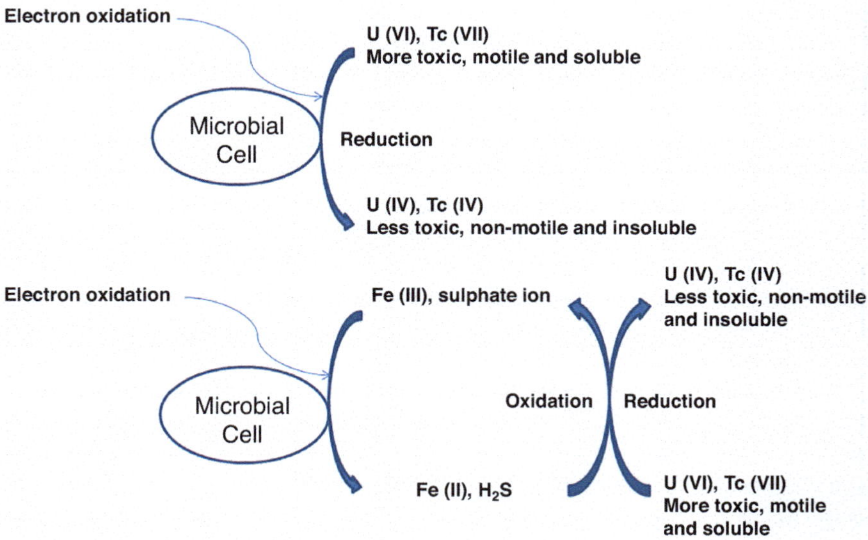

Fig. 5 Microbial approaches of dissolved radioactive (more toxic) metals into insoluble and less hazardous species (Roh et al. 2015)

Moreover, "indirect" methods of reduction can be essential in the immobilization of Tc (VII) as deposits, while the microbially reduced Fe (II) is able to transmit the electrons expeditiously to Tc (VII). Lovely and co-workers first reported that Fe (III)-reducing bacterial species *Geobacter metallireducens* and *Shewanella oneidensis* can be capable of acquiring energy for anaerobic growth by this reduction process. Other microorganisms such as *Clostridium* sp., *Desulfovibrio desulfuricans*, and *Desulfovibrio vulgaris* are also able to do this reduction process but cannot utilize this energy for growth purposes. The first reported microbial species in which enzyme system is accountable for U (VI) reduction was *Desulfovibrio vulgaris*. Payne and co-workers (2001) utilized cytochrome c3 mutant of the close relative *Desulfovibrio desulfuricans* G20 to affirm the function for this electron transfer protein in H_2-dependent reduction process of U (VI), but the electron donation pathway bypassed the use of cytochrome. It was suggested that c-type cytochromes in outer surface of *Geobacter sulfurreducens* might be responsible in U (VI) reduction, but it has now been reported efficient U (VI) reduction can take place at surfaces of electricity conducting pili or microbial nanowires. Orellana described that wild-type species did not precipitate uranium along pili as discussed in earlier studies but that U(IV) was precipitated at the outer layer of bacterial cells. These outcomes are consistent with earlier findings that have reported that *Geobacter sulfurreducens* does not need pili for reducing the extracellular electron acceptors but surface c-type cytochromes for this purpose. *Shewanella oneidensis* MR-1 has 42 putative c-type cytochromes which are necessary for reduction of metal (Roh et al. 2015).

10 Conclusions

Heavy metals and radionuclides are the emerging contaminants of today's world having highly toxic effect on living organisms including human beings. Though natural sources contribute to this contamination, exposure to heavy metals and radionuclides has increased considerably due to anthropogenic activities. We urgently need to curb these activities in order to save mankind from these harmful effects and also work toward removal of the existing contaminants as heavy metal and radionuclides are generally cannot be removed from the environment easily. They can be converted to a lesser toxic form, or contained in remedial organisms like plants and microbes or utilized by microorganisms in their metabolic activities. Modern biotechnological tools of genetic engineering are used nowadays to enhance the remedial properties of the microorganisms. Studies on various microbes and their pathways on one hand will lead to better research on better removal strategies, and also using bioinformatics tools will assist in storing of this information and using them accurately. The use of bioremedial measures over the previously used conventional ones will surely be the path forward in achieving a green, sustainable route for removal/containment of heavy metal and radionuclides from the environment and saving the living organisms from their exposure.

References

Abdel RO, Rahman HA, Hung YT (2011) Liquid radioactive wastes treatment: a review. Water 3:551–565

Aleksakhin RM (2009) Radioactive contamination as a type of soil degradation. Eurasian Soil Sci 42:1386–1396

Aleksakhin RM, Buldakov LA, Gubanov VA (2001) In: Il'in LA, Gubanov VA (eds) Large radiation accidents: consequences and control measures. IzdAT, Moscow, p 752

Al-Zoughool M, Krewski D (2009) Health effects of radon: a review of the literature. Int J Radiat Biol 85:57–69

Amachi S, Minami K, Miyasaka I, Fukunaga S (2010) Ability of anaerobic microorganisms to associate with iodine: 125I tracer experiments using laboratory strains and enriched microbial communities from subsurface formation water. Chemosphere 79:349–354

ATSDR (Agency for Toxic Substances and Disease Registry) (1999) Toxicological profile for ionizing radiation. ATSDR, Atlanta, p 438

Berradi H, Bertho JM, Dudoignon N, Mazur A, Grandcolas L, Baudelin C (2008) Renal anemia induced by chronic ingestion of depleted uranium in rats. Toxicol Sci 103(2):397–408

Bogutskaya NG, Zuykov MA, Naseka AM, Anderson EB (2011) Normal axial skeleton structure in common roach *Rutilus rutilus* (Actinopterygii: Cyprinidae) and malformations due to radiation contamination in the area of the Mayak (Chelyabinsk Province, Russia) nuclear plant. J Fish Biol 79:991–1016

Bonavigo L, Zucchetti M, Mankolli H (2009) Water radioactive pollution and related environmental aspects. J Int Environ Appl Sci 4(3):357–363

Brennwald MS, Dorp FV (2009) Radiological risk assessment and biosphere modelling for radioactive waste disposal in Switzerland. J Environ Radioact 100(12):1058–1061

California Base Closure Environmental Committee (1994) Treatment technologies applications matrix for base closure activities, Rev 1. Technology Matching Process Action Team, November 1994

Canu IG, Laurent O, Pires N, Laurier D, Dublineau I (2011) Health effects of naturally radioactive water ingestion: the need for enhanced studies. Environ Health Perspect 119(12):1676–1680
Carvalho FP, Oliveira JM (2010) Uranium isotopes in the Balkan's environment and foods following the use of depleted uranium in the war. Environ Int 36(4):352–360
Chang HO (2001) Hazardous and radioactive waste treatment technologies handbook. CRP Press, Boca Raton
Das A, Osborne JW (2018) Bioremediation of heavy metals. In: Environmental chemistry for a sustainable world 11. Springer, New York, p 9
Demey H, Vincent T, Guibal E (2018) A novel algal-based sorbent for heavy metal removal. Chem Eng J 332:582–595
Fang L, Huang Q, Wei X, Liang W, Rong X, Chen W, Cai P (2010) Microcalorimetric and potentiometric titration studies on the adsorption of copper by extracellular polymeric substances (EPS), minerals and their composites. Bioresour Technol 101:5774–5779
Gerber MA, Fayer MJ (1994) In situ remediation integrated program evaluation and assessment of containment technology. US DOE, Report
Ginneken LV, Meers E, Guisson R (2007) Phytoremediation for heavy metal-contaminated soils combined with bioenergy production. J Environ Eng Landsc Manag 15(4):227–236
Goodby J (2005) The limited test ban negotiations, 1954-63: how a negotiator viewed the proceedings. Int Negot 10:381–404
Green SJ, Prakash O, Jasrotia P, Overholt WA, Cardenas E, Hubbard D (2012) Denitrifying bacteria from the genus *Rhodanobacter* dominate bacterial communities in the highly contaminated subsurface of a nuclear legacy waste site. Appl Environ Microbiol 78:1039–1047
Hazra G (2018) Radioactive pollution: an overview. Holist Approach Environ 8(2):48–65
Hegazy AK, Emam MH (2011) Accumulation and soil-to-plant transfer of radionuclides in the Nile Delta coastal black sand habitats. Int J Phytoremediation 13:140–155
Hejabi AT, Basavarajappa HT, Karbassi AR, Monavari SM (2011) Heavy metal pollution in water and sediments in the Kabini River, Karnataka, India. Environ Monit Assess 182:1–13
Holker U, Schmiers H, Grosse S, Winkelhofer M, Polsakiewicz M, Ludwig S (2002) Solubilization of low-rank coal by *Trichoderma atroviride*: evidence for the involvement of hydrolytic and oxidative enzymes by using 14C-labelled lignite. J Ind Microbiol Biotechnol 28:207–212
Hu QH, Weng JQ, Wang JS (2010) Sources of anthropogenic radionuclides in the environment: a review. J Environ Radioact 101:426–437
Huertas MJ, Lopez-Maury L, Giner-Lamia J, Sanchez-Riego AM, Florencio FJ (2014) Metals in cyanobacteria: analysis of the copper, nickel, cobalt and arsenic homeostasis mechanisms. Life 4:865–886
IAEA (International Atomic Energy Agency) (2001) Handling and processing of radioactive waste from nuclear applications. Technical reports series no. 402. IAEA, Vienna
IAEA (International Atomic Energy Agency) (2004) IAEA safety glossary-terminology used in nuclear safety and radiation protection. IAEA, Vienna
IAEA (International Atomic Energy Agency) (2007) IAEA safety glossary-terminology used in nuclear safety and radiation protection. IAEA, Vienna, p 227
Isaksson M, Raaf CL (2016) Environmental radioactivity and emergency preparedness. CRC Press Taylor and Francis, Boca Raton
Izrael YA, Vakulovskii SM, Vetrov VA (1990) Chernobyl: radioactive contamination of natural media. Gidrometeoizdat, Leningrad, p 296
Izrael YA, Kvasnikova EV, Nazarov IM, Fridman SD (1994) Global and regional radioactive Cs-137 pollution in the European part of the former USSR. Meteorol Gidrol 5:5–9
Khan MF, Wesley SG (2011) Assessment of health safety from ingestion of natural radionuclides in seafoods from a tropical coast, India. Mar Pollut Bull 62(2):399–404
Khan R, Israili SH, Ahmad H, Mohan A (2005) Heavy metal pollution assessment in surface water bodies and its suitability for irrigation around the Neyveli lignite mines and associated industrial complex, Tamil Nadu, India. Mine Water Environ 24:155–161

Kinoshitaa N, Suekia K, Sasaa K, Kitagawaa J, Ikarashia S, Nishimuraa T, Wonga Y, Satoua Y, Handaa K, Takahashia T, Satob M, Yamagatab T (2011) Assessment of individual radionuclide distributions from the Fukushima nuclear accident covering central-East Japan. PNAS 108 (49):19526–19529

Lacerda LD (1997) Global mercury emissions from gold and silver mining. Water Air Soil Pollution 97:209–221

Law GT, Geissler A, Lloyd JR, Livens FR, Boothman C, Begg JD (2010) Geomicrobiological redox cycling of the transuranic element neptunium. Environ Sci Technol 44:8924–8929

Li C, Xu Y, Wei J, Dong X, Wang D, Liu B (2013) Effect of NaCl on the heavy metal tolerance and bioaccumulation of Zygosaccharomyces rouxii and Saccharomyces cerevisiae. Chem Eng J 171:1060–1068

Lloyd JR, Renshaw JC (2005) Bioremediation of radioactive waste: radionuclide-microbe interactions in laboratory and field-scale studies. Curr Opin Biotechnol 16:254

Mane P, Bhosle A, Jangam C, Vishwakarma C (2011) Bioadsorption of selenium by pretreated algal biomass. Adv Appl Sci Res 2:202–207

Mao J, Won SW, Yun Y-S (2013) Development of poly (acrylic acid)-modified bacterial biomass as a high-performance biosorbent for removal of Cd (II) from aqueous solution. Ind Eng Chem Res 52:6446–6452

Mohner M, Lindtner M, Otten H, Gille HG (2006) Leukemia and exposure to ionizing radiation among German uranium miners. Am J Ind Med 49:238–248

Nagajyoti PC, Lee KD, Sreekanth TVM (2010) Heavy metals, occurrence and toxicity for plants: a review. Environ Chem Lett 8:199–216

Pacyna JM (1986) Atmospheric trace elements from natural and anthropogenic sources. Toxic Met Atmos 17:33–52

Paul D (2017) Research on heavy metal pollution of river Ganga: a review. Ann Agrar Sci 15:278–286

Payne TE, Edis R, Fenton BR, Waite TD (2001) Comparison of laboratory uranium sorption data with 'in situ distribution coefficients' at the Koongarra uranium deposit, Northern Australia. J Environ Radioact 57(1):35–55

Prat O, Berenguer F, Steinmetz G, Ruat S, Sage N, Quemeneur E (2010) Alterations in gene expression in cultured human cells after acute exposure to uranium salt: involvement of a mineralization regulator. Toxicol In Vitro 24(1):160–168

Pravalie R (2014) Nuclear weapons tests and environmental consequences: a global perspective. Ambio 43:729–744

Prohl G, Olyslaegers G, Kanyar B (2005) Development and comparison of five site-specific biosphere models for safety assessment of radioactive waste disposal. J Radiol Prot 25 (4):343–373

Quiton KG, Doma B Jr, Futalan CM, Wan M-W (2018) Removal of chromium (VI) and zinc (II) from aqueous solution using kaolin-supported bacterial biofilms of Gram-negative E. coli and Gram-positive Staphylococcus epidermidis. Sustain Environ Res 28:206–213

Rajaganapathy V, Xavier F, Sreekumar D, Mandal PK (2011) Heavy metal contamination in soil, water and fodder and their presence in livestock and products: a review. J Environ Sci Technol 4 (3):234–249

Ramrakhiani L, Majumdar R, Khowala S (2011) Removal of hexavalent chromium by heat inactivated fungal biomass of Termitomyces clypeatus: surface characterization and mechanism of biosorption. Chem Eng J 171:1060–1068

Roh C, Kang C, Jonathan RL (2015) Microbial bioremediation processes for radioactive waste. Korean J Chem Eng 32(9):1720–1726

Rungini L, Costa G, Congestri R, Antonaroli S, di Toppi LS, Bruno L (2018) Phosphorus and metal removal combined with lipid production by the green microalga Desmodesmus sp.: an integrated approach. Plant Physiol Biochem 125:45–51

Sengupta S, Das P, Mukhopadhyay A, Datta S (2017) Microbial biosorption and improved/ genetically modified biosorbents for toxic metal removal and thermodynamics. In: Das S,

Dash HR (eds) Handbook of metal–microbe interactions and bioremediation. CRC Press, pp 267–280. Chapter 16. https://doi.org/10.1201/9781315153353
Smičiklas I, Šljivić-Ivanović M (2016) Radioactive contamination of the soil: assessments of pollutants mobility with implication to remediation strategies. In: Soil contamination-current consequences and further solutions, vol 13. InTechOpen, London
Souidi M, Gueguen Y, Linard C, Dudoignon N, Grison S, Baudelin C (2005) In vivo effects of chronic contamination with depleted uranium on CYP3A and associated nuclear receptors PXR and CAR in the rat. Toxicology 214(1–2):113–122
Su C, Jiang LQ, Zhang WJ (2014) A review on heavy metal contamination in the soil worldwide: situation, impact and remediation techniques. Environ Skeptics Crit 3(2):24–38
Szarlowicz K, Stobinski M, Hamerlik L, Bitusik P (2019) Origin and behavior of radionuclides in sediment core: a case study of the sediments collected from man-made reservoirs located in the past mining region in Central Slovakia. Environ Sci Pollut Res 26:7115–7122
Tangahu BV, Abdullah SRS, Basri H, Idris M, Anuar N, Mukhlisin M (2011) A review on heavy metals (As, Pb, and Hg) uptake by plants through phytoremediation. Int J Chem Eng 2011:939161
Tasat DR, Orona NS, Mandalunis PM, Cabrini RL, Ubios AM (2007) Ultrastructural and metabolic changes in osteoblasts exposed to uranyl nitrate. Arch Toxicol 81:319–326
Taulan M, Paquet F, Maubert C, Delissen O, Demaille J, Romey MC (2004) Renal toxicogenomic response to chronic uranyl nitrate insult in mice. Environ Health Perspect 112:1628–1635
Tchounwou PB, Yedjou CG, Patlolla AK, Sutton DJ (2012) Heavy metal toxicity and the environment. In: Molecular, clinical and environmental toxicology. Experientia Supplementum 101. Springer, Basel
Thompson LA, Darwish WS (2019) Environmental chemical contaminants in food: review of a global problem. Journal of Toxicology 2019:2345283
Tissandié E, Gueguen Y, Lobaccaro JM, Grandcolas L, Voisin P, Aigueperse J (2007) In vivo effects of chronic contamination with depleted uranium on vitamin D3 metabolism in rat. Biochim Biophys Acta 1770(2):266–272
UNSCEAR (2000) Sources and effects of ionizing radiation-report of the United Nations Scientific Committee on the effects of atomic radiation to the General Assembly. United Nations, New York, p 654
Valdovinos V, Monroy-Guzman F, Bustos E (2014) Treatment methods for radioactive wastes and its electrochemical applications. In: Environmental risk assessment of soil contamination. InTechOpen, London, p 14
Walker JS, Don GW (2013) Mathematics and music. CRC Press, Boca Raton
Wang J, Li Q, Li M-M, Chen T-H, Zhou Y-F, Yue Z-B (2014) Competitive adsorption of heavy metal by extracellular polymeric substances (EPS) extracted from sulfate reducing bacteria. Bioresour Technol 163:374–376
Wijnhoven S, Leuven RSEW, Velde GVD, Jungheim G, Koelemij EI, Vries FTD, Eijsackers HJP, Smits AJM (2007) Heavy-metal concentrations in small mammals from a diffusely polluted floodplain: importance of species- and location-specific characteristics. Arch Environ Contam Toxicol 52:603–613
Wuana RA, Okieimen FE (2011) Heavy metals in contaminated soils: a review of sources, chemistry, risks and best available strategies for remediation. International Scholarly Research Network Ecology 2011:402647
Yan G, Viraraghavan T (2000) Effect of pretreatment on the bioadsorption of heavy metals on Mucor rouxii. Water SA-Pretoria 26:119–124
Yin K, Lv M, Wang Q, Wu Y, Liao C, Zhang W, Chen L (2016) Simultaneous bioremediation and biodetection of mercury ion through surface display of carboxylesterase E2 from *pseudomonas aeruginosa* PA1. Water Res 103:383–390
Yin K, Wang Q, Lv M, Chen L (2019) Microorganism remediation strategies towards heavy metals. Chem Eng J 360:1553–1563

Yu H, Ni SJ, He ZW, Zhang CJ, Nan X, Kong B, Weng ZY (2014) Analysis of the spatial relationship between heavy metals in soil and human activities based on landscape geochemical interpretation. J Geochem Explor 146:136–148

Yue Z-B, Li Q, Li C-C, Chen T-H, Wang J (2015) Component analysis and heavy metal adsorption ability of extracellular polymeric substances (EPS) from sulfate reducing bacteria. Bioresour Technol 101:5774–5779

Removal of Heavy Metals Using Bentonite Clay and Inorganic Coagulants

Amit K. Thakur, Rahul Kumar, Pranava Chaudhari, and Ravi Shankar

Abstract

The cost-effective removal of heavy metals from aqueous solution is a major challenge for the scientific community. Heavy metal reduction from aqueous solution using bentonite clay, inorganic coagulants, and their mixture is a cost-effective technique. They are easily available and cheaper compared to the activated carbon, which is widely used but an expensive adsorbent. Bentonite is a clay mineral composed of very fine particles with high opening volume and high specific active site. Thus, it has a significantly high absorption capacity. The metal ion properties, adsorbent dosage, initial concentration, and operating conditions (temperature, pH, contacting time, etc.) are the major parameters for the effective application of raw and modified bentonite. Coagulation is also a traditional technique to minimize metal ions from the bulk solution. Excessive use of inorganic coagulants may result in corrosion and erosion effect on the equipment and excessive generation of sludge volume during coagulation process. Effectiveness of bentonite clay can be enhanced by mixing inorganic coagulants such as sodium carbonate, sulfates and chlorides of aluminum, iron, etc. These coagulants are mixed with the bentonite clay in an optimum ratio to reduce the turbidity and COD of solution. This chapter includes a detailed review of recent work on heavy metal reduction in aqueous system using bentonite clay, inorganic coagulants, and their mixtures.

A. K. Thakur · R. Kumar · P. Chaudhari
Department of Chemical Engineering, University of Petroleum and Energy Studies, Dehradun, Uttarakhand, India

R. Shankar (✉)
Department of Chemical Engineering, Madan Mohan Malaviya University of Technology, Gorakhpur, Uttar Pradesh, India
e-mail: rsch@mmmut.ac.in

M. P. Shah (ed.), *Removal of Emerging Contaminants Through Microbial Processes*,
https://doi.org/10.1007/978-981-15-5901-3_3

Keywords

Heavy metals · Bentonite clay · Adsorption · Low-cost adsorbents · Inorganic coagulants

1 Introduction

Elements with density above 5 g/cm^3 are categorized as the heavy metals. These pollutions are major components of wastewater system. Wastewater emanating from mining, textiles, tannery, fertilizers, pesticides, metal refining, and smelting processes is a major source of these types of unwanted metal contaminants. Generally, metal pollutants in the aqueous phase are found in the ionic form. Metal pollutants such as lead, cadmium, cobalt, zinc, nickel, arsenic, chromium, mercury, and iron are detrimental to living organisms including human and aquatic animals. Many of these metallic ions are highly toxic, carcinogenic, and non-biodegradable that can cause liver damage, brain damage, and other health difficulties when deposited in internal organs beyond the standard limits. Therefore, the discharge of heavy metal-enriched effluent in the water reservoir is a major concern for the scientific community (Xu et al. 2009; Zou et al. 2013; Vodyanitskii 2014).

There are several techniques, applied for the reduction of heavy metal ion contaminants from contaminant effluents. These are physicochemical, electrochemical, membrane filtration, and nanotechnology treatment techniques, which are classified in some common techniques such as chemical precipitation (Ku and Jung 2001), adsorption (Kongsuwan et al. 2009; Guo et al. 2010), reverse osmosis (Mohsen-Nia et al. 2007), ultrafiltration, nanofiltration (Al-Rashdi et al. 2013), chemical coagulation (Shan et al. 2017), flocculation (Huang et al. 2016), ion exchange (Zewail and Yousef 2015), and electrodialysis (Ebbers et al. 2015).

The chemical precipitation processes need chemical agents to reduce the metal ions into insoluble solid precipitates. The solid precipitates can be separated by filtration or sedimentation. Limestone is a most widely used inorganic chemical coagulant in chemical precipitation processes (Malik et al. 2017). The chemical agents used in this process may result in further source of water pollution. Moreover, the high cost of sludge disposal and slow process of metal precipitation often make it an impractical technique in many water cleaning techniques (Azimi et al. 2017). The adsorption process uses highly porous and large specific surface adsorbents for the reduction of heavy metals. This process can run in a reversible cycle of adsorption and desorption. The used adsorbents can be renewed and reused by desorption. Efficient and cost-effective adsorbents can be derived from the natural sources (Bobade et al. 2015), agriculture/plant wastes (Wan Ngah and Hanafiah 2008), and industrial by-products (Azimi et al. 2017). Reverse osmosis (Aljendeel 2011), ultrafiltration (Vinodhini and Sudha 2017), and nanofiltration (Al-Rashdi et al. 2013; Mohammad 2013) are pressure-driven processes through a semipermeable membrane. These processes remove heavy metals, oils, and inorganic materials from the aqueous solution. In reverse osmosis, hydraulic pressure is applied to the contaminated solution, water, and other lighter molecules across micro-openings of the semipermeable membrane, whereas the impurities/metallic ions are retained in the membrane. Flexible design, installation, and efficient removal are the main advantages of the reverse osmosis process. High maintenance/operating cost due to high pumping power requirement and membrane restoration is the disadvantage of the membrane process. The coagulation technique uses a chemical or coagulant to form aggregates of colloids called flocs. Polyaluminum chloride and ferric chloride are common types of coagulant, used for the reduction for impurities/heavy metals. The flocs attract the suspended materials to form lump of aggregates. In the flocculation technique, the solution is gently mixed so as to form an aggregate of flocs. The ion exchange method is used to exchange the metallic ions present in bulk solution, between liquid and solid. Alumina, zeolite, and silicates are used in the ion exchange process. This technique has high efficiency, low cost, & high selectivity and its performance is strong function of solution pH. In electrodialysis technique, dissolved ions are separated across the membrane under an electric field. This process has an ion-selective exchange membrane which does not allow simultaneous transport of opposite charge ions (Azimi et al. 2017). Among all the techniques discussed above, the adsorption process using low-cost adsorbents is suitable for wide applications. Adsorption techniques are simple, efficient, cost-effective, easily available, and easy to operate. This technique is a popular choice because the adsorbents can be regenerated and reused.

Activated carbon, carbon nanotubes, cellulosic materials, biochar, agriculture and plant wastes, industrial by-products, bioadsorbents, and natural clays are the commonly used adsorbents in water purification. Among these adsorbents, activated carbon is extensively used due to its high pore volume and high specific surface area. However, depletion of the coal-based commercial activated carbon is making the activated carbon adsorption process costlier. In past several years, agricultural waste and biomasses such as bagasse, coconut shell, saw dust, tree bark, banana pith, etc. are used for the activated carbon-based adsorbent preparation. The effectiveness of cellulosic material is enhanced by modification of its surface. It is also economically expensive at the commercial scale. Therefore, low-cost adsorbents are being explored for the reduction of heavy metal ions from bulk solution. These materials are locally available in abundance as natural clays (bentonite, kaolinite, smectite, ball clay, montmorillonite, etc.) (Uddin 2017), agricultural wastes (grape stalks, tea and coffee wastes, plant leaves, fruit peels, nut shells, rice husk, seaweed, etc.) (Pyrzynska 2019), industrial by-products and wastes (fly ash, blast furnace slag, lignin, paper waste, battery industry waste, etc.) (Tripathi and Rawat Ranjan 2015), marine materials (red and green algae, algal biomass, etc.) (Pyrzynska 2019), and biopolymers (chitosan, cellulosic biopolymers, etc.) (Kanmani et al. 2017). Kyzas and Kostoglou (2014) named these materials, *green adsorbents*.

Adsorbent preparation from organic waste is costlier in comparison with inorganic adsorbents. Bentonite is a natural clay mineral belonging to the smectite group. It has large active surface area, unique hydration characteristics, and good cation exchange capability. It is cheaper, abundantly available, and efficient for many impurities/metal ions. These characteristics of bentonite clay make it a strong option for the heavy metal ion reduction from bulk solution. Heavy metal reduction by coagulation technique using inorganic coagulant has also been a cost-effective choice in the water purification process. Effectiveness of coagulation can be enhanced by mixing bentonite clay with inorganic coagulants in an optimum volume ratio.

This chapter presented an extended literature study of heavy metal reduction from polluted effluent using bentonite and inorganic coagulants. This study is explained into two sections where the role of natural/modified bentonite and role of coagulants for the reduction of heavy metals from the aqueous solution are discussed. Here the effectiveness of natural/activated bentonite as adsorbent and various coagulants are presented in detail, and the various factors affecting the removal of heavy metals are also discussed. The combined effect of bentonite and inorganic coagulant for heavy metal reduction is also discussed.

2 Bentonite as Adsorbent

Bentonite is a natural clay which mainly consists of Al_2O_3, SiO_2, MgO, CaO, K_2O, and Fe_2O_3 (Wan Ngah et al. 2011). It is a three-layer type crystalline clay mineral in which an octahedral sheet of alumina is arranged between two tetrahedral sheets of silica giving a 2:1 layered structure (Uddin 2017). The length of the layer is in the order of several micrometers, and the thickness is in the order of few nanometers. The layers are arranged in stack with the gap filled with water-solvated metal cations and organic complexes. Its small crystal structure, chemical composition, and excellent ion exchange capability lead to the higher surface area, high porosity, and, consequently, high adsorption capacity toward various organic and inorganic water contaminants. There are two types of bentonite clay, namely, sodium bentonite and calcium bentonite. Both are abundantly available and widely used. Although the swelling tendency of sodium bentonite is relatively high compared to the calcium bentonite (Gleason et al. 1997), both have reasonably good adsorption capacity toward heavy metal ions.

There are various methods to activate the natural bentonite. Which includes activation by acids, thermal activation and treatment with organic surfactants, composites with clay and rubber, lyophilization, adsorption with organic and inorganic cations, etc. The modification of natural bentonite using polymer is done by insertion of polymer into the interlayer space of the bentonite. The modification using this technique improves the physical and chemical properties of the bentonite without changing its structure.

The surface characteristics of natural and modified bentonite can be studied by various techniques such as Fourier transform-infrared spectroscopy (FTIR), Brunauer-Emmett-Teller (BET) technique, X-ray diffraction (XRD), and scanning electron microscopy (SEM). FTIR spectroscopy technique is used to determine the change in structure and bonding of molecules by modification and activation process. BET technique determines the pore size and specific surface area. XRD is used to investigate the mineral composition, lattice structure, and crystallinity of the samples. SEM is used to evaluate the morphological and elemental characteristics of the sample.

The use of modified bentonite for the reduction of metal ions has been studied by several workers in the recent past. Sen and Khoo (2013) studied the removal of Zn^{2+} using bentonite as adsorbent. They reported that the adsorption of Zn^{2+} on bentonite is a two-step process. The first step is the rapid adsorption of metal ion at the surface of the adsorbent, and the second step is a rate-limiting step in which the metal ion diffuses into the interior of the adsorbent. The adsorption isotherm best fits the Langmuir model with a maximum adsorption capacity of 62.5 mg/g (Kanti Sen and Khoo 2013). Araujo et al. (2013) investigated the adsorption capacity of Zn^{2+} using bentonite clay. The bentonite clay was thermally treated for enhancement of adsorption capacity. The Langmuir model was found the best fit for isotherms with a maximum adsorption capacity of 4.95 mg/g (de Araujo et al. 2013). Guerra et al. (2013) used Brazilian bentonite and modified Brazilian bentonite to investigate the removal of Pb^{2+} from aqueous solution. The Brazilian bentonite was modified using 3-aminopropyltriethoxysilane and 3, 2-aminoethylamino propyltrimethoxysilane. The adsorption isotherm was reported to fit the Langmuir model with maximum adsorption capacity (mg/g) of 20.6843, 27.6524, and 29.5413 for Brazilian bentonite, Brazilian bentonite modified using 3-aminopropyltriethoxysilane, and Brazilian bentonite modified using 3, 2-aminoethylamino propyltrimethoxysilane, respectively (Guerra et al. 2013).

Aljlil and Alsewailem (2014) used natural bentonite to investigate the reduction of copper and nickel from wastewater. They reported the maximum adsorption capacity of copper as 13.22 mg/g while that of nickel as 9.29 mg/g. The heavy metal adsorption was found to increase with increase in temperature (Aljlil and Alsewailem 2014). El-Maghrabi and Mikhail (2016) also used natural bentonite for removal of Fe^{2+}, Zn^{2+}, and Ni^{2+} ions from aqueous system. They reported that the heavy metal adsorption from the single solution using bentonite clay follows the order $Fe^{2+} > Zn^{2+} > Ni^{2+}$ (El-maghrabi 2016). Choudhury et al. (2015) synthesize a nanocomposite of bentonite and hydroxyapatite cross-linked by glutaraldehyde. This nanocomposite was used to study the removal of Pb^{2+} ion from wastewater. Langmuir isotherm best described the adsorption of metal ion on the nanocomposite with maximum adsorption capacity (mg/g) of 346 at 303 K (Choudhury et al. 2015).

Olekhnovich et al. (2015) studied the reduction of heavy metal ions such as Cu^{2+}, Co^{2+}, and Ni^{2+} using a composite of poly (acrylic acid)-co-acrylamide/bentonite. Their observations reveal the high affinity of the bentonite-polymer composite toward the heavy metal ions (Olekhnovich et al. 2015). Ahmad and Hasan (2016) synthesized a nanocomposite of L-cysteine-modified bentonite-cellulose for adsorption of heavy metal ions (Cu^{2+}, Pb^{2+}, and Cd^{2+}) (Ahmad and Hasan 2016). They reported the values of maximum adsorption capacity (mg/g) of Cu^{2+}, Pb^{2+}, and Cd^{2+} as 32.36, 18.52, and 16.12, respectively. Their experimental data fitted better to Langmuir adsorption isotherm and a pseudo-second-order kinetic model compared to other isotherms and kinetic models (Ahmad and Hasan 2016). Green tea extracts based on nanoscale zero-valent ions supported on bentonite can be used for the removal of Cr^{6+} from aqueous system. It is found that, there is 34-fold increase in the adsorption capacity using zero-valent ions supported on bentonite compared to that of the natural bentonite (Soliemanzadeh and Fekri 2017). Table 1 summarizes the maximum adsorption capacity of some adsorption studies using natural bentonite, modified bentonite, and bentonite composites.

2.1 Parameters Affecting Adsorption of Heavy Metal Using Adsorption Process

The various factors affecting the adsorption of heavy metal are the contacting time, adsorbent dosage, pH, adsorbate concentration, and the temperature.

Effect of contacting time: Contact time of adsorbent and adsorbate is an important parameter affecting the adsorption. The adsorption first increases with an increase in the contact time, and maximum adsorption is achieved after a certain time, after that equilibrium is reached and there is no change in adsorption (Vhahangwele and Mugera 2015). This is because during initial stage, large amounts of adsorption sites are available, and as time progresses, the binding sites are occupied, and adsorption equilibrium is attained (Gandhi et al. 2014). In almost all studies Uddin (2017) using bentonite as adsorbent, it is observed that the rate of adsorption is initially fast and equilibrium is established in few hours. This observation suggests the strong affinity of bentonite to the metal ions.

Effect of adsorbent dosage: An increase in adsorbent dosage increases the percentage adsorption. In the case of multi-ion adsorption (Taha et al. 2016) at lower dose, the metal ions compete for available adsorption sites, but at higher adsorbent dose, enough vacant sites are available for adsorption of various metal ions and result in increase in the adsorption percentage (Liu and Zhou 2010; Tomul and Basoglu 2010). However, an increase in adsorbent dosage also results in decrease in the adsorption capacity due to reduction in the fraction of covered site per unit mass of the adsorbent. Therefore, an optimum dosage of adsorbent is required for optimum percentage removal and adsorption capacity.

Effect of pH: The pH is a key factor during adsorption process as it is related to the surface characteristics of adsorbent. Studies reported an optimum range of pH for maximum adsorption efficiency. Taha and coworkers (2016) conducted an adsorption study on Egyptian Na-activated bentonite for the reduction of Pb^{2+}, Cd^{2+}, and Ni^{2+} ions. The pH was varied from 2.0 to 7.0 at 298 K and 100 rpm. The adsorption percentage was found to increase with increase in pH (Taha et al. 2016). In an another study (Inglezakis et al. 2007) using pure bentonite for the removal Pb^{2+}, it was reported that the percentage removal of Pb^{2+} increases with increase in pH from 1 to 3, and at pH value of 4, there is a decrease in the percentage removal. A recent study (Wahab et al. 2019) using a silk-bentonite composite for the removal of Cd^{2+}, Cr^{6+}, Hg^{2+}, and Pb^{2+} from aqueous system varied the solution pH from 3.0 to 9.0. This study reported a decrease in adsorption efficiency at higher pH. An optimum pH

Table 1 Maximum adsorption capacity of natural, modified bentonite, and bentonite composites

Adsorbent	Metal ion removed	Maximum adsorption capacity (mg/g)	Remark	References
Natural bentonite	Cd^{2+}, Cu^{2+}	63.29		Karapinar and Donat (2009)
Sodium-activated bentonite	Pb^{2}, Cd^{2+}, Ni^{2+}, Cu^{2+}, Mn^{2+}, Cr^{3+}, and Zn^{2+}		>92% removal of heavy metal ions	Taha et al. (2016)
Natrified bentonite	Cd^{2+}	72.99		Schütz et al. (2016)
Quartz and bentonite mixture	Cd^{2+}	116.28		Schütz et al. (2016)
P4V4-bentonite	Cr^{6+}	45.25	Natural bentonite is a weak adsorbent for Cr^{6+}	Mansri et al. (2009)
Manganese oxide onto sodium bentonite	Cd^{2+}	108.69		Schütz et al. (2016)
GMZ-bentonite	Cu^{2+}	44.25	Adsorption of Cu^{2+} strongly depends on $pH < 6.5$	Li et al. (2009)
Cross-linked chitosan-bentonite beads Non-cross-linked chitosan-bentonite beads	Cu^{2+} Cu^{2+}	9.43 12.41	94.27% removal of Cu^{2+}	Aljlil and Alsewailem (2014) Ba et al. (2014)
Silk-bentonite composite	Pb^{2+} Cd^{2+} Hg^{2+} Cr^{6+}	11.1 11.35 10.5 10.2	>90% removal of metal ions	Wahab et al. (2019)
Humic acid-immobilized amine modified polyacrylamide bentonite	Cu^{2+}	112.5	>99% removal of metal ion at the pH range 5.0–6.0	Anirudhan and Suchithra (2010)
8-Hydroxy quinine bentonite	Cu^{2+} Cd^{2+} Pb^{2+}	56.55 67.35 142.94		Gök et al. (2008) Bentouami and Ouali (2006) Özcan et al. (2009)
Polyacrylamide composite bentonite	Cu^{2+}	33		Zhao et al. (2010)
4-Methybenzo-15-crown-5 bentonite	Cu^{2+} Pb^{2+}	34.5 47.5		Liu et al. (2006)

(continued)

Table 1 (continued)

Adsorbent	Metal ion removed	Maximum adsorption capacity (mg/g)	Remark	References
Manganese oxide-coated bentonite	Cu^{2+}	58.88		Eren et al. (2010)
HCl-activated bentonite	Zn^{2+} Cu^{2+} Pb^{2+} Ni^{2+}	8.87 7.01 9.6 0.99	0.05 M HCl is used to activate bentonite	Mohammed-Azizi et al. (2013)
Calcined biomass (pine cone)-modified bentonite	Cd^{2+} Cu^{2+} Pb^{2+}	16.45 20 23.2		Olu-Owolabi et al. (2016)
Calcined biomass (Carica papaya seeds)-modified bentonite	Cd^{2+} Cu^{2+} Pb^{2+}	13.61 32.79 17.76		Olu-Owolabi (et al. 2016)

of 5 is reported for the maximum adsorption efficiency. In several studies using bentonite as adsorbent, an optimum pH for many heavy metals is reported in the range of 5.0–6.0 after precipitation. For reduction of chromium ion, an optimum pH of 2.0 is reported in literature (Uddin 2017).

Effect of adsorbate concentration: Increase in the concentration of metal ions in adsorbate increases the adsorption capacity and decreases the adsorption percentage (El-maghrabi 2016). An increase in the metal ion concentration in adsorbate leads to a decrease in pH of the final solution, and subsequently the adsorption percentage decreases (Vhahangwele and Mugera 2015). The decrease in adsorption percentage with an increase in the adsorbate concentration is profound in the case of multicomponent adsorption and attributed to the competition between the metal ions for the available active sites (Taha et al. 2016).

Effect of temperature: Increase in temperature decreases the adsorption efficiency because high temperature favors desorption. However, the maximum adsorption capacity (mg/g of adsorbent) and the percentage removal of heavy metal increase with increase in temperature (Vieira et al. 2010). This is because a higher temperature leads to the activation of metal ions at the coordinating sites of the adsorbent (Inglezakis et al. 2007). Moreover, a higher temperature also accelerates the movement of cations to the active site leading to a higher percentage removal.

2.2 Equilibrium Adsorption Isotherms and Kinetics

The adsorption isotherm study is used to understand the interaction between the adsorbates and the adsorbent. Many equilibrium adsorption models are used to correlate the adsorption capacity and residual adsorbate concentration in liquid phase. Langmuir (1918), Freundlich (1906), Temkin and Pyzhev (1940), and Dubinin-Radushkevich (D-R) (Nguyen and Do 2001) models are widely used to obtain the relationship between the metal ions in solid and liquid phase at equilibrium.

Langmuir adsorption model (Langmuir 1918) assumes the formation of monolayer of metal ions on the homogeneous surface of adsorbent containing large number of active sites. The Langmuir equation in linear form can be given as:

$$\frac{C_e}{q_e} = \frac{C_e}{q_m} + \frac{1}{k_L q_m} \tag{1}$$

Here C_e is the concentration of the metal ion (mg/L) at equilibrium, q_e is the adsorption capacity at equilibrium (mg of metal ions adsorbed per gram of adsorbent), and q_m and k_L are the maximum adsorption capacity (mg/g) and the Langmuir constant, respectively. Further analysis of Langmuir model provides the dimensionless adsorption intensity (R_L), which determines the nature of the adsorption process (Ho et al. 2002).

Freundlich isotherm model (Freundlich 1906) is based on the assumption of formation of multilayer on the heterogeneous adsorbent surface. The Freundlich model in linear form is given as:

$$\log q_e = \log k_F + \frac{1}{n} \log C_e \tag{2}$$

Here k_F is the Freundlich adsorption capacity.

The Temkin isotherm (Temkin and Pyzhev 1940) in linear form is also used to model the adsorption process. The linear form of the Temkin model is given as:

$$q_e = \frac{RT}{b} \ln A + \frac{RT}{b} \ln C_e \tag{3}$$

Here b (kJ/mol) is heat of sorption. T is the temperature and R is universal gas constant. b and A are Temkin isotherm parameters.

The D-R isotherm (Nguyen and Do 2001) is often used to study the adsorption equilibrium data. This model is given as:

$$\ln q_e = \ln q_{DR} - \beta \varepsilon^2 \tag{4}$$

Here q_{DR} is the adsorption capacity corresponding to monolayer formation (mg/g), and β is a constant of adsorption energy.

The mechanism of adsorption involves the process of mass transfer by diffusion and chemical reactions. In order to examine the controlling mechanism of the adsorption process, various kinetic models are used such as Lagergren pseudo-first-order kinetics, pseudo-second-order kinetics, and intraparticle diffusion kinetics. Table 2 presents the various operating parameters, isotherms, and kinetics used in recent studies on adsorption of heavy metal contaminants using natural and modified bentonite.

Table 2 Operating conditions, isotherms, and kinetics of heavy metal adsorption using bentonite

Adsorbent	Metal ion removed	Operating conditions (contact time, θ (min), pH, temperature, T (K), adsorbent dose, C_{ads})	Adsorption isotherm	Adsorption kinetics	References
Natural bentonite	Cd^{2+} Cu^{2+}	θ: 5–60 min pH: 1–5 T: 298–348 K C_{ads}: 50–300 mg/l	Freundlich Langmuir D-R isotherm		Karapinar and Donat (2009)
Sodium-activated bentonite	Pb^{2+}, Cd^{2+}, Ni^{2+}, Cu^{2+}, Mn^{2+}, Cr^{3+}, and Zn^{2+}	θ: 1–60 min pH: 2–7 T: 290–328 K C_{ads}: 0.125–0.5 mg/l	Freundlich Langmuir	Pseudo-second-order Intraparticle pore diffusion is the rate-limiting step	Taha et al. (2016)
N-2-hydroxypropyl trimethyl ammonium chloride chitosan, HACC-bentonite	Cd^{2+}	θ: 5–80 min pH: 1–9 T: 293–323 K C_{ads}: 0.1–1.0 g	Langmuir Freundlich D-R isotherm Best fit: Langmuir	Pseudo-second-order	Huang et al. (2011)
Natrified bentonite, quartz, and bentonite mixture	Cd^{2+}	θ: 0–1440 min pH: 5 T: – C_{ads}: –	Langmuir	Pseudo-second-order	Schütz et al. (2016)
P4V4-bentonite	Cr^{6+}	θ: 0–180 min pH: 1–11 T: 308–333 K	Freundlich Langmuir		Mansri et al. (2009)
GMZ-bentonite	Cu^{2+}	θ: 0–1440 min pH: 5–6.5 T: 293.15–333.15 K C_{ads}: 1–3 mg/l	Freundlich Langmuir D-R isotherm		Li et al. (2009)
Cross-linked chitosan-bentonite beads Non-cross-linked chitosan-bentonite beads	Cu^{2+}	pH: 1–4	Freundlich Langmuir	Pseudo-first-order pseudo-second-order Best fit: Pseudo-second-order	Aljlil and Alsewailem (2014)

(continued)

Table 2 (continued)

Adsorbent	Metal ion removed	Operating conditions (contact time, θ (min), pH, temperature, T (K), adsorbent dose, C_{ads})	Adsorption isotherm	Adsorption kinetics	References
Silk-bentonite composite	Pb^{2+}, Cd^{2+}, Hg^{2+}, Cr^{6+}	θ: 0–100 min pH: 3–9 T: 293–353 K C_{ads}: 0.01–0.15 mg	Langmuir Freundlich Best fit: Langmuir	Pseudo-first-order pseudo-second-order Best fit: Pseudo-second-order	Wahab et al. (2019)
Humic acid-immobilized amine modified polyacrylamide bentonite	Cu^{2+}	θ: 0–300 min pH: 4–7 T: 283–333 K C_{ads}: 25–100 mg/l	Langmuir Freundlich D-R isotherm Best fit: Langmuir	Pseudo-first-order pseudo-second order Best fit: Pseudo-second-order	Anirudhan and Suchithra (2010)
Polyacrylamide composite bentonite	Cu^{2+}	θ: 0–2880 min pH: 2.4–7 T: 293–328 K C_{ads}: 0.08–0.65 g/l	Langmuir Freundlich Best fit: Langmuir		Zhao et al. (2010)
HCl-activated bentonite	Zn^{2+}, Cu^{2+}, Pb^{2+}, Ni^{2+}	θ: 10–120 min pH: 2–8 T: Room temperature	Langmuir Freundlich	Pseudo-second-order	Mohammed-Azizi et al. (2013)
Na-bentonite	Pb^{2+}	θ: 180–1440 min pH: 2–10 T: 298–338 K	Langmuir Freundlich D-R isotherm Best fit: Langmuir		Yang et al. (2010)

3 Removal of Heavy Metals Using Coagulation

The use of sustainable treatment technology is the major challenge for any industrial effluents. Operating conditions, types of effluents, environmental conditions, etc. are the major parameters during the selection of any wastewater treatment technology.

Coagulation is a traditional method to clean wastewater using chemicals, called as coagulants or flocculants. This technique plays a key role during reduction of heavy metals from aqueous solution. The coagulants can be roughly classified as inorganic coagulants, organic coagulants, and natural polymers based on their chemical composition and source. Figure 1 shows a classification of coagulants used in wastewater treatment.

Compared to the organic coagulants, inorganic coagulants are inexpensive and very effective in many cases especially with the low-turbid aqueous solution where the organic coagulants are not so effective. Every coagulant has its advantages and disadvantages in its application. Some major advantages and disadvantages of commonly used inorganic flocculants/coagulants are given in Table 3.

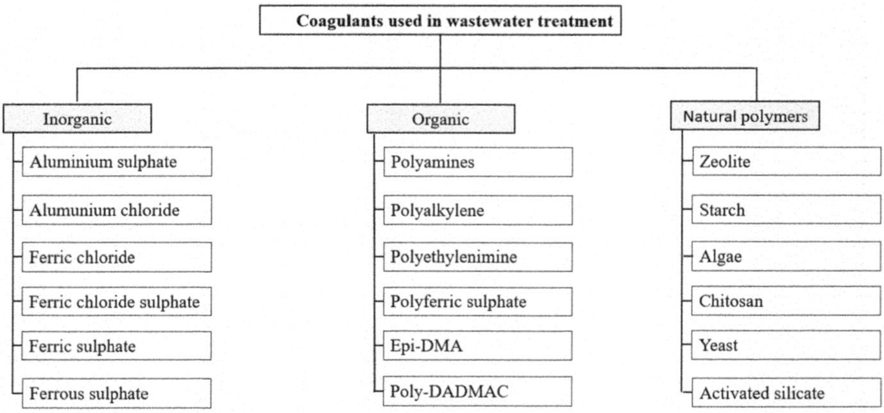

Fig. 1 Classification of coagulants used in wastewater treatment

Table 3 Advantages and disadvantages of commonly used inorganic coagulants

Coagulant/flocculant	Advantages	Disadvantages
Aluminum sulfate (alum)	1. Easily available 2. Most commonly used 3. Low sludge volume index 4. Effective within pH 5.5–7.5	1. Add dissolved salt to water 2. Effective in limited pH range
Ferric sulfate	Effective within pH 4–6 and 8.8–9.2	Add dissolved salt to water
Sodium aluminate	1. Effective in hard water 2. Low dosage of coagulant	1. Ineffective in soft water 2. High cost
Lime	1. Commonly used 2. Do not add dissolved salt	1. pH sensitive 2. More sludge generation

3.1 Parameters Affecting Removal Efficiency of Heavy Metal in Coagulation Process

The efficiency of heavy metal reduction often depends on the coagulation condition, nature of raw water, and type of coagulants (Hu et al. 2012). The rate of reduction of heavy metals in an aqueous solution mostly depends on the rate of adsorption, complexation, and coprecipitation. Concentration of metal ions, valence state of heavy metal, pH, and temperature are the main factors affecting the removal efficiency of heavy metal ions during coagulation.

Effect of metal ion concentration: Low concentration of metal ions results in lower removal efficiency. At low metal ion concentration, a small number of large flocks of low density are formed. Consequently, a relatively high coagulant dose is required to achieve high removal efficiency for low concentration of metal ion.

Effect of valence state of heavy metal: The valence state of metal ions also affects its efficiency of removal. It has been reported (Bordoloi et al. 2013; Lee et al. 2003) that As^{5+} is easily adsorbed compared to As^{3+} using inorganic coagulants in a coagulation process. It is also reported that the coagulant alone cannot reduce the concentration of As^{3+} below 10 μg/L (Bordoloi et al. 2013; Lee et al. 2003). Therefore, As^{3+} is pre-oxidized to As^{5+} before coagulation process to enhance its efficiency of removal.

Effect of pH: pH is the most important factor affecting efficiency of pollutant removal. As indicated in Table 3, the coagulant is effective if coagulation is carried out within an optimum range of pH moreover optimum range of pH is different for different coagulant and aqueous system. Therefore, the removal efficiency remains almost unaffected within the optimum range of pH for coagulation process and it shows significant variation in other pH range.

Effect of temperature: High temperature favors the coagulation efficiency because at high-temperature rate of hydrolysis, metal salt is high. Moreover, an increase in temperature results in increase in collision probability and Brownian motion of the metal ions and coagulants. Hence, the efficiency of heavy metal removal is favored at high temperature. A very few studies specifically reported the effect of temperature in a coagulation process, and further investigation is needed.

3.2 Role of Inorganic Coagulants for the Removal of Heavy Metals

There are various techniques which use chemical coagulants for heavy metal reduction. Some of these techniques are briefly described below.

Chemical precipitation: It is a common precipitation technique employed in the reduction of heavy metals/unused inorganic from industry/processes. During this process, coagulants react with the impurities/heavy metals and generate bulky/insoluble precipitates in the aqueous solution. Further, the insoluble precipitates are separated by the mechanism of sedimentation. Generally, heavy metals present in wastewater are separated in terms of carbonates, sulfides, phosphates, hydroxides, etc. (depending upon the nature of coagulant and other impurities properties).

Mirbagheri and Hosseini (2005) studied the removal of Cu^{2+} and Cr^{6+} ions from pilot plant petrochemical wastewater effluent using $Ca(OH)_2$ and NaOH as coagulants. The hexavalent ion of chromium was converted to trivalent chromium ion using ferrous sulfate. They reported that the maximum precipitation of chromate ion occurred at pH 8.7, and the concentration of chromate ion reduced from 30 to 0.01 mg/L (Mirbagheri and Hosseini 2005). Similarly, the copper ion was reduced from 48.51 to 0.694 mg/L at optimum pH of 12 (Mirbagheri and Hosseini 2005). Matlock et al. (2002) used 1,3-benzenediamidoethanethiol dianion for the reduction of heavy metals. They reported above 90% reduction of toxic metals from acid mine drainage (AMD) samples, and iron concentration was reduced from 194 to 0.009 ppm at pH 4.5 (Matlock et al. 2002).

Calcium hydroxide was used as a coagulant for an aqueous solution containing Cd(II), Zn(II), and Mn(II) (Charerntanyarak 1999). The initial concentrations of Cd (II), Zn(II), and Mn(II) were 150, 450, and 1085 mg/L, respectively. This study reported the complete removal of metal ions (up to 99.5%) at optimum pH of 11 (Charerntanyarak 1999). Chen et al. successfully removed the heavy metal ions, Cu (II), Zn(II), Cr(III), and Pb (II) using calcium hydroxide as a coagulant (Chen et al. 2009). In order to enhance the precipitation of metal ions, calcium hydroxide was mixed with fly ash, and the resultant suspension was carbonated in a carbon dioxide atmosphere (Chen et al. 2009). They reported the removal efficiency of up to 99.37–99.6% at optimum pH between 7 and 11 using carbonated fly ash-calcium hydroxide treatment (Chen et al. 2009).

Biogenic Mn oxides (BMO) have excellent heavy metal reduction capacity. Biologically *Pseudomonas putida*-generated BMO were investigated (Zhou et al. 2015) for the reduction of Cd^{2+}, Pb^{2+}, and Zn^{2+}. Based on kinetic and isotherm study, it was reported the BMO has seven to eight times higher adsorption capacity for Cd^{2+}, Pb^{2+}, and Zn^{2+} than the abiotic Mn oxide (Zhou et al. 2015). Effects of parameters like pH, ionic strength, and temperature were also reported in this study. The adsorption capacity was found to increase with increase in either temperature or pH, whereas it decreases with increase in ionic strength (Zhou et al. 2015).

Coagulation and flocculation: In the coagulation technique, heavy metals/inorganic impurities are reduced by suitable coagulants. Flocculation is a commonly used method to reduce the toxic elements from the effluents. During the flocculation process, agglomerates of fine particles are formed and clumped into flocs using suitable flocculants. Flocculants such as sodium dodecyl sulfate, polyferric sulfate, and polyacrylamide are commonly used for wastewater treatment.

Electrocoagulation (EC)-based treatment techniques are the ones in which metal electrodes dissolved and generate hydroxides of metal electrodes. The hydroxides react with pollutants/heavy metals and are separated by the mechanism of bridge and sweep coagulation (Shankar et al. 2014, 2015; Mahmad et al. 2016).

Leather industry-generated wastewater is a major source of Cr ion pollutants. Abdalhadi Deghles and Ugur Kurt (2016) studied the color, chemical oxygen demand (COD), NH_3-N, and Cr ion reduction of tannery wastewater using a hybrid process, EC integrated with electrodialysis (ED). Moreover, EC-treated solution using aluminum electrodes results in COD, color NH_3-N, and Cr ion percentage removal efficiency of 92, 100, 100, and 100%, respectively, whereas in the case of EC with iron electrodes, ED process also gives efficiency of removal for COD, color NH_3-N, and Cr as 87, 100, 100, and 100%, respectively (Deghles and Kurt 2016).

Chemically enhanced primary treatment (CEPT) is also an attractive option for the reduction of toxic metals such as, copper, chromium, nickel, zinc, and lead. CEPT using $FeCl_3$ and polymer-based coagulants is a more efficient option compared to CEPT using alum as coagulant. Johnson et al. (2008) reported that a mixture of 40 mg/L of $FeCl_3$ and 0.5 mg/L of anionic polymer results in enhanced metal reduction efficiency of 200% for Cr, Zn, and Ni ions and 475% for Pb ions compared to the traditional primary treatment (Johnson et al. 2008).

Coagulation-based combined/hybrid processes: The coagulation process combined with other suitable treatment technologies is an effective way to enhance the efficiency of heavy metal reduction. Other treatment technologies such as filtration, flocculation, membrane, adsorption, oxidation, etc. can be integrated with the coagulation process (Tang et al. 2016). Oxidation method can be used to destabilize the stable metals ions present in aqueous phase. Zotter and Licskó (1992) developed coagulation combined with the oxidation-reduction technique to reduce the dissolved Cr and Ni compounds from aqueous solution with the help of $FeSO_4 \cdot 7H_2O$ in both alkaline and neutral media. The partly soluble hydroxides of Cr are generated during the oxidation-reduction processes, and the Ni compounds precipitated are coagulated using iron hydroxides (Zotter and Licsko 1992).

Heavy metals in the aqueous phase can easily react with iminodiacetic acid (IDA), ethylenediaminetetraacetic acid (EDTA), and organic matters and result in reduction of rate of removal of heavy metals from aqueous phase. In this case, the presence of the strong oxidant (Fenton oxidation reagent) facilitates the oxidation and mineralization of the organic pollutants. This reaction results in transformation of the dissolved compounds into precipitates and leads to the simultaneous reduction of heavy metals and organic compounds. So, the integrating Fenton oxidation and coagulation technique lead to an effective wastewater treatment containing organic and heavy metal contaminants (Pachuau et al. 2013; Lan et al. 2012).

The membrane technology integrated with coagulation is also an interesting hybrid technology for the heavy metal reduction from aqueous phase. However, decrease in permeability of membrane due to fouling leads to regular cleaning or replacement of membrane. Due to this an increased operating cost becomes a major issue in the sustainable operation of hybrid method (Fu and Wang 2011).

Car wash wastewater contains oil grease and heavy contaminants, which can be coagulated by ferrous sulfate ($FeSO_4{\cdot}7H_2O$) and *Moringa oleifera* (a natural coagulant). The coagulation and flocculation process were integrated with sedimentation and filtration to treat wastewater from car wash. The concentration of *Moringa oleifera* and $FeSO_4{\cdot}7H_2O$ was taken as 35, 70, 105, and 140 mg/L (Al-Gheethi et al. 2016). The hybrid treatment using *Moringa oleifera* and filtration were successful in treatment of carwash wastewater (Al-Gheethi et al. 2016).

Industrial wastewater after primary treatment may contain small amounts of heavy metals/organic compounds in the aqueous phase which can be treated with the adsorption process for clean water production. Sahu et al. (2015) integrated EC process operation aeration and used the activated carbon as an adsorbent to treat biodigester effluent which contains organic compound as major pollutants and heavy metals as minor pollutants. They reported that this hybrid system gives improved results (Sahu et al. 2015).

Activated carbon (powder) can also be used to enhance the effectiveness of coagulants for the reduction of surface water pollutants. Activated carbon adsorption process can be integrated with aluminum chloride, aluminum sulfate, and polyaluminum chloride coagulant for removal of Cu, Cd, Ni, Pb, and indicator polychlorinated biphenyls (PCBs). It was reported that the integrated process gives 64, 39, 18, and 85% reduction for Cu, Cd, Ni, and Pb, respectively (Rosińska and Dąbrowska 2016).

Table 4 summarizes studies on coagulation and coagulation-based hybrid techniques for the reduction of toxic metals from aqueous system.

4 Role of Bentonite-Coagulant Mixture as Adsorbent and Coagulant

The main disadvantages associated with the chemical coagulants are corrosion and fouling of the equipments and generation of the large volume of sludge. Moreover, chemical-based commercial coagulants may create an adverse effect on surroundings and lead to health problems in human beings. Therefore, the combination of bentonite or other clay minerals and coagulants can be an eco-friendly choice for reduction of turbidity, COD, and heavy metals from aqueous solutions. Efforts have been made in the recent past to investigate the effect of combining the bentonite clay and coagulants to exploit the simultaneous advantages of adsorption and coagulation in wastewater treatment processes.

Syafalni et al. (2013) investigated the combination of bentonite and inorganic coagulants (alum, limestone, and zeolite) as coagulant for the reduction of iron, turbidity, and COD. An optimum value of both pH and coagulant dosage was reported for pure bentonite as well as thecombinations of bentonite-inorganic coagulants. They showed that the optimum volume ratio of bentonite-alum mixture to achieve 93.09% COD removal was 1:1 at a pH equal to 8 (Syafalni et al. 2013).

Table 4 Studies on coagulation and coagulation-based hybrid techniques

Source	Pollutant	Coagulant	Process	Removal efficiency	References
Simulated groundwater	Arsenic: 0.28 mg/L	Aluminum chloride: 6 mg/L	Coagulation	>96%	Hu et al. (2012)
Drainage water	Methylmercury: 1.20 pmol/L	Polyaluminum chloride: 1.61 mmol/L Ferric chloride: 1.05 mmol/L	Coagulation	70.83% 75%	Henneberry et al. (2011)
Simulated water	Manganese: 1 mg/L	Amorphous $Al(OH)_3$: 0.83 g/L	Coagulation	>92%	Wang et al. (2012)
Simulated water	Chromium: 10 mg/L	Ferric chloride: 0.7 g/L	Coagulation	90%	Golbaz et al. (2014)
Tap water	Copper: 15 mg/L Zinc: 17 mg/L Cobalt: 12 mg/L	Alum: 100 mg/L	Coagulation	>88% >64% 00%	Nilsson (1971)
Simulated wastewater	Nickel: 0.002 M	Ferric sulfate: 0.1–0.2 Mol	Coagulation	To 0.389 mg/L	Tu et al. (2012)
Simulated water	Arsenic: 25 mg/L	Ferric salts: 250 mg/L	Coagulation + pre-oxidation+ nanofiltration	98.4%	Pal et al. (2014)
Groundwater	Manganese: 0.235 mg/L	FeCl3 + FeSO4: 180 g/L	Coagulation +aeration + biological	>97%	Katsoyiannis et al. (2006)
Landfill leachate	Manganese: 0.77 mg/L Nickel: 0.77 mg/L	Polyferric sulfate	Coagulation +SBR+ Fenton +BAF	98.7% 98.2%	Wu et al. (2011)
Oil sand process	Barium: 0.35 mg/L	Alum: 250 mg/L	Coagulation +flocculation (PolyDADMAC)	42–63%	Pourrezaei et al. (2011)

The best volume ratio of bentonite-limestone for maximum COD removal was obtained as 6:4. The sludge generated using pure bentonite was higher (68.92 mL/g) compared to both bentonite-limestone (40.54 mL/g) and bentonite-alum mixture (44.59 mL/g) (Syafalni et al. 2013). Faghihian and Bahramian (2014) synthesized composites of bentonite and inorganic polymeric coagulants (polyaluminum chloride (PACl) and poly iron chloride (PICl)) for removal of heavy metals and turbidity from aqueous waste. Adsorption of Sr^{2+} and Pb^{2+} was studied to investigate the effect of pH, contact time, initial ion concentration, and the adsorbent dosage. The comparative performance study of four coagulants, composite bentonite-PACl, composite bentonite-PICl, $Al_2O_4Na_2$, and $Fe_2Cl_2{\cdot}H_2O$, toward percentage removal of metal ions (Sr^{2+} and Pb^{2+}) showed that the composite bentonite performs much better than the rest (Faghihian and Bahramian 2014). A recent study (Ntwampe and Moothi 2019) on acid mine drainage sample using combination of bentonite-inorganic coagulants ($FeCl_3$, $Al_2(SO_4)_3$, and Na_2CO_3) also reported that a combination of bentonite-$FeCl_3$ results in better turbidity removal as compared to the pure bentonite clay, $FeCl_3$ and $AlCl_3$ (Ntwampe and Moothi 2019). A flocculent synthesized using bentonite and Na_2CO_3 showed an optimal heavy metal reduction potential (Ntwampe and Moothi 2019).

5 Conclusions

Heavy metal contamination in different water bodies is continuously increasing because of rapid industrialization, population growth, and deforestation. The use of chemical-based treatment technologies and biological-based treatment technologies either consumes excess time or generates another pollutant in the aqueous phase. Adsorption/coagulation-based technologies are the best suited option for heavy metal reduction. Separation effectiveness of heavy metals from aqueous solution and cost of adsorption are the major parameters for the sustainable metal reduction. Bentonite clay-based coagulants and many inorganic coagulants are the better options for the heavy metal reduction from aqueous solution because of its availability and good separation properties. Inorganic-based coagulants generate less sludge and less preparation cost in comparison with organic coagulants. Metal removal efficiency from aqueous solution can be improved by increasing active surface site by different treatment methods such as acid-base, thermal, and hydro treatment. Many heavy metals from complex wastewater are not considerably reduced by single treatment technologies. The presence of organic and other impurities' content in the wastewater may affect the coagulation and adsorption properties. So, the use of hybrid technology could be the better option for the considerable reduction of heavy metals and other pollutants. The use of adsorption/filtration/electrocoagulation/membrane, etc. integrated with coagulation techniques could provide better metal-reducing options. The performance of bentonite clay and inorganic coagulants is the strong factor of operating condition and nature of wastewater. Chemical coagulant membrane-based hybrid technologies are prone to suffer corrosion and fouling, which makes the process less efficient and costlier. Hence, there is a need to address this issue in coagulant membrane-based hybrid techniques. Overall, the use of suitable materials, processes, and operating condition could be used to clean the heavy metal-containing wastewater.

References

Ahmad R, Hasan I (2016) L-cystein modified bentonite-cellulose nanocomposite (cellu/cys-bent) for adsorption of Cu 2+, Pb 2+, and Cd 2+ ions from aqueous solution. Sep Sci Technol 51:381–394

Al-Gheethi AA, Mohamed RMSR, Rahman MAA et al (2016) Treatment of wastewater from Car washes using natural coagulation and filtration system - IOPscience. IOP Conf Ser Mater Sci Eng 136:012046

Aljendeel HA (2011) Removal of heavy metals using reverse osmosis. J Eng 17:647–658

Aljlil SA, Alsewailem FD (2014) Adsorption of Cu & Ni on bentonite clay from waste water. Athens J Sci 1:21–30

Al-Rashdi BAM, Johnson DJ, Hilal N (2013) Removal of heavy metal ions by nanofiltration. Desalination 315:2–17

Anirudhan TS, Suchithra PS (2010) Humic acid-immobilized polymer/bentonite composite as an adsorbent for the removal of copper(II) ions from aqueous solutions and electroplating industry wastewater. J Ind Eng Chem 16:130–139

de Araujo ALP, Bertagnolli C, da Silva MGC et al (2013) Zinc adsorption in bentonite clay: influence of pH and initial concentration. Acta Sci Technol 35:325–332

Azimi A, Azari A, Rezakazemi M, Ansarpour M (2017) Removal of heavy metals from industrial wastewaters: a review. ChemBioEng Rev 4:37–59

Ba K, He LL, Tang H et al (2014) Use of chitosan-modified bentonite for removal of Cu2+, Cl - and 2,4-dichlorophenoxyacetic acid (2,4-D) from aqueous solution. Kem u Ind J Chem Chem Eng 63:253–258

Bentouami A, Ouali MS (2006) Cadmium removal from aqueous solutions by hydroxy-8 quinoleine intercalated bentonite. J Colloid Interface Sci 293:270–277

Bobade, Veena and Eshtiagi N (2015) Heavy metals removal from wastewater by adsorption process: a review. In: Asia Pacific Confederation of Chemical Engineering Congress 2015: APCChE 2015, incorporating CHEMECA 2015, pp 312–317

Bordoloi S, Nath SK, Gogoi S, Dutta RK (2013) Arsenic and iron removal from groundwater by oxidation-coagulation at optimized pH: laboratory and field studies. J Hazard Mater 260:618–626

Charerntanyarak L (1999) Heavy metals removal by chemical coagulation and precipitation. Water Sci Technol 39:135–138

Chen Q, Luo Z, Hills C et al (2009) Precipitation of heavy metals from wastewater using simulated flue gas: sequent additions of fly ash, lime and carbon dioxide. Water Res 43:2605–2614

Choudhury PR, Mondal P, Majumdar S (2015) Synthesis of bentonite clay based hydroxyapatite nanocomposites cross-linked by glutaraldehyde and optimization by response surface methodology for lead removal from aqueous solution. RSC Adv 5:100838–100848

Deghles A, Kurt U (2016) Treatment of tannery wastewater by a hybrid electrocoagulation/electrodialysis process. Chem Eng Process Process Intensif 104:43–50

Ebbers B, Ottosen LM, Jensen PE (2015) Electrodialytic treatment of municipal wastewater and sludge for the removal of heavy metals and recovery of phosphorus. Electrochim Acta 181:90–99

El-maghrabi HH (2016) Removal of heavy metals via adsorption using natural clay material removal of heavy metals via adsorption using natural clay material. J Environ Earth Sci 4:38–47

Eren E, Tabak A, Eren B (2010) Performance of magnesium oxide-coated bentonite in removal process of copper ions from aqueous solution. Desalination 257:163–169

Faghihian H, Bahramian E (2014) Synthesis of polymeric Al-bentonite and polymeric Fe-bentonite for simultaneous removal of turbidity, Sr+2, and Pb+2From colloidal aqueous wastes. Chem Eng Commun 201:1417–1430

Freundlich HMF (1906) Over the adsorption in solution. J Phys Chem 57:385–470

Fu F, Wang Q (2011) Removal of heavy metal ions from wastewaters: a review. J Environ Manag 92:407–418

Gandhi N, Sirisha D, Sekhar KBC (2014) Adsorption of chromium (Vi) from aqueous solution by using Multani Miti. Int J Res Pharm Chem 4:168–180

Gleason MH, Daniel DE, Eykholt GR (1997) Calcium and sodium bentonite for hydraulic containment applications. J Geotech Geoenviron Eng 123:438–445

Gök Ö, Özcan A, Erdem B, Özcan AS (2008) Prediction of the kinetics, equilibrium and thermodynamic parameters of adsorption of copper(II) ions onto 8-hydroxy quinoline immobilized bentonite. Colloids Surfaces A Physicochem Eng Asp 317:174–185

Golbaz S, Jafari AJ, Rafiee M, Kalantary RR (2014) Separate and simultaneous removal of phenol, chromium, and cyanide from aqueous solution by coagulation/precipitation: mechanisms and theory. Chem Eng J 253:251–257

Guerra DJL, Mello I, Resende R, Silva R (2013) Application as absorbents of natural and functionalized Brazilian bentonite in Pb2+ adsorption: equilibrium, kinetic, pH, and thermodynamic effects. Water Resour Ind 4:32–50

Guo M, Qiu G, Song W (2010) Poultry litter-based activated carbon for removing heavy metal ions in water. Waste Manag 30:308–315

Henneberry YK, Kraus TEC, Fleck JA et al (2011) Removal of inorganic mercury and methylmercury from surface waters following coagulation of dissolved organic matter with metal-based salts. Sci Total Environ 409:631–637

Ho YS, Huang CT, Huang HW (2002) Equilibrium sorption isotherm for metal ions on tree fern. Process Biochem 37:1421–1430

Hu C, Liu H, Chen G, Qu J (2012) Effect of aluminum speciation on arsenic removal during coagulation process. Sep Purif Technol 86:35–40

Huang R, Wang B, Yang B et al (2011) Equilibrium, kinetic and thermodynamic studies of adsorption of Cd (II) from aqueous solution onto HACC – bentonite. Desalination 280:297–304

Huang Y, Wu D, Wang X et al (2016) Removal of heavy metals from water using polyvinylamine by polymer-enhanced ultrafiltration and flocculation. Sep Purif Technol 158:124–136

Inglezakis VJ, Stylianou MA, Gkantzou D, Loizidou MD (2007) Removal of Pb(II) from aqueous solutions by using clinoptilolite and bentonite as adsorbents. Desalination 210:248–256

Johnson PD, Girinathannair P, Ohlinger KN et al (2008) Enhanced removal of heavy metals in primary treatment using coagulation and flocculation. Water Environ Res 80:472–479

Kanmani P, Aravind J, Kamaraj M et al (2017) Environmental applications of chitosan and cellulosic biopolymers: a comprehensive outlook. Bioresour Technol 242:295–303

Kanti Sen T, Khoo C (2013) Adsorption characteristics of zinc (Zn^{2+}) from aqueous solution by natural bentonite and kaolin clay minerals: a comparative study. Comput Water Energy Environ Eng 02:1–6

Karapinar N, Donat R (2009) Adsorption behaviour of Cu2+ and Cd2+ onto natural bentonite. Desalination 249:123–129

Katsoyiannis IA, Zikoudi A, Hug SJ (2006) Arsenic removal from groundwaters containing iron, ammonium, manganese and phosphate: a case study from a treatment unit in northern Greece. Desalination 224:330–339

Kongsuwan A, Patnukao P, Pavasant P (2009) Binary component sorption of Cu(II) and Pb(II) with activated carbon from Eucalyptus camaldulensis Dehn bark. J Ind Eng Chem 15:465–470

Ku Y, Jung IL (2001) Photocatalytic reduction of Cr(VI) in aqueous solutions by UV irradiation with the presence of titanium dioxide. Water Res 35:135–142

Kyzas G, Kostoglou M (2014) Green adsorbents for wastewaters: a critical review. Materials (Basel) 7:333–364

Lan S, Ju F, Wu X (2012) Treatment of wastewater containing EDTA-Cu(II) using the combined process of interior microelectrolysis and Fenton oxidation-coagulation. Sep Purif Technol 89:117–124

Langmuir I (1918) The adsorption of gases on plane surfaces of glass, mica and platinum. J Am Chem Soc 40:1361–1403

Lee Y, Um IH, Yoon J (2003) Arsenic (III) oxidation by iron (VI)(ferrate) and subsequent removal of arsenic (V) by iron (III) coagulation. Environ Sci Technol 37(24):5750–5756

Li J, Hu J, Sheng G et al (2009) Effect of pH, ionic strength, foreign ions and temperature on the adsorption of Cu(II) from aqueous solution to GMZ bentonite. Colloids Surfaces A Physicochem Eng Asp 349:195–201

Liu ZR, Zhou SQ (2010) Adsorption of copper and nickel on Na-bentonite. Process Saf Environ Prot 88:62–66

Liu Y, Shen X, Xian Q et al (2006) Adsorption of copper and lead in aqueous solution onto bentonite modified by 4′-methylbenzo-15-crown-5. J Hazard Mater 137:1149–1155

Mahmad MKN, Rozainy MAZMR, Abustan I, Baharun N (2016) Electrocoagulation process by using aluminium and stainless steel electrodes to treat total chromium, colour and turbidity. Proc Chem 19:681–686

Malik DS, Jain CK, Yadav AK (2017) Removal of heavy metals from emerging cellulosic low-cost adsorbents: a review. Appl Water Sci 7:2113–2136

Mansri A, Benabadji KI, Desbrières J, François J (2009) Chromium removal using modified poly (4-vinylpyridinium) bentonite salts. Desalination 245:95–107

Matlock MM, Howerton BS, Atwood DA (2002) Chemical precipitation of heavy metals from acid mine drainage. Water Res 36:4757–4764. https://doi.org/10.1016/S0043-1354(02)00149-5

Mirbagheri SA, Hosseini SN (2005) Pilot plant investigation on petrochemical wastewater treatment for the removal of copper and chromium with the objective of reuse. Desalination 171:85–93

Mohammad AW (2013) Editorial: Nanofiltration membranes. Desalination 315:1

Mohammed-Azizi F, Dib S, Boufatit M (2013) Removal of heavy metals from aqueous solutions by Algerian bentonite. Desalin Water Treat 51:4447–4458

Mohsen-Nia M, Montazeri P, Modarress H (2007) Removal of Cu2+ and Ni2+ from wastewater with a chelating agent and reverse osmosis processes. Desalination 217:217–281

Nguyen C, Do DD (2001) The Dubinin-Radushkevich equation and the underlying microscopic adsorption description. Carbon 39:1327–1336

Nilsson R (1971) Removal of metals by chemical treatment of municipal waste water. Water Res 5:51–60

Ntwampe IO, Moothi K (2019) Reaction dynamics of bentonite clay, FeCl3, Al2(SO4)3 and Na2CO3 dosage in AMD using varying dispersion techniques. J Environ Manag 231:552–561

Olekhnovich, RO, Volkova KV, Uspenskii AA, Slobodov AA, Uspenskaya MV (2015) Synthesis of poly (acrylic acid)-co-acrylamide/bentonite polymer nanocomposite as an absorbent for removal of heavy metal ions from water. Int Multidiscip Sci GeoConf Surv Geol Min Ecol Manag SGEM 2(5):477–484

Olu-Owolabi BI, Alabi AH, Unuabonah EI et al (2016) Calcined biomass-modified bentonite clay for removal of aqueous metal ions. J Environ Chem Eng 4:1376–1382

Özcan AS, Gök Ö, Özcan A (2009) Adsorption of lead(II) ions onto 8-hydroxy quinoline-immobilized bentonite. J Hazard Mater 161:499–509

Pachuau L, Lee SM, Tiwari D (2013) Ferrate(VI) in wastewater treatment contaminated with metal (II)-iminodiacetic acid complexed species. Chem Eng J 230:141–148

Pal P, Chakrabortty S, Linnanen L (2014) A nanofiltration-coagulation integrated system for separation and stabilization of arsenic from groundwater. Sci Total Environ 476–477:601–610

Pourrezaei P, Drzewicz P, Wang Y et al (2011) The impact of metallic coagulants on the removal of organic compounds from oil sands process-affected water. Environ Sci Technol 45:8452–8459

Pyrzynska K (2019) Removal of cadmium from wastewaters with low-cost adsorbents. J Environ Chem Eng 7:102795

Rosińska A, Dąbrowska L (2016) Enhancement of coagulation process with powdered activated carbon in PCB and heavy metal ions removal from drinking water. Desalin Water Treat 57:26336–26344
Sahu RK, Shankar R, Mondal P, Chand S (2015) Treatment potential of EC towards bio-digester effluent: effects of process parameters, aeration, and adsorbent. Desalin Water Treat 54:1912–1924
Schütz T, Dolinská S, Hudec P et al (2016) Cadmium adsorption on manganese modified bentonite and bentonite-quartz sand blend. Int J Miner Process 150:32–38
Shan TC, Matar A, Makky EA, Ali EN (2017) The use of Moringa oleifera seed as a natural coagulant for wastewater treatment and heavy metals removal. Appl Water Sci 7:1369–1376
Shankar R, Singh L, Mondal P, Chand S (2014) Removal of COD, TOC, and color from pulp and paper industry wastewater through electrocoagulation. Desalin Water Treat 52:7711–7722
Shankar R, Mondal P, Chand S (2015) Electrocoagulation as a process for the treatment of high organics bearing industrial waste water. In: Kanti Sen T (ed) Physical, chemical and biological treatment processes for water and wastewater. Nova Science, New York, pp 29–56
Soliemanzadeh A, Fekri M (2017) The application of green tea extract to prepare bentonite-supported nanoscale zero-valent iron and its performance on removal of Cr(VI): effect of relative parameters and soil experiments. Microporous Mesoporous Mater 239:60–69
Syafalni AR, Abustan I et al (2013) Wastewater treatment using bentonite, the combinations of bentonite-zeolite, bentonite-alum, and bentonite-limestone as adsorbent and coagulant. Int J Environ Sci 4:379–391
Taha AA, Shreadah MA, Ahmed AM, Heiba HF (2016) Multi-component adsorption of Pb(II), Cd (II), and Ni(II) onto Egyptian Na-activated bentonite; equilibrium, kinetics, thermodynamics, and application for seawater desalination. J Environ Chem Eng 4:1166–1180
Tang X, Zheng H, Teng H et al (2016) Chemical coagulation process for the removal of heavy metals from water: a review. Desalin Water Treat 57:1733–1748. https://doi.org/10.1080/19443994.2014.977959
Temkin MJ, Pyzhev V (1940) Kinetics of the synthesis of ammonia on promoted Iron catalysts. Acta Physicochim 12:217–222
Tomul F, Basoglu FT (2010) Adsorption of copper and zinc from aqueous solutions by bentonite. Asian J Chem 22:615–628
Tripathi A, Rawat Ranjan M (2015) Heavy metal removal from wastewater using low cost adsorbents. J Bioremed Biodegr 06:100315
Tu YJ, Chang CK, You CF, Wang SL (2012) Treatment of complex heavy metal wastewater using a multi-staged ferrite process. J Hazard Mater 209–210:379–384
Uddin MK (2017) A review on the adsorption of heavy metals by clay minerals, with special focus on the past decade. Chem Eng J 308:438–462
Vhahangwele M, Mugera GW (2015) The potential of ball-milled South African bentonite clay for attenuation of heavy metals from acidic wastewaters: simultaneous sorption of Co2+, Cu2+, Ni2+, Pb2+, and Zn2+ ions. J Environ Chem Eng 3:2416–2425
Vieira MGA, Neto AFA, Gimenes ML, da Silva MGC (2010) Sorption kinetics and equilibrium for the removal of nickel ions from aqueous phase on calcined Bofe bentonite clay. J Hazard Mater 177:362–371
Vinodhini PA, Sudha PN (2017) Removal of heavy metal chromium from tannery effluent using ultrafiltration membrane. Text Cloth Sustain 2:5
Vodyanitskii YN (2014) Natural and technogenic compounds of heavy metals in soils. Eurasian Soil Sci 47:255–265
Wahab N, Saeed M, Ibrahim M et al (2019) Synthesis, characterization, and applications of silk/bentonite clay composite for heavy metal removal from aqueous solution. Front Chem 7:1–12
Wan Ngah WS, Hanafiah MAKM (2008) Removal of heavy metal ions from wastewater by chemically modified plant wastes as adsorbents: a review. Bioresour Technol 99:3935–3948
Wan Ngah WS, Teong LC, Hanafiah MAKM (2011) Adsorption of dyes and heavy metal ions by chitosan composites: a review. Carbohydr Polym 83:1446–1456

Wang W, Zhang X, Wang H et al (2012) Laboratory study on the adsorption of Mn2+ on suspended and deposited amorphous Al(OH)3 in drinking water distribution systems. Water Res 46:4063–4070

Wu Y, Zhou S, Ye X et al (2011) Transformation of pollutants in landfill leachate treated by a combined sequence batch reactor, coagulation, Fenton oxidation and biological aerated filter technology. Process Saf Environ Prot 89:112–120

Xu C, Xia BC, Wu HN et al (2009) Speciation and bioavailability of heavy metals in paddy soil irrigated by acid mine drainage. Huanjing Kexue/Environmental Sci 30:900–906

Yang S, Zhao D, Zhang H et al (2010) Impact of environmental conditions on the sorption behavior of Pb(II) in Na-bentonite suspensions. J Hazard Mater 183:632–640

Zewail TM, Yousef NS (2015) Kinetic study of heavy metal ions removal by ion exchange in batch conical air spouted bed. Alexandria Eng J 54:83–90

Zhao G, Zhang H, Fan Q et al (2010) Sorption of copper(II) onto super-adsorbent of bentonite-polyacrylamide composites. J Hazard Mater 173:661–668

Zhou D, Kim DG, Ko SO (2015) Heavy metal adsorption with biogenic manganese oxides generated by Pseudomonas putida strain MnB1. J Ind Eng Chem 24:132–139

Zotter K, Licsko I (1992) Removal of chromium (VI) and other heavy metals from groundwaters in neutral and alkaline media. In: Water science and technology. IWA Publishing, London, pp 207–216

Zou G, Ylinen A, Di Capua FD et al (2013) Impact of heavy metals on denitrification of simulated mining wastewaters. In: Advanced materials research. Trans Tech Publications Ltd, Zurich, pp 500–503

Advanced Oxidation Processes for Wastewater Remediation: An Overview

Samreen Heena Khan and Virendra Kumar Yadav

Abstract

Water pollution has become the global scientific challenge with the emergence of organic/inorganic pollutant levels that raise to an alarming rate. There has been an increasing demand to develop green and sustainable wastewater treatment technology for the treatment of a wide range of pollutants. In the past few decades, the advanced oxidation process (AOP) has been emerged as the most widely anticipated treatment technology. It involves the chain of oxidative and reductive reactions by the production of the most powerful reactive oxidation species. These hydroxyl radicals as an oxidant degrade the range of persistent pollutants into environmental friendly compounds. AOP has been used for the treatment of recalcitrant pollutants that are resistant to conventional treatment technologies. In the present chapter, the fundamentals of advanced oxidation process along with the principle, type, and its applications for the treatment of wastewater have been presented in detail. The chapter also highlights the advantage and disadvantages of the various advanced oxidation technologies.

Keywords

Advanced oxidation · Wastewater · Oxidants · OH radicals · Remediation

S. H. Khan (✉)
School of Nanoscience, Central University of Gujarat, Gandhinagar, India

V. K. Yadav
School of Life sciences, Jaipur National University, Jaipur, India

M. P. Shah (ed.), *Removal of Emerging Contaminants Through Microbial Processes*,
https://doi.org/10.1007/978-981-15-5901-3_4

1 Introduction

Since the past few decades, the term "emerging contaminants" has been used frequently in many research papers, as the existence of new chemical substances in wastewater systems (Rodriguez-Narvaez et al. 2017). These contaminants can enter the water bodies through various points and nonpoint sources such as domestic sewage, industrial effluents, landfill leachate, agricultural runoff, etc. (Lima 2018). The presence of emerging contaminants in water systems causes a serious threat to humans and the environment (Richardson and Ternes 2018). Also, they do not have acceptable regulatory standards for the drinking water, and their long-term impacts on human health and environment are yet to be evaluated further. Studies are underway for the impact of these emerging contaminants on humans and ecosystems as majorities of these contaminants are persistent in nature and conventional treatment technologies are not sufficient for the removal of such contaminants from the wastewater (Bolong et al. 2009). Even at the low concentration such as ng/L of these contaminants poses serious hazards to the environment (Petrović et al. 2003; Murray et al. 2010).

To overcome the drawbacks associated with the conventional treatment technologies, the alternative method for complete treatment of recalcitrant toxicants from wastewater into the environmental friendly compounds has led to the research in the field of "advanced oxidation processes" (AOPs) as the advanced water treatment technologies (Buthiyappan et al. 2016, Wols and Hofman-Caris 2012). AOP utilizes powerful hydroxyl radicals as the main oxidizing agents which were proposed for the first time in 1980 for the treatment of potable water; later AOPs are being applied in broader ways for the treatment of different types of wastewaters due to the potential of strong oxidizing agents which can degrade recalcitrant a hazardous organic/inorganic pollutants from the wastewater (Xie et al. 2017; Huang et al. 2017).

The main aim of these AOPs is based on the in situ generation of highly reactive species for rapid degradation of unmanageable organic compounds and pollutants, water pathogens, toxicants, pesticides, and disinfection by-products (Kanakaraju et al. 2018). The major advantage of using AOP is that it doesn't produce any sludge and toxic end products that require further handling or secondary treatment. It has many advantages over other techniques such as low cost, complete degradation of the pollutant, less or no waste disposal problem, and less dependence on temperature and pressure (Arzate et al. 2019). All AOPs are identified by the same chemical feature, i.e., the production of hydroxyl radicals "OH" (Korpe et al. 2019). These radicals are extremely reactive species and attack every organic molecule with rate constants usually in the order of 10^6–10^9 mol l^{-1} s^{-1} (Tichonovas et al. 2017). Table 1 shows the relative oxidation potential of various oxidizing agents.

Table 1 Oxidation potential of various oxidizing species

S. no.	Oxidizing species	Oxidation potential
1.	Fluorine	3.06
2.	Hydroxyl radical	2.80
3.	Oxygen (atomic)	2.82
4.	Hole on TiO_2	2.35
5.	Ozone	2.08
6.	Hydrogen peroxide	1.78
7.	Permanganate	1.24
8.	Oxygen (molecular)	1.23
9.	Hypochlorous acid	1.10
10.	Chlorine	1.00

2 Present Treatment Technologies

Various conventional treatment technologies have been employed for the treatment of wastewater such as physical, chemical, and biological process or the combination of all the processes (Rodriguez-Narvaez et al. 2017). As we know that the main aim of wastewater treatment is the elimination of solids, organic matter, color, and odor from the wastewater, however, some of the contaminants are recalcitrant in nature, and they cannot be treated completely by any of the abovementioned technologies (Ahmed et al. 2017; Visanji et al. 2018). The increased public concern toward environmental pollutants has prompted the need to develop novel treatment methods for the treatment of wastewater, as existing technologies just convert the contaminant from one phase to another rather than complete degradation from the environment (Petrović et al. 2003; Grassi et al. 2012).

The present treatment technologies for the removal of hazardous contaminants and their intermediates demand high capital investment, operation cost, high maintenance cost, and large area (Swami and Buddhi 2006). The abovementioned methods are efficient but have certain limitations in terms of effectiveness, applicability, time, and cost.

The complete degradation of various organic pollutants is not possible by conventional methods, i.e., physicochemical treatment, activated sludge digestion, anaerobic digestion, and air stripping (Oh et al. 2012). The other technologies such as adsorption and coagulation simply concentrate the pollutants present by transferring them from one phase to another but still persist and not being completely degraded from the system. Chemical oxidation is one such method, but it is expensive and leads to the contamination of water with other toxic pollutants (Ksibi 2006). Further, methods such as sedimentation, filtration, chemical, and membrane technologies involve high operating costs and could generate toxic secondary pollutants into the ecosystem (Gogate and Pandit 2004). These findings demand the necessity of advance research for the removal of these contaminants to reduce the risk of their accumulation in the water bodies for direct or indirect usage.

At present, AOP has a particular advantage over conventional technologies because of its ability to eliminate recalcitrant compounds, and it is based on high oxidation strength of free radicals and highly reactive oxidation species (Zhou and Smith 2002). The main aim of the present chapter is to investigate the various AOPs used for the treatment of wastewater.

3 Advanced Oxidation Process (AOP)

The ability of advanced oxidation technology has been extensively demonstrated to remove persistent organic compounds (Tijani et al. 2014), toxicants (Ebbs 2004), pesticides (Sanches et al. 2010; Khan et al. 2018), and dye effluents (Suriyaprabha et al. 2016; Khan et al. 2017) from the wastewater. AOPs are defined as the oxidation process that involves the rapid generation of hydroxyl radicals (OH) in enough quantity to support the treatment of wastewater. The AOPs also explored sulfate radicals (SO4$^{\cdot -}$) later on (Méndez-Díaz et al. 2010; Deng and Ezyske 2011). These radicals are different from common oxidants such as ozone and chlorine as they have a dual role as disinfectants and decontaminants; on the other hand, AOPs are primarily employed for the destruction of organic/inorganic contaminants from the wastewaters (Lopez et al. 2000; Olmez-Hanci and Arslan-Alaton 2013). In the treatment of wastewater, these powerful oxidizing agents are expected to destruct the toxic pollutants and transform them into nontoxic end products, this way providing a solution to the remediation of wastewater (Liu et al. 2016). AOPs are characterized by a free radical mechanism initiated by the interactions of photons of a proper energy level with the molecules of chemical species present in the solution or with a catalyst (Wang and Xu 2012). Major factors affecting these AO processes are the initial concentration of the pollutant, the quantity of oxidizing agents and catalysts, irradiation time, light intensity, and the nature of the wastewater's solution such as pH, presence of solids, and other ions (Stasinakis 2008; Wang et al. 2003). At first, Glaze et al. (1987) defined the AOP as the near ambient temperature and pressure water treatment process that involves the generation of hydroxyl radicals in sufficient quantity to initiate water purification. The OH is the most powerful and nonselective chemical oxidant which acts effectively and rapidly on most of the organic compounds. These ·OH react unselectively once formed, and contaminants will be efficiently fragmented and converted into smaller inorganic molecules (Oppenländer 2007). In general terms, AOPs could proceed into three parts, such as (Faouzi et al. 2006):

1. Formation of ·OH radicals.
2. Attacks on targeted contaminant by ·OH radical and their breakdown into the fragments.
3. Lastly attacks by ·OH until complete mineralization of the contaminant.

3.1 Types of AOP

There are a few different ways by which AOPs can generate hydroxyl and other radicals; it clearly indicates that the highly reactive HO· is the main driver for all AOPs and the effectiveness of the method depends on the efficiency by which HO· is produced. From the many AOP systems, ozonation, ultraviolet (UV)/ozone, UV/hydrogen peroxide, and photocatalysis are most frequently studied and utilized for many applications (Klavarioti et al. 2009). The most used AOP technologies can be divided into the following groups (Poyatos et al. 2010) as irradiation based and non-irradiation based as shown in Fig. 1.

1. *Homogeneous AOP*: Homogeneous processes use energy in the form of light. Homogeneous AOP employs UV radiation for the degradation of pollutants which absorbs UV radiation within the corresponding range of the electromagnetic spectrum (Muruganandham et al. 2014a, b). Ultraviolet (UV) radiations and other sources such as ultrasound (sonolysis), electrical energy, and microwave and electron beam irradiation are generally utilized for the degradation of compounds (Cieśla et al. 2004). Moreover, the compounds that absorb UV light at lower wavelengths are suitable for such type of photodegradation reactions. The process-based on this approach includes UV/H_2O_2, UV/ozone, UV/photolysis, and photo-Fenton (Espinoza and Frimmel 2009).
2. *Heterogeneous AOP:* Heterogeneous processes generally use catalysts to enhance the degradation of the contaminants. The term heterogeneous refers to the *dual phase* such as the catalyst is in the solid phase, while the contaminants are present in the aqueous phase (Muruganandham et al. 2014a, b). In the heterogeneous process, two or more phases are used along with the light source (UV/solar radiation) in the presence semiconductor catalyst (TiO_2, ZnO, ZnS, CdS, etc.) (Kim and Ihm 2011). It accelerates the chemical reaction due to the presence of electron-hole pairs. The AOP-generated holes and electrons lead to redox reaction (oxidation and reduction) for the degradation of contaminants

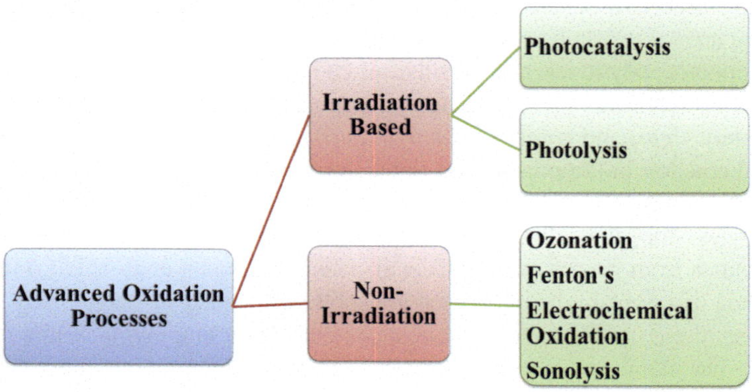

Fig. 1 General classification of advanced oxidation processes

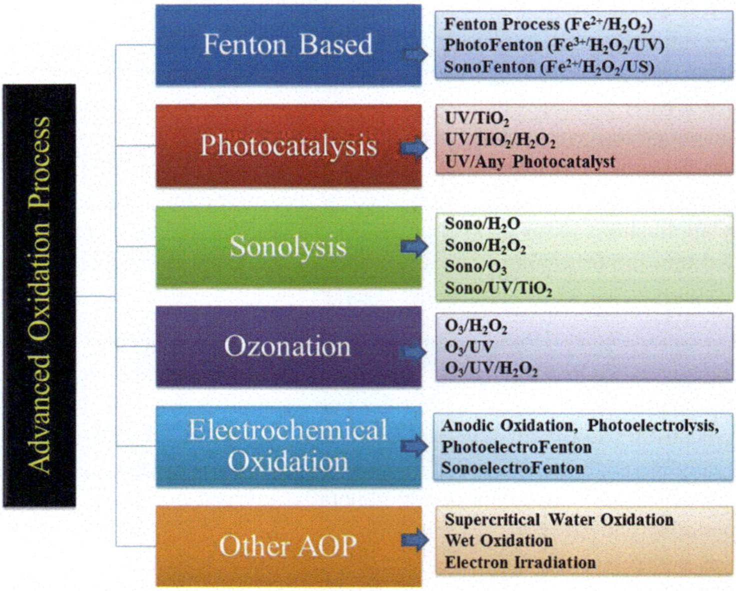

Fig. 2 Classification of various advanced oxidation processes

(Pinto et al. 2001). The AOP processes are further subdivided into processes that use energy and processes that do not use energy as shown in Fig. 2.

This section discusses the detailed account of various advanced oxidation processes for the treatment of wastewater.

3.1.1 Photocatalysis/UV-Based AOP

Mankind has received the gift of solar energy and had tried to explore the utilization of this vast and abundant source of energy. The utilization of solar energy is a very interesting aspect of science. Solar photocatalysis has, therefore, become a very important area of research, wherein sunlight is the source of illumination to perform different photocatalytic reactions (Mills and Le Hunte 1997). The term catalysis was first used by Berzelius in 1836. The word "photocatalysis" is the combination of *photo (phos, light)* and *catalysis (katalyo, decompose)*. It is the acceleration of the chemical reaction in the presence of light. In recent years, photocatalytic reactions gained much attention and emerged as one of the most successful and convenient technologies for the elimination and treatment of inorganic and organic contaminants from wastewater (Mills et al. 1993; Hoffmann et al. 1995). Fujishima and Honda first explain the photocatalytic splitting of water in 1972.

Among these AOPs, heterogeneous photocatalysis employing semiconductor catalysts has demonstrated its efficiency in degrading a wide range of toxic contaminants into biodegradable compounds and finally mineralized them to carbon dioxide and water (Ibhadon and Fitzpatrick 2013). The semiconductor-assisted photocatalyst is one of the most extensively investigated AOPs, because of its capacity to degrade a high number of recalcitrant chemicals in gaseous or aqueous systems, through comparatively inexpensive procedures (Pelizzetti and Serpone 2012).

Photocatalysis is defined as the “acceleration of a photoreaction by the action of a catalyst” (Fujishima et al. 2007). It occupies a central place in ecological equilibrium. The overall process of photocatalysis is very simple; a photon of light with energy equal to or higher than the bandgap energy produces a charge-carrier pair. As a result of this energy change, the formation of electron and holes takes place to produce hydroxide and other radicals for the degradation of organic contaminants (Litter 1999). The photocatalytic activity of the catalyst depends on the ability to create electron-hole pairs, which generate free radicals (e.g., hydroxyl radicals, ·OH).

Heterogeneous photocatalysis is currently being considered as a promising technique for water purification in comparison to other conventional methods, as it can break up complex long-chained organic molecules into simpler fragments (Martínez et al. 2011). Heterogeneous photocatalysis is a process in which two active phases solid and liquid are present; the solid phase is a catalyst, usually a semiconductor, based on the irradiation of a catalyst, which may be photoexcited to form electron-donor sites (reducing sites) and electron-acceptor sites (oxidizing sites), providing great scope as redox reagents (Doll and Frimmel 2005). Several researchers have typically used TiO_2 and ZnO; both are perfect photocatalyst, because of high chemical stability and highly oxidizing photo-generated holes. So, most of the investigations have focused on TiO_2 specially Degussa P25 and zinc oxide nanoparticles (Ilisz et al. 2004). However, other materials such as cadmium selenide (CdS) and zinc selenide (ZnS) and colloids were also investigated for their photochemical properties (Joshi and Shrivastava 2011; Guo et al. 2006); though, they are toxic, but they can be used after certain surface modifications (Sadegh et al. 2019; Wen et al. 2015).

Heterogeneous AOP using TiO_2 is an efficient advanced oxidation method for the treatment of various industrial effluents and other wastewaters, where photocatalyst is illuminated by artificial irradiation; as a result, electrons are being generated in conduction band and holes in the valence band, and these electrons interact with surface adsorbed oxygen to produce superoxide radicals anions; the hydroxyl radicals are produced when the holes interact with the water molecules (Lim et al. 2018).

It is a known fact that most of the organic compounds can be degraded by the oxidation under UV/photocatalyst and the degradation of organic compounds takes place by reacting with the hydroxyl radicals at the surface of photocatalysts. Besides, TiO_2 nanoparticles are extensively used and found most suitable for wastewater applications, because they are nontoxic, inexpensive, and chemically and biologically inert (Lamas Samanamud et al. 2012). Also, TiO_2 has higher oxidative power as compared to other oxidizing species (Arslan et al. 2000). Suspended TiO_2 nanoparticles were used in most studies to the photocatalytic oxidation of contaminants (Senthilkumaar and Porkodi 2005).

Principle and Mechanism of Photocatalysis: The mechanism and theory of semiconductor-based photocatalysis have been previously discussed in various literatures. The photocatalyst gets activated by the illumination with the UV radiations establishing a redox environment in the reaction medium. It is a process of degradation of pollutants by accelerating the rate of chemical reaction in the presence of a semiconductor catalyst without itself being consumed (Khan 2020; Nosaka et al. 2014). The process principally depends on the generation of hydroxyl radicals which are capable of converting various toxicants/nonbiodegradable pollutants into nontoxic end products such as CO_2, H_2O, and other environmental friendly compounds (Wang et al. 2013). The highest occupied valence band (VB) and the lowest unoccupied conduction band (CB) are separated by the energy bandgap. Holes in the valence band are the most powerful oxidizing agents, and the electrons in the conduction band are the most powerful reducing agents (Ismail and Bahnemann 2014). All the organic photodegradation processes utilize the oxidizing power of the holes. When a semiconductor absorbs the photon of energy equal or higher than the bandgap energy, electron from the valence band moves toward the conduction band creating a hole (h_{vb}^+) in the valence band; this is called generation of an "electron-hole (e^-/h^+) pair" (Kabra et al. 2004). Over the surface of the catalyst, electron and hole combine for the redox reaction; the photo-generated electron reacts with the oxygen to produce the superoxide anion radicals. The hole (h_{vb}^+) oxidizes the pollutant directly and produces the hydroxyl radical OH; on the other hand, electron (e_{cb}^-) efficiently reduces the oxygen adsorbed by the catalyst over its surface (Wenderich and Mul 2016). The basic reaction involved in photocatalysis is as follows:

$$\begin{aligned}\text{Photocatalyst} + \text{hv}\ (\geq \text{Energy band gap}) \rightarrow\ & \text{Photocatalyst}\ (e^-{}_{\text{CB}}) \\ & + \text{Photocatalyst}\ (h^+{}_{\text{VB}})\end{aligned}$$

$$\text{Photocatalyst}\ (h^+{}_{\text{VB}}) + \text{H}_2\text{O}/-\text{OH} \rightarrow \text{H}^+ + \text{OH}\cdot$$

$$\text{Photocatalyst}\ (e^-{}_{\text{CB}}) + \text{O}_2 \rightarrow \text{O}_2^{\cdot -}$$

$$\text{OH} + \text{O}_2^{\cdot -} + \text{Pollutant} \rightarrow \text{Intermediate} \rightarrow \text{CO}_2 + \text{H}_2\text{O} + \text{Inorganic Ions}$$

In the above steps, the generation of hydroxyl radical (·OH) and superoxide anion ($O_2^{-\circ}$) radical is the most important oxidizing species for the photocatalytic oxidation reaction.

In a photocatalytic process, hydroxyl radical generates due to the adsorption of water molecule over the surface as OH^-, oxygen molecule prevents the electron-hole pair recombination need no other oxidants, as atmospheric oxygen can be utilized as oxidants. When this ·OH radical attacks the organic contaminant, it involves the generation of various intermediate compounds; these intermediate again reacts with ·OH radical to produce final reaction products such as CO_2 and water (Herrmann 1999). The photocatalysis process is generally referred to as a thermodynamically downhill reaction that occurs spontaneously with the negative

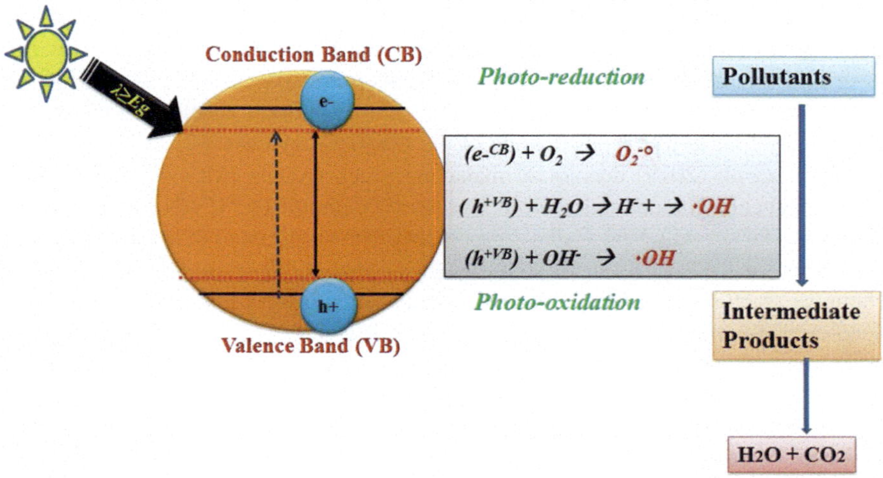

Fig. 3 Mechanism of photocatalytic degradation process

Gibbs energy change ($\Delta G < 0$). One most important point for the photocatalysis is that the oxidation of contaminants and reduction of oxygen should proceed simultaneously just to avoid the accumulation of electron in the CB which reduces the electron and hole recombination rate (Augugliaro et al. 2012); otherwise the energy will eliminate in the form of heat, and no reaction will happen. The complete processes of semiconductor-mediated photocatalysis during the light illumination are shown in Fig. 3.

Semiconductors as Photocatalyst: Most commonly used heterogeneous catalysts are semiconductors (Wang et al. 2014). The wide interest in small semiconducting nanoparticles originates due to their novel photophysical and photocatalytic properties. Semiconductors are basically important as a photocatalyst because of a unique combination of its electronic structures, charge transport characteristics, light absorption properties, and life of excitation state (Hisatomi et al. 2014). The main advantage of using semiconductors as photocatalysts is that they are comparatively inexpensive, nontoxic, high surface area, having broad absorption spectra with high absorption coefficients, exhibiting tunable properties, affording facility for the multi-electron transfer process, and capable of prolonged use without substantial loss of photocatalytic activity (Zou et al. 2011).

Jamshidi et al. (2009) investigated advanced photochemical oxidation (APO) processes (UV, UV/H_2O_2, UV/H_2O_2/Fe(II), and UV/H_2O_2/Fe(III)) in lab-scale experiments for the degradation of phenol (initial concentration = 0.5 mmol/L) in an aqueous medium. A medium-pressure 300 watts (UV-C) mercury ultraviolet lamp was used as the source of radiation and 30% H_2O_2 as the oxidant. In this study, parameters such as pH, H_2O_2 input concentration, the type of iron salt, iron catalyst concentration, and duration of UV radiation were studied based on the standard methods. The results indicated that the photo-Fenton process was the most effective treatment under acidic conditions producing a higher rate of phenol degradation over a very short radiation time, and the process accelerated the oxidation rate by four to five times the rate of the UV/H_2O_2 process.

Ghime et al. (2019) decolorize malachite green oxalate (MGO) dye by using photochemical advanced oxidation processes (UV photolysis, UV/H_2O_2, UV/TiO_2, UV/H_2O_2/TiO_2, and UV/H_2O_2/Fe^{2+} processes). The influence of oxidant (H_2O_2) and Fe^{2+} ions for the oxidation of MGO was also studied. A laboratory-scale photochemical reactor was used for the experimental runs. It was found that about 98% decolorization was obtained at initial 100 ppm MGO dye in photo-Fenton process under optimal conditions (60 ppm Fe^{2+} concentration, 12 mM of oxidant concentration, at pH 3.0 for 60 min) and 97% decolorization was obtained using UV/H_2O_2/TiO_2 process. The percentage decolorization of MGO was in the range of 94–95% for both UV/TiO_2 and UV/H_2O_2 processes. Sulfate radical-based AOPs were proved to be more effective in treating MGO dye with irradiation. Pseudo-first-order kinetics model was observed to be the best model fit to explain the decolorization of MGO dye solution.

Haddad et al. (2019) evaluated the degradation of high concentrations of formal dehyde in the chemical-less UVC/VUV photoreactor. It was observed that 99.5% degradation and 94% chemical oxygen demand (COD) removal of 200 mg/L formaldehyde was achieved in the UVC/VUV photoreactor at a reaction time of 60 min and solution pH of 7. The effect of water anions such as bicarbonate, carbonate, chloride, nitrate, sulfate, and phosphate was studied on degradation and COD removal of formaldehyde; nitrate and carbonate exhibited the highest inhibitory effects on the process. Moreover, the treatment of formaldehyde-contaminated tap water was also explored, and formaldehyde removal was decreased from 99.5% in aqueous solution to 86.2% in tap water. The findings of radical scavenging tests revealed that hydroxyl radical was the most predominant oxidizing agent contributed to the degradation of formaldehyde.

3.1.2 Ozone-Based AOP

Ozone (O_3) is one of the most powerful oxidants that have been used thoroughly in the treatment of wastewater. Ozone-based methods have been employing for the management of wastewater form marine aquaria, textiles, petroleum refineries, electroplating wastes, distilleries, mills, and molasses (Wu et al. 2007). A recent work on different wastewater treatments using ozone-based AOPs such as ozone/hydrogen peroxide, ozone/UV, ozone/ultrasound, and ozone/titanium dioxide photocatalysis is used frequently nowadays. In ozone-based processes, the formation of radicals such as the hydroxyl radical (·HO) and superoxide radical ($O_2^{\cdot-}$) occurs when ozone is added to water, through a complex series of reactions (Ried et al. 2007). The disintegration rate of ozone in water is better at higher pH levels (Ikehata and Gamal El-Din 2005). Moreover, the alkalinity of water, contact time, and the dose of ozone determine the performance of the process (Liu et al. 2019a, b).

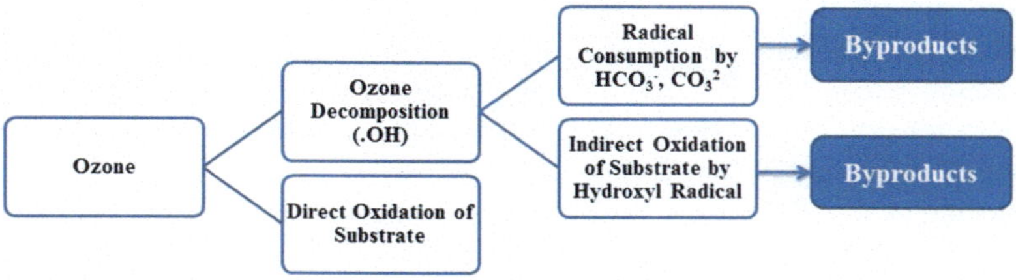

Fig. 4 Oxidation reaction of pollutants during ozonation

Ozone has been used as an oxidant and disinfectant in water treatment since long; as an oxidant, ozone is very selective and attacks primarily electron-rich functional groups like double bonds, amines, and activated aromatic rings (Chávez et al. 2019). Subsequently, ozone reactions in real aqueous solutions involve the formation of OH; ozonation itself is considered an AOP or AOP-like process. OH can be formed from the reaction of ozone with hydroxide ions. The process can be initiated by the introduction of hydrogen peroxide; the oxidation of highly reactive species with ozone took place which eventually enhances the degradation (Wang et al. 2020). The process of ozonation by H_2O_2/O_3 systems is the most established AOP in the treatment of wastewater as compared to other AOPs. Another way to speed up ozonation is by the use of homogenous/heterogeneous catalysts. Studies have been carried out with various metal ions/metal oxides that have shown significant changes in decomposition (Li et al. 2019). The concentration of ozone and ozonation time is the predominant factor for the enhancement of degradation rate (Liu et al. 2019a, b). Figure 4 shows the oxidation reaction during the process of ozonation.

Oxidation studies have been carried out on various organic contaminants using Fe_2O_3, MnO_2, TiO_2–Me, Fe^{2+}, Fe^{3+}, and Mn^{2+}. The ozone/catalyst system appears to be more effective for the reduction of chemical oxygen demand (COD) and total organic carbon (TOC) than oxidation with ozone alone at higher pH values (Lee et al. 2011). There are several oxidation mechanisms that result in the destruction of organic contaminants, but HO· is the predominant removal mechanism. The ozone reaction involves the following reactions (Ikehata and Gamal El-Din 2005):

$$O_3 + H_2O + h\upsilon \rightarrow O_2 + H_2O_2 \text{ at } (h\upsilon : \lambda < 300\text{ nm})$$

$$2O_3 + H_2O_2 \rightarrow 2HO \cdot + 3O_2$$

Figueredo et al. (2019) studied ozone-based advanced oxidation processes for primidone removal in water using simulated solar radiation and TiO_2 or WO_3 as photocatalyst. Primidone is a highly persistent pharmacological drug typically found in urban wastewaters. The comparison of processes, kinetics, nature of transformation products, ecotoxicity of treated water samples and influence of the water matrix were also evaluated. It was observed that in the presence of ozone, primidone is rapidly eliminated while hydroxyl radicals being the main species involved. TiO_2 was the most active catalyst regardless of the water matrix and the type of solar radiation applied. The synergy between ozone and photocatalysis (photocatalytic ozonation) for TOC removal was more evident at low O_3 doses. In spite of having a lower bandgap than TiO_2 P25, WO_3 did not bring any beneficial effects compared to TiO_2 P25 regarding TOC removal. The ozonation is found very effective for the removal of primidone.

Mansouri et al. (2019) investigated several AOPs such as O_3/H_2O_2, O_3/TiO_2, O_3/activated carbon (AC), O_3/Al_2O_3, $O_3/Fe^{2+}/H_2O_2$, and UV/TiO_2 and compared for the removal of diethyl phthalate (DEP), an endocrine-disrupting compound, in aqueous solutions. Hydroxyl radicals were the main species responsible for DEP degradation, and this was supported by computational chemistry calculation, scavenger experiments, and LC/MS/MS analysis. Operating parameters such as pH, ozone concentration, and catalyst dosage were also studied to achieve an enhanced performance of the combined processes. O_3/AC process was found to reduce the oxidation efficiency of O_3 at high ozone concentrations. Heterogeneous catalytic ozonation with Al_2O_3 was found to be the most effective process for DEP removal (~100% removal in about 15 min).

Mestankova et al. (2016) studied the removal of emerging contaminants during water treatment like UV- and ozone-based advanced oxidation processes (AOPs). In this study, AOPs were explored for their degradation capabilities of around 25 chemical contaminants. Twenty-three of these were found to be responsive to hydroxyl radical-based treatment, with second-order rate constants for their reactions with hydroxyl radicals (·OH) in the range of $3–8 \times 10^9\ M^{-1}\ s^{-1}$. Overall, this study provides fundamental and practical information on the AOP-based treatment of specific compounds.

3.1.3 Fenton-Based AOP

The combination of ferrous iron (Fe(II) and H_2O_2 in acidic condition results in the formation of ·OH termed as Fenton reaction. The Fenton reaction is restricted to acidic conditions to prevent the formation of precipitation of iron (Hartmann et al. 2010). In this process, iron acts as a catalyst with maximum catalytic activity at pH 3, especially due to the precipitation of ferric oxyhydroxide at higher pH value (Salimi et al. 2017). Also the addition of H_2O_2 in excess leads to the reduction of Fe(III) to Fe(II) in the reaction. It is also observed that the substitution of iron oxide with other transition metals can enhance total performance. The main benefit of the Fenton process is operation at low costs and the possibility of easy magnetic separation of residual iron without generating the secondary pollutant (Bach et al. 2010). The Fenton process is therefore established in several industrial pilot-scale applications.

Mechanism of Fenton Reaction: Like other AOPs, this process is also initiated by the hydroxyl radical for the degradation of organic compounds, mainly by oxidation reactions (Muruganandham et al. 2014a, b). But, the Fenton process is strict and requires absolute pH control to prevent precipitation of the iron. The reactor must be designed and developed to allow proper mixing of the Fe(II) and H_2O_2 to allow optimum hydroxyl radical formation and degradation of the contaminants (Kornweitz et al. 2015). Moreover, the Fenton process is a simple method to produce

HO· without the requirement of special apparatus and chemicals, and the reaction took place at ambient pressure and temperature; therefore, no special reaction condition is required for the process (Schrank et al. 2005). The Fenton method is an attractive way for oxidation, as H_2O_2 and iron salts are easily available, easy to handle, and environmentally safe as compared to other chemical-dependent reactions (Rodríguez et al. 2016). The destruction of organic compounds takes place by the reaction of the HO·. The formation of HO· radicals are shown in equation:

$$Fe^{2+} + H_2O_2 \rightarrow Fe^{3+} + OH^- + HO\cdot$$

It is important to note that the rate of degradation of organic pollutants with Fenton-Fenton-like reagents is strongly accelerated by irradiation with UV-VIS light at wavelengths greater than 300 nm (Rusevova et al. 2012). The photolysis of Fe^{3+} complexes allows Fe^{2+} to be regenerated. The occurrence of Fenton reactions in the presence of hydrogen peroxide is shown in equation:

$$Fe(OH)^{2+}\ h\nu \rightarrow Fe^{2+} + HO \cdot + OH^-$$

The generalized mechanism of the Fenton process is that HO radicals are produced during the activation of hydrogen peroxide (H_2O_2) by Fe^{2+} ion and form Fe^{3+} ion as shown in Fig. 5. The oxidized Fe^{3+} ion further reacts with H_2O_2 and forms Fe^{2+} ion. Until now, Fenton and photo-Fenton processes have been used to treat a wide range of toxic compounds such as herbicides, pesticides, phenols, dye effluents, pulp bleaching, and chemical manufacturing (Pouran et al. 2014). Additionally, Fenton pretreatment can be used to enhance biological wastewater treatment. The Fenton, photo-Fenton, and Fenton-like processes are popular methods for

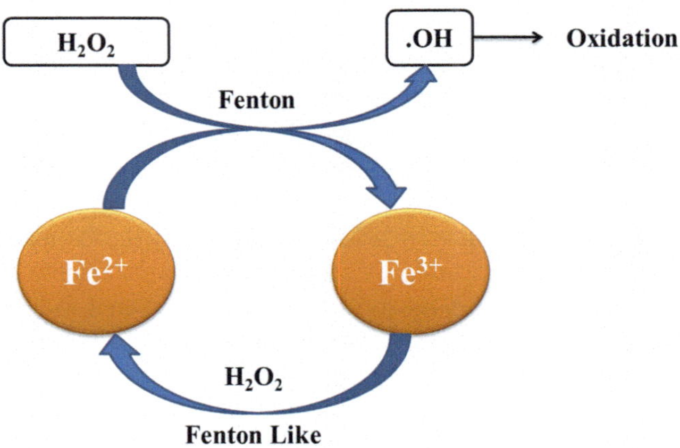

Fig. 5 The Fenton process

AOP due to their simplicity, flexibility, and integration into existing water treatment methods such as coagulation (Chacón et al. 2006). The major drawback of using the Fenton process is the production of sludge during the process that contains iron hydroxide as a by-product; also periodic addition of H_2O_2 is one of the drawbacks of the Fenton process, which considerably increases the operation cost (Jiang and Zhang 2007). The disposal of the sludge formed has to be included when estimating the cost of this process which increases the overall cost of the process (Pliego et al. 2015; Barbusiński 2005).

Zárate-Guzmán et al. (2019) investigated an iron-supported catalyst on granular activated carbon for its use in heterogeneous Fenton reaction coupled to an in situ H_2O_2 electrogeneration. An electrolysis cell was employed for this reaction using carbon felt as cathode and graphite as an anode. A solution of H_2O_2 (electrogenerated at a rate of 30 mg L^{-1} h^{-1}) was obtained using a current intensity of 12 mA. A carbon-Fe catalyst was used in order to promote the decomposition of H_2O_2 to ·OH. The catalyst was prepared by an incipient wet impregnation method using $FeSO_4$ as precursor salt to obtain samples with 9% wt of iron. Samples were characterized by EDX, FTIR, and XPS spectroscopy before and after wastewater treatment using phenol as a model molecule. Two iron oxidation states on the samples were found, Fe^{2+} and Fe^{3+}. The ratio between Fe^{2+} and Fe^{3+} was 1.29 which was later reduced to 0.92 after the Fenton process; this might be associated with the metal oxidation (Fe^{2+} to Fe^{+3}) occurring during Fenton reaction, thus indicating that H_2O_2 decomposition was carried out by Fe^{2+} on the carbon surface. Detection and quantification of hydroxyl radicals were carried out by fluorescence spectroscopy, obtaining a radical concentration of 3.5 μM in solution. Results showed an environmentally friendly process that can generate reagents in situ, with high efficiencies in the degradation of pollutants and minimizing the formation of toxic by-products, which are common in conventional treatments.

3.1.4 Sonolysis

Sonolysis is one of the best green chemistry-based environmental friendly AOPs. Sonolysis of water by ultrasound (US 20–5 kHz) leads to the rapid formation and collapse of fine bubbles formed due to the acoustic cavitation by induced compression and rarefaction (Stock et al. 2000). These bubbles explode rigorously after reaching the critical resonance size and generate high temperature (>5000 K) and high pressure (>1000 bar) and highly reactive oxidative reactive species (Ince and Tezcanlı 2001).

The destruction of contaminants in wastewater occurs by the thermal decomposition and reactions of various radicals; the ultrasonic cavitation exhibits low interference from the water matrix and less heat transfer as compared to UV irradiation (Olson and Barbier 1994). Though the process of ultrasonic is extensively energy-intensive, therefore, the coupling of ultrasound with UV irradiation (sonophotolysis), oxidants (O_3, H_2O_2), catalysts (TiO_2) (sonocatalysis), or both (sonophotocatalysis) receives the utmost attention (Isariebel et al. 2009; Chakma and Moholkar 2015). These hybrid processes can yield additional advantages, in such conditions that the development of advanced hybrid technologies has resulted in the improvement of the degradation efficiency of organic pollutants in waters and of the reduction of the sonochemical treatment time (Hoseini et al. 2013).

Ultrasonication is an aqueous medium involving particular AOP which can process via two different types of actions, either chemical or a physical mechanism. In chemical or indirect action, at high-energy frequency, water and dioxygen molecules undergo hemolytic fragment and yield ·OH, HO_2·, and ·O radicals (Cubillana-Aguilera et al. 2011). On the other hand, direct action, called sonication, involves the formation by the ultrasounds of cavitation bubbles which grow and then collapse, creating powerful breaking forces with extremely high temperatures ($T = 2000–5000$ K) and pressures (about 6 × 104 kPa) (Maezawa et al. 2007). At these extreme conditions, a sonolysis of water molecules occurs, which produces very reactive radicals that are able to react with organic chemical species present in the aqueous medium, and/or a pyrolysis degradation of organic compounds is taking place (Gogate et al. 2003). The main reaction involved during the process of ultrasonication is shown in equations:

$$H_2O + \text{Ultrasound} \rightarrow \cdot OH + \cdot H$$

$$\cdot OH + \text{Organic compound} \rightarrow \text{Products}$$

$$\text{Organic compound} + \Delta H \rightarrow \text{Products}$$

Thus, in the recent past, ultrasonication has been used widely for the oxidation and degradation of various organic pollutants in water and wastewaters. But, the main drawback of the water treatment by ultrasounds is that the number of generated ·OH radicals is generally insufficient (Serna-Galvis et al. 2019). Sonolysis is considered one of the cost-effective and environmentally safe processes and treats wastes at comparatively mild reaction conditions.

3.1.5 Electrochemical Oxidation Processes

Electrochemical AOP is based on the transfer of electrons which makes it interesting for environmental applications as it employs a clean way to generate hydroxyl radicals (in situ), which are able to destroy a wide range of toxic wastes from wastewater (Dewil et al. 2017). These ·OH radicals can be electrochemically produced either directly (AOP) or indirectly via electrocatalytically (in situ) generated electro-Fenton's reagent. The efficiency of the process can be further increased by combining both electrochemical processes and the classical EF process (Flores et al. 2018). The electrochemical system generates oxidizing agents such as the hydroxyl radical (˙OH) in the solution, resulting in the destruction of organic contamination up to their mineralization (Canizares et al. 2009).

In electrochemical AOP, the main electrodes which are commonly used are boron-doped diamond (BDD), doped PbO_2, SnO_2, RuO_2, and sub-stoichiometric and doped TiO_2 (Cañizares et al. 2008; Scialdone 2009). However, BDD electrodes are the most applied electrochemical AOP method due to their relatively low production costs compared to other electrodes and high stability of the diamond layer under anodic polarization (Rodrigo et al. 2001; Martinez-Huitle and Ferro 2006). The electrochemical AOP of contaminated wastewater with BDD electrodes proceeds with the generation of OH radical directly via O_2 evolution from the oxidation of water (Ganzenko et al. 2014). Thus, BDD electrode treatment gains interest as an eco-friendly and efficient method for the removal of various organic and inorganic pollutants (Alfaro et al. 2006).

3.1.6 Other AOPs

Microwave: The application of highly energetic radiation in the microwave range (30 MHz–300 GHz) has been investigated for the oxidation of contaminated water. Microwave irradiation has been used for the combination with oxidants (H_2O_2) and catalyst to assist the degradation of organic pollutants (Remya and Lin 2011). Microwave can enhance the rate of reaction and induce selective heating of the contaminants through internal molecular vibrations (Mudhoo and Sharma 2011).

Electron Beam

The utilization of ionizing radiation from an electron source for the treatment of wastewater is used since long ago. The electron penetrates the water surface and leads to the formation of electronically excited species in the water which includes various free radicals and ionic species (Arslan-Alaton 2003). This process exhibits a high oxidizing power and little interference by the water matrix, and the electrical efficiency is within the feasibility range for most contaminants (Wojnarovits and Takacs 2008).

The development of AOP applications has been motivated by increasingly strict regulations; the contamination of water resources through various sources and the regulatory requirements should meet effluent discharge standards. The above-discussed AOPs are very helpful in the remediation of wastewater and provide the better solution to the existing problems as compared to the conventional ones.

4 Conclusion

The present chapter provides the detail of the advanced oxidation process for wastewater treatment. AOP emerges as the most efficient green chemistry-based technology for the remediation of a wide range of emerging contaminants from water and wastewater. The AOPs provide environmental friendly solutions to the problem of wastewater in a sustainable manner. AOPs are based on a basic principle that entails the generation and use of a hydroxyl free radical as a strong oxidant for the destruction of compounds that cannot be oxidized using conventional oxidants and technologies. AOPs have shown premises in various water treatment sectors due to its superior efficiency in pollutant elimination in a sustainable way. Various combinations of AOP processes like ozone, UV, H_2O_2, etc. may provide efficient treatment of wastewater depending upon the characteristics of wastewater. Further, the economic optimization of these processes must be worked out for treating a huge amount of wastewater by AOPs. However, they need to be evaluated for effectiveness, costs, and possible side effects. In the general comparison, the effects of different inorganic/organic compounds in the water were not considered. There are still many research needs in the field of AOPs for water, wastewater, and polluted air treatment, including further R&D to provide a better understanding of the mechanisms of candidate AOPs, costs, efficiency, evaluation of by-products, and their toxicity. Moreover, the improvement of AOP could ensure economic solutions to the problem of water treatment and recovery.

References

Ahmed MB, Zhou JL, Ngo HH, Guo W, Thomaidis NS, Xu J (2017) Progress in the biological and chemical treatment technologies for emerging contaminant removal from wastewater: a critical review. J Hazard Mater 323:274–298

Alfaro MAQ, Ferro S, Martínez-Huitle CA, Vong YM (2006) Boron doped diamond electrode for the wastewater treatment. J Braz Chem Soc 17:227–236

Arslan I, Balcioglu IA, Bahnemann DW (2000) Heterogeneous photocatalytic treatment of simulated dyehouse effluents using novel TiO2-photocatalysts. Appl Catal B Environ 26 (3):193–206

Arslan-Alaton I (2003) A review of the effects of dye-assisting chemicals on advanced oxidation of reactive dyes in wastewater. Color Technol 119(6):345–353

Arzate S, Pfister S, Oberschelp C, Sánchez-Pérez JA (2019) Environmental impacts of an advanced oxidation process as tertiary treatment in a wastewater treatment plant. Sci Total Environ 694:133572

Augugliaro V, Bellardita M, Loddo V, Palmisano G, Palmisano L, Yurdakal S (2012) Overview on oxidation mechanisms of organic compounds by TiO2 in heterogeneous photocatalysis. J Photochem Photobiol C: Photochem Rev 13(3):224–245

Bach A, Shemer H, Semiat R (2010) Kinetics of phenol mineralization by Fenton-like oxidation. Desalination 264(3):188–192

Barbusiński K (2005) The modified Fenton process for decolorization of dye wastewater. Pol J Environ Stud 14(3):281–285

Bolong N, Ismail AF, Salim MR, Matsuura T (2009) A review of the effects of emerging contaminants in wastewater and options for their removal. Desalination 239(1–3):229–246

Buthiyappan A, Aziz ARA, Daud WMAW (2016) Recent advances and prospects of catalytic advanced oxidation process in treating textile effluents. Rev Chem Eng 32(1):1–47

Cañizares P, Paz R, Sáez C, Rodrigo MA (2008) Electrochemical oxidation of alcohols and carboxylic acids with diamond anodes: a comparison with other advanced oxidation processes. Electrochim Acta 53(5):2144–2153

Canizares P, Paz R, Sáez C, Rodrigo MA (2009) Costs of the electrochemical oxidation of wastewaters: a comparison with ozonation and Fenton oxidation processes. J Environ Manag 90(1):410–420

Chacón JM, Leal MT, Sanchez M, Bandala ER (2006) Solar photocatalytic degradation of azo-dyes by photo-Fenton process. Dyes Pigments 69(3):144–150

Chakma S, Moholkar VS (2015) Investigation in mechanistic issues of sonocatalysis and sonophotocatalysis using pure and doped photocatalysts. Ultrason Sonochem 22:287–299

Chávez AM, Gimeno O, Rey A, Pliego G, Oropesa AL, Álvarez PM, Beltrán FJ (2019) Treatment of highly polluted industrial wastewater by means of sequential aerobic biological oxidation-ozone based AOPs. Chem Eng J 361:89–98

Cieśla P, Kocot P, Mytych P, Stasicka Z (2004) Homogeneous photocatalysis by transition metal complexes in the environment. J Mol Catal A Chem 224(1–2):17–33

Cubillana-Aguilera LM, Franco-Romano M, Gil MLA, Naranjo-Rodríguez I, De Cisneros JHH, Palacios-Santander JM (2011) New, fast and green procedure for the synthesis of gold nanoparticles based on sonocatalysis. Ultrason Sonochem 18(3):789–794

Deng Y, Ezyske CM (2011) Sulfate radical-advanced oxidation process (SR-AOP) for simultaneous removal of refractory organic contaminants and ammonia in landfill leachate. Water Res 45(18):6189–6194

Dewil R, Mantzavinos D, Poulios I, Rodrigo MA (2017) New perspectives for advanced oxidation processes. J Environ Manag 195:93–99

Doll TE, Frimmel FH (2005) Removal of selected persistent organic pollutants by heterogeneous photocatalysis in water. Catal Today 101(3–4):195–202

Ebbs S (2004) Biological degradation of cyanide compounds. Curr Opin Biotechnol 15(3):231–236

Espinoza LAT, Frimmel FH (2009) A simple simulation of the degradation of natural organic matter in homogeneous and heterogeneous advanced oxidation processes. Water Res 43 (16):3902–3909

Faouzi M, Cañizares P, Gadri A, Lobato J, Nasr B, Paz R, Rodrigo MA, Saez C (2006) Advanced oxidation processes for the treatment of wastes polluted with azoic dyes. Electrochim Acta 52 (1):325–331

Figueredo MA, Rodríguez EM, Checa M, Beltran FJ (2019) Ozone-based advanced oxidation processes for primidone removal in water using simulated solar radiation and TiO_2 or WO_3 as photocatalyst. Molecules 24(9):1728

Flores N, Sharif F, Yasri N, Brillas E, Sirés I, Roberts EP (2018) Removal of tyrosol from water by adsorption on carbonaceous materials and electrochemical advanced oxidation processes. Chemosphere 201:807–815

Fujishima A, Honda K (1972) Electrochemical photolysis of water at a semiconductor electrode. Nature 238(5358):37–38

Fujishima A, Zhang X, Tryk DA (2007) Heterogeneous photocatalysis: from water photolysis to applications in environmental cleanup. Int J Hydrog Energy 32(14):2664–2672

Ganzenko O, Huguenot D, Van Hullebusch ED, Esposito G, Oturan MA (2014) Electrochemical advanced oxidation and biological processes for wastewater treatment: a review of the combined approaches. Environ Sci Pollut Res 21(14):8493–8524

Ghime D, Goru P, Ojha S, Ghosh P (2019) Oxidative decolorization of a malachite green oxalate dye through the photochemical advanced oxidation processes. Global NEST J 21(2):195–203

Glaze W, Kang J-W, Chapin DH (1987) The chemistry of water treatment processes involving ozone, hydrogen peroxide and ultraviolet radiation. Ozone Sci Eng 9(4):335–352. https://doi.org/10.1080/01919518708552148

Gogate PR, Pandit AB (2004) A review of imperative technologies for wastewater treatment I: oxidation technologies at ambient conditions. Adv Environ Res 8(3–4):501–551

Gogate PR, Mujumdar S, Pandit AB (2003) Sonochemical reactors for waste water treatment: comparison using formic acid degradation as a model reaction. Adv Environ Res 7(2):283–299

Grassi M, Kaykioglu G, Belgiorno V, Lofrano G (2012) Removal of emerging contaminants from water and wastewater by adsorption process. In: Emerging compounds removal from wastewater. Springer, Dordrecht, pp 15–37

Guo Z, Ma R, Li G (2006) Degradation of phenol by nanomaterial TiO2 in wastewater. Chem Eng J 119(1):55–59

Haddad FA, Moussavi G, Moradi M (2019) Advanced oxidation of formaldehyde in aqueous solution using the chemical-less UVC/VUV process: kinetics and mechanism evaluation. J Water Process Eng 27:120–125

Hartmann M, Kullmann S, Keller H (2010) Wastewater treatment with heterogeneous Fenton-type catalysts based on porous materials. J Mater Chem 20(41):9002–9017

Herrmann JM (1999) Heterogeneous photocatalysis: fundamentals and applications to the removal of various types of aqueous pollutants. Catal Today 53(1):115–129

Hisatomi T, Kubota J, Domen K (2014) Recent advances in semiconductors for photocatalytic and photoelectrochemical water splitting. Chem Soc Rev 43(22):7520–7535

Hoffmann MR, Martin ST, Choi W, Bahnemann DW (1995) Environmental applications of semiconductor photocatalysis. Chem Rev 95(1):69–96

Hoseini M, Safari GH, Kamani H, Jaafari J, Ghanbarain M, Mahvi AH (2013) Sonocatalytic degradation of tetracycline antibiotic in aqueous solution by sonocatalysis. Toxicol Environ Chem 95(10):1680–1689

Huang N, Wang T, Wang WL, Wu QY, Li A, Hu HY (2017) UV/chlorine as an advanced oxidation process for the degradation of benzalkonium chloride: synergistic effect, transformation products and toxicity evaluation. Water Res 114:246–253

Ibhadon AO, Fitzpatrick P (2013) Heterogeneous photocatalysis: recent advances and applications. Catalysts 3(1):189–218

Ikehata K, Gamal El-Din M (2005) Aqueous pesticide degradation by ozonation and ozone-based advanced oxidation processes: a review (part I). Ozone Sci Eng 27(2):83–114

Ilisz I, Bokros A, Dombi A (2004) TiO2-based heterogeneous photocatalytic water treatment combined with ozonation. Ozone Sci Eng 26(6):585–594

Ince NH, Tezcanlí G (2001) Reactive dyestuff degradation by combined sonolysis and ozonation. Dyes Pigments 49(3):145–153

Isariebel QP, Carine JL, Ulises-Javier JH, Anne-Marie W, Henri D (2009) Sonolysis of levodopa and paracetamol in aqueous solutions. Ultrason Sonochem 16(5):610–616

Ismail AA, Bahnemann DW (2014) Photochemical splitting of water for hydrogen production by photocatalysis: a review. Sol Energy Mater Sol Cells 128:85–101

Jamshidi N, Torabian A, Azimi A, Nabi Bidhendi GR, Jafarzadeh MT (2009) Investigation of phenol removal in aqueous solutions using advanced photochemical oxidation (APO). J Water Wastewater 72:24–29

Jiang CC, Zhang JF (2007) Progress and prospect in electro-Fenton process for wastewater treatment. J Zhejiang Univ Sci A 8(7):1118–1125

Joshi KM, Shrivastava VS (2011) Photocatalytic degradation of chromium (VI) from wastewater using nanomaterials like TiO 2, ZnO, and CdS. Appl Nanosci 1(3):147–155

Kabra K, Chaudhary R, Sawhney RL (2004) Treatment of hazardous organic and inorganic compounds through aqueous-phase photocatalysis: a review. Ind Eng Chem Res 43 (24):7683–7696

Kanakaraju D, Glass BD, Oelgemöller M (2018) Advanced oxidation process-mediated removal of pharmaceuticals from water: a review. J Environ Manag 219:189–207

Khan SH (2020) Green nanotechnology for the environment and sustainable development. In: Green materials for wastewater treatment. Springer, Cham, pp 13–46

Khan SH, Pathak B, Fulekar MH (2017) Spherical surfaced magnetic (Fe3O4) nanoparticles as nano adsorbent material for treatment of industrial dye effluents. Int J Nanosci Nanotechnol 13 (2):169–175

Khan SH, Pathak B, Fulekar MH (2018) Synthesis, characterization and photocatalytic degradation of chlorpyrifos by novel Fe: ZnO nanocomposite material. Nanotechnol Environ Eng 3(1):13

Kim KH, Ihm SK (2011) Heterogeneous catalytic wet air oxidation of refractory organic pollutants in industrial wastewaters: a review. J Hazard Mater 186(1):16–34

Klavarioti M, Mantzavinos D, Kassinos D (2009) Removal of residual pharmaceuticals from aqueous systems by advanced oxidation processes. Environ Int 35(2):402–417

Kornweitz H, Burg A, Meyerstein D (2015) Plausible mechanisms of the Fenton-like reactions, M= Fe (II) and Co (II), in the presence of RCO2–substrates: are OH· radicals formed in the process? Chem A Eur J 119(18):4200–4206

Korpe S, Bethi B, Sonawane SH, Jayakumar KV (2019) Tannery wastewater treatment by cavitation combined with advanced oxidation process (AOP). Ultrason Sonochem 59:104723

Ksibi M (2006) Chemical oxidation with hydrogen peroxide for domestic wastewater treatment. Chem Eng J 119(2–3):161–165

Lamas Samanamud GR, Loures CC, Souza AL, Salazar RF, Oliveira IS, Silva MB, Izário Filho HJ (2012) Heterogeneous photocatalytic degradation of dairy wastewater using immobilized ZnO. ISRN Chem Eng 2012:275371

Lee E, Lee H, Kim YK, Sohn K, Lee K (2011) Hydrogen peroxide interference in chemical oxygen demand during ozone based advanced oxidation of anaerobically digested livestock wastewater. Int J Environ Sci Technol 8(2):381–388

Li K, Xu L, Zhang Y, Cao A, Wang Y, Huang H, Wang J (2019) A novel electro-catalytic membrane contactor for improving the efficiency of ozone on wastewater treatment. Appl Catal B Environ 249:316–321

Lim J, Kwak DY, Sieland F, Kim C, Bahnemann DW, Choi W (2018) Visible light-induced catalytic activation of peroxymonosulfate using heterogeneous surface complexes of amino acids on TiO2. Appl Catal B Environ 225:406–414

Lima EC (2018) Removal of emerging contaminants from the environment by adsorption. Ecotoxicol Environ Saf 150:1–17

Litter MI (1999) Heterogeneous photocatalysis: transition metal ions in photocatalytic systems. Appl Catal B Environ 23(2–3):89–114

Liu Y, He X, Fu Y, Dionysiou DD (2016) Degradation kinetics and mechanism of oxytetracycline by hydroxyl radical-based advanced oxidation processes. Chem Eng J 284:1317–1327

Liu Z, Chys M, Yang Y, Demeestere K, Van Hulle S (2019a) Oxidation of trace organic contaminants (TrOCs) in wastewater effluent with different ozone-based AOPs: comparison of ozone exposure and ·OH formation. Ind Eng Chem Res 58(20):8896–8902

Liu Z, Hosseinzadeh S, Wardenier N, Verheust Y, Chys M, Hulle SV (2019b) Combining ozone with UV and H2O2 for the degradation of micropollutants from different origins: lab-scale analysis and optimization. Environ Technol 40(28):3773–3782

Lopez JL, Einschlag FSG, González MC, Capparelli AL, Oliveros E, Hashem TM, Braun AM (2000) Hydroxyl radical initiated photodegradation of 4-chloro-3, 5-dinitrobenzoic acid in aqueous solution. J Photochem Photobiol A Chem 137(2–3):177–184

Maezawa A, Nakadoi H, Suzuki K, Furusawa T, Suzuki Y, Uchida S (2007) Treatment of dye wastewater by using photo-catalytic oxidation with sonication. Ultrason Sonochem 14 (5):615–620

Mansouri L, Tizaoui C, Geissen SU, Bousselmi L (2019) A comparative study on ozone, hydrogen peroxide and UV based advanced oxidation processes for efficient removal of diethyl phthalate in water. J Hazard Mater 363:401–411

Martínez C, Fernández MI, Santaballa JA, Faria J (2011) Aqueous degradation of diclofenac by heterogeneous photocatalysis using nanostructured materials. Appl Catal B Environ 107 (1–2):110–118

Martinez-Huitle CA, Ferro S (2006) Electrochemical oxidation of organic pollutants for the wastewater treatment: direct and indirect processes. Chem Soc Rev 35(12):1324–1340

Méndez-Díaz J, Sánchez-Polo M, Rivera-Utrilla J, Canonica S, Von Gunten U (2010) Advanced oxidation of the surfactant SDBS by means of hydroxyl and sulphate radicals. Chem Eng J 163 (3):300–306

Mestankova H, Parker AM, Bramaz N, Canonica S, Schirmer K, Von Gunten U, Linden KG (2016) Transformation of contaminant candidate list (CCL3) compounds during ozonation and advanced oxidation processes in drinking water: assessment of biological effects. Water Res 93:110–120

Mills A, Le Hunte S (1997) An overview of semiconductor photocatalysis. J Photochem Photobiol A Chem 108(1):1–35

Mills A, Davies RH, Worsley D (1993) Water purification by semiconductor photocatalysis. Chem Soc Rev 22(6):417–425

Mudhoo A, Sharma SK (2011) Microwave irradiation technology in waste sludge and wastewater treatment research. Crit Rev Environ Sci Technol 41(11):999–1066

Murray KE, Thomas SM, Bodour AA (2010) Prioritizing research for trace pollutants and emerging contaminants in the freshwater environment. Environ Pollut 158(12):3462–3471

Muruganandham M, Suri RPS, Jafari S, Sillanpää M, Lee GJ, Wu JJ, Swaminathan M (2014a) Recent developments in homogeneous advanced oxidation processes for water and wastewater treatment. Int J Photoenergy 2014:821674

Muruganandham M, Suri RP, Sillanpää M, Wu JJ, Ahmmad B, Balachandran S, Swaminathan M (2014b) Recent developments in heterogeneous catalyzed environmental remediation processes. J Nanosci Nanotechnol 14(2):1898–1910

Nosaka Y, Nishikawa M, Nosaka AY (2014) Spectroscopic investigation of the mechanism of photocatalysis. Molecules 19(11):18248–18267

Oh PP, Lau HLN, Chen J, Chong MF, Choo YM (2012) A review on conventional technologies and emerging process intensification (PI) methods for biodiesel production. Renew Sust Energ Rev 16(7):5131–5145

Olmez-Hanci T, Arslan-Alaton I (2013) Comparison of sulfate and hydroxyl radical based advanced oxidation of phenol. Chem Eng J 224:10–16

Olson TM, Barbier PF (1994) Oxidation kinetics of natural organic matter by sonolysis and ozone. Water Res 28(6):1383–1391

Oppenländer T (2007) Photochemical purification of water and air: advanced oxidation processes (AOPs)-principles, reaction mechanisms, reactor concepts. Wiley, Chichester

Pelizzetti E, Serpone N (eds) (2012) Homogeneous and heterogeneous photocatalysis, vol 174. Springer, New York

Petrović M, Gonzalez S, Barceló D (2003) Analysis and removal of emerging contaminants in wastewater and drinking water. TrAC Trends Anal Chem 22(10):685–696

Pinto M, Amor M, Fuentes L, Troya JM (2001) Supporting heterogeneous users in collaborative virtual environments using AOP. In: International conference on cooperative information systems, Sept 2011. Springer, Berlin, pp 226–238

Pliego G, Zazo JA, Garcia-Muñoz P, Munoz M, Casas JA, Rodriguez JJ (2015) Trends in the intensification of the Fenton process for wastewater treatment: an overview. Crit Rev Environ Sci Technol 45(24):2611–2692

Pouran SR, Raman AAA, Daud WMAW (2014) Review on the application of modified iron oxides as heterogeneous catalysts in Fenton reactions. J Clean Prod 64:24–35

Poyatos JM, Muñio MM, Almecija MC, Torres JC, Hontoria E, Osorio F (2010) Advanced oxidation processes for wastewater treatment: state of the art. Water Air Soil Pollut 205 (1–4):187

Remya N, Lin JG (2011) Current status of microwave application in wastewater treatment—a review. Chem Eng J 166(3):797–813

Richardson SD, Ternes TA (2018) Water analysis: emerging contaminants and current issues. Anal Chem 90(1):398–428

Ried A, Mielcke J, Wieland A, Schaefer S, Sievers M (2007) An overview of the integration of ozone systems in biological treatment steps. Water Sci Technol 55(12):253–258

Rodrigo MA, Michaud PA, Duo I, Panizza M, Cerisola G, Comninellis C (2001) Oxidation of 4-chlorophenol at boron-doped diamond electrode for wastewater treatment. J Electrochem Soc 148(5):D60–D64

Rodríguez R, Espada JJ, Pariente MI, Melero JA, Martínez F, Molina R (2016) Comparative life cycle assessment (LCA) study of heterogeneous and homogenous Fenton processes for the treatment of pharmaceutical wastewater. J Clean Prod 124:21–29

Rodriguez-Narvaez OM, Peralta-Hernandez JM, Goonetilleke A, Bandala ER (2017) Treatment technologies for emerging contaminants in water: a review. Chem Eng J 323:361–380

Rusevova K, Kopinke FD, Georgi A (2012) Nano-sized magnetic iron oxides as catalysts for heterogeneous Fenton-like reactions—influence of Fe (II)/Fe (III) ratio on catalytic performance. J Hazard Mater 241:433–440

Sadegh H, Ali GA, Nia HJ, Mahmoodi Z (2019) Nanomaterial surface modifications for enhancement of the pollutant adsorption from wastewater: adsorption of nanomaterials. In: Nanotechnology applications in environmental engineering. IGI Global, Pennsylvania, pp 143–170

Salimi M, Esrafili A, Gholami M, Jafari AJ, Kalantary RR, Farzadkia M et al (2017) Contaminants of emerging concern: a review of new approach in AOP technologies. Environ Monit Assess 189(8):414

Sanches S, Crespo MTB, Pereira VJ (2010) Drinking water treatment of priority pesticides using low pressure UV photolysis and advanced oxidation processes. Water Res 44(6):1809–1818

Schrank SG, Jose HJ, Moreira RFPM, Schröder HF (2005) Applicability of Fenton and H2O2/UV reactions in the treatment of tannery wastewaters. Chemosphere 60(5):644–655

Scialdone O (2009) Electrochemical oxidation of organic pollutants in water at metal oxide electrodes: a simple theoretical model including direct and indirect oxidation processes at the anodic surface. Electrochim Acta 54(26):6140–6147

Senthilkumaar S, Porkodi K (2005) Heterogeneous photocatalytic decomposition of crystal violet in UV-illuminated sol–gel derived nanocrystalline TiO2 suspensions. J Colloid Interface Sci 288 (1):184–189

Serna-Galvis EA, Botero-Coy AM, Martínez-Pachón D, Moncayo-Lasso A, Ibáñez M, Hernández F, Torres-Palma RA (2019) Degradation of seventeen contaminants of emerging concern in municipal wastewater effluents by sonochemical advanced oxidation processes. Water Res 154:349–360

Stasinakis AS (2008) Use of selected advanced oxidation processes (AOPs) for wastewater treatment–a mini review. Global NEST J 10(3):376–385

Stock NL, Peller J, Vinodgopal K, Kamat PV (2000) Combinative sonolysis and photocatalysis for textile dye degradation. Environ Sci Technol 34(9):1747–1750

Suriyaprabha R, Khan SH, Pathak B, Fulekar MH (2016) Effective decolorization and adsorption of contaminant from industrial dye effluents using spherical surfaced magnetic (Fe3O4) nanoparticles. In: American Institute of Physics conference series, vol 1724, no 2, April 2016

Swami D, Buddhi D (2006) Removal of contaminants from industrial wastewater through various non-conventional technologies: a review. Int J Environ Pollut 27(4):324–346

Tichonovas M, Krugly E, Jankunaite D, Racys V, Martuzevicius D (2017) Ozone-UV-catalysis based advanced oxidation process for wastewater treatment. Environ Sci Pollut Res 24 (21):17584–17597

Tijani JO, Fatoba OO, Madzivire G, Petrik LF (2014) A review of combined advanced oxidation technologies for the removal of organic pollutants from water. Water Air Soil Pollut 225 (9):2102

Visanji Z, Sadr SM, Memon FA (2018) An implementation of a decision support tool to assess treatment of emerging contaminants in India. J Water Resour Protect 10:422

Wang JL, Xu LJ (2012) Advanced oxidation processes for wastewater treatment: formation of hydroxyl radical and application. Crit Rev Environ Sci Technol 42(3):251–325

Wang F, Smith DW, El-Din MG (2003) Application of advanced oxidation methods for landfill leachate treatment–a review. J Environ Eng Sci 2(6):413–427

Wang Y, Wang Q, Zhan X, Wang F, Safdar M, He J (2013) Visible light driven type II heterostructures and their enhanced photocatalysis properties: a review. Nanoscale 5 (18):8326–8339

Wang H, Zhang L, Chen Z, Hu J, Li S, Wang Z, Wang X (2014) Semiconductor heterojunction photocatalysts: design, construction, and photocatalytic performances. Chem Soc Rev 43 (15):5234–5244

Wang H, Zhan J, Gao L, Yu G, Komarneni S, Wang Y (2020) Kinetics and mechanism of thiamethoxam abatement by ozonation and ozone-based advanced oxidation processes. J Hazard Mater 390:122180

Wen J, Li X, Liu W, Fang Y, Xie J, Xu Y (2015) Photocatalysis fundamentals and surface modification of TiO2 nanomaterials. Chin J Catal 36(12):2049–2070

Wenderich K, Mul G (2016) Methods, mechanism, and applications of photodeposition in photocatalysis: a review. Chem Rev 116(23):14587–14619

Wojnarovits L, Takacs E (2008) Irradiation treatment of azo dye containing wastewater: an overview. Radiat Phys Chem 77(3):225–244

Wols BA, Hofman-Caris CHM (2012) Review of photochemical reaction constants of organic micropollutants required for UV advanced oxidation processes in water. Water Res 46 (9):2815–2827

Wu JJ, Muruganandham M, Chen SH (2007) Degradation of DMSO by ozone-based advanced oxidation processes. J Hazard Mater 149(1):218–225

Xie P, Guo Y, Chen Y, Wang Z, Shang R, Wang S et al (2017) Application of a novel advanced oxidation process using sulfite and zero-valent iron in treatment of organic pollutants. Chem Eng J 314:240–248

Zárate-Guzmán AI, González-Gutiérrez LV, Godínez LA, Medel-Reyes A, Carrasco-Marín F, Romero-Cano LA (2019) Towards understanding of heterogeneous Fenton reaction using carbon-Fe catalysts coupled to in-situ H2O2 electro-generation as clean technology for wastewater treatment. Chemosphere 224:698–706

Zhou H, Smith DW (2002) Advanced technologies in water and wastewater treatment. J Environ Eng Sci 1(4):247–264

Zou Z, Ye J, Sayama K, Arakawa H (2011) Direct splitting of water under visible light irradiation with an oxide semiconductor photocatalyst. In: Materials for sustainable energy. Nature Publishing Group, Berlin, pp 293–295

Rhizoremediation of Heavy Metal- and Xenobiotic-Contaminated Soil: An Eco-Friendly Approach

Pramila Verma and Seema Rawat

Abstract

In the last few decades, staggeredly increasing human population, indiscriminate use of pesticides, polycyclic aromatic hydrocarbons, and discharge of effluents containing toxic heavy metals, dyes, etc. have negatively impacted the soil fertility and biodiversity to the extent of converting the agricultural land into barren land. These pollutants can be removed or altered by conventional and physical methods. However, these methods are neither cost-effective nor eco-friendly as they generate large amount of toxic intermediates. The most powerful eco-friendly approach is bioremediation. Microorganisms play a pivotal role in the maintenance and sustainability of any ecosystem. They are versatile and capable to adapt to any challenges posed by the environment. The biotic activities of a diverse group of bacterial populations residing in rhizosphere affect the dynamics of soil which in turn affect the crop productivity. They stimulate plant growth by making soil nutrients available to them, stimulating the production of various plant growth hormones, protecting the plants from pathogens by their antagonistic activity and remediation of the pollutants by sequestering toxic heavy metals, and degrading xenobiotic compounds such as pesticides, PAHs, etc. This chapter highlights the role of PGPR in remediation of heavy metal- and xenobiotic-contaminated soil for sustainable crop production.

P. Verma
Department of Botany and Microbiology, H.N.B. Garhwal University, Srinagar, Uttarakhand, India

S. Rawat (✉)
Microbiology Lab, School of Life Sciences, Central University of Gujarat, Gandhinagar, Gujarat, India
e-mail: seema.rawat@cug.ac.in

M. P. Shah (ed.), *Removal of Emerging Contaminants Through Microbial Processes*,
https://doi.org/10.1007/978-981-15-5901-3_5

Keywords

Bioremediation · Pesticides · Polycyclic aromatic hydrocarbons · Rhizosphere · PGPR

1 Introduction

The soil fertility and biodiversity have been adversely affected, in the last few decades, due to continuously increasing human population, indiscriminate use of pesticides, polycyclic aromatic hydrocarbons, and discharge of effluents containing toxic heavy metals, dyes, etc. Industries such as plating, mining, petroleum, paint, automotive, and battery manufacturing industries, printing circuits and chemical manufacturing industries, etc. are the major sources of pollutants in the environment. As per 2009 estimates, approximately 33,900 million liter per day (MLD) urban waste water and 23,500 MLD industrial waste water were produced in India which contaminated the water and soil resources (Saha et al. 2017). According to CEPI (Comprehensive Environmental Pollution Index) rating, 43 critically polluted industrial areas or clusters or zones have been declared in 16 states by CPCB (Central Pollution Control Board) that have more than 70 CEPI rating. Among the 43 zones, 21 sites have been reported in four states, i.e., Gujarat, Uttar Pradesh, Maharashtra, and Tamil Nadu (Saha et al. 2017).

These pollutants have adversely affected human health, plants, and associated soil microorganisms. These toxic compounds accumulate in the agricultural soils in which they are applied, threaten food security, and pose health risks to living organisms by their transfer within the food chain. These pollutants accumulate in different parts of the body such as the muscle, bones, brain, and kidney and cause several harmful diseases like headache, nausea, vomiting, gastrointestinal tract disorders, asthma, allergy, cancer, heart attack, renal failure, nervous disorder, liver toxicity, Alzheimer's diseases, and muscle weakness.

These toxic pollutants can be removed or altered by conventional and physical methods such as membrane separation, chemical precipitation, reverse osmosis, solvent extraction, filtration, flocculation, electrodialysis, electrochemical treatments, ion-exchange, advanced oxidation process, and adsorption by using nanocomposite materials (Saravanan et al. 2019). However, the conventional (physical and chemical) methods are complicated and expensive, require large energy consumption, and generate large amount of toxic intermediates responsible for secondary pollution. Moreover some techniques are not useful in removing the high concentration of contaminants. This has led to search for a cost-effective alternative method. Bioremediation is the most powerful eco-friendly approach. It is the biological process that detoxifies, removes, and immobilizes the hazardous pollutants from the environment by the use of naturally occurring bacteria, fungi, or plants. It is a very useful technique for restoration of contaminated soil. The concept of using plants for the treatment of pollutants has emerged a few years ago. When plants remove, destroy, and stabilize the contaminants from soil, it is referred to as phytoremediation. This is a cost-effective and eco-friendly method, firstly investigated by Chaney (1983). This process removes the pollutants from soil by plants without affecting their biological activity, fertility, and structure of soil. It is also known as green remediation and agro-remediation (Pivetz 2001). The process of phytoremediation includes phytoextraction, phytoaccumulation, phytostabilization, phytostimulation/rhizostimulation, phytovolatilization, and rhizofiltration (Akpor and Muchie 2010).

Though phytoremediation reduces the contamination of many hazardous pollutants, the large-scale field application requires long time period. It suffers serious limitations by pollutant-induced stress which affects seed germination, plant development, and plant biomass. The use of rhizospheric microbes is the alternative way for reducing deleterious effect of pollutants to plants (Burd et al. 2000). The remediation of pollutants by rhizospheric microbes is called rhizoremediation (Kuiper et al. 2004). The rhizospheric bacteria have the extraordinary ability to promote the growth of plant by various mechanisms such as nitrogen fixation, solubilization of minerals, production of plant growth regulators and siderophores, and transformation of nutrient elements. Generally, the plant is not directly involved in the bioremediation process. The rhizobacteria can directly carry out the remediation process. Both endophytic and rhizospheric microbes have the potential to carry out the remediation of pollutants. Thus, among the various methods for the removal of pollutants, rhizoremediation is the most potential eco-friendly approach in which pollutants are removed and degraded by the interaction of microbes and plant roots. Plants provide the essential nutrients to microbes, while microbes convert the toxic compound into nontoxic minerals and make them available for plants. So, microbes enhance the soil health by addition of nutrients. These rhizobacteria play an important role in biological restoration and reclamation of contaminated sites. This process completely destroys pollutants and converts hazardous compounds into harmless products, which also eliminates the possibility of future liability associated with the treatment and disposal of contaminated materials (Abtenh 2017). A consortium of microbes detoxifies the pollutants more efficiently rather than single strain or species. This chapter focuses upon the rhizoremediation of heavy metals and xenobiotics from environment.

2 Phytoremediation

It is a Latin word in which *phyto* means plant and the *remedium* means restoring balance which maintains a permissible concentration of toxic material in the environment. It is a promising and emerging area of interest because it is an environmentally safe method for restoration of barren land. 0.2% of angiosperms tolerate and accumulated high concentration of metals referred as hyperaccumulaters (Baker and Brooks 1989). There are several examples of plants which act as hyperaccumulaters like *Alyssum* species, *Arabidopsis halleri*, *Astragalus racemosus*, *Brassica juncea*, *Noccaea* sp., and *Viola calaminaria* (Ganesan 2012). Plants use various mechanisms to remove pollutants from soil (Table 1, Fig. 1).

Table 1 Mechanisms used for bioremediation of for heavy metals and xenobiotics

Mechanism	Pollutants	Plants	References
Phytovolatization	Hg	*Chara canescens* (musk grass) and *Arabidopsis thaliana*	Ghosh and Singh (2005)
Phytoextraction	Pb As	*Glycine max* L. *Pteris vittata*	Aransiola et al. (2013), Xie et al. (2009)
Phytodegradation/ phytotransformation	Total Petrolium carbon (TPH)	*Scirpus grossus*	Al-Baldawi et al. (2015)
Phytostimulation/ rhizodegradation	Biodiesel Phenanthrene and pyrene	*Pisum sativum* Kandelia candel	Hawrot-Paw et al. (2019) Zhang et al. (2011)
Phytostabilization	As	*Piricum sativum*	Jonnalagadda and Nenzou (1997)
Rhizoremediation	Cr Lindane PHCs	*Helianthus annuus* *Withania somnifera* *Lolium perenne*	Gupta et al. (2018) Abhilash et al. (2011) Kukla et al. (2014)

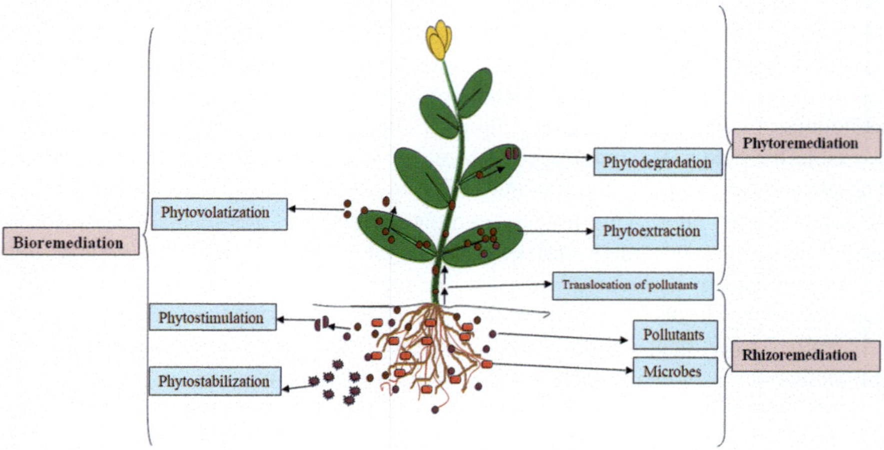

Fig. 1 Bioremediation of pollutants

These are the following. (1) Phytoremediation—It is the uptake of pollutants from soil by root system which are then converted into volatile form and released out in the atmosphere. (2) Phytoextraction—It is the translocation of pollutant by root from soil which gets accumulates in as hyperaccumulaters, and once the pollutant has reached the saturation limit, then plants are cut and disposed of. (3) Phytostabilization—It is the immobilization of pollutant in soil by absorption and precipitation. (4) Phytodegradation—It is the degradation and mineralization of pollutants via the metabolic processes of plants after uptake from the soils.

(5) Phytostimulation—It is the degradation of contaminants by microbial activity. Various types of xenobiotics are degraded by the process of phytoremediation. Although phytoremediation is a promising approach, it has some limitations which include the following: (1) long time period is required for remediation; (2) rate of plant growth and accumulation should be fast for reduction of time period; and (3) efficiency of phytoextraction is dependent upon translocation and bioconcentration; some plants do not allow translocation; and (4) the accumulation of higher concentration of pollutants by plant affects the growth and development of plants. The transgenic plants have been used to improve remediation, but they inhibit the growth of native plants by absorbing excess nutrients from soil.

3 Rhizoremediation

The word, "rhizoremediation" is composed of two words, *rhizo* (a latin word which means roots) and remediation (which means the process of removal of pollutants). It is a natural process and cost-effective and highly efficient method in which pollutants of soil are remediated by rhizospheric bacteria (Kala 2014). Rhizoremediation is a mutual relation between plant roots and soil microbes in the rhizosphere in which plants provide essential nutrients in the form of root exudates to microbes, and in return plants get protection from pathogens and help in nitrogen fixation. The plant roots secrete some substances such as sugars, organic acids, sterols, and amino acids which are utilized by the microbes for degradation of pollutants (Kuiper et al. 2004; Saravanan et al. 2019). These root exudates make a nutrient-rich environment in the rhizosphere where microbial metabolic activity and number are too high. The root exudates play an important role in rhizoremediation because these root exudates attract a specific microbial group. Rhizoremediation is enhanced by expanded root system of plants which provide large surface area for microbes which help in effective uptake and removal of contaminants (Praveen et al. 2019). This process is the combination of bioaugumentation and phytoremediation, which solves the problems that arise during the use of individual methods (Bhatia and Malik 2011). Thus, this process benefits the plants to enhance plant growth by making soil nutrients available to them, stimulating the production of various plant growth hormones, and protecting the plants from pathogens by their antagonistic activity and remediation of the pollutants by sequestering toxic heavy metals and degrading xenobiotic compounds such as pesticides, PAHs, etc. In the process of rhizoremediation, the rhizospheric microbes change the bioavailability of metals by changing the pH of soil, secretion of chelating agents (organic acids, siderophores, etc.), and oxidation/reduction reactions (Rajkumar et al. 2012). Plant growth-promoting rhizobacteria (PGPR) along with mycorrhizal and arbuscular mycorrhizal fungi (AMF) carry out remediation. The fungi spread around the plant roots increase competition for uptake of pollutants and reduce the uptake of pollutants by plants (Praveen et al. 2019). The microbial diversity and activity in the rhizosphere of plants are much more than that in bulk soil. Microbes residing in the rhizosphere of plant are more potent to degrade or transform pollutants than the microbes of bulk soil (Kamaludeen and Ramasamy 2008). The microbes of rhizosphere form a mutual association with root which has potential for detoxification of hazardous pollutants (De Souza et al. 1999).

Thus, there are many advantages of rhizoremediation like it is a cost-effective and eco-friendly approach, no need to add any chemical for degradation, and negligible side effects while chemical degradation adversely affect the environment. The efficient microbial strain and consortium has to be provided, and no investment is required (Saravanan et al. 2019). In addition to phytoremediation, rhizoremediation can be a better alternative for the removal of pollutants such as heavy metals, pesticides, polycyclic aromatic hydrocarbons (PAHs), etc.

3.1 Heavy Metals

Metals are the natural components of soil. Some of them are essential and required for various metabolic activities. Zn, Fe, Mg, Mn, Cu, Cr, Ni, Na, and Ca are essential micronutrients. Fe^{3+} is essential for all bacteria, while Fe^{2+} is required for only anaerobic bacteria and other metals such as Cd, Hg, Au, and Ag are nonessential and very toxic to soil microbes and other organisms. Heavy metals occur in soil as free cations in the form of complex with organic/inorganic ligands and soil colloids. Now pollution of heavy metals has become a severe problem for the environment. The increasing anthropogenic activities like industrial waste disposal, waste incineration, vehicle exhausts, energy and fuel production, mining and smelting of metals, long-term application of urban sewage *sludge*, and excessive use of agrochemicals and pesticides, etc. are the major source of heavy metal contamination which poses serious threat to soil ecosystem and human health (Khan 2005).

According to the report of Central Pollution Control Board (CPCB), in India, the states of Gujarat, Maharashtra, and Andhra Pradesh have 80% contribution in the generation of toxic pollutants including heavy metals. The report of Ministry of Mines, Government of India (2018), suggested that India is the major source of different types of metallic and nonmetallic minerals. Maximum numbers of mines are present in Tamil Nadu and Madhya Pradesh (Kumar et al. 2019). High concentration of heavy metal in soil are absorbed and accumulated by plant which ultimately reaches into the human body (Mishra et al. 2017) via food chain. The nonbiodegradable nature of these toxic heavy metals results in their persistence for long period in the environment. Even a very low concentration of these metals enters into the human body via contaminated food and contaminated drinking water and gets accumulated in animals and plants which can have lethal effects.

Various physical and chemical methods such as excavation and stabilization or solidification are used for removal of heavy metals, but these technologies have some demerits like less cost-effective, generation of hazardous products, difficult to operate, and cannot remove heavy metals permanently (Akshata et al. 2014). The remediation of heavy metals by phytoremediation takes a long time period while the addition of suitable microbes increases the rate of remediation. Microbes reduce the toxicity of metals by converting the metals from toxic to nontoxic form, by stabilizing the metal in rhizosphere so that pollutants cannot be used by plants and enhance the degradation of metals by increasing bioavailability.

The rhizosphere region of plants which is growing in heavy metal-contaminated soil harbors a large population of microbes (Idris et al. 2004) which are able to tolerate high concentration of metals. They convert toxic metals into bioavailable form in soil by mechanisms like precipitation, chelation, acidification, redox reactions, and complexation (Ma et al. 2016). These microbes have developed various resistance mechanisms for survival in the presence of heavy metal-contaminated soil such as bioabsorption, bioaccumulation, bioleaching, biotransformation, biomineralization, and plant-microbe interaction. Besides this PGPRs also act as biofertilizer which enhance the yield of crops by efficient way of heavy metals remediation in soil (Ganesan 2012).

Madhaiyan et al. (2007) reported that *Methylobacterium oryzae* strain CBMB20 and *Burkholderia* sp. strain CBMB40 reduce the toxicity of nickel (Ni) and cadmium (Cd) in tomato by reducing uptake and translocation of heavy metals to shoot in plant and also promote the plant growth in contaminated soil. Fatnassi et al. (2015) reported that at high concentration of copper (Cu), the growth of plant *Vicia faba* (fava bean) was inhibited, but when plant was inoculated with *Rhizobium* and *Pseudomonas*, the inhibitory effect was decreased. Wu et al. (2006) observed that *Azotobacter chroococcum* HKN-5 promoted the growth of *Brassica juncea* by phytoextraction from the soil which was highly contaminated by lead (Pb). Dary et al. (2010) reported that the consortium of *Bradyrhizobium* sp.750, *Ochrobactrum cytisi*, and *Pseudomonas* sp. protected and promoted the growth of *Lupinus luteus* in multi-metal-contaminated soil Cu, Cd, and Pb by the process of phytostabilization and also helped in the reclamation of soil. The rhizobacteria *Pseudomonas aeruginosa* and *Burkholderia gladioli* were reported to facilitate the plant growth of *Lycopersicon esculentum* seedlings by reducing the accumulation of Cd metal due to immobilization of metal on the root that prevented the translocation of metal towards shoot and reduce the toxicity of Cd by decreasing bioavailability (Khanna et al. 2019). Wani et al. (2008) reported that *Bacillus* sp. PSB10 improved the growth, nodulation, chlorophyll seed yield, etc. of chickpea grown in chromium-contaminated soil by reducing the uptake and toxicity of contaminant from soil. Another study was performed by Gupta et al. (2018) on sunflower (*Helianthus annuus* L.) in which inoculation of strain *Pseudomonas* sp. CPSB21 reduced the $Cr6+$ toxicity and enhanced the plant growth and nutrient uptake. Some examples of microbes involved in rhizoremediation of heavy metals are listed in Table 2.

3.2 Pesticides

Pesticides are organic chemical which were introduced in agriculture for increasing agricultural yield and soil productivity, to protect agricultural products from crop pests, and to control insect vectors for prevention of outbreaks of human and animal epidemics. Pesticides are used for many purposes such as livestock farming,

Table 2 Remediation of heavy metals by rhizobacteria

Plant	Pollutant	Microbes	References
Brassica juncea	Pb	*Azotobacter chroococcum* HKN-5	Wu et al. (2006)
Glycine max	Mn	*Glomus etunicatum* or *G. macrocarpum*	Nogueira et al. (2007)
Pteris vittata	As	*Pseudomonas*	Huang et al. (2007)
Zea mays	Pb, Zn, cu	*Brevibacterium haloterans*	Abou-Shanab et al. (2008)
Brassica juncea	Ni	*Pseudomonas* sp., *Bacillus megaterium*	Rajkumar and Freitas (2008)
Brassica juncea	Cu	*Achromobacter xylosoxidans* strain Ax10	Ma et al. (2009)
Cicer arietinum	Cr	*Bacillus* sp. PSB10	Wani and Khan (2010)
Helianthus annuus, *Ricinus communis*	Ni	*Psychrobacter* sp. SRS8	Ma et al. (2010)
Lupinus luteus	Cu, cd, Pb	*Bradyrhizobium* sp.750, *Pseudomonas* sp., and *Ochrobactrum cytisi*	Dary et al. (2010)
Orychophragmus violaceus	Zn	*Bacillus subtilis*, *Bacillus cereus*, *Flavobacterium* sp., *Pseudomonas aeruginosa*	He et al. (2010)
Crotalaria juncea	Cd	*Achromobacter* sp. AO22	Stanbrough et al. (2013)
Oryza sativa	Ni	*Bacillus licheniformis* NCCP-59	Jamil et al. (2014)
Triticum aestivum (wheat)	Zn	*Pseudomonas fluorescens*	Sirohi et al. (2015)
Vicia faba	Cu	*Rhizobium*, Enterobacter *cloacae*, and *Pseudomonas* sp.	Fatnassi et al. (2015)
Brassica nigra	Cu	*Kocuria* sp. *CRB15*	Hansda et al. (2017)
Leucaena leucocephala	Zn, cd	Rhizobia	Rangel et al. (2017)
Suaeda nudiflora	Zn, Pb	*Bacillus megaterium*, *Pseudomonas aeruginosa*	Jha et al. (2017)
Helianthus annuus	Cr	*Pseudomonas* sp. CPSB21	Gupta et al. (2018)

cropping, horticulture, forestry, home gardening, homes, hospitals, kitchens, roadsides, and industrial areas. Pesticides are a cost-effective method to control pest and increase agriculture productivity. But now the use of pesticides has become a serious threat as they cause many adverse effects upon environment and the human health. Residues of pesticides remain in the environment (air, soil, groundwater, and surface water) for long periods. Pesticides pollute the rivers, ponds, and lakes and also cause acid rain, which is hazardous to animals and especially amphibians

(Gilden et al. 2010). Every year around two million tonnes of pesticides are consumed worldwide, of which 24% is consumed by the United States, 45% in Europe, and the remaining 25% in the world (Abhilash and Singh 2009). Thus, three quarters of the total pesticide used worldwide (USEPA 2009) are consumed by the developed countries such as North America, Western Europe, and Japan (Yadav et al. 2015). In India, the use of pesticides started from 1948 where dichloro diphenyl trichloroethane (DDT) was used for the control of malaria and benzene hexachloride (BHC) for locusts. Now, India accounts for 3% of the world consumption of pesticide. After China, India is the second largest pesticide manufacturing country in Asia. The most common pesticides used in India are organophosphates, organochlorine, neonicotinoids, etc. The pesticides which are mostly used in India include monocrotophos, endosulfan, phorate, chlorpyrifos, methyl parathion, quinalphos, mancozeb, paraquat, butachlor, isoproturon, and phosphamidon (Bhushan et al. 2013). Ideally, according to definition a pesticide should be lethal only for targeted pests, but unfortunately, it has become lethal for nontargeted species including humans also. The persistent nature of pesticides is making the agriculture land barren.

Phytoremediation is a cost-effective technology that removes organic and inorganic contaminants used in polluted areas. But it is not an efficient technology without the contribution of rhizobacteria. Plant microbial interaction is very useful in remediation of contaminants. Romeh and Hendawi (2014) reported the degradation of organophosphorus insecticides, chlorpyrifos, chlorpyrifos-methyl, cyanophos, and malathion degraded by some phosphate-solubilizing bacteria like *Azospirillum lipoferum* and *Paenibacillus polymyxa* on mineral salt media as a carbon and phosphorus source. Various pesticides like acibenzolar-*S*-methyl, metribuzin, napropamide, propamocarb hydrochloride, and thiamethoxam have been reported to be degraded by *Bacillus subtilis* GB03, *Bacillus subtilis* FZB24, *Bacillus amyloliquefaciens* IN937a, and *Bacillus pumilus* SE34 in liquid culture and soil microcosm (Myresiotis et al. 2012). Some examples of microbes involved in rhizoremediation of pesticides are listed in Table 3.

The bacterial strain *Achromobacter xylosoxidans JCp4 and Ochrobactrum sp. FCp1* was found to increase crop productivity by degrading chlorpyrifos from soil and promote the plant growth of *Vigna unguiculata* (Akbar and Sultan 2016). *Fulekar (2014) studied the effect of consortium Stenotrophomonas maltophilia* MHF ENV20, *Stenotrophomonas maltophilia* MHF ENV22, and *Sphingobacterium thalpophilum* MHF ENV23 on plant growth of *Pennisetum pedicellatum* and *Cenchrus setigerus*. The consortium was found to degrade pesticides chlorpyrifos, cypermethrin, and fenvalerate in soil. Jacobsen (1997) suggested that colonization of the plant roots by the herbicide-degrading *Burkholderia cepacia* DBO1 (pRO101) can protect the plant *Hordeum vulgare* by degradation of the herbicide 2,4-dichlorophenoxyacetic acid (2,4-D) in the rhizosphere soil. Abhilash et al. (2011) reported that the integrated use of pesticide-tolerant plant *Withania somnifera* and rhizospheric microbe *Staphylococcus cohnii* decreases the concentration of lindane from soil and has practical application for in situ remediation of contaminated soils.

Table 3 Remediation of pesticides by rhizobacteria

Plant	Pollutant	Microbes	References
Zinnia angustifolia	Mefenoxam	*Pseudomonas fluorescens* and *Chryseobacterium indologenes*	Pai et al. (2001)
Greengram (*Vigna radiata*)	Fipronil and pyriproxyfen	*Pseudomonas aeruginosa* PS1	Ahemad and Khan (2011)
Withania somnifera	Lindane	*Staphylococcus cohnii*	Abhilash et al. (2011)
Grapevines (*Vitis vinifera*)	Profenofos	*Bacillus subtilis*	Salunkhe et al. (2013)
Cauliflower (*Brassica oleracea*)	Cypermethrin	*Serratia nematodiphila*	Tyagi and Prashar (2015)
Maize (*Zea mays*)	Thiamethoxam	*Bacillus subtilis* GB03) and *B. subtilis* FZB24	Myresiotis et al. (2015)
Vigna unguiculata	Chlorpyrifos	*Achromobacter xylosoxidans* JCp4, *Ochrobactrum* sp. FCp1	Akbar and Sultan (2016)

3.3 Polycyclic Aromatic Hydrocarbons

Polycyclic aromatic hydrocarbons (PAHs) are ubiquitous environmental pollutants generated from both natural and anthropogenic processes and pose a serious concern to the health of aquatic life and humans. They are released naturally in the environment due to incomplete combustion of organic material. Due their toxic, carcinogenic, and mutagenic effects on human health, 16 PAHs have been identified as priority pollutants by USEPA (United States Environmental Protection Agency) and monitored regularly in industrial wastes (Bisht et al. 2015; Gupte et al. 2016). Out of 16, seven priority pollutants are considered as probable human carcinogens by USEPA and IARC (International Agency for Research on Cancer) (Wang et al. 2018). There are various types of chemical methods used for removal of PAHs; however as mentioned earlier, due to their various demerits, phytoremediation and rhizoremediation are the most preferred method. Bisht et al. (2010, 2014) reported the rhizoremediation of PAHs by rhizospheric and endophytic bacteria in *Populus* sp. Some examples of microbes involved in PAH rhizoremediation are listed in Table 4.

Golubev et al. (2009) reported that *Sinorhizobium meliloti* was quite effective in degradation of PAHs besides its ability to produce surfactant and indole acetic acid (IAA). Kukla et al. (2014) reported that *Rhodococcus fascians* strain (L11) was able to degrade petroleum hydrocarbon and promoted the growth of ryegrass (*Lolium perenne* L.). Khan et al. (2014) reported that inoculation of *Pseudomonas putida* PD 1 to willow plant and ryegrass protected the plant from toxic effect by degrading phenanthrene and promoted plant growth. *Burkholderia fungorum* DBT1 along with hybrid poplar plant was reported to degrade PAHs from contaminated soil by the process of phytoextraction or rhizodegradation (Andreolli et al. 2013).

Table 4 Remediation of PAHs by rhizobacteria

Plant	Pollutant	Microbes	References
Grasses (*Lolium multiflorum*)	Napthalene	*Pseudomonas putida* PCL1444	Kuiper et al. (2001)
Festuca arundinacea	PAHs	*Pseudomonas fluorescens*	Ho et al. (2007)
Zea mays	Phenanthrene	*Pseudomonas putida* MUB1	Chouychai et al. (2009)
Populus deltoides	Naphthalene and anthracene	*Kurthia* sp. SBA4, *Micrococcus varians* SBA8, *Bacillus circulans* SBA12	Bisht et al. (2010)
Populus deltoides	PAHs	*Bacillus* sp. SBER3	Bisht et al. (2014)
Oryza sativa	Phenanthrene	*Phomopsis liquidambaris*	Fu et al. (2020)

3.4 Factors Affecting Rhizoremediation

The physicochemical properties of soil play an important role in successful rhizoremediation. The microbial activity as well as diffusion of the chemicals in the soil is affected by moisture, redox conditions, temperature, pH, organic matter, nutrients and nature, and amount of clay. The effect of soil moisture on microbial mineralization of three aerobically degradable pesticides (benzolin-ethyl, isoproturon, and glyphosate) was quantified by Schroll et al. (2006). They observed a linear correlation ($p < 0.0001$). The increasing soil moisture (within a soil water potential range of −20 and − 0.015 MPa) was found to increase the mineralization of pesticides.

3.4.1 Soil pH

pH of soil is an important factor that control the microbial growth and activity. Most of the microbes have optimum pH at 6.5–7.5. At particular pH, microbes secrete enzymes and organic acids that enhance the removal of pollutants. The catalytic activity of enzymes is dependent upon pH. The degradation process takes more time at acidic pH rather than alkaline and neutral pH. Singh et al. (2006) reported the biodegradation of organophosphate pesticide was found to be slow at low pH soil as compared to neutral and high pH soils.

3.4.2 Temperature

The removal and degradation rate of pollutants are affected by soil as the metabolic activity of microbes depends on temperature. Bioavailability of pollutants is increased with increasing temperature of soil. However the native microbes of low temperature soil like Arctic and Antarctic soils will degrade pollutants at low temperature. *Rhodococcus* isolated from an Antarctic soil was found to degrade various alkanes at −2 °C, while its activity was inhibited at high temperature (Oberai and Khanna 2018).

3.4.3 Soil Organic Matter

The soil organic matter is a rich source of nutrients which enhance microbial growth and thus affect the degradation of pollutants. The soil organic matters control the adsorption and desorption process. Sewage sludge which is a source of organic matter increases the biodegradation of isoproturon herbicide (Perrin-Ganier et al. 2001).

3.4.4 Plant Root Exudates

Rhizoremediation is mainly dependent on the quality of root exudates. The age of plants and presence of pollutants affect the quality and nature of root exudates which determine the microflora of the rhizosphere. It is a mixture of soluble organic compounds including sugars, amino acids, organic acids, and enzymes that influence the colonization of specific microbial community which promote the plant growth and protect the plant from soil pollutants, and microbes use root exudates for their growth and development.

3.4.5 Microbial Population

The composition of microbial community of rhizosphere can affect the rate and efficiency of rhizoremediation. There are many microbes like *Azotobacter*, *Bacillus* sp., *Pseudomonas* sp., etc. which are active colonizers of rhizosphere and have been reported to be efficient degraders of pollutants.

3.4.6 Pollutants

The nature of pollutants in soil can affect the composition of microbial population. The hydrophobic pollutants are degraded in aqueous medium such as the biodegradation of phenanthrene (Bouchez et al. 1995). The type and concentration of pollutants affect the process of rhizoremediation.

3.5 Methods to Improve Rhizoremediation

There are several limiting factors such as pH, temperature, organic matter, pollutant type, soil properties, source of energy, and microbes which slow the process of rhizoremediation. This process also depends on the capacity of plants to tolerate the maximum limit of pollutants which can be accumulated in the plants without any adverse effects to them. The efficiency of rhizoremediation can be enhanced by several techniques.

3.5.1 Biostimulation

It is a remediation method that refers to the addition of rate-limiting nutrients such as phosphorous, nitrogen, oxygen, carbon, and electron donors in the polluted soils to stimulate existing bacteria for degradation of pollutants (Tribedi et al. 2018). The activity of rhizobacteria is enhanced by the addition of fertilizers or minerals (N, P, and K) and compost in soil as they improve the root exudates. Biosurfactants improve the physicochemical property of soil and increase the bioavailability of pollutants in soil. The use of additive such as biochar, cattle manure, and fly ash enhances the soil fertility. Biochar is known as a universal sorbent due to its strong ability to bind pollutants and has long been used as a sorbent for both organic and inorganic pollutants (Ahmad et al. 2014). Biochar increases soil fertility moisture content and nutrient retention and remediate heavy metals from soil by surface adsorption, precipitation, partitioning, and sequestration (Tang et al. 2013; Saravanan et al. 2019). Recently, biochar has been implemented to enhance phytoremediation efficiency by improving plant growth, water holding capacity, and soil structure. Qin et al. (2013) observed the total petroleum hydrocarbon (TPH) removal after applying biochar as a soil amendment.

3.5.2 Bioaugumentation

Bioaugumentation is the addition of bacterial culture to increase the rate of degradation of pollutants. The native microbes of contaminated sites are able to detoxify the contaminants. The process of rhizoremediation is also enhanced by introducing specific pollutant-degrading microbes or consortium in the soil by various methods such as seed coating, root dipping, and soil drenching. The ability of microbe to remove contaminants is limited because the catabolic pathway for some pollutants like polychlorinated biphenyls (PCBs) has not evolved in microbes. To combat this problem, genetically modified microbes are used to increase efficiency of rhizoremediation by modification of a gene (Lorito et al. 1998). The genes of siderophore receptor introduced into *Pseudomonas* sp. made it more useful in rhizosphere region (Praveen et al. 2019). The selection of suitable plant-microbe combination for efficient rhizoremediation is also quite important (Kuiper et al. 2004).

3.5.3 Transgenic Plants

For efficient rhizoremediation, transgenic plants are now used. Transgenic plants have increased ability to uptake, transport, and degrade pollutants. The plant is improved by either gene transfer or breeding. Janssen et al. (2015) have improved *Salix* sp. (Willow tree) into *S. viminalis* and *S. alba* by metal extraction. Both species have improved ability to accumulate Cd and Zn. Plants are genetically modified to improve stabilization and accumulation of the pollutants. Doty et al. (2000) reported enhanced degradation of the halogenated hydrocarbons by the introduction of cytochrome P450 2E1 gene into plants.

3.5.4 Rhizoengineering

The use of transgenic plants increases the root exudates that improve the rhizoremediation, but they are not effective for soil which is contaminated by multiple contaminants. Wu et al. (2006) utilized the advantages of plant-microbe symbiosis within the plant rhizosphere for effective cleanup technology. They demonstrated the expression of a metal-binding peptide (EC20) in *Pseudomonas putida* 06909 that not only improved cadmium binding but also decreased the cellular toxicity of cadmium and also improved the plant growth. Thus rhizoengineering becomes very important to reclaim the soil.

3.5.5 Nanotechnology

Nanotechnology is characterized by the use of very small manufactured particles (<100 nm) called as nanoparticles. The use of nanoparticles for the removal of pollutants for environment cleanup is an emerging technique that has gained attention recently. Bacteria have the ability to mobilize and immobilize the metals, and they can reduce the metals by precipitation. The removal of environmental pollutants such as heavy metals, organic and inorganic, from polluted sites by using nanoparticles/nanomaterial produced by plants, fungi, and bacteria is called nanobioremediation which has been reported to enhance the efficiency of phytoremediation or rhizoremediation (Yadav et al. 2017). Recently some studies have reported the degradation of organic contaminants such as atrazine, molinate, and chlorpyrifos by nanosized zerovalent ions (Zhang 2003; Ghormade et al. 2011). Liu and Zhao (2007) used iron phosphate (vivianite) nanoparticles for in situ immobilization of Pb^{+2} in soils and reported the reduction of bioavailability and mobility of Pb^{+2} from soil effectively. Some organic pollutants like hydrocarbons and organochlorines cannot be degraded by plant as well as microbes, but the combined use of nanotechnology and biotechnology solved this problem (Yadav et al. 2017).

4 Conclusions

The rhizoremediation is a cost-effective and eco-friendly approach that is the emerging technique for remediation of pollutants. It is the mutual interaction of microbes and plant roots that enhances the degradation of contaminants. The versatility and adaptability of microbes to any challenges posed by the environment make them perfect for remediation of environmental pollutants. In rhizoremediation the degradation of pollutant is higher than that of phytoremediation in plant and microbe alone. So rhizoremediation is the best research topic for understanding the interaction of soil-root-microbe for bioremediation of pollutants. Plants and microbes can be genetically modified for improvement of degradation ability. Thus, rhizoremediation can be an environment-friendly cleaning technique.

References

Abhilash PC, Singh N (2009) Pesticide use and application: an Indian scenario. J Hazard Mater 165 (1–3):1–12

Abhilash PC, Srivastava S, Srivastava P, Singh B, Jafri A, Singh N (2011) Influence of rhizospheric microbial inoculation and tolerant plant species on the rhizoremediation of lindane. Environ Exp Bot 74:127–130

Abou-Shanab RA, Ghanem K, Ghanem N, Al-Kolaibe A (2008) The role of bacteria on heavy-metal extraction and uptake by plants growing on multi-metal-contaminated soils. World J Microbiol Biotechnol 24(2):253–262

Abtenh EG (2017) Application of microorganisms in bioremediation-review. J Environ Microbiol 1 (1):2–9

Ahemad M, Khan MS (2011) Pseudomonas aeruginosa strain PS1 enhances growth parameters of greengram [Vignaradiata (L.) Wilczek] in insecticide-stressed soils. J Pest Sci 84(1):123–131

Ahmad M, Rajapaksha AU, Lim JE, Zhang M, Bolan N, Mohan D, Vithanage M, Lee SS, Ok YS (2014) Biochar as a sorbent for contaminant management in soil and water: a review. Chemosphere 99:19–33

Akbar S, Sultan S (2016) Soil bacteria showing a potential of chlorpyrifos degradation and plant growth enhancement. Braz J Microbiol 47(3):563–570

Akpor OB, Muchie M (2010) Remediation of heavy metals in drinking water and wastewater treatment systems: processes and applications. Int J Phys Sci 5(12):1807–1817

Akshata JN, Udayashankara TH, Lokesh K (2014) Review on bioremediation of heavy metals with microbial isolates and amendments on soil residue. Int J Sci Res 3:118–123

Al-Baldawi IA, Abdullah SRS, Anuar N, Suja F, Mushrifah I (2015) Phytodegradation of total petroleum hydrocarbon (TPH) in diesel-contaminated water using *Scirpus grossus*. Ecol Eng 74:463–473

Andreolli M, Lampis S, Poli M, Gullner G, Biro B, Vallini G (2013) Endophytic *Burkholderia fungorum* DBT1 can improve phytoremediation efficiency of polycyclic aromatic hydrocarbons. Chemosphere 92(6):688–694

Aransiola SA, UJJ I, Abioye OP (2013) Phytoremediation of lead polluted soil by *Glycine max* L. Appl Environ Soil Sci 2013:1–7

Baker AJM, Brooks RR (1989) Terrestrial higher plants which hyperaccumulate metallic elements-a review of their distribution, ecology and phytochemistry. Biorecovery 1:81–126

Bhatia D, Malik DK (2011) Plant-microbe interaction with enhanced bioremediation. Res J Biotechnol 6(4):72–79

Bhushan C, Bhardwaj A, Misra SS (2013) State of pesticide regulations in India. Centre for Science and Environment, New Delhi

Bisht S, Pandey P, Sood A, Sharma S, Bisht NS (2010) Biodegradation of naphthalene and anthracene by chemo-tactically active rhizobacteria of *Populus deltoides*. Braz J Microbiol 41 (4):922–930

Bisht S, Pandey P, Kaur G, Aggarwal H, Sood A, Sharma S, Bisht NS (2014) Utilization of endophytic strain Bacillus sp. SBER3 for biodegradation of polyaromatic hydrocarbons (PAH) in soil model system. Eur J Soil Biol 60:67–76

Bisht S, Pandey P, Bhargava B, Sharma S, Kumar V, Sharma KD (2015) Bioremediation of polyaromatic hydrocarbons (PAHs) using rhizosphere technology. Braz J Microbiol 46(1):7–21

Bouchez M, Blanchet D, Vandecasteele JP (1995) Substrate availability in phenanthrene biodegradation: transfer mechanism and influence on metabolism. Appl Microbiol Biotechnol 43 (5):952–960

Burd GI, Dixon DG, Glick BR (2000) Plant growth-promoting bacteria that decrease heavy metal toxicity in plants. Can J Microbiol 46(3):237–245

Chaney RL (1983) Plant uptake of inorganic waste constituents. In: Parr JF, Marsh PB, Kla JM (eds) Land treatment of hazardous wastes. Noyes Data Corporation, Park Ridge, pp 50–76

Chouychai W, Thongkukiatkul A, Upatham S, Lee H, Pokethitiyook P, Kruatrachue M (2009) Plant-enhanced phenanthrene and pyrene biodegradation in acidic soil. J Environ Biol 30 (1):139–144

Dary M, Chamber-Pérez MA, Palomares AJ, Pajuelo E (2010) "In situ" phytostabilisation of heavy metal polluted soils using *Lupinus luteus* inoculated with metal resistant plant-growth promoting rhizobacteria. J Hazard Mater 177(1–3):323–330

De Souza MP, Huang CPA, Chee N, Terry N (1999) Rhizosphere bacteria enhance the accumulation of selenium and mercury in wetland plants. Planta 209(2):259–263

Doty SL, Shang TQ, Wilson AM, Tangen J, Westergreen AD, Newman LA, Gordon MP (2000) Enhanced metabolism of halogenated hydrocarbons in transgenic plants containing mammalian cytochrome P450 2E1. Proc Natl Acad Sci 97(12):6287–6291

Fatnassi IC, Chiboub M, Saadani O, Jebara M, Jebara SH (2015) Impact of dual inoculation with rhizobium and PGPR on growth and antioxidant status of *Vicia faba* L. under copper stress. C R Biol 338(4):241–254

Fu WQ, Xu M, Sun K, Chen XL, Dai CC, Jia Y (2020) Remediation mechanism of endophytic fungus *Phomopsis liquidambaris* on phenanthrene *in vivo*. Chemosphere 243:125305

Fulekar MH (2014) Rhizosphere bioremediation of pesticides by microbial consortium and potential microorganism. Int J Curr Microbiol Appl Sci 3(7):235–248

Ganesan V (2012) Rhizoremediation: a pragmatic approach for remediation of heavy metal-contaminated soil. In: Toxicity of heavy metals to legumes and bioremediation. Springer, Vienna, pp 147–161

Ghormade V, Deshpande MV, Paknikar KM (2011) Perspectives for nano-biotechnology enabled protection and nutrition of plants. Biotechnol Adv 29(6):792–803

Ghosh M, Singh SP (2005) A review on phytoremediation of heavy metals and utilization of it's by products. Asian J Energy Environ 6(4):18

Gilden R, Huffling K, Sattler B (2010) Pesticide and health risks. Obstet Gynecol Neonatal Nurs 39:103–110

Golubev SN, Schelud'ko AV, Muratova AY, Makarov OE, Turkovskaya OV (2009) Assessing the potential of rhizobacteria to survive under phenanthrene pollution. Water Air Soil Pollut 198 (1–4):5–16

Gupta P, Rani R, Chandra A, Kumar V (2018) Potential applications of *Pseudomonas* sp. (strain CPSB21) to ameliorate Cr^{6+} stress and phytoremediation of tannery effluent contaminated agricultural soils. Sci Rep 8(1):1–10

Gupte A, Tripathi A, Patel H, Rudakiya D, Gupte S (2016) Bioremediation of polycyclic aromatic hydrocarbon (PAHs): a perspective. Open Biotechnol J 10(1):363

Hansda A, Kumar V, Anshumali (2017) Cu-resistant *Kocuria* sp. CRB15: A potential PGPR isolated from the dry tailing of Rakha copper mine. 3 Biotech 7(2):132

Hawrot-Paw M, Ratomski P, Mikiciuk M, Staniewski J, Koniuszy A, Ptak P, Golimowski W (2019) Pea cultivar Blauwschokker for the phytostimulation of biodiesel degradation in agricultural soil. Environ Sci Pollut Res 26(33):34594–34602

He CQ, Tan GE, Liang X, Du W, Chen YL, Zhi GY, Zhu Y (2010) Effect of Zn-tolerant bacterial strains on growth and Zn accumulation in Orychophragmus violaceus. Appl Soil Ecol 44(1):1–5

Ho CH, Applegate B, Banks MK (2007) Impact of microbial/plant interactions on the transformation of polycyclic aromatic hydrocarbons in rhizosphere of *Festuca arundinacea*. Int J Phytoremediation 9(2):107–114

Huang A, Teplitski M, Rathinasabapathi B, Ma L (2007) Characterization of arsenic-resistant bacteria from the rhizosphere of arsenic hyperaccumulator. Can J Microbiol 56(3):236–246

Idris R, Trifonova R, Puschenreiter M, Wenzel WW, Sessitsch A (2004) Bacterial communities associated with flowering plants of the Ni hyperaccumulator *Thlaspi goesingense*. Appl Environ Microbiol 70(5):2667–2677

Jacobsen CS (1997) Plant protection and rhizosphere colonization of barley by seed inoculated herbicide degrading *Burkholderia* (*Pseudomonas*) *cepacia* DBO1 (pRO101) in 2, 4-D contaminated soil. Plant Soil 189(1):139–144

Jamil M, Zeb S, Anees M, Roohi A, Ahmed I, Rehman S, Rha ES (2014) Role of *Bacillus licheniformis* in phytoremediation of nickel contaminated soil cultivated with rice. Int J Phytoremediation 16(6):554–571

Janssen J, Weyens N, Croes S, Beckers B, Meiresonne L, Van Peteghem P, Vangronsveld J (2015) Phytoremediation of metal contaminated soil using willow: exploiting plant-associated bacteria to improve biomass production and metal uptake. Int J Phytoremediation 17(11):1123–1136

Jha Y, Subrmanian RB, Mishra KK (2017) Role of plant growth promoting rhizobacteria in accumulation of heavy metal in metal contaminated soil. Emergent Life Sci Res 3:48–56

Jonnalagadda SB, Nenzou G (1997) Studies on arsenic rich mine dumps. II The heavy element uptake by vegetation. J Environ Sci Health Part A 32(2):455–464

Kala S (2014) Rhizoremediation: a promising rhizosphere technology. IOSR J Environ Sci Toxicol Food Technol 8:23–27

Kamaludeen SPB, Ramasamy K (2008) Rhizoremediation of metals: harnessing microbial communities. Indian J Microbiol 48(1):80–88

Khan AG (2005) Role of soil microbes in the rhizospheres of plants growing on trace metal contaminated soils in phytoremediation. J Trace Elem Med Biol 18:355–364

Khan Z, Roman D, Kintz T, de las Alas M, Yap R, Doty S (2014) Degradation, phytoprotection and phytoremediation of phenanthrene by endophyte Pseudomonas putida, PD1. Environ Sci Technol 48(20):12221–12228

Khanna K, Jamwal VL, Gandhi SG, Ohri P, Bhardwaj R (2019) Metal resistant PGPR lowered cd uptake and expression of metal transporter genes with improved growth and photosynthetic pigments in Lycopersicon esculentum under metal toxicity. Sci Rep 9(1):1–14

Kuiper I, Bloemberg GV, Lugtenberg BJ (2001) Selection of a plant-bacterium pair as a novel tool for rhizostimulation of polycyclic aromatic hydrocarbon-degrading bacteria. Mol Plant-Microbe Interact 14(10):1197–1205

Kuiper I, Lagendijk Ellen L, Bloemberg Guido V, Lugtenberg Ben JJ (2004) Rhizoremediation: a beneficial plant-microbe interaction. Mol Plant Microbe Interact 17:6–15

Kukla M, Płociniczak T, Piotrowska-Seget Z (2014) Diversity of endophytic bacteria in Lolium perenne and their potential to degrade petroleum hydrocarbons and promote plant growth. Chemosphere 117:40–46

Kumar V, Sharma A, Kaur P, Sidhu GPS, Bali AS, Bhardwaj R, Thukral AK, Cerda A (2019) Pollution assessment of heavy metals in soils of India and ecological risk assessment: a state-of-the-art. Chemosphere 216:449–462

Liu R, Zhao D (2007) Reducing leachability and bioaccessibility of lead in soils using a new class of stabilized iron phosphate nanoparticles. Water Res 41(12):2491–2502

Lorito M, Woo SL, Fernandez IG, Colucci G, Harman GE, Pintor-Toro JA, Filippone E, Muccifora S, Lawrence CB, Zoina A, Tuzun S, Scala F (1998) Genes from mycoparasitic fungi as a source for improving plant resistance to fungal pathogens. Proc Natl Acad Sci 95(14):7860–7865

Ma Y, Rajkumar M, Freitas H (2009) Improvement of plant growth and nickel uptake by nickel resistant-plant-growth promoting bacteria. J Hazard Mater 166(2–3):1154–1161

Ma Y, Rajkumar M, Vicente JAF, Freitas H (2010) Inoculation of Ni-resistant plant growth promoting bacterium Psychrobacter sp. strain SRS8 for the improvement of nickel phytoextraction by energy crops. Int J Phytoremediation 13(2):126–139

Ma Y, Oliveira RS, Freitas H, Zhang C (2016) Biochemical and molecular mechanisms of plant-microbe-metal interactions: relevance for phytoremediation. Front Plant Sci 7:918

Madhaiyan M, Poonguzhali S, Sa T (2007) Metal tolerating methylotrophic bacteria reduces nickel and cadmium toxicity and promotes plant growth of tomato (*Lycopersicon esculentum* L.). Chemosphere 69(2):220–228

Mishra J, Singh R, Arora NK (2017) Alleviation of heavy metal stress in plants and remediation of soil by rhizosphere microorganisms. Front Microbiol 8:1706

Myresiotis CK, Vryzas Z, Papadopoulou-Mourkidou E (2012) Biodegradation of soil-applied pesticides by selected strains of plant growth-promoting rhizobacteria (PGPR) and their effects on bacterial growth. Biodegradation 23(2):297–310

Myresiotis CK, Vryzas Z, Papadopoulou-Mourkidou E (2015) Effect of specific plant-growth-promoting rhizobacteria (PGPR) on growth and uptake of neonicotinoid insecticide thiamethoxam in corn (Zea mays L.) seedlings. Pest Manag Sci 71(9):1258–1266

Nogueira MA, Nehls U, Hampp R, Poralla K, Cardoso EJBN (2007) Mycorrhiza and soil bacteria influence extractable iron and manganese in soil and uptake by soybean. Plant Soil 298 (1–2):273–284

Oberai M, Khanna V (2018) Rhizoremediation–plant microbe interactions in the removal of pollutants. Int J Curr Microbiol App Sci 7(1):2280–2287

Pai SG, Riley MB, Camper ND (2001) Microbial degradation of mefenoxam in rhizosphere of Zinnia angustifolia. Chemosphere 44(4):577–582

Perrin-Ganier C, Schiavon F, Morel JL, Schiavon M (2001) Effect of sludge-amendment or nutrient addition on the biodegradation of the herbicide isoproturon in soil. Chemosphere 44(4):887–892
Pivetz BE (2001) Phytoremediation of contaminated soil and ground water at hazardous waste sites. US Environmental Protection Agency, Office of Research and Development, Office of Solid Waste and Emergency Response, Washington, DC
Praveen A, Pandey VC, Marwa N, Singh DP (2019) Rhizoremediation of polluted sites: harnessing plant–microbe interactions. In: Phytomanagement of polluted sites. Elsevier, Amsterdam, pp 389–407
Qin G, Gong D, Fan MY (2013) Bioremediation of petroleum-contaminated soil by biostimulation amended with biochar. Int Biodeterior Biodegradation 85:150–155
Rajkumar M, Freitas H (2008) Effects of inoculation of plant-growth promoting bacteria on Ni uptake by Indian mustard. Bioresour Technol 99(9):3491–3498
Rajkumar M, Sandhya S, Prasad MNV, Freitas H (2012) Perspectives of plant-associated microbes in heavy metal phytoremediation. Biotechnol Adv 30(6):1562–1574
Rangel WM, Thijs S, Janssen J, Oliveira Longatti SM, Bonaldi DS, Ribeiro PRA, Jambon I, Eevers N, Weyens N, Vangronsveld J, Moreira FMS (2017) Native rhizobia from Zn mining soil promote the growth of *Leucaena leucocephala* on contaminated soil. Int J Phytorem 19(2):142–156
Romeh AA, Hendawi MY (2014) Bioremediation of certain organophosphorus pesticides by two biofertilizers, *Paenibacillus* (*Bacillus*) *polymyxa* (Prazmowski) and *Azospirillum lipoferum* (Beijerinck). J Agric Sci Technol 16(2):265–276
Saha JK, Selladurai R, Coumar MV, Dotaniya ML, Kundu S, Patra AK (2017) Status of soil pollution in India. In: Soil pollution-an emerging threat to agriculture. Springer, Singapore, pp 271–315
Salunkhe VP, Sawant IS, Banerjee K, Rajguru YR, Wadkar PN, Oulkar DP, Sawant SD (2013) Biodegradation of profenofos by Bacillus subtilis isolated from grapevines (*Vitis vinifera*). J Agric Food Chem 61(30):7195–7202
Saravanan A, Jeevanantham S, Narayanan VA, Kumar PS, Yaashikaa PR, Muthu CM (2019) Rhizoremediation–a promising tool for the removal of soil contaminants: a review. J Environ Chem Eng 8:103543
Schroll R, Becher HH, Dörfler U, Gayler S, Grundmann S, Hartmann HP, Ruoss J (2006) Quantifying the effect of soil moisture on the aerobic microbial mineralization of selected pesticides in different soils. Environ Sci Technol 40(10):3305–3312
Singh BK, Walker A, Wright DJ (2006) Bioremedial potential of fenamiphos and chlorpyrifos degrading isolates: influence of different environmental conditions. Soil Biol Biochem 38 (9):2682–2693
Sirohi G, Upadhyay A, Srivastava PS, Srivastava S (2015) PGPR mediated zinc biofertilization of soil and its impact on growth and productivity of wheat. J Soil Sci Plant Nutr 15(1):202–216
Stanbrough R, Chuaboonmee S, Palombo EA, Malherbe F, Bhave M (2013) Heavy metal phytoremediation potential of a heavy metal resistant soil bacterial isolate, Achromobacter sp. strain AO22. Apcbee Proc 5:502–507
Tang J, Zhu W, Kookana R, Katayama A (2013) Characteristics of biochar and its application in remediation of contaminated soil. J Biosci Bioeng 116(6):653–659
Tribedi P, Goswami M, Chakraborty P, Mukherjee K, Mitra G, Bhattacharyya P, Dey S (2018) Bioaugmentation and biostimulation: a potential strategy for environmental remediation. J Microbiol Exp 6:223–231
Tyagi H, Prashar P (2015) Isolation and identification of cypermethrin degrading Serratia nematodiphila from cauliflower rhizosphere. Int J Pharm Tech Res 7:64–71
USEPA (2009) The value of country working together to regulate pesticides and food safety—achieving public health and environmental protection through international collaboration. http://www.epa.gov/oppfead1/international/2009/workingtogether.pdf. Accessed 29 July 2014

Wang D, Ma J, Li H, Zhang X (2018) Concentration and potential ecological risk of PAHs in different layers of soil in the petroleum-contaminated areas of the loess plateau, China. Int J Environ Res Public Health 15(8):1785

Wani PA, Khan MS, Zaidi A (2008) Chromium-reducing and plant growth-promoting Mesorhizobium improves chickpea growth in chromium-amended soil. Biotechnol Lett 30 (1):159–163

Wani PA, Khan MS (2010) *Bacillus* species enhance growth parameters of chickpea (*Cicer arietinum* L.) in chromium stressed soils. Food Chem Toxicol 48(11):3262–3267

Wu CH, Wood TK, Mulchandani A, Chen W (2006) Engineering plant-microbe symbiosis for rhizoremediation of heavy metals. Appl Environ Microbiol 72(2):1129–1134

Xie QE, Yan XL, Liao XY, Li X (2009) The arsenic hyperaccumulator fern Pteris vittata L. Environ Sci Technol 43(22):8488–8495

Yadav IC, Devi NL, Syed JH, Cheng Z, Li J, Zhang G, Jones KC (2015) Current status of persistent organic pesticides residues in air, water, and soil, and their possible effect on neighboring countries: a comprehensive review of India. Sci Total Environ 511:123–137

Yadav KK, Singh JK, Gupta N, Kumar V (2017) A review of nanobioremediation technologies for environmental cleanup: a novel biological approach. J Mater Environ Sci 8(2):740–757

Zhang WX (2003) Nanoscale iron particles for environmental remediation: an overview. J Nanopart Res 5(3–4):323–332

Zhang Y, He L, Chen Z, Wang Q, Qian M, Sheng X (2011) Characterization of ACC deaminase-producing endophytic bacteria isolated from copper-tolerant plants and their potential in promoting the growth and copper accumulation of *Brassica napus*. Chemosphere 83(1):57–62

Biotreatment of High-Salinity Wastewater: Current Methods and Future Directions

Yiyi Zhao, Xuming Zhuang, Shakeel Ahmad, Shihwu Sung, and Shou-Qing Ni

Abstract

Saline wastewaters are usually generated by various industries, including the chemical, pharmaceutical, agricultural, and aquacultural industries. The discharge of untreated high-salinity wastewater may cause serious environmental pollution and damage the aquatic, terrestrial, and wetland ecosystems. For many countries, the treatment of saline wastewater has become an important task. Generally, saline wastewaters are treated through physical and chemical methods. However, these traditional techniques are associated with higher treatment costs and the generation of byproducts. In contrast, biotreatment techniques are environmentally friendly and inexpensive. This review highlights the sources and environmental concerns of high-salinity wastewater and illustrates the latest problems and solutions to the use of biological approaches for treating saline wastewater. Although high salinity may inhibit the effectiveness of aerobic and anaerobic biological wastewater treatment methods, such strategies as selecting salt-adapted microorganisms capable of degrading pollutants with tolerance to high salinity and optimizing operating conditions can be effective. This

Y. Zhao · S.-Q. Ni (✉)
Shandong Provincial Key Laboratory of Water Pollution Control and Resource Reuse, School of Environmental Science and Engineering, Shandong University, Qingdao, Shandong, P. R. China
e-mail: sqni@sdu.edu.cn

X. Zhuang
College of Chemistry and Chemical Engineering, Yantai University, Yantai, Shandong, P. R. China

S. Ahmad
Department of Soil and Environmental Sciences, Muhammad Nawaz Shareef University of Agriculture, Multan, Pakistan

S. Sung
College of Agriculture, Forestry and Natural Resource Management, University of Hawaii at Hilo, Hilo, HI, USA

M. P. Shah (ed.), *Removal of Emerging Contaminants Through Microbial Processes*,
https://doi.org/10.1007/978-981-15-5901-3_6

mini-review may serve as a reference for future efforts to treat high-salinity wastewater.

Keywords

Biotechnology · High-salinity wastewater · Biological reactors · Halophilic microorganisms

1 Introduction

Salinity is an important parameter for wastewater treatment (Kartal et al. 2006). Wastewater is defined as "high-salinity" or brine when the inorganic salt contents range from 1 to 3.5% w/w, and seawater usually contains 3.5% w/w sodium chloride (Pernetti and Di Palma 2005). High-strength wastewater with variable salinity and nutrient loads was produced in a number of industrial processes (Jesus et al. 2017), such as aquaculture in coastal areas, the nuclear industry, agriculture and food-processing, petroleum and natural gas extraction, and leather manufacturing (Calheiros et al. 2012; Lutz et al. 2013; Jesus et al. 2014; Lee et al. 2016; Zhang et al. 2019a).

There are some documented cases where biotechnology has been used successfully for treating saline wastewater from multiple sources. However, a comprehensive assessment regarding the feasibility of using biotreatment for the treatment of high-salinity wastewater is not available. Therefore, the review aimed to (1) outline the major sources of high-salinity wastewaters and their environmental influences; (2) summarize the traditional treatment technologies for high-salinity wastewater; and (3) emphasize the functions and mechanisms of biotreatment technologies for high-salinity wastewater.

2 Sources and Environmental Concerns of High-Salinity Wastewater

The composition and concentration of saline wastewater depend on their sources (Liang et al. 2017). There were three main sources of high-salinity wastewater: wastewater discharged during direct seawater use, agricultural runoff from saline-alkali lands, and industrial processes.

Due to the shortage of global freshwater resources, the direct use of seawater has become a feasible solution (Voutchkov 2018). Some coastal cities, such as Hong Kong and Qingdao, use seawater for fire control, road flushing, toilet flushing, and other nondirected contacts with human beings, and it is considered an important option to relieve the pressure on freshwater (Chen et al. 2012; Li et al. 2018a). High-salinity wastewater discharged from the direct use of seawater drains into the sewer affects the subsequent wastewater treatment process (Liu et al. 2016). In addition, China is the world's largest producer of aquaculture products, but the high level of mariculture is associated with many environmental issues (Xiang 2007). Inorganic nitrogen, active phosphate phosphorus, organics, and salts are the main pollutants in maricultural pollution areas (Liang et al. 2018). Seawater soda industries also produce a large quantity of high-salinity wastewater. For example, high concentrations of salts (NO_3^-: 188 g/L; TDS:1.63 $\times$ 10^5 mg/L) in wastewater from some soda ash factories were reported by Jadeja and Tewari (2007). Most desalination plants use reverse osmosis (RO) to desalinate seawater and brackish water; thus, these processes inevitably produce RO concentrate. The salinity of concentrate from seawater RO (SWRO) desalination facilities was up to 6.5–8.5 $\times$ 10^3 mg/L (Missimer and Maliva 2018).

The accumulation of soluble salts in the soil has caused land degradation, water quality deterioration, and serious problems related to agricultural development (Cassel and Sharma 2018; Fang et al. 2005). More than 20% of agricultural land is threatened by salinization globally (Raheja 1966). The salinized soils covered approximately one-tenth of the Chinese total land area, accounting for 100 million ha (Li 2010). However, the area of saline-alkali land is still expanding (Wei and Zhang 2018): saline-alkali soil area increased from 401.48 $\times$ 10^3 ha in 1954 to 1097.45 $\times$ 10^3 ha in 2005 in the western Songnen Plain, China (Yang et al. 2010), and continued desertification and salinization of the grasslands in the source regions of the Yangtze and Yellow Rivers were observed from 1968 to 2008 (Na et al. 2013). When the saline-alkali soil area was over-irrigated or rainfall occurred, it became a source of saline wastewater (Liang et al. 2017). Approximately 1.3–2 $\times$ 10^8 m^3 saline-alkali farmland drainage flows into the south of Chagan Lake from the Qianguo irrigation area every year (Yang et al. 2015). Generally, the soluble salt concentration is characterized by the conductivity (EC) value or by measuring the total dissolved solids (TDS) (Liang et al. 2017). In the Aksu Oasis area of Northwest China, the TDS in irrigation water increased from 1200 to 9.01 $\times$ 10^3 mg/L in drainage (Hu et al. 2019). In the Arys Turkestan Canal zone (Southern Kazakhstan), the TDS value of irrigated agricultural drainage exceeds 1200 mg/L, and Na^+ and HCO_3^- are the main constituents of dissolving salts (Karimov et al. 2009). An average TDS of 1191.33 mg/L was detected in drainage water of the Fayoum watershed, Egypt (Abdel Wahed et al. 2015).

Such industries as printing, dyeing, refining, chemicals, mining, currieries, pharmaceuticals, and nuclear energy may produce highly saline wastewater. The wastewater from printing and dyeing practices had high pH, high turbidity, poor biodegradability, complex composition, and high chrominance and contained inorganic contaminants, such as chloride, heavy metals, sulfate, sulfide, and nitrogen (Xu et al. 2018). Hossain et al. (2018) reported that the TDS values of the knit dyeing and woven dyeing industries in Bangladesh were as high as 2000–3000 and 5000–6000 mg/L, respectively. In leather manufacturing, after tanning processes, approximately 40% of the chromium amount remains in the solid and liquid wastes, especially spent tanning solutions (Fabiani et al. 1996). In addition, saline wastewater is discharged during soaking, liming, deashing, pickling, chrome tanning, and finishing operations (Xiao and Roberts 2010). The TDS concentration in tanneries ranges from 65.4 to 1281.1 mg/L in Nigeria (Akan et al. 2007). In the

Lokpaukwu-Ishiagu mining areas of southeastern Nigeria, it is conservatively estimated that nearly 3.3×10^4 m^3 of untreated drainage of abandoned mines, including approximately 710,000 kg of dissolved solids and 586 kg of potentially toxic metals, ran out to the Ivo River watershed each year (Ezekwe et al. 2013). Pharmaceutical wastewater carries not only COD, ammonia, and suspended solids but also organic and inorganic constituents (e.g., spent solvents, catalysts, and reactants) (Fent et al. 2006; Lefebvre and Moletta 2006). For example, an average TDS of 22,168 mg/L with Cl^-, PO_4^{3-}, Na^+, and K^+ as the major salts were reported in pharmaceutical wastewater of a Singapore pharmaceutical factory (producing penicillin family antibiotics) (Ng et al. 2014).

A large amount of high-salinity untreated wastewater that is discharged directly is the source of high-strength wastewater, which causes great damage to the environment. High-salinity wastewater strongly reduces soil productivity, worsens the water environment, hinders economic development, and threatens food production.

3 Treatment Technologies for High-Salinity Wastewater

The treatment methods and technologies of high-salinity wastewater are mainly divided into physical, chemical, biological, ecological engineering, and a combination of these technologies. Some examples of these methods are listed in Table 1. Physical and chemical technologies are widely used in high-salinity wastewater treatment, primarily including evaporation, membrane techniques, such as RO and nanofiltration (NF); ion exchange; advanced oxidation processes; and electrochemical techniques (Hou et al. 2019; Liang et al. 2017). However, physical and chemical techniques are associated with some disadvantages, such as high operational costs and difficulty in achieving the expected treatment results. Biological treatment technologies have the advantages of low operational cost, obvious treatment effect, and no secondary pollution. Researchers usually use molecular biology methods to change the environment of bacteria and enhance their salt tolerance and halophilism to improve high-salinity wastewater treatment.

4 Use of Biological Technologies for Treating High-Salinity Wastewater

Biological technologies eliminate pollutants from wastewater through the reproduction and metabolism of microorganisms that are economical, highly effective, stable, and environmentally friendly (Huang et al. 2019). Saline wastewater from industrial and aquaculture activities is frequently contaminated with nutrients and organics; thus, microorganisms are particularly important for the removal of COD and NH_4^+-N under high-salinity conditions (Liang et al. 2017). However, high salinity of wastewater leads to osmotic pressure in microbial cells to exceed normal living conditions, and a massive die-off can change the sludge microbial communities (Pollice et al. 2010; Zhang et al. 2019b), which considerably reduces treatment

Table 1 Summary of high-salinity wastewater treatment methods

	Type of wastewater	Major contaminants	Techniques	Effectiveness	References
Physical technologies	Simulated chromium tannery wastewater	Na^+, Cl^-, SO_4^{2-}	NF membrane	–	Yan et al. (2016)
Chemical technologies	Vinyl chloride monomer and polyvinyl chloride manufacturing plant	COD, TDS	Heterogeneous UV-assisted sono-Fenton	COD: 87%	Kakavandi and Ahmadi (2019)
Biological technologies	Mustard tuber wastewater	COD, Na^+, Cl^-	Microbial fuel cell	COD: 89.0% ± 1.5% BOD: 98.6% ± 2.0%	Zhang et al. (2019a)
	Synthetic wastewater	COD, TN, TP, NH_4^+-N, NaCl	SBR system	COD: 75% TN, NH_4^+-N: 98.5%	She et al. (2016)
	Landfill site leachate	NH_4^+-N, Cl^-	FBRs (plastic carriers; clay beads)	NH_4^+-N: 97%; 70%	Ramaswami et al. (2019)
Ecological engineering	Tannery wastewater	COD, TDS, TSS	CW	COD: 58–67% BOD: 60–77% TSS: 52–82% NH_4^+-N: 60–86%	Calheiros et al. (2010)
	Synthetic wastewater	Na^+, Cl^-, NO_3^--N, NH_4^+-N, PO_4^{3-}-P	CW	The best of NH_4^+-N: 85% The best of NO_3-N: 68% PO_4^{3-}-P: 100%	Jesus et al. (2017)
Combination technologies	Chemical plant	4,4′-Oxybis	Capacitive deionization-photocatalysis	EC: 55%	Ye et al. (2019)

(continued)

Table 1 (continued)

	Type of wastewater	Major contaminants	Techniques	Effectiveness	References
	Synthetic wastewater	COD, Na^+, Cl^-	CW-microbial fuel cell	COD: 64.79 ± 1.15% TP: 86.12 ± 0.38% TN: 70.86 ± 0.49% NH_4^+-N: 79.67 ± 0.45%	Xu et al. (2019)

efficiency (Ludzack and Noran 1965) and limits traditional biological treatment technologies. Biological treatment technologies of high-salinity wastewater generally include aerobic sludge plants (such as the traditional aerobic activated sludge process, aerobic granular sludge, the sequencing batch reaction (SBR) system, biofilms, and biofilters), anaerobic sludge plants, and cultivation and domestication of salt-tolerant and halophilic bacteria from saltwater lakes or oceans. Some examples of these methods are listed in Table 2. These reports have shown the potential use of biological technologies for saline wastewater treatment. For example, Ramaswami et al. (2019) found that fixed-bed reactor (FBR) with plastic media fed with NF permeate of landfill leachate yielded NH_4^+-N removals >97%. Muñoz Sierra et al. (2019) reported that anaerobic membrane bioreactors (AnMBR) exhibited a phenol removal of 96% at 26 g Na^+/L. In this mini-review, we reviewed the use of biotechnology to treat high-salinity wastewater, as well as the function and purification mechanisms of aerobic and anaerobic microorganisms under salt stress.

4.1 Aerobic Sludge Plants

The traditional activated sludge process has been used in wastewater biological treatment for more than a century (Bengtsson et al. 2019). In the mid-1960s, Ludzack and Noran (1965) studied the effect of salinity on an activated sludge treatment system. These researchers found that sludge focusability and microbial respiration decreased with increasing salinity and that the effluent became turbid. Yucel Tokuz and Wesley Eckenfelder (1979) found little effect of salinity on the activated sludge system when the inflow sodium chloride concentration was less than 35 g/L. The properly domesticated microbes can adapt to high-salt conditions (Hamoda and Al-Attar 1995). Aloui et al. (2009) described that under NaCl concentrations up to 4% (w/v) and organic loading rates up to 855 mg COD/L/d, the fish processing saline wastewater efficiently treated with acclimatizable activated sludge. At 2.0 wt% salinity, the SBR system inoculated with activated sludge achieved a 95% removal rate of COD, BOD, NH_4^+-N, and TP with *Candidate_division_TM7* as the dominant bacterial genus (Zhao et al. 2016). Some studies also reported the removal of heavy metals under high-salinity conditions by the aerobic activated sludge process. Industrial saline wastewaters contain heavy metals, such as tannery effluents (Zhuang et al. 2010). Zhang et al. (2019b) studied the removal of selenite (Se^{4+}) in artificial wastewater under high-salinity conditions of 70 g/L by using activated sludge in aerobic SBRs. The reactor removed soluble Se with relatively high efficiency in the beginning of the experiment. However, the experiment failed because the activated sludge was not adapted to high salinity, and the removal efficiency was recovered from the 20th batch.

Compared with the traditional activated sludge process, aerobic granular sludge (AGS) is denser, more compact and spherical, thereby occupying a lower footprint (Bengtsson et al. 2019; Beun et al. 1999). The effectiveness and stability of AGS in treating salt-containing wastewater have been evaluated in some studies. Pronk et al. (2014) evaluated the effect of stepwise increased salinity levels on nitrification,

Table 2 Use of biological technologies for high-salinity wastewater treatment

Treatment methods	Type of wastewater	Operation parameters	Removal efficiency	Comments	References
SBR system	Synthetic wastewater	HRT: 14.9 h	COD: 75% TN, NH_4^+-N: 98.5%	In the salt concentration range of 5 to 37.7 g/L, the increase of salinity did not inhibit ammonium oxidation and nitrite denitrification	She et al. (2016)
FBRs: Plastic carriers; clay beads	Landfill site leachate	Upflow velocities: 8 or 12 m/h	NH_4^+-N: 97%; 70%	Increased in chloride content did not have any observable detrimental effects on the performance of the reactors	Ramaswami et al. (2019)
AnMBRs	Phenolic wastewater	HRT: 7 days SRT: 40 ± 2 days	Phenol: 96% (26 g Na^+/L)	AnMBR shows higher stability than UASB under high-salinity conditions	Muñoz Sierra et al. (2019)
Aerobic SBR	Tanneries effluents	HRT: 2.5–5 days	The best of COD: 95%, PO_4^{3-}: 93%, TKN: 96%, SS: 92%	The halophilic bacteria responsible for nitrogen removal were most sensitive to modification of HRT, OLR, and salinity	Lefebvre et al. (2005)
Marine purple phototrophic bacteria cultivation	High-salinity domestic wastewater	HRT: 22.4–25 h SRT: 2.3–3.9 days	COD: 86 ± 1.7% TN: 62 ± 2.0% TP: 51 ± 2.6%	The adapted bacterial community was halophilic, with batch tests having a broad optimum between 20 and 70 mS/cm	Hulsen et al. (2019)

(continued)

Table 2 (continued)

Treatment methods	Type of wastewater	Operation parameters	Removal efficiency	Comments	References
Airlift SBR	Synthetic wastewater	OLR: 0.18–0.36 g/(L/day) HRT: 8 h SRT: 20 days	COD: 90.9 ± 0.8% NH_4^+-N: 72.6 ± 4.0%	Halophile sediment granulation in estuaries resolved the problem of sludge loss in high-salinity environment	Huang et al. (2019)

denitrification, nitrous oxide emissions, phosphate, and chemical oxygen demand (COD) removal from synthetic wastewater. Ammonia oxidation was not affected at any salt concentration, but nitrite oxidation and phosphate removal were severely inhibited at 20 g Cl^-/L. Hou et al. (2019) followed the performance and microbial characteristics of AGS under different salinities and alternating salinities. These researchers found that alternating salinity not only increased the COD removal efficiency but also generated a high concentration of granular biomass with good settling ability.

The effectiveness of biofilms and biofilters to treat high-salt wastewater has been discussed in several studies. Aslan and Simsek (2012) investigated the NO_2-N/NO_x-N ratio and NH_4^+-N removal efficiencies under various NaCl concentrations (0–40 g/L) under constant environmental conditions in a submerged biofilter reactor. These researchers found that the removal rate of NH_4^+-N (from 92 to 95%) increased with a small increase in salt concentration (1 g/L NaCl), and over this concentration, each NaCl addition induced NH_4^+–N oxidation. Navada et al. (2019) found that the ammonia oxidation capacity of moving bed biofilm reactors (MBBRs) was weakly influenced by the salinity increase rate, and the microbial community composition changed least for the largest salinity increment.

4.2 Anaerobic Sludge Plants

Anaerobic processes seem more attractive than aerobic processes because the processes consume little energy without aeration and can generate new energy in the form of methane or fuel alcohols (Xiao and Roberts 2010). Lefebvre et al. (2006) used an upflow sludge blanket reactor (UASB) reactor to study the anaerobic digestion of tannery soak liquor, and 78% COD removal efficiency was achieved at a TDS concentration of 71 g/L. Chung et al. (2009) observed approximately 100% removal efficiency of perchlorate and nitrate after 35 days of operation with denitrifying upflow packed-bed bioreactor (10% w/v NaCl concentration). Carrera et al. (2019) studied the development and stability of AGS in two SBR reactors treating fish canning wastewater. Their results demonstrated that the presence of anaerobic feeding/reaction phase increased the removal efficiency (80–90%) of organics, which was higher than with complete aerobic phase (75–85%).

In 1995, anaerobic ammonia oxidation (anammox) was first discovered unexpectedly in a denitrification fluidized bed reactor (Mulder et al. 1995). Anammox refers to the microbial anaerobic oxidation of ammonium by nitrite (NO_2^-), which forms N_2; it is a chemoautotrophic process (Reitner and Thiel 2011; Strous et al. 1998). The anammox process has the potential to treat saline wastewater, as studies have been conducted in the Black Sea, Golfo Dulce, and the tropical east Pacific, which showed that anammox was the main pathway to remove active nitrogen from the ocean (Dalsgaard et al. 2003; Kuypers et al. 2003; Lam et al. 2009). Numerous studies have indicated that the processes related to anammox have been successfully applied to treat saline wastewater. We summarized two strategies to treat saline wastewater with high ammonia nitrogen through the anammox process: (1) acclimation of freshwater-derived anammox bacteria (FAB) and (2) enrichment of marine anammox bacteria (MAB) from salt lakes or marine sediments.

Acclimation of FAB to the saline environment is a practical strategy to treat saline wastewater with high ammonia nitrogen (Li et al. 2018b). Jin et al. (2011) demonstrated that the anammox upflow anaerobic sludge blanket (UASB) reactor performed well under 30 g NaCl /L in the continuous flow. Gonzalez-Silva et al. (2017) found that stepwise adaptation of FAB from 0 to 3 g NaCl/L took 153 days while taking only 40 days from 3 to 30 g NaCl/L, and the dominant genera were shifted from *Candidatus Brocadia fulgida* to Ca. *Kuenenia stuttgartiensis* observed at 3 g NaCl/L. The nitritation-anammox process (SNAP) system was reported to be able to treat saline (3%) ammonium rich (185 mg/L) wastewater after gradual adaptation, and *Kuenenia* (anammox), *Nitrosomonas* (AOB), and *Nitrosovibrio* (AOB) bacteria were salt-adaptable microbes (Ge et al. 2019).

Nakajima et al. (2008) first successfully established an enrichment culture of MAB in a column-type reactor. Li et al. (2018c) reported that the nitrogen removal performance of MAB in SBR was enhanced with Mn^{2+} and Ni^{2+} addition. Rios-Del Toro et al. (2017) raised a novel upflow anaerobic sediment trapped (UAST) reactor for enrichment of MAB and realized high nitrogen removal efficiencies (>95%). The results obtained by these researchers indicated that *Candidatus Kuenenia* and *Candidatus Anammoximicrobium* had great potential for removing NH_4^+-N. These studies suggest that MAB can be applied to the treatment and recovery of high-salinity wastewater in the future.

4.3 Cultivation and Domestication of Salt-Tolerant and Halophilic Bacteria

According to the microorganisms' requirements for salt concentration classification, true halophilic organisms must grow with salt concentrations >3 M NaCl, whereas salt-tolerant organisms do not depend on salt for growth but can tolerate appreciable salt concentrations (<1 M NaCl) (Kushner 1993; Mokashe et al. 2018).

The physiological and biochemical mechanisms of salt-tolerant and halophilic bacteria to salt stress are complex. We reviewed three aspects. (1) maintain the balance of osmotic pressure. Extremely halophilic archaea and halophilic bacteria accumulate compatible solutes (organic solutes: trehalose, alanine, proline, glycine betaine), maintain high inorganic solute concentrations (K^+, Mg^{2+}), or elevate the anionic phospholipid proportion of the cellular membrane to maintain the hydration state of the cytoplasmic membrane (Mokashe et al. 2018; Mukhtar et al. 2019; Nath 2016). (2) Maintain protease activities by salinity. Most enzymes of halobacteria have robust activity and stability under hypersaline conditions and lose activity at salt concentrations lower than 2 M (Lanyi. 1974; Zhuang et al. 2010). (3) Produce exopolysaccharides (EPS) appearing to prevent injury from free radicals. Boujida et al. (2018) isolated ten halophilic strains from different hypersaline environments, and all the halophilic bacteria EPS had high emulsifying and antioxidant activities.

The treatment of high-salinity wastewater in a biological reactor acclimates common activated sludge by gradually increasing the salinity of wastewater, which undoubtedly increases the treatment time and cost. Halophiles prefer salt and thrive on saline or hypersaline environments. The cultivation and domestication of salt-tolerant and halophilic bacteria growing in extreme high-salinity environments (such as salt lakes, bay salt fields, saline-alkali soils, and pickled food factories) used directly to treat high-salinity wastewater are currently the research hotspot. The bacterial consortium was identified as *Ochrobactrum* sp., *Enterobacter cloacae*, and *Stenotrophomonas maltophilia* isolated from petroleum- or coal-contaminated sites from the seaport of Chennai, India, and were reported to have the ability to degrade PAHs present in crude oil-contaminated saline wastewater (Arulazhagan and Vasudevan 2009). Abou-Elela et al. (2010) isolated a salt-tolerant microorganism (*Staphylococcus xylosus*) from a vegetable pickled plant containing approximately 7.2% salts and used it as an inoculum for biodegradation. These researchers found that when the NaCl concentration was higher than 2%, the COD removal rate of *S. xylosus* alone was higher than that obtained using a *Staphylococcus*-supplemented mixture of activated sludge and activated bacterial culture alone. In addition, *S. xylosus* alone was applied to the case study of a vegetable acid wastewater with 7.2% salinity, and the COD removal efficiency was up to 88%. Wu et al. (2013) used halophilic microorganisms (*Bacillus* sp. strain) in the aerobic process to treat pretreated wastewater in ethyl chloride production with 4% NaCl and achieved 58.3% COD removal. Halophilic purple phototrophic bacteria (PPB) isolated from Brisbane River sediments were reported to be able to remove COD, nitrogen, and phosphorus from high-salinity domestic wastewater (Hulsen et al. 2019). These observations indicate that biological treatment techniques inoculated with specific halophilic microorganisms provide a promising avenue for the treatment of saline wastewater.

In conclusion, biological technologies are resource-saving and environmentally friendly and have been widely employed in different types of wastewater treatment. Saline wastewaters from different sources usually contain complex pollutants, such as NaCl, and high concentrations of various salts, organic compounds, heavy metals, pesticides, antibiotics, and so on. These pollutants can cause land degradation, water quality deterioration, and serious problems related to agricultural development. The current status and potential of biological treatment of high-salinity wastewaters were reviewed in this paper. Although high-salinity inhibits the growth of microbes, prior studies have shown that the acclimated microorganisms can gradually adapt to the environment of high salinity and can be used to treat saline wastewater. However, more studies are needed to investigate the different types of target contaminants and the interactions between salt and specific contaminants in activated sludge. In addition, it is suggested to carry out the purification culture and domestication of salt-tolerant and halophilic bacteria under practical conditions in future research.

Acknowledgments The authors gratefully acknowledge the support from the National Natural Science Foundation of China (21777086), Taishan Scholar Youth Expert Program of Shandong Province (tsqn201909005), Key Research & Developmental Program of Shandong Province (2019JZZY020308), Natural Science Foundation for Distinguished Young Scholars of Shandong Province (JQ201809), Young Scholars Program of Shandong University (2016WLJH16, 2020QNQT012), and Qingdao Science and Technology Huimin Demonstration Guide Project (20-3-4-4-nsh).

Conflict of Interest Authors declare no conflict of interest.

References

Abdel Wahed MSM, Mohamed EA, Wolkersdorfer C, El-Sayed MI, M'nif A, Sillanpää M (2015) Assessment of water quality in surface waters of the Fayoum watershed, Egypt. Environ Earth Sci 74:1765–1783. https://doi.org/10.1007/s12665-015-4186-0

Abou-Elela SI, Kamel MM, Fawzy ME (2010) Biological treatment of saline wastewater using a salt-tolerant microorganism. Desalination 250:1–5. https://doi.org/10.1016/j.desal.2009.03.022

Akan JC, Moses VOO, Abah J (2007) Assessment of tannery industrial effluents from Kano metropolis, Kano state, Nigeria. J Appl Sci 7:2788–2793. https://doi.org/10.3923/jas.2007.2788.2793

Aloui F, Khoufi S, Loukil S, Sayadi S (2009) Performances of an activated sludege process for the treatment of fish processing saline wastewater. Desalination 246:389–396. https://doi.org/10.1016/j.desal

Arulazhagan P, Vasudevan N (2009) Role of a moderately halophilic bacterial consortium in the biodegradation of polyaromatic hydrocarbons. Mar Pollut Bull 58:256–262. https://doi.org/10.1016/j.marpolbul.2008.09.017

Aslan S, Simsek E (2012) Influence of salinity on partial nitrification in a submerged biofilter. Bioresour Technol 118:24–29. https://doi.org/10.1016/j.biortech.2012.05.057

Bengtsson S, de Blois M, Wilen BM, Gustavsson D (2019) A comparison of aerobic granular sludge with conventional and compact biological treatment technologies. Environ Technol 40:2769–2778. https://doi.org/10.1080/09593330.2018.1452985

Beun JJ, Hendriks A, Van Loosdrecht MC, Morgenroth E, Wilderer PA, Heijnen JJ (1999) Aerobic granulation in a sequencing batch reactor. Water Res 33:2283–2290. https://doi.org/10.1016/S0043-1354(98)00463-1

Boujida N, Palau M, Charfi S, El Moussaoui N, Manresa A et al (2018) Isolation and characterization of halophilic bacteria producing exopolymers with emulsifying and antioxidant activities. Biocatal Agric Biotechnol 16:631–637. https://doi.org/10.1016/j.bcab.2018.10.015

Calheiros CS, Teixeira A, Pires C, Franco AR, Duque AF et al (2010) Bacterial community dynamics in horizontal flow constructed wetlands with different plants for high salinity industrial wastewater polishing. Water Res 44:5032–5038. https://doi.org/10.1016/j.watres.2010.07.017

Calheiros CS, Quiterio PV, Silva G, Crispim LF, Brix H, Moura SC, Castro PM (2012) Use of constructed wetland systems with Arundo and Sarcocornia for polishing high salinity tannery wastewater. J Environ Manage 95:66–71. https://doi.org/10.1016/j.jenvman.2011.10.003

Carrera P, Campo R, Mendez R, Di Bella G, Campos JL, Mosquera-Corral A, Val Del Rio A (2019) Does the feeding strategy enhance the aerobic granular sludge stability treating saline effluents? Chemosphere 226:865–873. https://doi.org/10.1016/j.chemosphere.2019.03.127

Cassel F, Sharma S (2018) Delocalization of salt solution in a semiarid farmland topsoil. Water Air Soil Pollut 229:254. https://doi.org/10.1007/s11270-018-3905-z

Chen GH, Chui HK, Wong CL, Tang DTW, Lu H, Jiang F, van Loosdrecht MC (2012) An innovative triple water supply system and a Novel SANI® process to alleviate water shortage and pollution problem for water-scarce coastal areas in China. J Water Sustain 2:121–129

Chung J, Shin S, Oh J (2009) Characterization of a microbial community capable of reducing perchlorate and nitrate in high salinity. Biotechnol Lett 31:959–966. https://doi.org/10.1007/s10529-009-9960-1

Dalsgaard T, Canfield DE, Petersen J, Thamdrup B, Acuna-Gonzalez J (2003) N2 production by the anammox reaction in the anoxic water column of Golfo Dulce, Costa Rica. Nature 422:606–608. https://doi.org/10.1038/nature01526

Ezekwe IC, Ezekwe AS, Chima GN (2013) Metal loadings and alkaline mine drainage from active and abandoned mines in the Ivo River Basin Area of Southeastern Nigeria. Mine Water Environ 32:97–107. https://doi.org/10.1007/s10230-013-0220-1

Fabiani C, Ruscio F, Spadoni M, Pizzichini M (1996) Chromium(III) salts recovery process from tannery wastewaters. Desalination 108:183–191

Fang H, Liu G, Kearney M (2005) Georelational analysis of soil type, soil salt content, landform, and land use in the Yellow River Delta, China. Environ Manag 35:72–83. https://doi.org/10.1007/s00267-004-3066-2

Fent K, Weston AA, Caminada D (2006) Ecotoxicology of human pharmaceuticals. Aquat Toxicol 76:122–159. https://doi.org/10.1016/j.aquatox.2005.09.009

Ge CH, Dong Y, Li H, Li Q, Ni SQ et al (2019) Nitritation-anammox process - a realizable and satisfactory way to remove nitrogen from high saline wastewater. Bioresour Technol 275:86–93. https://doi.org/10.1016/j.biortech.2018.12.032

Gonzalez-Silva BM, Ronning AJ, Andreassen IK, Bakke I, Cervantes FJ, Ostgaard K, Vadstein O (2017) Changes in the microbial community of an anammox consortium during adaptation to marine conditions revealed by 454 pyrosequencing. Appl Microbiol Biotechnol 101:5149–5162. https://doi.org/10.1007/s00253-017-8160-5

Hamoda MF, Al-Attar IMS (1995) Effects of high sodium chloride concentrations on activated sludge treatment. Water Sci Technol 31:61–72. https://doi.org/10.1016/0273-1223(95)00407-e

Hossain L, Sarker SK, Khan MS (2018) Evaluation of present and future wastewater impacts of textile dyeing industries in Bangladesh. Environ Dev 26:23–33. https://doi.org/10.1016/j.envdev.2018.03.005

Hou M, Li W, Li H, Li C, Wu X, Y-d L (2019) Performance and bacterial characteristics of aerobic granular sludge in response to alternating salinity. Int Biodeter Biodegr 142:211–217. https://doi.org/10.1016/j.ibiod.2019.05.007

Hu Q, Yang Y, Han S, Wang J (2019) Degradation of agricultural drainage water quantity and quality due to farmland expansion and water-saving operations in arid basins. Agric Water Manage 213:185–192. https://doi.org/10.1016/j.agwat.2018.10.019

Huang J-L, Wang H-H, Alam F, Cui Y-W (2019) Granulation of halophilic sludge inoculated with estuarine sediments for saline wastewater treatment. Sci Total Environ 682:532–540. https://doi.org/10.1016/j.scitotenv.2019.05.197

Hulsen T, Hsieh K, Batstone DJ (2019) Saline wastewater treatment with purple phototrophic bacteria. Water Res 160:259–267. https://doi.org/10.1016/j.watres.2019.05.060

Jadeja RN, Tewari A (2007) Effect of soda ash industry effluent on bioaccumulation of metals by seaweeds of coastal region of Gujarat, India. J Hazard Mater 147:148–154. https://doi.org/10.1016/j.jhazmat.2006.12.060

Jesus JM, Calheiros CS, Castro PM, Borges MT (2014) Feasibility of Typha latifolia for high salinity effluent treatment in constructed wetlands for integration in resource management systems. Int J Phytoremediation 16:334–346. https://doi.org/10.1080/15226514.2013.773284

Jesus JM, Cassoni AC, Danko AS, Fiuza A, Borges MT (2017) Role of three different plants on simultaneous salt and nutrient reduction from saline synthetic wastewater in lab-scale constructed wetlands. Sci Total Environ 579:447–455. https://doi.org/10.1016/j.scitotenv.2016.11.074

Jin R-C, Ma C, Mahmood Q, Yang G-F, Zheng P (2011) Anammox in a UASB reactor treating saline wastewater. Process Saf Environ Prot 89:342–348. https://doi.org/10.1016/j.psep.2011.05.001

Kakavandi B, Ahmadi M (2019) Efficient treatment of saline recalcitrant petrochemical wastewater using heterogeneous UV-assisted sono-Fenton process. Ultrason Sonochem 56:25–36. https://doi.org/10.1016/j.ultsonch.2019.03.005

Karimov A, Qadir M, Noble A, Vyshpolsky F, Anzelm K (2009) Development of magnesium-dominant soils under irrigated agriculture in Southern Kazakhstan. Pedosphere 19:331–343. https://doi.org/10.1016/S1002-0160(09)60124-7

Kartal B, Koleva M, Arsov R, van der Star W, Jetten MS, Strous M (2006) Adaptation of a freshwater anammox population to high salinity wastewater. J Biotechnol 126:546–553. https://doi.org/10.1016/j.jbiotec.2006.05.012

Kushner DJ (1993) Growth and nutrition of halophilic bacteria. In: Vreeland RH, Hochstein LI (eds) The biology of halophilic bacteria. CRC Press, Boca Raton, pp 87–103

Kuypers M, Sliekers AO, Lavik G, Schmid M, Jorgensen BB et al (2003) Anaerobic ammonium oxidation by anammox bacteria in the Black Sea. Nature 422:608–611. https://doi.org/10.1038/nature01472

Lam P, Lavik G, Jensen MM, van de Vossenberg J, Schmid M et al (2009) Revising the nitrogen cycle in the peruvian oxygen minimum zone. Proc Natl Acad Sci U S A 106:4752–4757. https://doi.org/10.1073/pnas.0812444106

Lanyi JK (1974) Salt-dependent properties of proteins from extremely halophilic bacteria. Am Soc Microbiol 38:272–290

Lee KY, Kim KW, Park M, Kim J, Oh M et al (2016) Novel application of nanozeolite for radioactive cesium removal from high-salt wastewater. Water Res 95:134–141. https://doi.org/10.1016/j.watres.2016.02.052

Lefebvre O, Moletta R (2006) Treatment of organic pollution in industrial saline wastewater: a literature review. Water Res 40:3671–3682. https://doi.org/10.1016/j.watres.2006.08.027

Lefebvre O, Vasudevan N, Torrijos M, Thanasekaran K, Moletta R (2005) Halophilic biological treatment of tannery soak liquor in a sequencing batch reactor. Water Res 39:1471–1480. https://doi.org/10.1016/j.watres.2004.12.038

Lefebvre O, Vasudevan N, Torrijos M, Thanasekaran K, Moletta R (2006) Anaerobic digestion of tannery soak liquor with an aerobic post-treatment. Water Res 40:1492–1500. https://doi.org/10.1016/j.watres.2006.02.004

Li B (2010) Soil salinization. In: Desertification and its control in China. Springer, Berlin, pp 263–298

Li Y, Wang Y, Gao Y, Zhao H, Zhou W (2018a) Seawater toilet flushing sewage treatment and nutrients recovery by marine bacterial-algal mutualistic system. Chemosphere 195:70–79. https://doi.org/10.1016/j.chemosphere.2017.12.076

Li J, Qi P, Qiang Z, Dong H, Gao D, Wang D (2018b) Is anammox a promising treatment process for nitrogen removal from nitrogen-rich saline wastewater? Bioresour Technol 270:722–731. https://doi.org/10.1016/j.biortech.2018.08.115

Li J, Feng L, Qiang Z, Dong H, Wang D (2018c) Enhanced performance and kinetics of marine anammox bacteria (MAB) treating nitrogen-rich saline wastewater with Mn(II) and Ni

(II) addition. Bioresour Technol 249:1085–1091. https://doi.org/10.1016/j.biortech.2017.10.101

Liang Y, Zhu H, Bañuelos G, Yan B, Zhou Q, Yu X, Cheng X (2017) Constructed wetlands for saline wastewater treatment: a review. Ecol Eng 98:275–285. https://doi.org/10.1016/j.ecoleng.2016.11.005

Liang Y, Cheng X, Zhu H, Shutes B, Yan B, Zhou Q, Yu X (2018) Historical evolution of mariculture in China during past 40 years and its impacts on eco-environment. Chin Geogr Sci 28:363–373. https://doi.org/10.1007/s11769-018-0940-z

Liu X, Dai J, Wu D, Jiang F, Chen G, Chui H-K, van Loosdrecht MCM (2016) Sustainable application of a novel water cycle using seawater for toilet flushing. Engineering 2:460–469. https://doi.org/10.1016/j.eng.2016.04.013

Ludzack FJ, Noran DK (1965) Tolerance of high salinities by conventional wastewater treatment processes. Water Pollut Control Fed 37:1404–1416

Lutz BD, Lewis AN, Doyle MW (2013) Generation, transport, and disposal of wastewater associated with Marcellus Shale gas development. Water Resour Res 49:647–656. https://doi.org/10.1002/wrcr.20096

Missimer TM, Maliva RG (2018) Environmental issues in seawater reverse osmosis desalination: intakes and outfalls. Desalination 434:198–215. https://doi.org/10.1016/j.desal.2017.07.012

Mokashe N, Chaudhari B, Patil U (2018) Operative utility of salt-stable proteases of halophilic and halotolerant bacteria in the biotechnology sector. Int J Biol Macromol 117:493–522. https://doi.org/10.1016/j.ijbiomac.2018.05.217

Mukhtar S, Ahmad S, Bashir A, Mehnaz S, Mirza MS, Malik KA (2019) Identification of plasmid encoded osmoregulatory genes from halophilic bacteria isolated from the rhizosphere of halophytes. Microbiol Res 228:126307. https://doi.org/10.1016/j.micres.2019.126307

Mulder A, van de Graaf AA, Robertson LA, Kuenen JG (1995) Anaerobic ammonium oxidation discovered in a denitrifying fluidized bed reactor. FEMS Microbiol Ecol 16:177–184. https://doi.org/10.1016/0168-6496(94)00081-7

Muñoz Sierra JD, Oosterkamp MJ, Wang W, Spanjers H, van Lier JB (2019) Comparative performance of upflow anaerobic sludge blanket reactor and anaerobic membrane bioreactor treating phenolic wastewater: overcoming high salinity. Chem Eng J 366:480–490. https://doi.org/10.1016/j.cej.2019.02.097

Na L, Genxu W, Guangsheng L, Yun L, Xiangyang S (2013) The ecological implications of land use change in the Source Regions of the Yangtze and Yellow Rivers, China. Reg Environ Change 13:1099–1108. https://doi.org/10.1007/s10113-013-0419-5

Nakajima J, Sakka M, Kimura T, Furukawa K, Sakka K (2008) Enrichment of anammox bacteria from marine environment for the construction of a bioremediation reactor. Appl Microbiol Biotechnol 77:1159–1166. https://doi.org/10.1007/s00253-007-1247-7

Nath A (2016) Insights into the sequence parameters for halophilic adaptation. Amino Acids 48:751–762. https://doi.org/10.1007/s00726-015-2123-x

Navada S, Vadstein O, Tveten A-K, Verstege GC, Terjesen BF et al (2019) Influence of rate of salinity increase on nitrifying biofilms. J Clean Prod 238:117835. https://doi.org/10.1016/j.jclepro.2019.117835

Ng KK, Shi X, Tang MKY, Ng HY (2014) A novel application of anaerobic bio-entrapped membrane reactor for the treatment of chemical synthesis-based pharmaceutical wastewater. Sep Purif Technol 132:634–643. https://doi.org/10.1016/j.seppur.2014.06.021

Pernetti M, Di Palma L (2005) Experimental evaluation of inhibition effects of saline wastewater on activated sludge. Environ Technol 26:695–703. https://doi.org/10.1080/09593330.2001.9619509

Pollice A, Rozzi A, Tomei MC, Di Pinto AC, Limoni N (2010) Monitoring the inhibitory effect of NaCl on anaerobic wastewater treatment processes by the Rantox biosensor. Environ Technol 21:535–544. https://doi.org/10.1080/09593332408618095

Pronk M, Bassin JP, de Kreuk MK, Kleerebezem R, van Loosdrecht MC (2014) Evaluating the main and side effects of high salinity on aerobic granular sludge. Appl Microbiol Biotechnol 98:1339–1348. https://doi.org/10.1007/s00253-013-4912-z

Raheja PC (1966) Aridity and salinity. In: Boyko H (ed) Salinity and aridity. Monographiae Biologicae. Springer, Dordrecht, pp 43–44

Ramaswami S, Jalal Uddin FM, Behrendt J, Otterpohl R (2019) High-rate nitrification of saline wastewaters using fixed-bed reactors. J Environ Manage 243:444–452. https://doi.org/10.1016/j.jenvman.2019.05.020

Reitner J, Thiel V (2011) Anammox. In: Reitner J, Thiel V (eds) Encyclopedia of geobiology. Springer, Dordrecht, pp 53–53. https://doi.org/10.1007/978-1-4020-9212-1_245

Rios-Del Toro EE, López-Lozano NE, Cervantes FJ (2017) Up-flow anaerobic sediment trapped (UAST) reactor as a new configuration for the enrichment of anammox bacteria from marine sediments. Bioresour Technol 238:528–533. https://doi.org/10.1016/j.biortech.2017.04.087

She Z, Zhao L, Zhang X, Jin C, Guo L et al (2016) Partial nitrification and denitrification in a sequencing batch reactor treating high-salinity wastewater. Chem Eng J 288:207–215. https://doi.org/10.1016/j.cej.2015.11.102

Strous M, Heijnen JJ, Kuenen JG, Jetten MSM (1998) The sequencing batch reactor as a powerful tool for the study of slowly growing anaerobic ammonium-oxidizing microorganisms. Appl Microbiol Biotechnol 50:589–596. https://doi.org/10.1007/s002530051340

Voutchkov N (2018) Energy use for membrane seawater desalination – current status and trends. Desalination 431:2–14. https://doi.org/10.1016/j.desal.2017.10.033

Wei Y, Zhang SH (2018) Abiostress resistance and cellulose degradation abilities of haloalkaliphilic fungi: applications for saline-alkali remediation. Extremophiles 22:155–164. https://doi.org/10.1007/s00792-017-0986-3

Wu X, Du YG, Qu Y, Du DY (2013) Ternary cycle treatment of high saline wastewater from pesticide production using a salt-tolerant microorganism. Water Sci Technol 67:1960–1966. https://doi.org/10.2166/wst.2013.072

Xiang JH (2007) Mariculture-related environmental concerns in the People's Republic of China. In: Bert TM (ed) Ecological and genetic implications of aquaculture activities. Methods and technologies in fish biology and fisheries, vol 6. Springer, Dordrecht, pp 219–228

Xiao Y, Roberts DJ (2010) A review of anaerobic treatment of saline wastewater. Environ Technol 31:1025–1043. https://doi.org/10.1080/09593331003734202

Xu H, Yang B, Liu Y, Li F, Shen C et al (2018) Recent advances in anaerobic biological processes for textile printing and dyeing wastewater treatment: a mini-review. World J Microbiol Biotechnol 34:165. https://doi.org/10.1007/s11274-018-2548-y

Xu F, Ouyang DL, Rene ER, Ng HY, Guo LL et al (2019) Electricity production enhancement in a constructed wetland-microbial fuel cell system for treating saline wastewater. Bioresour Technol 288:121462. https://doi.org/10.1016/j.biortech.2019.121462

Yan Z-Q, Zeng L-M, Li Q, Liu T-Y, Matsuyama H, Wang X-L (2016) Selective separation of chloride and sulfate by nanofiltration for high saline wastewater recycling. Sep Purif Technol 166:135–141. https://doi.org/10.1016/j.seppur.2016.04.009

Yang J, Zhang S, Li Y, Bu K, Zhang Y, Chang L, Zhang Y (2010) Dynamics of saline-alkali land and its ecological regionalization in western Songnen Plain, China. Chin Geogr Sci 20:159–166. https://doi.org/10.1007/s11769-010-0159-0

Yang YN, Sheng Q, Zhang L, Kang HQ, Liu Y (2015) Desalination of saline farmland drainage water through wetland plants. Agric Water Manage 156:19–29. https://doi.org/10.1016/j.agwat.2015.03.001

Ye G, Yu Z, Li Y, Li L, Song L, Gu L, Cao X (2019) Efficient treatment of brine wastewater through a flow-through technology integrating desalination and photocatalysis. Water Res 157:134–144. https://doi.org/10.1016/j.watres.2019.03.058

Yucel Tokuz R, Wesley Eckenfelder W (1979) The effect of inorganic salts on the activated sludge process performance. Water Res 13:99–104. https://doi.org/10.1016/0043-1354(79)90260-4

Zhang L, Fu G, Zhang Z (2019a) Electricity generation and microbial community in long-running microbial fuel cell for high-salinity mustard tuber wastewater treatment. Bioelectrochemistry 126:20–28. https://doi.org/10.1016/j.bioelechem.2018.11.002

Zhang Y, Kuroda M, Arai S, Kato F, Inoue D, Ike M (2019b) Biological treatment of selenate-containing saline wastewater by activated sludge under oxygen-limiting conditions. Water Res 154:327–335. https://doi.org/10.1016/j.watres.2019.01.059

Zhao Y, Park HD, Park JH, Zhang F, Chen C et al (2016) Effect of different salinity adaptation on the performance and microbial community in a sequencing batch reactor. Bioresour Technol 216:808–816. https://doi.org/10.1016/j.biortech.2016.06.032

Zhuang X, Han Z, Bai Z, Zhuang G, Shim H (2010) Progress in decontamination by halophilic microorganisms in saline wastewater and soil. Environ Pollut 158:1119–1126. https://doi.org/10.1016/j.envpol.2010.01.007

Facing Lethal Impacts of Industrialization via Green and Sustainable Microbial Removal of Hazardous Pollutants and Nanobioremediation

Basma A. Omran

Abstract

Industrialization exert a negative impact to the surrounding environment by causing pollution to soil, air, and water supplies in addition to its hazardous effect on human health because of the generated toxic by-products. Environment became polluted with heavy metals, pathogenic microorganisms, toxic industrial effluents, pharmaceutical compounds, and dyestuffs. Currently, several strategies have been employed for the removal of these hazardous substances involving physical and chemical technologies. Yet, these technologies have high operation costs as well as high consumption of energy and chemicals. Hence, development of green, low-cost, and sustainable treatment procedures is the need of the hour. Employing biological entities such as fungal, bacterial, and algal isolates is an interesting bioremediation approach which has proved its high efficacy. This bio-process is affected by diverse biotic and abiotic parameters. This chapter reviews the most prevailing treatment methodologies employed for removing heavy metals and other contaminants such as physical, chemical, physiochemical, and biological methodologies. It mainly focuses on bioremediation, its mechanism, and the employment of nanotechnology in bioremediation as an emerging strategy to reduce the hazardous effects of toxic pollutants.

Keywords

Conventional treatment technologies · Fungi · Bacteria · Algae · Bioremediation · Nanomaterials · Nanobioremediation

B. A. Omran (✉)
Researcher in Microbiology, Petroleum Biotechnology Laboratory, Processes Design and Development Department, Egyptian Petroleum Research Institute (EPRI), Cairo, Egypt
e-mail: basmaomran@epri.sci.eg

M. P. Shah (ed.), *Removal of Emerging Contaminants Through Microbial Processes*,
https://doi.org/10.1007/978-981-15-5901-3_7

Abbreviations

4-AP	4-Aminophenol
4-NP	4-Nitrophenol
AFM	Atomic force microscope
AgNPs	Silver nanoparticles
AOP	Advanced oxidation process
AuNPs	Gold nanoparticles
AYR	Alizarin Yellow R
BET	Brunauer-Emmett-Teller
CLC	*Cnidium monnieri* (L.) *Cuss*
CMC	Carboxymethyl cellulose
CNTs	Carbon nanotubes
CuNPs	Copper nanoparticles
CuO NPs	Copper oxide nanoparticles
DNA	Deoxyribonucleic acid
DTPA	Diethylenetriaminepentaacetic acid
EDTA	Ethylenediaminetetraacetic acid
EDX	Energy dispersive X-ray
Eh	Redox potential
EPA	Environmental protection agency
FESEM	Field emission scanning electron microscopy
IONPs	Iron oxide nanoparticles
MISFNPs	Magnetic inverse spinel iron oxide nanoparticles
MNMs	Magnetic nanomaterials
MNPs	Magnetic nanoparticles
MWCNTs	Multi-walled carbon nanotubes
NB	Nile blue
NBR	Nanobioremediation
nm	Nanometer
NMs	Nanomaterials
NPs	Nanoparticles
PCBs	Polyvinyl biphenyls
PRB	Permeable reactive barriers
QDs	Quantum dots
RhB	Rhodamine B
RO	Reverse osmosis
RY160	Reactive Yellow 160
SWCNTs	Single-walled carbon nanotubes
TCE	Tetrachloroethylene
TEM	Transmission electron microscopy
UV/Vis	Ultraviolet-visible
VSM	Vibrating sample magnetometry
WHO	World Health Organization

XPS	X-ray photoelectron spectroscopy
XRD	X-ray diffraction
ZVI NPs	Zerovalent iron nanoparticles

1 Introduction

Recently, the world is facing an unparalleled growth in population and an accelerated jump in industrialization in order to cope with our daily life demands (Kang and So 2016; Mary et al. 2018). However, with the high improvement in the quality of human life over the past few years, it is essential not to ignore that such industrial developments had a major influence over the quality of the surrounding environment. The increase in human activities particularly in the recent years all over the world has led to the elevation of environmental pollution levels via the discharge of massive amounts of hazardous pollutants (Raghunandan et al. 2018). Industrialization resulted in negative consequences which in turn affected human health by causing severe health problems as well as to the environment, such as, limitation of natural resources, increases in carbon emissions, and pollution (Ahuti 2015; Mgbemene et al. 2016). Several inorganic and organic chemicals are dumped onto earth's surface, either through industrial processes or accidentally by spilling. Among the serious threats that concern the whole world is water pollution. It takes place due to plenty of reasons including the rapid increase in worldwide population, developments of global industries, urbanization, etc. Different pollutants cause contamination of water, for instance, presence of dyes, toxic heavy metals, and organic and inorganic pollutants that represent a major threat to the surrounding ecosystems (Gupta and Mishra 2011). During the last century, versatile industries have witnessed major progress including mining, petroleum, gas electroplating, smelting, fertilizer, pesticides, tanneries, paper, and electronic industries. These industries resulted in the release and the discharge of large quantities of heavy metals and petroleum hydrocarbons within the natural ecosystem (Arivalagan et al. 2014; Taiwo et al. 2016). Other procedures which cause environmental contamination involve leaching of heavy metals, metal corrosion, atmospheric deposition, sediment resuspension to groundwater and soil, as well as metal evaporation from different water resources (Weerasundara et al. 2017; Francová et al. 2017).

Among the hazardous contaminants, heavy metals such as arsenic (As), cadmium (Cd), lead (Pb), copper (Cu), chromium (Cr), nickel (Ni), zinc (Zn), aluminum (Al), and manganese (Mn) are responsible for major threats to the environment (Ullah et al. 2015; Dhanarani et al. 2016; Karthik et al. 2017a). Additionally, they cause sever health issues to humans (Zeraatkar et al. 2016; Chen et al. 2015; Ullah et al. 2015). Different industries have been reported to release heavy metals (e.g., Zn, Ni, Cr, Cu, Mn, Pb, and Cd) including mining, textile, petrochemical, and pharmaceutical industries (Schaider et al. 2014). Fertilizers are sources for Cu, Co, Fe, Mn, Ni, Zn, and Mo (Roberts 2014). Pesticides release heavy metals such as Hg, Cu, Mn, Zn, and Pb (Tchounwou et al. 2012). Usually, heavy metals get into the environment via either natural or anthropogenic routes and hence are deposited in the air, soil, and water systems (Kuppusamy et al. 2017). Humus is the organic material that exists in soil. It has a high affinity towards absorption of heavy metal cations, and then they are taken up by extracting these metals from the water that passes through the soil (Mary et al. 2018). Moreover, plant roots, shoots, and stems pick up these elements along with water and diffuse into plants which in turn pass to animals (Manafi et al. 2012). On the other hand, particles bearing heavy metals settle down to the bottom sediments of water bodies and accumulate there (Kang and So 2016). When aquatic organisms consume these particles, heavy metals can easily enter the food chain, and hence their adverse effect is magnified (Mary et al. 2018). The entrance of the hazardous heavy metals into the food web leads to multiple health effects such as damage to lungs, kidneys, liver, pancreatic and nervous disorders (Arivalagan et al. 2014; Kumar et al. 2017). For instance, accumulation of Cr (VI) and its reduced form Cr (III) in cells cause severe damage in DNA. Therefore, the US Environmental Protection Agency (EPA) in addition to the World Health Organization (WHO) set a particular limit for each heavy metal (Kumar et al. 2017). Usually, salts containing heavy metals are easily dissolved in wastewater and, thus, cannot be easily purified by physical purification methodologies (Hussein et al. 2003). Contamination of soil with heavy metals results in persistent alterations in microbial population (Xu et al. 2019). Heavy metals have been proved to be lethal at low concentrations as they can harm living organisms via their accumulation. Different techniques have been employed earlier for the removal of heavy metals such as membrane filtration, ion exchange, adsorption, coagulation, electrochemical treatment, reduction, oxidation, evaporation, and chemical precipitation (Rajasulochana and Preethy 2016; Carolin et al. 2017). However, each process has its own disadvantages. Hence, designing strategies to tackle this issue remains a distressing problem. In order to create a pollution-free ecosystem, remediation of water and soil is urgently required, and as more attention is directed towards the development of green, environmentally benign and clean technologies, different alternative technologies are being investigated. Remediation via green technologies are categorized into two categories: firstly, treatment based upon using plants, animals, and microorganisms which is referred to as "bioremediation" and, secondly, treatment with modified and unmodified natural products (Gupta and Mishra 2011). Bioremediation is one of the cleaning strategies which are employed to protect human health as well as the environment (Raghunandan et al. 2018). This chapter reviews the conventional treatment technologies, bioremediation, and the intersection of nanotechnology with bioremediation as a new and innovative strategy to get rid of the hazardous contaminants.

2 Types of Treatment Technologies Employed for Removal of Contaminants

Multiple conventional methodologies exist to overcome the lethal impacts of industrialization. Listed below are the physical, chemical, and biological methods which are applied to get rid of heavy metals and other contaminants/pollutants.

2.1 Physical Treatment Technologies

Physical methods involve precipitation, ion exchange, electrowinning, electrocoagulation, cementation, membrane filtration, and electrodialysis. Precipitation is the process which involves the addition of suitable anion in order to precipitate the metal salt. Generally, some common chemicals are used in this technique, for instance, manganese sulfate, copper sulfate, ammonium sulfate, ferric salts, etc. The process efficacy is influenced by pH. The lower the pH, the higher the efficiency of the process (Mondal et al. 2006). It is worth mentioning that using bisulfide or lime in the precipitation process lacks the specificity, and as a result it becomes ineffective in heavy metal removal even at low concentrations. Ion exchange is a treatment technique which is usually used in industries to recover heavy metals from industrial effluents. This process involves the presence of an ion-exchange material (solid phase) which is capable of exchanging anions or cations. The synthetic ion-exchange resin is the most commonly used ion-exchange matrix material. However, it is somehow expensive, and it becomes ineffective when it is dealing with high metal concentrations due to foul formation. Furthermore, it is non-selective and highly sensitive to any variations in solution pH (Ahluwalia and Goyal 2007). Electrowinning is a physical process which is mainly used in mining and metallurgical industries as well as electronics, metal transformation, acid drainage, electrical industries, and metal leaching (Ahluwalia and Goyal 2007). Electrocoagulation is an electrochemical-based approach in which electric current is directed to remove metals from the targeted solution. In this method, contaminants in wastewater are retained in solution via the application of external voltage. Ions precipitate by the neutralization of charged particles with the ions of opposite electrical charges under the effect of the electrocoagulation system. Cementation is one type of the precipitation methods, and it is an electrochemical approach. In this approach, metals with higher oxidation capacity flow into the solution. For instance, copper (Cu), antimony (Sb), gallium (Ga), lead (Pb), silver (Ag), gold (Au), cadmium (Cd), tin (Sn), and arsenic (As) are usually separated and recovered by cementation. Adsorption is another physical process in which the metal species become adsorbed through physisorption and chemisorption processes. In this process, the efficiency of metal ion removal is affected by different factors, such as surface area of the adsorbent, its surface energy as well as pH. Generally, the most frequently used adsorbents are activated carbon, activated alumina, potassium permanganate ($KMnO_4$)-coated glauconite, granular ferric hydroxide, sand coated with iron oxide and copper–zinc granules, etc. Another important physical method is membrane filtration. In this method, metals are separated from water by passing through a semipermeable membrane with the help of pressure gradient which plays a role as the driving force. Fouling formation is the biggest drawback in this method. Fouling takes place because of the coprecipitation of both Fe^{2+} and Mn^{2+} ions. In addition, it is somehow expensive as it requires monitoring of pressure difference and water pretreatment (Ahluwalia and Goyal 2007). Electrodialysis is much related to the reverse osmosis (RO) process, though the driving force is the difference between both methods. In electrodialysis, metal ions are separated from contaminated water by the effect of an electric field which is applied across a semipermeable membrane. This method has shown high efficacy in removing heavy metals from groundwater. This method is mainly dependent upon porosity, pH, flow rate of groundwater, texture, water content, and ionic conductivity.

2.2 Chemical Treatment Technologies

Some chemical treatment methods are applied in order to remove heavy metals and other contaminants. These chemical methodologies involve reduction, chemical washing, and chelate flushing. Reduction depends upon using reducing agents such as gaseous H_2S and dithionites. Both are deeply injected into the polluted sites leading to degradation of these pollutants in the contaminated regions. However, the main drawback of this technique is the formation of toxic intermediates. Chemical washing depends on using strong extractants such as acids for the removal of contaminants. Unfortunately, this method affects the soil quality and consequently becomes hazardous to the surrounding environment. Henceforth, ex-situ treatment is risky, and it requires severe management procedures to overcome its negative impacts. Chelate flushing is a method which has the ability to extract large amounts of heavy metals. Yet, among the main drawbacks of this method is that chelating agents are expensive and carcinogenic. Ethylenediaminetetraacetic acid (EDTA) and diethylenetriaminepentaacetic acid (DTPA) are among the mostly used chelating agents.

2.3 Biological Treatment Technologies

In this section, several biotechnology-based treatment methodologies are overviewed and highlighted as the following. Biological treatment is mainly dependent upon the use of biological entities in a process referred to as bioremediation. Bioremediation is the process in which living systems such as bacteria, fungi, algae, and some plants are utilized to degrade, detoxify, transform, immobilize, or stabilize hazardous environmental pollutants to less toxic products. Bioremediation involves different mechanisms such as bioaugmentation, in which indigenous microorganisms are employed and stimulated by the addition of nutrients (biostimulation). It also includes the use of genetically modified microorganisms, phytoremediation via using particular plants, and biomineralization in which organic substances become biodegradable into inorganic constituents (Joutey et al. 2013). Biofiltration depends on using a biofilter made up of a porous medium at which the surface is covered with water and microorganisms. In this process, the contaminants are adsorbed by means of the porous medium and are eventually biotransformed into metabolic by-products, biomass, CO_2, and water. Three main stages take place in this process, and they are growth, degradation, and detachment of microorganisms (Srivastava and Majumder 2008; Joutey et al. 2013). Another important biological treatment technology is biosorption. In biosorption dead or living bacteria are used. This technology is cost-effective but highly efficient in metal recovery (Hashim et al. 2011). Different biosorbents are used like bacteria, fungi, and algae. These biosorbents are cost-effective, thus making this process more economical and more competitive particularly in in-situ environmental applications. It has been postulated that the immobilized biomass is more recommended for large-scale applications than the native biomass (Wang and Chen 2009). Yet, investigations regarding its efficacy, cost, and ease of application require much more research. Biophysiochemical method is another biotechnological treatment methodology. It is a biological process which is coupled with a coagulation or an adsorption technique. It possesses several advantages over the other physiochemical remediation techniques (Asia et al. 2012). In a study performed by Yan et al. (2010), *Acidithiobacillus ferrooxidans* BY-3, a chemolithotrophic bacterium was isolated from mines and was used as a natural biosorbent for the biological removal of toxic organic and inorganic arsenic compounds in aqueous solutions. Srivastava et al. (2011a) managed to isolate five different fungal strains and tested their potential for the removal of toxic arsenic contaminants in polluted sites. Another relevant methodology is the use of novel biosorbents. Novel biosorbents enhance the microbial selectivity and accumulating characteristics of the used microorganisms in the bioremediation process. It is mainly a genetic engineering-based biosorption process which helps to increase the remediating activities of the used microorganisms. Studies showed that genetically engineered microorganisms have high adsorption capacity and specificity for the remediation of toxic metal ions. However, it is somehow costly, and extensive research is required regarding pilot and full-scale processes (Vijayaraghavan and Yun 2008). Bioaugmentation is an in situ bioremediation technique in which genetic engineering has been used to increase the metabolizing potential of microorganisms. Kostal et al. (2004) investigated the genetic manipulation of *E. coli* strain and overexpressed ArsR genes which resulted in the accumulation of As. It proved to be an effective method to increase the accumulation and binding of arsenic in selective ligands for its removal. Chauhan et al. (2009) developed a collection of metagenomes from the sludge of industrial effluent treatment plants and discovered a new As (V) resistance gene (arsN) which encodes a protein similar to acetyl transferase. The overexpression of this protein led to high arsenic resistance in *E. coli*. Phytoremediation technology involves the use of different plant species for soil remediation to aid in reducing the contaminants in soil, surface water, groundwater, and sediments. Phyto-extraction is mainly dependent on using plant roots for the uptake of contaminants. Phyto-degradation or phyto-transformation deals with the breakdown of contaminants via the plant metabolic processes (Srivastava et al. 2011b; Ali et al. 2013).

3 Bioremediation

Microbes are prevalent in nature. They possess the capability to remediate heavy metals in polluted sites via mineralization of organic pollutants into end products such as CO_2 and H_2O or to intermediate metabolites that can be consumed as primary substrates for microbial cell growth during the bioremediation process (Verma and Kuila 2019). Bioremediation is a biological treatment process in which biological entities and their enzymes are involved to convert lethal and xenobiotic pollutants to less toxic forms (Silva et al. 2019). The process of bioremediation is affected by environmental, physical, and chemical parameters such as stereochemistry, toxicity, concentration of the pollutant, efficiency of the microbial strain, and reaction parameters during degradation (e.g., temperature, pH, retention time, presence of other toxic compounds, and their concentration (Misal et al. 2011)). Bioremediation has several advantages, as it is considered a safe process which transforms contaminants to less toxic form and it is a cheap process when compared with the other conventional methodologies (Silva et al. 2019). However, bioremediation has shown some drawbacks, for instance, the incomplete transformation, limitation to biodegradable compounds, and the need to select different microorganisms with specific features in order to exclude the use of genetically modified microorganisms. Figure 1 represents the different mechanisms of bioremediation process. Dixit et al. (2015) illustrated that microorganisms are extremely helpful in dissolving metals and also in oxidation and reduction of transition metals. Microbial cell membranes could be damaged by the contamination caused by some organic solvents. However, the defense mechanisms are sometimes developed by the cells via the formation of hydrophobic or solvent efflux pumps which play a role as an outer cell-membrane-protective material (Dixit et al. 2015). In a study presented by Ahemad (2014), authors found that systems like energy-dependent and plasmid-encoded metal efflux systems such as chemiosmotic ion/proton pumps and ATPases were resistible towards As, Cd, and Cr in many bacterial species.

3.1 Role of Microbes in Contaminant Removal

As mentioned earlier, bioremediation appears in different mechanisms involving bioaccumulation, bioleaching, biosorption, biotransformation, and biomineralization using microorganisms (Mary et al. 2018). Contrary, phytoremediation techniques are dependent upon using plants and different plant parts for heavy metal removal. Biosorption which is considered a very promising methodology as it exerts a major role in removing heavy metals. Biosorption is an arising technology which is mainly reversible and metabolism-independent, and it involves adsorption

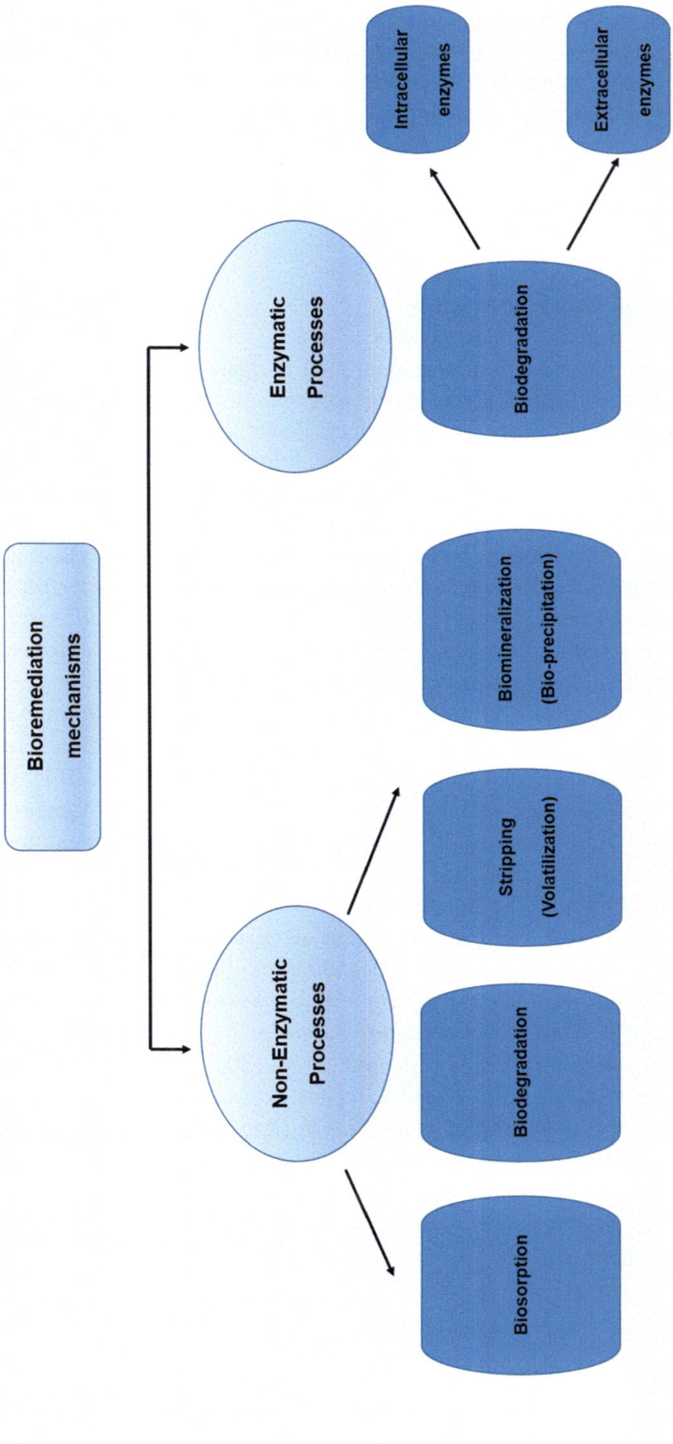

Fig. 1 An illustration showing the different bioremediation mechanisms

of contaminants on the cell surface of biological materials (Arivalagan et al. 2014). Biosorption possesses several advantages such as low operational cost, simplicity, high efficacy, as well as being an environment-friendly methodology (Karthik et al. 2017a). Nevertheless, when the concentration of these metals exceeds a certain level, metals induce numerous biochemical, physiological, and genotoxic effects (Emamverdian et al. 2015). Heavy metals highly influence the microbial communities in two different ways: (1) inhibition of metabolic function and (2) modulation in genetic materials.

3.1.1 Bacterial Bioremediation

Microbial biomass serves as potential candidates for heavy metal uptake from several polluted environments. Using microbial biomass in bioremediation can be easily acquired as leftover products from fermentation industries (Dhanarani et al. 2016). Among these microorganisms, bacteria have particular and unique genetic mechanisms which play an important role in mitigation of environmental pollution. *Pseudomonas* sp. and *Bacillus* sp. are among the mostly employed bacterial strains which are extensively used for bioremediation of heavy metals from soil and waste water. This might be attributed to the high metal binding affinities (Bachate et al. 2013). Huang and Liu (2013) and Abdi and Kazemi (2015) attributed the presence of functional groups in bacterial cell walls like hydroxyl, carboxyl, sulfonate, amide, and phosphonate groups to be mainly responsible for the metal uptake from aqueous solutions. Several studies highlighted that heavy metal bioremediation arises from the self-defense mechanism of bacterial cells. Generally, metal ions bind to the aforementioned reactive groups that exist on bacterial cell wall and then is followed by internalization of metal ions into the bacterial cell (Abbas et al. 2014). Usually, more heavy metals are uptaken by Gram-positive bacteria than Gram-negative bacteria owing to the presence of glycoproteins. Contrary, metal uptake by Gram-negative bacteria is less than Gram-positive species due to the presence of phospholipids and lipopolysaccharides (LPS) (Das et al. 2008). Removal of arsenic has been reported by different bacterial strains such as *Kocuria* sp. (Banerjee et al. 2011), *Bacillus* sp. (Majumder et al. 2013), and *Bacillus* sp. KM02 (Dey et al. 2016), while cadmium was reported to be removed by *Stenotrophomonas maltophilia* (Congeevaram et al. 2007), *Pseudomonas putida* mt2 (Shamim et al. 2014), and *Cupriavidus metallidurans* CH34 (Shamim et al. 2014). *Bacillus licheniformis* (Congeevaram et al. 2007), *Staphylococcus saprophyticus* (Jencarova and Luptakova 2012), and *Pseudomonas aeruginosa* (Babák et al. 2013) were reported to have the potential to remove chromium. Cobalt was reported to be removed by *Rhodopseudomonas palustris* (Gao et al. 2017). *Bacillus licheniformis* (Samarth et al. 2012), *Geobacillus thermodenitrificans* (Karakagh et al. 2012), *Bacillus cereus*, and *Pseudomonas aeruginosa* (Babák et al. 2013) had the capability to remove copper from contaminate sites.

3.1.2 Fungal Bioremediation (Mycoremediation)

Fungi are referred to as "mikes" in Greek. They are eukaryotic microorganisms and involve molds (fungi), yeasts, and mushrooms. Fungi are chemoheterotrophic organisms, and they are either parasitic or saprophytic. Most fungi are multicellular organisms, however; unicellular fungi are referred to as yeasts. The kingdom *Fungi* comprises phyla *Chytridiomycota* (the chytrids), *Zygomycota* (the conjugated fungi), *Ascomycota* (the sac fungi), *Basidiomycota* (the club fungi), *Deuteromycota* (the imperfect fungi), and *Glomeromycota*. This classification is based upon their mode of sexual reproduction and their molecular data (Silva et al. 2019). Fungi possess an efficient capability to transform a broad range of hazardous compounds to less toxic forms via non-specific intra- and extra-cellular oxidative enzymes (Morel et al. 2013; Durairaj et al. 2015). Fungal mycelia help the fungal strains to withstand sudden changes in pH humidity and play an important role in degrading complex organic compounds (Anastasi et al. 2013). Filamentous fungi are prevalent microorganisms. They can be observed in all types of ecological niches as they have the potential to cope their metabolism with variable carbon and nitrogen sources (Saratale et al. 2007). It is well-known that fungi possess a high tolerance and ability to detoxify heavy metal-contaminated effluents. Fungal cell wall is made up of chitin and other polysaccharides as well as proteins, polyphosphates, lipids, and inorganic ions which altogether cement the fungal cell wall. Because fungal biomass can be easily cultured in large scales, they can serve as good sorption materials (Congeevaram et al. 2007). Furthermore, fungal biomass is readily available and inexpensive. Thatoi et al. (2014) illustrated different mechanisms which might be involved in fungal detoxification of heavy metal-contaminated environments such as valence transformation, intra- and extra-cellular precipitations in addition to active uptake. Raja et al. (2015) referred that the high content of carboxyl groups in guluronic and mannuronic acids of the cell wall polysaccharides boosts heavy metal biosorption. Mamisahebei et al. (2007) showed that *Penicillium chrysogenum* was capable of removing arsenic. *Aspergillus niger*, *Aspergillus cristatus*, and *Hydrilla verticillata* were reported to have the potential to remove cadmium mercury, arsenic as shown by Bunluesin et al. (2007), Martínez-Juárez et al. (2012), and Acosta Rodríguez et al. (2013). Additionally, chromium was reported to be bioremediated by *Trichoderma viride* (Javaid and Bajwa 2007) and *Penicillium canescens* (Sethi et al. 2010). Furthermore, copper was demonstrated to be removed by *Pleurotus ostreatus* (Shipra et al. 2012) and *Aspergillus lentulus* (Sutherland and Venkobachar 2010).

3.1.3 Algal Bioremediation

Algae are photosynthetic organisms which are capable of transforming sunlight into biochemical energy (Mary et al. 2018). Algae are classified into micro- and macro-algae on the basis of size. Microalgae are unicellular, microscopic photosynthetic species, while macroalgae are multicellular organisms (Kathiraven et al. 2015). Algae are capable of thriving in all types of aquatic habitats including fresh and marine water and in moist soils. Algae have not been very well documented as biosorbents like fungi and bacteria. Bioaccumulation by algae is supported by various metabolic processes (Zeraatkar et al. 2016). Marine algae contain high quantities of biopolymers that have the potential to bind with heavy metals. Red and brown algae are among the most promising seaweeds which can be employed for biosorption (Flores-Chaparro et al. 2017). The high affinity of algal cell walls towards metal ions occurs due to the presence of certain polysaccharides. Algal cell wall is composed of different polysaccharides which possess functional groups such as carboxyl and amino groups that could serve as a platform to coordinate bonds along with metal ions. Proteins constitute approximately 10–70% and 37–50% in green and red algae, respectively (Mary et al. 2018). Kumar and Oommen (2012) demonstrated the capability of *Spirogyra hyaline* to remove arsenic. *Pithophora* spp. had the potential to remove chromium as proved by Subhashini et al. (2011). Subhashini et al. (2011) demonstrated the ability of *Sargassum* sp. to remove copper.

3.2 Factors Affecting Process of Bioremediation

Bioremediation represents a complex biological process in which microbes are capable of reducing heavy metals via the release of organic and inorganic acids (e.g., sulfuric and citric acids) and production of complexing agents (e.g., cyanide) (Gadd 2010). Several biotic and abiotic factors are involved in microbial bioremediation (Govarthanan et al. 2016). It is essential to determine and optimize such factors in order to reach the best conditions to assure the best removal of contaminants (Karthik et al. 2017b). Factors affecting the efficacy of microbial bioremediation are categorized into two main groups including (i) biotic and (ii) abiotic factors (i.e., physicochemical and climatic conditions).

3.2.1 Biotic Factors

Biotic factors exert a major influence upon the efficacy of microbial bioremediation via the changes in cell composition, size, and the production of extracellular by-products. The biomass concentration is one of the important biological factors for microbial bioremediation (Gokhale et al. 2008; Finocchio et al. 2010). According to Park and Choi (2001), low biomass concentrations increased the efficacy of metal removal. When the biomass concentration reaches the optimum level, heavy metals become adsorbed to the biomass surface and diffuse into the intracellular matrix through the concentration gradient of the metal (Mary et al. 2018). Contrary, high concentration of microbial biomass restrains the interaction between the heavy metals and the binding sites of microbial surfaces (Abbas et al. 2014). Another important biotic factor is the electrostatic interaction between cells (Ballester et al. 2007).

3.2.2 Abiotic Factors

Same as the biotic factors, the abiotic factors also affect the efficiency of heavy metal bioremediation efficiency by microorganisms. The various abiotic factors include the physicochemical parameters (e.g., temperature, pH, metal ion concentration, O_2, redox potential (Eh), nutrient, and organic carbon) and climatic conditions.

Bioremediation of heavy metals is highly influenced by the temperature of the reaction medium via affecting the physicochemical state of the targeted pollutant and through its effect on the metabolic activities of microorganisms. Thomson et al. (2017) demonstrated that the increase in temperature increases the metabolic activity of microorganisms till reaching the optimum temperature. However, further increment in temperature can lessen microbial metabolic activities via enzyme denaturation which makes the microorganism less active or being dead eventually. In addition, further increases in temperature may result in modifications in ribosomal conformations, and thus, protein synthesis is reduced (Mary et al. 2018). Contrary, when temperature is lower than that of the optimum standard one, fluidity of membranes becomes highly affected, and consequently functions of the transport systems become inhibited (Barria et al. 2013). Generally, the optimum temperature for in situ bioremediation process is approximately between 20 and 40°C. Another important factor which affects bioremediation is pH. It exerts a major influence upon microbial growth, enzymatic activities, and chemistry of metal complexation in addition to affecting the behavior of microbial cell surface functional groups (Gupta and Rastogi 2008). Bioremediation mainly takes place enzymatically, i.e., the unfavorable pH will directly affect enzyme ionization, which in turn will affect the efficiency of metal remediation.

3.3 Enzymes Involved in Microbial Bioremediation

The efficiency of biological entities to remediate heavy metals and other pollutants is enormously dependent upon their enzymatic system. Enzymes help to neutralize such contaminants to non-toxic counterparts (Karigar and Rao 2011). Table 1 reviews the major enzymes involved in the bioremediation process.

4 Nanobioremediation (NBR)

One of the interesting fields of science that have witnessed extensive research within the past few years is the science of nanotechnology. Nanotechnology is the result of the intersection of several science branches (Baker 2012). Nanotechnology is the science which deals with the manipulation and engineering of functional materials at different scales, i.e., atomic, molecular, and supramolecular scales, where significant upgrade in their properties is evident when compared to their bulk counterparts. The origin of nanotechnology was first postulated by the physicist Professor Richard Feynman in 1959 (El-Gendy and Omran 2019) via his historic lecture at the California Institute of Technology, titled with "there is a plenty of room at the bottom." In this lecture, he illustrated the capability of building up objects from the bottom-up approach (Feynman 1960). Later on, the term nanotechnology was firstly presented by the Japanese professor Norio Taniguchi in Tokyo Science University (Taniguchi et al. 1974). However, this important suggestion has drawn attention in the mid-1980s, when the American scientist Eric Drexler published a

Table 1 Enzymes involved during the bioremediation process

Enzyme	Function	Mechanism of action	Type of contaminant	Examples	References
Oxidoreductases	Responsible for oxidative coupling reactions	– Bond cleavage and consumption of energy produced from biochemical reactions – Electron transfer from the reduced organic compounds to other compounds; thereby harmful pollutants are eventually oxidized to harmless ones	Phenolic compounds, xenobiotics (e.g., anilinic compounds)	Oxygenases, laccases, peroxidases	Saratale et al. (2016)
Hydrolytic enzymes	Disruption of chemical bonds	Catalyze reactions such as condensations and alcoholysis	Organophosphates, oil spills, carbamate insecticides	Lipases, cellulases, proteases	Mary et al. (2018)

book, namely, "engines of creation" in 1986. In this book, he illustrated the potentiality of nanotechnology. Nanomaterials possess a plenty of unique features that are superior when compared with their bulk counterparts. These outstanding characteristics are originated from their high surface area to volume ratio. Nanotechnology offers a number of potential environmental applications involving treatment, remediation, sensing, detection, pollution reduction and prevention.

Products of nanotechnology are categorized into four generations as were identified by the US National Nanotechnology (Rao et al. 2007). The first generation arose since 2001, and this generation involved passive nanostructures like nanoparticle dispersions and nanostructured coatings. During this period the main focus was spotted upon the fabrication and control of processes dealing with the manufacturing of nanoscale particles. Second generation took place in 2005 and involved the development in active nanostructures and were successfully used in transistors, amplifiers, and drugs. It also witnessed the manufacture of novel devices and nanobiosensors. The third generation in 2010 included nanosystems with three dimensions and involved major advances in the assembling techniques for the fabrication of heterogeneous nanostructures as well as supramolecular engineering systems. Lately, the fourth generation has arisen since 2015 which involved heterogeneous molecular nanosystems containing molecules with specific structure and different roles to play.

Lately, human life has witnessed a major surge in nanotechnology and nanobiotechnology disciplines. Nanomaterials have been employed in different applications including water treatment, biosensing, catalysis, agriculture, and pollutant degradation. These advancements encouraged scientists working on bioremediation to integrate nanotechnology with bioremediation to upgrade the efficiency of bioremediation (Mukherjee et al. 2017). Hence, a new term arose due to the combination of both nanotechnology and bioremediation and was referred to as "nanobioremediation". Nanobioremediation endorses the involvement of nanoparticles for remediation of toxic pollutants. Generally, it has been found that microbial bioremediation was not effective for high metal concentrations, as these high concentrations were found to be lethal to the remediating organisms (Rizwan et al. 2014). Nanobioremediation refers to the removal of environmental pollutants/contaminants such as heavy metals and organic and inorganic compounds from contaminated sites using nanomaterials. This discipline combines the basics of bioremediation, nanotechnology and nanobiotechnology. Nanobiotechnology represents the fusion point between the disciplines of biotechnology and nanotechnology (El-Gendy and Omran 2019). Nanoparticles (NPs) can be employed for the remediation of water and soil polluted with heavy metals and organic and inorganic contaminants (Yadav et al. 2017). Ghormade et al. (2011) reported that zerovalent iron NPs had the potential to degrade organic contaminants such as atrazine, molinate, and chlorpyrifos.

4.1 Remediation of Contaminated Sites Using Nanomaterials

Nanomaterials (NMs) refer to materials that have a dimension of 100 nm or less (Omran et al. 2018). A nanometer (nm) is a billion part of meter, i.e., 10^{-9} m. Presence of a large proportion of atoms around the NPs surface helps in facilitating a high number of adsorption reactions with the contaminants. This unique feature candidates NPs as they are being more reactive than materials in the macro-scale (El-Gendy and Omran 2019). NMs can be categorized into naturally occurring, accidentally, or NMs manufactured in laboratories. Among the naturally occurring NMs, clay, organic matter, and iron oxide and all of them are considered as parts of the soil composition. The accidental NMs are those particles released/generated via air emissions, wastes (e.g., solid or liquid wastes) produced from facilities manufacturing nanoscale materials, weathering, fuel burning, and agricultural processes (Thomé and Reddy 2015). The manufactured NMs are generally produced via either a top to bottom (top-down) approach in which bulk materials are converted into smaller particles. The other approach is the bottom to top (bottom-up), where smaller building blocks like atoms and molecules fit together and become self-assembled to produce NPs (El-Gendy and Omran 2019). NMs differ according to the physicochemical composition, and they may be either organic or inorganic. Organic NMs are made up of carbon atoms and exist in the form of ellipsoids, spheres, or hollow tubes. Hollow spheres and ellipsoids are referred to as "fullerenes," while the hollow tubes are known as "nanotubes," and they are either single or multiple walls. On the other hand, inorganic NMs are classified into quantum dots, metal, and metal oxide. Quantum dots (QDs) are composed of two chemical elements with a size of 10 nm (e.g., CdS, CdSe, and CdTe). Metal oxides in the nanoscale appear in the form of oxides of metals (zinc oxide, iron oxide, etc.). Metal NPs include metals in pure form such as nanogold, nanosilver, and nanoiron (O'Carroll et al. 2013).

Nanoscale zerovalent metals such as iron, palladium, and nickel have been reported by several authors to have the ability to decontaminate sites from different types of toxic pollutants, for instance, chromium and arsenic as well as dehalogenation of persistent lethal organic compounds (Murugesan et al. 2011; Tosco et al. 2014; Thomé and Reddy 2015). Thomé and Reddy (2015) assumed that different features of NPs such as morphology, particle size distribution, specific surface area, surface charge, and crystallographic characterization help to determine the behavior of the used NPs (Fig. 2).

4.1.1 Zerovalent Iron Nanoparticles (ZVI NPs)

In the 1990s, water decontamination using nanomaterials started to begin and continued to arise during the following years (Thomé and Reddy 2015). The potential ouse of ZVI NPs in permeable reactive barriers (PRB) for water decontamination was first introduced by Gillham and co-colleagues in 1994 (Gillham et al. 1994). ZVI NPs have been applied for the remediation of lethal organic pollutants such as tetrachloroethylene (TCE) and polyvinyl biphenyls (PCBs). ZVI NPs have been found to be a rapid and an efficient alternative to degrade these contaminants

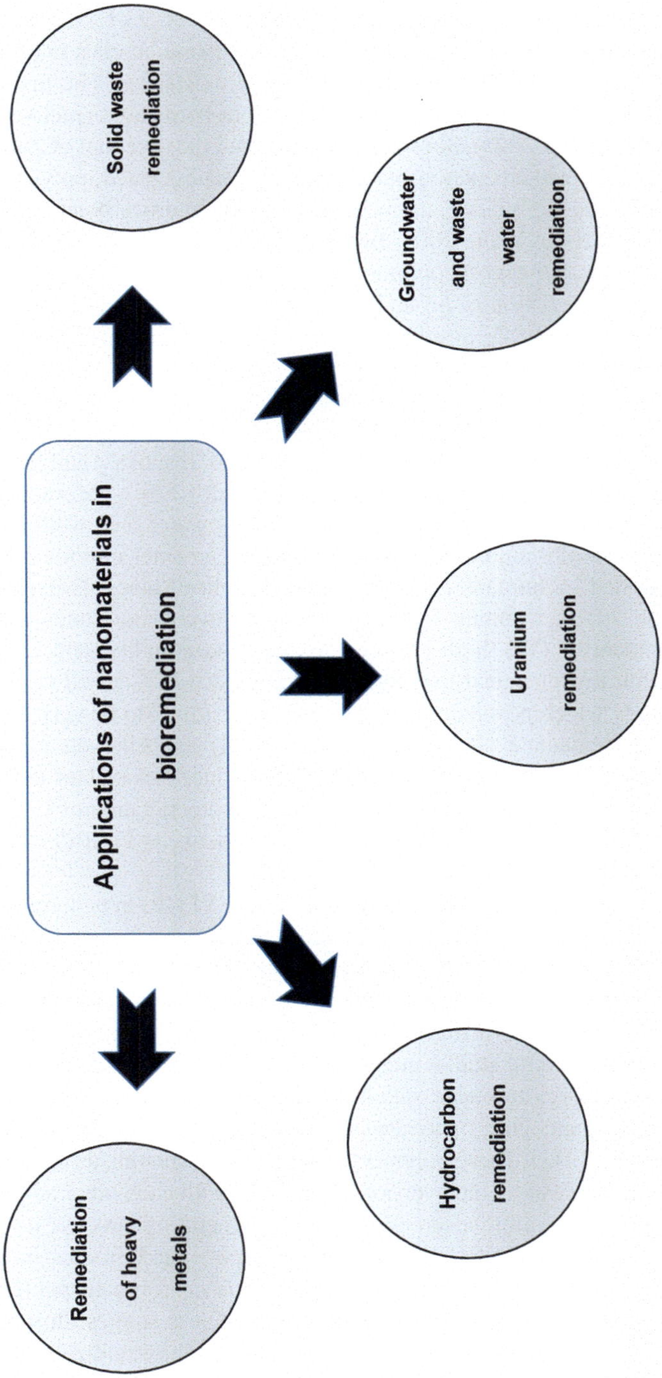

Fig. 2 Different applications of nanomaterials in bioremediation

(Fernanda et al. 2017). Moreover, ZVI NPs are characterized by their safety as well as low cost of production. Recently, the applications of ZVI NPs have been increased remarkably, becoming one of the most used nanomaterials in-situ for the remediation of contaminated regions with hazardous substances. The first trial for using ZVI NPs was by using it as a filling material in permeable reactive barriers (PRB) (Zanetti and Fiore 2005). Several reports showed the potential of ZVI NPs as reducing agents to remediate sites contaminated with halogenated, polychlorinated biphenyls, pesticides, herbicides, aromatic polycyclic hydrocarbons, and metals (Ponder and Darab 2000; Waite 2004; Berge and Ramsburg 2010). ZVI NPs have been proposed as one of the most commonly used nanomaterials in Europe and in the United States for remediation of groundwater and soil and recently in waste water remediation (Fu et al. 2014), in addition to enhancement of anaerobic digestion processes (Carpenter et al. 2015).

Uranium (U) is one of the most common contaminants existing in soil and groundwater. It is usually found in sites near uranium mining and processing, nuclear power plants, weapon-related production and disposal, and agriculture sites (Schnug and Lottermoser 2013; Sheng et al. 2014). Moreover, high concentrations of uranium have been found in drinking water (i.e., wells) (Stoliker et al. 2013). Release of uranium into the environment can enter into the food chain and can be ingested by humans and, consequently, cause kidney or liver damage (Ilaiyaraja et al. 2013). Uranium is usually found in two major states within the natural environment: U (VI) (high chemical toxicity and mobility) and U(IV) (less soluble form with low chemical toxicity) (Chen et al. 2017). Cantrell et al. (1995) introduced a study which postulated that ZVI NPs were found to be very promising nanomaterials to remediate sites polluted with U(VI) as being an inexpensive remediating material (Liang et al. 2013). ZVI NPs reduce hexavalent Uranium U (VI) through different mechanisms including reductive precipitation of U(IV) oxide (UO_2). As a consequence, electron transfer reactions between Fe (0)/Fe(II) and U (VI) take place at the material's surface (Huang et al. 2012).

However, some challenges face the employment of ZVI NPs in nanoremediation. Among these problems are reactivity, mobility, and in situ transport processes (Tosco et al. 2014). Fernanda et al. (2017) postulated that when NPs are applied in situ, they tend to have a clustering affinity and, hence, become deposited in soil particles, making it hard to diffuse through the soil pores towards the targeted contaminated zones. Several studies suggested the use of inert organic polymers as an effective methodology to create stabilized NPs such as sodium carboxymethyl cellulose (CMC), guar gum, and lactate, among others (Wei et al. 2010; Jiemvarangkul et al. 2011). These polymers cover the NPs, creating an electronically negative film which enhances the aversion of particles with each other and with the target soil particles, therefore decreasing the agglomeration process and hence increasing the mobility of NPs in soil. Several reports and studies have been introduced to illustrate the toxicity of ZVI NPs towards microorganisms (Diao and Yao 2009; Kirschling et al. 2010; Ševců et al. 2017). These studies illustrated that ZVI NPs exerted a high potential toxicity when used with no stabilizers. However, the behavior of ZVI NPs is extremely preferred in remediation (Tosco et al. 2014). It was observed that using ZVI NPs with organic stabilizers aids in enhancing the bioremediation process. The stabilizer acts as biostimulants.

4.1.2 Metal-Based Nano-adsorbents (Metal and Metal Oxide NPs) and Photocatalytic NMs

Several metallic NPs such as mercury, copper, chromium, nickel, arsenic, cadmium, lead, have been extensively investigated for their potential to remove heavy metals (Khan et al. 2016b). Photocatalysis is considered a very promising oxidation process which plays a major role in the deterioration of organic contaminants in an efficient way. Photocatalysis is an innovative advanced oxidation process (AOP) that can be employed to mitigate microbial pathogens as well (Kumar and Packirisamy 2016). Yet, this process suffers from certain obstacles, among which, slow kinetic reactions and the photocatalytic activity is diminished by time (Qu et al. 2013). Zinc oxide (ZnO), titanium dioxide (TiO), tungsten oxide (WO_3), and cadmium sulfide (CdS) are among the most common photocatalytic NMs (Nakata and Fujishima 2012). These photocatalytic NMs are cost-effective, readily available, and almost non-toxic.

4.1.3 Nanobiomaterials for Remediation of Heavy Metals and Other Contaminants

Mahmoud and co-authors performed a study in which heat-inactivated *Aspergillus ustus* and silicon dioxide nanopowder combined together for the biosorption of Cd (II) from aqueous solutions using batch equilibrium technique (Mahmoud et al. 2011). Scanning electron microscope (SEM) and Fourier-transform infrared spectroscopy (FTIR) were performed to examine immobilization of fungal cells on silicon dioxide-nanopowder. Cadmium biosorption was affected by different parameters including pH, reaction time, initial metal concentration, and sorbent dosage. Silicon dioxide nanopowders combined with heat-inactivated *Aspergillus ustus* exhibited a maximum sorption cadmium capacity. In as study performed by Ahmed et al. (2014), gold nanoparticles (AuNPs) were biologically fabricated using aqueous leaf extract of *Salicornia brachiate*. The biogenic AuNPs were evaluated for their catalytic potential to reduce methylene blue and nitrophenol. Ultraviolet-visible (UV/Vis) spectrophotometer was employed to track the bioreduction of 4-nitrophenol (4-NP) into 4-aminophenol (4-AP). Change in color took place from yellowish green to colorless. Additionally, the green synthesized AuNPs had the potential to reduce methylene blue into leucomethylene blue in presence of sodium borohydride ($NaBH_4$). The characteristic methylene blue color diminished. Water extract of *Dimocarpus longan* was used for the biological fabrication of silver nanoparticles (AgNPs) as reported by Khan et al. (2016a). The prepared AgNPs were assessed for its photocatalytic degradation potential of methylene blue and reduction of 4-nitrophenol (4-NP). Results showed that the prepared AgNPs exhibited a strong chemo catalytic activity. Nasrollahzadeh et al. (2014) introduced a study in which copper nanoparticles (CuNPs) were biologically fabricated via the hot water leaf extract of *Euphorbia esula*. The synthesized CuNPs exerted a green catalytic effect regarding the degradation of 4-NP. Lingamdinne and co-workers managed to synthesize magnetic inverse spinel iron oxide nanoparticles (MISFNPs) through a biogenic route (Lingamdinne et al. 2017). The seed extract of *Cnidium monnieri* (L.) *Cuss* (CLC) was investigated for the biosynthesis of MISFNPs. The synthesized MISFNPs were characterized via PXRD, FTIR, SEM-EDX, SEM, Brunauer-Emmett-Teller (BET), atomic force microscope (AFM), and X-ray photoelectron spectroscopy (XPS). The biogenic MISFNPs exhibited the potential to remove Pb (II) and Cr (III) from aqueous solutions. Batch adsorption experiments showed that Pb (II) and Cr (III) followed pseudo second order type of kinetics during adsorption onto the surface of MISFNPs. Further, authors suggested that adsorption of both Pb (II) and Cr(III) on MISFNPs was an endothermic process. Additionally, MISFNPs did not lose its stability during removal of the tested heavy metals.

Psidium guajava leaf extract was used to biologically synthesize copper oxide nanoparticles (CuO NPs) as a simple, eco-friendly, and low-cost synthesis procedure (Singh et al. 2019). The leaf extract acted as a reducing and a capping agent for synthesizing CuO NPs. The resultant data showed that the biologically fabricated CuO NPs exhibited an excellent degradation efficiency towards two commonly used industrial dyes, e.g., Nile blue (NB) (93% removal within 120 min) and Reactive Yellow 160 (RY160) (81% removal within 120 min). Interestingly, authors assumed that *Psidium guajava*-derived CuO NPs were reusable for the photocatalytic dye degradation after five successive cycles. Usman et al. (2019) demonstrated the biological fabrication of CuNPs using the extract of *Ficus carica*. The prepared CuNPs were used in treating water polluted with lethal organic dye, e.g., Alizarin Yellow R (AYR). The size of the prepared CuNPs was approximately 61 nm. Degradation of AYR dye reached a high rate of 89.71% within 36 h. Authors studied the different parameters that would affect the degradation process including pH, catalyst concentration, reaction time, and temperature. It was observed that solution pH greatly affected the degradation efficiency. The more acidic, the better the degradation kinetics. Additionally, increase in temperature accelerated the reaction rate because of the endothermic nature of the degradation reaction. The degradation rate was also influenced by the concentration of the tested CuNPs. A high rate of degradation was attained by increasing the concentration of CuNPs. Rafique et al. (2020) demonstrated an environmentally friendly and a green route for the synthesis of CuO NPs using the leaf extract of *Citrofortunella microcarpa*. UV/Vis spectroscopy, XRD, SEM, and FTIR were employed to study the optical, structural, and morphological features of the prepared CuO NPs. EDX and SEM affirmed the high purity of the spherical shaped CuO-NPs with a size of 54–68 nm. UV/Vis spectroscopy revealed the characteristic surface plasmon resonance (SPR) peak at 270–300 nm. The prepared CuO NPs displayed a high potential to degrade rhodamine B (RhB) dye in polluted water.

Chatterjee et al. (2020) conducted a study in which superparamagnetic 20–40 nm sized iron oxide nanoparticles (IONPs) (Fe_3O_4) of 20–40 were successfully prepared using *Aspergillus niger* BSC-1. The biogenic IONPs were investigated for their potential to remove hexavalent chromium from aqueous solution. The mycosynthesized IONPs were characterized via UV/Vis spectroscopy, attenuated total reflectance (ATR-FTIR) spectroscopy, Raman spectroscopy, XRD, transmission electron microscopy (TEM) and field emission scanning electron microscopy (FESEM), Zetasizer, and vibrating sample magnetometry (VSM). Results revealed that removal of Cr (VI) was attained at 40°C, pH 3 and by using 2.5 g/L of IONPs. Leaf extract of *Prosopis juliflora* was used for the green synthesis of AgNPs as reported by Malini et al. (2020). The biologically fabricated AgNPs were successfully encapsulated with chitosan nanocapsules. The resultant encapsulated product possessed antibacterial, photocatalytic, and adsorption activities. The produced biosorbent (green synthesized AgNPs encapsulated in chitosan nanocapsules) had a photocatalytic degradation efficiency of approximately 83% against rose Bengal dye. In addition, the prepared biosorbent managed to absorb 81% of heavy metals (e.g., copper) from a contaminated solution.

4.1.4 Carbon-Based Nanoparticles

Carbon-based NMs include fullerenes, graphenes, and carbon nanotubes (CNTs). There are two main types of CNTs, single-walled carbon nanotubes (SWCNTs) and multi-walled carbon nanotubes (MWCNTs). Interestingly, SWCNTs possess a high antimicrobial potential, while MWCNTs have both strong antimicrobial properties and high adsorption potential for the removal of heavy metals (Kumar and Packirisamy 2016). CNTs are extensively used in wastewater treatment for heavy metal removal.

4.1.5 Magnetic Nanomaterials (MNMs)

Recently magnetic nanomaterials (MNMs) have attracted major attention because of their distinctive features which make them applicable in different disciplines. MNMs can be manipulated under the effect of an external magnetic field (Gómez-Pastora et al. 2014). Water contaminated with organic and inorganic pollutants, heavy metals, and dyes has become a severe problem due to their tremendously hazardous impacts on humans and the surrounding ecological systems. Magnetic NPs (MNPs) act as adsorbent materials which have been extensively employed in solving environmental problems because of their distinctive physicochemical properties, which make them preferable when compared with conventional adsorbents (Reddy et al. 2012). However, in order to guarantee the successful implementation of MNPs during adsorption of contaminants, optimization of the MNPs during the recovery stages and regeneration stages is required. Several synthetic methods have been employed for the fabrication of superparamagnetic NPs over the past few decades (Xu and Sun 2013). Generally, MNPs are made up of magnetic elements, like cobalt, nickel, iron, and their oxides such as magnetite (Fe_3O_4), maghemite (γ-Fe_2O_3), nickel ferrite ($NiFe2O_4$), cobalt ferrite ($CoFe_2O_4$), etc. (Behrens and Appel 2016). Commonly, MNPs are coated with organic layers such as surfactants and polymers (e.g., dextran and polyethylene glycol) or inorganic components like metallic elements (e.g., gold, silver or platinum), metal oxides (zinc oxide, cobalt oxide, etc.), activated carbon, silica, etc. (Karimi et al. 2013; Xu and Sun 2013).

5 Conclusion

Enormous amounts of hazardous pollutants/contaminants are increasing day by day because of the increase in human activities and needs and as a consequence of industrialization. Henceforth, the surrounding environment suffers from severe pollution problem which in turn affects human health. These hazardous contaminants include heavy metals, inorganic and organic contaminants, dyes, etc. As a result, several research studies postulated different treatment technologies that can be employed in order to get rid of these contaminants. Physical, chemical, and biological methodologies are among the employed treatment technologies. Bioremediation has emerged as a relatively economical process, which proved its efficiency as an effective tool to counter the side effects of pollutant spreading and to render the release of toxic contaminants. Microorganisms like bacteria, yeast, fungi, and algae are powerful candidates to remove toxic contaminants. Different parameters affect the bioremediation process including biotic and abiotic factors Bioremediation can be applied either in-situ (on-site contamination) or ex-situ in which contamination is transferred away from its original site. Nanotechnology plays a major role in developing new products which can substitute available products, resulting in improved performance, prospective environmental protection, and cost savings. Nanobioremediation arose as a potential remediation technology in which nanomaterials have been used to remediate the effect of toxic contaminants. Nanobioremediation provides an eco-friendly alternative for the conventional environmental management technologies. Zerovalent iron nanoparticles, metal and metal oxide nanoparticles, magnetic nanomaterials, nanobiomaterials, and carbon-based nanomaterials have been used in remediation of heavy metals and other contaminants.

References

Abbas SH, Ismail IM, Mostafa TM, Sulaymon AH (2014) Biosorption of heavy metals: a review. J Chem Sci Technol 3:74–102

Abdi O, Kazemi M (2015) A review study of biosorption of heavy metals and comparison between different biosorbents. J Mater Environ Sci 6:1386–1399

Acosta Rodríguez I, Martínez-Juárez VM, Cárdenas-González JF et al (2013) Biosorption of arsenic (III) from aqueous solutions by modified fungal biomass of *Paecilomyces* sp. Bioinorg Chem Appl 2013:376780, 5 p. https://doi.org/10.1155/2013/376780

Ahemad M (2014) Remediation of metalliferous soils through the heavy metal resistant plant growth promoting bacteria: paradigms and prospects. Arab J Chem 12:1365–1377

Ahluwalia SS, Goyal D (2007) Microbial and plant derived biomass for removal of heavy metals from wastewater. Bioresour Technol 98:2243–2257

Ahmed KBA, Subramanian S, Sivasubramanian A et al (2014) Preparation of gold nanoparticles using *Salicornia brachiata* plant extract and evaluation of catalytic and antibacterial activity. Spectrochim Acta Part A Mol Biomol Spectrosc 130:54–58

Ahuti S (2015) Industrial growth and environmental degradation. Int Educ Res J 1:5–7

Ali H, Khan E, Anwar M (2013) Chemosphere phytoremediation of heavy metals — concepts and applications. Chemosphere 91:869–881

Anastasi A, Tigini V, Varese GC (2013) The bioremediation potential of different ecophysiological groups of fungi. In: Fungi as bioremediators. Springer, Berlin, pp 29–49

Arivalagan P, Singaraj D, Haridass V, Kaliannan T (2014) Removal of cadmium from aqueous solution by batch studies using *Bacillus cereus*. Ecol Eng 71:728–735

Asia E, Jain CK, Singh RD (2012) Technological options for the removal of arsenic with special reference to South. J Environ Manage 107:1–18

Babák L, Šupinova P, Zichova M et al (2013) Biosorption of Cu, Zn and Pb by thermophilic bacteria–effect of biomass concentration on biosorption capacity. Acta Univ Agric Silvic Mendelianae Brun 60:9–18

Bachate SP, Nandre VS, Ghatpande NS, Kodam KM (2013) Simultaneous reduction of Cr (VI) and oxidation of As (III) by *Bacillus firmus* TE7 isolated from tannery effluent. Chemosphere 90:2273–2278

Baker S (2012) Endophytes : toward a vision in synthesis of nanoparticle for future therapeutic agents. Int J Bioinorg Hybrid Nanomater 1:67–77

Ballester A, Bla ML, Mun JA et al (2007) Comparative study of biosorption of heavy metals using different types of algae. Bioresour Technol 98:3344–3353

Banerjee S, Datta S, Chattyopadhyay D et al (2011) Arsenic accumulating and transforming bacteria isolated from contaminated soil for potential use in bioremediation. J Environ Health Sci Eng 46:1736–1747

Barria C, Malecki M, Arraiano CM (2013) Bacterial adaptation to cold. *Microbiology* 159 (12):2437–2443

Behrens S, Appel I (2016) Magnetic nanocomposites. Curr Opin Biotechnol 39:89–96

Berge ND, Ramsburg CA (2010) Iron-mediated trichloroethene reduction within nonaqueous phase liquid. J Contam Hydrol 118:105–116

Bunluesin S, Kruatrachue M, Pokethitiyook P (2007) Batch and continuous packed column studies of cadmium biosorption by *Hydrilla verticillata* biomass. J Biosci Bioeng 103:509–513

Cantrell KJ, Kaplan DI, Wietsma W (1995) Zero-valent iron for the in situ remediation of selected metals in groundwater. J Hazard Mater 42:201–212

Carolin CF, Kumar PS, Saravanan A et al (2017) Efficient techniques for the removal of toxic heavy metals from aquatic environment: a review. J Environ Chem Eng 5:2782–2799

Carpenter AW, Laughton SN, Wiesner MR (2015) Enhanced biogas production from nanoscale zero valent iron-amended anaerobic bioreactors. Environ Eng Sci 32:647–655

Chatterjee S, Mahanty S, Das P et al (2020) Biofabrication of iron oxide nanoparticles using manglicolous fungus *Aspergillus niger* BSC-1 and removal of Cr (VI) from aqueous solution. Chem Eng J 385:123790

Chauhan NS, Ranjan R, Purohit HJ et al (2009) Identification of genes conferring arsenic resistance to *Escherichia coli* from an effluent treatment plant sludge metagenomic library. FEMS Microbiol Ecol 67:130–139

Chen M, Xu P, Zeng G et al (2015) Bioremediation of soils contaminated with polycyclic aromatic hydrocarbons, petroleum, pesticides, chlorophenols and heavy metals by composting: applications, microbes and future research needs. Biotechnol Adv 33:745–755

Chen A, Shang C, Shao J et al (2017) Science of the total environment the application of iron- based technologies in uranium remediation : a review. Sci Total Environ 575:1291–1306

Congeevaram S, Dhanarani S, Park J et al (2007) Biosorption of chromium and nickel by heavy metal resistant fungal and bacterial isolates. J Hazard Mater 146:270–277

Das N, Vimala R, Karthika P (2008) Biosorption of heavy metals–an overview

Dey U, Chatterjee S, Mondal NK (2016) Isolation and characterization of arsenic-resistant bacteria and possible application in bioremediation. Biotechnol Rep 10:1–7

Dhanarani S, Viswanathan E, Piruthiviraj P et al (2016) Comparative study on the biosorption of aluminum by free and immobilized cells of *Bacillus safensis* KTSMBNL 26 isolated from explosive contaminated soil. J Taiwan Inst Chem Eng 69:61–67

Diao M, Yao M (2009) Use of zero-valent iron nanoparticles in inactivating microbes. Water Res 43:5243–5251

Dixit R, Malaviya D, Pandiyan K et al (2015) Bioremediation of heavy metals from soil and aquatic environment: an overview of principles and criteria of fundamental processes. Sustainability 7:2189–2212

Durairaj P, Malla S, Nadarajan SP et al (2015) Fungal cytochrome P450 monooxygenases of *Fusarium oxysporum* for the synthesis of ω -hydroxy fatty acids in engineered *Saccharomyces cerevisiae*. Microb Cell Fact 14:45–61

El-Gendy NS, Omran BA (2019) Green synthesis of nanoparticles for water treatment. In: Nano Bio-based technology wastewater treatment: prediction and control tools for the dispersion of pollutants in the environment. Scrivener Publishing, Beverly, pp 205–263

Emamverdian A, Ding Y, Mokhberdoran F, Xie Y (2015) Heavy metal stress and some mechanisms of plant defense response. Sci World J 2015:1–19

Fernanda E, Cecchin I, Reddy KR, Schnaid F (2017) Nanobioremediation: integration of nanoparticles and bioremediation for sustainable remediation of chlorinated organic contaminants in soils. Int Biodeter Biodegr 119:419–428

Feynman RP (1960) There's plenty of room at the bottom. Calif Inst Technol Eng Sci Mag 23 (5):22–36

Finocchio E, Lodi A, Solisio C, Converti A (2010) Chromium (VI) removal by methylated biomass of Spirulina platensis: the effect of methylation process. Chem Eng J 156:264–269

Flores-Chaparro CE, Ruiz LFC, de la Torre MCA et al (2017) Biosorption removal of benzene and toluene by three dried macroalgae at different ionic strength and temperatures: algae biochemical composition and kinetics. J Environ Manage 193:126–135

Francová A, Chrastný V, Šillerová H et al (2017) Evaluating the suitability of different environmental samples for tracing atmospheric pollution in industrial areas. Environ Pollut 220:286–297

Fu F, Dionysiou DD, Liu H (2014) The use of zero-valent iron for groundwater remediation and wastewater treatment: a review. J Hazard Mater 267:194–205

Gadd GM (2010) Metals, minerals and microbes: geomicrobiology and bioremediation. Microbiology 156:609–643

Gao R, Wang Y, Zhang Y et al (2017) Cobalt (II) bioaccumulation and distribution in *Rhodopseudomonas palustris*. Biotechnol Biotechnol Equip 31:527–534

Ghormade V, Deshpande MV, Paknikar KM (2011) Perspectives for nano-biotechnology enabled protection and nutrition of plants. Biotechnol Adv 29:792–803

Gillham RW, Robert W, Stephanie F (1994) Enhanced degradation of halogenated aliphatic by zero-valent iron. Groundwater 32:958–967

Gokhale SV, Jyoti KK, Lele SS (2008) Kinetic and equilibrium modeling of chromium (VI) biosorption on fresh and spent *Spirulina platensis/Chlorella vulgaris* biomass. Bioresour Technol 99:3600–3608

Gómez-Pastora J, Bringas E, Ortiz I (2014) Recent progress and future challenges on the use of high performance magnetic nano-adsorbents in environmental applications. Chem Eng J 256:187–204

Govarthanan M, Mythili R, Selvankumar T et al (2016) Bioremediation of heavy metals using an endophytic bacterium *Paenibacillus* sp. RM isolated from the roots of *Tridax procumbens*. 3 Biotech 6:242–249

Gupta A, Mishra A (2011) Sustainable water remediation. Encycl Inorg Bioinorg Chem 21:1-16

Gupta VK, Rastogi A (2008) Equilibrium and kinetic modelling of cadmium (II) biosorption by nonliving algal biomass *Oedogonium s*p. from aqueous phase. J Hazard Mater 153:759–766

Hashim MA, Mukhopadhyay S, Narayan J, Sengupta B (2011) Remediation technologies for heavy metal contaminated groundwater. J Environ Manage 92:2355–2388

Huang W, Liu Z (2013) Biosorption of Cd (II)/Pb (II) from aqueous solution by biosurfactant-producing bacteria: isotherm kinetic characteristic and mechanism studies. Colloids Surf B Biointerfaces 105:113–119

Huang K, Park Y, Ewh A et al (2012) Interdiffusion and reaction between uranium and iron. J Nucl Mater 424:82–88

Hussein H, Farag S, Moawad H (2003) Isolation and characterization of *Pseudomonas* resistant to heavy metals contaminants. Arab J Biotechnol 7:13–22
Ilaiyaraja P, Kumar A, Deb S et al (2013) Adsorption of uranium from aqueous solution by PAMAM dendron functionalized styrene divinylbenzene. J Hazard Mater 250–251:155–166
Javaid A, Bajwa R (2007) Biosorption of Cr (III) ions from tannery wastewater by *Pleurotus ostreatus*. Mycopathologia 5:71–79
Jencarova J, Luptakova A (2012) The elimination of heavy metal ions from waters by biogenic iron sulphides. Chem Eng 28:205–2010
Jiemvarangkul P, Zhang W, Lien H (2011) Enhanced transport of polyelectrolyte stabilized nanoscale zero-valent iron (nZVI) in porous media. Chem Eng J 170:482–491
Joutey NT, Bahafid W, Sayel H, Ghachtouli N (2013) Biodegradation: involved microorganisms and genetically engineered microorganisms. In: Biodegradation-life of science. InTech, Rijeka, pp 289–320
Kang C-H, So J-S (2016) Heavy metal and antibiotic resistance of ureolytic bacteria and their immobilization of heavy metals. Ecol Eng 97:304–312
Karakagh RM, Chorom M, Motamedi H (2012) Biosorption of Cd and Ni by inactivated bacteria isolated from agricultural soil treated with sewage sludge. Ecohydrol Hydrobiol 12:191–198
Karigar CS, Rao SS (2011) Role of microbial enzymes in the bioremediation of pollutants: a review. Enzyme Res 2011:805187
Karimi Z, Karimi L, Shokrollahi H (2013) Nano-magnetic particles used in biomedicine: core and coating materials. Mater Sci Eng C 33:2465–2475
Karthik C, Barathi S, Pugazhendhi A et al (2017a) Evaluation of Cr (VI) reduction mechanism and removal by *Cellulosimicrobium funkei* strain AR8, a novel haloalkaliphilic bacterium. J Hazard Mater 333:42–53
Karthik C, Ramkumar VS, Pugazhendhi A et al (2017b) Biosorption and biotransformation of Cr (VI) by novel *Cellulosimicrobium funkei* strain AR6. J Taiwan Inst Chem Eng 70:282–290
Kathiraven T, Sundaramanickam A, Shanmugam N, Balasubramanian T (2015) Green synthesis of silver nanoparticles using marine algae *Caulerpa racemosa* and their antibacterial activity against some human pathogens. Appl Nanosci 5:499–504
Khan FU, Chen Y, Khan NU et al (2016a) Antioxidant and catalytic applications of silver nanoparticles using *Dimocarpus longan* seed extract as a reducing and stabilizing agent. J Photochem Photobiol B Biol 164:344–351
Khan SB, Marwani HM, Asiri AM, Bakhsh EM (2016b) Exploration of calcium doped zinc oxide nanoparticles as selective adsorbent for extraction of lead ion. Desalination Water Treat 57:19311–19320
Kirschling TL, Gregory KB, Minkley EG et al (2010) Impact of nanoscale zero valent iron on geochemistry and microbial populations in trichloroethylene contaminated aquifer materials. Environ Sci Technol 44:3474–3480
Kostal J, Yang R, Nu CH et al (2004) Enhanced arsenic accumulation in engineered bacterial cells expressing ArsR. Appl Environ Microbiol 70:4582–4587
Kumar JIN, Oommen C (2012) Removal of heavy metals by biosorption using freshwater alga *Spirogyra hyalina*. J Environ Biol 33:27–31
Kumar SR, Packirisamy G (2016) Applications of nanotechnology for bioremediation. In: Handbook of advanced industrial and hazardous wastes management. CRC Press, pp 27–48
Kumar P, Kim K-H, Bansal V et al (2017) Progress in the sensing techniques for heavy metal ions using nanomaterials. J Ind Eng Chem 54:30–43
Kuppusamy S, Thavamani P, Venkateswarlu K et al (2017) Remediation approaches for polycyclic aromatic hydrocarbons (PAHs) contaminated soils: technological constraints, emerging trends and future directions. Chemosphere 168:944–968
Liang L, Yang W, Guan X et al (2013) Kinetics and mechanisms of pH-dependent selenite removal by zero valent iron. Water Res 47:5846–5855
Lingamdinne LP, Chang YY, Yang JK et al (2017) Biogenic reductive preparation of magnetic inverse spinel iron oxide nanoparticles for the adsorption removal of heavy metals. Chem Eng J 307:74–84

Mahmoud ME, Yakout AA, Abdel-Aal H, Osman MM (2011) Enhanced biosorptive removal of cadmium from aqueous solutions by silicon dioxide nano-powder, heat inactivated and immobilized *Aspergillus ustus*. Desalination 279:291–297

Majumder A, Ghosh S, Saha N et al (2013) Arsenic accumulating bacteria isolated from soil for possible application in bioremediation. J Environ Biol 34:841–846

Malini S, Vignesh Kumar S, Hariharan R et al (2020) Antibacterial, photocatalytic and biosorption activity of chitosan nanocapsules embedded with *Prosopis juliflora* leaf extract synthesized silver nanoparticles. Mater Today Proc 21:828–832

Mamisahebei S, Khaniki GRJ, Torabian A et al (2007) Removal of arsenic from an aqueous solution by pretreated waste tea fungal biomass. J Environ Health Sci Eng 4:85–92

Manafi Z, Abdollahi H, Majd MZ et al (2012) Isolation and identification of microorganisms causing microbial degradation of organic phase of the solvent extraction unit in the copper industries. Int J Miner Process 112:43–48

Mary J, Karthik C, Ganesh R et al (2018) Biological approaches to tackle heavy metal pollution: a survey of literature. J Environ Manage 217:56–70

Martinez-Juarez VM, Cárdenas-González JF, Torre-Bouscoulet ME et al (2012) Biosorption of mercury (II) from aqueous solutions onto fungal biomass. Bioinorg Chem Appl 2012:156190, 5 p. https://doi.org/10.1155/2012/156190

Mgbemene CA, Nnaji CC, Nwozor C (2016) Industrialization and its backlash: focus on climate change and its consequences. J Environ Sci Technol 9:301–316

Misal SA, Lingojwar DP, Shinde RM, Gawai KR (2011) Purification and characterization of azoreductase from alkaliphilic strain *Bacillus badius*. Process Biochem 46:1264–1269

Mondal P, Majumder CB, Mohanty B (2006) Laboratory based approaches for arsenic remediation from contaminated water: recent developments. J Hazard Mater 137:464–479

Morel M, Meux E, Mathieu Y et al (2013) Xenomic networks variability and adaptation traits in wood decaying fungi. J Microbial Biotechnol 6:248–263

Mukherjee T, Chakraborty S, Ali A, Kumar T (2017) Bioremediation potential of arsenic by non-enzymatically biofabricated silver nanoparticles adhered to the mesoporous carbonized fungal cell surface of *Aspergillus foetidus* MTCC8876. J Environ Manage 201:435–446

Murugesan K, Bokare V, Jeon J et al (2011) Bioresource technology effect of Fe – Pd bimetallic nanoparticles on *Sphingomonas* sp. PH-07 and a nano-bio hybrid process for triclosan degradation. Bioresour Technol 102:6019–6025

Nakata K, Fujishima A (2012) TiO2 photocatalysis: design and applications. J Photochem Photobiol C Photochem Rev 13:169–189

Nasrollahzadeh M, Sajadi SM, Khalaj M (2014) Green synthesis of copper nanoparticles using aqueous extract of the leaves of *Euphorbia esula* L and their catalytic activity for ligand-free Ullmann-coupling reaction and reduction of 4-nitrophenol. RSC Adv 4:47313–47318

O'Carroll D, Sleep B, Krol M et al (2013) Nanoscale zero valent iron and bimetallic particles for contaminated site remediation. Adv Water Resour 51:104–122

Omran BA, Nassar HN, Fatthallah NA et al (2018) Characterization and antimicrobial activity of silver nanoparticles mycosynthesized by *Aspergillus brasiliensis*. J Appl Microbiol 125:370–382

Park JK, Choi SB (2001) Metal recovery using immobilized cell suspension from a brewery. Korean J Chem Eng 19:68–74

Ponder SM, Darab JG (2000) Remediation of Cr (VI) and Pb (II) aqueous solutions using supported, nanoscale zero-valent iron. Environ Sci Technol 34:2564–2569

Qu X, Alvarez PJJ, Li Q (2013) Applications of nanotechnology in water and wastewater treatment. Water Res 47:3931–3946

Rafique M, Shafiq F, Ali Gillani SS et al (2020) Eco-friendly green and biosynthesis of copper oxide nanoparticles using *Citrofortunella microcarpa* leaves extract for efficient photocatalytic degradation of Rhodamin B dye form textile wastewater. Optik 208:164053–164061

Raghunandan K, Kumar A, Kumar S et al (2018) Production of gellan gum, an exopolysaccharide, from biodiesel-derived waste glycerol by *Sphingomonas* spp. 3 Biotech 8:71–84

Raja CP, Jacob JM, Balakrishnan RM (2015) Selenium biosorption and recovery by marine *Aspergillus terreus* in an upflow bioreactor. J Environ Eng 142:C4015008

Rajasulochana P, Preethy V (2016) Comparison on efficiency of various techniques in treatment of waste and sewage water–a comprehensive review. Resour Technol 2:175–184

Rao CNR, Thomas PJ, Kulkarni GU (2007) Basics of nanocrystals. In: Nanocrystals synthesis, properties and applications. Springer, Berlin, pp 1–23

Reddy LH, Arias JL, Nicolas J, Couvreur P (2012) Magnetic nanoparticles: design and characterization, toxicity and biocompatibility, pharmaceutical and biomedical applications. Chem Rev 112:5818–5878

Rizwan M, Singh M, Mitra CK, Morve RK (2014) Ecofriendly application of nanomaterials: nanobioremediation. J Nanoparticles 2014:1–7

Roberts TL (2014) Cadmium and phosphorous fertilizers: the issues and the science. Proc Eng 83:52–59

Samarth DP, Chandekar CJ, Bhadekar RK (2012) Biosorption of heavy metals from aqueous solution using *Bacillus Licheniformis*. Int J Pure Appl Sci Technol 10:12–19

Saratale G, Kalme S, Bhosale S, Govindwar S (2007) Biodegradation of kerosene by *Aspergillus ochraceus* NCIM-1146. J Basic Microbiol 47:400–405

Saratale RG, Sivapathan SS, Woo JJ et al (2016) Preparation of activated carbons from peach stone by H4P2O7 activation and its application for the removal of Acid Red 18 and dye containing wastewater. J Environ Sci Heal Part A 51:164–177

Schaider LA, Senn DB, Estes ER et al (2014) Sources and fates of heavy metals in a mining-impacted stream: temporal variability and the role of iron oxides. Sci Total Environ 490:456–466

Schnug E, Lottermoser BG (2013) Fertilizer derived uranium and its threat to human health. Environ Sci Technol 47: 2433–2434

Sethi BK, Kanungo S, Rout JR et al (2010) Effect of chromium on *Mucor* species and optimization of growth conditions. Nat Sci 8:29–32

Ševců A, El-Temsah YS, Filip J et al (2017) Zero-valent iron particles for PCB degradation and an evaluation of their effects on bacteria, plants, and soil organisms. Environ Sci Pollut Res 24:21191–21202

Sheng G, Shao X, Li Y et al (2014) Enhanced removal of Uranium (VI) by nanoscale zerovalent iron supported on na – bentonite and an investigation of mechanism. J Phys Chem A 16:2952–2958

Shamim S, Rehman A, Qazi MH (2014) Cadmium-resistance mechanism in the bacteria *Cupriavidus metallidurans* CH34 and *Pseudomonas putida* mt2. Arch Environ Contam Toxicol 67:149–157

Shipra J, Dikshit SN, Pandey G (2012) Comparative study of growing/immobilized biomass verses resting biomass of *A. lentulus* for the effect of pH on Cu metal removal. Res J Pharm Biol Chem Sci 3:421–427

Silva A, Delerue-Matos C, Figueiredo SA, Freitas OM (2019) The use of algae and fungi for removal of pharmaceuticals by bioremediation and biosorption processes: a review. Water (Switzerland) 11:1555–1591

Singh J, Kumar V, Kim K, Rawat M (2019) Biogenic synthesis of copper oxide nanoparticles using plant extract and its prodigious potential for photocatalytic degradation of dyes. Environ Res 177:108569–108581

Srivastava NK, Majumder CB (2008) Novel biofiltration methods for the treatment of heavy metals from industrial wastewater. J Hazard Mater 151:1–8

Srivastava PK, Vaish A, Dwivedi S et al (2011a) Science of the total environment biological removal of arsenic pollution by soil fungi. Sci Total Environ 409:2430–2442

Srivastava S, Shrivastava M, Suprasanna P, Souza SFD (2011b) Phytofiltration of arsenic from simulated contaminated water using *Hydrilla verticillata* in field conditions. Ecol Eng 37:1937–1941

Stoliker DL, Campbell KM, Fox PM et al (2013) Evaluating chemical extraction techniques for the determination of uranium oxidation state in reduced aquifer sediments. Environ Sci Technol 16:9225–9232

Sutherland C, Venkobachar C (2010) A diffusion-chemisorption kinetic model for simulating biosorption using forest macro-fungus, fomes fasciatus. Int Res J Plant Sci 1:107–117

Subhashini SS, Kaliappan S, Velan M (2011) Removal of heavy metal from aqueous solution using *Schizosaccharomyces pombe* in free and alginate immobilized cells. In: 2nd international conference on environmental science and technology, pp 108–111
Taiwo AM, Gbadebo AM, Oyedepo JA et al (2016) Bioremediation of industrially contaminated soil using compost and plant technology. J Hazard Mater 304:166–172
Taniguchi N, Arakawa C, Kobayashi T (1974) On the basic concept of 'nano-technology'. In: Proceedings of the international conference on production engineering, pp 18–23
Tchounwou PB, Yedjou CG, Patlolla AK, Sutton DJ (2012) Heavy metals toxicity and the environment. EXS 101:133–164
Thatoi H, Das S, Mishra J et al (2014) Bacterial chromate reductase, a potential enzyme for bioremediation of hexavalent chromium: a review. J Environ Manage 146:383–399
Thomé A, Reddy KR (2015) Review of nanotechnology for soil and groundwater remediation: Brazilian Perspectives. Water Air Soil Pollut 226:121–141
Thomson B, Hepburn CD, Lamare M, Baltar F (2017) Temperature and UV light affect the activity of marine cell-free enzymes. Biogeosciences 14:3971–3977
Tosco T, Petrangeli M, Cruz C, Sethi R (2014) Nanoscale zerovalent iron particles for groundwater remediation : a review. J Clean Prod 77:10–21
Ullah A, Heng S, Munis MFH et al (2015) Phytoremediation of heavy metals assisted by plant growth promoting (PGP) bacteria: a review. Environ Exp Bot 117:28–40
Usman M, Ahmed A, Yu B et al (2019) Photocatalytic potential of bio-engineered copper nanoparticles synthesized from *Ficus carica* extract for the degradation of toxic organic dye from waste water: Growth mechanism and study of parameter affecting the degradation performance. Mater Res Bull 120:110583–110594
Verma S, Kuila A (2019) Bioremediation of heavy metals by microbial process. Environ Technol Innov 14:100369–100380
Vijayaraghavan K, Yun Y (2008) Bacterial biosorbents and biosorption. Biotechnol Adv 26:266–291
Waite TD (2004) Oxidative degradation of the carbothioate herbicide, molinate, using nanoscale zero-valent iron. Environ Sci Technol 38:2242–2247
Wang J, Chen C (2009) Biosorbents for heavy metals removal and their future. Biotechnol Adv 27:195–226
Weerasundara L, Amarasekara RWK, Magana-Arachchi DN et al (2017) Microorganisms and heavy metals associated with atmospheric deposition in a congested urban environment of a developing country: Sri Lanka. Sci Total Environ 584:803–812
Wei Y, Wu S, Chou C et al (2010) Influence of nanoscale zero-valent iron on geochemical properties of groundwater and vinyl chloride degradation: a field case study. Water Res 44:131–140
Xu C, Sun S (2013) New forms of superparamagnetic nanoparticles for biomedical applications. Adv Drug Deliv Rev 65:732–743
Xu Y, Seshadri B, Bolan N et al (2019) Microbial functional diversity and carbon use feedback in soils as affected by heavy metals. Environ Int 125:478–488
Yadav KK, Singh JK, Gupta N, Kumar V (2017) A review of nanobioremediation technologies for environmental cleanup: a novel biological approach. J Mater Environ Sci 8:740–757
Yan L, Yin H, Zhang S et al (2010) Biosorption of inorganic and organic arsenic from aqueous solution by *Acidithiobacillus ferrooxidans* BY-3. J Hazard Mater 178:209–217
Zanetti MC, Fiore S (2005) Evaluation of mutual connections between zero-valent iron reactivity and groundwater composition in trichloroethylene degradation. Ann Chim 95:779–789
Zeraatkar AK, Ahmadzadeh H, Talebi AF et al (2016) Potential use of algae for heavy metal bioremediation, a critical review. J Environ Manage 181:817–831

Wastewater Treatment Techniques: An Introduction

Mukesh Meena, Priyankaraj Sonigra, Garima Yadav,
and Tansukh Barupal

Abstract

Currently, the world is facing water quality crisis resulting from land use changes, industrialization, agricultural practices, sewerage, and poor wastewater management practices. Wastewater comprises a combination of one or more of domestic effluent, industrial effluent, agricultural runoff, and water from commercial establishment and institution including hospitals. Untreated wastewater released into the environment can cause health and environmental problem and can have an economic impact. Wastewater treatment process is used worldwide for removal of existing contamination before reuse or disposal into the environment. In this article, various wastewater treatment techniques, basic fundamentals, and their consequent pollutants removal are discussed briefly.

Keywords

Wastewater · Sewerage · Industrial effluent · Contamination

1 Introduction

Only 3% of the total water on earth is clean water, of which only 1% is accessible. The major portion of the clean water present in the form of glaciers and snow; therefore, there is always a higher demand for fresh and clean water (Gonçalves et al. 2017). Availability of water resources is becoming more and more scarce all around the world because of the imbalance between freshwater availability and its consumption; thus nowadays one of the major challenge for modern society is how to access freshwater (Aslan and Kapdan 2006). Demand for the fresh waters keeps increasing

M. Meena (✉) · P. Sonigra · G. Yadav · T. Barupal
Department of Botany, Mohanlal Sukhadia University, Udaipur, India
e-mail: drmukeshmeena321@mlsu.ac.in

M. P. Shah (ed.), *Removal of Emerging Contaminants Through Microbial Processes*,
https://doi.org/10.1007/978-981-15-5901-3_8

day by day, and there are several reasons behind this. The major reasons are climatic changes, industrial development, increasing water use per capita, and increased migrating population to drought regions. On the other hand, the water quality is also very important to us which is also threatened due to the release of wastewaters in water bodies in large amount without going through any treatment processes which are mainly produced by industrialization, agricultural and domestic practices, and urbanization (Rawat et al. 2011; Gonçalves et al. 2017). Different industrial activities such as mining, textile, oil industries, oil refining, storage, transportation and different chemicals are used for the pesticides, fertilizers, and other products manufacturing also produced unwanted by-products (waste material) (Yu et al. 2017; Crini and Lichtfouse 2019). Due to these types of wastewater discharge activities, a large number of anthropogenic chemicals (Schwarzenbach et al. 2006), and pathogens entered into the urban and rural water bodies (Rizz et al. 2013). These reasons have been recognized as one of the major causes of water pollution all around the world (Reemtsma et al. 2006; Seow et al. 2016).

The continuous disposal of wastewaters in water courses without proper treatment can cause serious pollution problems and diseases. Eutrophication is a deleterious problem that comes from the disposal of untreated wastewater containing a higher amount of organic matters, specially a higher amount of nitrogen (N) and phosphorus (P) in water bodies. The organic matters lead to the overgrowth of the aquatic plants (algal bloom), which cause increased biological oxygen demand (BOD), low oxygen availability and result in loss of keystone species and complete degradation of aquatic ecosystems. Wastewater also contains a large number of xenobiotic compounds, toxic metals, and disease-causing pathogenic microorganisms which can affect human health (Renuka et al. 2013; Ruiz et al. 2013). Another one is the textile industry wastewater which contains highly colored synthetic dyes, which severely affects the aquatic plants. Dark-colored dye has an impact on aquatic life such that low light penetration into the water, decreases photosynthesis rate and decreases oxygen consumption which results in the death of aquatic animals. It may also be toxic to certain types of aquatic life due to the presence of metals and chlorine. So, this textile wastewater must be treated before its discharge (Holkar et al. 2016). Furthermore, a large number of harmful gases are also produced during the decomposition of the organic matters present in wastewater (Topare et al. 2011; Seow et al. 2016); thus these problems clearly show the need of an effective method for wastewater treatment before it enters into the water resources, to save water quality, aquatic ecosystems and human health from harmful effects.

During the past decades, different types of physical, chemical and biological treatment methods have been reported for the wastewater treatment. Some commonly used methods are as follows: coagulation, flotation, flocculation, chemical precipitation, ion exchange, oxidation, absorption, adsorption, reverse osmosis, screening, membrane filtration, sedimentation, ultra-filtration, electrolysis, phytoremediation, aerobic and anaerobic biological degradation, etc. All the methods have its own performance characteristics and impacts on the environment (Berefield et al. 1982; Liu and Liptak 2000; Harvey et al. 2002; Chen 2004; Forgacs et al. 2004; Crini and Badot 2007; Cox et al. 2007; Hai et al. 2007; Bhattacharjee et al. 2007; El-Ashtoukhy et al. 2009; Kishimoto et al. 2010; Barakat 2011; Rathoure and Dhatwalia 2016; Morin-Crini and Crini 2017; Seow et al. 2016; Crini and Lichtfouse 2019).

The treatment methods which are used for wastewater treatment depend upon the final use of the water such as drinking, recreation, irrigation, etc. Each technology is different from the other not only in terms of cost but also in terms of efficiency, feasibility, reliability, practicability, impact on the environment, sludge production, requirements of pretreatment, formation of harmful by-products, and operation difficulty. Normally, elimination of pollutants is processed by biological, chemical, and physical means because not even a single method is able of appropriate wastewater treatment; the only reason behind this is the complex nature of industrial waste material. Table 1 shows different wastewater treatment techniques. These are basically in three types: physical waste treatment, chemical waste treatment, and biological waste treatment.

The legislation has become more aware in context to liquid industrial effluent, especially in the developed countries (Crini and Lichtfouse 2019). Among all the methods, biological wastewater treatment methods rely on the use of microorganisms. Due to its unique abilities to degrade organic matters, in a simple form, they remove nutrients (N, P) and neutralize or transform toxic compounds into harmless products (Barupal et al. 2019). These qualities make them the essential players in wastewater treatment. Biological methods can be further classified into two methods on the basis of oxygen presence and absence. One is the aerobic method; this method uses microbes for the treatment of the wastewater in the presence of oxygen, while an anaerobic method uses microbes in the absence of oxygen for treatment. Generally, the combination of aerobic and anaerobic treatment is used in real practices, in which anaerobic process is used to treat wastewater of chemical oxygen demand (COD), followed by the aerobic treatment to treat the resulting wastewater of low COD (Wang et al. 2011; Holkar et al. 2016).

In the beginning, the main requirement was the only removal of organic matters and suspended soils from the wastewater, but in present days there are additional

Table 1 Wastewater treatment techniques

Physical water treatment	Chemical water treatment	Biological water treatment
Screening	pH regulation	Aerobic treatment
Gravity separation	Precipitation	Anaerobic treatment
Dissolved air flotation	Oxidation	
Filtration	Reduction	
Adsorption	Resin-based process	
Desorption	Disinfection	
Membrane-based processes		
Coagulation and flocculation		
De-emulsification		
Sterilization		

objectives as follows: removal of toxic metals, odors, nutrients (N, P), and pathogens (Bitton 2005; Ferrera and Sanchez 2016). The biological methods have many benefits such as eco-friendly, cost-effective, giving non-hazardous compound, less sludge production, and low consumption of water (Holkar et al. 2016).

This article includes some important biological wastewater treatment techniques and some chemical and physical techniques also. Table 2 showed different treatment techniques, their basic purposes, and consequent pollutant(s) removal.

2 Chemical Methods

2.1 Chemical Oxidation

This is one of the newer treatment technology in which an oxidizing agent is introduced in wastewater during treatment and electrons move from the electron donor (oxidant) to an electron acceptor (pollutants). The most commonly used chemical oxidants are O_3, H_2O_2, ClO_2, K_2MnO_4, K_2FeO_4, and so on. This leads to the structural modification in pollutants and transforms them into the less harmful or less toxic components (Zheng et al. 2013).

2.2 Chemical Precipitation

Chemical precipitation method is most commonly used for removal of toxic metals that dissolved in wastewater. In this method, a precipitation reagent is mixed with wastewater solution to convert the dissolved toxic metals into solid particles. A chemical reaction initiated by the precipitation reagent causes the dissolved metals to form solid particles. To remove the particles from the mixture, a filtration step can be included. The efficiency of the process is depend upon the type of metal and concentration present in wastewater, and also the type of precipitation reagent is used for treatment. The most commonly used reagent for this treatment is calcium or sodium hydroxide, to create solid metal hydroxides (Wang et al. 2005).

2.3 Ion Exchange

An ion exchange process is used to soften the hard water. When water is too hard, it often leaves gray spots or residue. This process is similar to the reverse osmosis. The most common ions which are responsible for water hardness are calcium and magnesium (Clifford 1991). This treatment introduces dissolve sodium chloride salts which give rise to positively charged sodium ions. Hard water ions such as calcium and magnesium exchanged by sodium ions, and free sodium ions are simply released in the water (Samer 2015).

Table 2 Treatment techniques, basic purpose, and consequent pollutant(s) removal

Technique	Basic purpose	Pollutant(s) removed
1. Chemical techniques		
pH regulation	Neutralization, pH adjustment	Adjustment of H^{+}/OH^{-} concentration, precipitation of heavy metal ions
Precipitation	Dissolved inorganic substance removal	Some inorganic cations and anions
Oxidation and reduction	Specific dissolved substances removal	Specific toxic substances and some recalcitrant organic compounds
Resin-based processes	Removal of some dissolved inorganic and organic substances	Dissolved ionized substances and non-biodegradable organic substances
Disinfection	Destruction of microorganisms	Destruction of pathogenic microorganisms
2. Physical techniques		
Screening	Large particulate removal	Light and heavy particles of inorganic and organic substances
Gravity separation	Heavy particles, free oil, and grease removal	Finer suspended particles, oil, grease, and biological oxygen demand (BOD)
Dissolved air flotation	Both heavy and light particles, free oil, and grease removal	Finer suspended particles, free oil, grease, and BOD
Filtration	Both heavy and light particles	Suspended solid particles (organic as well as inorganic) and BOD
Adsorption	Volatile organic compounds (VOCs) dissolved/suspended (non-particulate) non-biodegradable organic substances, color, and odor removal	Recalcitrant organic substances, recoverable organic compounds, color, and odor
Desorption	Removal of ammonia, VOCs	Free ammonia, VOCs
Membrane-based processes	Dissolved inorganic and organic substances removal	Dissolved inorganic and organic substances and BOD
Coagulation and flocculation	Colloidal particles removal	Turbidity (due to suspended fine solid particles) and BOD
De-emulsification	Dispersed phase separation	Turbidity (due to suspended fine liquid droplets) and BOD
Sterilization	Complete destruction of microorganisms	Microorganisms
3. Biological techniques		
Aerobic treatment	Biodegradable dissolved and suspended organic substances	Biodegradable organic, BOD, some nitrogenous, and phosphorous compounds
Anaerobic treatment	Biodegradable dissolved and suspended organic substances	Biodegradable organic, BOD, some nitrogenous and phosphorous compounds

3 Physical Methods

3.1 Adsorption

Adsorption is a physical method where adsorbate that should be soluble particles of wastewater is separated by attachment of the adsorbate to the surface of an adsorbent (solid substrate with extremely high surface area). Examples of adsorbents involve clay colloids, activated alumina, hydroxides, resins, and activated carbon. The adsorbent should be activated before use. A broad range of organic materials can be removed by adsorption, including detergents and toxic compounds (Samer 2015). One of the most important characteristics of an adsorbent is the adsorption capacity that can be calculated as follows:

$$\text{Adsorption Capacity} = \text{Adsorbate}/\text{Adsorbent}$$

3.2 Emulsification (ELM)

Emulsion liquid membrane techniques have been getting acceptance over conventional separation methods (Saki and Uzal 2018), since their invention by Li (1968), who used this method for hydrometallurgical recovery of heavy metals. Emulsion liquid membrane process is a straight way to remove chemical contaminants from wastewater. Unlike, permeable and semi-permeable conventional membranes such as reverse osmosis, ultrafiltration, and microfiltration methods, emulsion liquid membrane techniques do not have constrains such as low efficiency, high operating and maintenance costs related to large sludge quantities, sensitive and difficult operating conditions, and elevated power consumption. Emulsion liquid membrane technique has huge advantages, comprising high removal and recovery efficiency, low operating and capital costs, less power consumption, a simple operating process, concentration factors, simultaneous extraction and re-extraction (stripping), and high fluxes. Apart from these advantages, ELM has higher extraction efficacy in the removal and recovery of heavy metal ion from wastewater over conventional methods (Goyal et al. 2011). Due to coalescence or swelling during the extraction process, emulsion liquid membrane techniques also have limitations, such as emulsion stability and breakage/leakage of the membrane phase, and these can decrease overall ELM efficiency.

3.3 De-emulsification

Although, the emulsion liquid membrane techniques have a number of advantages over other physicochemical wastewater treatment methods. There are some issues that need to be solved so that this technique would be applied in a practical way. The most serious problem related to emulsion liquid membrane technique is the stability of the emulsion and partition of the two immiscible liquids (Alsabagh et al. 2016). A membrane rupture can lead to a continuous loss in the internal phase due to the breakdown of the emulsion (Djenouhat et al. 2008; Hussein et al. 2019).

3.4 Dissolved Air Flotation

Flotation is the result of pouring of fine bubbles into the water. These tiny air bubbles became suspended in the water with the adhesion of oil particles. A scum layer is made during flotation which became separated from the water because the floating density of oil is less than that of water (Moosai and Dawe 2003). Recently, the most frequently used flotation methods are dissolved air flotation and jet impeller flotation (Yu et al. 2017).

3.5 Membrane-Based Processes

Forward osmosis (FO) is proving to be more advantageous over traditional membrane techniques such as nanofiltration, reverse osmosis, and ultrafiltration. Forward osmosis is another emerging technology for both wastewater treatment and desalination. An FO process involves the movement of molecules of water from a low-concentration solution (feed) to a high-concentration solution (draw) through a semi-permeable membrane under an osmotic pressure gradient. Application of FO technique is greatly reduces wastewater's volume because of the relatively low osmotic pressure difference results in extraction of clean water by osmotic pressure (Yuan et al. 2015).

3.6 Coagulation and Flocculation

The coagulation-flocculation method is one of the frequently used methods for the treatment of urban and industrial wastewater. Nowadays, the coagulation–flocculation of water is carried out to reduce turbidity, natural organic matter, as well as other soluble inorganic and organic pollutants in the wastewater where this technique involves agglomeration of fine particles and colloids into larger particles. This process contains two distinct phases. In the first phase, dispersed coagulant is rapidly mixed into water/wastewater to be treated via violent agitation. Later in the second phase of flocculation, small particles are agglomerated into well-defined flocs via gentle agitation. Finally, the flocs are left to settle down and then removed in the form of sludge, whereas the treated water/wastewater (supernatant) is transported into subsequent treatment process or for release into a watercourse (Teh et al. 2016).

3.7 Aerobic Treatment

The biological removal of organic matters from settled wastewater is conducted by microorganisms, mainly heterotrophic bacteria but also occasionally fungi. The microorganisms are able to decompose the organic matters through two different biological processes: biological oxidation and biosynthesis (Rudakova and Sakaeva 2019). The biological oxidation forms some end products, such as minerals, that remain in the solution and are discharged with the effluent. The biosynthesis transforms the colloidal and dissolved organic matter into new cells that form in turn the dense biomass that can be then removed by sedimentation (Samer 2015).

Under the aerobic or anaerobic conditions, microorganisms such as bacteria (aerobically or anaerobically), algae, and fungi (aerobically) utilize the organic matters and reproduce energy for themselves. Fundamentally, the wastewater is purifying by the bioconversion of dissolved organic matters into thick bacterial biomass. Eventually, it becomes crucial to the removal of microbial biomass from the treated wastewater through secondary sedimentation. This secondary sedimentation is generally similar to primary sedimentation except that the sludge contains bacterial cells rather than fecal solids (Samer 2015). An aerobic treatment technique is about 100 years old. It is a commonly used technique for the treatment of sewage wastewater spread on land. The treatment technologies were evolved from trickling filter treatment to "activated sludge" as the volume of sewage wastewater increased (Derco and Vrana 2018). Currently, the use of particular microbiological consortium in aerobic wastewater treatment has been increased as the technologies evolved to decompose the waste of discrete characteristic to obtain desired results. Aerobic treatment can be not only termed as economical in design but also having end product substantial which is able to manage all kinds of wastewater with variable composition and toxic pulses.

3.8 Activated Sludge Method

Traditional activated sludge process is widely adopted for treating low strength wastewater like municipal wastewater, etc. In the activated sludge process, a thick mixture of the bacterial population is suspended in the wastewater under aerobic conditions. The bacterial population shows high growth rate and respiration activity upon consumption of unlimited nutrients and oxygen-rich available organic matters that results in the production of oxidized end products (e.g., CO_2, NO^{3-}, SO_4^{2-}, and PO_4^{3-}) and the biosynthesis of new microorganisms too. The activated sludge method is based on five interdependent components; those are aeration, bioreactor, mixing system, sedimentation tank, activated sludge, and returned sludge. The biological process employing activated sludge is a frequently used method for wastewater treatment, where the running costs are inexpensive (Samer 2015). This process comprises autotrophic bacteria, heterotrophic bacteria, protozoa, and fungi too. Apart from cultural techniques, molecular biological techniques such as PCR-DGGE, FISH also assists in understanding microbial consortia in wastewater treatment systems (Liu et al. 2007). Approximately 20% of aerobic treatment systems do not meet the desired discharge standards aerobic treatment can be improved by using more diverse microbial communities (immobilized or free floating organisms) with an increased in metabolic activity. Due to the poor settling process, aerobic wastewater treatment produces relatively low density of sludge in the reactor, which can be boosted by enabling the biomass to adhere to the carrier like sand particles and operate in upflow fluidized bed. The excess amount of sludge consists of a lot of moisture, and it's also become tough to treat polluted water. Then by-products of wastewater treatment are dewatered, dried, and finally burnt into ashes that can be used in farm lands as compost fertilizer (Dhoble and Ahmed 2018).

4 Biological Techniques

4.1 Biological Filters

There are various types of biological filters, one of them is submerged aerated filters that are commonly known as biological aerated filters (BAFs), and these are the commonly executed design and the trickling filter. The BAFs execute either the sunken granular media with upward or download flows or floating granular media with upward flow. This is the most common design of biological aerated filters. The prime systems of operation of biological filters are (a) single filtration and (b) recirculation (Samer 2015).

4.2 Aerobic Granular Sludge Technology

Aerobic granular sludge (AGS) technique was employed primarily for anaerobic systems during the 1980s. Aerobic granular sludge can be defined as a type of self-immobilized microbial communities, comprising mainly of aerobic and facultative bacteria. Aerobic granules have been cultured mostly in sequenced batch reactors known as granulated sequenced batch reactor under hydraulic selection pressure. Granulated sequenced batch reactor gives stability in the process, removal of toxins and better quality effluents, and treatment of large organic loads than traditional methods (Adav et al. 2008). Granulation is also greatly influenced by volume exchange ratio and settling time (Dhoble and Ahmed 2018).

Aerobic granulation technology is a relatively new technique for the treatment of wastewater that possesses a number of promising features. Aerobic granules are type of microbial self-immobilization. However, AGS technology makes the use of smaller reactors, but it has a high concentration of microorganisms and good settling capability of granules in their structure. Granule development is considered as the backbone of the technology, but it is the least explicated aspect. Currently, a four-step granulation process has been put forwarded. It includes (1) initial cell to cell attachment to initiate the process, (2) formation of micro aggregates by these self-attached cells, (3) a large amount of extra-cellular polymeric substance (EPS) biosynthesized by the aggregated microorganisms, and (4) granule maturation in response to the external hydrodynamic parameters applied by the reactor configuration and operating situations (Jackson et al. 2001). The presence of slow-growing bacteria and denitrification process by microbes promotes the aggregation and formation of granules in the biomass (Caluwé et al. 2017). In the case of biogranulation of fungi and filamentous bacteria, they can easily form mycelial pellets. Those mycelial pellets can be retained in the reactor and also settle very well. The pellets undergo a lysis process due to the oxygen starvation in the inner part of the pellet after that they grow out to a diameter of 5 ± 6 mm. Eventually, the bacteria can grow out to colonies on these mycelial pellets since these pellets serve as an immobilization matrix. Subsequently, the bacterial colonies growing on mycelial pellets became capable to maintain themselves because now they were large enough to settle, since the mycelial pellets fall apart due the process of tylosis which takes place in the inner part of the pellets. These micro colonies further grow out to become denser granular sludge, leading eventually to a bacterial dominated population in the reactor as the granulation proceeds (Beun et al. 2002; Seow et al. 2016).

Energy-rich substrates such as glucose or fructose trigger the growth of filaments which results in the formation of loosely structured granules. The size of granule is correlated with organic load and too high load induces the disintegration of granules. Aerobic microorganisms is inhabited at the outer layer of granule while facultative and obligatory anaerobes develop in deeper layers of granules (Deliyanni et al. 2017). The interchanges between quorum sensing and quorum quenching surely have a greater role in the process of granulation, because the cell surface of microorganism has a negative charge. Thus, repulsions among similar charges avert the cells from attaching to each other without the support of another mechanism such as neutralization of the microbial cell surface charges by using divalent cations like Ca^{2+}, etc. It could be considered as a possible mechanism to enable initial cell–cell attachment (Henze and Comeau 2008). The van der Waals force may also help in this cell–cell attachment. Aeration is requisite for the development of aerobic granules, but it decreases the phosphorus and nitrogen removal potential of the technology. The elimination of emerging contaminants is one of the domains where this technology might be predominant to the traditional activated-sludge methods (Sarma et al. 2017).

Biogranulation can generate two types of granular sludge that is aerobic granular sludge and anaerobic granular sludge. Both of them are developed in a precise sequencing cycle of feeding, reacting, settling, and decanting under a single sequencing batch reactor structure (Ibrahim et al. 2010). The granular sludge produced via biogranulation processes has an elevated level of biomass retention and renewable and higher density of microorganisms with millions of bacteria cells per gram of biomass (Liu et al. 2015). The aerobic granules manifest characteristics of regular, nearly round in shape, compact, smooth, capability to cope with high organic loading rate or shock loadings, tolerance to toxicity and simultaneous COD, excellent settle capability, dense microbial structure, etc. (Seow et al. 2016).

4.3 Aerobic Systems Using High Rate Bioreactors

The perfect combination of the treatment process is the key to the successful management of municipal and industrial wastewater. The fusion of different aerobic and anaerobic bioreactors has been employed to treat a wide range of industrial wastewater containing pulp and paper industry wastewater, textile industry wastewater, mixture of olive oil mill wastewater, food solid waste, pharmaceutical industry wastewater, and primary municipal wastewater, starch industry wastewater, and slaughter house wastewater (Ashrafi et al. 2015). The anaerobic–aerobic treatment is an effective process to treat municipal and industrial wastewater.

4.4 Aerobic Fluidized Bed (AFB)

The particles are enveloped with biofilm fluidized by the recirculation of liquid. Fluidized bed reactors are packed with mobile supports. Substrate diffusion limitations are eliminated by them. It also eliminates nitrogen and organic carbon from municipal wastewater. AFB is fabricated with a cylindrical fluidized bed with four cylindrical fine bubble membrane diffusers that provide aeration. A pulverized pumice-stone is also present as support material for microbes to attach. In spite of variations in organic load transfers and short start-up time for the operation. AFB provides good stability, although, AFB reactor manifests various advantages like high organic loading rates (OLR), no bed clogging, excessive biomass concentration, large surface area for mass transfer, short hydraulic retention time (HRT), and small external mass deliver resistance (Seow et al. 2016).

But on the other hand, there are some limitations like high-energy expenditure due to the high liquid recirculation ratio, thickness of the biofilm, control of the bed expansion, and oxygen distribution system which constrain their applicability on a huge industrial scale (Rezaei et al. 2009).

4.5 Sequencing Batch Reactor (SBR)

The aerobic sequencing batch reactor (SBR) is an advanced version of the fill and draw activated sludge system. It consists of one or more tanks; each of the tanks is capable of solid separation and waste stabilization (Hasan et al. 2016). The SBR process provides flexibility in the treatment of variable flows, good removal efficiency, good oxygen availability to microorganisms and substrate, small floor space, minimum operator interaction, and the option for aerobic or anaerobic conditions in the same tank (Kim et al. 2008). These advantages make it an adequate process for industrial, and municipal, wastewater treatment (Yu and Chao 2008).

4.6 Anaerobic–Aerobic Fixed Film Bioreactor (FFB)

Anaerobic–aerobic fixed film bioreactor system immobilized cells on the surface (fixed-film) of the media. Use of media to immobilization of cell provides various advantages over culture suspension like less sensitivity to environmental variations (pH, temperature, and toxic substances); greater changes in population; higher growth rate; and faster consumption of the substrate in relation to free biomass. This is assigned to the fixed cells that undergo physiological modification due to either the increase in the concentration of nutrients and enzymes or the collective effect of the extracellular polymeric matrix related to toxic substances; two fixed-film bioreactors were integrated with both anaerobic and aerobic media, those connected in series with the recirculation system. It gives some advantages such as to treat wastewater that have high content of oil and grease (Del Pozo and Diez 2003), less sensitivity to environmental changes, and higher growth rate due to the used of immobilized cells on the surface of the media (Czarnitzki et al. 2015).

4.7 Aerobic Membrane Bioreactors (MBR)

Aerobic membrane bioreactors are functions as a dual mechanism in which membrane filtration occurs along with biodegradation processes. When water and small solution molecules are allowed to pass through the membrane at the same time, biomass, solid materials, pathogenic bacteria, and even macromolecules are retained in the water so that a very high value effluent is obtained (Dhaouadi and Marrot 2008; Ueda and Horan 2000). Aqueous and particulate-based enzymes were retained by the membrane in MBR which are otherwise lost at the sedimentation–clarification step of the conventional methods. These enzymes are also able to enhance the metabolic rate in the MBR which makes it an adequate choice to treat high-strength synthetic wastewater. Aerobic membrane bioreactor provides various advantages comprising the reduced sludge production, the high quality of the effluent, and the separation of solid retention time (SRT) from HRT. While there are various advantages, one of the major constraints in the application of aerobic membrane bioreactor is the fouling of the membrane. Cross-flow filtration through the membrane is being the most frequently used method to alleviate this problem (Chan et al. 2009).

4.8 Oxidation Ponds

Oxidation ponds are aerobic systems where the oxygen is provided by not only the photosynthetic activity of algae but also transferred from the atmosphere by the heterotrophic bacteria. The dark green ponds are building up to a depth of between 1.2 and 1.8 m to make sure of maximum penetration of sunlight. This dark green color is due to dense algal development in the euphotic zone, which is often only a few centimeters deep. In oxidation ponds, the bloom of algal biomass depends upon the sunlight for energy and the inorganic compounds (N, P, CO_2) released by aerobic bacteria. Further, this symbiotic cycle is completed by the consumption of oxygen by bacteria that is produced by algal biomass. There are two well-defined zones that are present in facultative ponds: upper and lower zones. Where bacterial (facultative) activity and anaerobically degradation of sludge produced from the settle out of suspension take place in the upper aerobic zone and the lower anaerobic zone, respectively (Zheng et al. 2013; Samer 2015).

4.9 Aeration Lagoons

Aeration lagoons are deeper than oxidation ponds. Aeration in the aeration lagoons is provided by aerators. Where the suspended biomass of microbial community is retained by the aerators that also provide sufficient dissolved oxygen that enables maximal aerobic activity. On the other hand, bubble aeration is commonly used to provide aeration in aeration lagoons. A plastic tubing present at the base of the lagoon pumped compressed air which results in the generation of bubbles to provide aeration. A bacterial biomass blooms in large amount without causing sedimentation and sludge return also. That is why this method is considered as suitable mixed liquor formed in the lagoon. Hence, the aeration lagoons are counted more adequate for degradable wastewater like wastewaters of food industries. Natural treatment systems such as facultative ponds, lagoons and constructed wetland, and wastewater stabilization ponds have long retention time and a low socioeconomic condition that is why these methods are adapted by developing countries for treating municipal and industrial wastewater (Samer 2015).

4.10 Anaerobic Wastewater Treatment

There is a different traditional treatment such as conventional aerobic activated sludge which generates large residual by-products and also required high amounts of energy. Currently, anaerobic technology is widely accepted as a wastewater treatment strategy because it saves energy sources and also helps in the generation of methanogenic biogas by effectively converting organic chemicals into methane (CH_4) gas. Therefore, wastewater is now recommended as a vital resource for energy (Visvanathan and Abeynayaka 2012; Gude 2016). Wastewater treatment by anaerobic degradation follows a process, in which microbes break down biodegradable waste material in the absence of oxygen. After degradation soluble organic pollutants are used by microorganism, acidogenic bacteria play a role and convert sugar and amino acids into carbon dioxide gas, ammonia, hydrogen, and organic acid. In the next step, organic acid is converted into acetic acid. At the end, methanogenic bacteria convert acetic acid into methane, carbon dioxide, and hydrogen (Alvarez and Illman 2005). This energy-rich methane or biogas can be used further for boiler feed or combined heat and for the production of green electricity and heat. Anaerobic treatment processes have been considered to be time-consuming and inefficient, as compared to aerobic degradation. However, it not only decreases the COD and BOD in the wastewater but also produces renewable energy. Moreover, it includes some anaerobic bacteria which could break down some persistent organic material present in wastewater, such as lignin and some other high molecular weight compound which is not possible in the case of aerobic degradation. Generally, anaerobic treatment reactors are of two types. One is anaerobic-activated sludge process which includes a conventional stirred anaerobic reactor, upflow anaerobic sludge blanket reactor. The second one is anaerobic contact tank which includes fluidized bed reactor, anaerobic rotating biological contactor, and anaerobic filter reactor (Zheng et al. 2013).

4.11 Microbial Fuel Cells (MFCs)

Microbial fuel cells are devices in which bacterial cells worked as catalysts to oxidize organic and inorganic matter and produce electricity. It consists of two chambers: one is anodic and the other is cathodic chamber divided by a proton exchange membrane (PEM). In this process, microbes and substrates are added to the anodic chamber. Bacterial cells oxidized the substrate and produced electron and protons. CO_2 is produced as an oxidation product. MFCs oxidize added substrates and generate electrons and protons in the process. Carbon dioxide is produced as an oxidation product. This carbon dioxide is used in photosynthesis therefore here no net carbon dioxide emission. The bacteria decompose organic matter on the anode and produced free H^+ ions and electrons. Electrons generated in this process are transferred to the anode (−) and flow to the cathode (+). Both terminals are linked by a resistor containing a conductive material. This flow of electron generates electricity in the device (Hernández-Flores et al. 2015; Ramadan and Purwono 2017). The H^+ ions flow to the cathode through the semi-permeable membrane. Near the anode a high concentration of H^+ ions is presently known as electrochemical gradient which plays an important role in this process. These H^+ ions and dissolved oxygen combine with electrons that move from the cathode, to form pure water (Brown 2012).

Electrode reactions are shown below.

Anodic reaction:

$$CH_3COO^- + 2H_2O \rightarrow CO_2 + 7H^+ + 8e^-$$

Cathodic reaction:

$$O_2 + 4e- +4H^+ \rightarrow 2H_2O$$

The energy or electricity is a by-product of this overall reaction (Rahimnejad et al. 2015). Single compartment, continuous flow, and membrane-less MFCs are preferred in wastewater treatment (He et al. 2005; Moon et al. 2005; Do et al. 2018).

4.12 Rotating Biological Contactors

The rotating biological contactor (RBC) system can be used as equipment or tool to modify and improve the available treatment method used for the secondary or tertiary treatment processes. The RBC is successfully used for biological treatment in three steps such as reduction of biological oxygen, nitrification, and denitrification. It contains a fixed-biofilm of anaerobic microbes which help to remove carbonaceous and nitrogenous compounds from wastewater. It was designed to remove ammonia nitrogen (NH3-N) and BOD or both, from wastewater (Patwardhan 2003). The RBC consists of a cover, bearings, drive, shaft, and media. Its hardware consists of a closely spaced circular plastic media which is present on a horizontal shaft and supported by bearings, and for its slow rotation, an electric motor is attached. The media are made up of corrugated polyethylene or polystyrene material with different densities, dimensions, and designs. The designs of the model are based on increasing surface area and stability, allowing a windy wastewater flow path and inducing air turbulence (Samer 2015).

4.13 Anaerobic Contact Process Reactor

Anaerobic contact process reactor can also be considered as an anaerobic activated sludge because the sludge is recycled from a separator to the reactor. In this process, waste material contains a mixture of gas, liquid, solid. The presence of gas may cause floating sludge in the clarifier or separator, therefore, this system required a vacuum degasifier to separate the gas and avoid floating sludge production. A set of reactors are created in series with recycling. The recycled material is poured into the bottom of the first reactor, an upflow reactor. It is a large reactor which facilitates the material to flow up from the bottom and separates the materials into three zones. The first zone is covered by the biogas. The lower zone occupied by bacterial digested matter and the mid-zone is a clarifier zone and helps to exports the stabilized waste (Nähle 1991).

4.14 Anaerobic Biofilter

Anaerobic biofilter is also called anaerobic fixed film reactors. It is a type of highly efficient anaerobic treatment equipment which is developed in the 1960s. In these reactors, anaerobic bacteria are provided by support materials to give the surface for their growth and to reduce turbulence to retain unattached populations in the system. These bacteria degrade organic matter present in wastewater and generate methane gas; it will be released from the top of the pool (Kassab et al. 2010). Microorganisms grow on the filler and reduce the speed of run of the degraders. These features could increase the efficiency of the treatment and the water quality. The drawback of this system could be blocked when it deal with higher concentration organic containing wastewater, particularly in the water inlet parts (Zheng et al. 2013).

4.15 Upflow Anaerobic Sludge Blanket Reactor (UASB)

The system was developed in the 1970s. In this system no carrier is present, and wastewater moves upward through a blanket of anaerobic granular sludge suspended in the UASB. Mixing of wastewater and sludge is achieved by the production of methane gas within the blanket and by hydraulic flow too (Bal and Dhagat 2001), and the three phase separators such as gas, liquid, and sludge biomass could inhibit the biomass loss of the sludge by the water discharge and gas emission. There are many advantages provided by this system, such as high concentration of naturally immobilized bacteria with excellent capacity to remove the organic pollutants from wastewater, high concentrations of biomass availability without support materials, and reduced cost of construction (Samer 2015).

4.16 Fluidized Bed Reactor (FBR)

This reactor contains a sand bed on which the complete biomass is grown. Because the sand particles are small in size, therefore, a large biomass can be developed in a reactor with small volume. For the growth of healthy biomass sand bed must be wet (Seow et al. 2016). A fluid may be gas or liquid form is passed through the spherical shaped beds solid material with high velocities. After this process, the solid material becomes wet and looks like fluid. Thus process is known as fluidization which gives many important advantages to the FBR (Samer 2015).

4.17 Anaerobic Lagoons

Anaerobic lagoons are often used to treat animal wastes and industrial wastes. In this method, anaerobic condition is created in the deep lagoon by restricting oxygen. This process occurs in the deep ground ponds, and such basins are applied for anaerobic pretreatment (Pal 2017). These lagoons are not aerated, heated, or mixed. The depth of a typical lagoon should be deeper than 2.5 m, because extra deeper lagoons are more efficient in wastewater treatment. Such depths do not allow the oxygen diffused from the environment into the lagoon, creating the anaerobic conditions (Samer 2015).

4.18 Combination of the Aerobic and Anaerobic Biodegradation

As compared to the single anaerobic and aerobic reactors, the combination of the both reactor is more efficient in organic pollutant degradation and waste treatment. The combined systems have many advantages as follows: the organic matters and suspended solid could get rid of the wastewater with the anaerobic process and reduce the aerobic sludge and organic load and reduce the volume of the reactors; pretreatment of wastewater by anaerobic could lower the load of fluctuation of the wastewater, thus decreasing the oxygen requirement in the aerobic degradation; the anaerobic process could make the aerobic process more efficient by modifying the biochemical property of the waste. Investigation on a combination of the aerobic and anaerobic biodegradation showed that it is more stable and has a huge potential for treatment. The classical combined reactors include oxidation ditch, constructed wetland, and anaerobic–anoxic–aerobic (A (1)-A (2)-O) and an anoxic–aerobic (A/O) reactor (Li et al. 2003; Zheng et al. 2013).

4.19 Oxidation Ditch

The oxidation ditch is a modified activated sludge technique that is equipped with a circular basin which provides aeration and circulation of the wastewater. The wastewater moves through the ditch at 1 to 2 ft./s speed. This process uses the long solids times (SRTs) to remove organics (Shammas and Wang 2009). In oxidation ditch activated sludge is added so that the microbes will digest the organic pollutants present in the water. This mixture of raw waste and returned sludge is known as mixed liquor. The oxygen could add into the flowing mixed liquor by the rotating biological contractor and could also increase surface area and generate waves within the ditches. After the organic pollutant has been removed from the wastewater, the oxidation ditch flows out of the mixed liquor. Sludge is moved in the secondary settling tank, and part of the sludge is pumped into a sludge pumping room where it is thickened via the help of aerator pumps (Yongzhen et al. 2008). A small portion of the sludge is returned to the reactor, while the remaining sludge is sent to waste. The important characteristics of oxidation ditch are simple process, steady operation, low maintenance consumption, and high shock resistance. There are some problems associated with this technique such as rising sludge, foam production and higher expansion, therefore, it required some improvement (Zheng et al. 2013).

4.20 Constructed Wetland

A constructed wetland is a technique in which an artificial wetland is created. This wetland could act as a biofilter to remove sediments and pollutants such as organic pollutants and heavy metals from the water. Constructed wetland is a combination of plants, microorganisms, media, water, and other animals. This technique has two basic types: one is surface flow wetlands and another one is subsurface-flow wetlands (Mook et al. 2008). In this method, physical, chemical, and biological methods are combined to remove pollutants from wastewater. Besides, absorbing organic pollutants and heavy metals on the constructed wetland filler, plants can supply the required nutrient for the growth and reproduction of microorganism such as carbon, nitrogen, etc. by their roots. Plants not only remediate pollutants directly into their cells or tissues but also act as catalysts in the purification reactions and promote various chemical and biochemical reactions to participate in purification reactions (Maine et al. 2005). Plants could also help to create aerobic and anaerobic condition by pumping oxygen in the deep level of constructed wetland to facilitate the breaking down of organic materials. Here, approximately 90% of pollutant are removed by microorganisms and natural chemical processes, while about 7–10% of pollutants are removed by the plants. In addition to organic pollutants, the nitrogen and phosphorous could also be removed from the wastewater (Maine et al. 2006) and prevent eutrophication problem. Constructed wetland is economical easy to management and ecofriendly; therefore, developing countries use this as a promising technique for the wastewater treatment (Zheng et al. 2013).

5 Conclusion

Recently, evolution of novel, effective, and cheaper methods for purification of wastewater became a vital field of research, as shown by the various publications appearing each year. However, at present the numerous treatment methods currently have been cited for wastewater treatment, but only few of them are generally utilized by the industrial sector for technological and economic reasons. Commonly, the removal of pollutants from wastewater is performed either by physicochemical or by biological means or combined with both. The recent developments interpreted the use of the physical treatment methods as the primary treatment followed by the biological treatment methods as secondary treatment which can be further preceded to the chemical treatment methods. However, sometimes it is also hard to define a universal method that could be used for the removal of all contaminants from wastewaters.

References

Adav SS, Lee DJ, Show KY, Tay JH (2008) Aerobic granular sludge: recent advances. Biotechnol Adv 26(5):411–423

Alsabagh AM, Hassan ME, Desouky SEM, Nasser NM, Elsharaky EA, Abdelhamid MM (2016) Demulsification of W/O emulsion at petroleum field and reservoir conditions using some demulsifiers based on polyethylene and propylene oxides. Egypt J Pet 25(4):585–595

Alvarez PJ, Illman WA (2005) Bioremediation and natural attenuation: process fundamentals and mathematical models. Wiley, Hoboken

Ashrafi O, Yerushalmi L, Haghighat F (2015) Wastewater treatment in the pulp-and-paper industry: a review of treatment processes and the associated greenhouse gas emission. J Environ Manag 158:146–157

Aslan S, Kapdan IK (2006) Batch kinetics of nitrogen and phosphorus removal from synthetic wastewater by algae. Ecol Eng 28(1):64–70

Bal AS, Dhagat NN (2001) Upflow anaerobic sludge blanket reactor a review. Indian J Environ Health 43(2):1–82

Barakat MA (2011) New trends in removing heavy metals from industrial wastewater. Arab J Chem 4(4):361–377

Barupal T, Chittora D, Meena M (2019) Solid waste: characterization, assessment, monitoring, and remediation. In: Singh RP, Prasad V, Vaish B (eds) Advances in waste-to-energy technologies. Taylor & Francis/CRC Press, Boca Raton, FL, pp 01–19

Berefield LD, Judkins JF, Weand BL (1982) Process chemistry for water and wastewater treatment. Prentice-Hall, Upper Saddle River

Beun JJ, Van Loosdrecht MCM, Heijnen JJ (2002) Aerobic granulation in a sequencing batch airlift reactor. Water Res 36(3):702–712

Bhattacharjee S, Datta S, Bhattacharjee C (2007) Improvement of wastewater quality parameters by sedimentation followed by tertiary treatments. Desalination 212(1–3):92–102

Bitton G (2005) Wastewater microbiology. Wiley, Hoboken

Brown JL (2012) Fuel cells treat wastewater, generate electricity. Civ Eng Mag Arch 82(7):34–35

Caluwé M, Dobbeleers T, D'aes J, Miele S, Akkermans V, Daens D, Dries J (2017) Formation of aerobic granular sludge during the treatment of petrochemical wastewater. Bioresour Technol 238:559–567

Chan YJ, Chong MF, Law CL, Hassell DG (2009) A review on anaerobic–aerobic treatment of industrial and municipal wastewater. Chem Eng J 155(1–2):1–18

Chen G (2004) Electrochemical technologies in wastewater treatment. Sep Purif Technol 38 (1):11–41

Clifford DA (1991) Ion exchange and in organic adsorption. In: Pontius FW (ed) Water quality and treatment, 4th edn. American Water Works Association/McGraw Hill, New York

Cox M, Négré P, Yurramendi L (2007) Industrial liquid effluents. Inasmet Tecnalia, San Sebastian, p 283

Crini G, Badot PM (2007) Traitement et épuration des eaux industrielles polluées. PUFC, Besançon, p 353

Crini G, Lichtfouse E (2019) Advantages and disadvantages of techniques used for wastewater treatment. Environ Chem Lett 17(1):145–155

Czarnitzki D, Hussinger K, Schneider C (2015) R & D collaboration with uncertain intellectual property rights. Rev Ind Organ 46(2):183–204

Del Pozo R, Diez V (2003) Organic matter removal in combined anaerobic–aerobic fixed-film bioreactors. Water Res 37(15):3561–3568

Deliyanni EA, Kyzas GZ, Matis KA (2017) Various flotation techniques for metal ions removal. J Mol Liq 225:260–264

Derco J, Vrana B (2018) Introductory chapter: biosorption. In: Biosorption. InTechOpen, London. https://doi.org/10.5772/intechopen.78961

Dhaouadi H, Marrot B (2008) Olive mill wastewater treatment in a membrane bioreactor: process feasibility and performances. Chem Eng J 145(2):225–231

Dhoble YN, Ahmed S (2018) Sustainability of wastewater treatment in subtropical region: aerobic vs anaerobic process. Int J Eng Res Dev 14(1):2278–2267

Djenouhat M, Hamdaoui O, Chiha M, Samar MH (2008) Ultrasonication-assisted preparation of water-in-oil emulsions and application to the removal of cationic dyes from water by emulsion liquid membrane: part 1: membrane stability. Sep Purif Technol 62(3):636–641

Do MH, Ngo HH, Guo WS, Liu Y, Chang SW, Nguyen DD, Ni BJ (2018) Challenges in the application of microbial fuel cells to wastewater treatment and energy production: a mini review. Sci Total Environ 639:910–920

El-Ashtoukhy ES, Amin NK, Bdelwahab O (2009) Treatment of paper mill effluents in a batch-stirred electrochemical tank reactor. Chem Eng J 146(2):205–210

Ferrera I, Sanchez O (2016) Insights into microbial diversity in wastewater treatment systems: how far have we come? Biotechnol Adv 34(5):790–802

Forgacs E, Cserhati T, Oros G (2004) Removal of synthetic dyes from wastewaters: a review. Environ Int 30(7):953–971

Gonçalves AL, Pires JC, Simões M (2017) A review on the use of microalgal consortia for wastewater treatment. Algal Res 24:403–415

Goyal RK, Jayakumar NS, Hashim MA (2011) Chromium removal by emulsion liquid membrane using [BMIM]+[NTf2]− as stabilizer and TOMAC as extractant. Desalination 278(1–3):50–56

Gude VG (2016) Wastewater treatment in microbial fuel cells–an overview. J Clean Prod 122:287–307

Hai FI, Yamamoto K, Fukushi K (2007) Hybrid treatment systems for dye wastewater. Crit Rev Environ Sci Technol 37(4):315–377

Harvey PJ, Campanella BF, Castro PM, Harms H, Lichtfouse E, Schäffner AR, Werck-Reichhart D (2002) Phytoremediation of polyaromatic hydrocarbons, anilines and phenols. Environ Sci Pollut Res 9(1):29–47

Hasan HA, Abdullah SRS, Al-Attabi AWN, Nash DAH, Anuar N, Rahman NA, Titah HS (2016) Removal of ibuprofen, ketoprofen, COD and nitrogen compounds from pharmaceutical wastewater using aerobic suspension-sequencing batch reactor (ASSBR). Sep Purif Technol 157:215–221

He Z, Minteer SD, Angenent LT (2005) Electricity generation from artificial wastewater using an upflow microbial fuel cell. Environ Sci Technol 39(14):5262–5267

Henze M, Comeau Y (2008) Wastewater characterization. In: Biological wastewater treatment: principles modelling and design. IWA Publication, London, pp 33–52

Hernández-Flores G, Poggi-Varaldo HM, Solorza-Feria O, Ponce-Noyola MT, Romero-Castañón T, Rinderknecht-Seijas N, Galíndez-Maye J (2015) Characteristics of a single chamber microbial fuel cell equipped with a low cost membrane. Int J Hydrogen Energy 40 (48):17380–17387

Holkar CR, Jadhav AJ, Pinjari DV, Mahamuni NM, Pandit AB (2016) A critical review on textile wastewater treatments: possible approaches. J Environ Manag 182:351–366

Hussein MA, Mohammed AA, Atiya MA (2019) Application of emulsion and Pickering emulsion liquid membrane technique for wastewater treatment: an overview. Environ Sci Pollut Res 26:1–21. https://doi.org/10.1007/s11356-019-06652-3

Ibrahim Z, Amin MFM, Yahya A, Aris A, Muda K (2010) Characteristics of developed granules containing selected decolourising bacteria for the degradation of textile wastewater. Water Sci Technol 61(5):1279–1288

Jackson RB, Carpenter SR, Dahm CN, McKnight DM, Naiman RJ, Postel SL, Running SW (2001) Water in a changing world. Ecol Appl 11(4):1027–1045

Kassab G, Halalsheh M, Klapwijk A, Fayyad M, Van Lier JB (2010) Sequential anaerobic–aerobic treatment for domestic wastewater–a review. Bioresour Technol 101(10):3299–3310

Kim YH, Yoo C, Lee IB (2008) Optimization of biological nutrient removal in a SBR using simulation-based iterative dynamic programming. Chem Eng J 139(1):11–19

Kishimoto N, Nakagawa T, Okada H, Mizutani H (2010) Treatment of paper and pulp mill wastewater by ozonation combined with electrolysis. J Water Environ Technol 8(2):99–109

Li NN (1968) U.S. Patent No. 3,410,794. U.S. Patent and Trademark Office, Washington, DC

Li YM, Gu GW, Zhao JF, Yu HQ, Qiu YL, Peng YZ (2003) Treatment of coke-plant wastewater by biofilm systems for removal of organic compounds and nitrogen. Chemosphere 52(6):997–1005

Liu DHF, Liptak BG (eds) (2000) Wastewater treatment. CRC Press, Boca Raton

Liu XC, Yang M, Zhang Y, Yang XP, Gan YP (2007) Microbial community comparison of different biological processes for treating the same sewage. World J Microbiol Biotechnol 23 (1):135–143

Liu Y, Kang X, Li X, Yuan Y (2015) Performance of aerobic granular sludge in a sequencing batch bioreactor for slaughterhouse wastewater treatment. Bioresour Technol 190:487–491

Maine MA, Suñe N, Hadad H, Sánchez G, Bonetto C (2005) Phosphate and metal retention in a small-scale constructed wetland for waste-water treatment. In: Golterman HL, Serrano L (eds) Phosphate in sediment. Proceedings 4th international symposium on phosphate in sediments. Backhuys, Leiden, pp 21–32

Maine MA, Sune N, Hadad H, Sánchez G, Bonetto C (2006) Nutrient and metal removal in a constructed wetland for wastewater treatment from a metallurgic industry. Ecol Eng 26 (4):341–347

Mook WT, Chakrabarti MH, Aroua MK, Khan GMA, Ali BS, Islam MS, Yongzhen P, Hongxun H, Shuying W, Youwei C, Zhiguo Y (2008) Nitrogen and phosphorus removal in pilot-scale anaerobic-anoxic oxidation ditch system. J Environ Sci 20(4):398–403

Moon H, Chang IS, Jang JK, Kim BH (2005) Residence time distribution in microbial fuel cell and its influence on COD removal with electricity generation. Biochem Eng J 27(1):59–65
Moosai R, Dawe RA (2003) Gas attachment of oil droplets for gas flotation for oily wastewater cleanup. Sep Purif Technol 33(3):303–314
Morin-Crini N, Crini G (eds) (2017) Eaux industrielles contaminées. PUFC, Besançon
Nähle C (1991) The contact process for the anaerobic treatment of wastewater: technology, design and experiences. Water Sci Technol 24(8):179–191
Pal P (2017) Industrial water treatment process technology. Butterworth-Heinemann, Cambridge
Patwardhan AW (2003) Rotating biological contactors: a review. Ind Eng Chem Res 42 (10):2035–2051
Rahimnejad M, Adhami A, Darvari S, Zirepour A, Oh SE (2015) Microbial fuel cell as new technology for bioelectricity generation: a review. Alex Eng J 54(3):745–756
Ramadan BS, Purwono (2017) Challenges and opportunities of microbial fuel cells (MFCs) technology development in Indonesia. In: MATEC Web of Conferences, EDP Sciences, 101:02018, pp 1–5
Rathoure AK, Dhatwalia VK (eds) (2016) Toxicity and waste management using bioremediation. Hershey, PA: IGI Global, Engineering Science Reference, p 425
Rawat I, Kumar RR, Mutanda T, Bux F (2011) Dual role of microalgae: phycoremediation of domestic wastewater and biomass production for sustainable biofuels production. Appl Energy 88(10):3411–3424
Reemtsma T, Weiss S, Mueller J, Petrovic M, González S, Barcelo D, Knepper TP (2006) Polar pollutants entry into the water cycle by municipal wastewater: a European perspective. Environ Sci Technol 40(17):5451–5458
Renuka N, Sood A, Ratha SK, Prasanna R, Ahluwalia AS (2013) Evaluation of microalgal consortia for treatment of primary treated sewage effluent and biomass production. J Appl Phycol 25 (5):1529–1537
Rezaei F, Xing D, Wagner R, Regan JM, Richard TL, Logan BE (2009) Simultaneous cellulose degradation and electricity production by Enterobacter cloacae in a microbial fuel cell. Appl Environ Microbiol 75(11):3673–3678
Rizz L, Manaia C, Merlin C, Schwartz T, Dagot C, Ploy MC, Fatta-Kassinos D (2013) Urban wastewater treatment plants as hotspots for antibiotic resistant bacteria and genes spread into the environment: a review. Sci Total Environ 447:345–360
Rudakova LV, Sakaeva EH (2019) Quality assessment of activated sludge biological treatment pulp and paper plant with variable load. In: IOP conference series: earth and environmental science, vol 321. IOP Publishing, 1, p 012048
Ruiz J, Arbib Z, Álvarez-Díaz PD, Garrido-Pérez C, Barragán J, Perales JA (2013) Photobiotreatment model (PhBT): a kinetic model for microalgae biomass growth and nutrient removal in wastewater. Environ Technol 34(8):979–991
Saki S, Uzal N (2018) Preparation and characterization of PSF/PEI/$CaCO_3$ nanocomposite membranes for oil/water separation. Environ Sci Pollut Res 25(25):25315–25326
Samer M (ed) (2015) Biological and chemical wastewater treatment processes. In: Wastewater treatment engineering, pp 1–50
Sarma SJ, Tay JH, Chu A (2017) Finding knowledge gaps in aerobic granulation technology. Trends Biotechnol 35(1):66–78
Schwarzenbach RP, Escher BI, Fenner K, Hofstetter TB, Johnson CA, Von Gunten U, Wehrli B (2006) The challenge of micropollutants in aquatic systems. Science 313(5790):1072–1077
Seow TW, Lim CK, Nor MHM, Mubarak MFM, Lam CY, Yahya A, Ibrahim Z (2016) Review on wastewater treatment technologies. Int J Appl Environ Sci 11:111–126
Shammas NK, Wang LK (2009) Oxidation ditch. In: Biological treatment processes. Humana Press, Totowa, pp 513–538
Teh CY, Budiman PM, Shak KPY, Wu TY (2016) Recent advancement of coagulation–flocculation and its application in wastewater treatment. Ind Eng Chem Res 55(16):4363–4389

Topare NS, Attar SJ, Manfe MM (2011) Sewage/wastewater treatment technologies: a review. Sci Rev Chem Commun 1:18–24
Ueda T, Horan NJ (2000) Fate of indigenous bacteriophage in a membrane bioreactor. Water Res 34 (7):2151–2159
Visvanathan C, Abeynayaka A (2012) Developments and future potentials of anaerobic membrane bioreactors (AnMBRs). Membr Water Treat 3(1):1–23
Wang LK, Vaccari DA, Li Y, Shammas NK (2005) Chemical precipitation. In: Physicochemical treatment processes. Humana Press, Totowa, pp 141–197
Wang Z, Xue M, Huang K, Liu Z (2011) Textile dyeing wastewater treatment. In: Advances in treating textile effluent. InTech, London, pp 91–116
Yongzhen P, Hongxun H, Shuying W, Youwei C, Zhiguo Y (2008) Nitrogen and phosphorus removal in pilot-scale anaerobic-anoxic oxidation ditch system. J Environ Sci 20(4):398–403
Yu GAO, Chao LIU (2008) Municipal wastewater treatment using sequencing batch biofilm reactor (SBBR). In: The international conference on advances in chemical technologies for water and wastewater treatment, pp 617–624
Yu L, Han M, He F (2017) A review of treating oily wastewater. Arab J Chem 10:S1913–S1922. https://doi.org/10.1016/j.arabjc.2013.07.020
Yuan H, Abu-Reesh IM, He Z (2015) Enhancing desalination and wastewater treatment by coupling microbial desalination cells with forward osmosis. Chem Eng J 270:437–443
Zheng C, Zhao L, Zhou X, Fu Z, Li A (2013) Treatment technologies for organic wastewater. Water Treat 11:50–86

New Trends in Removing Heavy Metals from Industrial Wastewater Through Microbes

Mohammad Haris, Adnan Shakeel, Touseef Hussain, Gufran Ahmad, Moh. Sajid Ansari, and Abrar Ahmad Khan

Abstract

Heavy metal contaminations produced by industrial activities are one of the major important issues which are faced by many countries, typically developing economies. Heavy metals in wastewater are of particular major concern in recent times due to their persistence and recalcitrance in the aquatic environment. As a result of various consequences, wastewater treatment has reached a certain level which is becoming unmanageable nowadays. Previous studies have provided many innovative processes for the treatment of industrial wastewater containing heavy toxic metals, often involving toxicity reduction techniques to meet technology-based treatment standards. This chapter reviews the ability of various technologies to remove heavy metals from industrial wastewater. Ample natural materials such as agricultural waste, industrial by-products and microbial biomass have been suggested as a potential bio-absorbent for the removal of heavy metals due to the presence of functional metal-binding groups. Especially focus is given to innovative physicochemical elimination processes like adsorption in new adsorbents, membrane filtration, photocatalysis and electrodialysis. The investigation shows that new adsorbents and membrane filtration are the most frequently studied and widely applied techniques for the treatment of wastewater contaminated with toxic heavy metals. In general, the applicability, wastewater characteristics, profitability and simplicity of the plant are the major factors in choosing the appropriate method for contaminated wastewater treatment.

The original version of this chapter was revised. A correction to this chapter can be found at https://doi.org/10.1007/978-981-15-5901-3_26

M. Haris · A. Shakeel · T. Hussain (✉) · G. Ahmad · M. S. Ansari · A. A. Khan
Department of Botany, Aligarh Muslim University, Aligarh, Uttar Pradesh, India

M. P. Shah (ed.), *Removal of Emerging Contaminants Through Microbial Processes*,
https://doi.org/10.1007/978-981-15-5901-3_9

Keywords

Environment · Heavy metals · Microbes · Wastewater · Pollution

1 Introduction

Technological advancement and industrialization have put an increasing load on the environment through releasing large quantities of wastewater which contain hazardous substances including heavy metals (chromium, cadmium and lead), organic contaminants and metalloids (elements have intermediate characters between metal and non-metal, like antimony and arsenic) which cause serious injury to the environment. Similarly, many industries such as textile, chemical, food and metallurgy release high quantities of waste with toxic substances. Currently, industrialization is increasingly developing worldwide, which may affect water quality, food, animal feed and climatic conditions (Owolabi and Hekeu 2014). Most heavy metals are toxic at low concentrations and can enter the food chain, where they accumulate and cause damage to living organisms. All metal has the capacity to cause injuries at higher concentrations, and the harmful effects of each metal depend on its amount, duration of exposure, the route and the absorbed dose (Mani and Kumar 2014).

The treatment of wastewater effluents including heavy metal ions is difficult because it depends largely on environmental, technical, economic and social considerations. This complexity prevents the development of unique technologies capable of treating a large number of wastewater effluents, so several technologies must be implemented for the reduction of pollution from the water and remedy inherited sites in addition to their neighbouring aquatic ecosystems (Akcil et al. 2015; Oyewo et al. 2018). Examples of heavy metal elimination technologies currently use such as membrane filtration, chemical precipitation, coagulants/flocculants, ion exchange, photocatalysis and adsorption on inorganic materials. Among all the technologies, the utilization of living organisms for the absorption of pollutants and separate heavy metals from the environment is really interesting. Microorganisms can be used directly which have specific characteristics of catabolic potential, and/or their products like biosurfactants and enzymes are a new approach to improve and increase their remediation efficiency (Le et al. 2017; Schenk et al. 2012). The main advantage of using microorganisms to eliminate toxic elements is their safety in human aspects. Among these microorganisms, one of the most effective is yeast (*Saccharomyces cerevisiae*) which is highly used in bakeries and food industry (Wang and Chen 2006; Soares and Soares 2012). This chapter will provide an overview of the use of microbial bioabsorbents in the removal of heavy metals from industrial wastes and contaminated environments, as well as the sources and toxicity of these metals in humans.

2 Importance of Water in Life

The sustainability of water resources is a critical issue in the context of the growing demand for industrial, domestic and agricultural uses since the world needs around 60% more food (FAO 2013) to feed 9.5 billion peoples in 2050 (United Nations 2012). Quality of water plays a central role in the protection of habitat, industry, public health and agriculture (Akasaka et al. 2010; Lu et al. 2015). The activity that consumes the most water is irrigation for agriculture, which represents 70% of total water withdrawals and more than 90% of water consumption (Bates et al. 2008). Freshwater is also needed to sustain the landscaping and gardens of hotels for attractions and materializes in the development of tourism infrastructure, food and fuel production (Chapagain and Hoekstra 2008; Gossling 2001; Hoekstra and Hung 2002; Pigram 1995; Worldwatch Institute 2004). The FAO distinguishes the consumption of industrial, domestic and agricultural freshwater.

On a global an average, around 70% of the water is utilized in the agriculture sector, 20% for industrial sector and 10% for domestic purpose including homes, municipalities, public services, commercial establishments and also for the consumption of safe drinking water for future ahead (Bates et al. 2008; UNESCO 2009). The use of thermoelectric energy has a serious impact on resources of water, and the energy sector depends heavily on such water resources. Nationally, the United States Geological Survey (USGS) observed that 41% of the whole freshwater was used in the USA in 2005 primarily for cooling purpose in thermoelectric power plant (Kenny et al. 2009).

3 The Need for Treatments of Wastewater

Nowadays, a large scale of heavy metal wastewater has been discharged into the environment (O'Connell et al. 2008). Heavy metal is applied to a group of elements which have an atomic weight between 63.5 and 200.6 (O'Connell et al. 2008; Srivastava and Majumder 2008). Nickel (Ni), lead (Pb), chromium (Cr), cadmium (Cd), copper (Cu), etc. are the most hazardous pollutants of the freshwater reservoir. Moreover, heavy metals are teratogenic and carcinogenic and pose severe health problems that may be accumulated in the human bodies with the channel of the food chain (Lee et al. 2012). So, it is necessary to remediate contaminated sewage with heavy metals.

The rapidly increasing industrialization and urbanization are releasing large quantities of wastewater that is increasingly used as a precious resource for irrigation in agricultural areas. It stimulates valuable economic activity, particularly for poor farmers, supports countless livelihoods and significantly changes water quality in natural bodies of water (Marshall et al. 2007). Due to the rapid industrialization and urbanization water is becoming highly polluted day by day and the risk of the consumption of contaminated water and its sanitation problems increases day by day in most of the developing and undeveloped countries (Table 2). This growing water scarcity problem has a significant negative effect on economic development, quality of the environment and human livelihoods worldwide. Consequently, it has become an essential need for the current environment to protect water from contamination or develop a cost-effective repair method for its protection.

Therefore, strong growth in water requirement can be expected, but available water resources will decrease in many regions due to depletion of non-renewable fossil water resources (groundwater and glacial ice), pollution of water bodies, decrease in groundwater sources and climate change that lead to a reduction in precipitation and a greater frequency of droughts, greater evaporation and modification of runoff patterns (IPCC 2007; Parry et al. 2009a, b). The problem has become more difficult given the decrease in water resources (Singh and Panda 2012) due to urbanization, pollution and the impacts of climate change.

4 Sources of Industrial Wastewater

Wastewater is released in the environment by many industrial sources. It contains several harmful pollutants, including metals. The main sources of antimony are released in the environment from many industries such as lead storage batteries, welding equipment, lamination and power transmission, sheets and pipes of metals, ammunition, flame retardants, ceramics, foundry, tin, enamels and paintings (Iqbal et al. 2013) (Table 1), and the deadly disease caused by them is displayed in Fig. 1. The metals released from the wastewater of various industries are as follows:

- Wastewaters such as those generated during dye and pigment production, film and photography, galvanometry, metal cleaning, leather and mining, plating and electroplating may have an undesirable quantity of chromium(VI) anions (Vinodhini et al. 2010).

Table 1 Sources of various industrial toxic wastes

Hazardous waste component	Source
Heavy metals	
Arsenic	Mining, non-anthropogenic geochemical formation
Cadmium	Mining, fertilizer industry, battery waste
Chromium	Mining areas, tanneries
Lead	Lead-acid battery smelters
Manganese	Mining areas
Mercury	Chlor-alkali industries, healthcare institutes
Nickel	Mining, metal refining
Hydrocarbons	
Benzene	Petrochemical industries, solvents
Vinyl chloride	Plastics
Pesticides	Insecticides
Organic chemicals	
Dioxins	Waste incineration, herbicides
PCBs	Fluorescent lights, e-waste, hydraulic fluid

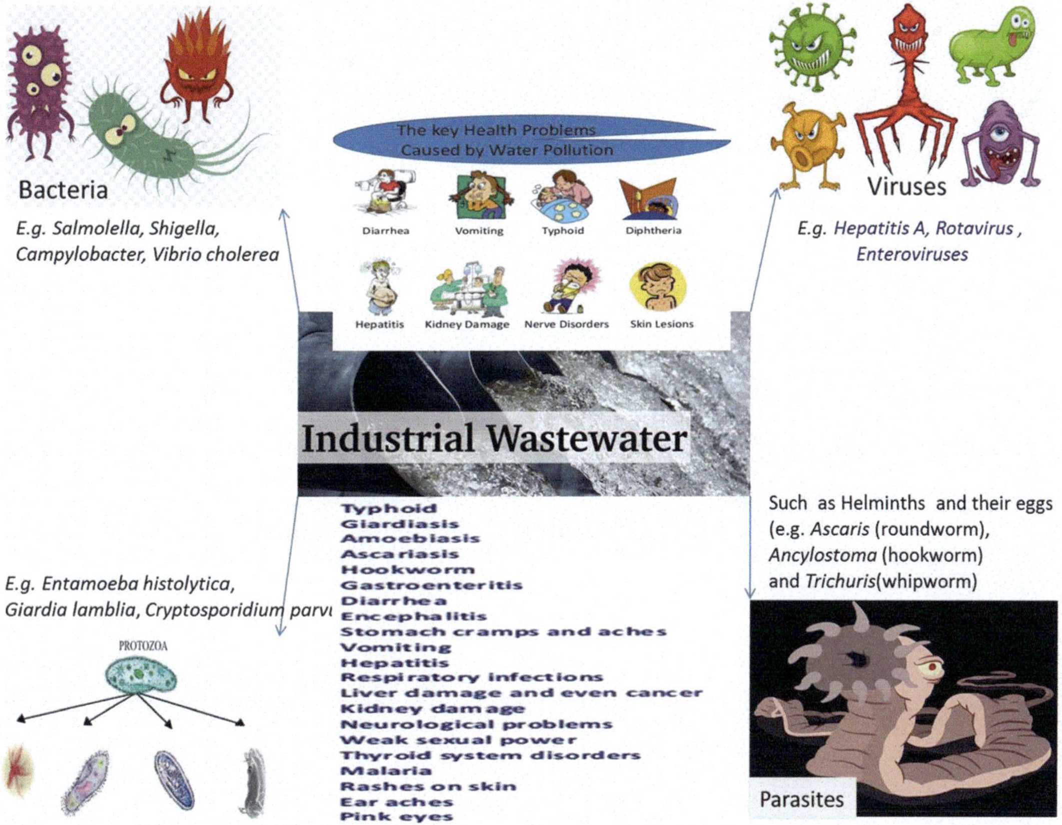

Fig. 1 Deadly disease-causing pathogens by wastewater in human beings

- Cobalt (Co) is generally found in polluted water because it is highly used in the electronics, porcelain, radioisotope therapy and alloys (especially stainless steels and magnetic steels) (Guo et al. 2013).
- Manganese (Mn) is released in the environment from the industries that involved in the manufacturing of fertilizer, electroplating, petrochemicals, metal processing, tanneries and mining (Abu Hasan et al. 2012).
- Mercury (Hg) can be present in wastewater which is discharged from paper and pulp, chlor-alkali, paint, oil refinery, fossil fuel burning, pharmaceutical, metallurgical processes and battery manufacturing (El-Shafey 2010).
- Effluent from the production of gasoline additives, batteries, pigments, sheets, alloys, etc. They have a high concentration of lead (Pb) ions (Tunali Akar et al. 2012).
- The metallurgy and extraction of nickel, aeronautics, stainless steel, nickel plating, manufacturing of batteries and pigment and ceramic industries release wastewater that contains large quantity of nickel (Ni) ions (Aloma et al. 2012).
- Zinc (Zn) is generally found in wastewater; it is released during the metallurgical processes and also from the galvanizing plants, stabilizers, pigment formation, thermoplastics, battery manufacturing and alloys as well as emissions during the municipal wastewater treatment (El-Shafey 2010).
- Arsenic (As) is introduced into water by anthropogenic and natural sources: release from the mineral, probably due to the long-term changes in geochemical and various industrial effluents such as ceramic industries, metallurgical industries, wood preservatives, pesticides and dye manufacturing industries (Sarý et al. 2011).

Cape Town is in the unenviable situation of being the first major city in the modern era to face the threat of running out of drinking water. However, the plight of the drought-hit South African city is just one extreme example of a problem that experts have long been warning about—water scarcity. Wordwide, majorities of the cities still facing water crises as well as going to face this serious issue in the near future, and this has been mentioned in Table 2.

5 Bioremediation of Heavy Metals by Microorganisms

Microbiology is an area of increasing interest for both environmental microbiologists and the general public. Nowadays, more and more people demonstrate that the magnitude of the problem of pollution in our soil and water requires immediate action. Among the achievement of toxins, hazardous levels are heavy metals. The threat of heavy metals is strongly intensified in the environment due to their absolute nature. To overcome the toxic effects of metals, microbes have progressed in various approaches and metalloids, using accumulation and resistance or, more interestingly, reducing their bioavailability or toxicity through biomethylation and transformation (Ojuederie and Babalola 2017; Ahamd et al. 2019). Bioremediation means that there are limitations, such as the conversion of toxic heavy metals into less toxic forms by microbes or their enzymes or, in other words, the production of toxic metabolites by microbes in bioremediation through

Table 2 Major cities worldwide that are going to face drinking water problem

S. No.	Cities	Reason
1	Bengaluru, India	Water pollution is a major problem being faced. Almost 85% of the lakes in Bengaluru had water that could neither be used for agriculture nor industrial use, the struggle of providing its inhabitants with freshwater
2	Sao Paulo, Brazil	Safe drinking water shortage problem back in 2015
3	Beijing, China	One of the major problems being faced is pollution. Studies showed that almost 40% of the surface water in Beijing was so polluted that it could not be used for agriculture or industrial use
4	Jakarta, Indonesia	The illegal system is draining the underground water reservoirs, leading the city to face a most probable water shortage trouble. Because of concrete and asphalt, the open cannot absorb the rainwater which in turn does not lead to replenishment of groundwater
5	Cairo, Egypt	Egypt is 97% dependent on river Nile as a source of drinking water. Due to urbanization and industrial development, river waters are getting contaminated with residential waste By 2025, the UN claims that there would a shortage of drinking water in Cairo
6	Istanbul, Turkey	Home to a pretty high population, the city is facing a shortage of water since 2016 when the supply of water lowered to almost 1700 m^3
7	Moscow, Russia	Russia has almost 1/4 of the world's freshwater reserves but is facing water crisis due to the industrial waste causing water pollution Since Moscow is 70% dependent on surface water, this is causing major problems in the city
8	Tokyo, Japan	Almost 70% of Tokyo depends on surface water. Both private and public have a proper water collection and utilization system in almost 750 buildings
9	Mexico City, Mexico	About 40% of the water is imported into the city from other sources (water losses occur due to a problem in the pipeline network). No proper system to recycle wastewater
10	Miami, USA	Water from the Atlantic Ocean polluted the Biscayne aquifer which was the city's main freshwater source due to a project conducted in the early twentieth century
11	London, Britain	London has an average rainfall of about 600 mm, which means that the city received almost 80% of its water supply from rivers Reports from the Greater London Authority show that the city would face supply problems by 2025 and would face severe water crisis by 2040

the cleaning of contaminated environments. Bioremediation through microorganism includes many ways like biosorption, bio-mineralization, biotransformation, absorption or active bioaccumulation (Ahamd et al. 2019). Bioremediation is gradually being accepted as standard practice for the restoration of contaminated soil with heavy metals, as it is more ecological and cost-effective than traditional chemical and physical methods that are often very expensive at times of metal concentrations and are disabled. In addition they also produces significant amounts of toxicity sludges (Ekperusi and Aigbodion 2015; Ayangbenro and Babalola 2017). The meaning of heavy metals concentrations varied over the years. However, it is collectively applied for metallic elements with a density above 4–5 g/cm^3 (Jennings 1993; Nies 1999) and those elements having atomic weight usually greater than 50 (Kennish 1992). These heavy metals are naturally occurs in environment and also contributed by many other sources like volcanoes, waste material of animals and plants, sewage, smoke of forest fire, acid rain (Kennish 1992; Johnson et al. 2019). In present world, rapid industrialization and municipal waste enriching the concentration and toxicity of hazardous heavy metals in environment (Girma 2015) cause heavy metal pollution. There is an increasing level of concentration of heavy metals in water bodies and crop lands, steadily entering into food chain.

For biological functions of plants and animals, heavy metals are necessary, but at high levels, they interfere with metabolic reactions in organism systems. Poisonous heavy metals such as lead (Pb), cadmium (Cd), mercury (Hg), chromium (Cr), zinc (Zn), uranium (Ur), selenium (Se), silver (Ag), gold (Au), nickel (Ni) and arsenic (S) which are not useful to plants are capable of reducing plant growth due to photosynthetic activities, mineral nutrition of plants and plant activity (Nematian and Kazemeini 2013; Kabata-Pendias 2010). These heavy metals are cytotoxic at low concentrations and can cause cancer in humans that can accumulate in the body when consumed in contaminated food through the food chain and become a health hazard to living organisms. It causes oxidative stress, an asymmetry that involves the ability of cells to produce free radicals and destroy them or repair damage (Chibuike and Obiora 2014; Dixit et al. 2015; Tak et al. 2013; Chandra et al. 2015; Mani 2015). Microorganisms through their natural inherently biological mechanisms allow them to survive under heavy metal stress and eliminate metals in the environment. Microorganisms, plants and fungi are of great importance in bioremediation as they work many tasks like improving soil quality, plant growth and detoxification. The ability of microorganisms to reduce pollution depends on the suitability of environmental conditions for their growth and metabolism, including adequate temperature, pH and humidity (Verma and Jaiswal 2016). It was reported that the biosorption capacity in the removal of pentachlorophenol (PCP) from aqueous solutions of *Aspergillus niger* (Mathialagan and Viraraghavan 2009) and *Mycobacterium chlorophenolicum* (Brandt et al. 1997; Bosso et al. 2015) depends on pH. Brandt et al. (1997) also assessed the effect of pH on the adsorption and de-adsorption behaviour of PCP by *M. chlorophenolicum* and reported that the pH value was an essential parameter affecting PCR adsorption, with the decrease being an increase in adsorption capacity. At pH 5.4, adsorption by bacteria was completely irreversible, while at pH 7.0 complete desorption was achieved. At pH 6–8, Jianlong et al. (2000) obtained better results on the adsorption behaviour of PCP by microbial biomass in aqueous solution. In alleviation of environments contaminated with heavy metals, microorganisms are necessary because they have different ways of coping with the toxicity of metals. The exploitation of microorganisms to sequence, form or alter the oxidation state of many heavy metals has been extensively studied (Gupta et al. 2016; Kang et al. 2016). Heavy metal bioremediation will be successful if an association of bacterial strains is used instead of using a single-strain culture. In the study by Wang and Chen (2009), synergistic effects of bacterial mixtures on bioremediation of a mixture of Pb, Cd and Cu from contaminated soil using four strains, *Viridibacillus arenosi* B-21, *Sporosarcina soli* B-22, *Enterobacter cloacae* KJ-46 and *E. cloacae* KJ-47, were investigated. They observed that the bacterial mixture had greater resistance and efficiency to heavy metal elimination, compared to the use of single-strain culture after 48 h, with a therapeutic efficiency of 98.3% for Pb, 85.4% for Cd and 5.6% for Cu recorded. *Penicillium* (a type of filamentous fungi) can remove a variety of heavy metal ions from aqueous solutions, such as Cu, Au, Zn, Cd, Mn, U and Th (Jianlong and Chen 2009). The fungi (ligninolytic) *Phanerochaete chrysosporium* have been reported as a best biodegradable agent and degrade a wide and diverse range of toxic pollutants (Novotny et al. 1999; Girma 2015). A number of microbes have been reported that are used for bioremediation. Some active members of aerobic and anaerobic group of bacterial species are *Pseudomonas*, *Sphingomonas*, *Rhodococcus*, *Bacillus*, etc. (Lata et al. 2019) that are highly useful in bioremediation. Powerful metal biosorbents in the bacteria class comprise the genus *Bacillus* (Hamer 1986; Hameed 2006), *Pseudomonas* (Huckle et al. 1993; Huston et al. 2002) and *Streptomyces* (Infante et al. 2014; Jain et al. 2012). Genetic engineering has improved the use and disposal of unwanted hazardous wastes under laboratory conditions by creating genetically modified organisms (Jain et al. 2011). Recombinant living organisms can be obtained by recombinant DNA techniques or by exchanging natural genetic material between organisms. They are currently able to insert genes suitable for the production of a particular enzyme that can degrade many contaminants (Jain et al. 2010). By developing an understanding of microbe's communities, their response to the natural environment and knowledge of the genetic sequences the expanding microbes can be utilized for their ability to reduce germs contaminations. There are field trials of new bioremediation techniques which are profitable and dedicated sites that are established apart from long-term research objectives; these opportunities provide the potential for significant progress.

6 Types of Heavy Metal Removal Methods

A huge amount and large volumes of effluent polluted with the ions of heavy metals and other organic compounds are produced by many anthropogenic sources like agricultural activities, industrial processes, traffic sources and living activities of human beings (Dinesh et al. 2014; Genç-Fuhrman et al. 2016). The pollution caused by contaminated effluents has become a great concern and receive increasing attention worldwide. The heavy metal pollution in wastewater could not be degraded by many degrading microorganisms and ultimately goes in water, soil, air and most importantly drinking water as well as the food chain. The exposure of heavy metals can be injurious because of it having mutagenic and carcinogenic nature. Many more studies and methods have been developed to remove the toxicity level of ions of heavy metals (Chang et al. 2018; Garrido-Baserba et al. 2018; Liu and Zhang 2018; Son et al. 2018; Thompson et al. 2016). The complexing agents like detergents, plastic wastages, pharmaceuticals wastages, oils, pesticides, fertilizers, etc. are making the constituents of heavy metals in wastewater and making it more complicated. The appearing substances like antibiotics, cyanide, humic acid, citrate, ethylenediaminetetraacetic acid (EDTA) and nitrilotriacetic acid (NTA) make the heavy metal ions to complex, increase stability with structures and toxicity levels (Wang et al. 2018; Wu et al. 2017; Xu et al. 2017). Moreover, a complex is defined as "a species formed by the consortium of two or more simpler species and each has a potential of independent existence" (Rossotti and Rossotti 1961). For example, the wastewater treated with EDTA and NTA, simple chemical neutralization and precipitation methods does not generally meet emission standards. Clearly, the treatment of heavy metal ion complex has become a condemnatory problem in the environmental protection act.

There are many methods of decontamination of heavy metals like adsorption, membrane filtration, electrolysis and photocatalysis (Gyliene et al. 2004; Ye et al. 2017; Zamariotto et al. 2010; Liu et al. 2016; Xu et al. 2015; Zhou et al. 2017) that are discussed in this part of chapter. The following are methods for the removal of heavy metals in wastewater.

6.1 Adsorption Methods

In recent decades, the removal of toxic heavy metal ions from wastewater by adsorption method has been applied effectively (Tong et al. 2011; Wang et al. 2014). There are two ways of adsorption method of heavy metal removal, that is, physisorption and chemisorptions. The physisorption method depends on the materials which have a suitable porous structure and are helpful in the removal of heavy metal ions. However, the physisorption method for the removal of heavy metals is not enough to receive the maximum removal, and therefore, the chemisorption method has received higher attention for wastewater treatment. The complexes of heavy metal can adsorb on the surface functional adsorbents by attracting electro-states and/or forcing chemical bonding to achieve the successful separation. For the moment the ion exchange and the sphere (inner/outer) are frequently involved in chemical adsorption of heavy metal ions (Jachuła et al. 2011; Zhou et al. 2019). Many methods of characterizations like density functional theory (DFT), Fourier transform infrared spectroscopy (FTIR), X-ray photoelectron spectroscopy (XPS) and X-ray absorption fine structure (EXAFS) calculations may help in the analysis of adsorption processes and mechanisms (Li et al. 2019; Luo et al. 2015).

Many mathematical representations have been used to explain the procedure of adsorption including the equations of kinetics like Elovich model, Ritchie second-order model, pseudo-first-order, pseudo-second-order and isothermal representation like Dubinin-Radushkevich, Freundlich and Langmuir. According to Gyliene et al. (2004), the Ni citrate complexes are removed from the solution of electroless nickel plating by adsorbing the NH_2 and OH groups onto chitosan. By using the equilibrium model, the adsorption of Cu-citrate complexes onto chitosan which makes it possible to understand the process of the mechanism of Cu adsorption at a varying ratio of Cu-citrate and pH conditions of solutions (Lu et al. 2017).

6.1.1 Types of Adsorbents

A wide range of adsorbents have been studied that have multiple characteristics features like proper pore and structure, surface area, easy to make and regenerate and better mechanical properties. The most commonly used adsorbents include biomaterials, carbon materials, the gel of silica, molecular sieves and naturally occurring clay (Abdolali et al. 2016; Cai et al. 2017; Habiba et al. 2017; Liu et al. 2014). Around 58% of adsorbents studies have reported the most common biological adsorbents that are applied for the removal of heavy metals which are presented in Table 3. Recently, many of the scientists use the biological product which has less toxicity, has chemical stability and is biodegradable like wastages of agriculture, waste crop materials, etc. These materials are used as absorbents to remove heavy metals from wastewater (Li et al. 2016; Tang et al. 2017).

6.2 Membrane Separation Method

It is one of the best methods and has an excellent capacity to remove heavy metal ions from contaminated water. The ultrafiltration membranes have many large pores with high permeability and do not trap the ions of heavy metal because of its small size. So, with the combination of water-soluble macromolecular polymers or other metal complexes, the ions of metals can be trapped easily. The association of

Table 3 Different types of adsorbent methods

S. No.	Adsorbents	Complexes of metal	pH	Isothermal models	Capacity of adsorption	References
1	Natural biofilms	Cd-EDTA	6	–	–	Hu et al. (2013)
2	MMIC-Fe (III)	Cd-tetracycline	8	Langmuir	194.3–516.3 mg/g	Chen et al. (2017)
3	Biochar	Cu-HA/FA	5–7	Freundlich	29.9/15.5 mg/g	He et al. (2017)
4	Sawdust biochar	Cu(II)/tetracycline	6	Langmuir	–	Zhou et al. (2017)
5	Chitosan	Cu-citrate	4–5	–	3–3.3 mmol/g	Guzman et al. (2003)
6	Chitosan	Cu-EDTA	3–5	–	–	Gylienė et al. (2006)
7	Chitosan	Cu-tartrate, Cu-glycine	>6	Freundlich, Langmuir	1.99/1.8/0.42 mmol/g	Gylienė et al. (2009)
8	Chitosan	Cd-tetracycline	3.5–6	Langmuir	1486.20 mmol/kg	Kang et al. (2010)
9	Bifunctional adsorbent	Cu-tetracycline	3–6	Langmuir	–	Ma et al. (2014)

polymer ultrafiltration processes have been widely used to treat the contaminated water with ions of metal, with citric acid chelation, and make a ternary polyethyleneimine (PEI)-Cu-citrate complex at the 5.5 pH which is allowing water purification and metal recovery because PEI as an N-rich artificial polymer have an ability to bind the ions of metal with a strong chelating effect (Molinari et al. 2006, 2007). Similarly, many other findings reported that metal removal rates of above 98% using PEI in the pH range of 4–9, from Cu(II) or Ni(II)-nitrilotriacetic acid and ethylenediaminetetraacetic acid complexes (Zamariotto et al. 2010). In the natural aquatic environment and contaminated water, the organic matter is more and combined with heavy metals that are increasing the much more serious problems to the environment. Moreover, very successful separation of heavy metals from wastewater using membrane ultrafiltration techniques has been studied (Munari and Schäfer 2010).

Recently, Gao et al. (2018) studied that use of membrane complexion ultrafiltration method for the recovery of Ni from metal-contaminated water, and for this study they used sodium polyacrylate as a complexation agent. It is very low cost and an appropriate way for the metal recovery that is also a highly ecologically standard method. The advantages of membrane ultrafiltration method are high efficiency and selectivity. However, various limitations have restricted its broad application such as high membrane preparation costs, ease of membrane blocking, inability to stretch membranes and the high requirement for additional complex agents.

6.3 Electrolysis Methods

Electrolysis is one of the earliest applied water treatment technologies and is widely used for the basic removal of heavy metals from industrial wastewater, due to its simple treatment principles, easy operation and ability to withstand the impact of large water volumes and high concentration wastewater.

Previous studies have comprehensively assessed the removal of heavy metals by electrolysis (Colantonio and Kim 2016; Tran et al. 2017). The basic mechanism of electrolysis in industrial wastewater utilizes the electrochemical properties of the contaminant metal. When exposed to an external direct current, heavy metal ions are separated from the high concentration solution and deposited on the cathode, resulting in the removal of harmful heavy metals from wastewater. Simultaneously, the reductive strong ions or anode material is discharged at the anode. Gyliené et al. (2004) reported a method using EDTA-metal (Ni(II), Zn(II), Co(II), Cd(II), Ca(II), Mg(II)) treatment, where metal ions in EDTA complexes are replaced by Cu(II), forming an insoluble $Cu_2EDTA{\cdot}4H_2O$ compound capable of being removed by electrolysis. Eivazihollagh et al. (2019) studied the simultaneous electrochemical recovery of copper, chelating agents (i.e. NTA, EDTA and DTPA) and a surface-active derivative of DTPA from alkaline solutions in an electrolysis membrane cell. The recovery mechanism was described with electrodeposition of metal ions occurring in the membrane electrolysis cell. Furthermore, the anode and cathode chambers were separated with a cation exchange membrane to prevent anodic oxidation (destruction) of chelating ligands during treatment, allowing recovery of chelating ligands for further reuse and simultaneous collection of metals by electrodeposition. However, to date, simplistic electrolysis methods have not been widely applied to the treatment of complex heavy metals.

The use of electrolysis methods for the removal of heavy metal complexes has the advantages of being convenient, easy to operate, effective with high concentration wastewater and having a large processing capacity. Nonetheless, the use of electrolysis to remove complex metal pollution remains an expensive approach because of the electrical energy required and the consumption of soluble anode materials. Furthermore, the electrodes are easily passivated, as this method is not suitable for the treatment of low-concentration complex heavy metal wastewater and the production of residual organic ligands leads to secondary pollution. Therefore, it is more feasible for use as a primary treatment. Based on the findings of recent studies, it appears that many combination methods using electrolysis and other means exhibit superior effects in the electrolytic removal of heavy metals. These combined methods include Fe-C micro-electrolysis, microbial electrolysis cells, electrocoagulation, bioelectrochemical systems and electrolysis as a clean technology option (Chen et al. 2012; Luo et al. 2014; Tao et al. 2014).

Alternatively, the removal of EDTA-chelated copper by interior micro-electrolysis has also been studied, which differs from electrolysis as it does not require an external power supply and exhibits excellent properties for the pretreatment of EDTA-chelated metals, due to its high efficiency and short reaction times (Gyliene et al. 2004). The combination of electrolysis with other methods will overcome some methodological shortcomings and further optimize removal efficiency. However, the high level of consumption of electrical energy and electrode materials, as well as the incomplete removal of organic matter in complex systems, may continue to limit the wider application of electrolysis.

6.4 Photocatalysis Methods

With the increasing attention on new energy sources, the use of renewable energy sources, such as solar energy, has become a key research focus. Photocatalytic oxidation has become a main research direction in recent years, in the remediation of heavy metal complexes from wastewater. Photocatalysis has the advantage of strong oxidation capability, allowing the destruction of heavy metal complexes and liberation of heavy metal ions, while organic complexes can be oxidized and degraded. Common photocatalysts include TiO_2, zinc oxide (ZnO) and tin (IV) oxide (SnO_2) (Anju et al. 2012; Marinho et al. 2017). TiO_2 photocatalysis has been extended over the last 10 years for the removal of organic pollutants and complex metals (Lin et al. 2016). Photocatalytic degradation techniques using TiO_2 are generally used for the treatment of wastewater containing heavy metal complexes, due to its ability to achieve complete mineralization of organic ligands under mild conditions (such as room temperature and atmospheric pressure). TiO_2 nanotubes have been developed for use as photoanodes to degrade Cu-cyanide and

Cu-EDTA complexes, at various [EDTA]:[CN-] ratios. The amount of Cu deposited onto the anode decreases in accordance with an increase in Cu deposition onto the cathode. Salama and Berk (2005) reported the use of a well-mixed reactor for the efficient degradation of Ni-EDTA by photocatalytic oxidation, which has significant potential for large-scale industrial application. With ongoing development, the focus of research has increasingly become the improvement of practical application prospects and reduction of operational costs. Therefore, Cho et al. (2007) studied the removal of the Cu(II)-EDTA complex using a combined TiO_2/solar light system, as well as investigating the effect of varying operational parameters and the feasibility of a solar light application. In the presence of H_2O_2 as an oxidant and OH source, this combined method was effective at increasing the rates of Cu(II)-EDTA removal. The results suggested that the TiO_2/solar system could be effectively applied in the treatment of metal-ligand contaminated wastewater in the natural environment.

Yang and Davis (2001) proposed the mechanism of photocatalytic oxidation of Cu(II)-EDTA with illuminated TiO_2 and then later compared the competitive photocatalytic oxidation of Cu(II)-EDTA and Cd(II)-EDTA with illuminated TiO_2. In these studies, it was demonstrated that TiO_2 can destroy the structure of metal complexes and further degrade EDTA, removing copper ions. The principle of TiO_2 photocatalysis is the use of a semiconductor to stimulate the generation of hydroxyl radicals ($^{\bullet}OH$) from photogenerated holes (h^+), to achieve oxidative degradation of heavy metal complexes. The mechanism is based on the example of Cu(II) organic matter (Saratale Rijuta et al. 2014; Xu et al. 2014), where photoexcitation generates electrons (e^-) and holes reacting with adsorbed electron acceptors and donors, respectively. The positive holes produced can then react with surface-bound hydroxyl groups, generating $^{\bullet}OH$ radicals that are involved in the oxidation of EDTA. Released metal ions can then be reduced by free electrons, which precipitate at the anode and can be recovered finally.

Currently, photocatalysis is an increasingly applied environmentally friendly technique that is highly suitable for the treatment of chelated heavy metal-containing wastewater. The popularity of this technique is due to its ability to achieve complete mineralization of organic ligands under ambient temperature and atmospheric pressure conditions while generating low-toxicity by-products (Chowdhury et al. 2018). Photocatalysis can degrade intricate organic complexes to nontoxic CO_2/NH_3 or other low molecular weight organic acids while assisting the decomplexation of metal ions and subsequent removal of free metal ions via adsorption or precipitation onto TiO_2. This highly efficient and rapid treatment method is both economical and has no secondary pollution implications. However, electrons and holes can recombine easily, limiting the practical use of TiO_2. The use of modified TiO_2 or other additional methods can significantly enhance the effectiveness and potential applications of photocatalytic systems. Therefore, wastewater treatment using TiO_2 photocatalytic reactor systems has been widely applied due to its simplicity and high efficiency degradation capability. Although current research into this method is still in the laboratory stage, practical and industrial applications are likely in the future with further method optimization.

7 New Emerging Technology in Wastewater Treatment

7.1 Biotechnological Improvements

Treatment of industrial wastewater uses processes similar to those described above. Therefore, any biotechnological improvements to these processes are likely to have immediate industrial application.

Biotechnological improvements may include the following

1. An increase in the capacity of treatment plants
2. Increased recovery of useful by-products
3. Replacement of synthetic chemical additives that are currently used
4. Removal of metals, recalcitrant compounds and odour

7.2 Advantages and Limitations in the Application of These Techniques

Physicochemical treatments generally offer different advantages, including fast phase, simple operation and control and adjustable temperature change. In physicochemical treatment, unlike in the biological system, the variable input and flow charges can be managed, for example, seasonal flows and complex discharge. Whenever it is required, chemical plants can be modified. In addition, the treatment system requires a lower space and installation cost. Their benefits, however, are outweighed by a number of drawbacks such as their high operational costs due to the chemicals used, high energy consumption and handling costs for sludge disposal. However, with reduced chemical costs (such as utilizing of low-cost adsorbents) and feasible sludge disposal, physicochemical treatments have been found as one of the most suitable treatments for inorganic effluent (Kurniawan et al. 2006).

In wastewater systems containing heavy metals with other organic pollutants, the presence of one species usually impedes the removal of the other. For instance, hydrometallurgy, a classical process to recover metals, is inhibited by the presence of organic compounds, and a pretreatment step, to remove or destroy organics, is generally required, pyrometallurgy which is able to decontaminate systems from lack of controllability, demanding extremely high temperatures. The most promising methods to treat such complex systems are the photocatalytic ones which consume cheap photons from the UV-near visible region. These photocatalysts serve as electron relays, from the organic substrates to metal ions. Thus, they induce both degradation of organic pollutants and recovery of metals in one-pot systems, operable at traces of the target compounds (less than ppm). Table 4 summarizes the main advantages and limitations of the various physicochemical treatments presented in this study.

Table 4 The key advantages and disadvantages of the different physicochemical approaches of wastewater treatment of heavy metal

S. No.	Treatment techniques	Advantages	Limitations	References
1.	Adsorption	Low cost, ease of use, large pH range and high metal connectivity capacity	Low selectivity, waste product generation	Babel and Kurniawan (2003), Aklil et al. (2004)
2.	Membrane filtration	Low pressure, high separation selectivity, small space requirement	High operating costs due to fouling of the membrane	Kurniawan et al. (2006)
3.	Electrodialysis	Selectivity for high separation	High operating costs due to fouling of membranes and consumption of energy	Mohammadi et al. (2005)
4.	Photocatalysis	Simultaneous, less harmful, removal of metals and organic pollutants	Long-term, minimal implementations	Kajitvichyanukula et al. (2005), Barakat et al. (2004).

8 Conclusion and Future Prospects

Although many techniques can be used to treat sewage loaded with heavy toxic metals, it is important to note that the selection of the most appropriate treatment methods for sewage contaminated with metals depends on some basic parameters such as pH, the initial concentration of the metal general performance treatment compared to economic parameters such as other technology, environmental impact as well as capital investment and operating costs. The selection and implementation of biosorbents for industrial waste management and soil remediation require much effort, as most of the reported adsorption studies have been limited to laboratory investigations in a batch system. It is necessary to develop a sustainable approach to select the most suitable biosorbent, operating conditions and efficient heavy metal removal systems in industrial flows, adequately addressing the main challenges involved. Furthermore, to develop a reliable biosorption process, more research is needed on the characterization of biosorbents in terms of surface and area morphology, zeta potential, functional groups and particle size, as these are important in bio-surgeon experiments, influenced by the hypocrisy of biosorbents. Similarly, further research is needed to develop microbial biomass with metal absorption capacity to explore the metabolic potential of this growing biomass and its application in the management of industrial wastewater. Biofilm mediation techniques, microbial gene transfer and microbial fuel cell-based techniques have emerged as strong contenders in recent years. The peptidoglycan and polysaccharide component of the biosorbent cell wall is an active binding site for greater metal absorption. In conclusion, technical applicability, plant simplicity and profitability and green technology are the major factors that play an important role in the selection of the most suitable treatment system for inorganic waste. All of the above factors should be taken into consideration when choosing the most effective and economical treatment to protect the environment. Treatment of heavy metals requires expanding the field of research on gene transfer approaches within biofuels. This will facilitate the development of improved techniques for bioremediation of heavy metals in the ecosystem and environmental research.

Conflict of Interest None of the authors has any conflict of interest.

References

Abdolali A, Ngo HH, Guo WS, Lu SY, Chen SS, Nguyen NC, Zhang XB, Wang J, Wu Y (2016) A breakthrough biosorbent in removing heavy metals: equilibrium, kinetic, thermodynamic and mechanism analyses in a lab-scale study. Sci Total Environ 542:603–611

Abu Hasan H, Abdullah SRS, Kofli NT, Kamarudin SK (2012) Isotherm equilibria of Mn2þ biosorption in drinking water treatment by locally isolated Bacillus species and sewage activated sludge. J Environ Manag 111:34–43

Ahamd G, Nishat Y, Haris M, Danish M, Hussain T (2019) Efficiency of soil, plant and microbes for the healthy plant immunity and sustainable agricultural system. In: Varma A, Tripathi S, Prasad R (eds) Plant-microbe interface. Springer, Cham., ISBN: 978-3-03019830-5. https://doi.org/10.1007/978-3-030-19831-2_15

Akasaka M, Takamura N, Mitsuhashi H, Kadono Y (2010) Effects of land use on aquatic macrophyte diversity and water quality of ponds. Freshw Biol 55(4):909–922

Akcil A, Erust C, Ozdemiroglu S, Fonti V, Beolchini F (2015) A review of approaches and techniques used in aquatic contaminated sediments: metal removal and stabilization by chemical and biotechnological processes. J Clean Prod 86:24–36

Aklil A, Mouflihb M, Sebti S (2004) Removal of heavy metal ions from water by using calcined phosphate as a new adsorbent. J Hazard Mater A112:183–190

Aloma I, Martín-Lara M, Rodríguez IL, Blazquez G, Calero M (2012) Removal of nickel(II) ions from aqueous solutions by biosorption on sugar cane bagasse. J Taiwan Inst Chem Eng 43:275–281

Anju S, Yesodharan S, Yesodharan E (2012) Zinc oxide mediated sonophotocatalytic degradation of phenol in water. Chem Eng J 189:84–93

Ayangbenro AS, Babalola OO (2017) A new strategy for heavy metal polluted environments: a review of microbial biosorbents. Int J Environ Res Public Health 14:94. https://doi.org/10.3390/ijerph14010094

Babel S, Kurniawan TA (2003) Low-cost adsorbents for heavy metals uptake from contaminated water: a review. J Hazard Mater B97:219–243

Barakat MA, Chen YT, Huang CP (2004) Removal of toxic cyanide and Cu(II) ions from water by illuminated TiO_2 catalyst. J Appl Catal B Environ 53:13–20

Bates BC, Kundzewicz ZW, Wu S, Palutikof JP (2008) Climate change and water. Technical paper of the Intergovernmental Panel on Climate Change. IPCC Secretariat, Geneva

Bosso L, Lacatena F, Cristinzio G, Cea M, Diez MC, Rubilar O (2015) Biosorption of pentachlorophenol by anthracophyllum discolor in the form of live fungal pellets. New Biotechnol 32:21–25. https://doi.org/10.1016/j.nbt.2014.08.001

Brandt S, Zeng AP, Deckwer WD (1997) Adsorption and desorption of pentachlorophenol on cells of mycobacterium chlorophenolicum PCP-1. Biotechnol Bioeng 55:480–489

Cai J, Lei M, Zhan Q, He JR (2017) Electrospun composite nanofiber mats of cellulose@ organically modified montmorillonite for heavy metal ion removal: design, characterization, evaluation of absorption performance. Composites A 92:10–16

Chandra K, Salman AS, Mohd A, Sweety R, Ali KN (2015) Protection against FCA induced oxidative stress induced DNA damage as a model of arthritis and in vitro anti-arthritic potential of Costus speciosus rhizome extract. Int J Pharm Phytopharmacol Res 7:383–389

Chang Y, Deng L, Meng XY, Zhang W, Wang CZ, Wang YX, Zhao S, Lin L, Crittenden JC (2018) Closed-loop electrochemical recycling of spent copper(II) from etchant wastewater using a carbon nanotube modified graphite felt anode. Environ Sci Technol 52:1262–1269

Chapagain AK, Hoekstra AY (2008) The global component of freshwater demand and supply: an assessment of virtual water flows between nations as a result of trade in agricultural and industrial products. Water Int 33(1):19e32

Chen RH, Chai LY, Wang YY, Liu H, Shu YD, Zhao J (2012) Degradation of organic wastewater containing Cu-EDTA by Fe-C micro-electrolysis. Trans Nonferr Met Soc China 22:983–990

Chen XW, Fu J, Shao JH, Nguyen U, Zhou SS, He YL (2017) Simultaneous removal of humic acid and heavy metal from aqueous solutions using charged ultrafiltration membranes. Sep Sci Technol 52(11):1913–1919

Chibuike G, Obiora S (2014) Heavy metal polluted soils: effect on plants and bioremediation methods. Appl Environ Soil Sci. https://doi.org/10.1155/2014/752708

Cho IH, Lee NH, Yang JK, Lee SM (2007) Treatment of wastewater containing Cu(II)-EDTA using immobilized TiO_2/solar light. J Environ Sci Health A 42:165–170

Chowdhury IH, Kundu S, Naskar MK (2018) Template-free hydrothermal synthesis of MgO-TiO_2 microcubes toward high potential removal of toxic water pollutants. J Phys Chem Solids 112:171–178

Colantonio N, Kim Y (2016) Cadmium(II) removal mechanisms in microbial electrolysis cells. J Hazard Mater 311:134–141

Dinesh M, Ankur S, Yong SO, Charles U, Pittman J (2014) Organic and inorganic contaminants removal from water with biochar, a renewable, low cost and sustainable adsorbent-a critical review. Bioresour Technol 160:191–202

Dixit R, Malaviya D, Pandiyan K, Singh UB, Sahu A, Shukla R, Singh BP, Rai JP, Sharma PK, Lade H (2015) Bioremediation of heavy metals from soil and aquatic environment: an overview of principles and criteria of fundamental processes. Sustainability 7:2189–2212. https://doi.org/10.3390/su7022189

Eivazihollagh A, Svanedal I, Edlund H, Norgren M (2019) On chelating surfactants: molecular perspectives and application prospects. J Mol Liq 278:688–705

Ekperusi O, Aigbodion F (2015) Bioremediation of petroleum hydrocarbons from crude oil-contaminated soil with the earthworm: Hyperiodrilus africanus. 3 Biotech 5:957–965. https://doi.org/10.1007/s13205-015-0298-1

El-Shafey EI (2010) Removal of Zn(II) and Hg(II) from aqueous solution on a carbonaceous sorbent chemically prepared from rice husk. J Hazard Mater 175:319–327

FAO (2013) FAO statistical year book 2013, World Food and Agriculture. Food and Agriculture Organization of the United Nations, Rome, p 289

Gao J, Qiu YR, Hou B, Zhang Q, Zhang XD (2018) Treatment of wastewater containing nickel by complexation-ultrafiltration using sodium polyacrylate and the stability of PAA-Ni complex in the shear field. Chem Eng J 334:1878–1885

Garrido-Baserba M, Vinardell S, Molinos-Senante M, Rosso D, Poch M (2018) The economics of wastewater treatment decentralization: a techno-economic evaluation. Environ Sci Technol 52:8965–8976

Genç-Fuhrman H, Mikkelsen PS, Ledin A (2016) Simultaneous removal of As, Cd, Cr, Cu, Ni and Zn from stormwater using high-efficiency industrial sorbents: effect of pH, contact time and humic acid. Sci Total Environ 566–567:76–85

Girma G (2015) Microbial bioremediation of some heavy metals in soils: an updated review. Ind J Sci Res 6:147–161

Gossling S (2001) The consequences of tourism for sustainable water use on a tropical island: Zanzibar, Tanzania. J Environ Manag 61(2):179–191
Guo W, Chen R, Liu Y, Meng M, Meng X, Hu Z et al (2013) Preparation of ion-imprinted meso porous silica SBA-15 functionalized with triglycine for selective adsorption of Co(II). Colloids Surf A Physicochem Eng Asp 436:693–703
Gupta A, Joia J, Sood A, Sood R, Sidhu C, Kaur G (2016) Microbes as potential tool for remediation of heavy metals: a review. J Microb Biochem Technol 8:364–372. https://doi.org/10.4172/1948-5948.1000310
Guzman J, Saucedo I, Revilla J, Navarro R, Guibal E (2003) Copper sorption by chitosan in the presence of citrate ions: influence of metal speciation on sorption mechanism and uptake capacities. Int J Biol Macromol 33:57–65
Gyliene O, Aikaite J, Nivinskiene O (2004) Recovery of EDTA from complex solution using Cu (II) as precipitant and Cu(II) subsequent removal by electrolysis. J Hazard Mater 116:119–124
Gyliené O, Tarozaité R, Nivinskiené O (2004) Sorption of Ni(II)-citrate complex from electroless nickel plating solutions onto chitosan. Trans Inst Met Finish 82:38–42
Gylienė O, Nivinskienė O, Razmutė I (2006) Copper(II)-EDTA sorption onto chitosan and its regeneration applying electrolysis. J Hazard Mater 137:1430–1437
Gylienė O, Binkienė R, Butkienė R (2009) Sorption of Cu(II) complexes with ligands tartrate, glycine and quadrol by chitosan. J HazardMater 171:133–139
Habiba U, Afifi AM, Salleh A, Ang BC (2017) Chitosan/(polyvinyl alcohol)/zeolite electrospun composite nanofibrous membrane for adsorption of Cr6+, Fe3+ and Ni2+. J Hazard Mater 322:182–194
Hameed MSA (2006) Continuous removal and recovery of lead by alginate beads free and alginate-immobilized Chlorella vulgaris. Afr J Biotechnol 5:1819–1823
Hamer DH (1986) Metallothionein. Annu Rev Biochem 55:913–951
He PJ, Yu QF, Zhang H, Shao LM (2017) Removal of copper(II) by biochar mediated by dissolved organic matter. Sci Rep 7:1–10
Hoekstra AY, Hung PQ (2002) Virtual water trade: a quantification of virtual water flows between nations in relation to international crop trade, Value of water research report series, vol 11. UNESCO-IHE, Delft, the Netherlands
Hu XY, Hu JR, Jiang X, Dong DM (2013) Adsorption of Cd to natural biofilms in the presence of EDTA: effect of pH, concentration, and component addition sequence. Environ Sci Pollut Res 20:1079–1088
Huckle JW, Morby AP, Turner JS, Robinson NJ (1993) Isolation of a prokaryotic metallothionein locus and analysis of transcriptional control by trace metal ions. Mol Microbiol 7:177–187
Huston WM, Jennings MP, McEwan AG (2002) The multi-copper oxidase of Pseudomonas aeruginosa is a ferroxidase with a central role in iron acquisition. Mol Microbiol 45:1741–1750
Infante JC, De Arco RD, Angulo ME (2014) Removal of lead, mercury and nickel using the yeast Saccharomyces cerevisiae. Rev MVZ Córdoba 19:4141–4149
IPCC (2007) Climate change 2007: impacts, adaptation and vulnerability. In: Parry ML, Canziani OF, Palutikof JP, van der Linden PJ, Hanson CE (eds) Contribution of working group II to the fourth assessment report of the Intergovernmental Panel on Climate Change. Cambridge University Press, Cambridge
Iqbal M, Saeed A, Edyvean RGJ (2013) Bioremoval of antimony(III) from contaminated water using several plant wastes: optimization of batch and dynamic flow conditions for absorption by green bean husk (Vigna radiata). Chem Eng J 225:192–201
Jachuła J, Kołodyńska D, Hubicki Z (2011) Sorption of Cu(II) and Ni(II) ions in the presence of the methylglycinediacetic acid by microporous ion exchangers and sorbents from aqueous solutions. Cent Eur J Chem 9:52–65
Jain PK, Gupta V, Gaur RK, Bajpai V, Gautama N, Modi DR (2010) Fungal enzymes: potential tools of environmental processes. In: Gupta VK, Tuohy M, Gaur RK (eds) Fungal biochemistry and biotechnology. LAP Lambert Academic Publishing AG and Co. KG, Germany, pp 44–56

Jain PK, Gupta VK, Bajpai V, Lowry M, Jaroli DP (2011) GMO's: perspective of bioremediation. In: Jain PK, Gupta VK, Bajpai V (eds) Recent advances in environmental biotechnology. LAP Lambert Academic Publishing AG and Co. KG, Germany, pp 6–23

Jain AN, Udayashankara TH, Lokesh KS (2012) Review on bioremediation of heavy metals with microbial isolates and amendments on soil residue. Int J Sci Res 6:2319–7064

Jennings D (1993) Stress tolerance of fungi, vol 10. M. Dekker, New York

Jianlong W, Chen C (2009) Biosorbents for heavy metals removal and their future. Biotechnol Adv 27:195–226

Jianlong W, Yi Q, Horan N, Stentiford E (2000) Bioadsorption of pentachlorophenol (PCP) from aqueous solution by activated sludge biomass. Bioresour Technol 75:157–161. https://doi.org/10.1016/S0960-8524(00)00041-9

Johnson H, Cho H, Choudhary M (2019) Bacterial heavy metal resistance genes and bioremediation potential. Comput Mol Biosci 9:1–12

Kabata-Pendias A (2010) Trace elements in soils and plants. CRC Press, New York, NY

Kajitvichyanukula P, Ananpattarachaia J, Pongpom S (2005) Sol–gel preparation and properties study of TiO_2 thin film for photocatalytic reduction of chromium(VI) in photocatalysis process. Sci Technol Adv Mater 6:352–358

Kang J, Liu HJ, Zheng YM (2010) Systematic study of synergistic and antagonistic effects on adsorption of tetracycline and copper onto a chitosan. J Colloid Interface Sci 344:117–125

Kang CH, Kwon YJ, So JS (2016) Bioremediation of heavy metals by using bacterial mixtures. Ecol Eng 89:64–69. https://doi.org/10.1016/j.ecoleng.2016.01.023

Kennish MJ (1992) Ecology of estuaries: anthropogenic effect. CRC Press, Boca Raton, FL

Kenny RF, Barber NL, Hutson SS, Linsey KS, Lovelace JK, Maupin MA (2009) Estimated use of water in the United States in 2005. US Geol Surv Circular 1344:52p

Kurniawan TA, Chan GYS, Lo WH, Babel S (2006) Physico-chemical treatment techniques for wastewater laden with heavy metals. Chem Eng J 118:83–98

Lata S, Kaur HP, Mishra T (2019) Cadmium bioremediation: a review. Int J Pharm Sci Res 10:4120–4128

Le TT, Son MH, Nam IH, Yoon H, Kang YG, Chang YS (2017) Transformation of hexabromocyclododecane in contaminated soil in association with microbial diversity. J Hazard Mater 325:82–89

Lee JC, Son YO, Pratheesh Kumar P, Shi XL (2012) Oxidative stress and metal carcinogenesis. Free Radic Biol Med 53(4):742–757

Li BB, Zhou F, Huang K, Wang YP (2016) Highly efficient removal of lead and cadmium during wastewater irrigation using a polyethylenimine grafted gelatin sponge. Sci Rep 6:1–9

Li QM, Song HO, Han RM, Wang GX, Li AM (2019) Efficient removal of Cu(II) and citrate complexes by combined permanent magnetic resin and its mechanistic insights. Chem Eng J 366:1–10

Lin YH, Hsueh HT, Chang CW, Chu H (2016) The visible light-driven photodegradation of dimethyl sulfide on S-doped TiO_2: characterization, kinetics, and reaction pathways. Appl Catal B Environ 199:1–10

Liu L, Zhang K (2018) Nanopore-based strategy for sequential separation of heavy-metal ions in water. Environ Sci Technol 52:2007–2015

Liu DG, Li ZH, Zhu Y, Li ZX (2014) Recycled chitosan nanofibril as an effective Cu(II), Pb(II) and Cd(II) ionic chelating agent: adsorption and desorption performance. Carbohydr Polym 111:469–476

Liu L, Li RZ, Liu Y, Zhang JD (2016) Simultaneous degradation of ofloxacin and recovery of Cu (II) by photoelectrocatalysis with highly ordered TiO_2 nanotubes. J Hazard Mater 308:264–275

Lu Y, Song S, Wang R, Liu Z, Meng J, Sweetman AJ, Jenkins A, Ferrier RC, Li H, Luo W, Wang T (2015) Impacts of soil and water pollution on food safety and health risks in China. Environ Int 77:5–15

Lu X, Zhang DX, Red AT, Liu C, Yang Z, Guo SS, Xiao ST, Ouyang YG (2017) Synthesis of amidoxime-grafted activated carbon fibers for efficient recovery of uranium(VI) from aqueous solution. Ind Eng Chem Res 56:11936–11947

Luo HP, Liu GL, Zhang RD, Bai YP, Fu SY, Hou YP (2014) Heavy metal recovery combined with H2 production from artificial acid mine drainage using the microbial electrolysis cell. J Hazard Mater 270:153–159

Luo JM, Luo XB, Crittenden J, Qu JH, Bai YH, Peng Y, Li JH (2015) Removal of antimonite (Sb (III)) and antimonate (Sb(V)) from aqueous solution using carbon nanofibers that are decorated with zirconium oxide (ZrO_2). Environ Sci Technol 49:11115–11124

Ma Y, Zhou Q, Zhou SC, Wang W (2014) A bifunctional adsorbent with high surface area and cation exchange property for synergistic removal of tetracycline and Cu_2^+. Chem Eng J 58:26–33

Mani S (2015) Production of reactive oxygen species and its implication in human diseases. In: Vibha R, Umesh CSY (eds) Free radicals in human health and disease. Springer, New Delhi, India, pp 3–15

Mani D, Kumar C (2014) Biotechnological advances in bioremediation of heavy metals contaminated ecosystems: an overview with special reference to phytoremediation. Int J Environ Sci Technol 11:843–872

Marinho BA, Cristóvão RO, Djellabi R, Loureiro JM, Boaventura RAR, Vilar VJP (2017) Photocatalytic reduction of Cr(VI) over TiO_2-coated cellulose acetate monolithic structures using solar light. Appl Catal B Environ 203:18–30

Marshall FM, Holden J, Ghose C, Chisala B, Kapungwe E, Volk J et al (2007) Contaminated irrigation water and food safety for the urban and peri-urban poor: appropriate measures for monitoring and control from field research in India and Zambia, Inception report DFID Enkar R8160. SPRU, University of Sussex, Falmer

Mathialagan T, Viraraghavan T (2009) Biosorption of pentachlorophenol from aqueous solutions by a fungal biomass. Bioresour Technol 100:549–558. https://doi.org/10.1016/j.biortech.2008.06.054

Mohammadi T, Mohebb A, Sadrzadeh M, Razmi A (2005) Modeling of metal ion removal from wastewater by electrodialysis. Sep Purif Technol 41:73–82

Molinari R, Argurio P, Poerio T (2006) Ultrafiltration of polymer-metal complexes for metal ion removal from wastewaters. Macromol Symp 235:206–214

Molinari R, Teresa P, Argurio P (2007) Chemical and operational aspects in running the polymer assisted ultrafiltration for separation of copper(II)-citrate complexes from aqueous media. J Membr Sci 295:139–147

Munari AD, Schäfer AI (2010) Impact of speciation on removal of manganese and organic matter by nanofiltration. J Water Supply Res Technol AQUA 59:152–163

Nematian MA, Kazemeini F (2013) Accumulation of Pb, Zn, C and Fe in plants and hyperaccumulator choice in galali iron mine area, Iran. Int J Agric Crop Sci 5:426

Nies DH (1999) Microbial heavy metal resistance. Appl Microbiol Technol 51:730–750. https://doi.org/10.1007/s002530051457

Novotny C, Erbanova P, Sasek V, Kubatova A, Cajthman T, Lang E, Krahl J, Zadrazil F (1999) Extracellular oxidative enzyme production and PAH removal in soil by exploratory mycelium of white rot fungi. Biodegradation 10:159–168

O'Connell DW, Birkinshaw C, O'Dwyer TF (2008) Heavy metal adsorbents prepared from the modification of cellulose: a review. Bioresour Technol 99(15):6709–6724

Ojuederie OB, Babalola OO (2017) Microbial and plant-assisted bioremediation of heavy metal polluted environments: a review. Int J Environ Res Public Health 14(12):1504. https://doi.org/10.3390/ijerph14121504

Owolabi JB, Hekeu MM (2014) Heavy metal resistance and antibiotic susceptibility pattern of bacteria isolated from selected polluted soils in Lagos and Ota, Nigeria. Int J Basic Appl Sci 4 (6):6–12

Oyewo OA, Agboola O, Onyango MS, Popoola P, Bobape MF (2018) Chapter 6 - Current methods for the remediation of acid mine drainage including continuous removal of metals from wastewater and mine dump. In: Prasad MNV, de Campos Favas PJ, Maiti SK (eds) Bio-geotechnologies for mine site rehabilitation. Elsevier, Amsterdam, pp 103–114

Parry M, Arnell N, Berry P, Dodman D, Fankhauser S, Hope C et al (2009a) Assessing the costs of adaptation to climate change: a review of the UNFCCC and other recent estimates. International Institute for Environment and Development and Grantham Institute for Climate Change, London

Parry M, Lowe J, Hanson C (2009b) Overshoot, adapt and recover. Nature 485:1102e1103

Pigram JJJ (1995) Resource constraints on tourism: water resources and sustainability. In: Butler RW, Pearce D (eds) Change in tourism: people, places, processes. Routledge, London, pp 208–228

Rossotti JC, Rossotti H (1961) The determination of stability constants. McGraw-Hill, New York

Salama P, Berk D (2005) Photocatalytic oxidation of Ni-EDTA in a well-mixed reactor. Int J Environ Res 44:7071–7077

Saratale Rijuta G, Noh HS, Song JY, Kim DS (2014) Influence of parameters on the photocatalytic degradation of phenolic contaminants in wastewater using TiO2/UV system. J Environ Sci Health A 49:1542–1552

Sarý A, Uluozlü OD, Tüzen M (2011) Equilibrium, thermodynamic and kinetic investigations on biosorption of arsenic from aqueous solution by algae (*Maugeotia genuflexa*) biomass. Chem Eng J 167:155–161

Schenk PM, Carvalhais LC, Kazan K (2012) Unraveling plant microbe interactions: can multi-species transcriptomics help? Trends Biotechnol 30(3):177–184

Singh A, Panda SN (2012) Effect of saline irrigation water on mustard (Brassica juncea) crop yield and soil salinity in a semi-arid area of North India. Exp Agric 48(1):99–110

Soares EV, Soares HM (2012) Bioremediation of industrial effluents containing heavy metals using brewing cells of Saccharomyces cerevisiae as a green technology: a review. Environ Sci Pollut Res 19:1066–1083

Son EB, Poo KM, Chang JS, Chae KJ (2018) Heavy metal removal from aqueous solutions using engineered magnetic biochars derived from waste marine macro-algal biomass. Sci Total Environ 615:161–168

Srivastava NK, Majumder CB (2008) Novel biofiltration methods for the treatment of heavy metals from industrial wastewater. J Hazard Mater 151(1):1–8

Tak HI, Ahmad F, Babalola OO (2013) Advances in the application of plant growth-promoting rhizobacteria in phytoremediation of heavy metals. In: Whitacre DM (ed) Reviews of environmental contamination and toxicology, vol 223. New York, Springer, pp 33–52

Tang J, He JG, Liu TT, Xin XD (2017) Removal of heavy metals with sequential sludge washing techniques using saponin: optimization conditions, kinetics, removal effectiveness, binding intensity, mobility and mechanism. RSC Adv 7:33385–33401

Tao HC, Lei T, Shi G, Sun XN, Wei XY, Zhang LJ, Wu WM (2014) Removal of heavy metals from fly ash leachate using combined bioelectrochemical systems and electrolysis. J Hazard Mater 264:1–7

Thompson KA, Shimabuku KK, Kearns JP, Knappe DRU, Scott Summers R, Cook SM (2016) Environmental comparison of biochar and activated carbon for tertiary wastewater treatment. Environ Sci Technol 50:11253–11262

Tong XJ, Li JY, Yuan JH, Xu RK (2011) Adsorption of Cu(II) by biochars generated from three crop straws. Chem Eng J 172:828–834

Tran TK, Leu HJ, Chiu KF, Lin CY (2017) Electrochemical treatment of heavy metal containing wastewater with the removal of COD and heavy metal ions. Chin Chem Soc 64:493–502

Tunali Akar S, Arslan S, Alp T, Arslan D, Akar T (2012) Biosorption potential of the waste bio material obtained from Cucumis melo for the removal of Pb2þ ions from aqueous media: equilibrium, kinetic, thermodynamic and mechanism analysis. Chem Eng J 185–186:82–90

UNESCO (2009) United Nations world water development report 3. Water in a changing world. UNESCO, Paris

United Nations (2012) World population prospects: 2012 Revision Population Database. http://www.un.org/esa/population/unpop.htm. Accessed 19 July 2013

Verma JP, Jaiswal DK (2016) Book review: advances in biodegradation and bioremediation of industrial waste. Front Microbiol 6:1555. https://doi.org/10.3389/fmicb.2015.01555

Vinodhini V, Das N, Vinodhini V (2010) Relevant approach to assess the performance of saw dust as adsorbent of chromium(VI) ions from aqueous solutions. Int J Environ Sci Technol 7:85–92

Wang J, Chen C (2006) Biosorption of heavy metals by Saccharomyces cerevisiae: a review. Biotechnol Adv 24:427–451

Wang J, Chen C (2009) Biosorbents for heavy metals removal and their future. Biotechnol Adv 27:195–226. https://doi.org/10.1016/j.biotechadv.2008.11.002

Wang XQ, Liu WX, Tian J, Zhao ZH, Hao P, Kang XL, Sang YH, Liu H (2014) Cr(VI), Pb(II), Cd (II) adsorption properties of nanostructured BiOBr microspheres and their application in a continuous filtering removal device for heavy metal ions. J Mater Chem A 2:2599–2608

Wang TC, Cao Y, Qu GZ, Sun QH, Xia TJ, Guo XT, Jia HZ, Zhu LY (2018) Novel Cu(II)-EDTA decomplexation by discharge plasma oxidation and coupled Cu removal by alkaline precipitation: underneath mechanisms. Environ Sci Technol 52:7884–7891

Worldwatch Institute (2004) Rising impacts of water use

Wu P, Jiang LY, He Z, Song Y (2017) Treatment of metallurgical industry wastewater for organic contaminant removal in China: status, challenges, and perspectives. Environ Sci Water Res Technol 3:1015–1031

Xu H, Ouyang SX, Liu LQ, Reunchan P, Umezawaace N, Ye JH (2014) Recent advances in TiO_2-based photocatalysis. J Mater Chem A 2:12642–12661

Xu Z, Gao GD, Pan BC (2015) A new combined process for efficient removal of Cu(II) organic complexes from wastewater: Fe(III) displacement/UV degradation/alkaline precipitation. Water Res 87:378–384

Xu Z, Shan C, Xie BH, Liu Y, Pan BC (2017) Decomplexation of Cu(II)-EDTA by UV/persulfate and UV/H_2O_2: efficiency and mechanism. Appl Catal B Environ 200:439–447

Yang JK, Davis AP (2001) Competitive photocatalytic oxidation of Cu(II)-EDTA and Cd(II)-EDTA with illuminated TiO_2. Environ Sci Technol 35:3566–3570

Ye MY, Li GJ, Yan PF, Ren J, Zheng L, Han DJ, Sun SY, Huang SS, Zhong YJ (2017) Removal of metals from lead-zinc mine tailings using bioleaching and followed by sulfide precipitation. Chemosphere 185:1189–1196

Zamariotto D, Lakard B, Fievet P, Fatin-Rouge N (2010) Retention of Cu(II)- and Ni(II)-polyaminocarboxylate complexes by ultrafiltration assisted with polyamines. Desalination 258:87–92

Zhou YY, Zhang FF, Tang L, Zhang JC, Zeng GM, Luo L, Liu YY, Wang P, Peng B, Liu XC (2017) Simultaneous removal of atrazine and copper using polyacrylic acid functionalized magnetic ordered mesoporous carbon from water: adsorption mechanism. Sci Rep 7:1–10

Zhou L, Li N, Owens G, Chen ZL (2019) Simultaneous removal of mixed contaminants, copper and norfloxacin, from aqueous solution by ZIF-8. Chem Eng J 362:628–637

Biological Wastewater Treatment Technology: Advancement and Drawback

Ifra Zoomi, Harbans Kaur Kehri, Ovaid Akhtar, Dheeraj Pandey, Uma Singh, Kanhaiya Lal Chaudhary, and Raghvendra Pratap Narayan

Abstract

Rapid population growth, industrialization and long-term droughts result in the occurrence of a wide range of contaminant in the surface and groundwater. These contaminants, when released in the environment cause problems to human beings and the environment. Therefore, there is immense need to remove these contaminants form wastewater. In this context, biological wastewater treatment system serves as a good option to combat the problem of water pollution. Biological wastewater treatment system is a cost-effective and environmental-friendly technology for the treatment of wastewater. Advances in technology have led to the evolution of several techniques for the removal of contaminant from wastewater. In this chapter, we discussed about the advancement and limitations of biological wastewater technology.

Keywords

Contaminant · Wastewater · Membrane bioreacter · Constructed wetland

---

I. Zoomi · H. K. Kehri · D. Pandey · U. Singh · K. L. Chaudhary
Sadasivan Mycopathology Laboratory, Department of Botany, University of Allahabad, Allahabad, UP, India

O. Akhtar
Department of Botany, Kamla Nehru Institute of Physical and Social Sciences, Sultanpur, UP, India

R. P. Narayan (✉)
Netaji Subhash Chandra Bose Government Girls P.G. College, Lucknow, UP, India

M. P. Shah (ed.), *Removal of Emerging Contaminants Through Microbial Processes*,
https://doi.org/10.1007/978-981-15-5901-3_10

1 Introduction

Rapid population growth, industrialization and long-term droughts result in the occurrence of a wide range of contaminants in the surface as well as in groundwater (Cundy et al. 2008; Chong et al. 2010; Zeng et al. 2011). There are various contaminants persisting in wastewater such as heavy metals and many other organic as well as inorganic complex compounds (O'Connor 1996; Fatta et al. 2011; Li et al. 2011a, b). These contaminants when released in the environment cause problem to human beings as well the ecological environment. Therefore, there is a need to remove these contaminants forms wastewater (Jiang et al. 2006; Pang et al. 2001). In this context, there are various methods for the removal of contaminant including bioremediation, biodegradation, photocatalytic oxidation, use of iron oxide nanomaterials, and adsorption/separation processing (Huang et al. 2006; Ahmed et al. 2017; Long et al. 2011; Pang et al. 2001; Xu et al. 2012; Garcia-Rodríguez et al. 2014). Among them, biological wastewater treatment system serves as a good option to combat the problem of water pollution. Biological wastewater treatment system is a cost-effective and environmental-friendly technology for the treatment of wastewater. Biological treatment is an integral part of any wastewater treatment system and is usually suggested for treating wastewater as they have economic advantages over chemical treatment technologies (Dogruel et al. 2006). Biological wastewater treatment technologies have been widely used for the removal of contaminants principally by biodegradation mechanisms. High-molecular-weight compounds are broken down by microorganisms (bacteria, algal, and fungi) into simple molecules, and the process is called as biodegradation (Garcia-Rodríguez et al. 2014). There have been an increasing number of studies on advanced biological treatment system of wastewater. This chapter presents the wastewater treatment technologies and drawback of biological wastewater treatment technologies.

2 Biological Wastewater Treatment

Biological treatment is an integral part of any wastewater treatment plant and is usually suggested for treating wastewater as they have economic advantages over chemical treatment technologies (Dogruel et al. 2006). Biological wastewater treatment technologies have been widely used for the removal of contaminants principally by biodegradation mechanisms. High-molecular-weight compounds are broken down by microorganisms (bacteria, algal, and fungi) into simple molecules, and the process is called as biodegradation (Garcia-Rodríguez et al. 2014). Microorganisms utilize large organic compounds as substrates for the their growth and metabolism (Tran et al. 2013). It has been reported that there are some contaminants which are toxic in nature (poorly biodegradable compounds as well as heavy metals), inhibiting the growth of the microorganisms, thereby inhibiting the process of biodegradation (Tran et al. 2013). Stasinakis et al. (2002) reported a substantial inhibition in the growth of heterotrophic microorganism in the presence of chromium (10 mg L^{-1}). It has also been reported that the chromium concentration had more impact on nitrification bacteria than on denitrification bacteria (Farabegoli et al. 2004). Biodegradation is basically divided into two processes: (1) aerobic and (2) anaerobic process. Furthermore, it has also been reported that the characteristics of wastewater play an important role in the selection of biological treatments (Raj and Anjaneyulu 2005; Deegan et al. 2011).

2.1 Aerobic Processes

Before we go in to the details of aerobic treatment technology, it is important to understand the term aerobic; as the word itself suggests, it means in the presence of air (oxygen). Thus, aerobic treatment processes utilize those microorganisms which grow in the presence of oxygen. Aerobic-activated sludge treatment has been widely used worldwide treatment technology (Gujer et al. 1995; Jeon et al. 2003; Chen and Gu 2005; Lopez et al. 2006; Polesel et al. 2016). Song et al. (2005) isolated two bacterial strains (*Arthrobacter* sp. 2AC and *Comamonas* sp. 4BC) from tannery-activated sludge and reported that these bacterial strains were capable to degrade naphthalene-2-sulfonic acid which is a main component of the naphthalene sulfonate. Similarly, Song and Burns (2005) reported the degradation of 2-naphthalenesulfonic acid and formaldehyde by *Cunninghamella polymorpha* (fungus). Senthilkumar et al. (2008) isolated four halotolerant bacteria, *viz*., *Pseudomonas aeruginosa*, *Bacillus flexus*, *Exiguobacterium homiense* and *Staphylococcus aureus* from different saline habitat and used these salt-tolerant bacteria in consortia for the biodegradation of tannery soak liquor. They reported that 80% COD removal efficiency was observed at 8% (w/v) salinity for halotolerant bacteria in consortia, but 10% (w/v) salinity reduced the COD removal. It has been reported that the temperature also plays a role in the removal of contaminants from the wastewater at large scale (Görgün et al. 2007). Insel et al. (2009) studied the functioning of intermittent aeration type of operation at 21 and 35 °C temperature in fluctuation and reported that the process of nitrification was improved by increasing the intensity of aeration that improves the efficiency total nitrogen removal.

2.2 Anaerobic Processes

The term anaerobic simply means in the absence of air (oxygen). The anaerobic treatment processes take place in the absence of oxygen by those microorganisms that do not require oxygen.

Anaerobic process has been attracting much attention in the scientific world due to its application in the treatment of wastewater with medium to high organic impurities (COD $>$ 1000 ppm). Mannucci et al. (2010) in their article reported 63 applications of anaerobic processes in tannery wastewater treatment. Lefebvre et al. (2006) and El-Sheikh Mahmoud et al. (2011) reported anaerobic treatment process mainly performed by using anaerobic filters composed of upflow anaerobic filters, downflow anaerobic filters as well as upflow anaerobic sludge blanket reactors. However, Zupančič and Jemec (2010) also reported expanded granular sludge bed and anaerobic baffled reactor. The filling materials used in the anaerobic filter were polyurethane foam cubes and polypropylene rasching rings (Daryapurkar et al. 2001).

3 Advancement in the Treatment Technologies

3.1 Microalgae-Based Treatment

Algae-based wastewater treatment can simulate the capacity of natural ecosystems as it remove the contaminant from the water in an eco-friendly way. Algae-based treatment systems have been proved to be effective to remove the contaminant from the water due to the use of fertilizer and resource recovery of algal biomass in algal ponds (Ahmed et al. 2017). It has also been reported that in algal pond, algae and bacteria also grow in symbiotic association (Ahmed et al. 2017). In a symbiotic association, bacteria uses oxygen which is produced during photosynthesis of algae and bacteria degrade the contaminants (Passos et al. 2014; Matamoros et al. 2015). Contaminants such as pharmaceutical compounds, anti-inflammatory drugs, and antibiotics can be removed by algal reactors (Ahmed et al. 2017). Algae-based treatment system can eliminate several types of contaminants such as endocrine-disrupting compounds and pharmaceuticals. However, it has also been reported that algal-based treatment system has lesser affinity toward pesticides removal (Ahmed et al. 2017).

3.2 Membrane Bioreactors

In recent years the use of membrane technologies has been focused on, particularly membrane bioreactors as it allow the operation at higher concentrations of biomass and sludge retention time. The application possibilities of membrane bioreactors are ever extending as the use of membrane bioreactor represents an economic advantage. Membranes used in this bioreactors can be categorized according to their hydrophilic (having a tendency to mix with or dissolve in water) or hydrophobic (tending to repel or fail to mix with water) nature depending on the properties of the construction material (Jermann et al. 2009; Alvarino et al. 2018).

3.3 Sequencing Batch Reactor

Sequencing batch reactor has been reported to remove the nutrients biologically. Kuba et al. (1993) reported the biological removal of phosphorus from wastewater by anaerobic-anoxic sequencing batch reactor. Similarly, Manning and Irvine (1985) reported the biological removal of phosphorus in a sequencing batch reactor. Lee et al. (2001) reported the biological nitrogen removal with enhanced phosphate uptake in a sequencing batch reactor using single-sludge system. It has also been reported that the biological phosphorus removal was enhanced due to the use of anaerobic-aerobic sequencing batch reactor system (Danesh and Oleszkiewicz 1997). Sequencing batch reactor has the ability to carry out nitrification and denitrification in the existence of inhibiting substances (Farabegoli et al. 2004; Ganesh et al. 2006). The nitrification and denitrification processes occur due to the selection and enrichment of particular microbial species (Lofrano et al. 2013). It has also been reported that the sequencing batch reactor treatment technology is also used for tannery wastewater because of the flexibility in its operation. Murat et al. (2006) studied the effect of a wide range of temperature (7–30 °C) on nitrogen removal from tannery wastewater and reported that the nitrification and denitrification was maintained by the adjustment of the sludge age for each temperature range.

3.4 Constructed Wetland

Constructed wetland is biological based treatment system that is designed to filter and remove contaminants by the use of natural methods including plants, soils and their associated microbes. According to Töre et al. (2012), constructed wetland treatment is an integrated combination of biological, physical and chemical processes. A number of studies have been undertaken for selecting the species of plant (Calheiros et al. 2007, 2008, 2012), selecting substrate (Calheiros et al. 2008) and for microbial dynamics (Calheiros et al. 2009a, b). Mant et al. (2004) reported the phytoremediation of chromium by interaction of *Pennisetum purpureum*, *Brachiaria decumbens* and *Phragmites australis* in constructed wetland. Similarly, Calheiros et al. (2008) studied the effect of tannery effluent on *Canna indica*, *Typha latifolia*, *P. australis*, *Stenotaphrum secundatum* and *Iris pseudacorus* in constructed wetland. They further reported that only two plants (*P. australis* and *T. latifolia*) were able to grow successfully. Moreover, Töre et al. (2012) reported the removal of steroid, estrogens, bisphenol and phthalates in constructed wetland.

3.5 Moving Bed Biofilm Reactor

Moving bed biofilm reactor developed in Norway in the late 1980s and early 1990s and has been used worldwide for the treatment of wastewater (Rusten et al. 2006). In moving bed biofilm reactor, the carrier elements are used which are attached to the biomass that moves freely along with the water in the reactor (Ødegaard et al. 1994). This technology was used by several researchers for the removal of unwanted materials from the wastewater (Pastorelli et al. 1997; Winkler et al. 2012; Barwal and Chaudhary 2014; Dvořák et al. 2014). This reactor was studied for the removal of nitrogen through simultaneous nitrification and denitrification processes (Wang et al. 2006). Li et al. (2011a, b) reported the removal of thiocyanate and ammonium from wastewater of coal gasification in moving bed biofilm reactor. Chen et al. (2008) reported the simultaneous exclusion of COD and ammonium from landfill leachate in moving bed biofilm reactor system. Additionally, this technology was also used for the treatment of pesticide and phenolic from the wastewater (Borghei and Hosseini 2004; Chen et al. 2007).

4 Drawback of Biological Wastewater Treatment

Biological wastewater treatment system displays a number of limitations over physical and chemical wastewater treatment systems. Ochoa-Herrera et al. (2009) reported the toxicity of fluoride to microbes in biological wastewater treatment systems. Similarly, copper(II) ions are reported to have toxic effect on microorganisms in biological wastewater treatment (Ochoa-Herrera et al. 2011). It has also been reported that algal-based treatment system has lesser affinity toward pesticide removal (Ahmed et al. 2017). Moreover, in membrane-based bioreactor, membrane fouling is the most serious issue (Meng et al. 2009; Lin et al. 2013) that diminished the hydraulic performance which limits the extensive use of membrane technology (Judd 2010; Yu et al. 2013). Various factors such as functioning conditions, characteristics of wastewater, membrane, biomass properties and their interactions are responsible for membrane fouling (Meng et al. 2007). As reported by Ahmed et al. (2017), there are still some unresolved problems in the removal of emerging contaminants from wastewater in moving bed bioreactor technology.

5 Conclusions

There are various contaminants persisting in wastewater such as heavy metals and many other organic and inorganic complex compounds (O'Connor 1996; Fatta et al. 2011; Li et al. 2011a, b). These contaminants when released in the environment cause problems to human beings as well ecological environments. Therefore, there is a need to remove these contaminants form wastewater (Jiang et al. 2006; Huang et al. 2006; Pang et al. 2001). In this context, biological wastewater treatment system serves as a good option to combat the problem of water pollution. Biological wastewater treatment system is a cost-effective and environmental-friendly technology for the treatment of wastewater. However, there are still some limitations in biological wastewater treatment system due to the variable efficiencies of different treatment systems.

Acknowledgments The authors gratefully acknowledge the Head of the Department of Botany, University of Allahabad, Allahabad, for providing the necessary facilities and are also grateful to acknowledge the University Grant Commission and CSIR for providing financial assistance.

References

Ahmed MB, Zhou JL, Ngo HH, Guo W, Thomaidis NS, Xu J (2017) Progress in the biological and chemical treatment technologies for emerging contaminant removal from wastewater: a critical review. J Hazard Mater 323:274–298

Alvarino T, Suarez S, Lema J, Omil F (2018) Understanding the sorption and biotransformation of organic micropollutants in innovative biological wastewater treatment technologies. Sci Total Environ 615:297–306

Barwal A, Chaudhary R (2014) To study the performance of biocarriers in moving bed biofilm reactor (MBBR) technology and kinetics of biofilm for retrofitting the existing aerobic treatment systems: a review. Rev Environ Sci Biotechnol 13(3):285–299

Borghei SM, Hosseini SH (2004) The treatment of phenolic wastewater using a moving bed biofilm reactor. Process Biochem 39(10):1177–1181

Calheiros CSC, Rangel AOSS, Castro PML (2007) Constructed wetland systems vegetated with different plants applied to the treatment of tannery wastewater. Water Res 41:1790–1798

Calheiros CSC, Rangel AOSS, Castro PML (2008) Evaluation of different substrates to support the growth of *Typha latifolia* in constructed wetlands treating tannery wastewater over long-term operation. Bioresour Technol 99:6866–6877

Calheiros CSC, Duque AF, Moura A, Henriques IS, Correia A, Rangel AOSS (2009a) Changes in the bacterial community structure in two-stage constructed wetlands with different plants for industrial wastewater treatment. Bioresour Technol 100:3228–3235

Calheiros CSC, Rangel AOSS, Castro PML (2009b) Treatment of industrial wastewater with two-stage constructed wetlands planted with *Typha latifolia* and *Phragmites australis*. Bioresour Technol 100:3205–3213

Calheiros CSC, Quitério PVB, Silva G, Crispim LFC, Brix H, Moura SC (2012) Use of constructed wetland systems with *Arundo* and *Sarcocornia* for polishing high salinity tannery wastewater. J Environ Manag 95:66–71

Chen Y, Gu G (2005) Preliminary studies on continuous chromium (VI) biological removal from wastewater by anaerobic–aerobic activated sludge process. Bioresour Technol 96 (15):1713–1721

Chen S, Sun D, Chung JS (2007) Treatment of pesticide wastewater by moving-bed biofilm reactor combined with Fenton-coagulation pretreatment. J Hazard Mater 144(1–2):577–584

Chen S, Sun D, Chung JS (2008) Simultaneous removal of COD and ammonium from landfill leachate using an anaerobic–aerobic moving-bed biofilm reactor system. Waste Manag 28 (2):339–346

Chong MN, Jin B, Chow CWK, Saint C (2010) Recent developments in photocatalytic water treatment technology: a review. Water Res 44(10):2997–3027

Cundy AB, Hopkinson L, Whitby RLD (2008) Use of iron-based technologies in contaminated land and groundwater remediation: a review. Sci Total Environ 400(1–3):42–51

Danesh S, Oleszkiewicz JA (1997) Use of a new anaerobic-aerobic sequencing batch reactor system to enhance biological phosphorus removal. Water Sci Technol 35(1):137–144

Daryapurkar RA, Nandy T, Kaul SN, Deshpande CV, Szpyrkowicz L (2001) Evaluation of kinetic constants for anaerobic fixed film fixed bed reactors treating tannery wastewater. Int J Environ Stud 58:835–860

Deegan AM, Shaik B, Nolan K, Urell K, Oelgemöller M, Tobin J, Morrissey A (2011) Treatment options for wastewater effluents from pharmaceutical companies. Int J Environ Sci Technol 8:649–666

Dogruel S, Genceli EA, Babuna FG, Orhon D (2006) An investigation on the optimal location of ozonation within biological treatment for a tannery wastewater. J Chem Technol Biotechnol 81:1877–1885

Dvořák L, Lederer T, Jirků V, Masak J, Novak L (2014) Removal of aniline, cyanides and diphenylguanidine from industrial wastewater using a full-scale moving bed biofilm reactor. Process Biochem 49(1):102–109

El-Sheikh Mahmoud A, Hazem I, Saleh J, Flora R, AbdEl-Ghany MR (2011) Biological tannery wastewater treatment using two stage UASB reactors. Desalination 276:253–259

Farabegoli G, Carucci A, Majone M, Rolle E (2004) Biological treatment of tannery wastewater in the presence of chromium. J Environ Manag 71:345–349

Fatta KD, Kalavrouziotis IK, Koukoulakis PH, Vasquez MI (2011) The risks associated with wastewater reuse and xenobiotics in the agroecological environment. Sci Total Environ 409 (19):3555–3563

Ganesh R, Balaji G, Ramanujam RA (2006) Biodegradation of tannery wastewater using sequencing batch reactor—respirometric assessment. Bioresour Technol 97:1815–1821

Garcia-Rodríguez A, Matamoros V, Fontàs C, Salvadó V (2014) The ability of biologically based wastewater treatment systems to remove emerging organic contaminants—a review. Environ Sci Pollut Res 21(20):11708–11728

Görgün E, Insel G, Artan N, Orhon D (2007) Model evaluation of temperature dependency for carbon and nitrogen removal in a full-scale activated sludge plant treating leather-tanning wastewater. J Environ Sci Health A Tox Hazard Subst Environ Eng 42:747–756

Gujer W, Henze M, Mino T, Matsuo T, Wentzel MC, Marais GV (1995) The activated sludge model No. 2: biological phosphorus removal. Water Sci Technol 31(2):1–11

Huang DL, Zeng GM, Jiang XY, Feng CL, Yu HY, Huang GH (2006) Bioremediation of Pb contaminated soil by incubating with *Phanerochaete chrysosporium* and straw. J Hazard Mater 134(1–3):268–276

Insel GH, Görgün E, Artan N, Orhon D (2009) Model based optimization of nitrogen removal in a full scale activated sludge plant. Environ Eng Sci 26:471–480

Jeon CO, Lee DS, Park JM (2003) Microbial communities in activated sludge performing enhanced biological phosphorus removal in a sequencing batch reactor. Water Res 37(9):2195–2205

Jermann D, Pronk W, Boller M, Schäfer AI (2009) The role of NOM fouling for the retention of estradiol and ibuprofen during ultrafiltration. J Membr Sci 329(1–2):75–84

Jiang XY, Zeng GM, Huang DL, Chen Y, Liu F, Huang GH (2006) Remediation of pentachlorophenol-contaminated soil by composting with immobilized *Phanerochaete chrysosporium*. World J Microbiol Biotechnol 22(9):909–913

Judd S (2010) The MBR book: principles and applications of membrane bioreactors for water and wastewater treatment. Elsevier, Amsterdam

Kuba T, Smolders G, Van Loosdrecht MC, Heijnen JJ (1993) Biological phosphorus removal from wastewater by anaerobic-anoxic sequencing batch reactor. Water Sci Technol 27(5–6):241–252

Lee DS, Jeon CO, Park JM (2001) Biological nitrogen removal with enhanced phosphate uptake in a sequencing batch reactor using single sludge system. Water Res 35(16):3968–3976

Lefebvre O, Vasudevan N, Torrijosa M, Thanasekaran K, Moletta R (2006) Anaerobic digestion of tannery soak liquor with an aerobic post-treatment. Water Res 40:1492–1500

Li HQ, Han HJ, Du MA, Wang W (2011a) Removal of phenols, thiocyanate and ammonium from coal gasification wastewater using moving bed biofilm reactor. Bioresour Technol 102 (7):4667–4673

Li X, Zeng GM, Huang JH, Zhang DM, Shi LJ, He SB (2011b) Simultaneous removal of cadmium ions and phenol with MEUF using SDS and mixed surfactants. Desalination 276:136–141

Lin H, Peng W, Zhang M, Chen J, Hong H, Zhang Y (2013) A review on anaerobic membrane bioreactors: applications, membrane fouling and future perspectives. Desalination 314:169–188

Lofrano G, Meriç S, Zengin GE, Orhon D (2013) Chemical and biological treatment technologies for leather tannery chemicals and wastewaters: a review. Sci Total Environ 461:265–281

Long F, Gong JL, Zeng GM, Chen L, Wang XY, Deng JH (2011) Removal of phosphate from aqueous solution by magnetic Fe–Zr binary oxide. Chem Eng J 171:448–455

Lopez C, Pons MN, Morgenroth E (2006) Endogenous processes during long-term starvation in activated sludge performing enhanced biological phosphorus removal. Water Res 40 (8):1519–1530

Manning JF, Irvine RL (1985) The biological removal of phosphorus in a sequencing batch reactor. J Water Pollut Control Fed 1:87–94

Mannucci A, Munz G, Mori G, Lubello C (2010) Anaerobic treatment of vegetable tannery wastewaters: a review. Desalination 264:1–8

Mant C, Costa S, Williams J, Tambourgi E (2004) Phytoremediation of chromium by model constructed wetland. Bioresour Technol 97:1767–1772

Matamoros V, Gutiérrez R, Ferrer I, García J, Bayona JM (2015) Capability of microalgae-based wastewater treatment systems to remove emerging organic contaminants: a pilot-scale study. J Hazard Mater 288:34–42

Meng F, Shi B, Yang F, Zhang H (2007) Effect of hydraulic retention time on membrane fouling and biomass characteristics in submerged membrane bioreactors. Bioproc Biosystems Eng 30 (5):359–367

Meng F, Chae SR, Drews A, Kraume M, Shin HS, Yang F (2009) Recent advances in membrane bioreactors (MBRs): membrane fouling and membrane material. Water Res 43(6):1489–1512

Murat S, Insel G, Artan N, Orhon D (2006) Performance evaluation of SBR treatment for nitrogen removal from tannery wastewater. Water Sci Technol 53:275–284

O'Connor GA (1996) Organic compounds in sludge-amended soils and their potential for uptake by crop plants. Sci Total Environ 185(1):71–81

Ochoa-Herrera V, Banihani Q, León G, Khatri C, Field JA, Sierra-Alvarez R (2009) Toxicity of fluoride to microorganisms in biological wastewater treatment systems. Water Res 43 (13):3177–3186

Ochoa-Herrera V, León G, Banihani Q, Field JA, Sierra-Alvarez R (2011) Toxicity of copper (II) ions to microorganisms in biological wastewater treatment systems. Sci Total Environ 412:380–385

Ødegaard H, Rusten B, Westrum T (1994) A new moving bed biofilm reactor-applications and results. Water Sci Technol 29(10–11):157

Pang Y, Zeng GM, Tang L, Zhang Y, Liu YY, Lei XX (2001) Cr(VI) reduction by *Pseudomonas aeruginosa* immobilized in a polyvinyl alcohol/sodium alginate matrix containing multi-walled carbon nanotubes. Bioresour Technol 102:10733–10736

Passos F, Uggetti E, Carrère H, Ferrer I (2014) Pretreatment of microalgae to improve biogas production: a review. Bioresour Technol 172:403–412

Pastorelli G, Andreottola G, Canziani R, Darriulat C, de Fraja Frangipane E, Rozzi A (1997) Organic carbon and nitrogen removal in moving-bed biofilm reactors. Water Sci Technol 35 (6):91–99

Polesel F, Andersen HR, Trapp S, Plósz BG (2016) Removal of antibiotics in biological wastewater treatment systems. A critical assessment using the activated sludge modeling framework for xenobiotics (ASM-X). Environ Sci Technol 50(19):10316–10334

Raj DS, Anjaneyulu Y (2005) Evaluation of biokinetic parameters for pharmaceutical wastewaters using aerobic oxidation integrated with chemical treatment. Process Biochem 40(1):165–175

Rusten B, Eikebrokk B, Ulgenes Y, Lygren E (2006) Design and operations of the Kaldnes moving bed biofilm reactors. Aquac Eng 34(3):322–331

Senthilkumar S, Surianarayanan M, Sudharshan S, Susheela R (2008) Biological treatment of tannery wastewater by using salt-tolerant bacterial strains. Microb Cell Factories 7:15

Song Z, Burns RG (2005) Depolymerisation and biodegradation of a synthetic tanning agent by activated sludges, the bacteria *Arthrobacter globiformis* and *Comamonas testosteroni*, and the fungus *Cunninghamella polymorpha*. Biodegradation 16:305–318

Song Z, Edwards SR et al (2005) Biodegradation of naphthalene-2-sulfonic acid present in tannery wastewater by bacterial isolates *Arthrobacter* sp. 2AC and *Comamonas* sp. 4BC. Biodegradation 16:237–252

Stasinakis AS, Mamais D, Thomaidis NS, Lekkas TD (2002) Effect of chromium (VI) on bacterial kinetics of heterotrophic biomass of activated sludge. Water Res 36:3342–3350

Töre GY, Meriç S, Lofrano G, De Feo G (2012) Removal of trace pollutants from wastewater in constructed wetlands. In: Emerging compounds removal from wastewater. Springer, Dordrecht, pp 39–58

Tran NH, Urase T, Ngo HH, Hu J, Ong SL (2013) Insight into metabolic and cometabolic activities of autotrophic and heterotrophic microorganisms in the biodegradation of emerging trace organic contaminants. Bioresour Technol 146:721–731

Wang XJ, Xia SQ, Chen L, Zhao JF, Renault NJ, Chovelon JM (2006) Nutrients removal from municipal wastewater by chemical precipitation in a moving bed biofilm reactor. Process Biochem 41(4):824–828

Winkler MK, Yang J, Kleerebezem R, Plaza E, Trela J, Hultman B, van Loosdrecht MC (2012) Nitrate reduction by organotrophic Anammox bacteria in a nitritation/anammox granular sludge and a moving bed biofilm reactor. Bioresour Technol 114:217–223

Xu P, Zeng GM, Huang DL, Feng CL, Hu S, Zhao MH, Lai C, Wei Z, Huang C, Xie GX, Liu ZF (2012) Use of iron oxide nanomaterials in wastewater treatment: a review. Sci Total Environ 424:1–10

Yu L, Zhang Y, Zhang B, Liu J, Zhang H, Song C (2013) Preparation and characterization of HPEI-GO/PES ultrafiltration membrane with antifouling and antibacterial properties. J Membr Sci 447:452–462

Zeng GM, Li X, Huang JH, Zhang C, Zhou CF, Niu J (2011) Micellar-enhanced ultrafiltration of cadmium and methylene blue in synthetic wastewater using SDS. J Hazard Mater 185 (2–3):1304–1310

Zupančič GD, Jemec A (2010) Anaerobic digestion of tannery waste: semi-continuous and anaerobic sequencing batch reactor processes. Bioresour Technol 101:26–33

Role of Extremophiles and Extremophilic Proteins in Industrial Waste Treatment

Sharmistha Tapadar, Deeksha Tripathi, Saurabh Pandey, Khyati Goswami, Arunima Bhattacharjee, Kunwali Das, Espita Palwan, Mamta Rani, and Ashutosh Kumar

Abstract

Majority of the industrial products are made in variety of extreme environments. These industrial processes generate by-products that are difficult to degrade, harmful for environment, and toxic to animals and humans. These industrial by-products are present in extreme conditions such as high salt and high or low temperature. It is an unfavorable condition for most of the waste-degrading enzymes as they work at ambient condition. The harmful industrial by-products, therefore, needed to shift at ambient condition for their degradation. Extremophiles grow at extreme conditions, and so their enzymes too work optimally under these extreme conditions. Extremophiles have enormous potential in biotechnological industry and waste remediation/management. Some halophilic microorganisms have great efficiency to remove petroleum, heavy metals, and dyes from the water polluted by industries. The chapter details the protein

Sharmistha Tapadar, Deeksha Tripathi, and Saurabh Pandey contributed equally to this work.

S. Tapadar · K. Goswami · A. Bhattacharjee · K. Das · E. Palwan · A. Kumar (✉)
Department of Microbiology, Tripura University (A Central University), Suryamaninagar, Agartala, Tripura, India
e-mail: ashutoshkumar@tripurauniv.in

D. Tripathi
Department of Microbiology, School of Life Sciences, Central University of Rajasthan, Ajmer, Rajasthan, India

S. Pandey
Department of Biochemistry, School of Chemical and Life Sciences, Jamia Hamdard, New Delhi, India

M. Rani
Plant-Microbe Interactions Laboratory, National Institute of Plant Genome Research, New Delhi, India

M. P. Shah (ed.), *Removal of Emerging Contaminants Through Microbial Processes*,
https://doi.org/10.1007/978-981-15-5901-3_11

adaptations in different extremophiles for their survival and uses of different extremophilic organisms/proteins in bioremediation.

Keywords

Extremophiles · Protein adaptations · Extremophilic proteins · Industrial waste treatment

1 Extremophiles

Extremophiles thrive in intolerably hostile or even lethal environments. These extreme conditions are extreme hot niches, arctic frost, and salty lakes. Some extremophiles can grow in presence of toxic waste and organic and heavy metal contaminants that are lethal to other organisms. Extremophiles are found in hydrothermal vents, in deep sea, as well as in the volcano. They are the Earth's most prolific group of organisms yet less studied. Extremophiles have members of all the three domains of life, bacteria, eukarya, and archaea. However, the members of the group archaea are least versatile; archaea are the major group that can survive in extreme environmental conditions. Among the bacteria, cyanobacteria are best adapted in extreme environments. Fungi are the most versatile among the eukaryotes. Example of most impressive eukaryotic polyextremophiles is the Tardigrade, a microscopic invertebrate. Tardigrades can survive in temperature of about −272 °C to 151 °C, pressure of about 6000 atmospheric pressure, extreme dehydration, and in the exposure to X-rays and gamma rays (Rampelotto 2013).

Extremophiles can be grouped based on different extreme growth conditions. Thermophiles can grow between 55 and 80 °C or higher temperature, whereas psychrophiles are between −20 and 10 °C. Halophiles are salinity tolerant and can grow at extremely high salt concentration, approximately 0.6 M to >5 M NaCl. Acidophiles and alkaliphiles are pH-tolerant microorganisms. Acidophiles have optimal growth at pH 3 or below 3, whereas alkaliphiles have optimal growth up to pH 10. Piezophiles are organisms that survive at high pressure (Dworkin et al. 2006). Radiation-resistant or radio-resistant extremophiles grow under extreme radiation such as gamma rays (GR), X-rays, UV radiation (UVR), and radio waves.

Recently, multicellular eukaryotic salmon parasite *Henneguya salminicola* is characterized with ability to grow in hypoxic environment. Evolution at hypoxic environment makes them lose the total of aerobic pathway. They have unusual characteristics of lacking mitochondrial genome, thus devoid of aerobic respiration. This makes them unique; otherwise, aerobic respiration is ubiquitous among eukaryotes (Yahalomi et al. 2020).

2 Industrial Waste and Bioremediation

Manufacturing, mining, and other industries emit/discharge unwanted industrial wastes. Due to complex nature, industrial waste management is more problematic and less predictable than management of municipal solid waste (MSW), particularly as industrial waste often contains hazardous pollutants. Aquatic organisms are continuously exposed to different industrial waste in contaminated water bodies. Industrial and household wastewater contains chemicals such as carbaryl, chlorpyrifos, diethyl phthalate, *p*-nonylphenol, tri(2-chloroethyl)phosphate, naphthalene, anthracene, 1,2,3-tricholoropropane, phenol, 1,4-dichlorobenzene, acetophenone, etc. (Petrie et al. 2015). Some chemicals, which have anthropogenic origin, such as dichlorodiphenyltrichloroethane (DDT), trichloroethylene, 1,2,3-trichloropropane, etc., are resistant to natural biodegradation (Janssen et al. 2005). 1,2,3-trichloropropane (TCP) is an industrial by-product of anthropogenic origin, which is a major groundwater contaminant (Samin and Janssen 2012). Chemicals in industrial waste have diverse physicochemical properties and occupy heterogeneous physical niches in the environment (Meckenstock et al. 2015).

Unregulated anthropogenic activities result in metal pollution. Bioremediation strategies have been implemented in many contaminated areas, using bacterial bioremediation approach. Some examples are as follows. In *Chromobacterium violaceum*, various genes and proteins are associated with the metabolism of metals, such as arsenic, iron, zinc, etc. (Alencar et al. 2016). *Pseudomonas aeruginosa* is mostly used for soluble cadmium removal (Sinha and Mukherjee 2009). *Pseudomonas putida* uses bioaccumulation and biosorptive mechanism for removal of metals such as cobalt, nickel, manganese, vanadium, lead, titanium, and copper (Kamika and Momba 2013). *Acinetobacter guillouiae* also uses the mechanism of biosorption for the removal of copper (Majumder et al. 2015). *Geobacter metallireducens* is used in the removal vanadium, in which the V(V) ions are precipitated after reduced to V (IV) ions (Ortiz-Bernad et al. 2004). *Paenibacillus polymyxa* is used for the removal of cadmium, copper, and zinc, by biosorption, adsorption, and bioaccumulation in the polymeric matrix or biomass (Martins et al. 2008). Intracellular accumulation and extracellular adsorption are used by the bacterium *Bacillus cereus*, for bioremediation of cadmium (Huang et al. 2014). Also, chromium ions are removed by biosorption of ions by immobilized *Bacillus cereus* (Maiti et al. 2009). *Rhodococcus erythropolis* is used for the removal of copper, cadmium, and lead, due to its capability to synthesize microbial surfactant (Pirog et al. 2013). *Bacillus licheniformis* uses bioaccumulation and biosorptive mechanism for the removal of zinc, copper, nickel, manganese, vanadium, lead, titanium, and copper (Kamika and Momba 2013). *Bacillus barbaricus* uses the mechanism of biosorption with bacterial consortium action, for the removal of cadmium and lead (Sen et al. 2014). *Chlamydomonas* spp., *Oscillatoria* spp., and *Chlorella vulgaris* are suggested to bioaccumulate heavy metals inside their tissues in higher concentration. *Aspergillus niger, Bacillus* spp., *Pseudomonas aeruginosa, Citrobacter* spp., *Chlorella vulgaris, Rhizopus arrhizus, Zooglea* spp., and *Volvariella volvacea* are some examples of microbes that utilize heavy metals (Mishra 2017).

Radionuclides are the radioactive wastes that are life-threatening upon exposure. Nuclear power plants contribute to about 95% of the total radioactive wastes generated (Ahier and Tracy 1995; Tamponnet and Declerck 2008). Bioremediation is the ecologically beneficial way by which radioactive wastes can be removed from the environment. Microorganism-mediated bioremediation can affect the solubility, bioavailability, and mobility of radionuclides (Prakash et al. 2013).

Bacteria such as *Shewanella putrefaciens* and *Geobacter sulfurreducens* are capable of U(VI) reduction (Wildung et al. 2000; Lloyd et al. 2003). ^{99}Tc, a radionuclide found in nuclear wastes, can be reduced by direct metabolic reduction by the bacteria *Shewanella putrefaciens* and *Geobacter metallireducens* (Lloyd and Macaskie 1996). Some biofilm-producing microorganisms serve as a platform for the precipitation of insoluble minerals (Prakash et al. 2013). *Citrobacter* sp. produces deposits of metal phosphate enzymatically (Prakash et al. 2013). Genetically modified *Pseudomonas aeruginosa* is be able to precipitate a complex containing phosphorus and uranium on their cell surface (Keasling et al. 2000). Several proteins of microorganisms such as *Deinococcus radiodurans*, *Sphingomonas* sp. BSAR-1, *Salmonella enterica* serovar Typhi, and *Desulfovibrio vulgaris* are utilized in the bioremediation of uranium (Appukuttan et al. 2006; Nilgiriwala et al. 2008; Misra et al. 2012). In the bioremediation of cobalt, genes and proteins of *Rhodopseudomonas palustris* and *Novosphingobium aromaticivorans* can be utilized (Raghu et al. 2008). Thus, huge progress had been made in the field of radionuclide bioremediation using microorganisms in recent years. However, many challenges still lie ahead in remediation of radioactive wastes.

3 Contaminants in Wastewater

3.1 Microbial Components

Wastewater contains numerous pathogenic as well as non-pathogenic bacteria, helminths, protozoa, and viruses. *Enterococcus faecalis*, *Escherichia coli*, *Staphylococcus aureus*, *Streptococcus pneumoniae*, *Vibrio cholerae*, *Bacillus anthracis*, *Clostridium botulinum*, *Clostridium perfringens*, *Pseudomonas aeruginosa*, *Clostridium difficile*, *Corynebacterium diphtheria*, *Mycobacterium tuberculosis*, *Streptococcus agalactiae*, *Yersinia pestis*, *Bacillus anthracis*, and *Salmonella enterica* are bacterial pathogens present in wastewater (Kumaraswamy et al. 2014). There is presence of different viruses such as noroviruses (Pouillot et al. 2015), rotaviruses (Baggi et al. 2001), adenoviruses (Osuolale and Okoh 2015), rhinoviruses, enteroviruses (Baggi et al. 2001), and herpes simplex viruses (Bibby and Peccia 2013).

3.2 Non-microbial Components

Nitrogen is abundantly found in wastewater. Both nitrogen and phosphorus are found in waste of agricultural industries. This increase of inorganic nitrogen enhances eutrophication of water bodies. Similarly, phosphorous-containing sludge from industries also causes eutrophication (Camargo and Alonso 2006).

Industrial wastewater is often heterogeneous mixture of aromatic amines, nitro-containing compounds, polynuclear aromatic hydrocarbons (PAHs), chlorinated organics, solvents, and heavy metals. Chemical industries are the largest producer of hazardous materials, such as hydrocarbons and coal tars (Houk 1992). Their wastes may contain carcinogens like acetonitrile and acetamide (Houk 1992). Extracts of wastewater also contain nitrobenzoic acids, nitrotoluenes, benzoic acids, and cresols (Sundvall et al. 1984). Organic chemical manufacturers release some of the most genotoxic discharges (McGeorge et al. 1985). Also plastics, resins, and rubber industries produce black tar, containing high levels (550 mg/g) of aniline, which is a mutagen in some organisms (DeMarini et al. 1987; DeMarini and Houk 1988). Textile industries produce carcinogens, such as benzidine and beta-naphthylamine. They also release heavy metals, especially chromium, copper, and zinc, and hazardous chemical intermediates, such as aromatic amines as waste (Houk 1992). In addition, these industries discharge several mutagens and carcinogens such as benzo-pyrene, fluoranthene, and phenanthrene/anthracene (Houk 1992). However, some chlorinated compounds or solvents, including dichlorobenzenes, carbon tetrachloride, and trichloroethylene, are identified in the wastes (DeMarini and Houk 1988). Also, pesticide manufacturers generate high alkylamines containing waste (DeMarini et al. 1989). Wood-preserving wastes contain phenols, cresols, carcinogenic PAHs (such as fluoranthene, pyrene), pentachlorophenol, and other chlorinated hydrocarbons (Donnelly et al. 1983, 1987a, b). Latex paint waste contains high levels of ethylbenzene, zinc, copper, and mercury and smaller amounts of some carcinogenic or mutagenic metals such as arsenic and selenium (Houk 1992). Polynuclear aromatic and aliphatic hydrocarbons and their derivatives were found to be the primary components of petrochemical wastes (Somani et al. 1980).

A mutagen, neoabietic acid, is also found in some effluents from pulp and paper mills (Houk 1992). The discharges from defense and munitions plants are trinitrotoluene (TNT), which is nonpolar, and the major organic component is dinitrotoluene (DNT) sulfonic acids (Houk 1992). TNT and DNT are mutagenic (Spanggord et al. 1982). Discharge from munition plants also contains small amounts of genotoxic compounds, such as 2,4,5-trinitrotoluene, 3,5-dinitroaniline, and 1,3,5-trinitrobenzene (Spanggord et al. 1982). Food processing industries produce large amounts of organic wastewater, containing carbon, ammonium, sodium hydroxide, etc. (Frenkel et al. 2017). Pharmaceutical industry contributes a lot of chemical waste to the water bodies.

Industrial wastewaters contain toxic heavy metals such as Zn, Cu, Cr, Ni, Cd, and Hg and may cause clinical manifestations (Zawierucha et al. 2016). Thus, all these wastes must be treated before discharging into water bodies (Frenkel et al. 2017). Radioactive contaminants can also be seen in wastewater from different industries that work with radioactive metals, mining industry, nuclear power station, and biomedical engineering.

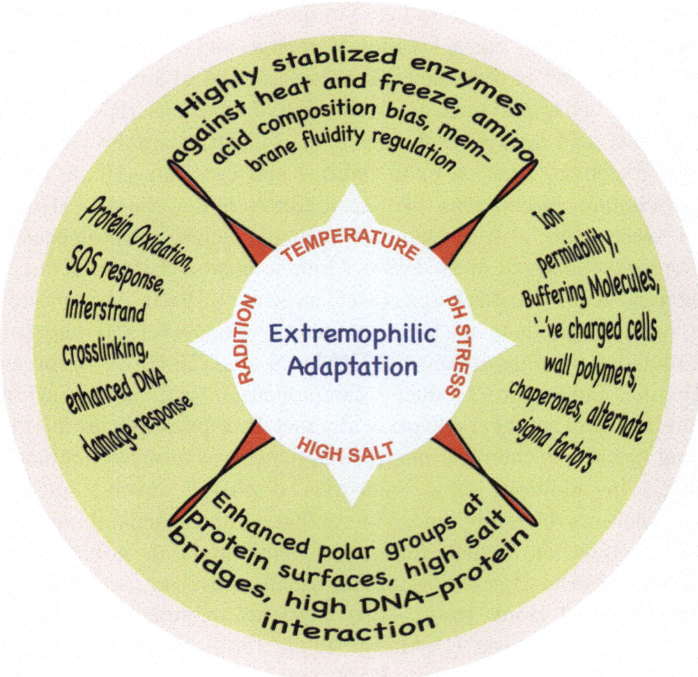

Fig. 1 Extremophilic adaptations for the survival in different stresses such as pH stress, temperature, high salt, and radiation. (Figure adapted, with permission, from Kumar et al. 2018 © 2018 Elsevier Ltd.)

4 Protein Adaptation of Extremophiles

Extremophiles can thrive in extreme environments that are intolerably hostile or even lethal for other life forms. Extremophiles have various adaptations that keep their cellular proteins stable and active which enables them to cope up with such harsh conditions. One such adaption is their enzymes known as "extremozymes" which perform the same enzymatic functions as their non-extreme counterparts. Extremozymes are capable to catalyze chemic activity inside the cell under harsh conditions (Fig. 1).

4.1 Protein Adaptation in Thermophiles

Though thermophiles can grow at very high temperature because of their thermostable enzymes and adaptation of the membrane, there is also high energy demand. Protons and sodium ions are used during energy transduction in bacteria; by increasing the rate of proton pumping, they sustain the proton's electrochemical gradient to increase permeability at high temperature; consequently, they use an increased fraction of metabolic energy for maintenance.

The proteins of thermophilic bacteria have different amino acid content than that of ordinary protein. Thermophile proteins undergo various modifications for adaptation. The most common are increase in the number of large hydrophobic residues, larger hydrophobic core, more number of disulfide bonds, and increased ionic interactions (Reed et al. 2013). Also, increased disulfide bridges lead to enhanced stability of the thermophilic proteins (Boutz et al. 2007; Cacciapuoti et al. 2012). Thermophilic proteins also have increased electrostatic interactions by replacement of uncharged residues with charged ones, increasing stability (Haney et al. 1999). Further, thermal stability also increases by subunit-subunit and subunit-cofactor interactions (Reed et al. 2013). Thermostable proteins have high amount of arginine and result in elevated incidence of salt bridge formation and ion pairing to stabilize thermophilic proteins. Lastly, extremophilic proteins have biotechnological applications. It can be used in many industries to reduce pollution by industrial waste treatment.

4.2 Protein Adaptation in Psychrophiles

Psychrophiles inhabit frozen lakes, polar regions, deep sea, arctic glaciers, and high altitudes. Since low temperature hampers bacterial cellular activity, they have adapted to low temperatures. Severe physicochemical constraints like increased water viscosity, decreased molecular diffusion rates, reduced biochemical reaction rates, increased solubility of gases, increased osmotic stress, desiccation, and ice formation are faced constantly by psychrophiles (D'Amico et al. 2006; Gerday and Glansdorff 2007). To overcome all these conditions, psychrophiles modify the fatty acid composition by increasing the content of unsaturated fatty acids of the lipid bilayer in the cell membrane (Chintalapati et al. 2004; Russell 2008). Psychrophile contains enzymes which have a high specific activity in the cold. In psychrophile proteome, there is an increased occurrence of glycine and decreased occurrence of proline. Glycine provides greater conformational mobility, whereas proline residues ensure conformational rigidity (Feller 2010). Psychrophilic proteins have less cysteine residues which correspond to decrease in the number of disulfide bridges (Parrilli et al. 2019). Psychrophilic enzymes have larger cavities that help to retain hydrophilic groups which in turn increases enzyme flexibility by enhancing the internal solvation (Paredes et al. 2011). Also, anti-freeze or ice-binding proteins (IBPs) inhibit the formation of ice crystals inside the cell (Bar Dolev et al. 2016).

4.3 Protein Adaptation in Halophiles

Halophiles thrive in salty oceans, sea, lakes, and coastal areas. Halophiles adapt themselves to prevent losing water or can shrink and ultimately die. Low hydrophobicity is considered as one of the protein adaptive mechanisms in halophiles. Lower water availability is responsible for dehydration of a protein as a result hydrophobic interactions strengthen. Also, halophilic proteins bind salt and water in solvent conditions like their environment. This binding ability is dependent on the acidic amino acid residues in the protein surface (Mevarech et al. 2000; Bergqvist et al. 2003; Tadeo et al. 2009). Halophilic proteins also have a high excess of negatively charged amino acids such as aspartate and glutamate over amino acids with a positive charge such as lysine and arginine (Gunde-Cimerman et al. 2018). Increased negative surface charge on the proteins makes them more soluble and provides them flexibility at high salt concentrations, whereas their non-halophilic counter-proteins tend to form a cluster making them non-flexible (Gunde-Cimerman et al. 2018).

4.4 Protein Adaptation in Acidophiles

Acidophiles maintain the cytoplasmic pH close to neutrality to protect their acid-labile cellular protein and other constituents, which needs a large pH gradient across cell membrane. Also, DNA becomes unstable in highly acidic condition. Acidophiles have special mechanisms to pump acid out of the cell cytoplasm to maintain neutral to weak acid conditions (Matin 1999). Three main mechanisms are involved in the adaptation of acidophiles: use of a reversed membrane potential, building an impermeable cell membrane, and cytoplasmic buffering (Baker-Austin and Dopson 2007). They have decreased permeability for the entry of protons into the cytoplasm. The inside positive membrane potential restricts the influx of protons by the formation of K^+ ions (Buetti-Dinh et al. 2016; Christel et al. 2018). Cytoplasmic buffering helps in maintaining intracellular pH because the cell cytoplasm possesses basic amino acids such as lysine, histidine, and arginine which are capable of sequestering protons (Baker-Austin and Dopson 2007).

4.5 Protein Adaptation in Alkaliphiles

Alkaliphiles are usually found in acidic environment of soda deserts and soda pans. Maintaining cell integrity and function of intracellular organelles is difficult but critical for survival. pH homeostasis in cells is achieved by lowering of cytoplasmic pH by proton uptake, reducing proton leakage, also production of organic acid, and inhibiting the entry of hydroxyl ions (Mamo 2019). Alkaliphilic *Bacillus* species uses the H^+ ions and also employs sodium ion solute uptake taken from extracellular environment for solute transport systems. The bacterial Na^+/H^+ antiporters help in uptake of cytoplasmic H^+. Alkaliphiles have more sodium proton antiporters, sodium-dependent flagellum rotor proteins, and F_1F_0-ATPase pump (Wernick et al. 2013).

5 Role of Extremophiles in Waste Treatment

Due to urbanization, the expansion of industrial activities has increased the various contaminations in the environment. Human activities create waste, and these wastes can pose risks to both the environment and public health. Extremophiles can produce different novel enzymes which are stable in harsh environments. They have potential to be used for waste degradation and bioremediation.

5.1 Bioremediations Using Thermophiles

For bioremediation, there are two widely used approaches: bio-absorption and bioaccumulation. Thermophiles are always preferred for bioremediation of heavy metals. Some thermophiles can tolerate high metal concentration which may increase metal solubilization by oxidation. Thermophiles are able to reduce a wide spectrum of metals such as Mn, U, Tc, Cr, Co, Mo, Au, and Hg. A thermophilic bacterium *Thermus thermophilus* is able to tolerate very high concentration of arsenate and arsenite. It has been reported that *Thermus scotoductus* and *Thermoanaerobacter* sp. are able to reduce enzymatic uranium and technetium. For the biodegradation of hydrocarbons, use of thermophiles is well known. Also, thermophiles have high potentiality for bioremediation of heavy metals of groundwater and surface water. They are also used for the removal of organic compounds such as aliphatic and aromatic hydrocarbons and synthetic dyes (Sar et al. 2013).

Thermophiles survive at temperature, like hot springs, volcanic environments, fumaroles, geysers, and deep-sea hydrothermal vents. Their enzymes remain active even at very high temperature. *Bacillus* sp. from hot springs have industrial applications (Derekova et al. 2008; Kumar et al. 2013; Panda et al. 2016). Metagenomic studies have shown great diversity of thermophiles present in the hot springs (Tekere 2011).

Anoxybacillus sp. produces hydrolytic enzymes and oxidoreductases that are useful for bioremediation of wastewater (Jardine et al. 2018). Also, thermophiles can reduce phenol, a constituent of many pollutants (Jardine et al. 2018). Enzymes produced by the *Anoxybacillus* sp. may reduce pollutants from food industries, viz., polyaromatic hydrocarbons dyes, antibiotic residues, phosphates, and heavy metals (Jardine et al. 2018). Thermophilic enzymes or thermozymes have maximum activity at elevated temperatures (Mehta et al. 2016).

Thermophilic molds also secrete thermostable enzymes capable of degradation of organic and other toxic contaminants and hence have potential applications in bioremediation of industrial wastes and effluents (Singh and Satyanarayana 2009; Singh et al. 2016). *Talaromyces emersonii*, *Mucor* sp., *Rhizopus* sp., and *Thermomucorindicae seudaticae* are some of the examples of thermophilic molds that are employed in the bioremediation of polluted water and decolorization of dyes (Singh et al. 2016). Enzymatic technetium reduction have been shown in thermophilic microorganisms, such as *Thermus scotoductus*, *Pyrobaculumis landicum*, *Thermoanaerobacter* sp., and *Thermoterrabacterium ferrireducens* (Chernyh et al. 2007). Considering the various advantages, bioremediation of wastes using thermophiles and thermophilic proteins (enzymes) seems to be a promising tool to increase the efficiency of the process of bioremediation of wastes (Urbieta et al. 2015).

5.2 Bioremediation Through Psychrophiles

Psychrophiles are known to be used for the bioremediation of polluted cold soils and wastewater. At low temperature, *Arthrobacter psychrolactophilus* degrades organic compounds and clarifies synthetic wastewater turbid medium (Margesin and Feller 2010). Some psychrophiles can act as a biofertilizer at low temperature. They degrade xenobiotic compounds which are man-made synthetic compounds. Enzymes produced by psychrophiles act on organic pollutants. Psychrophiles degrades toxic compounds into nontoxic substances (Kumar et al. 2019). Oil spills in marine water are one of the reasons of water pollution. During Deepwater Horizon spill, initial phases were dominated by *Oceanospirillales* which consume a variety of alkanes (Hazen et al. 2010), followed by dominance of *Colwellia* and *Cycloclasticus* that consume propane, ethane, and butane and BTEX (benzene, toluene, ethyl benzene, and xylenes), respectively (Redmond and Valentine 2012). Finally, there is dominance of *Flavobacteria* (*Tenacibaculum* and *Polaribacter*), *Alteromonadaceae*, and *Rhodobacteraceae* (Dubinsky et al. 2013) which degrade high molecular weight organics and dissolved organic matter in the marine water.

5.3 Bioremediation by Halophiles

Polyhydroxyalkanoates (PHAs) and polyhydroxybutyrate (PHB) are storage materials that are accumulated within bacteria as a source of energy and carbon reserve. The interest in these polymers is due to its unique characteristics of biodegradable, eco-friendly, and also biocompatible in nature. *Cupriavidus necator*, *Alcaligenes latus*, *Azotobacter vinelandii*, *Methylobacterium extorquens*, *Pseudomonas*, and recombinant *Escherichia coli* are some of the bacteria that accumulate PHAs (Lee 1996; Steinbüchel and Füchtenbusch 1998). Among halophiles, *Haloferax mediterranei* is so far the best PHA producer. Some other examples of haloarchaea that can synthesize PHA are *Haloterrigena hispanica*, *Haloquadratum walsbyi*, *Halorhabdus tiamatea*, *Halorhabdus utahensis*, *Halopiger aswanensis*, *Halobiforma haloterrestris*, and *Natrinema altunense* (Wainø et al. 2000; Xu et al. 2005; Burns et al. 2007; Romano et al. 2007; Antunes et al. 2008; Hezayen et al. 2010). Halophiles that produce poly-3′-hydroxybutyrate (PHB) are obtained from high salt sludge that are capable of producing bio-plastic (Hermann-Krauss et al. 2013; Legat et al. 2010). PHB biopolymer can be isolated from a halophilic strain

Halomonas boliviensis in its stationary phase (Quillaguamán et al. 2005). PHB biopolymer can be obtained by culturing with glucose, sucrose, volatile fatty acid, and hydrolyzed starch in fed-batch culture medium, making it cheapest and easily made PHB biopolymer (Quillaguamán et al. 2005). Biofuels are considered the best substitute for fossil fuel. Among all the biofuel, bioethanol is known as the best substitution. It has been shown that under aerobic and anaerobic conditions, *Nesterenkonia* sp., a moderately halophilic bacterium, is capable of producing butanol, ethanol, and acetone (Amiri et al. 2016).

Few strains of halophilic microorganisms can generate organophosphorus acid anhydrases (OPAA) which have a strong hydrolytic property to denature organophosphorus chemicals and its derivative. Decontamination of chemical substances can be done by these biocatalysts (DeFrank et al. 1993). By using recombinant DNA technology, cloning of OPAA compound facilitates the isolation and characterization of specific OPAA coding genetic segment which can help in detoxification of organophosphorus substances.

5.4 Bioremediation by Acidophiles

Acidophilic archaea and some thermophilic bacteria are capable of pollutant degradation from industrial wastewater, as it is hot and acidic in nature. *Sulfolobus solfataricus* can degrade phenol at 80 °C and pH 3.2 (Christen et al. 2011, 2012; Comte et al. 2013). *Thiomonas arsenitoxydans*, *Acidithiobacillus caldus*, and *Acidithiobacillus ferrooxidans* are some examples of acidophiles that can tolerate significantly high loads of heavy metals (Navarro et al. 2013). *Acidithiobacillus ferrooxidans* is widely used in mineral bioleaching (Ramos-Zúñiga et al. 2019). Similarly, acid mine drainage contributes a major portion of heavy metal waste by releasing highly acidic effluent. *Ferroplasma* spp. are acidophilic metal oxidizers which has optimal growth at very low pH (Edwards et al. 2000; Golyshina and Timmis 2005). Waste electric and electronic equipment generates huge amount of heavy metals like copper, lead, zinc, and nickel. Studies showed that a mixed culture of *A. ferrooxidans* and *A. thiooxidans* is more efficient in metal recovery than their pure culture (Wang et al. 2009; Liang et al. 2010). Use of mixotrophic acidophiles is shown to significantly increase cadmium removal from cadmium-contaminated soils (Hao et al. 2019). Low-pH iron oxidation is also used as a bioremediation strategy. Although sulfate is not considered as toxic, but it is still very harmful when present in wastewater. Sulfate can be reduced to zero-valent sulfur (ZVS) in order to remove it from wastewater. Autotrophic acidophilic and neutrophilic bacteria catalyze conversion of gaseous H_2S to ZVS under low redox conditions. Sulfur reducers usually grow at a broad spectrum of both temperature ranging from 2 to 110 °C and pH ranging from 1 to 10.5 (Johnson and Sánchez-Andrea 2019).

5.5 Bioremediation by Basophiles

Tributyltin (TBT) is present in waste affluent discharged from various industries such as wood, textiles, paper mill, and breweries. It is toxic to aquatic fauna and humans. Alkaliphilic bacteria *Stenotrophomonas chelatiphaga* decreases the TBT concentration (Hassan et al. 2018). Further, phenol toxic to aquatic terrestrial fauna and humans is degraded by *Arthrobacter* spp. which is obligate alkaliphilic bacteria (Kanekar et al. 1999). Hydrocarbon removal from high pH industrial wastes is also treated by alkaliphiles (Margesin and Schinner 2001). In textile industry, waste discharge contains various synthetic dyes. Synthetic dyes deplete the dissolved oxygen and hence affect the aquatic life. Biodegradation of azo dyes is made by alkaliphilic bacterial consortium in effluents of textile industry. Other bacterial species capable to remove dyes of textile effluents are *Bacteroides* spp., *Eubacterium* spp., *Clostridium* spp., *Proteus vulgaris*, *Streptococcus faecalis*, *Bacillus* spp., and *Sphingomonas* (Wuhrmann et al. 1980; Rafii et al. 1990; Bragger et al. 1997). Gold mining and jewelry industry generate large amount of cyanide waste. *Pseudomonas pseudoalcaligenes* uses cyanide, cyanate, and different metal-cyanide complexes as the sole nitrogen source (Luque-Almagro et al. 2005, 2008). Industrial effluent also contains nitrile as pollutant at high pH/salt conditions. *Comamonas* sp. (Manolov et al. 2005), *Pseudomonas putida* (Chapatwala et al. 1993), and *Rhodococcus* spp. (Blakey et al. 1995; Kohyama et al. 2006) use acetonitrile as source of carbon, nitrogen, and energy. They possess the nitrile hydratase/amidase enzymatic pathway. *Natronocella acetinitrilica* is high salt-tolerant, and obligate alkaliphile can be used for removal of nitriles as they use aliphatic nitriles as carbon and energy source (Sorokin et al. 2007).

6 Conclusion

Extremophiles have a huge potential in bioremediation as these organisms survive at extreme conditions and majority of pollutants are also present in extreme environment. Extremophilic proteins are functional in respective extreme conditions due to adaptations in protein structure. Many extremophilic organisms are being used in bioremediation. Genetic engineering techniques offer further opportunities to improve different strains or extremophilic proteins to make improvements in bioremediation.

Funding This project has been funded by "UGC-BSR Research Start-Up-Grant project No. F. 30-487/2019(BSR) sanctioned to Ashutosh Kumar."

References

Ahier BA, Tracy BL (1995) Radionuclides in the Great Lakes basin. Environ Health Perspect 103 (Suppl):89–101. https://doi.org/10.1289/ehp.95103s989

Alencar FLS, Araújo MFF, Do Nascimento ED (2016) Microbiology for environmental conservation: a systematic review of bioremediation of heavy metals by Chromobacterium violaceum. Gaia Sci 10:320–333

Amiri H, Azarbaijani R, Parsa Yeganeh L, Shahzadeh Fazeli A, Tabatabaei M, Hosseini Salekdeh G, Karimi K (2016) Nesterenkonia sp. strain F, a halophilic bacterium producing acetone, butanol, and ethanol under aerobic conditions. Sci Rep 6:18408. https://doi.org/10.1038/srep18408

Antunes A, Taborda M, Huber R, Moissl C, Nobre MF, da Costa MS (2008) Halorhabdus tiamatea sp. nov., a non-pigmented extremely halophilic archaeon from a deep-sea hypersaline anoxic basin of the Red Sea, and emended description of the genus Halorhabdus. Int J Syst Evol Microbiol 58:215–220. https://doi.org/10.1099/ijs.0.65316-0

Appukuttan D, Rao AS, Apte SK (2006) Engineering of Deinococcus radiodurans R1 for bioprecipitation of uranium from dilute nuclear waste. Appl Environ Microbiol 72:7873–7878. https://doi.org/10.1128/AEM.01362-06

Baggi F, Demarta A, Peduzzi R (2001) Persistence of viral pathogens and bacteriophages during sewage treatment: lack of correlation with indicator bacteria. Res Microbiol 152:743–751. https://doi.org/10.1016/S0923-2508(01)01255-4

Baker-Austin C, Dopson M (2007) Life in acid: pH homeostasis in acidophiles. Trends Microbiol 15:165–171

Bar Dolev M, Braslavsky I, Davies PL (2016) Ice-binding proteins and their function. Annu Rev Biochem 85:515–542. https://doi.org/10.1146/annurev-biochem-060815-014546

Bergqvist S, Williams MA, O'Brien R, Ladbury JE (2003) Halophilic adaptation of protein-DNA interactions. Biochem Soc Trans 31:677–680. https://doi.org/10.1042/bst0310677

Bibby K, Peccia J (2013) Identification of viral pathogen diversity in sewage sludge by metagenome analysis. Environ Sci Technol 47:1945–1951. https://doi.org/10.1021/es305181x

Blakey AJ, Colby J, Williams E, O'Reilly C (1995) Regio- and stereo-specific nitrile hydrolysis by the nitrile hydratase from Rhodococcus AJ270. FEMS Microbiol Lett 129:57–61. https://doi.org/10.1016/0378-1097(95)00135-R

Boutz DR, Cascio D, Whitelegge J, Perry LJ, Yeates TO (2007) Discovery of a thermophilic protein complex stabilized by topologically interlinked chains. J Mol Biol 368:1332–1344. https://doi.org/10.1016/j.jmb.2007.02.078

Bragger JL, Lloyd AW, Soozandehfar SH, Bloomfield SF, Marriott C, Martin GP (1997) Investigations into the azo reducing activity of a common colonic microorganism. Int J Pharm 157:61–71. https://doi.org/10.1016/S0378-5173(97)00214-7

Buetti-Dinh A, Dethlefsen O, Friedman R, Dopson M (2016) Transcriptomic analysis reveals how a lack of potassium ions increases Sulfolobus acidocaldarius sensitivity to pH changes. Microbiology 162(8):1422–1434. https://doi.org/10.1099/mic.0.000314

Burns DG, Janssen PH, Itoh T, Kamekura M, Li Z, Jensen G, Rodríguez-Valera F, Bolhuis H, Dyall-Smith ML (2007) Haloquadratum walsbyi gen. nov., sp. nov., the square haloarchaeon of Walsby, isolated from saltern crystallizers in Australia and Spain. Int J Syst Evol Microbiol 57:387–392. https://doi.org/10.1099/ijs.0.64690-0

Cacciapuoti G, Fuccio F, Petraccone L, Del Vecchio P, Porcelli M (2012) Role of disulfide bonds in conformational stability and folding of 5′-deoxy-5′-methylthioadenosine phosphorylase II from the hyperthermophilic archaeon Sulfolobus solfataricus. Biochim Biophys Acta Proteins Proteomics 1824:1136–1143. https://doi.org/10.1016/j.bbapap.2012.06.014

Camargo JA, Alonso Á (2006) Ecological and toxicological effects of inorganic nitrogen pollution in aquatic ecosystems: a global assessment. Environ Int 32:831–849

Chapatwala KD, Babu GRV, Dudley C, Williams R, Aremu K (1993) Degradative capability of Pseudomonas putida on acetonitrile. Appl Biochem Biotechnol 39–40:655–666. https://doi.org/10.1007/BF02919026

Chernyh NA, Gavrilov SN, Sorokin VV, German KE, Sergeant C, Simonoff M, Robb F, Slobodkin AI (2007) Characterization of technetium(VII) reduction by cell suspensions of thermophilic bacteria and archaea. Appl Microbiol Biotechnol 76:467–472. https://doi.org/10.1007/s00253-007-1034-5

Chintalapati S, Kiran MD, Shivaji S (2004) Role of membrane lipid fatty acids in cold adaptation. Cell Mol Biol 50:631–642

Christel S, Herold M, Bellenberg S, El Hajjami M, Buetti-Dinh A, Pivkin IV, Sand W, Wilmes P, Poetsch A, Dopson M (2018) Multiomics reveals the lifestyle of the acidophilic, mineral-oxidizing model species Leptospirillum ferriphilum^{T}. Appl Environ Microbiol 84. https://doi.org/10.1128/AEM.02091-17

Christen P, Davidson S, Combet-Blanc Y, Auria R (2011) Phenol biodegradation by the thermoacidophilic archaeon Sulfolobus solfataricus 98/2 in a fed-batch bioreactor. Biodegradation 22:475–484. https://doi.org/10.1007/s10532-010-9420-6

Christen P, Vega A, Casalot L, Simon G, Auria R (2012) Kinetics of aerobic phenol biodegradation by the acidophilic and hyperthermophilic archaeon Sulfolobus solfataricus 98/2. Biochem Eng J 62:56–61. https://doi.org/10.1016/j.bej.2011.12.012

Comte A, Christen P, Davidson S, Pophillat M, Lorquin J, Auria R, Simon G, Casalot L (2013) Biochemical, transcriptional and translational evidences of the phenol-meta-degradation pathway by the hyperthermophilic Sulfolobus solfataricus 98/2. PLoS One 8:e82397. https://doi.org/10.1371/journal.pone.0082397

D'Amico S, Collins T, Marx JC, Feller G, Gerday C (2006) Psychrophilic microorganisms: challenges for life. EMBO Rep 7:385–389

DeFrank JJ, Beaudry WT, Cheng TC, Harvey SP, Stroup AN, Szafraniec LL (1993) Screening of halophilic bacteria and Alteromonas species for organophosphorus hydrolyzing enzyme activity. Chem Biol Interact 87:141–148. https://doi.org/10.1016/0009-2797(93)90035-W

DeMarini DM, Houk VS (1988) Assessment of hazardous wastes for genotoxicity. In: Hazardous waste: detection, control, treatment. Elsevier, Amsterdam, pp 1107–1115

DeMarini DM, Inmon JP, Simmons JE, Berman E, Pasley TC, Warren SH, Williams RW (1987) Mutagenicity in Salmonella of hazardous wastes and urine from rats fed these wastes. Mutat Res Toxicol 189:205–216. https://doi.org/10.1016/0165-1218(87)90054-1

DeMarini DM, Gallagher JE, Houk VS, Simmons JE (1989) Toxicological evaluation of complex industrial wastes: implications for exposure assessment. Toxicol Lett 49:199–214. https://doi.org/10.1016/0378-4274(89)90033-7

Derekova A, Mandeva R, Kambourova M (2008) Phylogenetic diversity of thermophilic carbohydrate degrading bacilli from Bulgarian hot springs. World J Microbiol Biotechnol 24:1697–1702. https://doi.org/10.1007/s11274-008-9663-0

Donnelly KC, Brown KW, Scott BR (1983) The use of short-term bioassays to monitor the environmental impact of land treatment of hazardous wastes. In: Short-term bioassays in the analysis of complex environmental mixtures III. Springer, Boston, MA, pp 59–78

Donnelly KC, Brown KW, Kampbell D (1987a) Chemical and biological characterization of hazardous industrial waste. I. Prokaryotic bioassays and chemical analysis of a wood-preserving bottom-sediment waste. Mutat Res Fundam Mol Mech Mutagen 180:31–42. https://doi.org/10.1016/0027-5107(87)90064-9

Donnelly KC, Brown KW, Scott BR (1987b) Chemical and biological characterization of hazardous industrial waste. II. Eukaryotic bioassay of a wood-preserving of a wood-preserving bottom sediment. Mutat Res Fundam Mol Mech Mutagen 180:43–53. https://doi.org/10.1016/0027-5107(87)90065-0

Dubinsky EA, Conrad ME, Chakraborty R, Bill M, Borglin SE, Hollibaugh JT, Mason OU, M Piceno Y, Reid FC, Stringfellow WT, Tom LM, Hazen TC, Andersen GL (2013) Succession of hydrocarbon-degrading bacteria in the aftermath of the deepwater horizon oil spill in the gulf of Mexico. Environ Sci Technol 47:10860–10867. https://doi.org/10.1021/es401676y

Dworkin M, Falkow S, Rosenberg E, Schleifer K-H, Stackebrandt E (2006) The prokaryotes: a handbook on the biology of bacteria: archaea, Bacteria: Firmicutes, Actinomycetes. Springer, New York

Edwards KJ, Bond PL, Gihring TM, Banfield JF (2000) An archaeal iron-oxidizing extreme acidophile important in acid mine drainage. Science 287:1796–1799. https://doi.org/10.1126/science.287.5459.1796

Feller G (2010) Protein stability and enzyme activity at extreme biological temperatures. J Phys Condens Matter 22:323101. https://doi.org/10.1088/0953-8984/22/32/323101

Frenkel VS, Cummings GA, Maillacheruvu KY, Tang WZ (2017) Food-processing wastes. Water Environ Res 89:1360–1383. https://doi.org/10.2175/106143017X15023776270368

Gerday C, Glansdorff N (2007) Physiology and biochemistry of extremophiles. American Society of Microbiology, Washington, DC

Golyshina OV, Timmis KN (2005) Ferroplasma and relatives, recently discovered cell wall-lacking archaea making a living in extremely acid, heavy metal-rich environments. Environ Microbiol 7:1277–1288

Gunde-Cimerman N, Plemenitaš A, Oren A (2018) Strategies of adaptation of microorganisms of the three domains of life to high salt concentrations. FEMS Microbiol Rev 42:353–375

Hao X, Zhu P, Zhang H, Liang Y, Yin H, Liu X, Bai L, Liu H, Jiang H (2019) Mixotrophic acidophiles increase cadmium soluble fraction and phytoextraction efficiency from cadmium contaminated soils. Sci Total Environ 655:347–355. https://doi.org/10.1016/j.scitotenv.2018.11.221

Haney PJ, Badger JH, Buldak GL, Reich CI, Woese CR, Olsen GJ (1999) Thermal adaptation analyzed by comparison of protein sequences from mesophilic and extremely thermophilic Methanococcus species. Proc Natl Acad Sci U S A. 96(7):3578–3583. https://doi.org/10.1073/pnas.96.7.3578

Hassan HA, Dawah SE, El-Sheekh MM (2018) Monitoring the degradation capability of novel haloalkaliphilic tributyltin chloride (TBTCl) resistant bacteria from butyltin-polluted site. Rev Argent Microbiol 51:39–46. https://doi.org/10.1016/j.ram.2017.12.002

Hazen TC, Dubinsky EA, DeSantis TZ, Andersen GL, Piceno YM, Singh N, Jansson JK, Probst A, Borglin SE, Fortney JL, Stringfellow WT, Bill M, Conrad ME, Tom LM, Chavarria KL, Alusi TR, Lamendella R, Joyner DC, Spier C, Baelum J, Auer M, Zemla ML, Chakraborty R, Sonnenthal EL, D'Haeseleer P, Holman HYN, Osman S, Lu Z, Van Nostrand JD, Deng Y, Zhou J, Mason OU (2010) Deep-sea oil plume enriches indigenous oil-degrading bacteria. Science 330:204–208. https://doi.org/10.1126/science.1195979

Hermann-Krauss C, Koller M, Muhr A, Fasl H, Stelzer F, Braunegg G (2013) Archaeal production of polyhydroxyalkanoate (PHA) co- and terpolyesters from biodiesel industry-derived by-products. Archaea 2013:1–10. https://doi.org/10.1155/2013/129268

Hezayen FF, Gutiérrez MC, Steinbüchel A, Tindall BJ, Rehm BHA (2010) Halopiger aswanensis sp. nov., a polymer-producing and extremely halophilic archaeon isolated from hypersaline soil. Int J Syst Evol Microbiol 60:633–637. https://doi.org/10.1099/ijs.0.013078-0

Houk VS (1992) The genotoxicity of industrial wastes and effluents. Mutat Res Genet Toxicol 277:91–138. https://doi.org/10.1016/0165-1110(92)90001-P

Huang F, Guo CL, Lu GN, Yi XY, Zhu LD, Dang Z (2014) Bioaccumulation characterization of cadmium by growing Bacillus cereus RC-1 and its mechanism. Chemosphere 109:134–142. https://doi.org/10.1016/j.chemosphere.2014.01.066

Janssen DB, Dinkla IJT, Poelarends GJ, Terpstra P (2005) Bacterial degradation of xenobiotic compounds: evolution and distribution of novel enzyme activities. Environ Microbiol 7:1868–1882. https://doi.org/10.1111/j.1462-2920.2005.00966.x

Jardine JL, Stoychev S, Mavumengwana V, Ubomba-Jaswa E (2018) Screening of potential bioremediation enzymes from hot spring bacteria using conventional plate assays and liquid chromatography - tandem mass spectrometry (Lc-Ms/Ms). J Environ Manag 223:787–796. https://doi.org/10.1016/j.jenvman.2018.06.089

Johnson DB, Sánchez-Andrea I (2019) Dissimilatory reduction of sulfate and zero-valent sulfur at low pH and its significance for bioremediation and metal recovery. Adv Microb Physiol 75:205–231. https://doi.org/10.1016/bs.ampbs.2019.07.002

Kamika I, Momba MN (2013) Assessing the resistance and bioremediation ability of selected bacterial and protozoan species to heavy metals in metal-rich industrial wastewater. BMC Microbiol 13:28. https://doi.org/10.1186/1471-2180-13-28

Kanekar PP, Sarnaik SS, Kelkar AS (1999) Bioremediation of phenol by alkaliphilic bacteria isolated from alkaline lake of Lonar, India. J Appl Microbiol 85(Suppl 1):128S–133S

Keasling JD, Van Dien SJ, Trelstad P, Renninger N, McMahon K-A (2000) Application of polyphosphate metabolism to environmental and biotechnological problems. Biochemist 65:324–331

Kohyama E, Yoshimura A, Aoshima D, Yoshida T, Kawamoto H, Nagasawa T (2006) Convenient treatment of acetonitrile-containing wastes using the tandem combination of nitrile hydratase and amidase-producing microorganisms. Appl Microbiol Biotechnol 72:600–606. https://doi.org/10.1007/s00253-005-0298-x

Kumar P, Patel SKS, Lee JK, Kalia VC (2013) Extending the limits of Bacillus for novel biotechnological applications. Biotechnol Adv 31:1543–1561

Kumar A, Alam A, Tripathi D, Rani M, Khatoon H, Pandey S, Ehtesham NZ, Hasnain SE (2018) Protein adaptations in extremophiles: an insight into extremophilic connection of mycobacterial proteome. Semin Cell Dev Biol 84:147–157. https://doi.org/10.1016/j.semcdb.2018.01.003

Kumar R, Kumar P, Giri A (2019) Regional impact of psychrophilic bacteria on bioremediation. In: Smart bioremediation technologies. Elsevier, Amsterdam, pp 119–135

Kumaraswamy R, Amha YM, Anwar MZ, Henschel A, Rodríguez J, Ahmad F (2014) Molecular analysis for screening human bacterial pathogens in municipal wastewater treatment and reuse. Environ Sci Technol 48:11610–11619. https://doi.org/10.1021/es502546t

Lee SY (1996) Plastic bacteria? Progress and prospects for polyhydroxyalkanoate production in bacteria. Trends Biotechnol 14:431–438

Legat A, Gruber C, Zangger K, Wanner G, Stan-Lotter H (2010) Identification of polyhydroxyalkanoates in Halococcus and other haloarchaeal species. Appl Microbiol Biotechnol 87:1119–1127. https://doi.org/10.1007/s00253-010-2611-6

Liang G, Mo Y, Zhou Q (2010) Novel strategies of bioleaching metals from printed circuit boards (PCBs) in mixed cultivation of two acidophiles. Enzym Microb Technol 47:322–326. https://doi.org/10.1016/j.enzmictec.2010.08.002

Lloyd JR, Macaskie LE (1996) A novel PhosphorImager-based technique for monitoring the microbial reduction of technetium. Appl Environ Microbiol 62:578–582. https://doi.org/10.1128/aem.62.2.578-582.1996

Lloyd JR, Leang C, Hodges Myerson AL, Coppi MV, Cuifo S, Methe B, Sandler SJ, Lovley DR (2003) Biochemical and genetic characterization of PpcA, a periplasmic c-type cytochrome in Geobacter sulfurreducens. Biochem J 369:153–161. https://doi.org/10.1042/BJ20020597

Luque-Almagro VM, Blasco R, Huertas MJ, Martínez-Luque M, Moreno-Vivián C, Castillo F, Roldán MD (2005) Alkaline cyanide biodegradation by Pseudomonas pseudoalcaligenes CECT5344. Biochem Soc Trans 33(Pt 1):168–169

Luque-Almagro VM, Huertas MJ, Sáez LP, Luque-Romero MM, Moreno-Vivián C, Castillo F, Roldán MD, Blasco R (2008) Characterization of the Pseudomonas pseudoalcaligenes CECT5344 cyanase, an enzyme that is not essential for cyanide assimilation. Appl Environ Microbiol 74:6280–6288. https://doi.org/10.1128/AEM.00916-08

Maiti SK, Bera D, Chattopadhyay P, Ray L (2009) Determination of kinetic parameters in the biosorption of Cr(VI) on immobilized Bacillus cereus M116 in a continuous packed bed column reactor. Appl Biochem Biotechnol 159:488–504. https://doi.org/10.1007/s12010-008-8519-2

Majumder S, Gangadhar G, Raghuvanshi S, Gupta S (2015) A comprehensive study on the behavior of a novel bacterial strain Acinetobacter guillouiae for bioremediation of divalent copper. Bioprocess Biosyst Eng 38:1749–1760. https://doi.org/10.1007/s00449-015-1416-5

Mamo G (2019) Challenges and adaptations of life in alkaline habitats. Adv Biochem Eng Biotechnol. https://doi.org/10.1007/10_2019_97

Manolov T, Kristina H, Benoit G (2005) Continuous acetonitrile degradation in a packed-bed bioreactor. Appl Microbiol Biotechnol 66:567–574. https://doi.org/10.1007/s00253-004-1744-x

Margesin R, Feller G (2010) Biotechnological applications of psychrophiles. Environ Technol 31:835–844

Margesin R, Schinner F (2001) Biodegradation and bioremediation of hydrocarbons in extreme environments. Appl Microbiol Biotechnol 56:650–663

Martins PSDO, De Almeida NF, Leite SGF (2008) Application of a bacterial extracellular polymeric substance in heavy metal adsorption in a co-contaminated aqueous system. Braz J Microbiol 39:780–786. https://doi.org/10.1590/S1517-83822008000400034

Matin A (1999) pH homeostasis in acidophiles. Novartis Found Symp 221:152–163. https://doi.org/10.1002/9780470515631.ch10

McGeorge LJ, Louis JB, Atherholt TB, McGarrity GJ (1985) Mutagenicity analyses of industrial effluents: results and considerations for integration into water pollution control programs. In: Short-term bioassays in the analysis of complex environmental mixtures IV. Springer, Boston, MA, pp 247–268

Meckenstock RU, Elsner M, Griebler C, Lueders T, Stumpp C, Aamand J, Agathos SN, Albrechtsen HJ, Bastiaens L, Bjerg PL, Boon N, Dejonghe W, Huang WE, Schmidt SI, Smolders E, Sørensen SR, Springael D, Van Breukelen BM (2015) Biodegradation: updating the concepts of control for microbial cleanup in contaminated aquifers. Environ Sci Technol 49:7073–7081. https://doi.org/10.1021/acs.est.5b00715

Mehta R, Singhal P, Singh H, Damle D, Sharma AK (2016) Insight into thermophiles and their wide-spectrum applications. 3 Biotech 6:1–9

Mevarech M, Frolow F, Gloss LM (2000) Halophilic enzymes: proteins with a grain of salt. Biophys Chem 86(2–3):155–164

Mishra GK (2017) Microbes in heavy metal remediation: a review on current trends and patents. Recent Pat Biotechnol 11:188–196. https://doi.org/10.2174/1872208311666170120121025

Misra CS, Appukuttan D, Kantamreddi VSS, Rao AS, Apte SK (2012) Recombinant D. radiodurans cells for bioremediation of heavy metals from acidic/neutral aqueous wastes. Bioeng Bugs 3:44–48. https://doi.org/10.4161/bbug.3.1.18878

Navarro CA, von Bernath D, Jerez CA (2013) Heavy metal resistance strategies of acidophilic bacteria and their acquisition: importance for biomining and bioremediation. Biol Res 46:363–371

Nilgiriwala KS, Alahari A, Rao AS, Apte SK (2008) Cloning and overexpression of alkaline phosphatase PhoK from Sphingomonas sp. strain BSAR-1 for bioprecipitation of uranium from alkaline solutions. Appl Environ Microbiol 74:5516–5523. https://doi.org/10.1128/AEM.00107-08

Ortiz-Bernad I, Anderson RT, Vrionis HA, Lovley DR (2004) Vanadium respiration by Geobacter metallireducens: novel strategy for in situ removal of vanadium from groundwater. Appl Environ Microbiol 70:3091–3095. https://doi.org/10.1128/AEM.70.5.3091-3095.2004

Osuolale O, Okoh A (2015) Incidence of human adenoviruses and hepatitis A virus in the final effluent of selected wastewater treatment plants in Eastern Cape Province, South Africa. Virol J 12:98. https://doi.org/10.1186/s12985-015-0327-z

Panda AK, Bisht SS, De Mandal S, Kumar NS (2016) Bacterial and archeal community composition in hot springs from Indo-Burma region, North-east India. AMB Express 6:111. https://doi.org/10.1186/s13568-016-0284-y

Paredes DI, Watters K, Pitman DJ, Bystroff C, Dordick JS (2011) Comparative void-volume analysis of psychrophilic and mesophilic enzymes: structural bioinformatics of psychrophilic enzymes reveals sources of core flexibility. BMC Struct Biol 11:42. https://doi.org/10.1186/1472-6807-11-42

Parrilli E, Tedesco P, Fondi M, Tutino ML, Lo Giudice A, de Pascale D, Fani R (2019) The art of adapting to extreme environments: the model system Pseudoalteromonas. Phys Life Rev. https://doi.org/10.1016/j.plrev.2019.04.003

Petrie B, Barden R, Kasprzyk-Hordern B (2015) A review on emerging contaminants in wastewaters and the environment: Current knowledge, understudied areas and recommendations for future monitoring. Water Res., occurrence, fate, removal and assessment of emerging contaminants in water in the water cycle (from wastewater to drinking water) 72:3–27. https://doi.org/10.1016/j.watres.2014.08.053

Pirog T, Sofilkanych A, Shevchuk T, Shulyakova M (2013) Biosurfactants of rhodococcus erythropolis IMV Ac-5017: synthesis intensification and practical application. Appl Biochem Biotechnol 170:880–894. https://doi.org/10.1007/s12010-013-0246-7

Pouillot R, van Doren JM, Woods J, Plante D, Smith M, Goblick G, Roberts C, Locas A, Hajen W, Stobo J, White J, Holtzman J, Buenaventura E, Burkhardt W, Catford A, Edwards R,

DePaola A, Calci KR (2015) Meta-analysis of the reduction of norovirus and male-specific coliphage concentrations in wastewater treatment plants. Appl Environ Microbiol 81:4669–4681. https://doi.org/10.1128/AEM.00509-15

Prakash D, Gabani P, Chandel AK, Ronen Z, Singh OV (2013) Bioremediation: a genuine technology to remediate radionuclides from the environment. Microb Biotechnol 6:349–360. https://doi.org/10.1111/1751-7915.12059

Quillaguamán J, Hashim S, Bento F, Mattiasson B, Hatti-Kaul R (2005) Poly(β-hydroxybutyrate) production by a moderate halophile, Halomonas boliviensis LC1 using starch hydrolysate as substrate. J Appl Microbiol 99:151–157. https://doi.org/10.1111/j.1365-2672.2005.02589.x

Rafii F, Franklin W, Cerniglia CE (1990) Azoreductase activity of anaerobic bacteria isolated from human intestinal microflora. Appl Environ Microbiol 56:2146–2151. https://doi.org/10.1128/aem.56.7.2146-2151.1990

Raghu G, Balaji V, Venkateswaran G, Rodrigue A, Maruthi Mohan P (2008) Bioremediation of trace cobalt from simulated spent decontamination solutions of nuclear power reactors using E. coli expressing NiCoT genes. Appl Microbiol Biotechnol 81:571–578. https://doi.org/10.1007/s00253-008-1741-6

Ramos-Zúñiga J, Gallardo S, Martínez-Bussenius C, Norambuena R, Navarro CA, Paradela A, Jerez CA (2019) Response of the biomining Acidithiobacillus ferrooxidans to high cadmium concentrations. J Proteome 198:132–144. https://doi.org/10.1016/j.jprot.2018.12.013

Rampelotto PH (2013) Extremophiles and extreme environments. Life 3:482–485

Redmond MC, Valentine DL (2012) Natural gas and temperature structured a microbial community response to the deepwater horizon oil spill. Proc Natl Acad Sci U S A 109:20292–20297. https://doi.org/10.1073/pnas.1108756108

Reed CJ, Lewis H, Trejo E, Winston V, Evilia C (2013) Protein adaptations in archaeal extremophiles. Archaea 2013:1–14

Romano I, Poli A, Finore I, Huertas FJ, Gambacorta A, Pelliccione S, Nicolaus G, Lama L, Nicolaus B (2007) Haloterrigena hispanica sp. nov., an extremely halophilic archaeon from Fuente de Piedra, southern Spain. Int J Syst Evol Microbiol 57:1499–1503. https://doi.org/10.1099/ijs.0.64895-0

Russell NJ (2008) Membrane components and cold sensing. In: Psychrophiles: from biodiversity to biotechnology. Springer, Berlin, pp 177–190

Samin G, Janssen DB (2012) Transformation and biodegradation of 1,2,3-trichloropropane (TCP). Environ Sci Pollut Res Int 19(8):3067–3078. https://doi.org/10.1007/s11356-012-0859-3

Sar P, Kazy SK, Paul D, Sarkar A (2013) Metal bioremediation by thermophilic microorganisms. In: Thermophilic microbes in environmental and industrial biotechnology: biotechnology of thermophiles. Springer, Dordrecht, pp 171–201

Sen SK, Raut S, Dora TK, Mohapatra PK (2014) Contribution of hot spring bacterial consortium in cadmium and lead bioremediation through quadratic programming model. J Hazard Mater 265:47–60. https://doi.org/10.1016/j.jhazmat.2013.11.036

Singh B, Satyanarayana T (2009) Thermophilic molds in environmental management. In: Fungi from different environments. Science Publishers, Lucknow, India, pp 355–379

Singh B, Poças-Fonseca MJ, Johri BN, Satyanarayana T (2016) Thermophilic molds: biology and applications. Crit Rev Microbiol 42:985–1006

Sinha S, Mukherjee SK (2009) Pseudomonas aeruginosa KUCd1, a possible candidate for cadmium bioremediation. Braz J Microbiol 40:655–662. https://doi.org/10.1590/s1517-83822009000300030

Somani SM, Teece RG, Schaeffer DJ (1980) Identification of cocarcinogens and promoters in industrial discharges into and in the Illinois river. J Toxicol Environ Health 6:315–331. https://doi.org/10.1080/15287398009529854

Sorokin DY, van Pett S, Tourova TP, Takaichi S, Muyzer G (2007) Acetonitrile degradation under haloalkaline conditions by Natronocella acetinitrilica gen. nov., sp. nov. Microbiology 153:1157–1164. https://doi.org/10.1099/mic.0.2006/004150-0

Spanggord RJ, Mortelmans KE, Griffin AF, Simmon VF (1982) Mutagenicity in Salmonella typhimurium and structure-activity relationships of wastewater components emanating from the manufacture of trinitrotoluene. Environ Mutagen 4:163–179. https://doi.org/10.1002/em.2860040207

Steinbüchel A, Füchtenbusch B (1998) Bacterial and other biological systems for polyester production. Trends Biotechnol 16:419–427

Sundvall A, Marklund H, Rannug U (1984) The mutagenicity on Salmonella typhimurium of nitrobenzoic acids and other wastewater components generated in the production of nitrobenzoic acids and nitrotoluenes. Mutat Res Toxicol 137:71–78. https://doi.org/10.1016/0165-1218(84)90094-6

Tadeo X, López-Méndez B, Trigueros T, Laín A, Castaño D, Millet O (2009) Structural basis for the aminoacid composition of proteins from halophilic archaea. PLoS Biol 7:e1000257. https://doi.org/10.1371/journal.pbio.1000257

Tamponnet C, Declerck S (2008) Radionuclide (RN) pollution is a worldwide problem that arises from human activities. J Environ Radioact 99:773–774

Tekere M (2011) Metagenomic analysis of bacterial diversity of Siloam hot water spring, Limpopo, South Africa. African J Biotechnol 10. https://doi.org/10.5897/ajb11.899

Urbieta MS, Donati ER, Chan KG, Shahar S, Sin LL, Goh KM (2015) Thermophiles in the genomic era: biodiversity, science, and applications. Biotechnol Adv 33:633–647

Wainø M, Tindall BJ, Ingvorsen K (2000) Halorhabdus utahensis gen. nov., sp. nov., an aerobic, extremely halophilic member of the Archaea from Great Salt Lake, Utah. Int J Syst Evol Microbiol 50:183–190. https://doi.org/10.1099/00207713-50-1-183

Wang J, Bai J, Xu J, Liang B (2009) Bioleaching of metals from printed wire boards by Acidithiobacillus ferrooxidans and Acidithiobacillus thiooxidans and their mixture. J Hazard Mater 172:1100–1105. https://doi.org/10.1016/j.jhazmat.2009.07.102

Wernick DG, Choi KY, Tat CA, Lafontaine Rivera JG, Liao JC (2013) Genome sequence of the extreme obligate alkaliphile Bacillus marmarensis strain DSM 21297. Genome Announc 1. https://doi.org/10.1128/genomeA.00967-13

Wildung RE, Gorby YA, Krupka KM, Hess NJ, Li SW, Plymale AE, McKinley JP, Fredrickson JK (2000) Effect of electron donor and solution chemistry on products of dissimilatory reduction of technetium by Shewanella putrefaciens. Appl Environ Microbiol 66:2451–2460. https://doi.org/10.1128/AEM.66.6.2451-2460.2000

Wuhrmann K, Mechsner K, Kappeler T (1980) Investigation on rate - determining factors in the microbial reduction of azo dyes. Eur J Appl Microbiol Biotechnol 9:325–338. https://doi.org/10.1007/BF00508109

Xu XW, Ren PG, Liu SJ, Wu M, Zhou PJ (2005) Natrinema altunense sp. nov., an extremely halophilic archaeon isolated from a salt lake in Altun Mountain in Xinjiang, China. Int J Syst Evol Microbiol 55:1311–1314. https://doi.org/10.1099/ijs.0.63622-0

Yahalomi D, Atkinson SD, Neuhof M, Chang ES, Philippe H, Cartwright P, Bartholomew JL, Huchon D (2020) A cnidarian parasite of salmon (Myxozoa: Henneguya) lacks a mitochondrial genome. Proc Natl Acad Sci U S A 117:5358–5363. https://doi.org/10.1073/pnas.1909907117

Zawierucha I, Kozlowski C, Malina G (2016) Immobilized materials for removal of toxic metal ions from surface/groundwaters and aqueous waste streams. Environ Sci Process Impacts 18:429–444

Biosorption of Carcinogenic Heavy Metals by Bacteria: Role and Mechanism

Ravi Chauhan, Hardi Patel, and Seema Rawat

Abstract

The various kinds of anthropogenic activities, viz., modern farming, coal burning, mining, and industries, have led to the discharge of heavy metals in the environment which besides causing disaster in the environment may also cause acute and chronic diseases in human beings. The bioaccumulation and biomagnification of heavy metals in soil and water make it a major threat to biodiversity. International Agency for Research on Cancer has classified arsenic, beryllium, cadmium, chromium, and nickel as group 1 carcinogens. The conventional physicochemical technologies used for the removal of heavy metals are cost-ineffective and sometimes generate a huge amount of toxic by-products. Bioremediation is considered as one of the safer, cleaner, cost-effective, and environmentally friendly technology for decontaminating heavy metal-contaminated sites. Microbes possess several mechanisms for metal sequestration, one of which is biosorption. This chapter will focus on the role played by bacteria in metal sequestration and various mechanisms, like binding to EPS, metallothionein, cell surface adsorption, etc., used by them for biosorption of most carcinogenic heavy metals.

Keywords

Carcinogenic · Heavy metals · Bioremediation · Biosorption · Metallothionein

Ravi Chauhan and Hardi Patel have contributed equally to this work.

R. Chauhan · H. Patel · S. Rawat (✉)
Microbiology Lab, School of Life Sciences, Central University of Gujarat, Gandhinagar, Gujarat, India
e-mail: seema.rawat@cug.ac.in

M. P. Shah (ed.), *Removal of Emerging Contaminants Through Microbial Processes*,
https://doi.org/10.1007/978-981-15-5901-3_12

1 Introduction

Heavy metals are defined as naturally occurring elements with atomic weights usually above 50 and density above 5 g/cm^3 (Tchounwou et al. 2012; Jobby and Desai 2017). Weast (1984) has classified naturally occurring 90 elements as 21 non-metals, 16 light metals, and 53 heavy metals. Heavy metals, being naturally occurring elements, are present all over the earth's crust in the form of physical and chemical weathering of metamorphic and igneous rocks, volcanic eruptions, atmospheric deposition of airborne particles, wind erosion, forest fire smoke, decomposition of animal and plant waste matter, oceanic spray, etc. (Jobby and Desai 2017). However, the majority of environmental contamination and human exposure results due to metal corrosion, leaching of heavy metals, sediment resuspension and metal evaporation from metal-contaminated waterbodies to soil and ground waters, industrialization and technological development, anthropogenic activities like mining and smelting, domestic and agricultural use, refineries, coal burning in power plants, nuclear power plants, petroleum combustion, plastics, textiles, microelectronics, wood preservation and paper processing plants, etc. (Tchounwou et al. 2012).

Heavy metal pollution is one of the major environmental issues because the non-biodegradable nature of heavy metals leads to their prolonged prevalence in the environment. Most of the heavy metals are toxic even at low concentrations, and once entered in the food chain, they can bioaccumulate and biomagnify causing severe toxicity to humans, animals, plants, as well as microbes (Ayangbenro and Babalola 2017). Table 1 summarizes the sources of potent carcinogenic heavy metals and their hazardous health effects on humans, plants, and microorganisms.

Taking into account the toxicity and hazardous health concerns associated with heavy metal contamination, their removal or reduction from the environment is need of the hour in order to prevent the possibility of their uptake into the food chain (Ayangbenro and Babalola 2017). Physicochemical techniques routinely employed for heavy metal removal are not only inefficient and costly but also generate a large amount of toxic by-products and thus are not eco-friendly. Hence, a cheap, highly effective, and eco-friendly alternative of heavy metal removal is inevitable. One such promising and widely used alternative is bioremediation (Timková et al. 2018).

Bioremediation can be defined as use of bacteria, algae, fungi, plants, or their enzymes to either degrade or transform the environmental contaminants into harmless or less toxic forms (Jobby and Desai 2017). Microbes not only use diverse mechanisms, viz., biomineralization, for gaining resistance to heavy metals but also use them as a source of energy for growth and development via respiration, fermentation, and co-metabolism. Therefore, they are able to survive in heavy metal-contaminated environments. Biomineralization, bioaccumulation, biotransformation and biosorption are the mechansims used by microorganisms for metal remediation (Ayangbenro and Babalola 2017). This chapter focuses upon the various mechanisms used by bacteria for biosorption of carcinogenic heavy metal. It will also discuss the biosorption of carcinogenic heavy metals by bacteria with the aid of

Table 1 Sources and effects of carcinogenic heavy metals

Metal	Source	Effects on humans	Effects on plants	Effects on microorganisms	References
Arsenic	Atmospheric deposition, mining, pesticides, insecticides, fungicides, algicides, rock, sedimentation, smelting, thermal power plants, fuel burning, wood preservatives, dyestuff, pigments, car batteries manufacturing, alloyed semiconductor materials, contaminated water	Brain damage; cardiovascular and respiratory disorder; conjunctivitis; dermatitis; cancer of the bladder, kidney, prostrate, skin, and liver; diabetes; hearing loss; portal fibrosis; hematologic disorders (anemia, leukopenia, and eosinophilia)	Damage cell membrane, inhibition of growth, inhibits roots extension and proliferation, interferes with critical metabolic processes, loss of fertility, yield and fruit production, oxidative stress, physiological disorders	Deactivation of enzymes	Bissen and Frimmel (2003), Abdul-Wahab and Marikar (2012), Tchounwou et al. (2012), Kim et al. (2015), Jobby and Desai (2017)
Nickel	Electroplating, nonferrous metal, coins and stainless steel, production, paints, porcelain enameling, smelting, fossil fuel combustion, thermal power plants, battery industry, road runoff	Cardiovascular diseases, chest pain, dermatitis, allergy, dizziness, dry cough and shortness of breath, headache, kidney diseases, lung and nasal cancer, nausea, skin and oral epithelial damage	Decrease chlorophyll content, inhibit enzyme activities and growth, reduced nutrient uptake	Disrupt cell membrane, inhibit enzyme activities, oxidative stress	Malik (2004), Chibuike and Obiora (2014), Kim et al. (2015), Fashola et al. (2016), Jobby and Desai (2017)
Chromium	Dyes and pigments electroplating, industrial welding, paints production, steel fabrication, anticorrosive agent for utensils, metal processing, leather tanning, textile, mining,	Bronchopneumonia, chronic bronchitis, diarrhea, emphysema, headache, irritation of the skin, allergy and asthma, itching of respiratory tract, liver diseases, lung cancer, nausea, renal	Chlorosis, delayed, senescence, wilting, biochemical lesions, reduced biosynthesis germination, stunted growth, oxidative stress	Elongation of lag phase, growth inhibition, inhibition of oxygen uptake	Cervantes et al. (2001), Barakat (2011), Mohanty et al. (2012), Tchounwou et al. (2012), Kim et al. (2015), Jobby and Desai (2017)

(continued)

Table 1 (continued)

Metal	Source	Effects on humans	Effects on plants	Effects on microorganisms	References
	smelting, wood preservation, industrial coolants, road runoff	failure, reproductive toxicity, vomiting, nose ulcers, irritation in lining of nose, irritation and ulcers in stomach and small intestine, anemia, sperm damage, male reproductive system damage			
Cadmium	Fertilizer, mining, pesticide, plastic, refining, welding, alloys, pigments, electroplating, Ni/Cd batteries, e-waste, paint sludge, incinerations and fuel combustion, road runoff	Bone disease, coughing, emphysema, headache, hypertension, itai-itai, kidney diseases, lung, ovarian, breast and prostate cancer, lymphocytosis, microcytic hypochromic anemia, testicular atrophy, vomiting	Chlorosis, decrease in plant nutrient content, growth inhibition, reduced seed germination	Damage nucleic acid, denature protein, inhibit cell division and transcription, inhibits carbon and nitrogen mineralization	Nagajyoti et al. (2010), Sebogodi and Babalola (2011), Tchounwou et al. (2012), Chibuike and Obiora (2014), Sankarammal et al. (2014), Kim et al. (2015), Fashola et al. (2016), Jobby and Desai (2017)
Beryllium	Coal and oil combustion, volcanic dust	Allergic reactions, berylliosis, cancer, heart diseases, lung diseases	Inhibits seed germination	Chromosomal aberration, mutation	Gordon and Bowser (2003), Blais et al. (2008)

various mechanisms like binding to EPS, metallothionein, biofilm formation, cell surface adsorption, etc., as well as some aspects that can help bridge the gap between existing knowledge and challenges for a sustainable development of technology.

2 Conventional Methods for Heavy Metal Removal

Since last several decades, various conventional physicochemical methods have been employed for removal of heavy metal. These methods are as follows:

1. Chemical precipitation: This method is most commonly used for heavy metal removal by addition of anions, due to which metal ions get precipitated by forming metal hydroxides, carbonates, sulfides, phosphates, etc. which can be separated by sedimentation, filtration, or centrifugation (Hassan et al. 2010; Kanamarlapudi et al. 2018).
2. Ion exchange: It is based on reversible exchange of cations and anions between solid and aqueous phase. The ion-exchange resin exchanges ion from an electrolytic solution. Thus, they take up cationic heavy metals and in return leach out anions of similar charge in a chemically equivalent amount (Kanamarlapudi et al. 2018). However, this method is comparatively expensive, nonselective, and highly sensitive to pH fluctuations (Hassan et al. 2010).
3. Membrane filtration: It is a widely accepted technique because it can remove heavy metal as well as suspended solids and organic components from wastewater without generating any toxic by-products (Hassan et al. 2010; Kanamarlapudi et al. 2018). A membrane is a selective layer with porous or nonporous structure which makes contact between two homogenous phases (Kanamarlapudi et al. 2018).
4. Ultrafiltration: It uses a permeable membrane with pore size in the range of 0.1–0.001 μm which permits only water and low molecular weight solutes, whereas macromolecules, particles, and colloids are retained back (Kanamarlapudi et al. 2018).

   Qiu and Mao (2013) reported 98.8% removal of heavy metals like Cu^{2+}, Ni^{2+}, Zn^{2+}, and Mn^{2+} using ultrafiltration with a copolymer of maleic acid and acrylic acid in combination with polyvinyl butyral hollow fiber.
5. Microfiltration: It removes solutes larger in size than those retained back by ultrafiltration (Kanamarlapudi et al. 2018). Cross-flow microfiltration accompanied with *Saccharomyces cerevisiae*-based bioaccumulation removed 31% of Cu^{2+}, 7% of Cd^{2+}, 63% of Pb^{2+}, and 71% of Cr^{3+} from artificially contaminated tap water (Brady et al. 1994).
6. Nanofiltration: It uses a membrane with pore diameter of 0.5–2 nm which separates molecules within molecular weight range of 300–500 Da (Kanamarlapudi et al. 2018). Removal of heavy metal Cd^{2+}, Mn^{2+}, and Pb^{2+} with an efficiency of 99%, 89%, and 74%, respectively, was achieved using a commercially available nanofiltration membrane NF270 (Al-Rashdi et al. 2013).

In spite of its high efficiency, cost of operation, maintenance, and pre-treatment is comparatively higher (Hassan et al. 2010).

7. Reverse osmosis (RO): It employs pressure that elevates the passage of solution through semipermeable membrane. This method is extensively used for heavy metal removal from industries. A polyamide thin-film composite membrane TW30-1812-50 was reported to be successful in removal of heavy metals like Cu^{2+}, Ni^{2+}, and Zn^{2+} by reverse osmosis (Bakalár et al. 2009). One of the major disadvantages of this technique is fouling of membrane with highly suspended solids and slightly soluble compounds present in wastewater (Hassan et al. 2010).
8. Electrodialysis: This technique uses cation and anion membranes fitted between the electrodes in electrolytic cell which enables the migration of ions under continuous electric current which leads to recovery of heavy metals (Hassan et al. 2010). Electrodialysis system was used for removal of 73.9% of arsenic, 89.9% of lead, 98.9% of manganese, and 91.1% of nitrate nitrogen from groundwater of Korea (Choi et al. 2015).
9. Photocatalysis: It is used for removal of different types of pollutants due to their rapid and effective destruction with the help of nontoxic semiconductors. In this method, firstly, the pollutant is transferred and adsorbed on the surface of semiconductor, and later the photocatalytic reactions lead to decomposition and removal of pollutants from interface region (Kanamarlapudi et al. 2018). Selenium-doped ZnO nanocomposite semiconductor has been reported to remove 0.421 (Cu), 0.211 (Cr), 0.147 (Pb), and 0.097 (Cd) per 0.5 g of ZnO/Se nanocomposite from pharmaceutical wastewater by photocatalysis (Shyni et al. 2016).

3 Mechanisms Used by Bacterial Cell for Heavy Metal Bioremediation

1. Bioaccumulation
   It is a metabolic activity (requires input of energy) in which bacteria imports heavy metals (Diep et al. 2018), which either get bound to proteins or are transformed to less toxic state inside the bacteria, like arsenic +3 converted into arsenic +5 (Mishra and Malik 2013).
2. Biosorption
   It includes the binding of the metal to the bacterial cell which requires the physical interaction between metal and bacterial dead cell particulate. At neutral pH, the various anionic functional groups present on cell surface bind to the cationic heavy metals, but at acidic pH, this bonding is not favored (Michalak et al. 2013).
3. Metal Sequestration
   It is the binding of heavy metals to the cytoplasmic components like proteins. The metalloproteins which contain a high amount of cysteine are capable of binding with the heavy metals more efficiently (Higham et al. 1986). Glutathione has also

Table 2 Genes of efflux pumps involved in removal of carcinogenic heavy metals

Gene	Metal	Bacteria	References
ArsA and ArsB	Arsenic	*E. coli*, *Leptospirillum ferriphilum*	Chen et al. (1986), Yang et al. (2012)
cadA2R	Cadmium	*Pseudomonas*	Nowicki et al. (2015)
Czn C and Czn A	Nickel	*Helicobacter pylori*	Stähler et al. (2006)
Chromate ABC efflux pump	Chromium	*Campylobacter* sp.	Krawiec et al. (2017)

been reported to play role in intracellular sequestration of *Rhizobium leguminosarum* cells (Lima et al. 2006). Extracellular sequestration is the formation of insoluble complex of metal and organic matter of bacteria either on the surface of bacteria or on the periplasmic space (Igiri et al. 2018). Cha and Cooksey (1991) reported that copper-resistant bacteria *Pseudomonas syringae* produces three proteins, Cops A, B, and C, out of which A and B bind to the metal on periplasmic space while C binds on surface of bacteria.

4. Efflux Pumps
   These are used by living cell to throw out the toxic metal. The use of efflux pump is considered as the main mechanism used by the bacteria which makes them resistant to different metals. Nowicki et al. (2015) reported some genes, viz., cadA2R, czcCBA1, and colRS, in *Pseudomonas*. The various genes coding for efflux pump in different bacteria are summarized in Table 2.
5. Biosorption
   It is carried out by the biomolecules present on the surface of cell which includes EPS (exopolysaccharide), cell wall, carbohydrates, etc. The major biosorbent involved in the adsorption of heavy metal is the EPS which contains various biomolecules which are potent adsorbent (Diep et al. 2018).
6. Precipitation
   It occurs on the surface of cell due to the functional group present on cell surface. Extracellular components of cell like EPS are involved in the precipitation of heavy metal and form insoluble complexes (Kanamarlapudi et al. 2018). Mohamad et al. (2009) reported the precipitation of copper on the surface of *Mesorhizobium amorphae*.
7. Ion Exchange
   It involves either an anionic or cationic exchanger. Carboxyl and imidazole groups both act as cationic and anionic exchanger in case of bacteria. Phosphate, hydroxyl, and carboxyl groups are also involved as ion exchanger (Fig. 1). *Spirulina* has been studied for the phenomenon of ion exchange for carcinogenic heavy metals (Chojnacka et al. 2005).
8. Methylation
   It is one of the key processes used by bacteria to transform heavy metal to less toxic form (Rajkumar et al. 2013). Addition of methyl group to heavy metal results in transformation of state of metal to volatile nature. Since volatile compounds get diffused form the bacteria, the metal concentration decreases

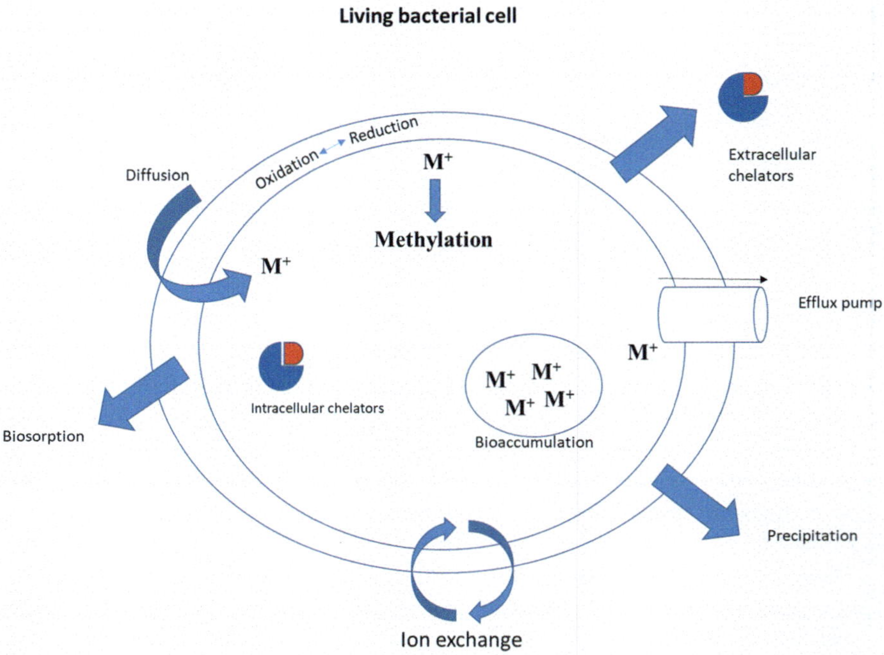

Fig. 1 Mechanisms of removal of heavy metals by microorganisms. This figure demonstrates all the possible mechanism used by the living cell for bioremediation, biotransformation, and bioaccumulation of heavy metal. Bioaccumulation and biosorption do not transform heavy metal, whereas methylation and oxidation reduction of the metal require the involvement of enzymes which changes the valency of metal and converts it into less toxic form or makes it volatile (methylation) so as it diffuses out of the cell and reduces the heavy metal toxicity inside the cell

inside the bacterial cell (Meyer et al. 2007). For the transfer of the methyl group, enzyme reductase plays a key role which reduces metal in the presence of an electron donor (Fig. 2).

4 Biosorption of Carcinogenic Heavy Metals by Bacteria

Various bacteria such as *Pseudomonas*, *Enterobacter*, *Bacillus*, and *Micrococcus* species have been extensively exploited for biosorption owing to their small size which provides high surface-area-to-volume ratio, ability to thrive under controlled as well as extreme environmental conditions, numerous chemisorption sites, presence of teichoic acid, etc. (Ayangbenro and Babalola 2017). Biosorption efficiency not only depends upon the type of heavy metal being adsorbed but also on the type of bacteria due to the variation in their cell wall components (Hassan et al. 2010). Some bacteria also produce slime or capsule like layer, i.e., EPS composed of

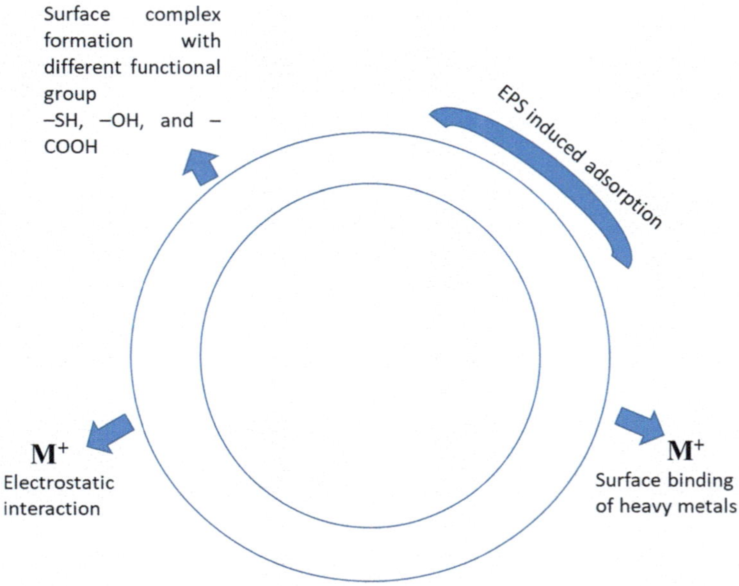

Fig. 2 Biosorption of heavy metals in bacterial cell

polysaccharide which also helps in biosorption of metal ions (Kanamarlapudi et al. 2018). The biosorption of various carcinogenic metals by bacteria is summarized in Table 3. Biosorption has various advantages over conventional physicochemical methods (Abbas et al. 2014; Shamim 2018). These are as follows:

1. Production of bacterial biomass is comparatively easy and cheap.
2. Multiple types of heavy metals can be bio adsorbed.
3. Effective for treating large volume of wastewater.
4. Functions effectively at wide range of temperature, pH, and also in presence of other metal ions.
5. Desorption of metals attached to biomass is easier and economical.
6. Less production of waste or toxic compounds.

However, there are some disadvantages of biosorption which are as follow (Abdi and Kazemi 2015):

1. Active sites of metal binding ligands get saturated.
2. Reversible sorption of metals on biomass can occur.

The biosorption of carcinogenic heavy metal in dead bacterial cell occurs in two ways: (1) via EPS (exopolysaccharides) and (2) via cell wall-mediated adsorption.

Table 3 Biosorption of various carcinogenic heavy metals by bacteria

Heavy metal	Bacteria	Temp (°C)	pH	Time (h)	Agitation speed (rpm)	Initial conc. (mg/L)	Biosorption efficiency (mg/g or %)	References
Arsenic (As)	*Bacillus* sp. KM02	–	–	–	–	–	–	Banerjee et al. (2011), Majumder et al. (2013), Dey et al. (2016)
	Kocuria sp.	–	–	–	–	–	–	
	Bacillus sp.	–	–	–	–	–	–	
	Sulfate-reducing bacteria	–	6.9	24	–	1	0.07	Teclu et al. (2008)
	Pseudomonas aeruginosa	–	–	2	–	–	98%	Tariq et al. (2019)
	Bacillus cereus	–	7.5	–	–	–	32.42	
	Bacillus sp.	40	6	–	–	6	–	Saravanan et al. (2012)
	Arthrobacter sp.	28	2–7	0.5	–	–	74–81	Prasad et al. (2013)
Cadmium (Cd)	*Pseudomonas putida* mt2	–	–	–	–	–	–	Congeevaram et al. (2007), Rani et al. (2010), Shamim et al. (2014a, b), Shamim and Rehman (2014, 2015)
	Cupriavidus metallidurans CH34	–	–	–	–	–	–	
	Enterobacter cloacae	25	5	2	240	100	58.9%	
	Stenotrophomonas maltophilia	28	5	2	140	20,000	0.12	
	Actinomycetes sp.	30	6	24	150	5000	32.63	
	Rhizobium sp.	–	–	–	–	–	135.3	Abd-Alla et al. (2012)
	Aeromonas caviae	–	7	–	–	–	155.3	Loukidou et al. (2004), Yilmaz and Ensari (2005), Lu et al. (2006), Quintelas et al. (2009)
	Bacillus circulans	–	7	–	–	–	26.5	
	Enterobacter sp. J1	–	6	–	–	–	46.5	
	B. laterosporus	25	7	2	–	1000	159.5	Zouboulis et al. (2004), Haq et al. (2016)
	B. licheniformis	25	7	2	–	1000	142.7	
	Kocuria rhizophila	35	8	1	–	150	9.07	
	Enterobacter cloacae	25	5	2	240	300	58.9%	Liu et al. (2004), Congeevaram et al. (2007), Rani et al. (2010)
	Pseudomonas sp.	30	5–9	24	100	25	90.41%	
	Stenotrophomonas maltophilia	25	5	2	140	10	0.12	
	Actinomycetes sp.	30	6	24	150	50–400	32.63	

Chromium	*Micrococcus* sp.	35	5	24	120	100	92%	Congeevaram et al. (2007), Rani et al. (2010), Sulaymon et al. (2013), Jencarova and Luptakova (2012), Babák et al. (2013)
	Bacillus licheniformis	28	3.5	48	120	–	95%	
	Staphylococcus saprophyticus	27	2	3	150	200	24.1	
	Enterobacter cloacae	25	4	2	240	100	55.8	
	Pseudomonas aeruginosa	25	–	–	–	–	1.07	
	Bacillus coagulans	–	–	–	–	–	39.9	Vijayaraghavan and Yun (2008)
	B. laterosporus	25	2.5	2	–	1000	72.6	Zouboulis et al. (2004), Haq et al. (2016), Kim et al. (2015), Kang et al. (2005)
	B. licheniformis	25	2.5	2	–	1000	62	
	Desulfovibrio desulfuricans (immobilize on zeolite)	37	7.8	168	–	100	99.8	
	Kocuria rhizophila	35	4	1	–	150	14.4	
	Pseudomonas aeruginosa	25	3.4	10	–	52	6.42	
	Pseudomonas fluorescence	–	2	–	–	–	40.8	Uzel and Ozdemir (2009)
	Staphylococcus saprophyticus	27	2	3	150	193.6	24.1%	Liu et al. (2004), Rani et al. (2010), Jencarova and Luptakova (2012), Babák et al. (2013), Karakagh et al. (2012), Sulaymon et al. (2013)
	Enterobacter cloacae	25	4	2	240	100	55.8%	
	Pseudomonas sp.	30	5.5	–	200	1–10 mmol/L	8.9–238	
	Pseudomonas aeruginosa	25	–	–	–	1000	1.07	
	Bacillus licheniformis	28	3.5	48	120	1200	95%	
Nickel	*Actinomycetes* sp.	30	5	24	150	5000	36.55	Congeevaram et al. (2007), Sulaymon et al. (2013), Abbas et al. (2014)
	Micrococcus sp.	35	5	24	120	–	90%	
	Bacillus thuringiensis	–	–	–	–	–	15.7%	Öztürk (2007), Quintelas et al. (2009)
	E. coli	–	2.7–3.6	–	–	–	6.9	
	Micrococcus sp.	–	–	–	–	50	55%	Congeevaram et al. (2007), Kumar et al. (2011), Bhattacharya and Gupta (2013), Kumaran et al. (2011)
	Pseudomonas sp.	–	–	–	–	1	53%	
	Acinetobacter sp. B9	–	–	–	–	51	68.94%	

(continued)

Table 3 (continued)

Heavy metal	Bacteria	Temp (°C)	pH	Time (h)	Agitation speed (rpm)	Initial conc. (mg/L)	Biosorption efficiency (mg/g or %)	References
	Streptomyces coelicolor	–	8	–	–	–	11.1	Öztürk et al. (2004), Selatnia et al. (2004), Gabr et al. (2008), Tuzen et al. (2008), Kao et al. (2008), Morillo et al. (2009)
	Streptomyces rimosus	–	6.5	–	–	–	36	
	P. aeruginosa ASU 6a	–	7	–	–	–	113.6	
	Pseudomonas aeruginosa	–	8.5	–	–	–	5.25	
	Escherichia coli	–	7	–	–	–	34.3	
	Paenibacillus jamilae	–	6	–	–	–	17.6	
	Desulfovibrio desulfuricans (immobilize on zeolite)	37	7.8	168	–	100	90.1	Kang et al. (2005), Rajkumar and Freitas (2008), Kim et al. (2015)
	Pseudomonas aeruginosa	25	5.5	10	–	58.69	8.26	
	P. jessenii	25	–	6	–	275	1.36	
	Pseudomonas sp.	25	–	6	–	275	2.79	
	Pseudomonas sp.	30	5.5	–	200	1–10 mmol/L	556	Liu et al. (2004)
	Pseudomonas fluorescence	–	2	–	–	–	40.8	Uzel and Ozdemir (2009)

EPS-induced adsorption is mediated by bonding of positively charged heavy metals to the negatively charged EPS components on the surface of the bacteria (Gupta and Diwan 2017). Based on the composition, there are two types of EPS, i.e., homopolysaccharide and heteropolysaccharide. Both of them are synthesized intra-cellularly, but homopolysaccharides are predominantly synthesized on the surface of the cell when responsible enzymes are active. Dextransucrase and levansucrase are the enzymes responsible for the stereo structure of sucrose in the chain (Whitfield 1988). EPS mainly consists of protein and polysaccharide (Subramanian et al. 2010; More et al. 2014; Nouha et al. 2016a). Polysaccharide covers most of the part of EPS. The extracellular EPS of bacteria can be homo- or heterogeneous (Monsan et al. 2001). D-Glucose and L-fructose are some of the common examples of homopolysaccharide, whereas alginate which is produced by *Pseudomonas aeruginosa* is a type of heteropolysaccharide. The composition of EPS of the same bacteria can be different in different condition, for example, when bacteria is fed with more carbon source, then the composition of EPS will be different compared to when bacteria produce EPS in stress condition. The use of different carbon source also affects the concentration of EPS produced for the same bacteria (Ye et al. 2012). Sugars contain a lot of hydroxyl group which play a major role in binding of the metal to EPS (Abbas et al. 2014). Protein is also a crucial component of EPS of bacteria (Ton-That et al. 2004). As discussed above the protein concentration of bacteria is not hugely correlated with the carbon source used, as compared to the polysaccharide. It has been reported that the protein concentration is inversely proportional to the nitrogen content (Nouha et al. 2016b). The proteins are chains of amino acids which are diverse in nature with different functional groups like imidazole, hydroxyl, thiol, thioester, carboxyl etc., which play a major role in the binding of heavy metals to EPS (Abbas et al. 2014). Table 4 summarizes the functional groups involved in heavy metal biosorption by bacteria.

There are different types of EPS produced in different bacteria which are summarized in Table 5.

Table 4 Functional group present in bacterial cell for biosorption of heavy metals

Component	Functional group present	Bonding molecule	Reference
Polysaccharides	Hydroxyl group	O	Maddela et al. (2018)
Amino acid	Imidazole Thiol Carboxyl Hydroxyl Thioester Imine	N S O O S N	Abbas et al. (2014)
Phospholipid	Phosphonate	O	
Lipopolysaccharide	Phosphodiester	O	
Peptidoglycan	Secondary amine	N	
Teichoic acid	Phosphodiester	O	
Uronic acid	Carboxyl group Hydroxyl group	O O	

Table 5 Types of sugars in exopolysaccharide of bacteria

Homopolysaccharides	Monomers	Bacteria	References
Dextran Mutan Alternan Reuteran	Glucose	*Leuconostoc mesenteroides*, *Streptococcus*, and *Lactobacillus*	Monsan et al. (2001), Ruas-Madiedo et al. (2002), Kralj et al. (2004), Majumder et al. (2009), Mcintosh et al. (2005), Ruas-Sarwat et al. (2008), Freitas et al. (2009)
Curdlan		*Agrobacterium biovar*, *Cellulomonas* sp., *Rhizobium* sp., *Bradyrhizobium japonicum*, *Streptococcus pneumoniae*	Mcintosh et al. (2005), Czaczyk and Myszka (2007)
Levan Fructans	Fructose	*B. subtilis*, *Zymomonas mobilis*, *Streptococcus salivarius*, and *Streptococcus mutans*	Monsan et al. (2001), Czaczyk and Myszka (2007)
Heteropolysaccharide	*Monomers*		
Alginate	β-D-mannurosyl and α-L-glucuronosyl units	*Pseudomonas aeruginosa*, *Azotobacter vinelandii*	Czaczyk and Myszka (2007)
Xanthan	Glucose backbone, linked with trisaccharide side chain of glucuronic acid, mannose, pyruvil, and acetyl residues	*Xanthomonas campestris*	De Vuyst and Degeest (1999), Freitas et al. (2011)
Lactobacillus EPS	Glucose, galactose, rhamnose, fructose	*Lactobacillus* sp.	De Vuyst and Degeest (1999), Nishimura (2014)
Hyaluronan	Glucuronic acid and *N*-acetylglucosamine	*Pseudomonas aeruginosa* and *Streptococci attenuated strains*	Freitas et al. (2011)
Sphingans	Rhamnose, glucose, glucuronic acid	*Sphingomonas* sp.	Coleman et al. (2008)

Cell wall of bacteria contain many ionizable functional groups like amino, hydroxyl, phosphate, and carboxyl which play a major role in the binding of the heavy metal on the cell wall of the bacteria (El-Helow et al. 2000; Taniguchi et al. 2000). Both gram-positive and gram-negative bacterial cell walls are primarily made up of peptidoglycan. The teichoic acid monomers present in bacterial cell walls are linked by phosphodiester bonds which imparts an overall negative charge to the cell wall which facilitates the adsorption of metal ions (Kanamarlapudi et al. 2018). Gram-negative bacteria have comparatively thinner cell wall due to less amount of peptidoglycan, and hence their metal uptake capacity is less (Shamim 2018). However, the presence of an additional lipopolysaccharide (LPS) layer in gram-negative bacterial cell wall provides the negative charge which aids in metal uptake.

5 Kinetics of Biosorption

Biosorption of metals by bacteria has been explained by various adsorption isotherms (Hassan et al. 2010; Abbas et al. 2014). The isotherm equation describes the correlation between the amount of metal ion adsorbed by the cell and the amount of metal ion left in the solution. It also describes the equilibrium between the distribution of metal between liquid and solid phase (Hassan et al. 2010). Biosorption also largely depends upon the quality of biosorbent which is decided by two factors:

1. The amount of metal ion that can be adsorbed by biosorbent.
2. The amount of time for which metal ions remain bound to biosorbent in immobilized form.

The amount of metal uptake by the biosorbent can be calculated using the formula (Ahalya et al. 2003; Abbas et al. 2014; Abdi and Kazemi 2015):

$$q = \frac{V(C_{\mathrm{i}} - C_{\mathrm{e}})}{M}$$

q = amount of metal biosorbed by biomass (mg/g)
V = volume of metal solution (L)
C_{i} = initial concentration of metal (mg/L)
C_{e} = concentration of metal (mg/L) at equilibrium
M = mass of adsorbent

While studying the isotherm model for biosorption, parameters such as temperature, pH, ionic strength, etc. are kept constant, whereas the metal ion concentration is varied. Biosorption isotherm is usually explained by two models, i.e., (1) Freundlich and (2) Langmuir (Abbas et al. 2014; Shamim 2018).

5.1 Freundlich Model

It is the first mathematical equation describing an isotherm given by Freundlich and Kuster in 1907. It is a nonlinear sorption model that describes the sorption of heavy metal within the active site of biosorbent in a monolayer format on a heterogeneous surface. It can be represented by following equation (Jalali-Rad et al. 2004; Abdi and Kazemi 2015):

$$q_e = KC_e{}^{1/n}$$

$K = \text{mg/g} =$ Freundlich constant related to adsorption capacity
$1/n$ or $n =$ Freundlich constant related to adsorption intensity

The linearized form of this model can be given by the equation (Hassan et al. 2010):

$$\ln q_{eq} = \ln K_f + 1/n \ln C_{eq}$$

5.2 Langmuir Model

This model published by Langmuir in 1918 described the adsorption of gas or liquid on solid biosorbent. It is also a type of monolayer adsorption, but it describes the interaction between metal and active site and not between adsorbed molecules. It can be represented by equation (Abbas et al. 2014; Abdi and Kazemi 2015):

$$q_e = \frac{q_{max} bC_e}{1 + bC_e}$$

$q_e =$ Amount of metal ion removed (mg/g)
$C_e =$ Equilibrium concentration (mg/L)
$b =$ Langmuir constant related to affinity
$q_{max} =$ Maximum metal uptake (mg/g) under the given conditions

While the linear form of Langmuir equation is (Hassan et al. 2010):

$$C_{eq}/q = 1/q_{max} b + C_{eq}/q_{max}$$

Langmuir adsorption isotherm is based on various assumptions (Fourest et al. 1994; Volesky 2003):

1. Metal ions are chemically adsorbed at a fixed number of well-defined sites.
2. All sites are energetically equivalent.

3. The energy of adsorption is constant.
4. Each site can only hold one ion.
5. There is no interaction between the ions.
6. There is no migration of adsorbate molecules in the surface plane (localized adsorption).
7. The maximum adsorption occurs when a saturated monolayer of solute molecules is present on the adsorbent surface.

6 Factors Affecting Biosorption

The various physicochemical factors affect biosorption of carcinogenic heavy metal. These are:

1. Temperature—The normal temperature range for biosorption is 20–35 °C. Higher temperatures up to a certain extent can enhance biosorption, but further increasing the temperature can lead to protein denaturation affecting the metal uptake capacity of bacterial cell and can also cause structural damage to biosorbent (Abbas et al. 2014; Shamim 2018; Kanamarlapudi et al. 2018).
2. pH—It has been reported to affect binding site of biomass and solubility of metal ions. The normal pH range for biosorption is 2.5–6. Higher pH affects the biosorption ability of bacterial cell (Abbas et al. 2014; Shamim 2018).
3. Nature of biosorbents—Biosorption of metals can be carried out by various forms like biofilms, freely suspended cells, or immobilized cells. Age of bacterial cell as well as the growth media components also widely affects the biosorption due to alterations in cell wall components, cell size, and EPS formations (Abbas et al. 2014; Shamim 2018).
4. Surface-area-to-volume ratio—Surface-area-to-volume ratio plays a vital role in efficient removal of heavy metal. One of the most relevant examples is the preference of bacteria to uptake metal by formation of biofilm rather than binding to bacterial cell wall. Biofilm formation, though being an energy consuming process, is more preferred by bacteria due to the availability of higher surface-area-to-volume ratio (Abbas et al. 2014; Shamim 2018).
5. Concentration of biomass—Metal uptake capacity is directly proportional to biomass concentration. At a given concentration of metal, increasing the biosorbent concentration will elevate the metal uptake efficiency due to more available binding site. However, at increased biosorbent concentration, the amount of biomass adsorbed per unit weight is lower compared to lower concentration of biosorbent (Abbas et al. 2014; Shamim 2018; Kanamarlapudi et al. 2018).
6. Initial metal ion concentration— The metal uptake by bacterial cell will be more if the initial concentration of metal is high at a given concentration of biomass. This is because the initial concentration is a driving force to overcome mass transfer resistance of metal between aqueous and solid phase. Further increasing the metal ion concentration will reduce the biosorption efficiency due to

saturation of metal binding sites (Abbas et al. 2014; Shamim 2018; Kanamarlapudi et al. 2018).

7. Metal affinity to biosorbent—Physical (autoclaving, drying, boiling, sonication) and chemical (addition of acid or alkali) pre-treatment widely affects the surface charges and accessibility of metal binding groups of the biomass (Abbas et al. 2014; Shamim 2018).
8. Effect of contact time—Types of biosorbent and metal ion are important parameters that affect the time required for achieving maximum biosorption. Almost 90% biosorption of metals occur in initial hour of treatment because all the active sites of biosorbent are vacant and available for biosorption (Kanamarlapudi et al. 2018).
9. Effect of agitation speed— The biosorption of metals will be high with the increase in agitation due to minimum mass transfer resistance. Very high agitation speed can increase turbulence and result in vortex phenomenon that can disrupt the physical nature of biosorbent, thus decreasing its efficiency. Best homogeneity and high biosorption can be achieved at moderate agitation speed (Kanamarlapudi et al. 2018).

7 Commercialized Techniques for Biosorption of Heavy Metals

Biosorption is one of the potential and economical techniques for treatment of large volume of heavy metal-contaminated wastewater. Several attempts have been made for commercialization of biosorption technique in the past which is given in Table 6 (Wang and Chen 2009).

8 Desorption of Heavy Metals

Desorption is the mechanism of separating heavy metal from the adsorbent (Chatterjee & Abraham 2019). Desorption study is useful as it helps to regenerate the adsorbent as well as help in extraction of the metals that can be used in other industrial applications. There are several methods and chemical compounds which are used as desorbents like acid, bases, chelating agents, salts, etc. Desorption is basically ratio of the solid (adsorbent) and liquid (desorbent) (Wang and Chen 2009). Desorption agents are selected based on their desorption efficiency, non-hazardous and eco-friendly nature (Dhankhar and Hooda 2011).

8.1 Agents Used for Desorption of Heavy Metals

8.1.1 Acid as Desorbent

Certain organic as well as nonorganic acids are used for the purpose of extracting bound metal from the sorbent. There are some common examples which are used for

Table 6 Commercialized biosorption techniques for removal of heavy metals

Technology	Microorganism	Specifications	References
–	Specific microbial biomass	First patent of biosorption in 1980	Tsezos (2001)
B.V. SORBEX	*S. natans*, *A. nodosum*, *Halimeda opuntia*, *Palmyra pamata*, *Chondrus crispus*, and *C. vulgaris*	Biosorption was not affected by Ca^{2+}, Mg^{2+}, or organic matter	Volesky (1990)
AlgaSORB	Immobilized *Chlorella* in silica or polyacrylamide gel	Can reduce heavy metal concentration up to 1 mg/L or lower, resembles an ion exchange resin, and can undergo more than 100 biosorption-desorption cycles	Kuyucak (1990), Garnham and Wase (1997)
AMT-BIOCLAIM	*Bacillus subtilis*	Can accumulate 2.90 mmol Pb/g, 2.39 mmol Cu/g, 2.09 mmol Zn/g, 1.90 mmol Cd/g, or 0.8 mmol Ag/g metal cations with 99% efficiency	Kuyucak (1990), Brierley (1990), Garnham and Wase (1997), Veglio and Beolchini (1997), Vijayaraghavan and Yun (2008)
BIO-FIX	Mixture of biomasses of *Sphagnum* peat moss, algae, yeast, bacteria, and/or aquatic flora immobilized in high density polysulfone	Used for treating acid mine waste, has metal affinity for $Al^{3+} > Cd^{2+} > Zn^{2+} > Mn^{2+}$, and can be used for more than 120 extraction-elution cycles	Gupta et al. (2000)
–	Immobilized *R. arrhizus* biomass	Recovery of uranium from ore	Veglio and Beolchini (1997)
MetaGeneR and RACHO Bio-Beads	–	Effectively removes metal ions from mining and electroplating wastes	Atkinson et al. (1998)
Vitrokele	–	Used for mercury and iron removal	Huber et al. (1990)

the desorption of metal bound to the bacterial cell matter. Sulfuric acid was used for *Cladosporium cladosporioides* to remove bound copper (Carmo et al. 2013). Apart from sulfuric acid, other acids like nitric acid and citric acid are also used (Gong et al. 2005). Nitric acid acts as a far good desorbent compared to other nonorganic acids, as the high proton concentration in nitric acid is a key factor which disrupt the bond between heavy metal and sorbent and elute the heavy metal (Davila-Guzman et al. 2016). Nitric acid was used to remove metal from dead cell of *Spirulina* and

Spirulina maxima (Gong et al. 2005). Sulfuric acid was also found to be a good eluting agent at 1 M concentration; at this concentration, it can elute up to 31.1% of copper and 20.3% of cadmium (Kamaruzaman et al. 2017). However, sulfuric acid has one limitation that it cannot elute lead because lead gets precipitated to form lead sulfate when reacted with sulfuric acid (Puranik and Paknikar 1997).

8.1.2 Bases as Desorbent

There are several bases which are used as desorbent for their alkaline property which include sodium hydroxide, sodium bicarbonate, sodium carbonate, etc. These are mostly used for the removal of metal from bioorganic sorbent (Bai and Abraham 2003) like fungus *Penicillium digitatum* (Galun et al. 1983). Bases were found to be more effective than acid in elution of heavy metal. Sodium hydroxide was found to be effective in elution of chromium (Selvi et al. 2001). Sodium carbonate was used for the elution of chromium from *Rhizopus nigricans* immobilized on the matrix of polysulfone, and 91.9% of chromium was eluted (Ahalya et al. 2003).

8.1.3 Chelating Agents

EDTA (Ethylenediaminetetraacetic acid) is considered to be one of the best desorbents (Akhtar and Iram 2014). EDTA was used for the elution of zinc from dead cell of bacterium *Streptoverticillium cinnamoneum* (Puranik and Paknikar 1997). The unique property of EDTA which makes it an ideal candidate for the desorption of heavy metals is that it does not change the property of adsorbent (Pandey et al. 2007).

8.1.4 Salt

Sodium nitrate and calcium chloride are used as desorbents but none of them are used for the removal of the metals from dead bacterial cell. Calcium chloride along with HCl is used for the desorption of heavy metal from *Penicillium* sp. (Mishra 2014). Sodium nitrate can also be used as desorbent, but acid and alkaline are better than salts (Chatterjee and Abraham 2019). Table 7 summarizes various desorbents used for extracting heavy metals from bacterial cell.

Table 7 Desorbents used for recovery of heavy metals

Bacteria	Desorbent used	Metals recovered	Efficiency of desorbent (%)	Reference
Spirulina sp.	EDTA	Chromium Cadmium Copper	60 41 53	Chojnacka et al. (2005)
Spirulina maxima	EDTA	Lead	91	Gong et al. (2005)
Penicillium digitatum	$(NH_4)_2CO_D$	Uranium	98	Galun et al. (1983)
Cladosporium cladosporioides	Sulfuric acid	Copper	43	Carmo et al. (2013)
Spirulina maxima	Citric acid	Lead	81	Gong et al. (2005)
Spirulina sp.	Nitric acid	Chromium Cadmium	90 95	Chojnacka et al. (2005)

9 Future Prospects

Commercialization of biosorption is a challenge mainly because of nontechnical issues such as pre-treatment, source and type of biosorbent, capitalization, feasibility, etc. (Wang and Chen 2009). A hybrid technology involving the use of biotechnological-based processes like biosorption, bioprecipitation, bioreduction, biotransformation, etc. as well as non-biotechnological-based processes like electrochemical processes, chemical precipitation, flotation, membrane technology, etc. can be used for treating wastewater at a large scale for simultaneous removal of organic matter as well as heavy metals (Tsezos 2001).

The source and type of biosorbent utilized for biosorption also play a major role in making the large-scale process economical. If the biomass used for the sorption process needs to be specifically cultured, it will increase the production and maintenance cost for the industry. However, some of the potential bioadsorbents can be obtained as waste from various industries at almost zero cost, and if these wastes are used for metal removal, then their disposal issue also gets resolved. Yeasts (*Saccharomyces cerevisiae* from food and beverage industry; *Candida albicans*—clinical isolate), molds (*Rhizopus arrhizus* from food industry, *Penicillium chrysogenum* from antibiotics industry, and *Aspergillus niger* from citric acid and industrial enzyme production), and bacteria (*Bacillus* sp. from amino acid and antibiotic fermentation and *Streptomyces noursei* from pharmaceutical industries) are some of the candidates which can serve as ideal candidates for bioadsorbents. The only cost which the industry has to bear will be drying, if required, and transportation (Wang and Chen 2009).

However further studies to gain deeper knowledge regarding improvement in biomass immobilization, optimization of biosorption parameters and physicochemical conditions, reuse and recycling of biomass, and mechanisms of metal microbe interactions are essential for removal of heavy metals from industrial wastewater (Wang and Chen 2006, 2009).

References

Abbas SH, Ismail IM, Mostafa TM, Sulaymon AH (2014) Biosorption of heavy metals: a review. J Chem Sci Technol 3(4):74–102

Abd-Alla MH, Morsy FM, El-Enany AWE, Ohyama T (2012) Isolation and characterization of a heavy-metal-resistant isolate of *Rhizobium leguminosarum* bv. *viciae* potentially applicable for biosorption of Cd^{2+} and Co^{2+}. Int J Biodeterior Biodegrad 67:48–55

Abdi O, Kazemi M (2015) A review study of biosorption of heavy metals and comparison between different biosorbents. J Mater Environ Sci 6(5):1386–1399

Abdul-Wahab S, Marikar F (2012) The environmental impact of gold mines: pollution by heavy metals. Open Eng 2(2):304–313

Ahalya N, Ramachandra TV, Kanamadi RD (2003) Biosorption of heavy metals. Res J Chem Environ 7(4):71–79

Akhtar S, Iram S (2014) Effect of chelating agents on heavy metal extraction from contaminated soils. Res J Chem Sci 4(9):70–87

Al-Rashdi BAM, Johnson DJ, Hilal N (2013) Removal of heavy metal ions by nanofiltration. Desalination 315:2–17

Atkinson BW, Bux F, Kasan HC (1998) Considerations for application of biosorption technology to remediate metal-contaminated industrial effluents. Water SA 24(2):129–135

Ayangbenro AS, Babalola OO (2017) A new strategy for heavy metal polluted environments: a review of microbial biosorbents. Int J Environ Res Public Health 14(1):94

Babák L, Šupinova P, Zichova M, Burdychova R, Vitova E (2013) Biosorption of Cu, Zn and Pb by thermophilic bacteria–effect of biomass concentration on biosorption capacity. Acta Univ Agric Silvic Mendelianae Brun 60(5):9–18

Bakalár T, Búgel M, Gajdošová L (2009) Heavy metal removal using reverse osmosis. Acta Montan Slovaca 14(3):250

Bai RS, Abraham TE (2003) Studies on chromium (VI) adsorption – desorption using immobilized fungal biomass. Bioresour Technol 87(1):17–26

Banerjee S, Datta S, Chattyopadhyay D, Sarkar P (2011) Arsenic accumulating and transforming bacteria isolated from contaminated soil for potential use in bioremediation. J Environ Sci Health A 46(14):1736–1747

Barakat MA (2011) New trends in removing heavy metals from industrial wastewater. Arab J Chem 4(4):361–377

Bhattacharya A, Gupta A (2013) Evaluation of Acinetobacter sp. B9 for Cr(VI) resistance and detoxification with potential application in bioremediation of heavy-metals-rich industrial wastewater. Environ Sci Pollut Res 20(9):6628–6637

Bissen M, Frimmel FH (2003) Arsenic—a review. Part I: Occurrence, toxicity, speciation, mobility. Acta Hydrochim Hydrobiol 31(1):9–18

Blais JF, Djedidi Z, Cheikh RB, Tyagi RD, Mercier G (2008) Metals precipitation from effluents. Pract Period Hazard Toxic Radioactive Waste Manag 12(3):135–149

Brady D, Rose PD, Duncan JR (1994) The use of hollow fiber cross-flow microfiltration in bioaccumulation and continuous removal of heavy metals from solution by *Saccharomyces cerevisiae*. Biotechnol Bioeng 44(11):1362–1366

Brierley JA (1990) Production and application of a *Bacillus*-based product for use in metals biosorption. In: Biosorption of heavy metals. CRC Press, Boca Raton, FL, pp 305–312

Carmo JRD, Pimenta CJ, Silva JFD, Souza SMCD (2013) Recovery of copper (II) absorbed in biomass of Cladosporium cladosporioides. Sci Agric 70(3):147–151

Cervantes C, Campos-García J, Devars S, Gutiérrez-Corona F, Loza-Tavera H, Torres-Guzmán JC, Moreno-Sánchez R (2001) Interactions of chromium with microorganisms and plants. FEMS Microbiol Rev 25(3):335–347

Cha JS, Cooksey DA (1991) Copper resistance in *Pseudomonas syringae* mediated by periplasmic and outer membrane proteins. Proc Natl Acad Sci U S A 88(20):8915–8919

Chatterjee A, Abraham J (2019) Desorption of heavy metals from metal loaded sorbents and e-wastes: a review. Biotechnol Lett 41(3):319–333

Chen CM, Misra TK, Silver S, Rosen BP (1986) Nucleotide sequence of the structural genes for an anion pump. The plasmid-encoded arsenical resistance operon. J Biol Chem 261 (32):15030–15038

Chibuike GU, Obiora SC (2014) Heavy metal polluted soils: effect on plants and bioremediation methods. Appl Environ Soil Sci:1–14

Choi SY, Keun-Young P, Hee JK, Kweon JH (2015) Removal of heavy metal and nitrate nitrogen in polluted groundwater by electrodialysis process. J Water Resour Hydraulic Eng 4:412

Chojnacka K, Chojnacki A, Gorecka H (2005) Biosorption of Cr^{3+}, Cd^{2+} and Cu^{2+} ions by blue–green algae Spirulina sp.: kinetics, equilibrium and the mechanism of the process. Chemosphere 59(1):75–84

Coleman RJ, Patel YN, Harding NE (2008) Identification and organization of genes for diutan polysaccharide synthesis from *Sphingomonas* sp. ATCC 53159. J Ind Microbiol Biotechnol 35 (4):263–274

Congeevaram S, Dhanarani S, Park J, Dexilin M, Thamaraiselvi K (2007) Biosorption of chromium and nickel by heavy metal resistant fungal and bacterial isolates. J Hazard Mater 146 (1–2):270–277

Czaczyk K, Myszka K (2007) Biosynthesis of extracellular polymeric substances (EPS) and its role in microbial biofilm formation. Pol J Environ Stud 16(6):799–806

Davila-Guzman NE, Cerino-Córdova FDJ, Loredo-Cancino M, Rangel-Mendez JR, Gómez-González R, Soto-Regalado E (2016) Studies of adsorption of heavy metals onto spent coffee ground: equilibrium, regeneration, and dynamic performance in a fixed-bed column. Int J Chem Eng 2016:1

De Vuyst L, Degeest B (1999) Heteropolysaccharides from lactic acid bacteria. FEMS Microbiol Rev 23(2):153–177

Dey U, Chatterjee S, Mondal NK (2016) Isolation and characterization of arsenic-resistant bacteria and possible application in bioremediation. Biotechnol Rep 10:1–7

Dhankhar R, Hooda A (2011) Fungal biosorption–an alternative to meet the challenges of heavy metal pollution in aqueous solutions. Environ Technol 32(5):467–491

Diep P, Mahadevan R, Yakunin AF (2018) Heavy metal removal by bioaccumulation using genetically engineered microorganisms. Front Bioeng Biotechnol 6:157

El-Helow ER, Sabry SA, Amer RM (2000) Cadmium biosorption by a cadmium resistant strain of *Bacillus thuringiensis*: regulation and optimization of cell surface affinity for metal cations. Biometals 13(4):273–280

Fashola MO, Ngole-Jeme VM, Babalola OO (2016) Heavy metal pollution from gold mines: environmental effects and bacterial strategies for resistance. Int J Environ Res Public Health 13(11):1047

Fourest E, Canal C, Roux JC (1994) Improvement of heavy metal biosorption by mycelial dead biomasses (*Rhizopus arrhizus*, *Mucor miehei* and *Penicillium chrysogenum*): pH control and cationic activation. FEMS Microbiol Rev 14(4):325–332

Freitas F, Alves VD, Pais J, Costa N, Oliveira C, Mafra L et al (2009) Characterization of an extracellular polysaccharide produced by a Pseudomonas strain grown on glycerol. Bioresour Technol 100(2):859–865

Freitas F, Alves VD, Reis MA (2011) Advances in bacterial exopolysaccharides: from production to biotechnological applications. Trends Biotechnol 29(8):388–398

Gabr RM, Hassan SHA, Shoreit AAM (2008) Biosorption of lead and nickel by living and non-living cells of Pseudomonas aeruginosa ASU 6a. Int Biodeterior Biodegrad 62(2):195–203

Galun M, Keller P, Feldstein H, Galun E, Siegel S, Siegel B (1983) Recovery of uranium (VI) from solution using fungi II. Release from uranium-loaded *Penicillium* biomass. Water Air Soil Pollut 20(3):277–285

Garnham GW, Wase J (1997) The use of algae as metal biosorbents. In: Biosorbents for metal ions. Taylor & Francis, London, pp 11–37

Gong R, Ding Y, Liu H, Chen Q, Liu Z (2005) Lead biosorption and desorption by intact and pretreated *Spirulina maxima* biomass. Chemosphere 58(1):125–130

Gordon T, Bowser D (2003) Beryllium: genotoxicity and carcinogenicity. Mutat Res Fundam Mol Mech Mutagen 533:99–105

Gupta P, Diwan B (2017) Bacterial exopolysaccharide mediated heavy metal removal: a review on biosynthesis, mechanism and remediation strategies. Biotechnol Rep 13:58–71

Gupta R, Ahuja P, Khan S, Saxena RK, Mohapatra H (2000) Microbial biosorbents: meeting challenges of heavy metal pollution in aqueous solutions. Curr Sci 78:967–973

Haq F, Butt M, Ali H, Chaudhary HJ (2016) Biosorption of cadmium and chromium from water by endophytic *Kocuria rhizophila*: equilibrium and kinetic studies. Desalin Water Treat 57 (42):19946–19958

Hassan SH, Awad YM, Kabir MH, Oh SE, Joo JH (2010) Bacterial biosorption of heavy metals. In: Biotechnology cracking new pastures. MD Publications PVT LTD, New Delhi, pp 79–110

Higham DP, Sadler PJ, Scawen MD (1986) Cadmium-binding proteins in *Pseudomonas putida*: Pseudothioneins. Environ Health Perspect 65:5–11

Huber AL, Holbein BE, Kidby DK (1990) Metal uptake by synthetic and biosynthetic chemicals. In: Biosorption of heavy metals. CRC Press, Boca Raton, FL, pp 249–292

Igiri BE, Okoduwa SI, Idoko GO, Akabuogu EP, Adeyi AO, Ejiogu IK (2018) Toxicity and bioremediation of heavy metals contaminated ecosystem from tannery wastewater: a review. J Toxicol 2018:2568038

Jalali-Rad R, Ghafourian H, Asef Y, Dalir ST, Sahafipour MH, Gharanjik BM (2004) Biosorption of cesium by native and chemically modified biomass of marine algae: introduce the new biosorbents for biotechnology applications. J Hazard Mater 116(1–2):125–134

Jencarova J, Luptakova A (2012) The elimination of heavy metal ions from waters by biogenic iron sulphides. Chem Eng 28:205–210

Jobby R, Desai N (2017) Bioremediation of heavy metals. In: Biodegradation and bioremediation. Environmental science and engineering. Stadium Press, New Delhi, pp 201–220

Kamaruzaman S, Aris NF, Yahaya N, Hong LS, Razak MR (2017) Removal of Cu(II) and Cd (II) ions from environmental water samples by using cellulose acetate membrane. J Environ Anal Chem 4(4):2380–2391

Kanamarlapudi SLRK, Chintalpudi VK, Muddada S (2018) Application of biosorption for removal of heavy metals from wastewater. Biosorption 18:69–116

Kang SY, Lee JU, Kim KW (2005) Metal removal from wastewater by bacterial sorption: kinetics and competition studies. Environ Technol 26(6):615–624

Kao WC, Huang CC, Chang JS (2008) Biosorption of nickel, chromium and zinc by MerP-expressing recombinant Escherichia coli. J Hazard Mater 158(1):100–106

Karakagh RM, Chorom M, Motamedi H, Kalkhajeh YK, Oustan S (2012) Biosorption of Cd and Ni by inactivated bacteria isolated from agricultural soil treated with sewage sludge. Ecohydrol Hydrobiol 12(3):191–198

Kim HS, Kim YJ, Seo YR (2015) An overview of carcinogenic heavy metal: molecular toxicity mechanism and prevention. J Cancer Prev 20(4):232

Kralj S, van Geel-Schutten GH, Dondorff MMG, Kirsanovs S, Van Der Maarel MJEC, Dijkhuizen L (2004) Glucan synthesis in the genus *Lactobacillus*: isolation and characterization of glucansucrase genes, enzymes and glucan products from six different strains. Microbiology 150(11):3681–3690

Krawiec M, Woźniak-Biel A, Bednarski M, Wieliczko A (2017) Antimicrobial susceptibility and genotypic characteristic of *Campylobacter* spp. isolates from free-living birds in Poland. Vector Borne Zoon Dis 17(11):755–763

Kumar R, Bhatia D, Singh R, Rani S, Bishnoi NR (2011) Sorption of heavy metals from electroplating effluent using immobilized biomass Trichoderma viride in a continuous packed-bed column. Int Biodeterior Biodegrad 65(8):1133–1139

Kumaran N, Sundaramanickam A, Bragadeeswaran S (2011) Absorption studies on heavy metals by isolated bacterial strain (*Pseudomonas* sp.) from Uppanar estuarine water, southeast coast of India. J Appl Sci Environ Sanit 6(4)

Kuyucak N (1990) Feasibility of biosorbents application. In: Biosorption of heavy metals. CRC Press, Boca Raton, FL, pp 372–377

Lima AIG, Corticeiro SC, Figueira EMDAP (2006) Glutathione-mediated cadmium sequestration in *Rhizobium leguminosarum*. Enzym Microb Technol 39(4):763–769

Liu HL, Chen BY, Lan YW, Cheng YC (2004) Biosorption of Zn(II) and Cu(II) by the indigenous Thiobacillus thiooxidans. Chem Eng J 97(2–3):195–201

Loukidou MX, Zouboulis AI, Karapantsios TD, Matis KA (2004) Equilibrium and kinetic modeling of chromium (VI) biosorption by *Aeromonas caviae*. Colloids Surf A Physicochem Eng Asp 242(1–3):93–104

Lu WB, Shi JJ, Wang CH, Chang JS (2006) Biosorption of lead, copper and cadmium by an indigenous isolate *Enterobacter* sp. J1 possessing high heavy-metal resistance. J Hazard Mater 134(1–3):80–86

Maddela, NR, Zhou Z, Yu Z, Zhao S, Meng F (2018) Functional determinants of extracellular polymeric substances in membrane biofouling: Experimental evidence from pure-cultured sludge bacteria. Appl Environ Microbiol 84(15):e00756–18

Majumder A, Singh A, Goyal A (2009) Application of response surface methodology for glucan production from Leuconostoc dextranicum and its structural characterization. Carbohydr Polym 75(1):150–156

Majumder A, Ghosh S, Saha N, Kole SC, Sarkar S (2013) Arsenic accumulating bacteria isolated from soil for possible application in bioremediation. J Environ Biol 34(5):841–846

Malik A (2004) Metal bioremediation through growing cells. Environ Int 30(2):261–278

Mcintosh M, Stone BA, Stanisich VA (2005) Curdlan and other bacterial (1→3)-β-D-glucans. Appl Microbiol Biotechnol 68(2):163–173

Meyer J, Schmidt A, Michalke K, Hensel R (2007) Volatilisation of metals and metalloids by the microbial population of an alluvial soil. Syst Appl Microbiol 30(3):229–238

Michalak I, Chojnacka K, Witek-Krowiak A (2013) State of the art for the biosorption process—a review. Appl Biochem Biotechnol 170(6):1389–1416

Mishra SP (2014) Adsorption–desorption of heavy metal ions. Curr Sci 107(4):601–612

Mishra A, Malik A (2013) Recent advances in microbial metal bioaccumulation. Crit Rev Environ Sci Technol 43(11):1162–1222

Mohamad OA, Hao X, Xie P, Hatab S, Lin Y, Wei G (2009) Biosorption of copper (II) from aqueous solution using non-living *Mesorhizobium amorphae* strain CCNWGS0123. Microbes Environ 27(3):234–241

Mohanty M, Pattnaik MM, Mishra AK, Patra HK (2012) Bio-concentration of chromium—an in-situ phytoremediation study at South Kaliapani chromite mining area of Orissa, India. Environ Monit Assess 184(2):1015–1024

Monsan P, Bozonnet S, Albenne C, Joucla G, Willemot RM, Remaud-Siméon M (2001) Homopolysaccharides from lactic acid bacteria. Int Dairy J 11(9):675–685

More TT, Yadav JSS, Yan S, Tyagi RD, Surampalli RY (2014) Extracellular polymeric substances of bacteria and their potential environmental applications. J Environ Mgmt 144:1–25

Morillo JA, Antizar-Ladislao B, Monteoliva-Sánchez M, Ramos-Cormenzana A, Russell NJ (2009) Bioremediation and biovalorisation of olive-mill wastes. Appl Microbiol Biotechnol 82(1):25

Nagajyoti PC, Lee KD, Sreekanth TVM (2010) Heavy metals, occurrence and toxicity for plants: a review. Environ Chem Lett 8(3):199–216

Nishimura J (2014) Exopolysaccharides produced from Lactobacillus delbrueckii subsp. bulgaricus. Adv Microbiol 4(14):1017

Nouha K, Hoang NV, Song Y, Tyagi RD, Surampalli R (2016a) Characterization of extracellular polymeric substances (EPS) produced by *Cloacibacterium normanense* isolated from wastewater sludge for sludge settling and dewatering. Civil Environ Engg 5(6):191–200

Nouha K, Kumar RS, Tyagi RD (2016b) Heavy metals removal from wastewater using extracellular polymeric substances produced by *Cloacibacterium normanense* in wastewater sludge supplemented with crude glycerol and study of extracellular polymeric substances extraction by different methods. Bioresour Technol 212:120–129

Nowicki EM, O'Brien JP, Brodbelt JS, Trent MS (2015) Extracellular zinc induces phosphoethanolamine addition to Pseudomonas aeruginosa lipid A via the ColRS two-component system. Mol Microbiol 97(1):166–178

Öztürk A (2007) Removal of nickel from aqueous solution by the bacterium Bacillus thuringiensis. J Hazard Mater 147(1–2):518–523

Öztürk A, Artan T, Ayar A (2004) Biosorption of nickel (II) and copper (II) ions from aqueous solution by Streptomyces coelicolor A3 (2). Colloids Surf B Biointerfaces 34(2):105–111

Pandey A, Bera D, Shukla A, Ray L (2007) Studies on Cr (VI), Pb (II) and Cu (II) adsorption–desorption using calcium alginate as biopolymer. Chem Spec Bioavail 19(1):17–24

Prasad KS, Ramanathan AL, Paul J, Subramanian V, Prasad R (2013) Biosorption of arsenite (As+ 3) and arsenate (As+ 5) from aqueous solution by Arthrobacter sp. biomass. Environ Technol 34 (19):2701–2708

Puranik PR, Paknikar KM (1997) Biosorption of lead and zinc from solutions using *Streptoverticillium cinnamoneum* waste biomass. J Biotechnol 55(2):113–124

Qiu YR, Mao LJ (2013) Removal of heavy metal ions from aqueous solution by ultrafiltration assisted with copolymer of maleic acid and acrylic acid. Desalination 329:78–85

Quintelas C, Rocha Z, Silva B, Fonseca B, Figueiredo H, Tavares T (2009) Removal of Cd(II), Cr (VI), Fe(III) and Ni(II) from aqueous solutions by an *E. coli* biofilm supported on kaolin. Chem Eng J 149(1–3):319–324

Rajkumar M, Freitas H (2008) Influence of metal resistant-plant growth-promoting bacteria on the growth of Ricinus communis in soil contaminated with heavy metals. Chemosphere 71 (5):834–842

Rajkumar M, Prasad MNV, Swaminathan S, Freitas H (2013) Climate change driven plant–metal–microbe interactions. Environ Int 53:74–86

Rani MJ, Hemambika B, Hemapriya J, Kannan VR (2010) Comparative assessment of heavy metal removal by immobilized and dead bacterial cells: a biosorption approach. Afr J Environ Sci Technol 4(1):23–30

Ruas-Madiedo P, Hugenholtz J, Zoon P (2002) An overview of the functionality of exopolysaccharides produced by lactic acid bacteria. Int Dairy J 12(2–3):163–171

Sankarammal M, Thatheyus A, Ramya D (2014) Bioremoval of cadmium using Pseudomonas fluorescens. Open J Water Pollut Treat 1:92–100

Saravanan S, Anitha MCA, Venkadesan S, Annadurai G (2012) Optimization of biosorption of arsenic metal ions by using immobilized metal resistant Bacillus sp. Int J Res Environ Sci Technol 2(4):114–118

Sarwat F, Qader SAU, Aman A, Ahmed N (2008) Production and characterization of a unique dextran from an indigenous Leuconostoc mesenteroides CMG713. Int J Biol Sci 4(6):379

Sebogodi KM, Babalola OO (2011) Identification of soil bacteria from mining environments in Rustenburg, South Africa. Life Sci J 8(S2):25–32

Selatnia A, Boukazoula A, Kechid N, Bakhti MZ, Chergui A, Kerchich Y (2004) Biosorption of lead (II) from aqueous solution by a bacterial dead *Streptomyces rimosus* biomass. Biochem Eng J 19(2):127–135

Selvi K, Pattabhi S, Kadirvelu K (2001) Removal of Cr(VI) from aqueous solution by adsorption onto activated carbon. Bioresour Technol 80(1):87–89

Shamim S (2018) Biosorption of heavy metals. Biosorption 2:21–49

Shamim S, Rehman A (2014) Physicochemical surface properties of *Cupriavidus metallidurans* CH34 and Pseudomonas putida mt2 under cadmium stress. J Basic Microbiol 54(4):306–314

Shamim S, Rehman A (2015) Antioxidative enzyme profiling and biosorption ability of *Cupriavidus metallidurans* CH34 and *Pseudomonas putida* mt2 under cadmium stress. J Basic Microbiol 55(3):374–381

Shamim S, Rehman A, Qazi MH (2014a) Cadmium-resistance mechanism in the bacteria *Cupriavidus metallidurans* CH34 and Pseudomonas putida mt2. Arch Environ Contam Toxicol 67(2):149–157

Shamim S, Rehman A, Qazi MH (2014b) Swimming, swarming, twitching, and chemotactic responses of *Cupriavidus metallidurans* CH34 and *Pseudomonas putida* mt2 in the presence of cadmium. Arch Environ Contam Toxicol 66(3):407–414

Shyni LS, Jagadish K, Srikantaswamy S, Abhilash MR (2016) Photocatalytic degradation and removal of heavy metals in pharmaceutical waste by selenium doped ZnO nanocomposite semiconductor. J For Res 2(05):47–54

Stähler FN, Odenbreit S, Haas R, Wilrich J, Van Vliet AH, Kusters JG et al (2006) The novel *Helicobacter pylori* CznABC metal efflux pump is required for cadmium, zinc, and nickel resistance, urease modulation, and gastric colonization. Infect Immun 74(7):3845–3852

Subramanian SB, Yan S, Tyagi RD, Surampalli RY (2010) Extracellular polymeric substances (EPS) producing bacterial strains of municipal wastewater sludge: isolation, molecular identification, EPS characterization and performance for sludge settling and dewatering. Water Res 44 (7):2253–2266

Sulaymon AH, Mohammed AA, Al-Musawi TJ (2013) Competitive biosorption of lead, cadmium, copper, and arsenic ions using algae. Environ Sci Pollut Res 20(5):3011–3023

Taniguchi J, Hemmi H, Tanahashi K, Amano N, Nakayama T, Nishino T (2000) Zinc biosorption by a zinc-resistant bacterium, *Brevibacterium* sp. strain HZM-1. Appl Microbiol Biotechnol 54 (4):581–588

Tariq A, Ullah U, Asif M, Sadiq I (2019) Biosorption of arsenic through bacteria isolated from Pakistan. Int Microbiol 22(1):59–68

Tchounwou PB, Yedjou CG, Patlolla AK, Sutton DJ (2012) Heavy metal toxicity and the environment. Mol Clin Environ Toxicol 101:133–164

Teclu D, Tivchev G, Laing M, Wallis M (2008) Bioremoval of arsenic species from contaminated waters by sulphate-reducing bacteria. Water Res 42(19):4885–4893

Timková I, Sedláková-Kaduková J, Pristaš P (2018) Biosorption and bioaccumulation abilities of actinomycetes/streptomycetes isolated from metal contaminated sites. Separations 5(4):54

Ton-That H, Marraffini LA, Schneewind O (2004) Protein sorting to the cell wall envelope of Gram-positive bacteria. Biochim Biophys Acta Mol Cell Res 1694(1–3):269–278

Tsezos M (2001) Biosorption of metals. The experience accumulated and the outlook for technology development. Hydrometallurgy (Amsterdam) 59(2–3):241–243

Tuzen M, Saygi KO, Usta C, Soylak M (2008) *Pseudomonas aeruginosa* immobilized multiwalled carbon nanotubes as biosorbent for heavy metal ions. Bioresour Technol 99(6):1563–1570

Uzel A, Ozdemir G (2009) Metal biosorption capacity of the organic solvent tolerant *Pseudomonas fluorescens* TEM08. Bioresour Technol 100(2):542–548

Veglio F, Beolchini F (1997) Removal of metals by biosorption: a review. Hydrometallurgy 44 (3):301–316

Vijayaraghavan K, Yun YS (2008) Bacterial biosorbents and biosorption. Biotechnol Adv 26 (3):266–291

Volesky B (1990) Removal and recovery of heavy metals by biosorption. In: Biosorption of heavy metals. CRC Press, Boca Raton, FL, pp 7–43

Volesky B (2003) Sorption and biosorption. BV Sorbex, Montreal

Wang J, Chen C (2006) Biosorption of heavy metals by Saccharomyces cerevisiae: a review. Biotechnol Adv 24(5):427–451

Wang J, Chen C (2009) Biosorbents for heavy metals removal and their future. Biotechnol Adv 27 (2):195–226

Weast RC (1984) CRC Handbook of chemistry and physics, 6th edn. CRC Press, Boca Raton

Whitfield C (1988) Bacterial extracellular polysaccharides. Can J Microbiol 34(4):415–420

Yang HC, Fu HL, Lin YF, Rosen BP (2012) Pathways of arsenic uptake and efflux. Current topics in membranes, vol 69. Academic Press, London, pp 325–358

Ye S, Liu F, Wang J, Wang H, Zhang M (2012) Antioxidant activities of an exopolysaccharide isolated and purified from marine pseudomonas PF-6. Carbohydr Polym 87(1):764–770

Yilmaz EI, Ensari NY (2005) Cadmium biosorption by *Bacillus circulans* strain EB1. World J Microbiol Biotechnol 21(5):777–779

Zouboulis AI, Loukidou MX, Matis KA (2004) Biosorption of toxic metals from aqueous solutions by bacteria strains isolated from metal-polluted soils. Process Biochem 39(8):909–916

The Interest in Nanotechnology: A Step Towards Bioremediation

Komal Agrawal and Pradeep Verma

Abstract

Green technology, an approach for the removal/bioremediation of the pollutant, has been gaining interest over physical and chemical methods. The concept of nanotechnology for green and clean environment has expanded interest in various sectors due to its multidisciplinary approach. The nanotechnology allows to play with various aspects of processing such as size of the particle, optical properties and its shape which cannot be the case when other treatment methods are employed. The utility of nanotechnology has been extended for the treatment of pollutant from a broad range of sectors, e.g. dyes, pharmaceutical and medical. The main concern which arises while considering the present approach is the disposal of the nanoparticles after it has served its purpose in the treatment of the waste. Thus, the present chapter would deal with basic introduction on nanoparticles, metals used for synthesis, its potential for the removal of soil and water pollutant and lastly its future prospects.

Keywords

Nanotechnology · Bioremediation · Emerging pollutant · Multidisciplinary · Green technology

1 Introduction

The industrialization led to the development of technologies, increased financial independence as well as improvement of the living standards of the community. However, with time and overexploitation of resources to meet the demands and

K. Agrawal · P. Verma (✉)
Bioprocess and Bioenergy Laboratory, Department of Microbiology, Central University of Rajasthan, Ajmer, Rajasthan, India
e-mail: pradeepverma@curaj.ac.in

M. P. Shah (ed.), *Removal of Emerging Contaminants Through Microbial Processes*,
https://doi.org/10.1007/978-981-15-5901-3_13

release of effluents, undesired end-products have damaged the ecosystem globally. Thus, various physical, chemical and physiochemical methods were devised for the treatment of the pollutant, which after a certain period of time diverted towards the biological treatment processes involving microbes. Now with the development of technologies, availability of resources has diverted the focus towards nanotechnology (Sastry et al. 2004). Nanotechnology is a new and a promising technology which will help tremendously in the future to improve the lives and also contribute uniqueness in various fields of study. The integration of nanotechnology with environmental biotechnology will enable the development of nano-devices. The nano-devices along with biological potential can be effectively used for remediation (Singh and Tripathi 2007), as it leads to the synthesis/development of products with new abilities.

As per Singh and Tripathi (2007), many incidents support the importance and efficiency of nanotechnology, i.e. development of new nanotechnologies, superiority in products, integration of computational techniques and possible development of high-performance products (Tolles and Rath 2003) that have enabled researchers to design, manipulate and use for various applications.

Nanotechnology helps replace the toxic metals from the environment to non-/less toxic forms which act towards favouring the sustainable environment. In the study by Choi et al. (1999), carbon nanotubes were proposed for manufacturing monitors of computers instead of cathode ray tubes. It prevented pollution as the toxic heavy metal which was initially used was now replaced, thereby saving the ecosystem as well as energy (Choi et al. 1999). Pizzini et al. (2005) used nano-porous silica in the photovoltaic cells and later Grätzel (2003) proposed TiO_2 nanoparticles coated with an organic dye for the development of organic solar cells for the conversion of light to energy, a process similar to photosynthesis. Thus, the present chapter would discuss various types of metal used in nanotechnology, its remediation potential in soil and water, its limitation and future prospect.

2 Synthesis and Unique Properties of Nanoparticles

The material with bulk properties has been used for the development of various types of nanostructures with the reduction in size to nano-range. As the particle size changes from micrometre to nanometre, the ferromagnetic property changes which can be done either at low or high temperatures (Komarneni 1995). In low temperature treatment methods adapted include precipitation, hydrothermal synthesis, inverse micelle and sol-gel synthesis. The treatment at high temperatures includes condensation using gas, liquid aerosol thermodynamics and wire explosion. The conventional techniques involve using hydrothermal, microwave hydrothermal and microwave solvo-thermal methods for the preparation of nanophase (varied sizes and shapes) materials (Komarneni 2003). The synthesis of nano-powder or colloidal powders consists of ball milling, condensation or precipitation, glassy material drawing, biological assembly via self-assembly and growth of the second material on a crystalline lattice where the lattice parameters are not similar. Other methods used for the nanostructure fabrication include lithography, molecular beam epitaxy and growth of one material onto another. The chemical methods are also used which include self-assembly and spontaneous aggregation of molecular groups (Singh and Tripathi 2007; Poole Jr and Owens 2003).

The physical, chemical and electronic characteristic features of nanoparticles depends on its number, type of atoms and its interaction among the crystal atoms and atoms in grain boundaries. The van der Waals' forces, magnetism and electron resistance play a very important role in determining the factors of nanoparticles as compared to gravity and inertia. As the spectral properties can enable better understanding of the nanoparticle, the nanoparticle study is a cost-effective strategy and has been widely studied (Singh and Tripathi 2007; Bhat 2003).

3 Advances in Environmental Application of Nanobiotechnology

The main advantage of nanobiotechnology include certain characteristic features such as clean and green nanomaterial for the removal of contaminant (Tratnyek and Johnson 2006). The two main techniques applied in nano-remediation include the use of adsorption and reactive techniques for both in situ and ex situ treatment of pollutants. The development of various technologies and its application has enabled mitigation of the environment towards a more sustainable environment. The biogenic nanomaterials have been designed using metal oxides and noble metals of which the nanoscale zero-valent iron (ZVI) has a wider range of applications in remediation (Tripathi et al. 2018). The nanoparticles synthesized using the biogenic approach have its utility in the remediation of soil and water which has been discussed in the following sections of the chapter.

4 Green Synthesis of Nanoparticles

The green synthesis of nanoparticles has been done using vitamins, sugars, extracts of plants and polymers (biodegradable). In addition, microbes have been used as reductant and capping agent. The nanoparticle is also synthesized using inorganic nanoparticles of which the plant-based nanoparticles seem to be the best alternative towards a greener approach and can be used for large-scale production as well (Iravani 2011). The green synthesis of nanoparticles has been more preferred over other synthesis methods as it is cost-effective, a stable material and comparatively reproducible (Kalaiarasi et al. 2010). Overtime the focus diverted towards microorganism where the rate of synthesis is slow and only a limited number of nanoparticle can be synthesized. In spite of the drawbacks associated with the microorganisms, fungal system is gaining popularity for the synthesis of the nanoparticles (Dhillon et al. 2012) as they are environmentally friendly and do not require an extensive chemical for the synthesis (Parsons et al. 2007). Thus, the various methods used for the synthesis of green nanoparticles have been discussed.

4.1 Silver Nanoparticles

The silver nanoparticle (5–10 nm) synthesized using microwave irradiation conditions using glutathione is a benign antioxidant (Baruwati et al. 2009) and takes only 30–60 min for the synthesis. Similar nanoparticle synthesis can be applied using gold, palladium and platinum. The silver nanoparticle synthesis has been done using geraniol, which has the potential to inhibit (less than 30%) cancer cell lines, i.e. Fibrosarcoma-Wehi 164. Sugar along with silver salt has been used for the production of silver mirror as a part of Tollens' test; sugar has been used for the fabrication of silver nanoparticles, where it acts as a capping and a reducing agent. Utilization of sugar has advantages such as cost-effective, less/negligible toxicity and easy availability and adaptability (Nazir et al. 2011). The synthesis of silver nanoparticles has also been done using polymeric matrix (natural), silver nitrate, NaOH, $C_6H_{12}O_6$ and gelatin (Darroudi et al. 2010). The use of plant extract for the synthesis of nanoparticle is carried out at room temperatures, but under certain exception, heating below 100 °C is added (Kharissova et al. 2013). The synthesis of nanoparticles involves the use of extracts from tea and coffee (Moulton et al. 2010) (Table 1).

4.2 Gold Nanoparticles

The silver nanoparticle is more extensively studied as compared to the gold nanoparticle, where its size usually ranges from 20 to 300 nm (Kharissova et al. 2013). Gold nanoparticle has been synthesized using banana peel extract (Bankar et al. 2010). On the other hand, a nanocomposite of gold nanoparticles was synthesized using tea extract (Afzal et al. 2009). The concept of green synthesis has been adapted significantly in a study by Wu and Chen (2007); gold nanoparticles were synthesized using rice wine, soda and Au(III) ion and dissolved at pH 6.5 and 25–55 °C. In this experimental setup, rice wine was the solvent and reducing agent and soda as the protective agent and base catalyst resulting in highly stable resultant solution with no precipitation even after several months. In addition to the gold nanoparticles, gold alloys were also produced, wherein dextrose and ethanol extract from black tea, i.e. *Camellia sinensis*, were used for the synthesis of gold-silver alloy nanoparticle under the microwave irradiation. The synthesized gold-silver alloy was purified using agarose gel electrophoresis, helped infer the spherical shape with size range of approximately 200 nm with an average size of 92 nm (Nori et al. 2013). Thus, the study of nanotechnology has not only enabled the synthesis of gold or silver nanoparticles but also its alloy, thereby extending and broadening the study with numerous future possibilities (Table 2).

Table 1 Synthesis and application of silver nanoparticles

S. no.	Source for the synthesis	Application	References
1.	*Aspilia pluriseta* extracts	Antimicrobial and catalytic activity	Nyabola et al. (2020)
2.	Babassu mesocarp starch (*Attalea speciosa* Mart. ex Spreng.)	Antimicrobial activity	Bastos Araruna et al. (2020)
3.	Pomegranate seeds extract	Nanocomposite Ag-starch synthesis	Mohseni et al. (2020)
4.	Tea leaves (*Camellia sinensis*) leaves	–	
5.	Coconut inflorescence sap	Antimicrobial and cytotoxic	Rajesh et al. (2020)
6.	Secondary metabolites of *Acacia nilotica*	Antimicrobial and detoxification agent	Shah et al. (2020)
7.	*Mangifera indica* and *Prunus dulcis* extract	Antibacterial activity	Akujobi et al. (2020)
8.	*Serratia marcescens* ssp. *sakuensis* supernatant	Antibacterial activity	Akl et al. (2020)
9.	Endophytic *Trichoderma atroviride*	Antimicrobial activity	Abdel-Azeem et al. (2020)
10.	Rhizome extract of *Alpinia officinarum*	Potential biomaterial to remediate environmental pollution	Li et al. (2020)
11.	*Kitasatospora albolonga*	Antimicrobial against multidrug-resistant *Pseudomonas aeruginosa*	D'Lima et al. (2020)
12.	Leaf extract of *Bridelia stipularis*	Effect on mitotic chromosomes of *Pisum sativum*	Daphedar and Taranath (2019)
13.	*Phyllanthus emblica* fruit extract	Antimicrobial application	Renuka et al. (2020)
14.	*Bacillus methylotrophicus DC3*	*Bacillus methylotrophicus* DC3	Wang et al. (2016)
15.	Bacterial synthesis of silver nanoparticles	*Bioremediation of Congo red dye*	Modi et al. (2015)

4.3 Iron Nanoparticles

The nanoparticles of Fe_xO_y and FeOOH which consist of 16 polymorphic forms have a disinfectant and bioremediation potential of water and soil (Li et al. 2008; Kharisov et al. 2012). Thus, various methods were used for the synthesis of iron nanoparticles such as using tea extracts (polyphenol) and other plant extracts (Hoag et al. 2009). The pure iron nanoparticles synthesized using green tea leaves have been effectively used for the removal of methylene blue and methyl orange (Shahwan et al. 2011). The iron nanoparticles have been synthesized using soya

Table 2 Synthesis and application of silver nanoparticles

S. no.	Source for the synthesis	Application	References
1.	Bacteria cellulose nanofibres	Biosensing	Zhang et al. (2010)
2.	Guar gum	Ammonia sensor	Pandey et al. (2013)
3.	Leaf extracts of *Magnolia kobus* and *Diospyros kaki*	–	Song et al. (2009)
4.	Water-soluble methylimidazolium-based ionic polymer	Reduction of *p*-nitrophenol and in the hydrogenation of cinnamaldehyde	Biondi et al. (2011)
5.	*Enterobacteriaceae*	–	Honary et al. (2012)
6.	*Stenotrophomonas maltophilia* (AuRed02)	–	Nangia et al. (2009)
7.	*Padina gymnospora*	–	Singh et al. (2013)
8.	*Cinnamomum zeylanicum* leaf broth		Smitha et al. (2009)
9.	*Shewanella algae*	–	Ogi et al. (2010)
10.	*Coriander leaf extract*	–	Narayanan and Sakthivel (2008)

bean sprout templates (Cai et al. 2010). In the study by Smuleac et al. (2011), membranes containing reactive Fe and Fe/Pd nanoparticles immobilized on polyacrylic acid coated with polyvinylidene fluoride were synthesized using tea extract. This membrane was used for the removal of trichloroethylene.

4.4 Palladium and Platinum Nanoparticles

Palladium as an option for the synthesis of nanoparticle has been explored where they have been synthesized using various methodologies. In the study by Kanchana et al. (2010), palladium nanoparticles were phyto-synthesized using *Solanum trilobatum* in conjugation with lipoic acid and vitamins under moderate pH and room temperature. The palladium nanoparticles like the silver and gold nanoparticles have been synthesized using coffee and tea extracts (Nadagouda and Varma 2008), broth of *Cinnamomum camphora* leaf (Yang et al. 2010) and soybean (*Glycine max*) leaf extract (Petla et al. 2011). Among the various metals used for the synthesis of nanoparticles, platinum too has been used where the nanoparticles were synthesized using leaf extract of *Diospyros kaki* (Song et al. 2010).

4.5 Titanium(IV) Oxide and Zinc(II) Oxide Nanoparticles

Nanoparticles using titanium(IV) oxide have been synthesized using leaf extract of *Nyctanthes arbor-tristis* (Sundrarajan and Gowri 2011) and latex of *Jatropha curcas* L. (Hudlikar et al. 2012). In addition, zinc(II) oxide nanoparticles were synthesized using zinc nitrate and miller leaf extract of *Aloe barbadensis* (Sangeetha et al. 2011); they were also synthesized using gelatin media using sol-gel method (Zak et al. 2011).

4.6 Copper Oxides, Indium(III) Oxide and Lead(II) Sulphide Nanoparticles

Cuprous oxide (Cu_2O) nanoparticles were synthesized using a leaf waste of an agricultural waste, i.e. *Arachis hypogaea* L., by the reduction of Barfoed's solution (Ramesh et al. 2011). The synthesized nanoparticles had an antibacterial effect against *Escherichia coli*. In addition, the indium(III) oxide was synthesized using indium acetylacetonate and plant extract of aloe vera using a calcification step for 2 h at 400–600 °C (Maensiri et al. 2008). Lastly the cubic lead sulphide nanoparticle was synthesized using *Rhodosporidium diobovatum* (Seshadri et al. 2011).

5 Nano-remediation

The nanomaterial due to their small size has high surface area which allows its effective utilization in nano-bioremediation. Its size, morphology, surface area, distribution, surface charge and crystallographic characterization have been studied in details by various researchers. The detailed study enables a better understanding of the particles, thereby broadening its application in various sectors of bioremediation (Thomé et al. 2015). Degradation of the two pollutants, i.e. tetrachloroethylene and polyvinyl biphenyls, by the use of nanoparticles has been fast and effective as compared to the microbial remediation. The implementation of nanoparticles at the contaminated sites involves the use of nanoscale zero-valent iron/nano-iron (nZVI). nZVI has been injected into the environment subsurface of the contaminated sites for the removal of the pollutant as it is less toxic and economically feasible (Cecchin et al. 2016). The zero-valent iron, e.g. iron filing, is used for the construction of permeable reactive barriers (Gillham and O'Hannesin 1994). The various limitations associated with the ZVI in permeable reactive barriers were reactivity loss of ZVI, gas formation and precipitation at the application area and barrier and difficulty in digging in areas where bedrocks were present and in recycling of the reactive material. However later Zhang et al. (1998) addressed the limitations and designed a nanoscale ZVI which improved the reductive dichlorination of tetrachloroethylene (Zhang et al. 1998). The particle size was smaller than the soil size; as a result of which, it can be easily injected in the subsurface via aqueous suspension for the removal of the contamination (Berge and Ramsburg 2010; Varadhi et al. 2005).

Various types of materials have been used for the bioremediation including nanoscale zeolites; oxides of metals; noble metals, i.e. bimetallic nanoparticles; nanotubes of carbon; and titanium dioxide. Among all the explored materials, nanoscale zero-valent iron is the most widely used nanoparticle; the size of which ranges from 10 to 100 nm (Karn et al. 2009).

5.1 Nano-remediation of Water

The pollutant in the water can be converted in to nontoxic products via chemical reaction. The pollutant trichloroethane found in industrial pollution can be treated using nanoparticles (Li et al. 2017a, b). Nanotechnology utilization for the remediation of the water sources eases its treatment process as it does not require pumping of water from the contaminated sites for its treatment (Fig. 1). Nanoscale particle can be directly added to the contaminated sites for the bioremediation (Berge and Ramsburg 2010; Varadhi et al. 2005). The nano-sized fibres use for the deionization uses fibre as electrodes which is both cost and energy efficient (Ryoo et al. 2003). The traditional wastewater treatment involves semipermeable membrane reverse osmosis or electrodialysis. The pore size is an important parameter for both dialysis and osmosis, as too large-sized pores will not be able to filter bacteria and viruses. Thus, membrane with pore size in nanometre range was developed which will eventually increase the selectivity of the contaminants passing through the membrane (Auffan et al. 2008). Nano-adsorbents are also used for the treatment of water as they have high specific surface area; the diffusion distance is short among the intraparticle, tunable pore size and surface chemistry. All these features allow high adsorption capacity enabling better treatment. The surface of the nanomaterial has a significant

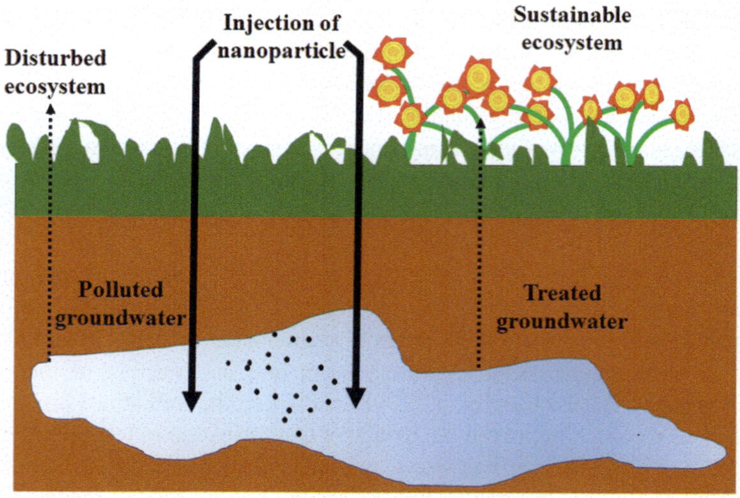

Fig. 1 Nano-remediation for the removal of pollutant from groundwater

impact on the pollutant to be acted upon, e.g. porous nanomaterial enables control of adsorption kinetics, and nano-adsorbents can be used in treatment processes of slurry reactors, adsorbers and filters. Carbo-nanotubes and graphite oxide nanosheets are capable of removing organic and metal pollutants (Yang and Xing 2010; Gao et al. 2011). Disinfection techniques involve chlorination and ozonation. The main limitation associated with such disinfection is the formation of toxic end-products. Thus, nanomaterial has been playing an important role in the removal of the pollutant and exhibits antimicrobial properties as well. The various antimicrobial nanoparticles include silver, titanium(IV) oxide and zinc(II) oxide. These nanomaterials compromise the cell membrane of the microorganism on direct contact or generate reactive oxygen species (Li et al. 2008; Klaine et al. 2008; Lee et al. 2011).

5.2 Nano-remediation of Soil

The various environmental remediation methods include in situ and ex situ, adsorptive and reactive methodologies (Tratnyek and Johnson 2006; Hodson 2010), and integration of nanomaterial in all these treatment technologies has been studied. In case of in situ treatment either reactive zone has to be created or reactive nanoparticle plume which has the potential to migrate to the polluted areas is required for remediation. Contamination of soil and water is closely associated with one another, as contaminated soil will pollute the water and vice versa. The contamination of both soil and water has raised concern globally. As a result of which focus has diverted the research towards more eco-friendly and sustainable treatment methods. The sources of contamination includes heavy metals, organic substances, pesticides and herbicides. The conventional treatment involves ex situ and in situ treatment by washing of soil, pump-and-treat operation, thermal treatment, chemical oxidation and reactive iron barriers (Hodson 2010). It has been stated that the pump-and-treat operation required 18 years on an average. Thus, nanoscale ZVI is more appropriate for the treatment of soil as it requires only 1–2 years (Fig. 2). The technique of adsorption has been effectively applied for the removal of the metal contaminant from the soil. The metals in the soil can be immobilized using nanoscale metal

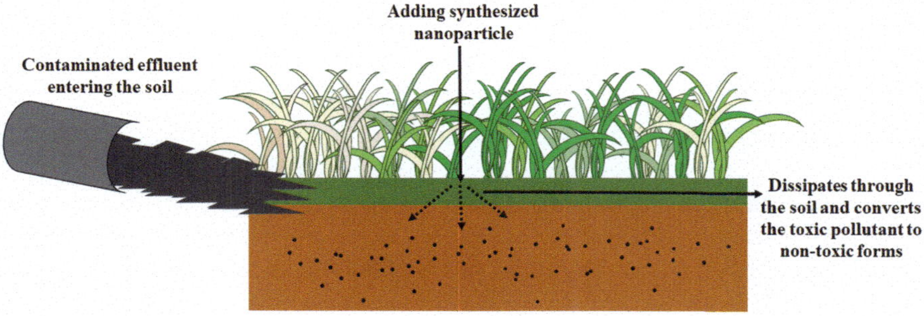

Fig. 2 Effect of nano-remediation on the removal of pollutant from soil

oxides where iron and iron oxide were used for the removal of phosphate (Schorr 2007). In the study by Rickerby and Morrison (2007), iron oxide (green rust) was used for the reduction of Cr(VI) to Cr(III). The study has extended further to carbon-based nanomaterials (Mueller and Nowack 2010), iron oxide minerals can also be used for the adsorption of arsenic (Rickerby and Morrison 2007) and nanoscale magnetite particles have been used for the removal of phosphates via adsorption in soil (Pan et al. 2010). The non-stabilized nano-magnetite could not pass through the column consisting of soil as they aggregated/agglomerated due to gravity into microparticles. Thus, coated nanoparticle due to its small size and negative charge due to carboxymethyl groups enable its mobility through the column as even the soil is negatively charged due to repulsion (O'Day and Vlassopoulos 2010; Mueller and Nowack 2010).

6 Difficulties Associated with the Use of Nanoparticle

Nanoparticles have numerous applications in the field of bioremediation of soil and water; however its limitations such as its useful life, reactivity problems and in situ transport process have been stated in various literature (Cundy et al. 2008; Tosco et al. 2012). As stated above agglomeration is also an issue which hinders its mobility in the required phase, e.g. soil repulsion (O'Day and Vlassopoulos 2010; Mueller and Nowack 2010). The study by Reddy (2010) showed the blocking effect of nanoscale ZVI in soil due to deposition which in turn attracts other particles present in the suspension and blocks the passage of the liquid or fluid through the soil. Nanoscale ZVI has numerous applications and has been studied tremendously; however the effect it has on the native microflora is in the budding stage; as a result of which, its implementation by the agencies has been sceptical causing the major hindrance factor for its implementation in the field. However, there are certain exceptions where in the USA nanoscale ZVI has been used extensively as compared to the European countries where its use is restricted (Wiesner et al. 2006; Mueller et al. 2012). Further the impact of nanoparticles on microflora has been studied which includes growth inhibition or production of enzyme under laboratory-maintained conditions wherein conflict in results have been observed. In certain studies inhibitory effect were observed, while others showed bio-stimulation (Reardon et al. 2008; Wilkin et al. 2003; Morones et al. 2005; Cecchin et al. 2016).

7 Nanoparticles and Its Toxicity

Nanoparticles have been used for the bioremediation of soil; the natural microflora of soil has the potential to degrade the organic pollutants as well as immobilize heavy metals (Bokare et al. 2012). Thus, hampering of soil microflora or decrease in the population of the native microorganisms will weaken the resilience of the soil to the contaminants (Dinesh et al. 2012). The study of nanoparticles on the other hand for the remediation of soil is in the budding stage; thus more study is required for its better and effective implementation. In certain studies, it has been observed that nanoscale ZVI has a bactericidal effect on the soil microorganisms, though the use of it is quite diverse where they effectively reduce the pollutant to either nontoxic or less toxic forms which can be easily biodegraded by the microbes. Thus, it becomes important to treat the soil using nanotechnology side by side to maintain soil microflora to enable it to maintain the viability of the soil (Wilkin et al. 2003). The mechanism of the nanoscale of ZVI is well studied; however the effect may vary depending upon the geochemistry, microflora, mineral and substrate content. In a study where zero-valent microscale iron has been used for the site remediation as a permeable reactive barrier, the formation of H^+ by the degradation of zero-valent microscale iron acted as a bio-stimulant for methanogens, sulphate-reducing and dehalogens where nanoscale ZVI is the primary electron donor (Weathers et al. 1997; Dinh et al. 2004; Liu and Lowry 2006). Cell membrane disruption and generation of oxidative stress due to the generation of Fe^{2+} and the reactive oxygen species lead to the bactericidal effect on the microorganisms of the soil and have been stated due to the direct contact with the cell wall of microorganism (Lefevre et al. 2016).

8 Aggregation of Nanoparticles and Its Implication

Nanoparticles released in the environment may aggregate to certain extent; however for its effective implementation, it has to be stable and exhibit dispersive properties for it to pass through the porous material/column which has been saturated with water. Thus methodologies and techniques need to be developed for the prevention of nanoparticle aggregation, thereby allowing it to remediate the pollutants effectively (Phenrat et al. 2007). Mobility of the nanoparticles in certain condition allows them to travel longer distance, resulting them to carry adsorbed contaminants from the surface. They have the potential to travel in the redox zones which can facilitate/inhibit its transport (Kersting et al. 1999; Novikov et al. 2006; Vilks et al. 1997; Waite et al. 1999). Though the formation of nanoscale minerals plays an important role in the environmental colloid, the aggregation and the transporting features have not been studied in details. The other aspect which is of serious concern is that nanoparticles can suspend with solid and sediment and bioaccumulate, thereby entering the food chain and the water resources (Boxall et al. 2007). The US EPA has raised a concern of bio-magnification, though data supporting the research or disapproving the concern has not been published as per Biswas and Wu (2005); USEPA (2007).

9 Nanoparticles: Safety Concerns

As the potential risk of nanoparticles is unknown and its implementation at field scale has unknown risk, certain nongovernmental organizations have raised queries resulting in the halt of practical implementation until the technology has been proven to be entirely safe. The ETC Group (2003) asked for the precautionary principles to be applied in nanotechnology. In the study by Drexler (1986) based on the concept of 'Eric Drexler's' raised a concern that the nanoscale machines can self-replicate and convert the matter into 'grey goo' which was later clarified by Phoenix and Drexler (2004). Later, UK Royal Society (2004) reported against the use of nanomaterial in bioremediation until, proper research has been carried out regarding the benefits of nanomaterials, which would then overpower or supress its risks. Studies were conducted wherein SCENIHR (2005) strictly prohibited the use of nanoparticles, and Commission de l'Éthique de la Science et la Technologie (2006) asked for more risk-related research to be conducted. Similar research was supported by other studies as well where Maynard et al. (2006) and USEPA (2007) both stated the negative effects of the nanomaterials and implications should be further studied in detail prior to its implementation in bioremediation.

10 Conclusion

Nanotechnology involves the use of nanoparticles for the conversion of toxic products in soil and water to less toxic or nontoxic forms. These forms can also be utilized by the existing microflora of the soil for further degradation, thereby enabling better efficiency of the process. In general, an argument exists with respect to the nanoparticles regarding its toxicity, aggregation and clogging effects. Further detailed study would be required for focusing on research considering the toxic effects of the nanoparticles if they enter the ecosystem via soil or water (Cecchin et al. 2016). Nano-remediation has the potential to reduce the contaminant level and also reduce the cost associated with the overall clean-up of the environment if implemented at a large scale. Thus, further research and detailed study about the toxicological study will allow its implementation in the field in the future to help achieve sustainable environment (Tripathi et al. 2018).

References

Abdel-Azeem A, Nada AA, O'Donovan A, Thakur VK, Elkelish A (2020) Mycogenic silver nanoparticles from endophytic *Trichoderma atroviride* with antimicrobial activity. J Renew Mater 8:2

Afzal AB, Akhtar MJ, Nadeem M, Hassan MM (2009) Investigation of structural and electrical properties of polyaniline/gold nanocomposites. J Phys Chem C 113(40):17560–17565

Akl BA, Nader MM, El-Saadony MT (2020) Biosynthesis of silver nanoparticles by Serratia marcescens ssp sakuensis and its antibacterial application against some pathogenic bacteria. J Agric Chem Biotechnol 11(1):1–8

Akujobi CO, Anuforo HU, Okereke JN, Ibeh C, Agbo CJ (2020) Parametric optimization of synthesis of silver nanoparticles from Mangifera indica and Prunus dulcis extracts and their antibacterial activity. Synthesis 15(30):21–26

Auffan M, Rose J, Proux O, Borschneck D, Masion A, Chaurand P, Hazemann JL, Chaneac C, Jolivet JP, Wiesner MR, Van Geen A (2008) Enhanced adsorption of arsenic onto maghemites nanoparticles: As(III) as a probe of the surface structure and heterogeneity. Langmuir 24 (7):3215–3222

Bankar A, Joshi B, Kumar AR, Zinjarde S (2010) Banana peel extract mediated synthesis of gold nanoparticles. Colloids Surf B Biointerfaces 80(1):45–50

Baruwati B, Polshettiwar V, Varma RS (2009) Glutathione promoted expeditious green synthesis of silver nanoparticles in water using microwaves. Green Chem 11(7):926–930

Bastos Araruna F, Oliveira Sousa Araruna F, Lima Alves Pereira LP, Aranha Brito MC, Veras Quelemes P, de Araújo-Nobre AR, de Oliveira TM, da Silva DA, de Souza de Almeida Leite J, Fernandes Coutinho D, da Rocha Borges MO (2020) Green syntheses of silver nanoparticles using babassu mesocarp starch (*Attalea speciosa* Mart. ex Spreng.) and their antimicrobial applications. Environ Nanotechnol Monit Manag 13:100281

Berge ND, Ramsburg CA (2010) Iron-mediated trichloroethene reduction within nonaqueous phase liquid. J Contam Hydrol 118(3–4):105–116

Bhat JSA (2003) Hearlding a new future-nanobiotechnology? Curr Sci 85(2):147–154

Biondi I, Laurenczy G, Dyson PJ (2011) Synthesis of gold nanoparticle catalysts based on a new water-soluble ionic polymer. Inorg Chem 50(17):8038–8045

Biswas P, Wu CY (2005) Nanoparticles and the environment. J Air Waste Manage Assoc 55 (6):708–746

Bokare V, Murugesan K, Kim JH, Kim EJ, Chang YS (2012) Integrated hybrid treatment for the remediation of 2,3,7,8-tetrachlorodibenzo-p-dioxin. Sci Total Environ 435:563–566

Boxall AB, Tiede K, Chaudhry Q (2007) Engineered nanomaterials in soils and water: how do they behave and could they pose a risk to human health? Nanomedicine (Lond) 2(6):919–927

Cai Y, Shen Y, Xie A, Li S, Wang X (2010) Green synthesis of soya bean sprouts-mediated superparamagnetic Fe_3O_4 nanoparticles. J Magn Magn Mater 322(19):2938–2943

Cecchin I, Reddy KR, Thomé A, Tessaro EF, Schnaid F (2016) Nanobioremediation: integration of nanoparticles and bioremediation for sustainable remediation of chlorinated organic contaminants in soils. Int Biodeterior Biodegradation 119:419–428

Choi WB, Chung DS, Kang JH, Kim HY, Jin YW, Han IT, Lee YH, Jung JE, Lee NS, Park GS, Kim JM (1999) Fully sealed, high-brightness carbon-nanotube field-emission display. Appl Phys Lett 75(20):3129–3131

Commission de l'Éthique de la Science et la Technologie (2006) Ethics and nanotechnology: a basis for action. Commission de l'Éthique de la Science et la Technologie, Québec

Cundy AB, Hopkinson L, Whitby RL (2008) Use of iron-based technologies in contaminated land and groundwater remediation: a review. Sci Total Environ 400(1–3):42–51

D'Lima L, Phadke M, Ashok VD (2020) Biogenic silver and silver oxide hybrid nanoparticles: a potential antimicrobial against multi drug-resistant Pseudomonas aeruginosa. New J Chem 44:4935

Daphedar A, Taranath CT (2019) Biomimetic synthesis and characterization of silver nanoparticles synthesized by leaf extract of *Bridellia stipularis* L Blume and evaluation of their effect on mitotic chromosomes of *Pisum Sativum* L. J Nanosci Curr Res 4(116):2

Darroudi M, Ahmad MB, Abdullah AH, Ibrahim NA, Shameli K (2010) Effect of accelerator in green synthesis of silver nanoparticles. Int J Mol Sci 11(10):3898–3905

Dhillon GS, Brar SK, Kaur S, Verma M (2012) Green approach for nanoparticle biosynthesis by fungi: current trends and applications. Crit Rev Biotechnol 32(1):49–73

Dinesh R, Anandaraj M, Srinivasan V, Hamza S (2012) Engineered nanoparticles in the soil and their potential implications to microbial activity. Geoderma 173:19–27

Dinh HT, Kuever J, Mußmann M, Hassel AW, Stratmann M, Widdel F (2004) Iron corrosion by novel anaerobic microorganisms. Nature 427(6977):829–832

Drexler, E (1986) Reprint. Engines of Creation. The Coming Era of Nanotechnology, 1st ed. Anchor Books, New York. Available from www.e-drex-ler.com/d/06/00/EOC/

ETC Group (2003) The big down: from genomes to atoms. ETC Group, Winnepeg, Manitoba

Gao W, Majumder M, Alemany LB, Narayanan TN, Ibarra MA, Pradhan BK, Ajayan PM (2011) Engineered graphite oxide materials for application in water purification. ACS Appl Mater Interfaces 3(6):1821–1826

Gillham RW, O'Hannesin SF (1994) Enhanced degradation of halogenated aliphatics by zero-valent iron. Groundwater 32(6):958–967

Grätzel M (2003) Dye-sensitized solar cells. J Photochem Photobiol C Photochem Rev 4 (2):145–153

Hoag GE, Collins JB, Varma RS, Nadagouda M (2009) Green synthesis of metal nanoparticles using plant extracts. PCT Int Appl, WO, 2009140694, A2

Hodson ME (2010) The need for sustainable soil remediation. Elements 6(6):363–368

Honary S, Gharaei-Fathabad E, Paji ZK, Eslamifar M (2012) A novel biological synthesis of gold nanoparticle by Enterobacteriaceae family. Trop J Pharm Res 11(6):887–891

Hudlikar M, Joglekar S, Dhaygude M, Kodam K (2012) Green synthesis of TiO_2 nanoparticles by using aqueous extract of *Jatropha curcas L.* latex. Mater Lett 75:196–199

Iravani S (2011) Green synthesis of metal nanoparticles using plants. Green Chem 13 (10):2638–2650

Kalaiarasi R, Jayallakshmi N, Venkatachalam P (2010) Phytosynthesis of nanoparticles and its applications. Plant Cell Biotechnol Mol Biol 11(1/4):1–16

Kanchana A, Devarajan S, Ayyappan SR (2010) Green synthesis and characterization of palladium nanoparticles and its conjugates from *Solanum trilobatum* leaf extract. Nano-Micro Lett 2 (3):169–176

Karn B, Kuiken T, Otto M (2009) Nanotechnology and in situ remediation: a review of the benefits and potential risks. Environ Health Perspect 117(12):1813–1831

Kersting AB, Efurd DW, Finnegan DL, Rokop DJ, Smith DK, Thompson JL (1999) Migration of plutonium in ground water at the Nevada Test Site. Nature 397(6714):56–59

Kharisov BI, Dias HR, Kharissova OV, Jiménez-Pérez VM, Pérez BO, Flores BM (2012) Iron-containing nanomaterials: synthesis, properties, and environmental applications. RSC Adv 2 (25):9325–9358

Kharissova OV, Dias HR, Kharisov BI, Pérez BO, Pérez VMJ (2013) The greener synthesis of nanoparticles. Trends Biotechnol 31(4):240–248

Klaine SJ, Alvarez PJ, Batley GE, Fernandes TF, Handy RD, Lyon DY, Mahendra S, McLaughlin MJ, Lead JR (2008) Nanomaterials in the environment: behavior, fate, bioavailability, and effects. Environ Toxicol Chem 27(9):1825–1851

Komarneni S (1995) Novel microwave-hydrothermal processing for materials synthesis leads to energy savings. Ionics 21(12):95–98

Komarneni S (2003) Nanophase materials by hydrothermal, microwave-hydrothermal and microwave-solvothermal methods. Curr Sci 85:1730–1734

Lee J, Hong S, Mackeyev Y, Lee C, Chung E, Wilson LJ, Kim JH, Alvarez PJ (2011) Photosensitized oxidation of emerging organic pollutants by tetrakis C60 aminofullerene-derivatized silica under visible light irradiation. Environ Sci Technol 45(24):10598–10604

Lefevre E, Bossa N, Wiesner MR, Gunsch CK (2016) A review of the environmental implications of in situ remediation by nanoscale zero valent iron (nZVI): behavior, transport and impacts on microbial communities. Sci Total Environ 565:889–901

Li Q, Mahendra S, Lyon DY, Brunet L, Liga MV, Li D, Alvarez PJ (2008) Antimicrobial nanomaterials for water disinfection and microbial control: potential applications and implications. Water Res 42(18):4591–4602

Li H, Qiu YF, Wang XL, Yang J, Yu YJ, Chen YQ (2017a) Biochar supported Ni/Fe bimetallic nanoparticles to remove 1,1,1-trichloroethane under various reaction conditions. Chemosphere 169:534–541

Li Y, Li S, Zhang K (2017b) Influence of hydrophilic carbon dots on polyamide thin film nanocomposite reverse osmosis membranes. J Membr Sci 537:42–53

Li JF, Liu YC, Chokkalingam M, Rupa EJ, Mathiyalagan R, Hurh J, Ahn JC, Park JK, Pu JY, Yang DC (2020) Phytosynthesis of silver nanoparticles using rhizome extract of Alpinia officinarum and their photocatalytic removal of dye under UV and visible light irradiation. Optik 208:164521

Liu Y, Lowry GV (2006) Effect of particle age (Fe0 content) and solution pH on NZVI reactivity: H2 evolution and TCE dechlorination. Environ Sci Technol 40(19):6085–6090

Maensiri S, Laokul P, Klinkaewnarong J, Phokha S, Promarak V, Seraphin S (2008) Indium oxide (In_2O_3) nanoparticles using Aloe vera plant extract: synthesis and optical properties. J Optoelectron Adv Mater 10(3):161–165

Maynard AD, Aitken RJ, Butz T, Colvin V, Donaldson K, Oberdörster G, Philbert MA, Ryan J, Seaton A, Stone V, Tinkle SS (2006) Safe handling of nanotechnology. Nature 444 (7117):267–269

Modi S, Pathak B, Fulekar MH (2015) Microbial synthesized silver nanoparticles for decolorization and biodegradation of azo dye compound. J Environ Nanotechnol 4(2):37–46

Mohseni MS, Khalilzadeh MA, Mohseni M, Hargalani FZ, Getso MI, Raissi V, Raiesi O (2020) Green synthesis of Ag nanoparticles from pomegranate seeds extract and synthesis of Ag-starch nanocomposite and characterization of mechanical properties of the films. Biocatal Agric Biotechnol 101569

Morones JR, Elechiguerra JL, Camacho A, Holt K, Kouri JB, Ramírez JT, Yacaman MJ (2005) The bactericidal effect of silver nanoparticles. Nanotechnology 16(10):2346

Moulton MC, Braydich-Stolle LK, Nadagouda MN, Kunzelman S, Hussain SM, Varma RS (2010) Synthesis, characterization and biocompatibility of "green" synthesized silver nanoparticles using tea polyphenols. Nanoscale 2(5):763–770

Mueller NC, Nowack B (2010) Nanoparticles for remediation: solving big problems with little particles. Elements 6(6):395–400

Mueller NC, Braun J, Bruns J, Černík M, Rissing P, Rickerby D, Nowack B (2012) Application of nanoscale zero valent iron (NZVI) for groundwater remediation in Europe. Environ Sci Pollut Res 19(2):550–558

Nadagouda MN, Varma RS (2008) Green synthesis of silver and palladium nanoparticles at room temperature using coffee and tea extract. Green Chem 10(8):859–862

Nangia Y, Wangoo N, Goyal N, Shekhawat G, Suri CR (2009) A novel bacterial isolate Stenotrophomonas maltophilia as living factory for synthesis of gold nanoparticles. Microb Cell Factories 8(1):39

Narayanan KB, Sakthivel N (2008) Coriander leaf mediated biosynthesis of gold nanoparticles. Mater Lett 62(30):4588–4590

Nazir S, Hussain T, Iqbal Md MK, Muazzam AG, Ismail JM (2011) Novel and cost-effective green synthesis of silver nano particles and their in vivo antitumor properties against human cancer cell lines. Biosci Tech 2(6):425–430

Nori NM, Abdi K, Khoshayand MR, Ahmadi SH, Lamei N, Shahverdi AR (2013) Microwave-assisted biosynthesis of gold–silver alloy nanoparticles and determination of their Au/Ag ratio by atomic absorption spectroscopy. J Exp Nanosci 8(4):442–450

Novikov AP, Kalmykov SN, Utsunomiya S, Ewing RC, Horreard F, Merkulov A, Clark SB, Tkachev VV, Myasoedov BF (2006) Colloid transport of plutonium in the far-field of the Mayak Production Association, Russia. Science 314(5799):638–641

Nyabola AO, Kareru PG, Madivoli ES, Wanakai SI, Maina EG (2020) Formation of silver nanoparticles via Aspilia pluriseta extracts their antimicrobial and catalytic activity. J Inorg Organomet Polym Mater. https://doi.org/10.1007/s10904-020-01497-7

O'Day PA, Vlassopoulos D (2010) Mineral-based amendments for remediation. Elements 6 (6):375–381

Ogi T, Saitoh N, Nomura T, Konishi Y (2010) Room-temperature synthesis of gold nanoparticles and nanoplates using Shewanella algae cell extract. J Nanopart Res 12(7):2531–2539

Pan G, Li L, Zhao D, Chen H (2010) Immobilization of non-point phosphorus using stabilized magnetite nanoparticles with enhanced transportability and reactivity in soils. Environ Pollut 158(1):35–40

Pandey S, Goswami GK, Nanda KK (2013) Green synthesis of polysaccharide/gold nanoparticle nanocomposite: an efficient ammonia sensor. Carbohydr Polym 94(1):229–234

Parsons JG, Peralta-Videa JR, Gardea-Torresdey JL (2007) Use of plants in biotechnology: synthesis of metal nanoparticles by inactivated plant tissues, plant extracts, and living plants. Dev Environ Sci 5:463–485

Petla RK, Vivekanandhan S, Misra M, Mohanty AK, Satyanarayana N (2011) Soybean (Glycine max) leaf extract based green synthesis of palladium nanoparticles. J Biomater Nanobiotechnol. https://doi.org/10.4236/jbnb.2012.31003

Phenrat T, Saleh N, Sirk K, Tilton RD, Lowry GV (2007) Aggregation and sedimentation of aqueous nanoscale zerovalent iron dispersions. Environ Sci Technol 41(1):284–290

Phoenix C, Drexler E (2004) Safe exponential manufacturing. Nanotechnology 15(8):869

Pizzini S, Acciarri M, Binetti S (2005) From electronic grade to solar grade silicon: chances and challenges in photovoltaics. Phys Status Solidi A 202(15):2928–2942

Poole CP Jr, Owens FJ (2003) Introduction to nanotechnology. Wiley, Hoboken, NJ

Rajesh MK, Muralikrishna KS, Nair SS, Kumar BK, Subrahmanya TM, Sonu KP, Subaharan K, Sweta H, Prasad TK, Chandran N, Karunasagar I (2020) Facile coconut inflorescence sap mediated synthesis of silver nanoparticles and its diverse antimicrobial and cytotoxic properties. Mater Sci Eng C 111:110834

Ramesh C, HariPrasad M, Ragunathan V (2011) Effect of *Arachis hypogaea* L. leaf extract on Barfoed's solution; Green synthesis of Cu_2O nanoparticles and its antibacterial effect. Curr Nanosci 7(6):995–999

Reardon EJ, Fagan R, Vogan JL, Przepiora A (2008) Anaerobic corrosion reaction kinetics of nanosized iron. Environ Sci Technol 42(7):2420–2425

Reddy KR (2010) Nanotechnology for site remediation: dehalogenation of organic pollutants in soils and groundwater by nanoscale iron particles. In Proceedings of the 6th International Congress on Environmental Geotecnics, November, 1, pp 163–180

Renuka R, Devi KR, Sivakami M, Thilagavathi T, Uthrakumar R, Kaviyarasu K (2020) Biosynthesis of silver nanoparticles using Phyllanthus emblica fruit extract for antimicrobial application. Biocatal Agric Biotechnol 24:101567

Rickerby D, Morrison M (2007) Report from the workshop on nanotechnologies for environmental remediation, JRC Ispra

Ryoo MW, Kim JH, Seo G (2003) Role of titania incorporated on activated carbon cloth for capacitive deionization of NaCl solution. J Colloid Interface Sci 264(2):414–419

Sangeetha G, Rajeshwari S, Venckatesh R (2011) Green synthesis of zinc oxide nanoparticles by Aloe barbadensis Miller leaf extract: structure and optical properties. Mater Res Bull 46 (12):2560–2566

Sastry M, Ahmad A, Khan MI, Kumar R (2004) Microbial nanoparticle production. In: Nanobiotechnology: concepts, applications and perspectives. WileyVCH, Weinheim, pp 126–135

SCENIHR (2005) Scientific Committee on Emerging and Newly Identified Health Risks (SCENIHR) opinion on the appropriateness of existing methodologies to assess the potential risks associated with engineered and adventitious products of nanotechnologies

Schorr JR (2007) Promise of nanomaterials for water cleanup. Water Conditioning & Purification

Seshadri S, Saranya K, Kowshik M (2011) Green synthesis of lead sulfide nanoparticles by the lead resistant marine yeast, *Rhodosporidium diobovatum*. Biotechnol Prog 27(5):1464–1469

Shah Z, Hassan S, Shaheen K, Khan SA, Gul T, Anwar Y, Al-Shaeri MM, Khan M, Khan R, Haleem MA, Suo H (2020) Synthesis of AgNPs coated with secondary metabolites of Acacia nilotica: an efficient antimicrobial and detoxification agent for environmental toxic organic pollutants. Mater Sci Eng C 111:110829

Shahwan T, Sirriah SA, Nairat M, Boyacı E, Eroğlu AE, Scott TB, Hallam KR (2011) Green synthesis of iron nanoparticles and their application as a Fenton-like catalyst for the degradation of aqueous cationic and anionic dyes. Chem Eng J 172(1):258–266

Singh SN, Tripathi RD (eds) (2007) Environmental bioremediation technologies. Springer Science & Business Media, Berlin

Singh M, Kalaivani R, Manikandan S, Sangeetha N, Kumaraguru AK (2013) Facile green synthesis of variable metallic gold nanoparticle using Padina gymnospora, a brown marine macroalga. Appl Nanosci 3(2):145–151

Smitha SL, Philip D, Gopchandran KG (2009) Green synthesis of gold nanoparticles using *Cinnamomum zeylanicum* leaf broth. Spectrochim Acta A Mol Biomol Spectrosc 74 (3):735–739

Smuleac V, Varma R, Sikdar S, Bhattacharyya D (2011) Green synthesis of Fe and Fe/Pd bimetallic nanoparticles in membranes for reductive degradation of chlorinated organics. J Membr Sci 379 (1–2):131–137

Song JY, Jang HK, Kim BS (2009) Biological synthesis of gold nanoparticles using Magnolia kobus and Diopyros kaki leaf extracts. Process Biochem 44(10):1133–1138

Song JY, Kwon EY, Kim BS (2010) Biological synthesis of platinum nanoparticles using Diopyros kaki leaf extract. Bioprocess Biosyst Eng 33(1):159

Sundrarajan M, Gowri S (2011) Green synthesis of titanium dioxide nanoparticles by *Nyctanthes arbor-tristis* leaves extract. Chalcogenide Lett 8(8):447–451

Thomé A, Reddy KR, Reginatto C, Cecchin I (2015) Review of nanotechnology for soil and groundwater remediation: Brazilian perspectives. Water Air Soil Pollut 226(4):121

Tolles WM, Rath BB (2003) Nanotechnology, a stimulus for innovation. Curr Sci 85:1746–1759

Tosco T, Coisson M, Xue D, Sethi R (2012) Zerovalent iron nanoparticles for groundwater remediation: surface and magnetic properties, colloidal stability, and perspectives for field application. Res Signpost, Kerala

Tratnyek PG, Johnson RL (2006) Nanotechnologies for environmental cleanup. Nano Today 1 (2):44–48

Tripathi S, Sanjeevi R, Anuradha J, Chauhan DS, Rathoure AK (2018) Nano-bioremediation: nanotechnology and bioremediation. In: Biostimulation remediation technologies for groundwater contaminants. IGI Global, Pennsylvania, pp 202–219

United Kingdom Royal Society (2004) Nanoscience and nanotechnologies: opportunities and uncertainties. United Kingdom Royal Society, London

USEPA (2007) Nanotechnology white paper. US Environmental Protection Agency, Washington, DC, p 20460

Varadhi SN, Gill H, Apoldo LJ, Liao K, Blackman RA, Wittman WK (2005) Full-scale nanoiron injection for treatment of groundwater contaminated with chlorinated hydrocarbons. In Natural Gas Technologies 2005 Conference, February, Orlando, FL

Vilks P, Frost LH, Bachinski DB (1997) Field-scale colloid migration experiments in a granite fracture. J Contam Hydrol 26(1–4):203–214

Waite TD, Schäfer AI, Fane AG, Heuer A (1999) Colloidal fouling of ultrafiltration membranes: impact of aggregate structure and size. J Colloid Interface Sci 212(2):264–274

Wang C, Kim YJ, Singh P, Mathiyalagan R, Jin Y, Yang DC (2016) Green synthesis of silver nanoparticles by *Bacillus methylotrophicus*, and their antimicrobial activity. Artif Cells Nanomed Biotechnol 44(4):1127–1132

Weathers LJ, Parkin GF, Alvarez PJ (1997) Utilization of cathodic hydrogen as electron donor for chloroform cometabolism by a mixed, methanogenic culture. Environ Sci Technol 31 (3):880–885

Wiesner MR, Lowry GV, Alvarez P, Dionysiou D, Biswas P (2006) Assessing the risks of manufactured nanomaterials. Environ. Sci. Technol 40:4336

Wilkin RT, Puls RW, Sewell GW (2003) Long-term performance of permeable reactive barriers using zero-valent iron: geochemical and microbiological effects. Groundwater 41(4):493–503

Wu CC, Chen DH (2007) A facile and completely green route for synthesizing gold nanoparticles by the use of drink additives. Gold Bull 40(3):206–212

Yang K, Xing B (2010) Adsorption of organic compounds by carbon nanomaterials in aqueous phase: Polanyi theory and its application. Chem Rev 110(10):5989–6008

Yang X, Li Q, Wang H, Huang J, Lin L, Wang W, Sun D, Su Y, Opiyo JB, Hong L, Wang Y (2010) Green synthesis of palladium nanoparticles using broth of *Cinnamomum camphora* leaf. J Nanopart Res 12(5):1589–1598

Zak AK, Majid WA, Darroudi M, Yousefi R (2011) Synthesis and characterization of ZnO nanoparticles prepared in gelatin media. Mater Lett 65(1):70–73

Zhang WX, Wang CB, Lien HL (1998) Treatment of chlorinated organic contaminants with nanoscale bimetallic particles. Catal Today 40(4):387–395

Zhang T, Wang W, Zhang D, Zhang X, Ma Y, Zhou Y, Qi L (2010) Biotemplated synthesis of gold nanoparticle–bacteria cellulose nanofiber nanocomposites and their application in biosensing. Adv Funct Mater 20(7):1152–1160

Sustainable Phycoremediation of Xenobiotics Polluted Water

Nour Sh. El-Gendy, Hussein N. Nassar, Khaled A. El-Tarabily, and Synan F. AbuQamar

Abstract

Xenobiotics pose serious risks to the environment and human health because of their carcinogenic, teratogenic, mutagenic, and other toxic properties. Biodegradation of different xenobiotics has been widely covered using bacteria, yeast, and filamentous fungi. Because of the efficient capabilities of microalgae in remediation of polluted water via potential enzymatic system and production of biosurfactants, relatively little attention has been paid to the phycodegradation of xenobiotics by microalgae. This chapter summarizes the opportunities and challenges of integrated sustainable application of microalgae in the remediation of xenobiotics polluted water.

N. S. El-Gendy (✉)
Processes Design and Development Department, Egyptian Petroleum Research Institute (EPRI), Cairo, Egypt

Center of Excellence, October University for Modern Sciences and Arts (MSA), 6th of October City, Egypt

H. N. Nassar
Processes Design and Development Department, Egyptian Petroleum Research Institute (EPRI), Cairo, Egypt

Faculty of Pharmacy, October University for Modern Sciences and Arts (MSA), 6th of October City, Egypt

K. A. El-Tarabily
Department of Biology, College of Science, United Arab Emirates University, Al-Ain, United Arab Emirates

Khalifa Center for Genetic Engineering and Biotechnology, United Arab Emirates University, Al-Ain, United Arab Emirates

S. F. AbuQamar
Department of Biology, College of Science, United Arab Emirates University, Al-Ain, United Arab Emirates

M. P. Shah (ed.), *Removal of Emerging Contaminants Through Microbial Processes*,
https://doi.org/10.1007/978-981-15-5901-3_14

Keywords

Algae · Integrated process · Phycodegradation · Xenobiotics

1 Introduction

Oil pollution is one of the major worldwide concerns, especially in fresh and marine water ecosystems (Kalhor et al. 2016). Crude oil is a main source of toxic xenobiotics contaminating the water ecosystem (Rodrigues et al. 2010; Das and Chandran 2011; El Mahdi et al. 2016). Oil spill can spread and harm hundreds of miles away from the leakage source. Moreover, its negative impact on ecosystem is highly persistent as it takes long time to recover (Drozdova et al. 2013; Demirel et al. 2017). It can be considered as an immediate fire hazard, releasing greenhouse gasses as well as an indirect source of air pollution via the volatile hydrocarbons. Consequently, oil spill can act as a source for global warming and adds to the problem of climate change. Once oil spill occurs, it disorders the food chain and destroys coral reefs, mangroves, and coastal habitats (El-Tarabily 2002; Mzoughi and Chouba 2011; Farrington 2014; Ouada et al. 2018). Oil pollution directly and indirectly affects human, animals, and birds; thus, it severely disturbs biodiversity. Such contamination has adverse effects on economy and tourism (El-Gendy et al. 2014). The US-Environmental Protection Agency (US-EPA) has recorded 16 polycyclic aromatic hydrocarbons (PAHs) as priority pollutants, based on their hydrophobic nature, low water solubility, recalcitrance, long persistence in ecosystem, xenobiotic properties, and their toxic, mutagenic, neurotoxic, and carcinogenic properties, in addition to their severe negative impact on human health and ecosystem (Perelo 2010; Luo et al. 2014a, b; Takáčová et al. 2014; Abdel-Shafy and Mansour 2016). Moreover, there are other xenobiotic organic pollutants which can be released into water ecosystem from petroleum refineries and other petrochemical industries, such as diaryl ethers, phenols, chlorophenols, nitrophenols, and other compounds which are also toxic, recalcitrant, and persistent for a long period of time (Todd et al. 2002; Lika and Papadakis 2009; Duan et al. 2018; Ouada et al. 2018). Moreover, phenols like PAHs have toxic, carcinogenic, mutagenic, and teratogenic properties and are listed on the EPA priority pollutants (Thakurta et al. 2018). Phenol has been also reported to have acute toxicity to human and fish even at low concentrations (9–25 mg/L), and its lethal concentration is reported to be 150 mg/100 mL (Kulkarni and Kaware 2013; Duan et al. 2017).

There are three major routes for remediation of polluted water: physical, chemical, and biological (Bohdziewicz et al. 2012; Dakhil 2013; Kaushik and Chatterjee 2015; Hosny et al. 2016; Saththasivam et al. 2016; Zheng 2017). The eco-friendly and cost-effective bioremediation process using microorganism (bacteria, yeast, and filamentous fungi) is recommended worldwide for remediation of polluted water (McNally et al. 1999; Dean-Ross et al. 2002; Todd et al. 2002; Silva et al. 2009; Karn and Chakrabarti 2015; Liu et al. 2016; Tian et al. 2017). Bioremediation of xenobiotics and organic pollutants using micro- and/or macro-algae is known as phycoremediation (Rawat et al. 2011). *Arthrobacter*, *Burkholderia*, *Mycobacterium*, *Pseudomonas*, *Sphingomonas*, *Bacillus*, *Alcaligenes*, *Acinetobacter*, *Corynebacterium*, *Nocardia*, and *Rhodococcus* are the most widely applied and efficient bacterial species, whereas biodegradation varied between 0.003% and 100% (Yakimov et al. 2007; El-Gendy et al. 2009; Seo et al. 2009; Saiyood et al. 2010; Sun et al. 2012; Deriase et al. 2013; Ali et al. 2014; El-Sheshtawy et al. 2014; Liu et al. 2016; Chen et al. 2017). However, limited reports are published on the application of algae for bioremediation of petroleum hydrocarbons as a source of toxic xenobiotics polluting the water ecosystem (Kumar et al. 2009; Subashchandrabose et al. 2013; Ben et al. 2014; Gao and Chi 2015; Ghosal et al. 2016; Wang et al. 2019).

One of the advantages of phycoremediation is the fixation of carbon dioxide by non-pathogenic algae via photosynthesis producing oxygen and improving effectively the water quality in a cost-effective manner (Xiong et al. 2017). Phycoremediation is eco-friendly and sustainable; since no secondary pollution occurs, the biomass is reused and efficient nutrient recycling is allowed (Pérez-Garcia et al. 2011). Waste biomass can be even used as a source of heating or electricity via the production of biogas. It has been reported by Delrue et al. (2012) that 10–20% of the cost of algal biomass is related to the nutritional components of the cultivation media. Moreover, due to the problems of water scarcity, climate change, deforestation, decline of non-renewable energy resources, and the evolved "food vs. fuel" upon the production of biofuels from food crops. There is a worldwide mandate for sustainable and fully integrated process for wastewater treatment via microalgae and biomass production for different valuable applications in energy and industrial sectors. For example, Quinn and Davis (2015) reported that the average estimated production cost of biodiesel from microalgae is about 2.5 €/L which is approximately four times higher than that of petro-diesel. Thus, to make it competitive for petrochemical industry, a reduction in the cost of the algal biomass production is a mandate. In another study related to the cost of wastewater treatment, Guo et al. (2015) also reported that the cost of the conventional wastewater treatment techniques can reach to up to 0.682 $\$/m^3$ upon the application of membrane bioreactors.

Muñoz and Guieysse (2006) reported the promising application of microalgae for wastewater treatment and attributed this to the suitableness of the aquatic ecology in wastewater for production of microalgae biomass and the efficient synergism between microalgae and bacteria (Borde et al. 2003; Liang et al. 2013). Taking into consideration that 50% of the energetic requirements in wastewater treatment plants comes from the oxygen aeration (Delrue et al. 2016), thus, upon the application of open ponds algae, there is an expected great reduction in energy demand, greenhouse gas (GHG) emissions, and electricity costs. Algae provide oxygen to bacteria required for biodegradation and bacteria provide carbon dioxide to algae required for photosynthesis. Moreover, Kumar and Singh (2019) estimated algae biomass production cost to be approximately 4.92 $/kg. However, upon the usage of phycoremediation of wastewater for biomass production, it offers an economy savings of approximately 170 $/ton, and upon the production of different valuable products of such biomass, it offers a profit of approximately 123 $/kg. Consequently, the overall biomass production cost would decrease to approximately 0.5 \$/kg. Wijffels et al. (2010) also reported the reduction of the overall algae biomass production cost to 90% upon the usage of wastewater and production of biofuel and valorization of the de-oiled biomass into different valuable products from the obtained algae biomass.

This chapter emphasizes on phycoremediation of petroleum hydrocarbons as a source of toxic xenobiotic organic pollutants in water ecosystems using microalgae (i.e., microphytes). As from the economic and environmental points of view, it is very promising and feasible. Microalgae can detoxify, transform, accumulate, and/or degrade such pollutants via its metabolic system, producing lipids for biodiesel production, in addition to biorefineries, bioethanol, and other different valuable by-products (Brennan and Owende 2010; Halim et al. 2012; Olguín 2012; Abdelaziz et al. 2014; Vassilev and Vassileva 2016; Kalhor et al. 2017), thus sustainably remediating polluted water and valorizing the wastes and pollutants into value-added products for pharmaceutical, food supplements, animal feed, feed in aquaculture, bio-fertilizer, cosmetics, etc. (Mulbry et al. 2005; Becker 2007; Del Campo et al. 2007; Eriksen 2008; Adarme-Vega et al. 2012; Sasso et al. 2012; Fathi et al. 2013; Renuka et al. 2015; Sutherland and Ralph 2019; Zhang et al. 2019). Moreover, for reaching to the point of zero waste, the waste biomass can be valorized into biogas, solid biofuel, biochar, and activated carbon which have different applications (Biller et al. 2012; Bird et al. 2012; Agrafioti et al. 2013; Hognon et al. 2015).

2 Microalgae for Phycoremediation

The unicellular photosynthetic microalgae are categorized into four main classes: diatoms, green algae, blue-green algae, and golden algae. They can also be categorized as prokaryotic and eukaryotic marine and freshwater algae. The oxygenic photosynthetic cyanobacteria is also a microalgae (Priyadarshani et al. 2011). There are four modes of microalgae growth and biomass production: (1) photoautotrophic, which requires light, CO_2, water, and nutrients to yield a biomass rich in lipids, protein, and sugars; (2) heterotrophic, which does not need light and depends mainly on organic carbon source (i.e., electron donor) for energy via oxidative phosphorylation convoyed with electron acceptor in the form of oxygen intake; (3) photoheterotrophic, in which microalgae have the ability to switch between photoautotrophic and heterotrophic growth modes and presence of light and organic carbon is obligatory, as energy and carbon source, respectively; and (4) mixotrophic, which is the integration of autotrophic and heterotrophic metabolism, in which the acclimatization of organic carbon, nutrients, and CO_2 concurrently, in the presence of light, achieves respiration and photosynthesis (Wang et al. 2014). Usually, mixotrophic and heterotrophic modes of cultivation are more flexible and feasible, producing as much biomass, lipids, and carbohydrates for biorefinery processes to produce different valuable products, chemicals, and biomass for fuel gas (Kröger and Müller-Langer 2011; Tabernero et al. 2012; Zheng et al. 2012; Leyva et al. 2015; Delrue et al. 2016). Mixotrophic mode is the uptake of inorganic and organic carbon at the same time, whereas CO_2 is taken in the presence of light via photosynthesis and organic carbon is taken via aerobic respiration. Thus, light is not an absolute limiting factor for mixotrophic growth mode. It should be noted also that the growth rate and biomass production for some microalgae via mixotrophic are much faster than those via photoautotrophic, and the metabolic products depend on the quality and quantity of the organic carbon in the medium. For example, lipid content is highly affected by carbon source, amount of light intensity, and photoperiod. Not only these but also other nutrients such as nitrogen and phosphorous are other important factors for algal growth, lipid content, and other by-products (Hu et al. 2012; Zhou et al. 2013a, b; Nawkarkar et al. 2019; Sipaúba-Tavares et al. 2020).

Mixotrophic mode can be feasibly applied for biotreatment of polluted water, and this mode of microalgae cultivation poses different advantages such as (1) flexibility upon switching between heterotrophic or photoautotrophic cultivation modes; (2) low and/or no photo-inhibitory impact, resulting in low biomass loss under darkness or photooxidative biomass loss due to the oxygen accumulation in photobioreactors; (3) higher growth rate, shorter growth cycles, higher biomass, cost-effectiveness, and higher efficiency in the removal of nitrogen, phosphorous, and organic pollutants under darkness; (4) higher lipid and/or carbohydrates and other biomass components; and (5) sustainability by integrating wastewater treatment with valorization into valuable products and bioenergy production, instead of using expensive organic carbon, nitrogen, and phosphorous as nutrient sources for algae biomass productivity (Li et al. 2014; Zhan et al. 2017).

Phycoremediation can be performed in waste stabilization ponds (WSPs) and/or high-rate algal ponds (HRAPs) by synergism between heterotrophic bacteria and microalgae. That is, microalgae produce oxygen via photosynthesis, which is utilized by bacteria for aerobic biodegradation of organic pollutants, producing carbon dioxide and other mineral compounds as the end-products which are nutritionally reused by microalgae, producing biomass. Thus, there is a decrease in nutrient load and total dissolved solids as they act as nutrient sources. Consequently, phycoremediation can act as a potential process for energy and nutrient recovery. Nevertheless, it should be taken into consideration the GHG emission reduction, declines in sludge formation, low energy requirement, simple operation and maintenance, and the disinfection effect of algal photosynthesis on most pathogenic bacteria (e.g., *Escherichia coli*) and viruses, via the pH increment with the assimilation of CO_2, H^+ ions, and bicarbonate by algae, adding also to the advantage of wastewater treatment via microalgae. Some microalgae which are generally used for the polluted water treatment are listed in Table 1. Concentration of the pollutant, presence or absence of alternative carbon source, light intensity, and oxygen availability are listed to be the major factors affecting phycoremediation potentials of microalgae (Papazi and Kotzabasis 2007; Lika and Papadakis 2009; Ji et al. 2014; Nazos et al. 2017). However, when nitrogen was flushed, replacing air in phycodegradation of phenol by *Chlamydomonas reinhardtii*, under illumination, the biodegradation efficiency was not affected since oxygen can be photosynthetically produced under light conditions (Nazos et al. 2017). Nevertheless, the type of

Table 1 Phycodegradation of different xenobiotic organic pollutants

Microalgae	Pollutant	Reference
Ankistrodesmus braunii *Scenedesmus quadricauda*	Phenols	Pinto et al. (2002)
Chlamydomonas mexicana	Bisphenol A	Ji et al. (2014)
Chlamydomonas mexicana *Chlorella vulgaris*	Bisphenol A	Ji et al. (2014)
Chlamydomonas reinhardtii	Phenol	Nazos et al. (2017)
Chlorella kessleri LARG/1	Benzo[a]pyrene	Takáčová et al. (2014)
Chlorella sorokiniana	Acetonitrile	Chia et al. (2019)
Chlorella vulgaris *Coenochloris pyrenoidosa*	*p*-Chlorophenol	Lima et al. (2004)
Cyclotella caspia	Fluoranthene	Liu et al. (2006)
Isochrysis galbana	Phenol	Wang et al. (2019)
Navicula sp.	4-Methylcyclohexaneacetic acid	Headley et al. (2008)
Nitzschia sp.	Fluoranthene, phenanthrene	Hong et al. (2008)
Oscillatoria sp.	Naphthalene	Cerniglia et al. (1980a)
Scenedesmus sp. *Chlorella* sp.	Crude oil	El-Sheekh et al. (2013)
Scenedesmus acutus	Benzo[a]pyrene	de Llasera et al. (2016)
Scenedesmus obliquus	Chlorophenols	Papazi and Kotzabasis (2013)
Scenedesmus obliquus	*p*-Cresol	Papazi et al. (2012)
Scenedesmus obliquus	1-Naphthalenesulfonic acid	Kneifel et al. (1997)
Selenastrum capricornutum	Benzene, naphthalene, pyrene, 2,6-dinitrotoluene	Semple et al. (1999)
Selenastrum capricornutum	Benzo[a]pyrene	de Llasera et al. (2016)
Selenastrum capricornutum	Phenanthrene, fluoranthene, pyrene	Tam et al. (2010)
Selenastrum capricornutum *Scenedesmus acutus*	Benzo[a]pyrene Benzo[a]anthracene	García de Llasera et al. (2018)
Spirulina maxima	Phenol	Klekner and Kosaric (1992b)

light has been also reported to affect the phycoremediation process (Schoeny et al. 1988).

Large numbers of enzymes produced by microalgae are involved in biodegradation of xenobiotic organic pollutants (Semple et al. 1999; Takáčová et al. 2014;

Sutherland and Ralph 2019). Moreover, some microalgae have been reported to produce biosurfactants (Radmann et al. 2015) and exopolysaccharides (Paniagua-Michel et al. 2014) that might accelerate the phycoremediation process.

3 Phycoremediation of Water Polluted by Crude Oil and Its Fractions

To our knowledge, the first publication for phycoremediation of petroleum compounds and its fractions was by Walker et al. (1975a). Walker et al. (1975b) reported 10–40% phycoremediation of water polluted by 1% (v/v) motor oil and crude oil, respectively, within 30 days, using *Prototheca zopfii*, the aerobic, unicellular, yeast-like, achlorophyllic (without chlorophyll) microalga. The recorded lower degradation percentage of motor oil-polluted water was accounted to the composition of the pollutant itself, as motor oil has lower percentage of readily biodegradable saturates and higher percentage of recalcitrant cycloalkanes and aromatics. Since then, several other algal species have been reported for crude oil degradation, for example, *Cellana*, *Volvox*, *Chlorella vulgaris*, *Scenedesmus obliquus*, *Synechococcus elongatus*, *Chlorella autotrophica*, *Coccochloris elabens*, and *Dunaliella tertiolecta* (Rajasulochana et al. 2009; Tang et al. 2010a; Cao et al. 2013; El-Sheekh et al. 2013).

Gamila and Ibrahim (2004) isolated freshwater green algae *Scenedesmus obliquus* and the diatom *Nitzschia linearis* from River Nile water and studied their biodegradation effect on 1 mL/L Egyptian light crude oil. The growth as chlorophyll (a) was enhanced by 183% and 228% than the control (without crude oil), respectively, with the prolongation of the growth phase. The phycodegradation of aliphatic hydrocarbons nC10–nC24 ranged between 81% to 90% and 79.5% to 86.1%, respectively, and decreased with the increase of carbon numbers. The two isolates expressed higher phycodegradation efficiencies on PAHs than aliphatics. However, the degradation efficiency decreased with the increase of numbers of aromatic rings and methylation of polyaromatic compounds and its position. For example, the biodegradation of naphthalene was higher than 1-methyl naphthalene. Similarly, the phycodegradation of 1-methyl naphthalene was higher than 2-methyl naphthalene. The degradation efficiencies of PAHs decreased in the following order: diaromatic ring compounds > three aromatic ring compounds > four aromatic ring compounds > five aromatic ring compounds (Gamila and Ibrahim 2004).

The mixotrophic phycoremediation of water polluted by different concentrations of crude oil (0.5, 1, and 1.55) was investigated using the cyanobacterium *Anabaena oryzae* and the microalgae *Chlorella kessleri* in individual and consortium culture (Hamouda et al. 2016), where the cyanobacteria/microalgae consortium enhanced the phycodegradation of crude oil than individual cultures. The maximum growth and productivity of chlorophyll (a) and carotenoids occurred at 1% crude oil, with efficient degradation of aliphatic and aromatic petroleum hydrocarbons. However, chlorophyll (b) decreased with the increase of crude oil concentration. Kalhor et al. (2016) investigated the utilization of 10 g/L and 20 g/L crude oil by *Chlorella vulgaris*. The biomass yield was increased by 12% and 39% with a concomitant increase in oil yield of approximately 35% and 67%, relative to control (i.e., without crude oil), respectively, and attributed this to the constituents of the crude oil which act as suitable growth medium for *C. vulgaris*. The produced lipid was successfully trans-esterified into biodiesel. However, Kalhor et al. (2017) noticed the decrease of biodegradation efficiency with increasing crude oil concentration and reported approximately complete degradation of light hydrocarbons (boiling point <350 °C) and 78% degradation of heavy fraction (boiling point >350 °C) using the green algae *C. vulgaris* within 14 days of incubation at 10 g/L initial crude oil concentration. On the other hand, the degradation was decreased to 90% and 58.8%, respectively, at 20 g/L initial crude oil concentration. This was related to the inhibition of degrading enzymes which might have occurred at high crude oil concentration.

Several synergetic algal and bacterial consortia have been reported for bioremediation of oil-polluted water. Tang et al. (2010b) reported that individual algal culture of *Scenedesmus obliquus* GH2 was capable of degrading crude oil. However, its axenic culture could not degrade crude oil. But such axenic culture synergistically promoted the efficiency of bacterial consortium for biodegradation of crude oil. The bacterial consortium consisted of *Sphingomonas* GY2B, *Burkholderia cepacia* GS3C, *Pseudomonas* GP3A, and *Pandoraea pnomenusa* GP3B. *S.* GY2B and *B. cepacia* GS3C degraded only PAHs and aliphatic hydrocarbons, respectively, while *Pseudomonas* GP3A and *P. pnomenusa* GP3B degraded both PAHs and aliphatic hydrocarbons. Further, Tang et al. (2010a) previously noticed that the individual algal culture was associated with two uncultured bacteria and two culturable ones: *Methylophilus methylotrophus* and *Hydrogenophaga* sp. Thus, Tang et al. (2010b) attributed the antagonistic effect of the individual algal culture and the four bacterial consortia on the crude oil degradation to such concurrent presence of other bacteria in the individual algal culture which might have produced metabolites, intermediates, and/or by-products which inhibited the activity of the inoculated bacterial consortia.

4 Phycoremediation of Water Polluted by PAHs

Cerniglia et al. (1980b) reported the biodegradation of naphthalene by five green algae, one red alga, and one brown alga. However, Cody et al. (1984) illustrated the effect of light type on phycodegradation of benzo[a]pyrene by *Selenastrum capricornutum*. Gold illumination enhanced the phycodegradation even at high benzo[a]pyrene (BaP) concentration of 1200 μg/L, while phycodegradation was inhibited by low concentration of BaP 160 μg/L under white illumination. Luther (1990) reported the biotransformation of 1-naphthalene sulfonate into 1-naphthol and sulfite by *Scenedesmus obliquus*. In another study by Warshawsky et al. (1990), white light illumination produced quinones and other toxic products during the phycodegradation of BaP by freshwater green alga *S. capricornutum*, which inhibited the phycodegradation process. Todd et al. (2002) reported the involvement of cytochrome P450 and oxygenases in the biotransformation of three xenobiotic pollutants, naphthalene, dibenzofuran, and dibenzo-*p*-dioxin, into more polar compounds, 1-naphthol; mixtures of 2-, 3-, and 4-hydroxydibenzofurans; and dihydroxylated dibenzofurans and 2-hydroxydibenzo-*p*-dioxin by three green algae, *C. vulgaris*, *Scenedesmus* SI1, and *Ankistrodesmus* SI2, respectively. Safonova et al. (2005) reported the phycodegradation of phenanthrene by *S. obliquus* ES-55, isolated from oil-polluted soil into different metabolites, dihydroxy-dihydro-phenanthrene, 3-*H*-benzo[f]chromene-2-carboxylic acid, 2-naphthalenepropanoic acid, (*Z*)-2-hydroxy-4-(1-hydroxynaphthalen-2-yl)-but-3enoic acid, 4-(1-hydroxynaphthalen-2-yl)-2-oxobutanoic acid, 4-(1-hydroxynaphthalen-2-yl)-4-oxobutanoic acid, and 2-((*E*)-5-carboxypent-3-enyl) benzoic acid. The diatoms *Skeletonema costatum* and *Nitzschia* sp. have been reported for phycodegradation of phenanthrene and fluoranthene, as representative model compounds for the three- and four-membered aromatic ring compounds, respectively (Hong et al. 2008). Gavrilescu (2010) reported the biodegradation of benzene, toluene, naphthalene, phenanthrene, and pyrene by *S. capricornutum*. Moreover, Warshawsky et al. (1988) reported the involvement of dioxygenase in the biodegradation of BaP by *S. capricornutum*. In another study by Schoeny et al. (1988), three microalgae *S. capricornutum*, *Scenedesmus acutus*, and *Ankistrodesmus braunii* biotransformed BaP into 3,6-quinones under white light illumination, which inhibited microbial growth. However, under gold light illumination, the microalgae growth was enhanced, and 98% of BaP was biotransformed into BaP diols. Lei et al. (2003) reported that the activity of the antioxidant protector and xenobiotic metabolic enzyme, glutathione *S*-transferase (GST), changes with algal species and its tolerance to the toxicity of PAHs, although it plays a vital role in pyrene phycodegradation. For example, the activity of GST in *Scenedesmus platydiscus* and *S. capricornutum* increased with the increase of pyrene concentration from 0.1 to 1.0 mg/L. On the contrary, *Scenedesmus quadricauda* was highly sensitive to pyrene and *C. vulgaris* did not metabolize pyrene. Olmos-Espejel et al. (2012) proved the involvement of dioxygenase in the biodegradation of BaP by *S. capricornutum* via the production of BaP-dihydrodiols. Ke et al. (2010) and SureshKumar et al. (2018) reported the involvement of cytochrome P450 monooxygenase in phycodegradation of low and high molecular weight PAHs by *S. capricornutum*, *Haematococcus pluvialis*, *Auxenochlorella protothecoides*, and *Parachlorella kessleri*. Gao and Chi (2015) reported the involvement of esterases in the phycodegradation of phthalic derivatives. Asghari et al. (2018) reported the increase of total phenol and flavonoid content and the antioxidant enzymes, ascorbate peroxidase, catalase, and superoxide dismutase, in phenanthrene algal culture of *C. vulgaris* to overcome the stress caused by reactive oxygen species (ROS) and consequently degrade phenanthrene.

It is worth to know that although microalgae have enzymatic system to degrade PAHs, PAHs express toxicity to microalgae and inhibit its growth, with different degrees depending on the algal species and the type of PAHs (Baun et al. 2002; Chan et al. 2006; Hong and Yuan 2008). For example, the four-membered rings fluoranthene expressed higher toxicity on diatoms *Skeletonema costatum* and

Nitzschia sp. than phenanthrene (Hong et al. 2008). However, *S. costatum* expressed more tolerance to PAHs than *Nitzschia* sp. (Hong et al. 2008). Moreover, sometimes the presence of PAHs in mixture stimulates the phycodegradation efficiency. For example, the phycodegradation of pyrene by *S. capricornutum* was enhanced in the presence of phenanthrene and fluoranthene (Chan et al. 2006). Moreover, the phycodegradation of phenanthrene and fluoranthene by four different algae strains, *C. vulgaris*, *Scenedesmus platydiscus*, *Scenedesmus quadricauda*, and *S. capricornutum*, was stimulatory enhanced when they are added to algal cultures as a mixture than when they were added individually (Lei et al. 2007). In another study, the phycodegradation of phenanthrene using *Skeletonema costatum* was nearly doubled when found in mixture with fluoranthene (Hong et al. 2008). It is worth to know that one of the advantages of microalgae is that the phycodegradation of high molecular weight PAHs is reported to be higher than the low molecular weight PAHs in some applied microalgae strains, for example, *S. capricornutum* (Ke et al. 2010). This phenomenon is attributed to the composition of the algal cell wall which is characterized by the abundance of polysaccharides. However, low molecular weight PAHs would lead to swelling in such amorphous polysaccharide cell wall, enhancing the accessibility and sorption affinity of high molecular weight PAHs (Wang et al. 2007). Not only this, but stress caused by presence of heavy metals can express synergetic positive effect on phycodegradation of PAHs (Ke et al. 2010), which also adds to the advantages of the application of microalgae in phycoremediation as usually oil pollution of water is concomitant with heavy metal pollution. Heavy metals and PAHs would act as inducers for the ROS in algal cells, producing antioxidant enzymatic defense systems, for example, superoxide dismutase, peroxidase, and catalase, and degrade the pollutant in a pathway nearly similar to that of photodegradation (Mittler 2002; Babu et al. 2003). Luo et al. (2014a) studied the phycodegradation of seven high molecular weight PAHs, benzo [a] anthracene (BaA), benzo[b]fluoranthene (BbF), benzo[k]fluoranthene BkF, BaP, dibenzo[a,h]anthracene (DA), benzo[g,h,i]perylene (BghiP), and indeno[c,d]pyrene (IP) by *S. capricornutum*, under dark, gold, and white illumination. The growth as chlorophyll (a) concentration was not affected by illumination type. However, gold illumination expressed better phycodegradation for the PAHs than white illumination, and no degradation occurred under darkness. Moreover, the phycodegradation of BaP as a single substrate was much better than its degradation percentage in the mixture PAH cultures, due to the competitive inhibition effect of PAHs on the microalgae enzymatic system. White light enhanced the photooxidation, thus increasing the inhibitor's oxidized metabolites. BaP and BaA have the highest degradation, due to high water solubility and lower molecular weight, respectively. On the contrary, the most recalcitrant ones were DA and BghiP, because of their high molecular weight, low water solubility, and structure, as bend PAHs are less photoreactive than linear ones (Zeppm and Schlotzhauer 1979).

Three marine algae *Cylindrotheca closterium*, *Dunaliella salina*, and *Chaetoceros muelleri* have been tested for the degradation of di-ethyl phthalate (DEP) and di-*n*-butyl phthalate (DBP) which are mainly used as additives and plasticizers for enhancing the flexibility and workability of the polymeric materials

(Gao and Chi 2015). The phycodegradation of DBP followed the first-order kinetic model equation with rate constants of 0.0169, 0.0035 and 0.0034 h^{-1}, respectively. Moreover, DBP decreased the biodegradation efficiency of DEP. However, the phycodegradation of DBP occurred via intra- and extracellular enzymes, while that of DEP was mainly via extracellular enzymes. The phycodegradation of DEP by the three algal strains decreased in the following order *C. closterium* > *C. muelleri* > *D. salina*. The high efficiency of *C. closterium* for phycodegradation of such phthalate derivatives was related to the ratio of surface area to volume and the cell/substrate interaction as it is a benthic diatom with more hydrophobic surface which has the capability of colony formation, while the other two planktonic algae do not have (Ozkan and Berberoglu 2013; Gao and Chi 2015).

In another study, BaP has been completely removed by *S. capricornutum* and *Scenedesmus acutus* mainly via degradation and sorption, respectively (García de Llasera et al. 2016). And the production of intermediates, 4,5-dihydrodiol-BaP, 7,8-dihydrodiol-BaP, and 9,10-dihydrodiol-BaP, was rapid and high applying *S. capricornutum*. However, Blanco and García de Llasera (2016) reported 79%, 69%, 84%, and 63% of BaP, benzo(a)anthracene (BaA), benzo(k)fluoranthene (BkF), and benzo(b)fluoranthene (BbF), respectively, with the production of dihydrodiols during the phycodegradation of such PAH mixture.

Subashchandrabose et al. (2017) reported the enhancement of adding non-ionic Tween 80 surfactant on pyrene phycodegradation via *Chlorella* sp. MM3 and proved the role of dihydrolipoamide acetyltransferase in pyrene degradation by MM3. In another study, marine algae *Rhodomonas baltica* reported to degrade mixture of phenanthrene (Phe), fluoranthene (Fla), and pyrene (Pyr) (Arias et al. 2017). Although the algal growth decreased in the presence of PAHs relative to reference cultures, *R. baltica* removed about 70% of such PAHs, and the bioaccumulation of such pollutants decreased in the following order Fla > Pyr > Phe, based on hydrophobicity, and the highest phycodegradation was recorded for Phe.

Aldaby and Mawad (2019) studied the effect of initial Pyr concentration on the phycodegradation capacity of *Oscillatoria* sp. and *Chlorella* sp., where complete degradation occurred up to 30 mg/L, within 30 days, and decreased to 95% and 78.71% at 50 mg/L Pyr and further decreased to 68% and 49% at 200 mg/L Pyr, applying the two algal strains, respectively, with a concomitant decease in algal growth. The phycodegradation process with both algal strains followed the zero-order kinetic model equation. Moreover, the highest antioxidant concentration (i.e., pigments: chlorophyll and carotenoids) was achieved at 50 mg/L Pyr by *Chlorella* sp. and that was attributed to the enhancement of stress-responsive antioxidant reactions.

5 Phycoremediation of Water Polluted by Phenols

To our knowledge, the first report for phenol phycodegradation was by Oswald et al. (1953). The green microalga *Ochromonas danica* has been reported for biodegradation of phenol via the *meta*-cleavage pathway (Semple and Cain 1996). Cytochrome

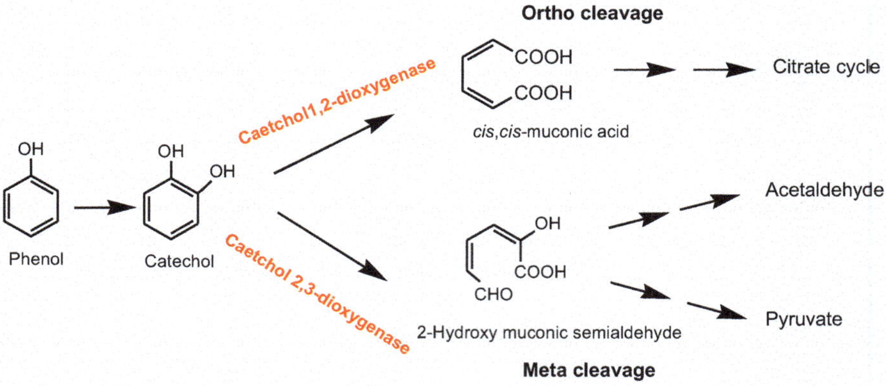

Scheme 1 Possible phenol phycodegradation pathways

P450 and NADH-dependent biodegradation pathways have been also reported for biodegradation of phenolic compounds by *Chlorella* sp. (Thies et al. 1996) and *Thalassiosira* sp. (Lovell et al. 2002). Moreover, Lika and Papadakis (2009) reported the involvement of phenol monooxygenase and catechol 2,3dioxygenase in the biodegradation of phenolic compounds by microalgae (Scheme 1). Whereas most of the eukaryotic microalgae degrade phenol through *ortho*-cleavage pathway, via catechol 1,2-dioxygenase, producing mainly *cis,cis*-muconic acid (Das et al. 2015), most of the green algae utilize both *ortho*- and *meta*-cleavage pathways with the predominance of the former (Surkatti and Al-Zuha 2018). However, Subashchandrabose et al. (2013) reported the involvement of intracellular polyphenol oxidase and laccase enzymes in phenol metabolism by some species of *Chlorella*.

Chlorella species have been reported for biodegradation of phenol, bisphenol A, 4-nitrophenol, 4-chlorophenol, 2,4-dinitrophenol, and 2,4-dimethylphenol (Klekner and Kosaric 1992a; Hirooka et al. 2003; Tsuji et al. 2003; Lima et al. 2004). *Scenedesmus* species have been also reported for biodegradation of acylated phenols, catechol, bisphenol A, tyrosol, hydroxytyrosol, and 4-hydroxybenzoic acid (Pinto et al. 2002; Nakajima et al. 2007; Zhou et al. 2013a, b). Lima et al. (2003) reported the complete degradation of 50 mg/L *p*-nitrophenol by a (3:1) consortium of *Coenochloris pyrenoidosa* and *C. vulgaris*, which were previously isolated from nitrophenol-, chlorophenol-, and fluorobenzene-polluted media. In a study performed by El-Sheekh et al. (2012), different algal strains have been isolated, and the effect of different xenobiotic organic pollutants on the phycodegradation efficiencies was studied. It was proved that the pollutant structure and the physiological metabolism of the applied microalgae affect the phycodegradation process. α-Naphthol, phenol, and catechol have been reported to be degraded by *Lyngbya lagerlerimi*. β-Naphthol has been reported to be degraded by *Volvox aureus*, *Nostoc linckia*, and *Oscillatoria rubescens*, while catechol has been degraded by *C. vulgaris* and *V. aureus*. But phenol was oxidized into catechol by *N. Linckia* and *O. rubescens*. Moreover, *L. lagerlerimi* expressed phycodegradation efficiency toward naphthalene and anthracene. It was worth noticing the higher degradation efficiency of *V. aureus*, *Elakatothrix viridis*, and *O. rubescens* toward anthracene than naphthalene. The phycodegradation efficiency of heterocyclic aromatic compounds was also found to be changed with the applied microalgae. *L. lagerlerimi* and *V. aureus* expressed higher biodegradation efficiency toward 2-methyl thie-3-phenyl quinazlin 4-(3*H*) one than that toward 2-phenyl-3,1-benzexazin-4-one and vice versa which occurred in case of *C. vulgaris*. However, *N. linckia*, *O. rubescens*, and *E. viridis* expressed the same phycodegradation efficiencies on such studied heterocyclic aromatic compounds. Complete mineralization of phenol by *Spirulina maxima* has been reported, with maximum specific growth rate at 50 mg/L phenol (Lee et al. 2015). Moreover, complete removal of phenol up to 400 mg/L has been also achieved within 6 days of incubation.

Two-level Plackett–Burman design (PBD) with five factors identified that incubation period and initial algal and phenol concentrations are the most significant factors affecting the phycodegradation of phenol by *Chlorella pyrenoidosa* (Priyadharshini and Bakthavatsalam 2016). But urea and potassium bicarbonate concentrations have non-significant effects. Further, response surface methodology based on central composite analysis proved the significance of the interactive effects of inoculum size with initial phenol concentration, incubation period with phenol concentration, and incubation period with inoculum size on phenol phycodegradation. Second-order polynomial equation was predicted to describe the phycodegradation process, and the maximum achieved degradation percentage was 97% using 4 g/L inoculum size and 0.8 g/L phenol within 4 days of incubation (Priyadharshini and Bakthavatsalam 2016).

Further, Wang et al. (2016) enhanced the phenol tolerance and phycodegradation capability of *Chlorella* sp. via its successive exposure to environmental stress throughout 31 cycles over 95 days to 500 mg/L phenol. And complete removal of 500 mg/L phenol has been achieved within 7 days with biomass production of approximately 3.4 g/L after 8 days. But the biomass decreased to 2.7 g/L when grown on 700 mg/L phenol, with biodegradation percentage of approximately 86.57%. To take the advantage of applying extremophilic algae that would have gained physiological adaptation and some kind of genetic modification from its survival under harsh physical conditions and severe pollution, Ouada et al. (2018) tested the ability of *Picocystis* sp. and *Graesiella* sp. previously isolated from alkaline household sewage water and hot water catchment basin highly polluted by plastics, respectively, to degrade bisphenol A. As was expected, the growth, exponential phase, and photosynthetic activity as oxygen evolution decreased with the increment of bisphenol A concentration. The tolerance of *Picocystis* sp. was very obvious, where the growth decreased by approximately 43% at 75 mg/L bisphenol A, while that of *Graesiella* sp. decreased by approximately 80%, with IC_{50} of >75 mg/L and 24 mg/L after 4 days, respectively. It was worth observing that the photosynthetic inhibition was not directly correlated to algal growth. The former sharply decreased to more than 80% and completely inhibited at 75 mg/L bisphenol A in case of *Picocystis* sp. and *Graesiella* sp., respectively, proving photo-heterotrophic metabolism. Bisphenol A expressed extremely lower lipid peroxidation levels and higher oxidative stress on *Picocystis* sp. than *Graesiella* sp., and the former proved to produce higher concentration of catalase, glutathione *S*-transferase, and ascorbate peroxidase as antioxidant enzymes that catalyzed reactions to overcome the excess stress of free radicles and reactive oxygen species induced by bisphenol A. The biodegradation percentage of bisphenol A by *Picocystis* sp. was nearly twice that of *Graesiella* sp. in all the studied concentrations, expressing maximum biodegradation percentage of 39.87% and 18.88% at 25 mg/L bisphenol A, respectively (Ouada et al. 2018).

In another study by López-Pacheco et al. (2019), a consortium of cyanobacterium *Arthrospira maxima* and green algae *C. vulgaris* was reported to efficiently degrade *p*-nitrophenol with the production of 4-(1-methyl-octyl)-4-hydroxy-cyclohex-2-enone, 4-nonyl-4-hydroxy-ciclohexa-2,5-dienone, 4-nonyl-4-hydroxy-ciclohex-2-enone, and 4-nonyl-(1-methyl,6,8-metoxy)-hydroxybenzene, presumptively indicating the involvement of cytochrome P-450 in phycodegradation of *p*-nitrophenol. However, cyanobacteria have been reported to efficiently degrade 2,4-dinitrophenol (Surkatti and Al-Zuha 2018).

In a study performed by Hao et al. (2019), 100 mg/L phenol has been reported to be completely removed by the marine golden algae *Isochrysis galbana* MACC/H59 at 20 °C, 180 μmol/m^2/s light intensity, and within only 24 h. However, phenol was also removed under darkness after 102 h. This added to the advantages of applying algae in phycoremediation in open seawater or oceans, as light intensity decreases with the increase of water depth. It is worth to mention that the phycodegradation process followed the pseudo-first-order and zero-order kinetic model equations under light and dark conditions, respectively.

6 Phycoremediation of Xenobiotic and Petroleum Hydrocarbons Polluted Wastewater Effluents

There are many industrial wastewater effluents (e.g., petroleum refinery, petrochemical industries, oil field formation water, produced water, coal coking, coal gas purification, etc.) containing toxic phenolic compounds and PAHs. Biotreatment of these effluents is suggested; however, the presence of heavy metals and other cyanide and nitrogenous compounds in such wastewater will act as inhibitors for bacterial activities. However, microalgae are reported as a promising candidate for biotreatment of such wastewater as they photoautotrophically and/or heterotrophically accumulate and metabolize nitrogenous compounds into valuable compounds and lipids to be transesterified to biodiesel (Matamoros et al. 2015; Ryu et al. 2017). Moreover, microalgae reported to be tolerable to heavy metals (Ke et al. 2010) and have the ability to detoxify and degrade many phenolic and PAH compounds (de Bashan and Bashan 2010; Luo et al. 2014a, b; Ryu et al. 2014a).

Freshwater microalgae *Microcystis aeruginosa* and *Scenedesmus intermedius* expressed some kind of adaptation toward exposure to high concentrations of petroleum and diesel oil (Romero-Lopez et al. 2012).

Sivasubramanian and Muthukumaran (2012) reported the isolation of *Chlorococcum humicola*, *Chlamydomonas pertusa*, and *Chlorococcum vitiosum* from oil drilling effluent, which were further efficiently utilized for phycoremediation of such effluent. And a consortium of all was applied for phycoremediation on laboratory scale, pilot scale, as well as large scale. In the laboratory scale, there was a significant reduction in total dissolved solids (TDS), total hardness, Ca, Mg, Na, K, and chlorides. Biochemical oxygen demand (BOD), chemical oxygen demand (COD), and nitrate were reduced by 96.17%, 52.19%, and 78.04%, respectively. Heavy metal reduction also occurred, recording a 92.57%, 88.17%, and 38.77% reduction for Cr, Cu, and Zn, respectively. In the pilot scale, pH increased from 6.45 to 7.48. Sludge, alkalinity, sulfate, and BOD reduced by 69.4%, 70.51%, 95.81%, and 70.27%, respectively. On the large scale, in a tank (12 m length × 12 m width × 1.9 m height) having a capacity of 44,000 L of oil drilling effluent, the sludge, turbidity, total solids, TDS, total suspended solids (TSS), electrical conductivity, and BOD reduced by 47.75%, 87.03%, 80.83%, 80.79%, 90.27%, 80.83%, and 93.20%, respectively. Uzoh et al. (2015) reported that the phycodegradation of Bonny light crude oil by *Closterium* sp. under dark condition was higher than that under sunlight, recording a percentage biodegradation of 98.84% and 88.15%, respectively.

In another study, Gods'gift and Fagade (2016) reported the synergetic effect of microalgae and bacteria in bioremediation of petroleum-based effluent, where a consortium of *Pseudomonas aeruginosa*, *Bacillus subtilis*, *Citrobacter* sp., and *Chlorella minutissima* degraded 92.07% of the total PAHs, followed by a consortium of *Pseudomonas aeruginosa*, *Bacillus subtilis*, *Citrobacter* sp., *Chlorella minutissima*, and *Aphanocapsa* sp. which degraded 67.76% of the total PAHs, and then finally a consortium of *Pseudomonas aeruginosa*, *Bacillus subtilis*, *Citrobacter* sp., and *Aphanocapsa* sp. which degraded 47.19% of the total PAHs.

The green alga *Scenedesmus quadricauda* has been reported for phycoremediation of phenols in coke wastewater (Ryu et al. 2017). But the meta-cleavage degradation of phenols was highly affected by the dilution of coke effluent. The highest phycodegradation with maximum production of fatty acids occurred at 20% coke wastewater concentration. However, upon the application of algal-bacterial cultures, complete degradation of phenols occurred even at 100% coke wastewater concentration. It has been also noted that, although it is known that NH_4^+N is the utmost energetically favored nitrogen source for microalgae and the least ones are NO_2 and NO_3 which are reduced by intracellular reductases, however, NH_4^+N which is important for chlorophyll synthesis was better removed at diluted concentrations of coke wastewater, and the algal-bacterial cultures expressed 2.3-fold greater removal of NH_4^+N than the algae alone. That proved the synergetic advantages of the application of algal-bacterial cultures over algal alone, as it overcomes the negative impact of the ROS produced by the oxidative stress of phenolic, PAHs, and other oxygen heterocyclic compounds in the coke wastewater (Ryu et al. 2017). In another study, Thakurta et al. (2018) reported the phycodegradation of phenol in coke-oven wastewater effluent after secondary biotreatment step using *Leptolyngbya* sp.

Budisa et al. (2019) reported the phycoremediation of 50% diluted oil refinery wastewater in a photobioreactor by marine microalgae *Microchloropsis gaditana* and *Pseudochloris wilhelmii* with lipid production of approximately 35.5% and 40.6%, respectively. Both have efficiently removed NH_4^+/NH_3, and the former had the advantage of low phosphorous requirement. That recommended *M. gaditana* for phycoremediation and biodiesel production. However, the percentage of unsaturated fatty acid methyl ester (FAME) increased after the lag phase with the depletion of nitrogen in the culture media, increasing the iodine value in the produced biodiesel.

Das and Deka (2019) reported a cost-effective mixotrophic phycoremediation oil field formation water that is produced during crude oil exploration by *C. vulgaris* previously isolated from formation water. Moreover, it expressed efficient phycoremediation for the crude oil of approximately 98.63% with biomass productivity of approximately 1.76 g/L/day, complete removal of SO_4, and 75% reduction of chemical oxygen demand (COD), within only 14 days, without dilution or nutrient addition. Thus, *C. vulgaris* depended on the main constituents of formation water from major constituents as SO_4, Fe, Cu, Zn, Na, Mn, Ca, N, and Mg, for growth and phycodegradation of petroleum hydrocarbons. The complete removal of SO_4 from the formation water is very beneficial in petroleum industry as it can omit the growth of sulfate-reducing bacteria and consequently decrease the use of expensive and non-eco-friendly biocides and the problem of toxic and corrosive H_2S production. The gas chromatography mass spectroscopy proved the complete degradation of aliphatics and PAHs with C12 to C60. In another study, Das et al. (2019) studied the phycoremediation of petroleum-derived produced water with initial low pH and salinity of 4.17 and 4.3 ppt, respectively, and total nitrogen of 52.5 ppm, total phosphorous of 0.21 ppm, and total organic content of 720 ppm after its pretreatment with NaOH to neutralize the pH. The pretreatment decreased the total nitrogen, total phosphorous, and total organic content by 38%, 19%, and 40%, respectively. Upon the application of different microalgae strains for biotreatment of the produced pretreated water, *Chlorella* sp. was the best grown isolate with and without the addition of supplementing nutrients, where its growth reached to 1.72 g/L on unsterilized pretreated produced water with an initial total organic content value of 428 mg/L. *Chlorella* sp. reduced the TOC to 115.4 mg/L after 15 days of incubation, with a concomitant significant reduction in the phenolic and aliphatic compounds, PAHs and heavy metals. Further 92% reduction in total nitrogen was also recorded after the phycoremediation process, with a synergetic concomitant growth in indigenous bacterial population. It is important to note that such high value of total organic content after phycoremediation was related to the degradation of such organic compounds into other organic compounds and/or the production of extracellular organic compounds. Fish bioassay study has been performed to guarantee the efficiency of the phycoremediation and the successfulness of producing water effluent which can be reused safely. All the tested zebrafish was killed within 48 h in the untreated water and attributed this to the high pollutant concentrations and low pH values. However, all the tested fishes survived and were healthy for 9 days in the phycoremediated water. The chlorophyll (a) content in the blue-green 11 (BG11) medium (i.e., −ve control) and the phycoremediation cultures was nearly the same. This also added to the value of the application of the concept of obtaining algal biomass via phycoremediation of polluted water. Another advantage of applying algae in the phycoremediation is the increase of pH during biotreatment operation and production of extracellular organic compounds, which together enhanced the floc formation and consequent removal of heavy metals (Gani et al. 2015; Das and Deka 2019).

7 Challenges and Opportunities

It is recommendable to invest in the double benefit, cost-effective, and eco-friendly phycoremediation of organic wastewater before disposing and/or reuse as it is a win–win process. Wastewater itself can be considered as a low-cost or, even better, costless source of carbon, nitrogen, and phosphorous for simultaneous polluted water treatment and algae biomass growth for further production of different valuable products, chemicals, and biofuels. Thus, it can be a promising management for the worldwide twin challenges: energy security and environmental pollution. Algae can help in solving the problem of climate change and global warming as it accumulates CO_2 during the sustainable phycoremediation of polluted water. Moreover, it can be applied in the in situ bioremediation in open seawater, oceans, lakes, shallow water, etc. as most of the algae have phototaxis properties and usually prefer to live in surface water (Hao et al. 2019). Moreover, researches proved that the prior acclimation of microalgae to toxic chemical stress enhances the algal biomass production, and the antioxidant enzyme activities, in addition to the lipid content, consequently increase the algal biodiesel production (Cho et al. 2016). Thus, it is important to focus on the researches concerning algal biomass produced from wastewater treatment, their enzymatic activity, and its biomass contents from pigments, lipids, and carbohydrates.

Algal-based organic wastewater treatment has been reported to be very promising and feasible. And it consumes approximately 30-fold less electricity than the conventional activated sludge process (ASP) (Matamoros et al. 2015). Moreover, the synergetic effect between bacteria and algae makes it a cost-efficient aeration process and also enhances the removal of organic and inorganic pollutants from wastewater effluents (Su et al. 2012; Posadas et al. 2013; Luo et al. 2014b; Zhao et al. 2014; Babatsouli et al. 2015; Ramanan et al. 2016; Gentili and Fick 2017). Not only this, but one of the synergetic effects is the usual formation of algal-bacterial flocs, which can omit the application of conventional flocculating agents in wastewater treatment process and facilitate the biomass harvesting via simple gravity sedimentation (Ryu et al. 2014b). Moreover, algal-bacterial mixed culture overcomes the ROS inhibitory effect on the algal photosynthesis as ROS are known to decrease the photosynthetic electron transfer and the formation of NADPH which consequently decrease the phycoremediation efficiency (Qian et al. 2009). Thus, algal-bacterial cultures can enhance the production of valuable compounds, e.g., chlorophylls (Ryu et al. 2017). Moreover, ROS is reported to highly affect the production of fatty acids and its type (i.e., long-chain saturated or unsaturated ones) (Paradies et al. 2002). Thus, the control of the concentration of ROS is important as it will affect the quality of the produced biodiesel. The percentage of unsaturation affects the cold flow properties of biodiesel and its oxidation stability (Mostafa and El-Gendy 2017; Ryu et al. 2017).

However, more researches are required to widen the range of xenobiotic organic pollutants that can be biodegraded by microalgae, through genetic engineering and/or screening for more new strains. Nevertheless, it should be taken into consideration the toxic effects of such xenobiotic organic pollutants and their inhibition on microalgae growth and reproduction. Thus, it is important to study well and predict the optimum initial inoculum size, the inoculated microalgae species, its cell wall composition, and its enzymatic system which is responsible for the pollutant's phycodegradation. It is worth to study also the toxic effect of the metabolites and take the precautions to overcome such obstacle. For example, isolating microalgae and designing a consortia and cultures to mixotrophically degrade phenolic compounds as pollutants and/or as metabolic intermediates is very vital for a successful phycoremediation process.

Usually, phycodegradation occurs throughout three successive steps: adsorption of the pollutant, absorption and accumulation into the algal cells, and finally degradation (Thakurta et al. 2018). Thus, the algal size, the surface area to bio-volume ratio of the microalgae, the cell wall composition, and morphology play a vital role in the successfulness of the phycoremediation process. Not only this, but the type of the pollutant; its chemical structure, molecular weight, water solubility, octanol–water partition coefficients (K_{ow}), and lipophilicity; and its chemical composition are also important factors to be known and studied to guarantee the successfulness of the phycoremediation process.

For the synergetic positive impact of the presence of heavy metals on the phycodegradation of xenobiotic organic pollutants (Ke et al. 2010), it would be very promising to use metal-contaminated outlet effluents of different industries to dilute concentrated xenobiotic organic and/or oil polluted water. However, further researches are required on; the benefits and feasibility of such strategy and the isolation of microalgae with efficient tolerance to high concentrations of heavy metals and organic pollutants, so as not to perform dilution step and other pretreatment processes.

Microalgae isolated form real xenobiotics organic and petroleum hydrocarbons polluted sites are usually reported to be efficient for phycoremediation (Lima et al. 2003; Ryu et al. 2017). However, further work is required on the phycodegradation metabolic pathway of such pollutants to assure its complete mineralization. Not only this, but the initial biomass concentration and water salinity are also essential parameters for the successfulness of the phycoremediation process (Kalhor et al. 2017). Moreover, the effect of illumination intensity, type of light (e.g., white and gold), and dark/light cycles are also important factors to be taken into consideration to achieve the highest phycoremediation with the production of sufficient biomass containing valuable components (e.g., natural dyes and lipids) (Luo et al. 2014a). The application of mathematical modeling and statistical optimization for the enhancement of the phycoremediation process of polluted water with concomitant algal biomass production will minimize the loss of time and effort and decrease the cost and energy consumption. The accumulation of PAH phycodegradation intermediates that might be not further degraded and their impact on the algal culture and aquatic ecosystem should be taken into consideration and accurately studied. Further, upon application on open ponds for polluted water treatment and algae biomass production, seasonal changes and contamination with other fast growing microorganisms are other problems to be solved. Since the overall cost will be increased, contamination will decrease the quality and quantity of the biomass and products of interest. It is also unrealistic to perform such phycoremediation and biomass production in expensive and energy-consuming self-sterilizer bioreactors. Thus, it is essential to design and integrate synergetic systems which depends on isolating new microalgae strains and/or applying microalgae consortium, extremophilic, genetically, and/or metabolically engineered microalgae, which can competitively survive in the presence of other possible microorganisms and work efficiently over a wide range of salinity and under different types and intensities of illumination in cost-effective means and lower energy consumption.

References

Abdelaziz AEM, Leite GB, Belhaj MA, Hallenbeck PC (2014) Screening microalgae native to Quebec for wastewater treatment and biodiesel production. Bioresour Technol 157:140–148

Abdel-Shafy HI, Mansour MSM (2016) A review on polycyclic aromatic hydrocarbons: source, environmental impact, effect on human health and remediation. Egypt J Pet 25:107–123

Adarme-Vega TC, Lim DKY, Timmins M, Vernen F, Li Y, Schenk PM (2012) Microalgal biofactories: a promising approach towards sustainable omega-3 fatty acid production. Microb Cell Factories 11:96–105

Agrafioti E, Bouras G, Kalderis D, Diamadopoulos E (2013) Biochar production by sewage sludge pyrolysis. J Anal Appl Pyrol 101:72–78

Aldaby ESE, Mawad AMM (2019) Pyrene biodegradation capability of two different microalgal strains. Global NEST J 21:291–296

Ali HR, Ismail DA, El-Gendy NS (2014) The biotreatment of oil polluted seawater by biosurfactant producer halotolerant *Pseudomonas aeruginosa* Asph2. Energy Source A 36:1429–1436

Arias AH, Souissi A, Glippa O, Roussin M, Dumoulin D, Net P, Ouddane B, Souissi S (2017) Removal and biodegradation of phenanthrene, fluoranthene and pyrene by the marine algae *Rhodomonas baltica* enriched from North Atlantic coasts. Bull Environ Contam Toxicol 98:392–399

Asghari S, Movafeghi A, Lisar SYS, Barar J, Omidi Y (2018) Effects of phenanthrene on growth parameters and antioxidant systems in the green microalga *Chlorella vulgaris*. Biointerface Res Appl Chem 8:3405–3411

Babatsouli P, Fodelianakis S, Paranychianakis N, Venieri D, Dialynas M, Kalogerakis N (2015) Single stage treatment of saline wastewater with marine bacterial–microalgae consortia in a fixed-bed photobioreactor. J Hazard Mater 292:155–163

Babu TS, Akhtar TA, Lampi MA, Tripuranthakam S, Dixon DG, Greenberg BM (2003) Similar stress responses are elicited by copper and ultraviolet radiation in the aquatic plant *Lemna gibba*: implication of reactive oxygen species as common signals. Plant Cell Physiol 44:1320–1329

Baun A, Justesen KB, Nyholm N (2002) Algal tests with soil suspensions and elutriates: a comparative evaluation for PAH-contaminated soils. Chemosphere 46:251–258

Becker EW (2007) Micro-algae as a source of protein. Biotechnol Adv 25:207–210

Ben CK, Sánchez E, Baghour M (2014) The role of algae in bioremediation of organic pollutants. Int Res J Public Environ Health 1:19–32

Biller P, Ross AB, Skill SC, Lea-Langton A, Balasundaram B, Hall C, Riley R, Llewellyn CA (2012) Nutrient recycling of aqueous phase for microalgae cultivation from the hydrothermal liquefaction process. Algal Res 1:70–76

Bird MI, Wurster CM, De Paula Silva PH, Paul NA, De Nys R (2012) Algal biochar: effects and applications. GCB Bioenergy 4:61–69

Blanco FJH, García de Llasera MP (2016) Monitoring dihydrodiol polyaromatic hydrocarbon metabolites produced by the freshwater microalgae *Selenastrum capricornutum*. Chemosphere 158:80–90

Bohdziewicz J, Kaminski G, Tytła M (2012) The removal of phenols from wastewater through sorption on activated carbon. Arch Civil Eng Environ 2:89–94

Borde X, Guieysse B, Delgado O, Muñoz R, Hatti-Kaul R, Nugier-Chauvin C, Patin H, Mattiasson B (2003) Synergistic relationships in algal–bacterial microcosms for the treatment of aromatic pollutants. Bioresour Technol 86:293–300

Brennan L, Owende P (2010) Biofuels from microalgae-a review of technologies for production, processing, and extractions of biofuels and co-products. Renew Sust Energ Rev 14:557–577

Budisa A, Haberle I, Konjevic L, Blazina M, Djakovac T, Spalj BL, Hrustic E (2019) Marine microalgae *Microchloropsis gaditana* and *Pseudochloris wilhelmii* cultivated in oil refinery wastewater – a perspective on remediation and biodiesel production. Fresenius Environ Bull 28:7888–7897

Cao X, Xiong Y, Lund J (2013) The effects of micro-algae characteristics on the bioremediation rate of Deepwater Horizon Crude Oil. J Emerg Invest:1–7

Cerniglia CE, Van Baalen C, Gibson DT (1980a) Metabolism of naphthalene by the cyanobacterium *Oscillatoria* sp., strain JCM. J Gen Microbiol 116:485–494

Cerniglia CE, Gibson DT, Van Baalen C (1980b) Oxidation of naphthalene by cyanobacteria and microalgae. J Gen Microbiol 116:495–500

Chan SMN, Luan TG, Wong MH (2006) Removal and biodegradation of polycyclic aromatic hydrocarbons by *Selenastrum capricornutum*. Environ Toxicol Chem 25:1772–1779

Chen Q, Li J, Liu M, Sun H, Bao M (2017) Study on the biodegradation of crude oil by free and immobilized bacterial consortium in marine environment. PLoS One 12(3):e0174445. https://doi.org/10.1371/journal.pone.0174445

Chia SR, Chew KW, Show PL, Sivakumar M, Ling TC, Tao Y (2019) Isolation of protein from *Chlorella sorokiniana* CY1 using liquid biphasic flotation assisted with sonication through sugaring-out effect. Chin J Oceanol Limn 37:898–908

Cho K, Lee C-H, Ko K, Lee Y-J, Kim K-N, Kim M-K, Chung Y-H, Kim D, Yeo I-K, Oda T (2016) Use of phenol-induced oxidative stress acclimation to stimulate cell growth and biodiesel production by the oceanic microalga *Dunaliella salina*. Algal Res 17:61–66

Cody TE, Radike MJ, Warshawsky D (1984) The phototoxicity of benzo[a]pyrene in green alga *Selenasrum capricornutum*. Environ Res 35:122–132

Dakhil IH (2013) Removal of phenol from industrial wastewater using sawdust. Int J Eng Sci 3:25–31

Das N, Chandran P (2011) Microbial degradation of petroleum hydrocarbon contaminants: an overview. Biotechnol Res Int 2011:941810. https://doi.org/10.4061/2011/941810

Das B, Deka S (2019) A cost-effective and environmentally sustainable process for phycoremediation of oil field formation water for its safe disposal and reuse. Sci Rep 9:15232. https://doi.org/10.1038/s41598-019-51806-5

Das B, Mandal TK, Patra S (2015) A comprehensive study on *Chlorella pyrenoidosa* for phenol degradation and its potential applicability as biodiesel feedstock and animal feed. Appl Biochem Biotechnol 176:1382–1401

Das P, Abdul Quadir M, Thaher M, Khan S, Chaudhary AK, Alghasal G, Al-Jabri HMSJ (2019) Microalgal bioremediation of petroleum-derived low salinity and low pH produced water. J Appl Phycol 31:435–444

de Bashan LE, Bashan Y (2010) Immobilized microalgae for removing pollutants: review of practical aspects. Bioresour Technol 101:1611–1627

de Llasera MPG, de Jesús Olmos-Espejel J, Díaz-Flores G, Montaño-Montiel A (2016) Biodegradation of benzo (a) pyrene by two freshwater microalgae *Selenastrum capricornutum* and *Scenedesmus acutus*: a comparative study useful for bioremediation. Environ Sci Pollut Res 23:3365–3375

Dean-Ross D, Moody J, Cerniglia CE (2002) Utilization of mixtures of polycyclic aromatic hydrocarbons by bacteria isolated from contaminated sediment. FEMS Microbiol Ecol 41:1–7

Del Campo JA, García-González M, Guerrero M (2007) Outdoor cultivation of microalgae for carotenoid production: current state and perspectives. Appl Microbiol Biotechnol 74:1163–1174

Delrue F, Setier P-A, Sahut C, Cournac L, Roubaud A, Peltier G (2012) An economic, sustainability, and energetic model of biodiesel production from microalgae. Bioresour Technol 111:191–200

Delrue F, Álvarez-Díaz PD, Fon-Sing S, Fleury G, Sassi J-F (2016) The environmental biorefinery: using microalgae to remediate wastewater, a win-win paradigm. Energies 9(132). https://doi.org/10.3390/en9030132

Demirel YT, Yati I, Donmez R, Sonmez HB (2017) Clean-up of oily liquids, fuels and organic solvents from the contaminated water fields using poly(propylene glycol) based organogels. Chem Eng J 312:126–135

Deriase SF, Younis SA, El-Gendy NS (2013) Kinetic evaluation and modeling for batch degradation of 2-hydroxybiphenyl and 2,2′-dihydroxybiphenyl by *Corynebacterium variabilis* Sh42. Desalin Water Treat 51:4719–4728

Drozdova S, Ritter W, Lendl B, Rosenberg E (2013) Challenges in the determination of petroleum hydrocarbons in water by gas chromatography (hydrocarbon index). Fuel 113:527–536

Duan WY, Meng FP, Lin YF, Wang GS (2017) Toxicological effects of phenol on four marine microalgae. Environ Toxicol Pharmacol 52:170–176

Duan WY, Meng FP, Cui HW, Lin YF, Wang GS, Wu JY (2018) Ecotoxicity of phenol and cresols to aquatic organisms: a review. Ecotoxicol Environ Saf 157:441–456

El Mahdi AM, Aziz HA, Abu Amr SS, El-Gendy NS, Nassar HN (2016) Isolation and characterization of *Pseudomonas* sp. NAF1 and its application in biodegradation of crude oil. Environ Earth Sci 75(380). https://doi.org/10.1007/s12665-016-5296-z

El-Gendy NS, Moustafa YM, Barakat MAK, Deriase SF (2009) Evaluation of a bioslurry remediation of petroleum hydrocarbons contaminated sediments using chemical, mathematical and microscopic analysis. Int J Environ Stud 66:563–579

El-Gendy NS, Ali HR, El-Nady MM, Deriase SF, Moustafa YM, Roushdy MI (2014) Effect of different bioremediation techniques on petroleum biomarkers and asphaltene fraction in oil polluted sea water. Desalin Water Treat 52:7484–7494

El-Sheekh MM, Ghareib MM, Abou-EL-Souod GW (2012) Biodegradation of phenolic and polycyclic aromatic compounds by some algae and *Cyanobacteria*. J Bioremed Biodegr 3 (133). https://doi.org/10.4172/2155-6199.1000133

El-Sheekh MM, Hamouda RA, Nizam AA (2013) Biodegradation of crude oil by *Scenedesmus obliquus* and *Chlorella vulgaris* growing under heterotrophic conditions. Int Biodeterior Biodegradation 82:67–72

El-Sheshtawy HS, Khalil NM, Ahmed W, Abdallah RI (2014) Monitoring of oil pollution at Gemsa Bay and bioremediation capacity of bacterial isolates with biosurfactants and nanoparticles. Mar Pollut Bull 87:191–200

El-Tarabily KA (2002) Total microbial activity and microbial composition of a mangrove sediment are reduced by oil pollution at a site in the Arabian Gulf. Can J Microbiol 48:176–182

Eriksen NT (2008) Production of phycocyanin—a pigment with applications in biology, biotechnology, foods and medicine. Appl Microbiol Biotechnol 80:1–14

Farrington JW (2014) Oil pollution in the marine environment II: fates and effects of oil spills. Environ Sci Policy Sustain Dev 65:16–31

Fathi AA, Azooz MM, Al-Fredan MA (2013) Phycoremediation and the potential of sustainable algal biofuel production using wastewater. Am J Appl Sci 10:189–194

Gamila HA, Ibrahim MBM (2004) Algal bioassay for evaluating the role of algae in bioremediation of crude oil: I—Isolated strains. Bull Environ Conatm Toxicol 73:883–889

Gani P, Sunar NM, Matias-Peralta H, Latiff AA, Parjo UK, Razak AR (2015) Phycoremediation of wastewaters and potential hydrocarbon from microalgae: a review. Adv Environ Biol 9:1–8

Gao J, Chi J (2015) Biodegradation of phthalate acid esters by different marine microalgal species. Mar Pollut Bull 99:70–75

García de Llasera MP, Olmos-Espejel JJ, Díaz-Flores G, Montaño-Montiel A (2016) Biodegradation of benzo(a)pyrene by two freshwater microalgae *Selenastrum capricornutum* and *Scenedesmus acutus*: a comparative study useful for bioremediation. Environ Sci Pollut Res 23:3365–3375

García de Llasera MP, León Santiago M, Loera Flores EJ, Bernal Toris DN, Covarrubias Herrera MR (2018) Mini-bioreactors with immobilized microalgae for the removal of benzo(a)anthracene and benzo(a)pyrene from water. Ecol Eng 121:89–98

Gavrilescu M (2010) Environmental biotechnology: achievements, opportunities and challenges. Dyn Biochem Process Biotechnol Mol Biol 4:1–36

Gentili FG, Fick J (2017) Algal cultivation in urban wastewater: an efficient way to reduce pharmaceutical pollutants. J Appl Phycol 29:255–262

Ghosal D, Ghosh S, Dutta TK, Ahn Y (2016) Current state of knowledge in microbial degradation of polycyclic aromatic hydrocarbons (PAHs): a review. Front Microbiol 7(1369). https://doi.org/10.3389/fmicb.2016.01369

Gods'gift OE, Fagade OE (2016) Microalgal-bacterial consortium in polyaromatic hydrocarbon degradation of petroleum – based effluent. J Bioremed Biodegr 7:359. https://doi.org/10.4172/2155-6199.1000359

Guo T, Englehardt J, Wu T (2015) Review of cost versus scale: water and wastewater treatment and reuse processes. Water Sci Technol 69:223–234

Halim R, Danquah MK, Webley PA (2012) Extraction of oil from microalgae for biodiesel production: a review. Biotechnol Adv 30:709–732

Hamouda RA, Sorour NM, Yeheia DS (2016) Biodegradation of crude oil by *Anabaena oryzae*, *Chlorella kessleri* and its consortium under mixotrophic conditions. Int Biodeterior Biodegradation 112:128–134

Hao L, Fanping M, Yuejie W, Yufei L (2019) Removal of phenol by *Isochrysis galbana* in seawater under varying temperature and light intensity. Chin J Oceanol Limn. https://doi.org/10.1007/s00343-019-9125-6

Headley JV, Du JL, Peru KM, Gurprasad N, McMartin DW (2008) Evaluation of algal phytodegradation of petroleum naphthenic acids. J Environ Sci Health A 43:227–232

Hirooka T, Akiyama Y, Tsuji N, Nakamura T, Nagase H, Hirata K, Miyamoto K (2003) Removal of hazardous phenols by microalgae under photoautotrophic conditions. J Biosci Bioeng 95:200–203

Hognon C, Delrue F, Texier J, Grateau M, Thiery S, Miller H, Roubeau A (2015) Comparison of pyrolysis and hydrothermal liquefaction of *Chlamydomonas reinhardtii*. Growth studies on the recovered hydrothermal aqueous phase. Biomass Bioenerg 73:23–31

Hong YW, Yuan DX (2008) Toxic effect of typical polycyclic aromatics hydrocarbons on diatoms in mangrove area. Chin J Mar Environ Sci 27:4–10

Hong YW, Yuan DX, Lin QM, Yang TL (2008) Accumulation and biodegradation of phenanthrene and fluoranthene by the algae enriched from a mangrove aquatic ecosystem. Mar Pollut Bull 56:1400–1405

Hosny R, Fathy M, Ramzi M, Abdel Moghny T, Desouky SEM, Shama SA (2016) Treatment of the oily produced water (OPW) using coagulant mixtures. Egypt J Pet 25:391–396

Hu B, Min M, Zhou W, Li Y, Mohr M, Cheng Y, Lei H, Liu Y, Lin X, Chen P, Ruan R (2012) Influence of exogenous CO_2 on biomass and lipid accumulation of microalgae *Auxenochlorella protothecoides* cultivated in concentrated municipal wastewater. Appl Biochem Biotechnol 166:1661–1673

Ji M-K, Kabra AN, Choi J, Hwang J-H, Kim JR, Abou-Shanab RAI, Oh Y-K, Jeon B-H (2014) Biodegradation of bisphenol A by the freshwater microalgae *Chlamydomonas mexicana* and *Chlorella vulgaris*. Ecol Eng 73:260–269

Kalhor AX, Mohammadi-Nassab AD, Abedi E, Bahrami A, Movafeghi A (2016) Biodiesel production in crude oil contaminated environment using *Chlorella vulgaris*. Bioresour Technol 222:190–194

Kalhor AX, Movafeghi A, Mohammadi-Nassab AD, Abedi E, Bahrami A (2017) Potential of the green alga *Chlorella vulgaris* for biodegradation of crude oil hydrocarbons. Mar Pollut Bull 123:286–290

Karn SK, Chakrabarti S (2015) Simultaneous biodegradation of organic (chlorophenols) and inorganic compounds from secondary sludge of pulp and paper mill by *Eisenia fetida*. Int J Recycl Org Waste Agric 4:53–62

Kaushik G, Chatterjee S (2015) Oil spill cleanup: role of environmental biotechnology. In: Applied environmental biotechnology: present scenario and future trends. Springer, India, pp 129–143

Ke L, Luo L, Wang P, Luan T, Tam NF-Y (2010) Effects of metals on biosorption and biodegradation of mixed polycyclic aromatic hydrocarbons by a freshwater green alga *Selenastrum capricornutum*. Bioresour Technol 101:6950–6961

Klekner V, Kosaric N (1992a) Degradation of phenolic mixtures by *Chlorella*. Environ Technol 13:503–506

Klekner V, Kosaric N (1992b) Degradation of phenols by algae. Environ Technol 13:493–501

Kneifel H, Elmendorff K, Hegewald E, Soeder CJ (1997) Biotransformation of 1-naphthalenesulfonic acid by the green alga *Scenedesmus obliquus*. Arch Microbiol 167:32–37

Kröger M, Müller-Langer F (2011) Impact of heterotrophic and mixotrophic growth of microalgae on the production of future biofuels. Biofuels 2:145–151

Kulkarni SJ, Kaware JP (2013) Review on research for removal of phenol from wastewater. Int J Sci Res Publ 3:1–5

Kumar D, Singh B (2019) Algal biorefinery: an integrated approach for sustainable biodiesel production. Biomass Bioenerg 131:105398

Kumar MS, Muralitharan G, Thajuddin N (2009) Screening of a hypersaline cyanobacterium, *Phormidium tenue*, for the degradation of aromatic hydrocarbons: naphthalene and anthracene. Biotechnol Lett 31:1863–1866

Lee H-C, Lee M, Den W (2015) *Spirulina maxima* for phenol removal: study on its tolerance, biodegradability and phenol-carbon assimilability. Water Air Soil Pollut 226(395). https://doi.org/10.1007/s11270-015-2664-3

Lei A-P, Wong Y-S, Tam NF-Y (2003) Pyrene-induced changes of glutathione-S-transferase activities in different microalgal species. Chemosphere 50:293–301

Lei A-P, Hu ZL, Wong Y-S, Tam NF-Y (2007) Removal of fluoranthene and pyrene by different microalgal species. Bioresour Technol 98:273–280

Leyva LA, Bashan Y, de Bashan LE (2015) Activity of acetyl-CoA carboxylase is not directly linked to accumulation of lipids when *Chlorella vulgaris* is co-immobilized with *Azospirillum brasilense* in alginate under autotrophic and heterotrophic conditions. Ann Microbiol 65:339–349

Li Y-R, Tsai W-T, Hsu Y-C, Xie M-Z, Chen J-J (2014) Comparison of autotrophic and mixotrophic cultivation of green microalgal for biodiesel production. Energy Procedia 52:371–376

Liang Z, Liu Y, Ge F, Xu Y, Tao N, Peng F, Wong M (2013) Efficiency assessment and pH effect in removing nitrogen and phosphorus by algae-bacteria combined system of *Chlorella vulgaris* and *Bacillus licheniformis*. Chemosphere 92:1383–1389

Lika K, Papadakis IA (2009) Modeling the biodegradation of phenolic compounds by microalgae. J Sea Res 62:135–146

Lima SAC, Castro PML, Morais R (2003) Biodegradation of *p*-nitrophenol by microalgae. J Appl Phycol 15:137–142

Lima SA, Raposo MF, Castro PM, Morais RM (2004) Biodegradation of p-chlorophenol by a microalgae consortium. Water Res 38:97–102

Liu Y, Luan TG, Lu NN, Lan CY (2006) Toxicity of fluoranthene and its biodegradation by *Cyclotella caspia* alga. J Integr Plant Biol 48:169–180

Liu Z, Wenyu X, Dehao L, Yang P, Zesheng L, Shusi L (2016) Biodegradation of phenol by bacteria strain *Acinetobacter calcoaceticus* PA isolated from phenolic wastewater. Int J Environ Res Public Health 13:300–308

López-Pacheco IY, Salinas-Salazar C, Silva-Núñez A, Rodas-Zuluaga LI, Donoso-Quezada J, Ayala-Mar S, Barceló D, Iqbal HMN, Parra-Saldívar R (2019) Removal and biotransformation of 4-nonylphenol by *Arthrospira maxima* and *Chlorella vulgaris* consortium. Environ Res 179:108848

Lovell CR, Eriksen NT, Lewitus AJ, Chen YP (2002) Resistance of the marine diatom *Thalassiosira* sp. to toxicity of phenolic compounds. Mar Ecol Prog Ser 229:11–18

Luo L, Wang P, Lin L, Luan T, Ke L, Tam NFY (2014a) Removal and transformation of high molecular weight polycyclic aromatic hydrocarbons in water by live and dead microalgae. Process Biochem 49:1723–1732

Luo S, Chen B, Lin L, Wang X, Tam NF-Y, Luan T (2014b) Pyrene degradation accelerated by constructed consortium of bacterium and microalga: effects of degradation products on the microalgal growth. Environ Sci Technol 48:13917–13924

Luther M (1990) Degradation of different substituted aromatic compounds as nutrient sources by the green alga *Scenedesmus obliquus*. Dechema Biotechnol Conf 4:613–615

Matamoros V, Gutiérrez R, Ferrer I, García J, Bayona JM (2015) Capability of microalgae-based wastewater treatment systems to remove emerging organic contaminants: a pilot-scale study. J Hazard Mater 288:34–42

McNally DL, Mihelcic JR, Lueking DR (1999) Biodegradation of mixtures of polycyclic aromatic hydrocarbons under aerobic and nitrate-reducing conditions. Chemosphere 38:1313–1321

Mittler R (2002) Oxidative stress, antioxidants and stress tolerance. Trends Plant Sci 7:405–410

Mostafa SSM, El-Gendy NS (2017) Evaluation of fuel properties for microalgae *Spirulina platensis* bio-diesel and its blends with Egyptian petro-diesel. Arab J Chem 10:S2040–S2050

Mulbry W, Westhead EK, Pizarro C, Sikora L (2005) Recycling of manure nutrients: use of algal biomass from dairy manure treatment as a slow release fertilizer. Bioresour Technol 96:451–458

Muñoz R, Guieysse B (2006) Algal–bacterial processes for the treatment of hazardous contaminants: a review. Water Res 40:2799–2815

Mzoughi N, Chouba L (2011) Distribution and partitioning of aliphatic hydrocarbons and polycyclic aromatic hydrocarbons between water, suspended particulate matter, and sediment in harbours of the West coastal of the Gulf of Tunis (Tunisia). J Environ Monit 13:689–698

Nakajima N, Teramotom T, Kasai F, Sano T, Tamaoki M, Aono M, Kubo A, Kamada H, Azumi Y, Saji H (2007) Glycosylation of bisphenol A by freshwater microalgae. Chemosphere 69:934–941

Nawkarkar P, Singh AK, Zainul Abdin M, Kumar S (2019) Life cycle assessment of *Chlorella* species producing biodiesel and remediating wastewater. J Biosci 44(89). https://doi.org/10.1007/s12038-019-9896-0

Nazos TT, Kokarakis EJ, Ghanotakis DF (2017) Metabolism of xenobiotics by *Chlamydomonas reinhardtii*: phenol degradation under conditions affecting photosynthesis. Photosynth Res 131:31–40

Olguín EJ (2012) Dual purpose microalgae-bacteria-based systems that treat wastewater and produce biodiesel and chemical products within a biorefinery. Biotechnol Adv 30:1031–1043

Olmos-Espejel JJ, García de Llasera MP, Velasco-Cruz M (2012) Extraction and analysis of polycyclic aromatic hydrocarbons and benzo[a]pyrene metabolites in microalgae cultures by

off-line/online methodology based on matrix solid-phase dispersion, solidphase extraction and high-performance liquid chromatography. J Chromatogr A 1262:138–147
Oswald WJ, Gotaas HB, Ludwig HF, Lynch V (1953) Algae symbiosis in oxidation ponds: III. Photosynthetic oxygenation. Sewage Ind Waste 25:692–705
Ouada SB, Ali RB, Leboulanger C, Zaghden H, Choura S, Ouada HB, Sayadi S (2018) Effect and removal of bisphenol A by two extremophilic microalgal strains (Chlorophyta). J Appl Phycol 30:1765–1776
Ozkan A, Berberoglu H (2013) Physico-chemical surface properties of microalgae. Colloids Surf B Biointerfaces 112:287–293
Paniagua-Michel JDJ, Olmos-Soto J, Morales-Guerrero ER (2014) Algal and microbial exopolysaccharides: new insights as biosurfactants and bioemulsifers. Adv Food Nutr Res 73:221–257
Papazi A, Kotzabasis K (2007) Bioenergetic strategy of microalgae for the biodegradation of phenolic compounds - exogenously supplied energy and carbon sources adjust the level of biodegradation. J Biotechnol 129:706–716
Papazi A, Kotzabasis K (2013) "Rational" management of dichlorophenols biodegradation by the microalga *Scenedesmus obliquus*. PLoS One 8(4):e61682. https://doi.org/10.1371/journal.pone.0061682
Papazi A, Assimakopoulos K, Kotzabasis K (2012) Bioenergetic strategy for the biodegradation of *p*-cresol by the unicellular green alga *Scenedesmus obliquus*. PLoS One 7(12):e51852. https://doi.org/10.1371/journal.pone.0051852
Paradies G, Petrosillo G, Pistolese M, Ruggiero FM (2002) Reactive oxygen species affect mitochondrial electron transport complex I activity through oxidative cardiolipin damage. Gene 286:135–141
Perelo LW (2010) Review: *In situ* and bioremediation of organic pollutants in aquatic sediments. J Hazard Mater 177:81–89
Pérez-Garcia O, Escalante FME, de Bashan LE, Bashan Y (2011) Heterotrophic cultures of microalgae: metabolism and potential products. Water Res 45:11–36
Pinto G, Pollio A, Previtera L, Temussi F (2002) Biodegradation of phenols by microalgae. Biotechnol Lett 24:2047–2051
Posadas E, García-Encina PA, Soltau A, Domínguez A, Díaz I, Muñoz R (2013) Carbon and nutrient removal from centrates and domestic wastewater using algal–bacterial biofilm bioreactors. Bioresour Technol 139:50–58
Priyadarshani I, Sahu D, Rath B (2011) Microalgal bioremediation: current practices and perspectives. J Biochem Technol 3:299–304
Priyadharshini SD, Bakthavatsalam AK (2016) Optimization of phenol degradation by the microalga *Chlorella pyrenoidosa* using Plackett–Burman design and response surface methodology. Bioresour Technol 207:150–156
Qian HF, Chen W, Sun LW, Jin YX, Liu WP, Fu ZW (2009) Inhibitory effects of paraquat on photosynthesis and the response to oxidative stress in *Chlorella vulgaris*. Ecotoxicology 18:537–543
Quinn JC, Davis R (2015) The potentials and challenges of algae based biofuels: a review of the techno-economic, life cycle, and resource assessment modeling. Bioresour Technol 184:444–452
Radmann EM, Greque de Morais E, Freitas de Oliveira C, Zanfonato K, Costa JAV (2015) Microalgae cultivation for biosurfactant production. Afr J Microbiol Res 9:2283–2289
Rajasulochana P, Dhamotharan R, Murgesan S, Ramachandra Murthy A (2009) Bioremediation of oil refinery effluent by using *Scenedesmus obliquus*. J Am Sci 5:17–22
Ramanan R, Kim B-H, Cho DH, Kim H-S (2016) Algae–bacteria interactions: evolution, ecology and emerging applications. Biotechnol Adv 34:14–29
Rawat I, Ranjith Kumar R, Mutanda T, Bux F (2011) Dual role of microalgae: phycoremediation of domestic wastewater and biomass production for sustainable biofuels production. Appl Energy 88:3411–3424

Renuka N, Sood A, Prasanna R, Ahluwalia AS (2015) Phycoremediation of wastewaters: a synergistic approach using microalgae for bioremediation and biomass generation. Int J Environ Sci Technol 12:1443–1460
Rodrigues FP, Angeli JPF, Mantovani MS, Guedes CLB, Jordão BQ (2010) Genotoxic evaluation of an industrial effluent from an oil refinery using plant and animal bioassays. Genet Mol Biol 33:169–175
Romero-Lopez J, Lopez-Rodaz V, Costas E (2012) Estimating the capability of microalgae to physiological acclimatization and genetic adaptation to petroleum and diesel oil contamination. Aquat Toxicol 124/125:227–237
Ryu BG, Kim J, Farooq W, Han JI, Yang JW, Kim W (2014a) Algal-bacterial process for the simultaneous detoxification of thiocyanate-containing wastewater and maximized lipid production under photoautotrophic/photoheterotrophic conditions. Bioresour Technol 162:70–79
Ryu BG, Kim J, Yoo G, Lim JT, Kim W, Han JI, Yang JW (2014b) Microalgae mediated simultaneous treatment of toxic thiocyanate and production of biodiesel. Bioresour Technol 158:166–173
Ryu BG, Kim W, Han JI, Yang JW (2017) Feasibility of using a microalgal-bacterial consortium for treatment of toxic coke wastewater with concomitant production of microbial lipids. Bioresour Technol 225:58–66
Safonova E, Kvitko K, Kuschk P, Möder M, Reisser W (2005) Biodegradation of Phenanthrene by the green alga *Scenedesmus obliquus* ES-55. Eng Life Sci 5:234–239
Saiyood S, Vangnai AS, Thiravetyan P, Inthorn D (2010) Bisphenol A removal by the Dracaena plant and the role of plant-associating bacteria. J Hazard Mater 178:777–785
Sasso S, Pohnert G, Lohr M, Mittag M, Hertweck C (2012) Microalgae in the postgenomic era: a blooming reservoir for new natural products. FEMS Microbiol Rev 36:761–785
Saththasivam J, Loganathan K, Sarp S (2016) An overview of oil–water separation using gas flotation systems. Chemosphere 144:671–680
Schoeny R, Cody T, Warshawsky D, Radike M (1988) Metabolism of mutagenic polycyclic aromatic hydrocarbons by photosynthetic algal species. Mutat Res 197:289–302
Semple KT, Cain RB (1996) Biodegradation of phenols by the alga *Ochromonas danica*. Appl Environ Microbiol 62:1265–1273
Semple KT, Cain RB, Schmidt S (1999) Biodegradation of aromatic compounds by microalgae. FEMS Microbiol Lett 170:291–300
Seo JS, Keum YS, Li QX (2009) Bacterial degradation of aromatic compounds. Int J Environ Res Public Health 6:278–309
Silva IS, Grossman M, Durrant LR (2009) Degradation of polycyclic aromatic hydrocarbons (2–7 rings) under microaerobic and very-low-oxygen conditions by soil fungi. Int Biodeterior Biodegrad 63:224–229
Sipaúba-Tavares LH, Scardoeli-Truzzi B, Fenerick DC, Tedesque MG (2020) Comparison of photoautotrophic and mixotrophic cultivation of microalgae *Messastrum gracile* (Chlorophyceae) in alternative culture media. Braz J Biol. https://doi.org/10.1590/1519-6984.226548
Sivasubramanian V, Muthukumaran M (2012) Large scale phycoremediation of oil drilling effluent. J Algal Biomass Utiliz 3:5–17
Su Y, Mennerich A, Urban B (2012) Synergistic cooperation between wastewater born algae and activated sludge for wastewater treatment: influence of algae and sludge inoculation ratios. Bioresour Technol 105:67–73
Subashchandrabose SR, Ramakrishnan B, Megharaj M, Venkateswarlu K, Naidu R (2013) Mixotrophic cyanobacteria and microalgae as distinctive biological agents for organic pollutant degradation. Environ Int 51:59–72
Subashchandrabose SR, Logeshwaran P, Venkateswarlu K, Naidu R, Megharaj M (2017) Pyrene degradation by *Chlorella* sp. MM3 in liquid medium and soil slurry: possible role of dihydrolipoamide acetyltransferase in pyrene biodegradation. Algal Res 23:223–232

Sun Q, Li Y, Chou P-H, Peng P-Y, Yu CP (2012) Transformation of bisphenol-A and alkylphenols by ammonia-oxidizing bacteria through nitration. Environ Sci Technol 46:4442–4448

SureshKumar P, Thomas J, Poornima V (2018) Structural insights on bioremediation of polycyclic aromatic hydrocarbons using microalgae: a modelling-based computational study. Environ Monit Assess 190(92). https://doi.org/10.1007/s10661-017-6459-4

Surkatti R, Al-Zuha S (2018) Microalgae cultivation for phenolic compounds removal. Environ Sci Pollut Res 25:33936–33956

Sutherland DL, Ralph PJ (2019) Microalgal bioremediation of emerging contaminants – opportunities and challenges. Water Res 164:114921

Tabernero A, Martín del Valle EM, Galán MA (2012) Evaluating the industrial potential of biodiesel from a microalgae heterotrophic culture: scale-up and economics. Biochem Eng J 63:104–115

Takáčová A, Smolinská M, Ryba J, Mackuľak T, Jokrllová J, Hronec P, Čík G (2014) Biodegradation of benzo[a]pyrene through the use of algae. Cent Eur J Chem 12:1133–1143

Tam NFY, Chan MN, Wong YS (2010) Removal and biodegradation of polycyclic aromatic hydrocarbons by immobilized microalgal beads. Waste Manage Environ 140:391–402

Tang X, He LY, Liu T, Yang DM, Dang Z, Lu GN (2010a) The axenic culture of *Scenedesmus obliquus* and its application in biodegradation of crude oil. Chin J Environ Eng 4:1079–1083

Tang X, He LY, Tao XQ, Dang Z, Guo CL, Lu GN, Yi XY (2010b) Construction of an artificial microalgal-bacterial consortium that efficiently degrades crude oil. J Hazard Mater 181:1158–1162

Thakurta SG, Aakula M, Chakrabarty J, Dutta S (2018) Bioremediation of phenol from synthetic and real wastewater using *Leptolyngbya* sp.: a comparison and assessment of lipid production. 3 Biotech 8(206). https://doi.org/10.1007/s13205-018-1229-8

Thies F, Backhaus T, Bossmann B, Grimme LH (1996) Xenobiotic biotransformation in unicellular green algae. Involvement of cytochrome P450 in the activation and selectivity of the pyridazinone pro-herbicide metflurazon. Plant Physiol 112:361–370

Tian M, Du D, Zhou W, Zeng X, Cheng G (2017) Phenol degradation and genotypic analysis of dioxygenase genes in bacteria isolated from sediments. Braz J Microbiol 48:305–313

Todd SJ, Cain RB, Schmidt S (2002) Biotransformation of naphthalene and diaryl ethers by green microalgae. Biodegradation 13:229–238

Tsuji N, Hirooka T, Nagase H, Hirata K, Miyamoto K (2003) Photosynthesis-dependent removal of 2,4-dichlorophenol by *Chlorella fusca* var. vacuolata. Biotechnol Lett 25:241–244

Uzoh CV, Ifeanyi VO, Okwuwe CI, Oranusi SU, Braide W, Iheukwumere IH, Anyanwuocha CE, Ntamzor BG (2015) Effect of light on the biodegradation of crude oil by the algae *Closterium* species. J Nat Sci Res 5:112–118

Vassilev SV, Vassileva CG (2016) Composition, properties and challenges of algae biomass for biofuel application: an overview. Fuel 181:1–33

Walker JD, Colwell RR, Vaituzis Z, Meyer SA (1975a) A petroleum-degrading achlorophyllous alga, *Prototheca zopfii*. Nature 254:423–424

Walker JD, Colwell RR, Petrakis L (1975b) Degradation of petroleum by an alga, *Prototheca zopfii*. Appl Microbiol 30:79–81

Wang X, Yang K, Tao S, Xing B (2007) Sorption of aromatic organic contaminants by biopolymers: effects of pH, copper(II) complexation, and cellulose coating. Environ Sci Technol 41:185–191

Wang J, Yang H, Wang F (2014) Mixotrophic cultivation of microalgae for biodiesel production: status and prospects. Appl Biochem Biotechnol 172:3307–3329

Wang L, Xue C, Wang L, Zhao Q, Wei W, Sun Y (2016) Strain improvement of *Chlorella* sp. for phenol biodegradation by adaptive laboratory evolution. Bioresour Technol 205:264–268

Wang YJ, Meng FP, Li H, Zhao SL, Liu QQ, Lin YF, Wang GS, Wu JY (2019) Biodegradation of phenol by *Isochrysis galbana* screened from eight species of marine microalgae: growth kinetic models, enzyme analysis and biodegradation pathway. J Appl Phycol 31:445–455

Warshawsky D, Radike M, Jayasimhulu K, Cody T (1988) Metabolism of benzo(a)pyrene by a dioxygenase enzyme system of the freshwater green alga *Selenastrum capricornutum*. Biochem Biophys Res Commun 152:540–544

Warshawsky D, Keenan TH, Reilman R, Cody TE, Radike MJ (1990) Conjugation of benzo[a] pyrene metabolites by freshwater green alga *Selenastrum capricornutum*. Chem Biol Interact 74:93–105

Wijffels RH, Barbosa MJ, Eppink MHM (2010) Microalgae for the production of bulk chemicals and biofuels. Biofuels Bioprod Biorefning Innov Sustain Econ 4:287–295

Xiong JQ, Kurade MB, Jeon BH (2017) Biodegradation of levofloxacin by an acclimated freshwater microalga, *Chlorella vulgaris*. Chem Eng J 313:1251–1257

Yakimov MM, Timmis KN, Golyshin PN (2007) Obligate oil-degrading marine bacteria. Curr Opin Biotechnol 18:257–266

Zeppm RG, Schlotzhauer PF (1979) Photoreactivity of selected aromatic hydrocarbons in water. In: Joncs PW, Leber P (eds) Polynuclear aromatic hydrocarbons. Ann Arbor Science Publishing, Ann Arbor, pp 141–158

Zhan J, Rong J, Wang Q (2017) Mixotrophic cultivation, a preferable microalgae cultivation mode for biomass/bioenergy production, and bioremediation, advances and prospect. Int J Hydrog Energy 42(12):8505–8517

Zhang F, Man YB, Mo WY, Won MH (2019) Application of *Spirulina* in aquaculture: a review on wastewater treatment and fish growth. Rev Aquacult 12:582. https://doi.org/10.1111/raq.12341

Zhao X, Zhou Y, Huang S, Qiu D, Schideman L, Chai X, Zhao Y (2014) Characterization of microalgae-bacteria consortium cultured in landfill leachate for carbon fixation and lipid production. Bioresour Technol 156:322–328

Zheng T (2017) Treatment of oilfield produced water with electrocoagulation: improving the process performance by using pulse current. J Water Reuse Desalin 7:378–386

Zheng Y, Chi Z, Lucker B, Chen S (2012) Two-stage heterotrophic and phototrophic culture strategy for algal biomass and lipid production. Bioresour Technol 103:484–488

Zhou GJ, Peng FQ, Yang B, Ying GG (2013a) Cellular responses and bioremoval of nonylphenol and octylphenol in the freshwater green microalga *Scenedesmus obliquus*. Ecotoxicol Environ Saf 87:10–16

Zhou Y, Schideman L, Yu G, Zhang Y (2013b) A synergistic combination of algal wastewater treatment and hydrothermal biofuel production maximized by nutrient and carbon recycling. Energy Environ Sci 6:3765–3779

Fundamentals of Biosensor Application in Environmental Pollutant Monitoring

Vinay Patel, Rishi Pramod, Namita Khanna, Prajakta Pawar, Abhilasha Singh Mathuriya, and Soumya Pandit

Abstract

Population explosion and rapid industrialization have led to increase in concentration of pollutants posing threat to the environment. Therefore, it is essential to detect and reduce the level of pollutants, to conserve our natural habitat and protect the environment from further degradation. Biosensors offer a solution as analytical devices that can be customized to detect a variety of pollutants. It comprises of three essential parts—bioreceptor, transducer, and data transfer and processing. The biosensing devices have a biological component: the bioreceptor, which is biological in origin and may comprise of whole cells, cell-based receptors, enzymes, antibodies, or nucleic acids. These biological components provide advantages like detection of pollutant in low amounts, high specificity, and precision. This chapter focuses on biosensors, their types, and principles as well as recent advancements in the field.

Keywords

Biosensors · Pollutant · PAH · Endocrine disruptive chemicals · Electrochemical sensors · Receptors

V. Patel
School of Biomedical Engineering, McMaster University, Hamilton, ON, Canada

R. Pramod · P. Pawar
Amity Institute of Biotechnology, Amity University, Mumbai, India

N. Khanna
Department of Biotechnology, Birla Institute of Technology and Science, Dubai, United Arab Emirates

A. S. Mathuriya · S. Pandit (✉)
Department of Life Sciences, School of Basic Sciences and Research, Sharda University, Greater Noida, India

M. P. Shah (ed.), *Removal of Emerging Contaminants Through Microbial Processes*,
https://doi.org/10.1007/978-981-15-5901-3_15

1 Introduction

Human activities have led to rapid destruction of the environment in the past five decades. The major activities resulting in the environment degradation includes discharge of industrial waste in streams, release of harmful chemicals into soil, and emission of polluting gases from industries and vehicles. Environmental pollution is defined as "the contamination of the physical and biological components of the earth/atmosphere system to such as extent that normal environmental processes are adversely affected". When the environment is unable to process or neutralize harmful by-products of human activities like poisonous gas or fluid emissions, it leads to the structural and functional damage to the ecosystem. In order to control the pollution, there is a need to first understand its type and the corresponding pollutants. Environmental pollution can be categorized into air, water, and soil pollution.

Air pollution includes introduction of substances like noxious gases, mainly oxides of carbon, sulphur, and nitrogen, into the atmosphere resulting in detrimental effects on the atmosphere. The primary causes of air pollution are burning of fossil fuels, volcanic eruptions, forest fires, and industrial and vehicular emissions. These pollutants can lead to health problems like asthma, bronchitis, emphysema, and possibly lung and skin cancer (Afsar et al. 2019).

Water pollution includes the contamination of water bodies such as lakes, rivers, streams, aquifers, oceans, and groundwater due to the presence of foreign materials like chemicals and biological contaminants. Water pollution can be further classified into point and non-point sources. When the source of pollution is discrete and identifiable, it is called a point source while the non-point sources are cumulative results of multiple types of contaminants. Groundwater pollution occurs through infiltration of pollutants which adversely affects the groundwater reserves. Few causes of water pollution are industrial wastes, agricultural run-off, oil spills, leaking landfills, and household wastewater. These pollutants can cause waterborne illnesses like diarrhoea, amoebiasis dysentery, typhoid, cholera, jaundice, and malaria (Mateo-Sagasta et al. 2017). There is an expanding need to develop regulations and policies for discharge and reuse of wastewaters. Many countries have included wastewater reuse as an important dimension of their water resource planning. In China, sewage use in agriculture has rapidly developed since 1958, and now over 1.33 million hectares are irrigated with sewage effluent. It is generally accepted that wastewater use in agriculture is justified on agronomic and economic grounds; however, care must be taken to minimize the adverse effects on health and environmental impacts (Lu et al. 2008).

Soil pollution includes contamination or degradation of the soil making it unsuitable for human use. It is mainly caused due to excessive use of chemicals and fertilizers in agriculture, untreated industrial and household waste disposal. Leaching of these toxic chemicals into the groundwater can further contribute to the water pollution (Steffan et al. 2018). Contaminated soil can impact human health causing sickness like food poisoning, reduced plant growth due to decreased soil fertility, and change in soil structure. These factors can lead to an unbalance in the ecological system.

1.1 Environmental Pollutants

Any pollution-causing agents are called pollutants, and these agents can be divided into two groups: conventional and nonconventional pollutants. Conventional pollutants include biodegradable elements, molecules, organic, or inorganic polymers that naturally occur in nature. However, due to human intervention, their rate of release is several folds higher than the rate of biodegradation leading to their accumulation in nature. Organic nutrients, heavy metals, inorganic carbon, and oil are some common examples of conventional pollutants usually found in domestic, commercial, or industrial wastes, while nonconventional pollutants include xenobiotic compounds, which are man-made and recalcitrant in nature. Industrialization has led to generation of non-native chemical compounds that have accumulated in nature due to their low biodegradability. These compounds include emerging pollutants like endocrine-disrupting compounds, pesticides, flame retardants, etc. (Bhushan et al. 2020).

1.2 Emerging Pollutants

Over the previous decade, emerging contaminants were detected in drinking water supplies, and this is an event that demands worldwide attention. Although these contaminants are present in low concentrations in the environment, even the low concentrations can be harmful to human and animal health. Further, over a period, these contaminants can accumulate to have severe impact on the living beings. Emerging contaminants are comprised of personal care products, pharmaceuticals, endocrine-disrupting compounds, industrial compounds, food additives, herbicides, and pesticides (Bhushan et al. 2020; Deblonde et al. 2011; Hernandez-Vargas et al. 2018). The majority of these contaminants enters the environment after getting consumed and flushed out as metabolites or chemically unaltered compounds. Current wastewater treatment plants are not fully equipped to deal with these compounds. Therefore, these contaminants can accumulate in the environment over a period.

1.2.1 Pharmaceuticals and Personal Care Products

Pharmaceuticals are the most common emerging contaminants detected in groundwater and surface water at a very low concentrations ranging from μg/L to ng/L (Deblonde et al. 2011). These are mixed into the wastewater through industrial and hospital effluents due to incomplete metabolism in the human body. These compounds can have the potential to gradually affect the ecological environment and human health even at low concentrations. Personal care, healthcare products and pharmaceuticals include sunscreen, transformation products, fragrances and their metabolites, cosmetics, food supplements, analgesics, antibiotics, diuretics, beta blockers, lipid regulators, non-steroidal anti-inflammatory drugs (NSAIDs), stimulant drugs, antiseptics, and antimicrobials. The bio accumulation of these contaminants in the environment could increase the incidence of diseases like breast and testosterone cancer, decrease fertility, cause persistent antibiotic resistance, and influence the hormone control in our body causing reproductive impairments (Deblonde et al. 2011).

1.2.2 Endocrine-Disrupting Compounds

Endocrine-disrupting compounds (EDCs) are the type of emerging contaminants which are detected in aquatic environments like wastewater, surface water, groundwater, and landfill leachates. EDCs are substances or mixtures that affect the function of the endocrine systems causing adverse health effects such as blocking a hormone, causing its overproduction or underproduction. With the higher concentration of EDCs, there is an increase in the incidence of diseases such as endometriosis and different types of cancers like testicular, prostate, and breast cancers. Common EDCs are 17β-ethinylestradiol(EE2), estrone(E1), and 17β-estradiol (Bhushan et al. 2020; Yang et al. 2015).

1.2.3 Pesticides

Pesticides are chemical and biological substances that can selectively control infestation of insects, weeds, and other pests in gardens or farms. At present, farmers rely intensively on the use of pesticides. Common pesticides used are 2,4-dichlorophenoxyacetic acid and omethoate to reduce the loss of agricultural products (Justino et al. 2017). Though these biochemicals have advantages in agriculture, however, an unregulated usage of these pesticides can lead to a significant environmental problem. Some of the environmental problems are immune-depressive effects in fishes and mammals. The pesticides can also modify haemopoietic tissue of anterior kidney (Bhushan et al. 2020).

1.2.4 Flame Retardants

Automotive industry, plastic polymers, coating textiles, and electronic equipment manufacturers are the largest consumers of brominated flame retardants (BFRs). BFRs prevent fire-related damage as their primary function along with protecting the materials from ignition. However, their low biodegradability leading to bioaccumulation and high toxicity raises environmental concerns. Flame retardants consist of hexabromocyclododecane (HBCD), polybrominated diphenyl ethers (PBDEs), tetrabromobisphenol A (TBBPA), decabrominated diphenyl ethers (DBDPE), phosphate flame retardants, and tris(chloroisopropyl) phosphate (TCIPP) (Segev et al. 2009). Previous studies showed that flame retardants have a significant effect on human health such as change in hormone levels and poor semen quality in males and exhibition of in vitro estrogenic and anti-androgenic effect on human osteosarcoma cell line and can be carcinogenic.

2 Biosensors

In order to monitor or screen the pollutants, there is a need for development of sensitive, rapid, and selective technologies. One such approach is to use biological components such as enzymes, whole cells, cellular organelles, cell-based receptors, antibodies, or nucleic acids for detection and quantification of certain specific or non-specific analytes (pollutants) (Justino et al. 2015). The sensing devices based on biological components are known as biosensors and can provide a promising platform for specific, reliable, and sensitive method to monitor environmental pollutants.

According to the International Union of Pure and Applied Chemistry (IUPAC), a biosensor is defined as "an independently integrated receptor transducer device, which is capable of providing selective quantitative or semi-quantitative analytical information using a biological recognition element". An ideal biosensor should have the following features:

1. The biorecognition element used should be highly specific for the purpose of analysis and stable under normal conditions.
2. The reaction should be independent of physical parameters as stirring, pH, and temperature.
3. The sensor response should be accurate and reproducible.
4. If the biosensor is used in invasive monitoring in clinical situations, the probe must be tiny and biocompatible.

2.1 Biosensors Classification and Components

Any biosensor has three essential parts: a bioreceptor, a transducer, and a data transfer and processing unit (Fig. 1). Bioreceptor is the primary tool for the sensing technology; this biological element exploits the biochemical mechanism for specific recognition and binding of analyte of interest resulting in a signal change. This signal change is converted to a readable format via a transducer which is processed by a signal processing unit to generate results for the user. However, biosensors have few drawbacks such as limited range of detection and interference from other compounds present in the sample. The choice of the bioreceptor molecule, immobilization technique (if needed), and transducer has profound impact on the selectivity, sensitivity, precision, and accuracy of the biosensor. The key characteristics analysed during biosensor development are accuracy, precision, speed, sensitivity, specificity, range of measurement, reliability, calibration and long-term stability, robustness, size, portability, cost of analysis, and acceptability by users (Badihi-Mossberg et al. 2007).

2.1.1 Bioreceptor

Bioreceptor or biological recognition element imparts the selectivity and specificity that allows the biosensor to respond to a specific target or group of analytes. The

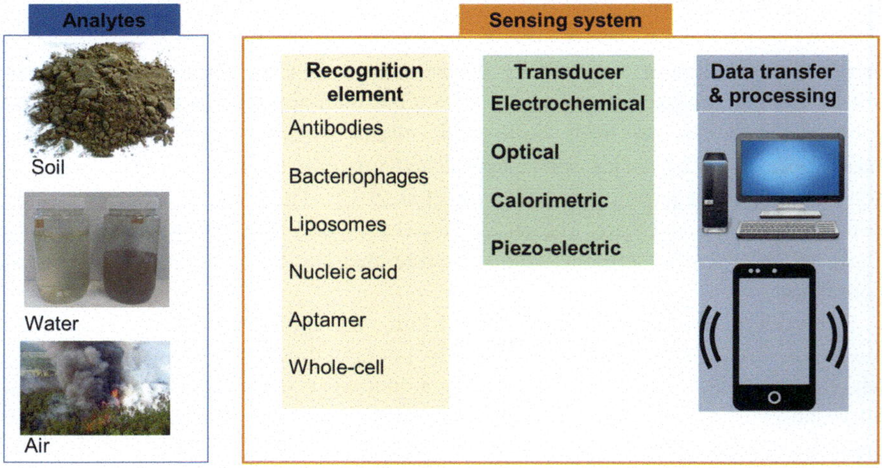

Fig. 1 Different analyte samples like soil, water, and air along with the sensing systems. Sensing systems have three components: recognition element, transducer, and data transfer and processing unit

analyte of interest (e.g. antibodies and aptamers are more suitable for the detection of bacteria or pathogens; however, in enzymes are more suitable for catalytic reactions) determines the selection criteria for the biological recognition element (Justino et al. 2015). In general, these recognition elements are immobilized using methods such as adsorption, covalent binding, entrapment, and membrane confinement. In general, covalent immobilization is the preferred method due to its stability and irreversibility which prevents loss of the biological elements from the support surface.

2.1.2 Transducer

Transducer plays a key role to convert the physical or chemical changes of biosensors into a readable signals such as (a) colour change due to the chemical reaction (colorimetric biosensors), (b) potential change due to charge redistribution (potentiometric biosensors), (c) electron movement due to the redox reaction (amperometric biosensors), (d) change in absorbance or transmittance due to detection products (optical biosensors), or (e) change in charges due to change in force (piezo-electric biosensors). For this chapter, we have limited the scope to electrochemical and optical biosensors.

2.1.3 Data Transfer and Processing Unit

This unit can amplify the signal to make it easier to analyse the generated signal. Further, the electrical interference generated within the electronic components of the transducer, the electrical signal from the transducer, is often low and superimposed with a high-frequency signal component which can be termed as noise. The signal processing normally involves subtracting a "reference" baseline signal, derived from a similar transducer without any biorecognition element. The sample signal is normalized with respect to the reference baseline resulting in a clean signal without any noise. The comparatively slow nature of the biosensor signal eases the problem of electrical noise filtration. After analysis and noise filtration, the signal is displayed or recorded in the sensing device (Kumar et al. 2013).

3 Biosensors Based on Different Bioreceptors

Based on bioreceptors, biosensors can be classified as enzyme, antibodies, bacteriophages, aptamers, whole cells, and nucleic acid-based biosensors (Justino et al. 2015). In this section, we will discuss briefly about bioreceptors and few examples of their application in sensing various environmental pollutants.

3.1 Enzyme Biosensors

Enzyme biosensors selectively covert the analyte to a chemical species which can be detected using various transducers like optical, electrochemical, and others (Justino et al. 2015). Enzymatic biosensors can use one enzyme or a multiple set of enzymes to cover the analyte of interest in detectable signal. Few commonly used enzymes for environmental pollutant detection are acetylcholinesterase used to detect paraoxon or parathion, horseradish peroxidase with antibodies used to detect pathogens like *Shigella*, tyrosinase used for catechol detection, and others (Justino et al. 2017, 2015). Acetylcholinesterase is used to detect multiple organophosphates, which are one of the most used pesticides. These compounds are widely used in modern agriculture to control various weeds, disease-spreading vector, and pests. Organophosphates inhibit the enzymatic activity of acetylcholinesterase which can be detected using transducers like electrochemical in the form of change in current and potential signals. Advantages of enzymatic sensors are the following: easy manufacturing and handling, have high specificity, and require simple apparatus and procedures for analyte detection. Few limitations are less tolerant to change in environmental factors like pH and temperature; costly to purify the enzymes; poor stability; and the need for optimum temperature for long-term storage. Further, enzymatic sensors are used to fabricate sensing platforms with multiple enzymes which can be used to detect various analytes present in the sample. A sensing platform was fabricated by using screen-printed electrodes to detect six organophosphates present in the water samples. The full assay can be performed in 6 min using the sensing platform to measure the organophosphates in the concentration range 10 μM to 1 nM (Crew et al. 2011).

3.2 Antibody Biosensors

Antibody biosensors use the antibody-antigen interaction to detect the analyte of interest. These sensors can be categorized into four types depending on the transducers used: thermometric, electrochemical, microgravimetric, and optical (Bhushan et al. 2020). Biosensors can use either polyclonal (secreted by B-cells of different lineages) or monoclonal (secreted by B-cells of single lineage) antibodies. The advantages of antibody sensors are high specificity and sensitivity. But these sensors have few limitations like polyclonal antibodies which can result in non-specific binding, while the more specific monoclonal antibodies are costly. These biosensors are used to detect various environmental pollutants, such as polycyclic aromatic hydrocarbons (PAHs), food pathogens, herbicides, etc. (Justino et al. 2015).

Polycyclic aromatic hydrocarbon (PAH) is one of the most widespread organic pollutants made of fused aromatic rings made of carbon and hydrogen. These compounds originate from oil and coal industrial processes during the incomplete combustion of carbon-containing fuels. Some PAHs which are present in non-processed foodstuff associated with environmental pollution can cause cancer and other health-related issues. Multiple sensing platforms based on antibodies are reported for PAH detection (Behera et al. 2018). These sensors are based on biosensing platforms like enzyme-linked immunosorbent assay (ELISA), electrochemical, and others. An electrochemical based on antibodies was proposed by Lin et al. to detect Benzo[a]pyrene (Lin et al. 2012). The sensor was fabricated using a ferrous-ferric oxide nanoparticle and Nafion film functionalized with multi-enzyme antibody complex. The sensor had a broad range of detection (8 pM–2 nM) with limit of detection of 4 pM which was better than the previous reported sensors.

3.2.1 Molecularly Imprinted Polymer (MIP)

Synthetic receptors that mimic the recognition properties of an antibody are called molecularly imprinted polymers (MIPs). MIPs are generally formed using self-assembly of functional monomers around a template placed around and the monomers are polymerized to form a polymer with cavities of specific shape, size, and functionality (Van Dorst et al. 2010). The imprinted polymer can recognize the analyte with high specificity and affinity. In the past decade, MIP are explored by numerous researchers to fabricate biosensors due their low-cost production and high stability as compared to antibodies and more tolerance to chemical, thermal, and mechanical changes. Therefore, these systems can be designed to use at high temperature, wide pH ranges, and multiple solvents like water, organic solvents. However, there are few challenges like the use of traditional bulk polymerization process which is time-consuming, full removal of the template particles from MIP, and application of MIP molecules for detection of high molecular weight analytes in complex matrices. MIP-based biosensors are used to detect numerous pollutants like pharmaceuticals, cosmetics, and phthalate esters.

Recent research of MIPs is moving towards nanosized MIPS due to their high surface area to volume ratio, higher affinity, easy sensor integration, and faster binding kinetics as compared to conventional MIPs (Justino et al. 2015). A voltammetric biosensor based on nanosized MIP was reported by Gholivand et al. (2014). The nanosized MIP was synthesized using polyparaben (a common cosmetics ingredient) as a template. The sensor reported a linear range of 1–100 nM with 0.32 nM as the lower limit of detection.

3.3 Bacteriophage Biosensors

Bacteriophages can bind naturally to specific bacterial cells which make them a suitable candidate to identify bacteria or pathogens in environmental samples. Phages identify the bacteria depending on the markers such as surface proteins, lipoproteins, etc. present on bacterial cells. Wild phages can be used as such to fabricate sensors (Shabani et al. 2008). However researchers have also developed engineered phage for specific sensing applications including expressing a marker protein (Gervais et al. 2007), attaching a fluorescence (Edgar et al. 2006) or luminescence (Blasco et al. 1998) tag on the phage, or inserting gene into viral genome. Along with bacterial cell infection, phages are also used to display certain peptides and proteins on the surface known as phage display. Phages offer a versatile platform which is used to detect various environmental pollutants such as trinitrobenzene, trinitrotoluene, toxins like aflatoxin B1, and common food and water pathogens like *E. coli*, *Salmonella*, and *Listeria*. For example, Bhardwaj et al.(2016) demonstrated a screen-printed graphene electrode which was covalently immobilized with lytic phages using 1-(3-dimethylaminopropyl)ethylcarbodiimide (EDC) hydrochloride. The measuring range of the sensor was from 2 to 2×10^3CFU/mL with a detection limit of 2 CFU/mL and a rapid response time of 2 min. The sensor was tested in river water and juice. The sensor was stable for 3 months in storage. Phage-based biosensors have mainly used two transducers: electrochemical and optical (Van Dorst et al. 2010). These biosensors are low-cost and more robust and tolerant to environmental parameter changes like solvents, pH, and temperature as compared to immunosensors. Phages are reported to be stable in a wide pH range of 3–11, and filamentous phages retain their binding affinity for 6 days at 63 °C. These biosensors are highly specific and sensitive to the bacterial species or the surface proteins. However, more research needs to be done for further optimization of the immobilization protocol of phages on sensor surface and to improve the limit of detection of these biosensors.

3.4 Nucleic Acid Biosensors

Nucleic acid-based probes work on the complementarity. A nucleic acid probe can be designed using deoxyribonucleic acid (DNA), ribonucleic acid (RNA), and peptide nucleic acid (PNA). DNA and RNA can be amplified using PCR and reverse transcription PCR (RT-PCR). PNAs are synthetic oligonucleotides which have high binding affinity to the complimentary DNA or RNA sequences. PNAs can be tailored to improve the binding affinity and enhance stability against various conditions including physical, biological, and chemical degradation (Singh et al. 2013). Nucleic acid-based probes have advantages such as cost-effective and rapid detection, stable probes, and a versatile process. There are multiple commercial kits based on nucleic acid-based probes for pathogen detection like BAX system for *E. coli* detection and Gen-Probe for *Mycobacterium avium complex* (Singh et al. 2013). But the probes have disadvantages including false negative and positive results due to nucleic acid degradation and contamination and inability to distinguish between live and dead cells and detect the toxins produced by the bacteria (Singh et al. 2013; Richter et al. 2018). These sensors are used to detect target analytes such as chlorinated benzenes like hexachlorobenzene; heavy metals like mercury, lead, and cadmium; and toxins like aflatoxin M1, cholera, and botulinum.

The most common transducers used for nucleic acid-based biosensors are optical and electrochemical (Van Dorst et al. 2010). An electrochemical DNA biosensor was used to detect hexachlorobenzene in real water samples (Wu et al. 2011). The sensor demonstrated a measuring range of 100 pM–100 nM, with 30 pM as lower limit of detection. The sensor can be used as a rapid screening tool for real water samples.

3.4.1 Aptamer-Based Biosensor

Aptamers are oligonucleotide (20–60 nucleotides) molecules designed to have a high selectivity, specificity, and affinity towards a target analyte ranging from small molecules to whole cells. Aptamer's long-term stability and their inexpensive and rapid synthesis and ability to be altered with labels with little effect on their binding site performance, stability, or binding characteristics are some of the advantages of aptamers as biological recognition elements as compared to the natural recognition elements like DNA antibodies. However, aptamers have few disadvantages like cross-reactivity with similar compounds; aptamer generation is a time-taking and laborious process (Hayat and Marty 2014; Van Dorst et al. 2010; Justino et al. 2015). These biosensors are used for detection of various compounds like antibiotics, toxins, pesticides, and heavy metals. A sensor for mercury detection was reported by Zhang and Guo (2012). The sensor obtained a linear measuring range between 0.1 nM and 10 nm of mercury (Hg^{2+}) concentration with 20 pM as the lower range of detection. The sensor was tested in spiked tap and lake water.

3.5 Whole-Cell Biosensors

Whole-cell biosensors act as the biorecognition agents that monitor general or specific toxicity caused by one or more pollutants. Pollutants can be monitored using these biosensors by quantifying the change in light, fluorescence, colour, electric potential, or current (Justino et al. 2015). Whole-cell biosensors have advantages like the ability to detect multiple chemicals due to the presence of multiple enzymes present in the cells, easy to manipulate, and conducive to use on various substrates. The versatility and flexibility of these biosensors are best demonstrated by Microtox system (Bereza-Malcolm et al. 2015). The system is a standard toxicity test which is easy to use, sensitive, rapid, and reproducible. The system uses an engineered bioluminescent *Vibrio fischeri* as the test organism. The bacterium is exposed to different concentrations of the pollutant, and the reduction in luminescence is correlated to the pollutant concentration. Whole-cell biosensors are used in pharmacology, medicine, cell biology, toxicology, and environmental monitoring. Few examples of whole-cell biosensors for detection of environmental pollutants are heavy metals, alkanes, and aromatic compounds in water. Unlike common analytical methods, whole cells can react to the presence of specific metals which are not present in biological systems. The heavy metals inhibit the biological enzymes that offer a simple and a sensitive way to detect their presence. Cerminati et al. reported a whole-cell biosensor to detect various toxic metals including mercury, lead, cadmium, and/or gold ions in water samples (Cerminati et al. 2015). Sensing system is an example of a rapid and highly sensitive sensor to detect multiple heavy metals.

4 Biosensors Based on Transducers

Based on the transducers used, biosensor can be classified in different groups like electrochemical-, optical-, or mass-based sensors (Justino et al. 2017). For this chapter, we will focus mostly on electrochemical- and optical-based biosensors.

4.1 Electrochemical Biosensors

These biosensors are based on a single or multiple chemical reactions to generate a readable signal in the form of either a current or potential change. Electrochemical sensors offer rapid detection, high sensitivity and selectivity, low-cost, and easy to miniaturization and integration biomolecules and synthetic molecules as recognition elements (Badihi-Mossberg et al. 2007; Hernandez-Vargas et al. 2018; Ronkainen et al. 2010). Further electrochemical reactions mainly occur on electrode surface or electrode-sample interfaces. So, these sensors do not strongly depend on the sample volume and can be used for very small sample volumes. Electrochemical biosensors can be categorized into three groups: potentiometric, amperometric/voltammetric, and impedimetric biosensors.

4.1.1 Potentiometric Biosensors

Potentiometric sensors measure change in potential over time. This change in potential is correlated to the analyte concentration. Permselective layer is designed for the selective transfer of the Ion selective electrodes are extensively researched for over 60 years. These electrodes measure potential changes with time, and the change in potential is correlated to the analyte concentration. The potential change is measured relative to a reference potential point which is provided by the reference electrode. So, ion-selective electrode is a two-electrode system: working electrode (ion-selective) and reference electrode (a constant potential electrode against which the potential variation in ion-selective electrode is measured). Ion-selective electrodes have advantages like small sensor size conducive to on-site measurements, small sample volume, and simple electronics system for measurement (De Marco et al. 2007). However, the sensors have challenges like membrane fouling in long-term monitoring due to chemical and biological components in real samples and electrode drift resulting in poor reproducibility. Few environment pollutants detected using ion-selective electrodes are copper, iron, cadmium, mercury, lead, phosphates, nitrates, sulphates, etc. (Badihi-Mossberg et al. 2007; Ibupoto et al. 2011). Table 1 summarizes few potentiometric sensors used for environment pollutant monitoring.

4.1.2 Amperometric/Voltammetric Biosensors

Voltammetric sensors measure the change in current, when potential is varied over a range, while amperometric sensors measure the current changes over a given time period. Amperometric sensors are commonly used in enzyme- and affinity-based sensors. The sensor response can be measured either under a constant potential bias or no potential bias. A voltammetric sensor for organophosphate detection was proposed using enzyme acetylcholinesterase recently by Bao et al. (2019). The sensor demonstrated a linear measuring range from 3 pM to 46.65 nM with 0.31 pM as limit of detection. The sensor was tested in spiked river and lake samples. A screen-printed carbon electrode coated with polyclonal antibodies was used for detected *Legionella pneumophila* (Martín et al. 2015). The current was measured by applying a potential bias of −0.15 V to measure the generated hydrogen peroxide in the presence of hydroquinone. The sensor demonstrated a limit of detection of 10^4 CFU/mL which can be lowered to 10 CFU/mL using a pre-concentration step. Table 1 summarizes few amperometric and voltammetric biosensors used for environmental monitoring.

4.2 Optical Biosensors

Optical biosensors detect changes due to the interaction between the target of interest and the biological recognition element and transform the signal into a measurable optical signal. The analytical technique has various applications in the field of medical diagnostics, food analysis, environmental applications, drug discovery, and security and defence. It is one of the most commonly used detection methods in the biosensor due to its advantages such as lower detection limit, non-destructive, and their ability to detect a wide variety of analytes or multiple analytes at the same time with fast signal monitoring and analysis. Optical biosensors can be categorized into direct and indirect detection. Direct detection of the analyte is done using techniques such as UV absorption, planar optical waveguide, fibre optics, surface-enhanced Raman spectroscopy (SERS), and surface plasmon resonance (SPR). Indirect detection includes optically labelled probes such as fluorescence and chemiluminescence. These biosensors are used to detect various environment pollutants like heavy metals (e.g. mercury, copper, arsenic), pesticides, nutrients (e.g. nitrates), endocrine-disrupting compounds, and microbial concentrations (Justino et al. 2017).

Table 1 Advantages and disadvantages of different sensing techniques

Analyte	Transducer	Biorecognition element	Limit of detection	Features	Matrix	Reference
Electrochemical sensors						
Penicillin	Potentiometry	Penicillinase	N/A	Working pH = 7.4 Calibration slope: ~121 mV/decade Measuring range: 0.1–10^{-4} M	Buffer solution	Ibupoto et al. (2011)
Paraoxon (organophosphate)	Potentiometry	*E. coli*-expressing organophosphorus hydrolase	2 μM	Stable for 2 months (at 4oC)optimum working pH = 8.5linear range: 0.055–1.8 mM	N/A	(Mulchandani et al. 1998)
Malathion (organophosphate)	Voltammetry	Acetylcholinesterase	0.31 pM	Linear range: 3 pM–46.65 nM Working pH = 7.4	Spiked river, lake water	Bao et al. (2019)
Legionella pneumophila	Amperometry	Polyclonal antibodies	10^4 CFU/mL	LOD after pre-concentration: 10 CFU/mL Potential bias: −0.15 V	Inoculated water	Martín et al. (2015)
Mercury (Hg^{2+})	Amperometry	Nucleic acid	3 fM	Optimum working pH = 7.5 Linear range: 10 fM–1 μM	Spiked lake water	Shi et al. (2017)
Bisphenol A	Amperometry	MIP and Au nanoparticles	13.8 μM	Monomer: 2-aminothiophenol Measuring range: 0.6 M–8 μM	Bottled water samples	Huang et al. (2011)
Lead (Pb^{2+})	Voltammetry	Algae (*Phormidium*)	0.025 μM	Optimum working pH = 8 Linear range: 0.05–20 μM	Aqueous media	Yüce et al. (2010)
Optical sensor						
Benzene	Colorimetry	Engineered *Escherichia coli*	1.35 mM	Measuring range: 0.4–7 mM	Oil refinery oil samples	Di Gennaro et al. (2011)
Escherichia coli	Surface plasma resonance	DNA aptamer	3 × 10^4 CFU/mL	Measuring range: 0.4–7 mM	PBS buffer	Wang et al. (2011)
Ochratoxin A	Colorimetry	DNA aptamer with Au nanoparticles	20 nM	Measuring range: 20–625 nM	Standard solution	Yang et al. (2011)

(continued)

Table 1 (continued)

Analyte	Transducer	Biorecognition element	Limit of detection	Features	Matrix	Reference
Copper	Colorimetry	Engineered *Saccharomyces*	1 μM	Measuring range: 1–100 μM Storage stability: 50 days	Well water	Vopálenská et al. (2015)
Arsenic (III)	Fluorescence	Engineered *Escherichia coli*	10 μg/L	Measuring range: 0–100 μg/L	Tap water	Truffer et al. (2014)
Nitrate	Fluorescence	Engineered *Escherichia coli*	0.05 mM	Measuring range: 0.05–5 mM	LB media	Taylor et al. (2004)
Methyl parathion	Colorimetry	Silica nanoparticles and *Sphingomonas*	0.1 ppm	Measuring range: 0.1–1 ppm (pH = 8) Storage stability: 180 days at 4 °C	Spiked water samples	Mishra et al. (2017)
Toluene	Phosphorescence	Toluene *ortho*-monooxygenase	3 μM	Measuring range: 3–100 μM	Wastewater	Zhong et al. (2011)

An engineered *E. coli* was demonstrated for nitrate detection in environment water without any interference (Taylor et al. 2004). The *E. coli* transformed the gene of a membrane-bound nitrate reductase enzyme coupled with a green fluorescence protein gene. The whole-cell biosensor was insensitive to interferents like chloride, phosphate, and nitrite. An enzymatic biosensor for detection of toluene in wastewater was reported by Zhong et al. (2011). The sensor was based on toluene *ortho*-monooxygenase enzyme with a phosphorescent dye as transducer. The sensor demonstrated a linear range of toluene measurement up to 100 μM, with a limit of detection 3 μM. The response time of the sensor was 1 h. The sensor was evaluated in spiked water samples from lake and reservoir water. A few optical biosensors are summarized in Table 1.

5 Practical Challenges with Environmental Samples

Sample matrix can drastically affect the application of biosensors in real-world applications. Matrix effects could be defined as an induced deviation from theoretically forecasted parameters, caused by the compounds present in the sample other than the analyte of interest. The exact composition of the matrix of environmental samples is usually unknown and may vary between samples. Therefore, quantifying analyte under such complex conditions makes the use of biosensors challenging. To address these challenges, steps such as pre-treatment can help to detect the analyte in this complex matrix. Further, matrix interferences in water samples can be prevented by simple methods such as dilution, buffering, and altering the pH and conductivity. Additionally, in the case of immunosensors, detergents or mimicking proteins are added to reduce matrix interferences. For seawater samples, ionic strength must also be considered while analysing the analyte concentration.

For biosensors to become a practical and reliable method, it must compete with the conventional analytical systems in terms of reliability, selectivity, sensitivity, and robustness. Biosensor measurements hereby need to be verified and validated before being accepted. The validation procedures can vary depending on the technology that is examined and include participation in inter-laboratory trials, the use of certified reference materials (CRM) or comparison with conventional chemical analysis. The credibility of biosensors may include some hurdles since the quality parameters such as stability of the biological element and or the element immobilized in the transducer being biological in origin may easily be compromised. Although several validation procedures are acceptable, there are no universal established guidelines for biosensor validation. Many analytical parameters like working and linear construction ranges, the limit of detection (LOD), limit of quantification (LOQ), selectivity, reliability, steady-state and transient response times, sample throughput, reproductivity, stability, and lifetime are some of the parameters that should be closely monitored during testing and validation of the biosensor. The overall commercial status and general acceptance of the technology will depend on the performance characteristics, sample throughput, associated costs, and acceptance by the regulatory authorities based on independent validation of data generated with the help of internationally recognized procedures like the International Organization for Standardization (ISO).

6 Outlook and Conclusion

Biosensing is a growing field of research that has rapidly transformed over the past decades through new discoveries related to novel micro- and nanofabrication techniques, improved electronics systems, improved biochemical techniques, and increased computational power. In spite of the ongoing research, there are still challenges to fabricate a robust, reliable, reproducible, and repeatable sensing device. In addition, pollutant concentrations in environment are constantly changing both as a result of pollutant input and changes in environmental factors like air or water flow. Establishment of multiple autonomous analytical stations which can extensively monitor the sites of interest in rivers, lakes, wells, or even water treatment has a huge demand. But there is still a dearth of continuous monitoring devices that can provide an easy and rapid on-site measurement. These devices would also be useful to obtain real-time data for accidental spill seasonal fluctuations or major pollution events. These continuous monitoring systems could facilitate the measurement of pollutant concentration in remote places.

Acknowledgement The financial support received from Sharda University, Greater Noida, India, is duly acknowledged.

References

Afsar B, Elsurer Afsar R, Kanbay A, Covic A, Ortiz A, Kanbay M (2019) Air pollution and kidney disease: review of current evidence. Clin Kidney J 12:19–32. https://doi.org/10.1093/ckj/sfy111

Badihi-Mossberg M, Buchner V, Rishpon J (2007) Electrochemical biosensors for pollutants in the environment. Electroanalysis 19:2015–2028. https://doi.org/10.1002/elan.200703946

Bao J, Huang T, Wang Z, Yang H, Geng X, Xu G, Samalo M, Sakinati M, Huo D, Hou C (2019) 3D graphene/copper oxide nano-flowers based acetylcholinesterase biosensor for sensitive detection of organophosphate pesticides. Sensors Actuators B Chem 279:95–101. https://doi.org/10.1016/j.snb.2018.09.118

Behera BK, Das A, Sarkar DJ, Weerathunge P, Parida PK, Das BK, Thavamani P, Ramanathan R, Bansal V (2018) Polycyclic aromatic hydrocarbons (PAHs) in inland aquatic ecosystems: perils and remedies through biosensors and bioremediation. Environ Pollut 241:212–233. https://doi.org/10.1016/j.envpol.2018.05.016

Bereza-Malcolm LT, Mann G, Franks AE (2015) Environmental sensing of heavy metals through whole cell microbial biosensors: a synthetic biology approach. ACS Synth Biol 4:535–546. https://doi.org/10.1021/sb500286r

Bhardwaj N, Bhardwaj SK, Mehta J, Mohanta GC, Deep A (2016) Bacteriophage immobilized graphene electrodes for impedimetric sensing of bacteria (Staphylococcus arlettae). Anal Biochem 505:18–25. https://doi.org/10.1016/j.ab.2016.04.008

Bhushan I, Singh VK, Tripathi DK (2020) Nanomaterials and environmental biotechnology. Springer, Cham

Blasco R, Murphy MJ, Sanders MF, Squirrell DJ (1998) Specific assays for bacteria using phage mediated release of adenylate kinase. J Appl Microbiol 84:661–666. https://doi.org/10.1046/j.1365-2672.1998.00393.x

Cerminati S, Soncini FC, Checa SK (2015) A sensitive whole-cell biosensor for the simultaneous detection of a broad-spectrum of toxic heavy metal ions. Chem Commun 51:5917–5920. https://doi.org/10.1039/c5cc00981b

Crew A, Lonsdale D, Byrd N, Pittson R, Hart JP (2011) A screen-printed, amperometric biosensor array incorporated into a novel automated system for the simultaneous determination of organophosphate pesticides. Biosens Bioelectron 26:2847–2851. https://doi.org/10.1016/j.bios.2010.11.018

De Marco R, Clarke G, Pejcic B (2007) Ion-selective electrode potentiometry in environmental analysis. Electroanalysis 19:1987–2001. https://doi.org/10.1002/elan.200703916

Deblonde T, Cossu-Leguille C, Hartemann P (2011) Emerging pollutants in wastewater: a review of the literature. Int J Hyg Environ Health 214:442–448. https://doi.org/10.1016/j.ijheh.2011.08.002

Di Gennaro P, Bruzzese N, Anderlini D, Aiossa M, Papacchini M, Campanella L, Bestetti G (2011) Development of microbial engineered whole-cell systems for environmental benzene determination. Ecotoxicol Environ Saf 74:542–549. https://doi.org/10.1016/j.ecoenv.2010.08.006

Edgar R, McKinstry M, Hwang J, Oppenheim AB, Fekete RA, Giulian G, Merril C, Nagashima K, Adhya S (2006) High-sensitivity bacterial detection using biotin-tagged phage and quantum-dot nanocomplexes. Proc Natl Acad Sci U S A 103:4841–4845. https://doi.org/10.1073/pnas.0601211103

Gervais L, Gel M, Allain B, Tolba M, Brovko L, Zourob M, Mandeville R, Griffiths M, Evoy S (2007) Immobilization of biotinylated bacteriophages on biosensor surfaces. Sensors Actuators B Chem 125:615–621. https://doi.org/10.1016/j.snb.2007.03.007

Gholivand MB, Shamsipur M, Dehdashtian S, Rajabi HR (2014) Development of a selective and sensitive voltammetric sensor for propylparaben based on a nanosized molecularly imprinted polymer-carbon paste electrode. Mater Sci Eng C 36:102–107. https://doi.org/10.1016/j.msec.2013.11.021

Hayat A, Marty JL (2014) Aptamer based electrochemical sensors for emerging environmental pollutants. Front Chem 2:1–9. https://doi.org/10.3389/fchem.2014.00041

Hernandez-Vargas G, Sosa-Hernández JE, Saldarriaga-Hernandez S, Villalba-Rodríguez AM, Parra-Saldivar R, Iqbal HMN (2018) Electrochemical biosensors: a solution to pollution detection with reference to environmental contaminants. Biosensors 8:1–21. https://doi.org/10.3390/bios8020029

Huang J, Zhang X, Liu S, Lin Q, He X, Xing X, Lian W (2011) Electrochemical sensor for bisphenol A detection based on molecularly imprinted polymers and gold nanoparticles. J Appl Electrochem 41:1323–1328. https://doi.org/10.1007/s10800-011-0350-8

Ibupoto ZH, Ali SMU, Khun K, Chey CO, Nur O, Willander M (2011) ZnO nanorods based enzymatic biosensor for selective determination of penicillin. Biosensors 1:153–163. https://doi.org/10.3390/bios1040153

Justino CIL, Freitas AC, Pereira R, Duarte AC, Rocha Santos TAP (2015) Recent developments in recognition elements for chemical sensors and biosensors. Trends Anal Chem 68:2–17. https://doi.org/10.1016/j.trac.2015.03.006

Justino CIL, Duarte AC, Rocha-Santos TAP (2017) Recent progress in biosensors for environmental monitoring: a review. Sensors (Switzerland) 17. https://doi.org/10.3390/s17122918

Kumar A, Kim H, Hancke GP (2013) Environmental monitoring systems: a review. IEEE Sensors J 13:1329–1339. https://doi.org/10.1109/JSEN.2012.2233469

Lin M, Liu Y, Sun Z, Zhang S, Yang Z, Ni C (2012) Electrochemical immunoassay of benzo[a]pyrene based on dual amplification strategy of electron-accelerated Fe_3O_4/polyaniline platform and multi-enzyme-functionalized carbon sphere label. Anal Chim Acta 722:100–106. https://doi.org/10.1016/j.aca.2012.01.059

Lu W-Q, Xie S-H, Zhou W-S, Zhang S-H, Liu A-L (2008) Water pollution and health impact in China: a mini review. Open Environ J 2:1–5. https://doi.org/10.2174/1874233500802010001

Martín M, Salazar P, Jiménez C, Lecuona M, Ramos MJ, Ode J, Alcoba J, Roche R, Villalonga R, Campuzano S, Pingarrón JM, González-Mora JL (2015) Rapid Legionella pneumophila determination based on a disposable core-shell Fe_3O_4@poly(dopamine) magnetic nanoparticles immunoplatform. Anal Chim Acta 887:51–58. https://doi.org/10.1016/j.aca.2015.05.048

Mateo-Sagasta J, Zadeh SM, Turral H (2017) Water pollution from agriculture: a global review. Food Agric. Organ. United Nation Rome, 2018 Int. Water Manag. Inst. behalf Water L. Ecosyst. Res. Progr. CGIAR Colombo, 2018. p 35. http://www.fao.org/3/a-i7754e.pdf

Mishra A, Kumar J, Melo JS (2017) An optical microplate biosensor for the detection of methyl parathion pesticide using a biohybrid of Sphingomonas sp. cells-silica nanoparticles. Biosens Bioelectron 87:332–338. https://doi.org/10.1016/j.bios.2016.08.048

Mulchandani A, Mulchandani P, Kaneva I, Chen W (1998) Biosensor for direct determination of organophosphate nerve agents using recombinant Escherichia coli with surface-expressed organophosphorus hydrolase. 1. Potentiometric microbial electrode. Anal Chem 70:4140–4145. https://doi.org/10.1021/ac9805201

Richter Ł, Janczuk-Richter M, Niedziółka-Jönsson J, Paczesny J, Hołyst R (2018) Recent advances in bacteriophage-based methods for bacteria detection. Drug Discov Today 23:448–455. https://doi.org/10.1016/j.drudis.2017.11.007

Ronkainen NJ, Halsall HB, Heineman WR (2010) Electrochemical biosensors. Chem Soc Rev 39:1747–1763. https://doi.org/10.1039/b714449k

Segev O, Kushmaro A, Brenner A (2009) Environmental impact of flame retardants (persistence and biodegradability). Int J Environ Res Public Health 6:478–491. https://doi.org/10.3390/ijerph6020478

Shabani A, Zourob M, Allain B, Marquette CA, Lawrence MF, Mandeville R (2008) Bacteriophage-modified microarrays for the direct impedimetric detection of bacteria. Anal Chem 80:9475–9482. https://doi.org/10.1021/ac801607w

Shi L, Wang Y, Chu Z, Yin Y, Jiang D, Luo J, Ding S, Jin W (2017) A highly sensitive and reusable electrochemical mercury biosensor based on tunable vertical single-walled carbon nanotubes and a target recycling strategy. J Mater Chem B 5:1073–1080. https://doi.org/10.1039/C6TB02658C

Singh A, Poshtiban S, Evoy S (2013) Recent advances in bacteriophage based biosensors for food-borne pathogen detection. Sensors (Switzerland) 13:1763–1786. https://doi.org/10.3390/s130201763

Steffan JJ, Brevik EC, Burgess LC, Cerdà A (2018) The effect of soil on human health: an overview. Eur J Soil Sci 69:159–171. https://doi.org/10.1111/ejss.12451

Taylor CJ, Bain LA, Richardson DJ, Spiro S, Russell DA (2004) Construction of a whole-cell gene reporter for the fluorescent bioassay of nitrate. Anal Biochem 328:60–66. https://doi.org/10.1016/j.ab.2004.01.013

Truffer F, Buffi N, Merulla D, Beggah S, Van Lintel H, Renaud P, Van Der Meer JR, Geiser M (2014) Compact portable biosensor for arsenic detection in aqueous samples with Escherichia coli bioreporter cells. Rev Sci Instrum 85:015120. https://doi.org/10.1063/1.4863333

Van Dorst B, Mehta J, Bekaert K, Rouah-Martin E, De Coen W, Dubruel P, Blust R, Robbens J (2010) Recent advances in recognition elements of food and environmental biosensors: a review. Biosens Bioelectron 26:1178–1194. https://doi.org/10.1016/j.bios.2010.07.033

Vopálenská I, Váchová L, Palková Z (2015) New biosensor for detection of copper ions in water based on immobilized genetically modified yeast cells. Biosens Bioelectron 72:160–167. https://doi.org/10.1016/j.bios.2015.05.006

Wang Y, Ye Z, Si C, Ying Y (2011) Subtractive inhibition assay for the detection of E. coli O157:H7 using surface plasmon resonance. Sensors 11:2728–2739. https://doi.org/10.3390/s110302728

Wu L, Lu X, Jin J, Zhang H, Chen J (2011) Electrochemical DNA biosensor for screening of chlorinated benzene pollutants. Biosens Bioelectron 26:4040–4045. https://doi.org/10.1016/j.bios.2011.03.027

Yang C, Wang Y, Marty JL, Yang X (2011) Aptamer-based colorimetric biosensing of Ochratoxin A using unmodified gold nanoparticles indicator. Biosens Bioelectron 26:2724–2727. https://doi.org/10.1016/j.bios.2010.09.032

Yang O, Kim HL, Weon J-I, Seo YR (2015) Endocrine-disrupting chemicals: review of toxicological mechanisms using molecular pathway analysis. J Cancer Prev 20:12–24. https://doi.org/10.15430/jcp.2015.20.1.12

Yüce M, Nazir H, Dönmez G (2010) An advanced investigation on a new algal sensor determining Pb(II) ions from aqueous media. Biosens Bioelectron 26:321–326. https://doi.org/10.1016/j.bios.2010.08.022

Zhang B, Guo LH (2012) Highly sensitive and selective photoelectrochemical DNA sensor for the detection of Hg^{2+} in aqueous solutions. Biosens Bioelectron 37:112–115. https://doi.org/10.1016/j.bios.2012.04.051

Zhong Z, Fritzsche M, Pieper SB, Wood TK, Lear KL, Dandy DS, Reardon KF (2011) Fiber optic monooxygenase biosensor for toluene concentration measurement in aqueous samples. Biosens Bioelectron 26:2407–2412. https://doi.org/10.1016/j.bios.2010.10.021

Sustainable Recovery of Resources from Industrial Wastewater: Trends in Biotechnology and Allied Industries

Muthusivaramapandian Muthuraj, Biswanath Bhunia, and Abhijit Chatterjee

Abstract

Various natural resources including freshwater reserve are depleting as a consequence of rapid increase in population. A sustainable wastewater system should aim at recovering energy and resource materials from waste discharge. Accordingly, this book chapter begins with a short introduction explaining the rationale behind the subject matter followed by a discussion on recent trends in the identification of resource materials from various types of wastewater. The major goal of the chapter, however, is to provide a comprehensive review summarizing recent advancement in various resource recovery technologies (bioelectrochemical systems, zero liquid discharge technologies, and adsorption-based recovery) commonly applied to sewage and effluent from biotechnology and allied industries. Finally, the chapter ends with a recommendation on development of economically feasible and environmentally benign resource recovery schemes for various types of wastewater.

Keywords

Sustainability · Bio-industries · Resource recovery · Biorefinery · ZLD · Bioelectrochemical systems

Abbreviations

AD	Anaerobic digestion
AEM	Anion exchange membrane
AGMD	Air gap membrane distillation
BECs	Bioelectrochemical systems

M. Muthuraj · B. Bhunia · A. Chatterjee (✉)
Department of Bioengineering, National Institute of Technology Agartala, Agartala, Tripura, India

M. P. Shah (ed.), *Removal of Emerging Contaminants Through Microbial Processes*,
https://doi.org/10.1007/978-981-15-5901-3_16

BMED	Bipolar membrane electrodialysis
CEM	Cation exchange membrane
DCMD	Direct contact membrane distillation
DC-MFC	Double chamber microbial fuel cell
DY	Drimarene Yellow K-2R
ED	Electrodialysis
EDM	Electrodialysis metathesis
EDR	Electrodialysis reversal
FO	Forward osmosis
MB	Maxilon Blue 5G
MD	Membrane distillation
MDC	Microbial desalination cell
MDCs	Microbial desalination cells
MEC	Microbial electrolysis cell
MFC	Microbial fuel cell
MLD	Minimal liquid discharge
NMCG	National Mission for Clean Ganga
PEM	Proton exchange membrane
PVDF	Polyvinylidene fluoride
RO	Reverse osmosis
SC-MFC	Single chamber microbial fuel cell
SF	Sodium fluorescein
SGMD	Sweeping gas membrane distillation
TDS	Total dissolved solids
UF	Ultrafiltration
VMD	Vacuum membrane distillation
ZLD	Zero liquid discharge

1 Introduction

The pressing issue that startled life on earth is the environmental hazards caused by anthropogenic activities (Meena et al. 2019). The limited potable water resources of the planet have been persistently polluted with emerging contaminants, due to the urban expansion and associated industrialization that raised concerns on environmental sustainability (Meena et al. 2019). For instance, over ten billion tons/year of organic-rich wastewater is reported to be generated globally (Lu et al. 2019). A significant portion of this wastewater is often released into the aquatic bodies without adequate treatment leading to adverse effects on the environment and ecology (Lu et al. 2019). Conventional physicochemical and biological methods are exploited for the remediation of the wastewaters at the expense of high energy inputs and other environmental resources (Luo and He 2020). These conventional wastewater treatment strategies target the generation of clean water as the only key objective. In order to meet the regulatory criteria, industries rely on conventional methods to mitigate pollution rather than converting the waste into resources attributed to augmented construction and operating investments. Thus, the waste remains a waste forever and not recovered as a resource by any of the conventional process. In contrast to that, the emerging wastewater treatment strategies should target to generate clean water while recovering various resources and converting them into several value-added products with possible revenue generation. This will attract the industries to invest in establishment of integrated closed-loop wastewater treatment strategies leading to circular economy (Pott et al. 2018). Thus, a paradigm shift is required to accentuate the ability of wastewater to generate revenue by exploiting the resources (energy, water, nutrients, heavy metals, etc.) therein. To that end, the present study is mainly focusing on the recent trends and advancements in various resource recovery strategies suitable for sustainable industrial wastewater treatment.

Wastewater can be broadly categorized as sewage and non-sewage. Sewage is the wastewater mainly produced through domestic activities in kitchen, toilets, and laundry. It is generally treated through municipal wastewater treatment plant which collects the wastewater from houses, restaurants, public toilets, hospitals, schools, and other similar public facilities. Non-sewage covers other types of wastewater such as stormwater, agricultural runoff, and industrial wastewater. It is not always practical, however, to distinguish between municipal and industrial wastewater systems. Often times, especially in developing countries, industrial wastewater is discharged into the municipal sewer either directly or after a minimal treatment. Only large-scale industries may have their fully developed effluent treatment plant. Besides, treatment trains for both systems are usually based on similar unit operations.

A typical list of wastewater resources comprises nutrients (phosphorus, nitrogen), water for reuse and energy. For biotechnology-based industrial wastewater, organic and inorganic chemicals of commercial importance may be included. The water recovered from an industrial discharge can be recycled back to the plant itself (zero liquid discharge approach) or may be utilized for other agricultural, environmental, and industrial applications. Water reuse is typically categorized as non-potable (fit for non-drinking purposes, such as toilet flushing, clothes washing, and irrigation), direct potable (fit for drinking water distribution system), and indirect potable (fit for disposal into an aquatic body which is used as a source of drinking water). The following table (Table 1) summarizes the recent technological advancement in resource recoveries from wastewater. It showed that major technologies for resource recovery can be broadly categorized as bioelectrochemical systems, membrane-based technologies leading to zero liquid discharge, adsorption-based recovery, and combination of all these technologies. A major part of this chapter will substantiate the recent progress in each of these areas followed by the development of a "wastewater biorefinery" concept which will illustrate different routes for wastewater valorization by exploiting different combinations of these technologies.

Table 1 Resource recovery from various wastewater resources and the corresponding technology employed for recovery (modified from Meena et al. 2019)

Resources recovered		Outcome	Wastewater source	Technology used	References
Nutrients	Nitrogen, phosphorus, and potassium are utilized for growth of microalgae	Biodiesel: 4060 L ha^{-1} $year^{-1}$	Carpet industry effluent	Microalgae-based biological treatment	Chinnasamy et al. (2010)
		Biomass: 29 tons ha^{-1} $year^{-1}$			
	Nitrogen and phosphorus	Ammonia nitrogen: 150%	Domestic wastewater	Advanced microbial nutrient recovery cell	Chen et al. (2017)
		Total phosphorus: 490% recovery in solid form			
	Phosphate	Phosphate: 27% as struvite recovery in solid form	Swine wastewater	Microbial fuel cell	Ichihashi and Hirooka (2012)
	Ammonia, phosphate	Ammonia: 45%	Synthetic	Microbial recovery cell coupled with osmotic membrane bioreactor	Hou et al. (2017)
		Phosphate: 65% recovery in solid form	wastewater		
	Phosphate	Orthophosphate recovery in solid form: 600 ppm	Digested sewage sludge	Microbial fuel cell	Fischer et al. (2011)
	Phosphate	Over 90% of phosphate was recovered in solid form	Municipal wastewater (seawater brine)	Forward osmosis-microfiltration	Qiu et al. (2015)
	Nitrogen and phosphate	Ammoniacal nitrogen >97.58%; phosphate >94.9%	Synthetic wastewater	Microbial fuel cell with cation exchange membrane	Ye et al. (2019a)
	Nitrogen	94% of ammonia recovered and used as fertilizer with 7.8% nitrogen and 21.6% phosphorus pentoxide	Urban wastewater: ammonia-rich stream from zeolite regeneration	Hollow fiber-liquid liquid membrane contactors	Vecino et al. (2019)
	Nitrogen, phosphorus	90.2% total nitrogen; 88.6% total phosphorus for microalgal biomass production	Industrial wastewater with molasses	Microalgae-based biological treatment	Ma et al. (2017)
	Nitrogen, phosphorus		Piggery-brewery waste	Microalgae-based biological treatment	Zheng et al. (2018)

		32–96% total nitrogen; 28–95% total phosphorus for microalgal biomass production			
	Phosphorus	Phosphorus: 98.7%	Domestic wastewater	Hollow fiber microfiltration membrane-biocathode microbial desalination cell	Zuo et al. (2018)
Acid/ alkalis	Sodium hydroxide and hydrochloric acid	Sodium hydroxide, 2.6 ± 0.9 mmol, and hydrochloric acid, 2.1 ± 0.7 mmol	Synthetic salt water	Microbial electrolysis desalination and chemical-production cell	Chen et al. (2012)
	Malic acid	L-Malic acid: 1.15–1.27 kWh kg^{-1}	Beverage industry wastewater	Electrodialysis with bipolar membrane	Lameloise and Lewandowski (2012)
	Citric acid	Citric acid: 97.1%	Fermentation industry liquid	Electrodialysis with bipolar membrane	Sun et al. (2017)
	Acetate and sodium hydroxide	0.491 mol L^{-1} (acetate); 0.556 mol L^{-1} NaOH	Solid waste residue from insecticide factories	Electrodialysis with bipolar membrane	Xue et al. (2015)
	Malic acid	Malic acid: 18.4 ± 0.6 mmol L^{-1} h	Effluent of matured anolyte	Microbial electrolysis desalination and chemical-production cell, microbial electrodialysis cell, modified microbial desalination cell	Lu et al. (2017)
	Formic Acid	Formic acid: 37.00 ± 1.15 mg L^{-1} h	Effluent of matured anolyte	Microbial electrolysis desalination and chemical-production cell	Luo et al. (2017)
Heavy metals	Gold	Gold: 99%	Tetrachloroaurate wastewater	Microbial fuel cell	Choi and Hu (2013)
	Chromium, cadmium, copper	Cr^6, Cu^2, and Cd^2	Influent with mixed metal solution	Microbial fuel cell-microbial electrochemical cell	Li et al. (2017)

(continued)

Table 1 (continued)

Resources recovered		Outcome	Wastewater source	Technology used	References
	Silver	Silver: 99%	Silver Ion containing wastewaters	Microbial fuel cell	Choi and Cui (2012)
	Copper	Copper	Metal containing waste streams	Microbial fuel cell	Heijne et al. (2010)
	Lead	Lead-226 mg g^{-1} of biomass	Lead zinc mine tailings	Microbial recovery through sorption	Chen et al. (2016)
High-value compounds	Bioplastics	Polyhydroxyalkanoates: 74% recovery	Synthetic wastewater similar to fermented oil mill wastewater	Mixed microbial culture in sequencing batch reactor (fermentation and extraction)	Mannina et al. (2019)
Water		Direct potable use	Municipal wastewater	Ozonation and reverse osmosis with disinfection	Leverenz et al. (2011)
		High-quality water	Raw sewage	Osmotic membrane bioreactor with reverse osmosis	Luo et al. (2016)
		Direct discharge	Synthetic basal anaerobic	Forward osmosis	Schneider et al. (2019)
		Direct discharge	Landfill leachate	Forward osmosis	Iskander et al. (2017)
Energy	Power per unit area	70 mW m^{-2}	Sugar cane bagasse wastes	Microbial fuel cell	Catal et al. (2019)
	Power per unit area	0.378 W m^{-2}	Synthetic wastewater	Microbial fuel cell	Dong et al. (2017)
	Power per unit area	4.38 W m^{-2}	Synthetic wastewater	Anaerobic acidification and forward osmosis membrane-microbial fuel cell	Liu et al. (2017)
	Current density	24 A m^{-2}	Effluent of matured anolyte	Microbial electrolysis desalination and chemical-production cell	Luo et al. (2017)
	Current	67.4 mA	Diluted industrial wastewater	Multistage membrane desalination cell	Zuo et al. (2017)

2 Resource Recovery Technologies

2.1 Bioelectrochemical Systems

Bioelectrochemical systems (BECs) are the systems that generate electricity from the organic compositions present in wastewater using biological components such as enzymes or microbes, thereby leading to energy and nutrient recovery. Unlike conventional electrochemical systems, the electrodes in BECs rely on the electron-synthesizing activity of organisms and/or biological enzymes (San-Martin et al. 2018). Thus, the electrons generated from the oxidation of complex organic compounds in the anode are transferred to the electron acceptor at the cathode where the electrons are reduced (Luo and He 2020). This movement of electron from anode to cathode through an external circuit is driven by the potential difference between the compartments. BECs are the most attractive systems attributed to their high efficiency in nutrient recovery and remediation of various wastewater types while generating electricity and less sludge as compared to other viable methods of recovery (Bajracharya et al. 2016). Thus, BECs are utilized as isolated units or in combination with other methods to obtain higher recovery efficiency coupled with electricity generation. Among the various systems enabled for BECs, microbial fuel cell (MFC), microbial electrolysis cell (MEC), microbial desalination cell (MDC), combination of different BEC systems, and combination of BEC systems with other membrane-based technologies are well exploited in many recovery processes (Li et al. 2017; Luo et al. 2017; Lu et al. 2017). To that end, this section will predominantly discuss the current trends and advancements in the nutrient recovery process by employing BECs.

Microbial fuel cell utilizes the metabolically active microbes that generate electrons from the degradation of organic and inorganic substances present in the wastewater (Zhao and Chen 2018). In general, anode comprises an electron secreting organism that metabolizes the complex organic compounds via oxidation and releases H^+ and e^- ions, which reach cathode via cation exchange membrane and external circuit, respectively. On the other hand, the cathode chamber comprises an electron acceptor, which accepts the electron generated in the anode and gets reduced by the H^+ ions produced and transferred from the anode. The flow of electrons from anode to cathode via an external connection generates electricity in the circuit (Zhao and Chen 2018). The anode comprises the electron-synthesizing microbes such as *Shewanella oneidensis* and *Geobacter sulfurreducens* which are the well-studied model organisms that exchange electrons with the electrode in the presence or absence of mediators (Logan et al. 2019). The other organisms predominantly used in BECs are *Thermincola ferriacetica*, *Sporomusa ovata*, *Moorella thermoacetica*, *Acetobacterium woodii*, *Clostridium ljungdahlii*, *Escherichia coli*, *Corynebacterium glutamicum*, etc. (Kracke et al. 2015). These prokaryotic microbes use their oxidative phosphorylation mechanism to transfer the electrons to an acceptor with high redox potential from a low potential donor, thereby leading to ATP biosynthesis (Kracke et al. 2015). In contrast to the other aerobic microorganisms which use oxygen as the electron acceptor, these anaerobic microorganisms rely on the surface metal ions as electron acceptors/donors that do not enter the cytosolic face or the periplasmic space. In such cases, these organisms excrete their redox-active molecules into the extracellular medium for electron exchange with metal ions available on the electrodes. Such electron shuttling with the extracellular metal ions leads to normal functioning and respiration in these microbes, which in turn assist in harnessing the electrons by the anode electrodes in the fuel cell (Kracke et al. 2015; Logan et al. 2019). Both the anode and cathode use graphite electrodes in general, whereas the efficiency in electron captures depends on the surface area of these electrodes. Thus, the anode usually runs under anaerobic condition with wastewater rich in organic and inorganic substrates which are used by the electron-secreting microbes as nutrient for their growth. On the other hand, the cathode comprises open-air cathode with oxygen as the electron acceptor or algal strains which synthesize and release oxygen via photosynthesis (Ye et al. 2019a, b). Based on the configuration of anode and cathode chambers, the MFC are classified (Fig. 1) into two types: (a) single chamber microbial fuel cell (SC-MFC) and (b) double chamber microbial fuel cell (DC-MFC). DC-MFC is the conventional design that mimics the electrochemistry-based galvanic cell with a separate anode and cathode electrode chambers connected by a proton exchange membrane or a salt bridge (Zhao and Chen 2018). The DC-MFC generates less power and current density attributed to high internal resistance offered by the proton exchange membrane and requires high cost for construction (Ye et al. 2019a). In contrast, the SC-MFC comprises both the anode and cathode in a single chamber with no additional separation by proton exchange membrane which makes them less resistant and economical. However, contamination with the oxidizing and reducing biological agents in the single chamber remains one of the toughest problems to resolve with SC-MFC. These MFC systems can offer green electricity generated from the wastewater which makes them sustainable. However, large-scale processes are still under experimentation due to low power density yields associated with these MFCs which have much scope for improvement in every aspect.

There are several factors that significantly influence the efficiency in sustainable recovery of nutrients with concomitant generation of high energy densities of the MFCs which includes the (a) microorganism and its metabolic activity; (b) substrate/nutrient availability; (c) operating conditions; (d) microbial fuel cell architecture: single or dual chamber, electrodes, surface area, mediators, and membrane configurations; (e) cathode configuration; and (f) modes of operation (continuous, self-circulation mode, fed-batch, or batch). For instance, Ye et al. (2019a) evaluated the performance of dual chamber microbial fuel cell and critically analyzed the effect of various parameters such as operation mode, aeration conditions, and type of membrane separators. The study evaluated the effect of three different modes of operation in which the anolyte was varied: (1) the batch mode with no addition or removal of anolyte in 24 h batch time, (2) self-circulation mode which involved a constant recirculation of anolyte in the anodic chamber at 10 mL min^{-1} flow rate for 24 h followed by complete replacement of anolyte with fresh one after every 24 h, and (3) continuous mode with a constant flow of fresh anolyte at 0.24 mL min^{-1} with 24 h retention time. The study showed a maximum voltage of 641 mV in self-

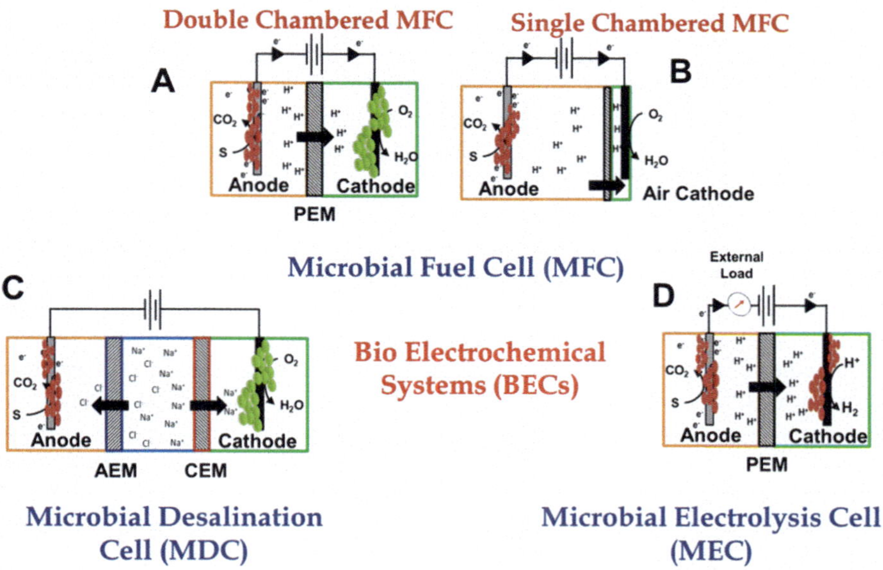

Fig. 1 Schematic figure on the operation of different bioelectrochemical systems used for wastewater treatment and recovery. (**a**) Double chamber microbial fuel cell; (**b**) single chamber microbial fuel cell; (**c**) microbial desalination cell; (**d**) microbial electrolysis cell. S represents the substrate. The picture has been modified from Luo and He (2020)

circulation mode with the dual chambered fuel cell with cathode comprising deionized water without aeration (Ye et al. 2019a). Influence of aeration in cathodic chamber and associated efficiency in nutrient removal was also evaluated in the dual chamber MFC which showed negligible recovery in the nutrients NH_4^+-N and phosphate ions from the anodic chamber. However, aeration in the cathodic chamber resulted in the precipitation of ammonia and phosphate ions in the cathode with catholyte pH ranging from 8 to 8.4 which assisted in increased recovery of the nutrients (Ye et al. 2019a). Similarly, increased pH and alkalinity in cathode chamber also reported to augment nutrient recovery through chemical precipitation (Liu et al. 2018). Further, the DC-MFC was evaluated with three membrane configurations: (a) cation exchange membrane, (b) non-woven membrane, and (c) forward osmosis. It was found that the MFC with cation exchange membrane resulted in maximum removal of both ammonia and phosphate ions as compared to other configurations (Ye et al. 2019a). In another study, a membrane less microbial fuel cell was used with a modified cup-shaped clay as cathode attributed to its augmented porosity, conductivity, and adsorption capability (Liu et al. 2020). This study demonstrated increased recovery of Cu^{2+}ions up to 96.5% from low concentration solutions while generating power densities up to 113.7 mW m^{-2} (Liu et al. 2020). The effect of organic loading rate (from 0.435 to 0.87 gCOD L^{-1} day^{-1}) on the performance of dual chamber microbial fuel cell and nutrient recovery efficiency was evaluated by Ye et al. (2019b) which depicted an increased power density with low organic loading rate. For instance, the maximum power density of 0.254 W m^{-2} was obtained for low organic loading rate of 0.435 gCOD L^{-1} day^{-1}, whereas the lowest power density of 0.071 W m^{-2} was obtained with 0.580 gCOD L^{-1} day^{-1} (Ye et al. 2019b). In addition to that, the increased organic loading rates from 0.435 to 0.870 gCOD L^{-1} day^{-1} resulted in reduced recovery of ammonia and phosphate ions (Ye et al. 2019b). Yang et al. (2019) used a novel photosynthetic microbial fuel cell coupled with photobioreactor which was operated in continuous mode for pollutant removal and algal biomass production from anaerobically digested food waste. The anode compartment comprised the anaerobically digested food waste as the influent. The cathode was continuously fed with 4% of anaerobically digested food waste and connected to a photobioreactor. The algae in the photobioreactor utilized the effluent from cathode and released necessary oxygen required as electron acceptor in the microbial fuel cell (Yang et al. 2019). The system resulted in maximum power density of 3.5 W m^{-3} and nitrogen removal rates up to 3 mg L^{-1} day^{-1} with 1.8 kWh m^{-3} of total energy produced from both the fuel cell and the algal biomass generated (Yang et al. 2019). Thus coupling of MFC technologies with microalgal cathode or by obtaining additional product output would result in feasibility and sustainability of the overall process.

Microbial electrolysis cell (MEC) is an electrolytic system that yields hydrogen gas from organic matters present in the wastewater through anaerobic digestion with an additional input of external voltage (Fig. 1). The MEC system has been reported to have high efficiency as compared to the conventional splitting of water for hydrogen synthesis (Kadier et al. 2014). Unlike the configuration of MFC, MEC comprises an additional external voltage with a sealed cathode under anaerobic condition. The anode compartment containing the electron-secreting microorganisms of the MEC generates the electrons, protons, and carbon dioxide. The generated protons and electrons pass through the cation exchange membrane and external circuit, respectively, to the cathode where the hydrogen gas is synthesized by the hydrogen-producing organisms. MEC is a potential technology that harnesses organic matters and pollutants in the wastewater and channelizes it toward hydrogen gas production. In several cases, the MEC systems are known to recover or remove chemical oxygen demands up to 75% and remove odor, with less sludge volume as output. The aeration-free cathode results in less energy consumption unlike MFC. However, MEC are known to accumulate high levels of methane attributed to the growth of methanogens in the strict anaerobic conditions that prevail in the cathode which reverses the chemical oxygen demand. Thus, controlling the cathodic chamber with high levels of hydrogen-synthesizing microbes is an essential task to attain high efficiencies in energy recovery. A study showed hydrogen production rates up to 300 mL L^{-1} day^{-1} of hydrogen gas from urban wastewater in a 100 mL reactor which was continuously operated for 24 h (Escapa et al. 2012). On the other hand, increased energy recovery up to 121% was attained with hydrogen production rates up to 4 L day^{-1} when urban wastewater was tested at a pilot-scale MEC with 130 L volume (Baeza et al. 2017). For further improvement in the recovery efficiencies with high hydrogen evolution rates, use of cost-effective metal catalysts such as nickel and molybdenum is reported to significantly augment the cathodic efficiencies (Lu et al. 2016).

Microbial desalination cells (MDCs) are known to generate electricity from wastewater utilizing the organic and inorganic compounds with simultaneous desalination of water, thus considered to be a potential BEC technology with sustainability (Al-Mamun et al. 2018). This MDC technology resolves the high energy requirements of conventional methods for seawater desalination. The MDC system is similar to the DC-MFC system with an additional chamber in between the anode and cathode chambers which contains the seawater to be desalinated. The intermediate chamber is separated from anionic chamber by an anion exchange membrane whereas the cationic chamber is separated by a cation exchange membrane. Similar to the anodic chamber of MFC, the anode chamber of the microbial desalination cell results in the generation of electrons which are further transferred to cathode via an external circuit where the electrons are captured by the electron acceptors leading to electricity generation. When the electron concentration in the anode is very high, it results in continuous augmentation of electron concentration in cathode. This leads to an ionic imbalance in the circuit which drives the movement of cations and anions to the respective chambers resulting in desalination of seawater in the intermediate chamber. The electrons are captured in the cathode, thereby resulting in the power generation. The major factors that influence the performance of MDCs and the recovery of different ions are the architecture of MDC, electrodes used, number of compartments, loading rates, salinity, and exchange membranes. In general, MDC is combined with different technologies such as osmosis to attain high efficiency in energy, nutrient recovery, and desalination. In certain conditions, the anodic chamber may accumulate increased concentrations of chlorine ions which may affect the pH of the anode, further affecting the efficiency in electron production. To that end, several studies have shown to improve the MDC system by utilizing an additional bipolar membrane between the anode and intermediate chamber which is known as bipolar MDC that resulted in avoiding the pH variations and associated drop in efficiency of MDC (Carmalin et al. 2016). In addition to that, a supercapacitive microbial desalination cell coupled with an additional capacitive electrode was demonstrated in which the anode and cathode act as capacitors with positive and negative charges, respectively, that discharge and self-recharge (Santoro et al. 2017). This supercapacitive microbial desalination cell showed maximum power of 3 W m^{-2} with reduction in solution conductivity more than 60% within 44 h of desalination (Santoro et al. 2017). Sequentially stacked chambers of MDC are usually employed for maximization of desalination efficiency with simultaneous generation of electricity which enhances the salt removal rates by 40% as compared to the three-chambered MDC (Al-Mamun et al. 2018). In addition to that, several modular MDCs such as recirculation MDC, MDC packed with ion exchange resins, upflow MDC, osmotic MDC, etc. are shown to have increased efficiency as compared to the simple MDC with three chambers. Further improvement in the desalination activity could be achieved by employing a biocathode comprising algal strains which generate oxygen continuously, thereby eliminating expensive electron acceptors such as ferricyanide. For instance, the biocathode microbial desalination cell has been reported to yield high salt removal efficiencies up to 92% with concomitant power generation up to 3.2 W m^{-3} (Al-Mamun et al. 2018). In a similar study with algal biocathode, sodium chloride concentration of

35 g L^{-1} resulted in power density up to 19.8 mW m^{-2} and salt removal rates up to 0.34 g L^{-1} day^{-1} (Zamanpour et al. 2017).

Even though several bioelectrochemical systems in combination with other membrane-based technologies are renowned for high-efficiency recovery of energy and nutrients, commercial-scale BECs are still awaited attributed to several techno-economic feasibility issues while considering large-scale operations. The major challenges extend from electrode design with increased surface area to spacing, membrane configuration, cathode configuration, associated aeration, recirculation frequency in cathode, wastewater types, and their compositions (Luo and He 2020). Thus, the future prospect is to develop a pilot-scale operation of BECs with better performance and low cost.

2.2 Zero Liquid Discharge (ZLD)

In ZLD approach, the wastewater generated is not released from the industrial process, rather purified and recycled back to the process. The basic treatment train typically includes reverse osmosis (RO) followed by thermal evaporation of RO reject stream (also termed as "RO brine") through concentrator and crystallizer. However, design of the treatment train varies on a case-by-case basis, and hence a number of variations are possible (Fig. 2). The clean water as produced by the RO and the evaporation is recycled, while the solids from the crystallizer are often recovered as by-products. For example, sodium sulfate was recovered from wastewater discharged by a SO_2 scrubber (Tunc and Groth 2011) and other industrial wastewaters (Quist-Jensen et al. 2017; Charisiadis 2018). Sodium chloride was recovered from coal mine wastewater (Charisiadis 2018) and textile wastewater (Lin et al. 2015). To remove natural organic matter from the salt (Van Linden et al. 2020), solids obtained in a ZLD treatment plant were packed in a fixed-bed reactor (bed volume = 51 mL), and saturated NaCl solution was used as an eluent. If recovery of salt is not feasible, then the solid is to be sent in an evaporation pond for disposal. A number of modifications to this basic system will be detailed below.

This basic zero liquid discharge model suffers from the following limitations:

- The external pressure of RO membrane module needs to overcome the osmotic pressure which depends on the salinity of the feed. But the maximum possible external pressure of RO membrane module is fixed by a given mechanical configuration. Hence RO membrane can only work within a limited salinity range (exit concentration = 70,000 mg L^{-1}) (Tong and Elimelech 2016). Recently a staged RO system has been developed (Wang et al. 2020b) that can concentrate high salinity wastewaters (>4.0 M NaCl or 234 g L^{-1} NaCl) at moderate hydraulic pressure (<70 bar). For a feedwater having 0.1–1 M NaCl, this RO system has four times lower specific energy consumption (approx. 2–8 kWh of electrical energy per m^3 of feedwater treated) compared to mechanical vapor compressor.
- RO membrane fouling issues may hamper the normal functionality of the system.

- Apparently, ZLD technology increases water reuse at the expense of high energy consumption as it relies on thermal processes such as evaporation and crystallization. The energy consumption (Yaqub and Lee 2019) by brine concentrator (20–25 kWh m^{-3} of treated feedwater) and brine crystallizer (52–66 kWh m^{-3}) is much higher than the energy requirement of RO. Since the brine concentrator works via mechanical vapor compression, very high salinity limit (2,50,000 mg L^{-1}) can be obtained in a brine concentrator (Tong and Elimelech 2016).
- Both of these equipment are also susceptible to scaling. RO reject water (also called RO brine) may be too concentrated to be directly used in the concentrator.

RO membrane fouling is mainly caused by hardness, alkalinity, silica precipitation, organic fouling, and biofouling. Some of these issues may be addressed by incorporating various water softening pretreatments (either by addition of chemicals or by using another membrane) for RO feedwater. For oil refinery effluent (Ronquim et al. 2020), it was found that a standard recovery of 59% can be increased to 89.2% by applying a $BaSO_4$ supersaturation step prior to the membrane equipment. By incorporating an intermediate water softening step using an additional RO membrane with CaO addition and calcite seeding, RO water recovery can be increased to 96.3% by removing scaling compounds.

A life cycle assessment showed that the energy consumption impact was decreased due to increase of RO recovery with consequent decrease of volumetric loading to thermal operations. In another pilot study using basal aquifer water (TDS = 21,300 mg L^{-1}) in the Athabasca oil sands in northern Alberta, Canada (Loganathan et al. 2016), RO recovery was improved using intermediate UF treatment according to the following sequence of unit operations: chemical oxidation,

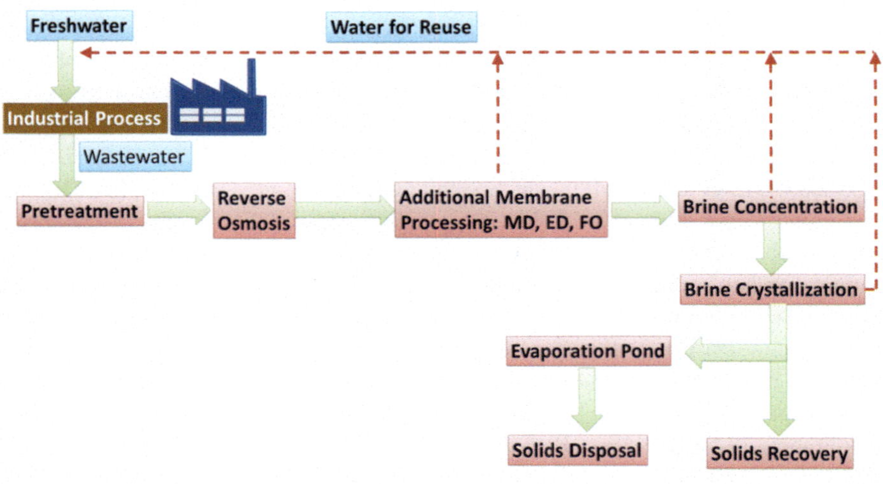

Fig. 2 Schematic diagram of generalized version of zero liquid discharge system. *RO* reverse osmosis, *MD* membrane distillation, *ED* electrodialysis, *FO* forward osmosis

polymeric ultrafiltration (UF), reverse osmosis, evaporation, and crystallization. A high-efficiency RO membrane system (trademarked as "HERO" membrane system), which is more resistant to fouling, has been developed by Toshiba Water Solutions Private Limited (TOSHIBA 2018) and has been commercially utilized in different full-scale ZLD applications (Tong and Elimelech 2016; Yaqub and Lee 2019).

To achieve zero liquid discharge in an economically feasible manner, a number of other membrane operations such as forward osmosis (FO), membrane distillation (MD), electrodialysis (ED), etc. may be incorporated to concentrate RO reject stream prior to the thermal treatment (Fig. 2). All three membrane operations (FO, MD, ED) have higher salinity limit. The energy demand for each of three membrane operations is higher than RO but much lower than thermal evaporation (Yaqub and Lee 2019). In forward osmosis, a highly concentrated salt solution called draw solution is kept separated from the RO brine by a semipermeable membrane. Unlike reverse osmosis in which external hydraulic pressure needs to overcome the osmotic pressure, forward osmosis is driven by the osmotic pressure difference between draw solution and RO brine solution. Water travels from the RO brine to the draw solution through osmotic membrane with gradual dilution of the draw solution. A draw solution containing 6 mol L^{-1} of ammonium ions resulted in high osmotic pressure up to 250 bar due to the mixing of highly soluble gases comprising ammonia and carbon dioxide (Lee et al. 2019). The draw solution can be regenerated by decomposing it at 60 °C (at atmospheric pressure) using the thermal energy available from industrial waste heat (Yaqub and Lee 2019). The RO reject from a textile wastewater (Wang et al. 2020a) was concentrated ten times using FO process. The FO process has been employed in the treatment of complex industrial wastewater produced by landfill leachate, oil and gas exploration, olive mill, digested sludge centrate, and molasses distillery (Singh et al. 2019).

Integration of membrane distillation with reverse osmosis is another emerging technique which is feasible if waste heat is available in the industry. The waste heat is then utilized to maintain a temperature difference (that leads into a partial vapor pressure difference) across the two sides of a hydrophobic microporous membrane which selectively permits water vapor to travel from higher to lower pressure. Membrane distillation can be done at a lower temperature compared to the conventional distillation. Three typical configurations of MD are direct contact (DCMD), air gap (AGMD), vacuum (VMD), and sweeping gas membrane distillation (SGMD). It was found (Schwantes et al. 2018) that membrane distillation is 40% more cost-effective than mechanical vapor compression. Laqbaqbi et al. (2019) studied the fouling mechanism in direct contact membrane distillation by using a flat-sheet polyvinylidene fluoride (PVDF) membrane and three different dyes: Maxilon Blue 5G (MB), Drimarene Yellow K-2R (DY), and sodium fluorescein (SF). The flux reduction factor was found to be lowest for the anionic SF dye solution but highest for the cationic DY solution. Membrane pore size reduction due to fouling was found in following order: PVDF > PVDF-SF > PVDF-DY > PVDF-MB. Leaper et al. (2019) demonstrated that air gap membrane distillation can be used to treat simulated dye wastewater containing salts, dyes, and surfactants. The main disadvantage of MD is lower permeate flux compared to other pressure-driven membrane systems.

In ED, electrodes are used to generate an electric field across a stack having alternate arrangement of cation and anion exchange membranes which are selectively permeable toward cations and anions, respectively. The feedwater, while passing through the system, generates a concentrated brine stream and a dilute salt-depleted stream. This operation removes ions from the water, so the neutral organic material and silica will remain in the deionized stream. Different configurations of ED system include (Tsai et al. 2017) electrodialysis reversal (EDR) providing an antifouling mechanism based on polarity reversal, electrodialysis metathesis (EDM) having four solution compartments, and bipolar membrane electrodialysis (BMED) having a bipolar membrane to split water into hydrogen and hydroxyl ions. The BMED is especially useful for clean production of HCl and NaOH (Herrero-Gonzalez et al. 2020; Lin et al. 2015) from wastewater. Incorporation of ED operation prior to the thermal evaporation of RO reject from textile wastewater led to a savings of 3.33 USD m^{-3} of RO reject (Praneeth et al. 2014). The ED system was found to be less susceptible to membrane fouling.

Since RO reject water can have a very high TDS (e.g., 47,000 mg L^{-1} for RO brine from a textile effluent (Praneeth et al. 2014)), additional treatment may be necessary to protect downstream equipment from scaling and membrane fouling. A number of pretreatment alternatives such as chemical precipitation, chemical coagulation, electrocoagulation, ion exchange, nanofiltration, and adsorption have been thoroughly studied (Semblante et al. 2018) and found suitable to remove hardness, silica, and organics from RO reject stream.

The energy consumption for brine concentrator and crystallizer is much higher than average membrane operations. A major share of total cost for ZLD is due to cost for removal of final 5–10% of water (Charisiadis 2018). Accordingly, industries (Frito-Lay North America, General Motors) often opted for minimal liquid discharge (MLD) approach to maximize water recovery at minimum operating cost (Desai et al. 2016). Here, 70–95% of water is recovered through membrane-based operations only.

Finally, Mansour et al. (2018) developed a robust cost assessment model where the end user can compare the cost of a number of alternative treatments in a versatile ZLD treatment train. For example, after primary chemical processing, feed can be treated through a number of thermal technologies (mechanical vapor compression, multi-effect distillation, brine concentrator) or membrane technologies (RO, ED) followed by different brine processing options (evaporation pond, brine crystallizers). Flow rates and compositions were taken as model variables. The model can determine the optimum choice for each of thermal, membrane, and brine processing technology under a fixed set of product water parameter values. The case study also shows the effect of model parameters (flow rates and compositions for each inlet) on optimized treatment train. Later Cham et al. (2020) refined this model by incorporating effect of other variables such as temperature, salinity, and pressure of feed in the ZLD networks.

Regulatory compliance and cost of wastewater treatment are the two main factors for an industry to choose this technology. This is especially important for industries in water-stressed regions and/or requiring high volume of freshwater input such as coal-to-chemical industries, cooling tower operation in power generation sector, and industries producing colored effluent (dye, paper, food color, carpet, and textile). For example, 50% of industrial wastewater in China was originated from coal-to-chemical-type industries having high COD (10.47 million tons of COD), high ammonia nitrogen, color, and other inorganic pollutants (Chen et al. 2020). The Chinese government has recently mandated application of ZLD technology in coal-fired power plants and coal-to-chemical refining facilities according to its 12th Five-Year Plan. Chen et al. (2020) have made a detailed economic analysis to compare the ZLD technology with conventional biological treatment technology for treatment of coal-to-chemical wastewater. It was found that ZLD's life cycle cost (0.94 US$ t^{-1} water) is a little higher than that of conventional approach. However, if the environmental impact of biologically treated effluent (high TDS and refractory pollutants after meeting regulatory standard) is incorporated within the life cycle economic assessment, ZLD treatment was found to be 1.07 US$ t^{-1} water less than the conventional one.

In India, the National Mission for Clean Ganga (NMCG) has recently sanctioned a project for treatment of tannery wastewater in a central effluent treatment plant of 20 MLD which includes the installation of a ZLD pilot plant (Yaqub and Lee 2019). The Chemplast Sanmar's PVC plant in Mettur, Tamil Nadu, has commissioned a ZLD plant to treat its wastewater having high COD and TDS of about 10,000 mg L^{-1} and rich in contaminants like oil and grease and volatile organic compounds. The membrane fouling issue was eliminated using HERO technology (Yaqub and Lee 2019).

2.3 Adsorption

Adsorption usually refers to the mass transfer operation in which the solute (termed as adsorbate) from the fluid phase is transferred into the surface of a solid phase (termed as adsorbent) and become bound through physical (physisorption) or chemical (chemisorption) interactions (da Costa et al. 2019). Adsorption technology is especially suitable for selective removal/recovery of substances such as rare earth metals (da Costa et al. 2019), dyes, pharmaceuticals, and personal care products (Yin et al. 2020) from dilute stream of wastewater because the process is relatively simple, cost-effective, and environment friendly. Adsorption experiments can be conducted in either batch or fixed-bed reactor. It is relatively easier to set up batch experiments, so screening of suitable adsorbent material is usually done through batch experiments, whereas fixed-bed reactor is more suitable for full-scale commercial application.

In batch experiment (da Costa et al. 2019), an aqueous solution containing the target adsorbate and adsorbents is incubated long enough to attain equilibrium. The remaining adsorbate concentration in aqueous phase is analyzed experimentally and corresponding solid-phase concentration is calculated by mass balance. The solid-phase concentration, i.e., concentration within the adsorbent, is also termed as the capacity of the adsorbent. An experimental data set comprising aqueous and solid phase equilibrium concentration of adsorbate is termed as isotherm which can be mathematically modeled by either using existing nonmechanistic isotherm functions (da Costa et al. 2019) such as Langmuir and Freundlich or developing a mechanistic model. An isotherm is a function of temperature, ionic strength, and pH of the solution. Therefore, either these parameters need to remain constant throughout the experiment, or the isotherm model used needs to include the effect of these parameters. A series of such experiments may be conducted to find the effect of various process parameters such as initial concentration of adsorbate, adsorbent dose, temperature, ionic strength and pH of the solution, and time of incubation. The time of incubation required to achieve equilibrium condition may be found by studying the adsorption kinetics. Once the adsorbent becomes fully saturated, it may be regenerated along with recovery of adsorbed species by eluting with proper solvent. Various organic and inorganic acids and salts have been utilized for this purpose.

Choice of adsorbent material is an important consideration for selective adsorption of a given target component and may be done on the basis of the following properties of the adsorbent: high uptake capacity, favorable kinetics, easy recovery, and regeneration. Adsorbents such as zeolites, synthetic polymeric resins, activated carbon, silica gel, and activated alumina have been extensively used commercially. Recently, graphene oxide derivatives (Huang et al. 2019), magnetic nanoparticles (Zhao et al. 2019), and metal-organic framework (Yin et al. 2020) are widely utilized to synthesize novel adsorbents. The adsorption process can be made cost-efficient by employing naturally abundant biomass (e.g., seaweeds) or waste biomass from other industries (agriculture, food industry, or biotechnology) as an adsorbent. Accordingly, a number of different biological materials have been explored as bio-adsorbent for removal of heavy metals and dyes. This may be broadly categorized as follows (Murnane et al. 2019):

(a) Agricultural residue: Cellulose and lignin-based agricultural residues such as rice husk, corn straw, fruit peels, and other similar materials contain oxygen, nitrogen, and sulfur-containing functional groups which impart them an affinity to bind metals. The waste material can be used in raw form or after a pretreatment.
(b) Biochar: This is the carbonaceous material obtained from the thermal conversion (gasification, pyrolysis) of biomass (wood type waste, agricultural waste, and sludge from biological treatment of wastewater). The biochar may further be "activated" by physical (steam) or chemical agents (acid/alkali).
(c) Microbial biomass: The outermost layer of various dead or inactive microbial biomass (bacteria, fungi, yeast, algae) contains a number of functional groups (carboxyl, hydroxyl, amine, amide) which can sequester the dissolved metal ions even from a dilute solution. The microbial biomass may be immobilized within suitable matrix to be used in a fixed-bed reactor.

The following table (Table 2) summarizes recent studies to find novel adsorbents for resource recoveries.

Table 2 Adsorption-based resource recovery

Adsorbent	Adsorbate	Maximum capacity ($mg\ g^{-1}$)	References
$Ni_{0.6}Fe_{2.4}O_4$	Au(III)	283.9	Zhao et al. (2019)
Rice husk silica	Nitrate	94.49	Suzaimi et al. (2019)
Mg-Fe LDHs	Uranium	710.0	Li et al. (2019)
Magnetic graphene oxide	Tetracycline	252.0	Huang et al. (2019)
Magnetic graphene oxide	Cd(II)	234 .0	Huang et al. (2019)
Magnetic graphene oxide	As(V)	14.0	Huang et al. (2019)
MIL-101(Cr)	Uranium	27.99	De Decker et al. (2017)
Micrococcus luteus	Cu(II)	408.0	Puyen et al. (2012)
Spirulina platensis	Cu(II)	817.7	Gunasundari and Senthil (2017)
Rhizopus sp.	Cr(VI)	9.95	Espinoza-Sánchez et al. (2019)
Pig manure	Methylene blue	48.30	Lonappan et al. (2016)
Zeolite A	Cu(II)	170.0	Hong et al. (2019)

3 Biorefinery

Biorefinery usually aims at conversion of different biomass types (such as oil or lignocellulosic-based crops and residues) into useful biobased product and biofuels. Similarly, wastewater biorefineries focus on treatment of wastewater with simultaneous recovery of other products or energy to build up a sustainable system. This concept works as a closed-loop system with simultaneous recovery of resources and reuse of water that can lead toward circular economy (Pott et al. 2018). The potential for biorefinery concept in any wastewater or resources can be evaluated by calculating the biorefinery complexity index (BCI) which in turn is governed by the major factors such as feedstock, platform/cell factory, process, and corresponding product (Chandra et al. 2018). Among them, the most significant factor is the type of wastewater to be processed through biorefinery or the feedstock. To employ biorefinery, the selected wastewater should have relatively high volumes, less complexity, and high concentration of major nutrients or compounds of interest. For instance, wastewater from sugar, dairy, food, paper, and brewery industries usually contain high concentrations of nutrients and organic compounds with high-volume outputs. Accordingly, this type of wastewater can be utilized directly for biorefinery. The second important factor is the product that can be obtained from the specific wastewater type. There are several products that are usually considered under biorefinery which predominantly involves the energy, biochemicals, enzymes, pigments, organic acids, volatile fatty acids, and polymer production utilizing the resources from wastewater. The major product outcomes that could be expected from biorefinery concept are categorized into three: (a) molecules that are derived from the complex molecules such as organic acids, volatile fatty acids, etc.; (b) molecules such as bioflocculants, surfactants, fertilizers, etc., which are in general a mixture of complex macromolecules; and (c) bioenergy molecules. The selection of a particular product will be based on whether the economic and energy goal of biorefinery concept is met or not, with a decent biorefinery complexity index. The third important factor is the cell factory or the platform that will be employed to execute the biorefinery process. The cell factory may be the photosynthetic organisms, microbes (bacteria or fungi), or phytoplanktons which are involved in the utilization of nutrients and resources available in the wastewater to produce biomass which can be further converted/valorized to produce value-added chemicals or bioproducts (Pott et al. 2018). In some cases, instead of living organisms, a physicochemical process such as liquefaction may be employed for the synthesis of energy products. Finally the process to be employed decides the feasibility of the wastewater biorefinery. The process chosen has to be economical and should not generate additional hazardous chemicals as wastes (Chandra et al. 2018). Figure 3 represents the schematic view of a potential wastewater biorefinery concept that may be designed for maximal utilization with various microbial systems and photosynthetic organisms. Such biorefinery model can be achieved by coupling the various

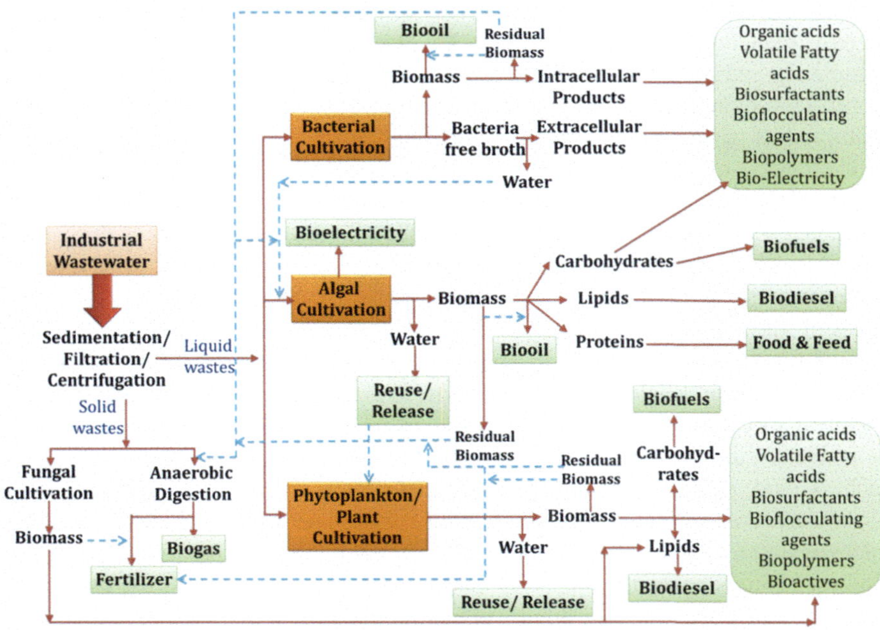

Fig. 3 Schematic representation of a potential wastewater biorefinery system. The green boxes represent the probable products, and the blue dotted lines represent the process flow in which output from one process acts as one of the inputs of the other processes. The red lines represent the main process flow toward biorefinery. The picture has been modified from Pott et al. (2018)

resource recovery technologies such as membrane-based operations, bioelectrochemical systems, ZLD technology, adsorption, etc.

A potential biorefinery model was proposed by Chandra et al. (2018) for the resource recovery from dairy wastes. This study proposed different strategies for utilization of whey proteins and the cattle excreta which are the major waste products in a dairy industry with concomitant revenue generation. The cattle excreta was shown to be utilized directly as a fertilizer attributed to their high content of ammonia (Tsai and Liu 2015) and rich organic carbon contents. On the other hand, the cattle dung is directly used in the biogas production through anaerobic digestion in which the rich organic carbon is converted into methane and carbon dioxide, thereby yielding revenue. There are novel technologies through which the methane is filtered out from the mixture of biogas comprising CO_2 and CO to produce methane as fuel for transportation (Chandra et al. 2018). The whey waste obtained from dairy industries is known to contain rich amounts of proteins, lactose, lipids, salts, and antibiotics with high biological oxygen demand and chemical oxygen demand. The high lactose content in whey supports the growth of various cell factories that can metabolize lactose, thereby paving way for production of lactic acid, polylactic acid, galacto-oligosaccharides, polysaccharides, polymers, and bioenergy molecules such as bioethanol, biobutanol, biodiesel, hydrogen, and methane (Chandra et al. 2018). The main protein content of the whey is recovered in the form of whey powder. The major issues are with the antibiotics that are present in the whey liquid which might inhibit the growth of the industrially important microbes, thereby reducing the yield potentials. The study also estimated the biorefinery complexity index to be 20 which ensures that the resource recovery through a closed-loop biorefinery process will be sustainable and economically feasible with multiproduct paradigm as output (Chandra et al. 2018).

Cultivation of microalgae in high BOD wastewater followed by its conversion into value-added products has been widely studied (Javed et al. 2019). In an integrated wastewater treatment system in a photobioreactor, algae act as the source of oxygen for bacterial culture while the bacterial culture supplies the carbon dioxide and nutrient (N and P through breakdown of organic matter) required for photosynthesis of microalgae.

Different types of cultivation (photoautotrophic, heterotrophic, mixotrophic) system have been developed for this purpose. Microalgae can be harvested through chemical, mechanical, electrical, and biological treatment. Harvested microalgae can be converted into biofuel or may be utilized on other various nonfuel applications such as food, pharmaceuticals, and nutraceuticals (Tang et al. 2020). Wang et al. (2019) have studied biotransformation of tofu wastewater by cultivating filamentous microalgae *Tribonema minus* mixotrophically to generate biorefinery feedstock. Highest biomass concentration was found to be 8.7 g L^{-1} using 100% tofu wastewater. Productivity of neutral lipid and palmitoleic acid was comparable to that in BG-11 control medium.

In another study (Shahid et al. 2019), primary treatment of city wastewater was done by mixed cultivation of *Chlorella* sp. and *Bracteacoccus* sp. (12 g $L^{-1}$of biomass) with simultaneous removal of nitrate (68%) and phosphate (75%). In a phytoremediation study (Khan et al. 2019), the consortium of microalgae *Chlorella minutissima*, *Scenedesmus* spp., and *Nostoc muscorum* was found to remove NH_4^+-N (92%), NO_3^--N (87%), PO_4^{3-}-P (85%), TDS (96%), BOD_5 (90%), and COD (81%) from sewage wastewater appreciably along with lipid production. Lipid production was highest for *Scenedesmus* (81.23 ± 4.5 mg L^{-1}) followed by *Nostoc muscorum* (14.29 ± 8.7 mg L^{-1}) and *C. minutissima* (11.33 ± 5.6 mg L^{-1}). A feasibility analysis has shown that using this biomass as manure can save 5584 USD worth of chemical fertilizer. In another wastewater biorefinery study (Bhowmick et al. 2019), three different sources of wastewater (kitchen waste, ethanol distillery waste, dairy waste) was evaluated to grow *Chlorella minutissima* with extraction of lutein (a healthcare product) and biodiesel. It was found that acidic pH, extremely high BOD and COD of distillery, and dairy wastewater were not suitable for the growth of this algae. The study was then continued with kitchen waste to evaluate the effect of different process parameters such as CO_2% supply, acetate concentration, poultry litter waste concentration, and light intensity on biomass cultivation. Guldhe et al. (2017) have treated aquaculture wastewater using *Chlorella sorokiniana* with high removal efficiency of COD, ammonium, nitrate, and phosphate with simultaneous production of biomass, lipid, carbohydrate, and protein. It was found that sodium nitrate can increase the metabolite yield.

Anaerobic treatment with biogas production has been proved to be economical compared to aerobic treatment (activated sludge process) for high strength wastewater. Secondary treatment of food processing wastewater (dairy, slaughterhouse, distillery, olive mill) is often done via anaerobic digestion (AD) as it contains a high amount of biodegradable organic carbon (carbohydrate, protein, and lipid). A significant number of research articles are available focusing on reactor stability (Shamurad et al. 2020) and optimization of different process parameters to maximize biogas recovery and COD removal in different types of reactors. Recently, environmental impact of VOC gases generated from AD plant has been studied by Zhang et al. (2019). However, AD effluent often needs tertiary treatment for nutrient (N and P) removal. Accordingly, Gupta et al. (2018) took an integrated approach for treatment of snacks processing industrial wastewater. Here effluent after anaerobic digestion (nutrient rich, carbon deficient) was mixed with raw wastewater (carbon rich) to produce microalgal biodiesel with simultaneous removal of total nitrogen, total phosphorus, and COD to regulatory standard. Similarly, piggery wastewater in China was found to contain high amount of nutrient after AD (Zheng et al. 2018). It was also found to have low yield of algal biomass for wastewater-based algal cultivation system due to low C:N ratio. Zheng et al. (2018) worked on this issue and found promising results by mixing piggery wastewater with brewery wastewater in 1:5 ratio.

4 Future Prospects

The manuscript critically reviews the major technical alternatives to integrate resource recovery with wastewater treatment. The current research trend shows that various types of resources (water, energy, nutrients such as nitrogen and phosphorus, other commodity chemicals, chemicals having pharmaceutical importance) from wastewater can be recovered through bioelectrochemical system, membrane-based minimal/zero liquid discharge system, and adsorption-based technologies in a sustainable manner. Finally a "biorefinery tree" has been developed for easy identification of various resource recovery tracts by utilizing those technologies. This "big picture" will be helpful to explore the different resource recovery routes almost exhaustively. Commercial success, however, requires development of systemic methodologies to identify the most sustainable solution in a given context. It needs to analyze the technical, economic, social, and environmental aspects on a case-by-case basis. The technology readiness level metric system, originally developed by NASA (Udugama et al. 2020), can be utilized to determine whether a technology is matured enough to be implemented under a given industrial setting. Similarly, life cycle analysis tools will be helpful to compare the environmental impact of a given technical solution. Overall, research in resource recovery from wastewater needs to be directed toward full-scale industrialization of existing methodologies.

References

Al-Mamun A, Ahmad W, Baawain MS, Khadem M, Dhar BR (2018) A review of microbial desalination cell technology: configurations, optimization and applications. J Clean Prod 183:458–480

Baeza JA, Martínez-Miró À, Guerrero J, Ruiz Y, Guisasola A (2017) Bioelectrochemical hydrogen production from urban wastewater on a pilot scale. J Power Sources 356:500–509

Bajracharya S, Sharma M, Mohanakrishna G, Benneton XD, Strick DPBTB, Sarma PM, Pant D (2016) An overview on emerging bioelectrochemical systems (BESs): technology for sustainable electricity, waste remediation, resource recovery, chemical production and beyond. Renew Energy 98:153–170

Bhowmick D, Sen GR, Sarmah AK (2019) Consolidated bioprocessing of wastewater cocktail in an algal biorefinery for enhanced biomass, lipid and lutein production coupled with efficient CO_2 capture: an advanced optimization approach. J Environ Manage 252:109696

Carmalin AS, Bhalambaal VM, Lima EC, Thirunavoukkarasu M (2016) Microbial desalination cell technology: contribution to sustainable waste water treatment process, current status and future applications. J Environ Chem Eng 4:3468–3478

Catal T, Liu H, Fan Y, Bermek H (2019) A clean technology to convert sucrose and lignocellulose in microbial electrochemical cells into electricity and hydrogen. Bioresour Technol Rep 5:331–334

Chandra R, Carlos CZ, Patricio D, Parra-Saldívar R (2018) A biorefinery approach for dairy wastewater treatment and product recovery towards establishing a biorefinery complexity index. J Clean Prod 183:1184–1196

Cham E, Alnouri S, Mansour F, Al-Hindi M (2020) Design of end-of-pipe zero liquid discharge systems under variable operating parameters. J Clean Prod 250:119569. https://doi.org/10.1016/j.jclepro.2019.119569

Charisiadis C (2018) Brine zero liquid discharge (ZLD) fundamentals and design. [WWW document] URL: https://doi.org/10.13140/RG.2.2.19645.31205/. Accessed 12 Mar 2020

Chen S, Liu G, Zhang R, Qin B, Luo Y (2012) Development of the microbial electrolysis desalination and chemical-production cell for desalination as well as acid and alkali productions. Environ Sci Tech 46:2467–2472

Chen Z, Pan X, Chen H, Guan X, Lin Z (2016) Biomineralization of Pb(II) into Pb-hydroxyapatite induced by *Bacillus cereus* 12-2 isolated from lead-zinc mine tailings. J Hazard Mater 301:531–537

Chen X, Zhou H, Zuo K, Zhou Y, Wang Q, Sun D, Gao Y, Liang P, Zhang X, Ren ZJ, Huang X (2017) Self-sustaining advanced wastewater purification and simultaneous in situ nutrient recovery in a novel bioelectrochemical system. Chem Eng J 330:692–697

Chen B, Yang S, Cao Q, Qian Y (2020) Life cycle economic assessment of coal chemical wastewater treatment facing the "Zero liquid discharge" industrial water policies in China: discharge or reuse. Energy Policy 137:111107

Chinnasamy S, Bhatnagar A, Hunt RW, Das KC (2010) Microalgae cultivation in a wastewater dominated by carpet mill effluents for biofuel applications. Bioresour Technol 101:3097–3105

Choi C, Cui Y (2012) Recovery of silver from wastewater coupled with power generation in a microbial fuel cell. Bioresour Technol 107:522–525

Choi C, Hu N (2013) The modeling of gold recovery from tetrachloroaurate wastewater using a microbial fuel cell. Bioresour Technol 133:589–598

da Costa TB, da Silva MGC, Vieira MGA (2019) Recovery of rare-earth metals from aqueous solutions by bio/adsorption using non-conventional materials: a review with recent studies and promising approaches in column applications. J Rare Earths 38(4):339–355

De Decker J, Folens K, De Clercq J, Meledina M, Van Tendeloo G, Du Laing G, Van Der Voort P (2017) Ship-in-a-bottle CMPO in MIL-101(Cr) for selective uranium recovery from aqueous streams through adsorption. J Hazard Mater 335:1–9

Desai S, Rosenberg S, Hermsen N (2016) Minimal liquid discharge: adopting a 'less is more' mindset. [WWW document] URL: https://www.wateronline.com/doc/minimal-liquid-discharge-adopting-a-less-is-more-mindset-0001. Accessed 12 Mar 2020

Dong H, Liu X, Xu T, Wang Q, Chen X, Chen S, Zhang H, Liang P, Huang X, Zhang X (2017) Hydrogen peroxide generation in microbial fuel cells using graphene-based air-cathodes. Bioresour Technol 247:684–689

Escapa A, Gil-Carrera L, García V, Morán A (2012) Performance of a continuous flow microbial electrolysis cell (MEC) fed with domestic wastewater. Bioresour Technol 117:55–62

Espinoza-Sánchez MA, Arévalo-Niño K, Quintero-Zapata I, Castro-González I, Almaguer-Cantú V (2019) Cr(VI) adsorption from aqueous solution by fungal bioremediation based using *Rhizopus* sp. J Environ Manage 251:109595

Fischer F, Bastian C, Happe M, Mabillard E, Schmidt N (2011) Microbial fuel cell enables phosphate recovery from digested sewage sludge as struvite. Bioresour Technol 102:5824–5830

Guldhe A, Ansari FA, Singh P, Bux F (2017) Heterotrophic cultivation of microalgae using aquaculture wastewater: a biorefinery concept for biomass production and nutrient remediation. Ecol Eng 99:47–53

Gunasundari E, Senthil Kumar P (2017) Adsorption isotherm, kinetics and thermodynamic analysis of Cu(II) ions onto the dried algal biomass (*Spirulina platensis*). J Ind Eng Chem 56:129–144

Gupta A, Dutta A, Sarkar J, Panigrahi MK, Sar P (2018) Low-abundance members of the firmicutes facilitate bioremediation of soil impacted by highly acidic mine drainage from the Malanjkhand copper project, India. Front Microbiol 9:1–18

Heijne AT, Liu F, van der Weijden R, Weijma J, Buisman CJN, Hamelers HVM (2010) Copper recovery combined with electricity production in a microbial fuel cell. Environ Sci Technol 44:4376–4381

Herrero-Gonzalez M, Diaz-Guridi P, Dominguez-Ramos A, Irabien A, Ibañez R (2020) Highly concentrated HCl and NaOH from brines using electrodialysis with bipolar membranes. Sep Purif Technol 242:116785

Hong M, Yu L, Wang Y, Zhang J, Chen Z, Dong L, Zan Q, Li R (2019) Heavy metal adsorption with zeolites: the role of hierarchical pore architecture. Elsevier B.V., Amsterdam
Hou D, Lu L, Sun D, Ge Z, Huang X, Cath TY, Ren ZJ (2017) Microbial electrochemical nutrient recovery in anaerobic osmotic membrane bioreactors. Water Res 114:181–188
Huang D, Wu J, Wang L, Liu X, Meng J, Tang X, Tang C, Xu J (2019) Novel insight into adsorption and co-adsorption of heavy metal ions and an organic pollutant by magnetic graphene nanomaterials in water. Chem Eng J 358:1399–1409
Ichihashi O, Hirooka K (2012) Removal and recovery of phosphorus as struvite from swine wastewater using microbial fuel cell. Bioresour Technol 114:303–307
Iskander SM, Zou S, Brazil B, Novak JT, He Z (2017) Energy consumption by forward osmosis treatment of landfill leachate for water recovery. Waste Manag 63:284–291
Javed F, Aslam M, Rashid N, Shamair Z, Khan AL, Yasin M, Fazal T, Hafeez A, Rehman F, Rehman MSU, Khan Z, Iqbal J, Bazmi AA (2019) Microalgae-based biofuels, resource recovery and wastewater treatment: a pathway towards sustainable biorefinery. Fuel 255:115826
Kadier A, Simayi Y, Kalil MS, Abdeshahian P, Hamid AA (2014) A review of the substrates used in microbial electrolysis cells (MECs) for producing sustainable and clean hydrogen gas. Renew Energy 71:466–472
Khan SA, Sharma GK, Malla FA, Kumar A, Gupta N (2019) Microalgae based biofertilizers: a biorefinery approach to phycoremediate wastewater and harvest biodiesel and manure. J Clean Prod 211:1412–1419
Kracke F, Vassilev I, Kromer JO (2015) Microbial electron transport and energy conservation—the foundation for optimizing bioelectrochemical systems. Front Microbiol 6:1–18
Lameloise M-L, Lewandowski R (2012) Recovering l-malic acid from a beverage industry waste water: experimental study of the conversion stage using bipolar membrane electrodialysis. J Membr Sci 403–404:196–202
Laqbaqbi M, García-Payo MC, Khayet M, El Kharraz J, Chaouch M (2019) Application of direct contact membrane distillation for textile wastewater treatment and fouling study. Sep Purif Technol 209:815–825
Leaper S, Abdel-Karim A, Gad-Allah TA, Gorgojo P (2019) Air-gap membrane distillation as a one-step process for textile wastewater treatment. Chem Eng J 360:1330–1340
Lee S, Choi J, Park YG, Shon H, Ahn CH, Kim SH (2019) Hybrid desalination processes for beneficial use of reverse osmosis brine: current status and future prospects. Desalination 454:104–111
Leverenz HL, Tchobanoglous G, Asano T (2011) Direct potable reuse: a future imperative. J Water Reuse Desalin 1:2–10
Li M, Pan Y, Huang L, Zhang Y, Yang J (2017) Continuous flow operation with appropriately adjusting composites in influent for recovery of Cr (VI), Cu (II) and Cd (II) in self-driven MFC–MEC system. Environ Technol 38:615–628
Li P, Chen P, Liu Z, Nie S, Wang X, Wang G, Zhang W, Chen H, Wang L (2019) Highly efficient elimination of uranium from wastewater with facilely synthesized Mg-Fe layered double hydroxides: Optimum preparation conditions and adsorption kinetics. Ann of Nuclear Energy 140:107140
Lin J, Ye W, Huang J, Ricard B, Baltaru MC, Greydanus B, Balta S, Shen J, Vlad M, Sotto A, Luis P, Van Der Bruggen B (2015) Toward resource recovery from textile wastewater: dye extraction, water and base/acid regeneration using a hybrid NF-BMED process. ACS Sustain Chem Eng 3:1993–2001
Liu J, Wang X, Wang Z, Lu Y, Li X, Ren Y (2017) Integrating microbial fuel cells with anaerobic acidification and forward osmosis membrane for enhancing bioelectricity and water recovery from low-strength wastewater. Water Resour 110:74–82
Liu J, Cheng X, Qi X, Li N, Tian J, Qiu B, Xu K, Qu D (2018) Recovery of phosphate from aqueous solutions via vivianite crystallization: thermodynamics and influence of pH. Chem Eng J 349:37–46

Liu SH, Lai CY, Chang PH, Lin CW, Chen YH (2020) Enhancing copper recovery and electricity generation from wastewater using low-cost membrane-less microbial fuel cell with a carbonized clay cup as cathode. J Clean Prod 247:119118

Logan BE, Rossi R, Ragab A, Saikaly PE (2019) Electroactive microorganisms in bioelectrochemical systems. Nat Rev Microbiol 17:307–319

Loganathan K, Chelme-Ayala P, El-Din MG (2016) Pilot-scale study on the treatment of basal aquifer water using ultrafiltration, reverse osmosis and evaporation/crystallization to achieve zero-liquid discharge. J Environ Manage 165:213–223

Lonappan L, Rouissi T, Das RK, Brar SK, Ramirez AA, Verma M, Surampalli RY, Valero JR (2016) Adsorption of methylene blue on biochar microparticles derived from different waste materials. Waste Manag 49:537–544

Lu L, Hou D, Fang Y, Huang Y, Ren ZJ (2016) Nickel based catalysts for highly efficient H_2 evolution from wastewater in microbial electrolysis cells. Electrochim Acta 206:381–387

Lu Y, Luo H, Yang K, Liu G, Zhang R, Li X, Ye B (2017) Formic acid production using a microbial electrolysis desalination and chemical-production cell. Bioresour Technol 243:118–125

Lu H, Peng M, Zhang G, Li B, Li Y (2019) Brewery wastewater treatment and resource recovery through long term continuous-mode operation in pilot photosynthetic bacteria-membrane bioreactor. Sci Total Environ 646:196–205

Luo S, He Z (2020) Resource recovery from wastewater by bioelectrochemical systems. In: Olivares JA, Puyol D, Melero JA, Dufour J (eds) Wastewater treatment residues as resources for biorefinery products and biofuels. Elsevier, Amsterdam

Luo W, Hai FI, Price WE, Guo W, Ngo HH, Yamamoto K, Nghiem LD (2016) Phosphorus and water recovery by a novel osmotic membrane bioreactor–reverse osmosis system. Bioresour Technol 200:297–304

Luo H, Cheng X, Liu G, Zhou Y, Lu Y, Zhang R, Li X, Teng W (2017) Citric acid production using a biological electrodialysis with bipolar membrane. J Membr Sci 523:122–128

Ma C, Wen H, Xing D, Pei X, Zhu J, Ren N, Liu B (2017) Molasses wastewater treatment and lipid production at low temperature conditions by a microalgal mutant *Scenedesmus* sp. Z-4. Biotechnol Biofuels 10:111

Mannina G, Presti D, Montiel-Jarillo G, Suárez-Ojedam ME (2019) Bioplastic recovery from wastewater: a new protocol for polyhydroxyalkanoates (PHA) extraction from mixed microbial cultures. Bioresour Technol 282:361–369

Mansour F, Alnouri SY, Al-Hindi M, Azizi F, Linke P (2018) Screening and cost assessment strategies for end-of-Pipe Zero Liquid Discharge systems. J Clean Prod 179:460–477

Meena RAA, Kannah KR, Sindhu J, Ragavi J, Kumar G, Gunasekaran M, Rajesh Banu J (2019) Trends and resource recovery in biological wastewater treatment system. Bioresour Technol Rep 7:100235

Murnane G, Ghanim B, O'Donoghue L, Courtney R, O'Dwyer TF, Pembroke JT (2019) Advances in metal recovery from wastewaters using selected biosorbent materials and constructed wetland systems. [WWW document] URL: https://doi.org/10.5772/intechopen.84335/. Accessed 12 Mar 2020.

Pott R, Johnstone-Robertson M, Verster B, Rumjeet S, Nkadimeng L, Raper T, Rademeyer S, Harrison STL (2018) Wastewater biorefineries: integrating water treatment and value recovery. In: The nexus: energy, environment and climate change, green energy and technology. Springer International Publishing, Berlin

Praneeth K, Manjunath D, Bhargava SK, Tardio J, Sridhar S (2014) Economical treatment of reverse osmosis reject of textile industry effluent by electrodialysis-evaporation integrated process. Desalination 333:82–91

Puyen ZM, Villagrasa E, Maldonado J, Diestra E, Esteve I, Solé A (2012) Biosorption of lead and copper by heavy-metal tolerant *Micrococcus luteus* DE2008. Bioresour Technol 126:233–237

Qiu G, Law Y-M, Das S, Ting Y-P (2015) Direct and complete phosphorus recovery from municipal wastewater using a hybrid microfiltration-forward osmosis membrane bioreactor process with seawater brine as draw solution. Environ Sci Technol 49:6156–6163

Quist-Jensen CA, Macedonio F, Horbez D, Drioli E (2017) Reclamation of sodium sulfate from industrial wastewater by using membrane distillation and membrane crystallization. Desalination 401:112–119

Ronquim FM, Sakamoto HM, Mierzwa JC, Kulay L, Seckler MM (2020) Eco-efficiency analysis of desalination by precipitation integrated with reverse osmosis for zero liquid discharge in oil refineries. J Clean Prod 250:119547

San-Martin MI, Leicester DD, Heidrich ES, Alonso RM, Mateos R, Escapa A (2018) Bioelectrochemical systems for energy valorization of waste streams. IntechOpen Limited, London

Santoro C, Abad FB, Serov A, Kodali M, Howe KJ, Soavi F, Atanassov P (2017) Supercapacitive microbial desalination cells: new class of power generating devices for reduction of salinity content. Appl Energy 208:25–36

Schneider C, Rajmohan RS, Zarebska A, Tsapekos P, Hélix-Nielsen C (2019) Treating anaerobic effluents using forward osmosis for combined water purification and biogas production. Sci Total Environ 647:1021–1030

Schwantes R, Chavan K, Winter D, Felsmann C, Pfafferott J (2018) Techno-economic comparison of membrane distillation and MVC in a zero liquid discharge application. Desalination 428:50–68

Semblante GU, Lee JZ, Lee LY, Ong SL, Ng HY (2018) Brine pre-treatment technologies for zero liquid discharge systems. Desalination 441:96–111

Shahid A, Ishfaq M, Ahmad MS, Malik S, Farooq M, Hui Z, Batawi AH, Shafi ME, Aloqbi AA, Gull M, Mehmood MA (2019) Bioenergy potential of the residual microalgal biomass produced in city wastewater assessed through pyrolysis, kinetics and thermodynamics study to design algal biorefinery. Bioresour Technol 289:121701

Shamurad B, Sallis P, Petropoulos E, Tabraiz S, Ospina C, Leary P, Dolfing J, Gray N (2020) Stable biogas production from single-stage anaerobic digestion of food waste. Appl Energy 263:114609

Singh N, Petrinic I, Hélix-Nielsen C, Basu S, Balakrishnan M (2019) Influence of forward osmosis (FO) membrane properties on dewatering of molasses distillery wastewater. J Water Process Eng 32:100921

Sun X, Lu H, Wang J (2017) Recovery of citric acid from fermented liquid by bipolar membrane electrodialysis. J Clean Prod 143:250–256

Suzaimi ND, Goh PS, Malek NANN, Lim JW, Ismail AF (2019) Performance of branched polyethyleneimine grafted porous rice husk silica in treating nitrate-rich wastewater via adsorption. J Environ Chem Eng 7:103–135

Tang DYY, Khoo KS, Chew KW, Tao Y, Ho S-H, Show PL (2020) Potential utilization of bioproducts from microalgae for the quality enhancement of natural products. Bioresour Technol 304:122997. https://doi.org/10.1016/j.biortech.2020.122997

Tong T, Elimelech M (2016) The global rise of zero liquid discharge for wastewater management: drivers, technologies, and future directions. Environ Sci Tech 50:6846–6855

TOSHIBA (2018) Solutions for industrial water treatment zero liquid discharge (ZLD) with HERO™. URL. https://www.toshiba.co.jp/sis/en/environment/solution/s2/zld.htm. Accessed 11 Mar 2020

Tsai W-T, Liu S-C (2015) Thermochemical characterization of cattle manure relevant to its energy conversion and environmental implications. Biomass Conversion Biorefinery 6:71–77

Tsai JH, Macedonio F, Drioli E, Giorno L, Chou CY, Hu FC, Li CL, Chuang CJ, Tung KL (2017) Membrane-based zero liquid discharge: myth or reality? J Taiwan Inst Chem Eng 80:192–202

Tunc CM, Groth AM (2011) Sustainable integrated membrane contactor process for water reclamation, sodium sulfate salt and energy recovery from industrial effluent. Desalination 283:187–192

Udugama IA, Petersen LAH, Falco FC, Junicke H, Mitic A, Alsina XF, Mansouri SS, Gernaey KV (2020) Resource recovery from waste streams in a water-energy-food nexus perspective: toward more sustainable food processing. Food Bioprod Process 119:133–147

Van Linden N, Shang R, Stockinger G, Heijman B, Spanjers H (2020) Separation of natural organic matter and sodium chloride for salt recovery purposes in zero liquid discharge. Water Resour Ind 23:100–117

Vecino X, Reig M, Bhushan B, Gibert O, Valderrama C, Cortina JL (2019) Liquid fertilizer production by ammonia recovery from treated ammonia-rich regenerated streams using liquid-liquid membrane contactors. Chem Eng J 360:890–899

Wang F, Gao B, Su M, Dai C, Huang L, Zhang C (2019) Integrated biorefinery strategy for tofu wastewater biotransformation and biomass valorization with the filamentous microalga Tribonema minus. Bioresour Technol 292:121938

Wang SY, Chen YF, Zhou H, Xu XH, Cheng LH (2020a) Calcium carbonate scaling in forward osmosis for textile reverse osmosis concentrate treatment. J Water Process Eng 35:101–181

Wang Z, Deshmukh A, Du Y, Elimelech M (2020b) Minimal and zero liquid discharge with reverse osmosis using low-salt-rejection membranes. Water Res 170:115–317

Xue S, Wu C, Wu Y, Chen J, Li Z (2015) Bipolar membrane electrodialysis for treatment of sodium acetate waste residue. Sep Purif Technol 154:193–203

Yang Z, Nie C, Hou Q, Zhang L, Zhang S, Yu Z, Pei H (2019) Coupling a photosynthetic microbial fuel cell (PMFC) with photobioreactors (PBRs) for pollutant removal and bioenergy recovery from anaerobically digested effluent. Chem Eng J 359:402–408

Yaqub M, Lee W (2019) Zero-liquid discharge (ZLD) technology for resource recovery from wastewater: a review. Sci Total Environ 681:551–563

Ye Y, Ngo HH, Guo W, Liu Y, Chang SW, Nguyen DD, Ren J, Liu Y, Zhang X (2019a) Feasibility study on a double chamber microbial fuel cell for nutrient recovery from municipal wastewater. Chem Eng J 358:236–242

Ye Y, Ngo HH, Guo W, Chang SW, Nguyen DD, Liu Y, Nghiem LD, Zhang X, Wang J (2019b) Effect of organic loading rate on the recovery of nutrients and energy in a dual-chamber microbial fuel cell. Bioresour Technol 281:367–373

Yin Y, Shi M, Ren Y, Wang S, Hua M, Lu J, Zhang W, Lv L (2020) Wrinkle structure on multifunctional MOFs to facilitate PPCPs adsorption in wastewater. Chem Eng J 387:124196

Zamanpour MK, Kariminia H-R, Vosoughi M (2017) Electricity generation, desalination and microalgae cultivation in a biocathode-microbial desalination cell. J Environ Chem Eng 5:843–848

Zhang Y, Zhu Z, Zheng Y, Chen Y, Yin F, Zhang W, Dong H, Xin H (2019) Characterization of volatile organic compound (VOC) emissions from swine manure biogas digestate storage. Atmosphere 10(7):411

Zhao W, Chen S (2018) Critical parameters selection in polarization behaviour analysis of microbial fuel cells. Bioresource Technol Rep 3:185–190

Zhao J, Wang C, Wang S, Zhang L, Zhang B (2019) Selective recovery of Au(III) from wastewater by a recyclable magnetic $Ni_{0.6}Fe_{2.4}O_4$ nanoparticles with mercaptothiadiazole: interaction models and adsorption mechanisms. J Clean Prod 236:117605

Zheng H, Liu M, Lu Q, Wu X, Ma Y, Cheng Y, Addy M, Liu Y, Ruan R (2018) Balancing carbon/nitrogen ratio to improve nutrients removal and algal biomass production in piggery and brewery wastewaters. Bioresour Technol 249:479–486

Zuo K, Chang J, Liu F, Zhang X, Liang P, Huang X (2017) Enhanced organics removal and partial desalination of high strength industrial wastewater with a multistage microbial desalination cell. Desalination 423:104–110

Zuo K, Chen M, Liu F, Xiao K, Zuo J, Cao X, Zhang X, Liang P, Huang X (2018) Coupling microfiltration membrane with biocathode microbial desalination cell enhances advanced purification and long-term stability for treatment of domestic wastewater. J Membr Sci 547:34–42

Accumulation and Detoxification of Metals by Plants and Microbes

Soma Nag and Swarup Biswas

Abstract

People and natural ecosystem deserve clean water and environment. Increased industrial and a number of anthropogenic activities have substantially increased the content of toxic metals and other pollutants in the wastewater. Approx. 132 lakh m^3 industrial effluents are generated per day in India. The contaminated water could cause harm to the aquatic animals and also pollute the soils affecting the agricultural crops and thus causing danger to the food security. In current years, bioremediation is treated as an efficient and eco-friendly biological method for detoxification of wastewater by researchers. The method has flexibility of use of adsorbents, and various kinds of waste biomass are used for the detoxification purpose. Agricultural wastes, residues of crops after processing, fruit peels and seeds, green plants, algae, fungi, bacteria, and a variety of modified biomass are preferred. Use of industrial biomass or seaweed plants may also be beneficial. Microbial degradation, bioremediation, biosorption, and bioaccumulation are new methods for biodegradation of toxic metal ions and are an effective technology for reducing the concentrations to a very low level. The negative charges of the surface functional groups of the microbes are often support the reclamation process. The modern biotechnology improves the ability of the microbial enzymes to decompose the toxins more rapidly. Microbial remediation and phytoremediation are the future techniques of bioremediation. In this chapter, the role of plants and microbes for degradation of the heavy metals from aqueous system has been discussed.

S. Nag (✉) · S. Biswas
Department of Chemical Engineering, National Institute of Technology Agartala, Agartala, Tripura, India

M. P. Shah (ed.), *Removal of Emerging Contaminants Through Microbial Processes*,
https://doi.org/10.1007/978-981-15-5901-3_17

Keywords

Detoxification · Heavy metals · Bioremediation · Green plants · Microbes · Algae · Fungi · Bacteria · Phytoremediation

1 Introduction

Water is a treasure resource on the earth for sustainability of life on earth. Getting clean water to the human population is the highest priority to the modern world. Millions of people die every year due several waterborne diseases, and a United Nations report discloses that more than a billion of people have no right to use safe drinking water in the third world countries (GLAAS 2012). Extension of industrial activity as well as population outburst has resulted serious degradation of surface, ground, and drinking water by contaminating it with a number of industrial chemicals, pharmaceuticals, pesticides, surfactants, and various personal care products. There are many endocrine disrupting compounds, food additives, etc. (Richardson and Ternes 2014) which are very difficult to eliminate. Expansion and development of nanotechnology and growth of chemical and pharmaceutical industry have raised the content of heavy metals, nanomaterials in the effluents. An estimated average industrial wastewater generation would be approximately 132×10^6 m^3 per day by 2050 in India (Kaur et al. 2012). The contaminated water has damaging effects to the human and aquatic life and is also a great concern for the soils. The affected crops are causing a danger to the food security. As a whole, water pollution is the major reason for drinking water crisis and for death and diseases throughout the globe (WWDR4 2012).

It is understood that this wastewater is 99% water with 1% toxicity and may be reusable by retrieval of contaminants. Among the toxic elements, heavy metals are viewed as most harmful, non-degradable pollutants which carry on for a prolonged period within the environment and cause serious hazards to mankind (Mehmet et al. 2007).

Lead, cadmium, chromium, mercury, uranium, zinc, silver, arsenic, nickel, and many other metals and heavy metals are potential contaminants to any life on earth. They are even toxic at low concentrations and carcinogenic in many occasions. So, the United States Environmental Protection Agency and World Health Organization have restricted the disposal of the heavy metals in the environment and laid down ceilings for each one.

In search of technological answers to this burning issue, a combination of physical, chemical, and biological processes are usually considered as the most effective solution. Bioremediation is useful to clean the infected environments, including soils, groundwater, and marine surroundings. In bioremediation, biomass, both dead and alive, is used to accumulate the contaminants for resolving the environmental issues (Vidali 2001). The bioaccumulation and bioremediation are treated as efficient and eco-friendly biological methods for detoxification of wastewater nowadays to have sustainable solution of the problem by using various kinds of waste biomass, bacteria, fungi, microbes, algae, and other plant materials. Researchers recommend the preferential use of microbes to degrade the contaminants and clean up the environment. The use of waste biomass can cover up the needs of the small scale industries in the developing countries, in a cheap and sustainable way. The key advantages of these methods are minimum production of secondary wastes thus less harmful to the environment. Microbial degradation, bioremediation, etc. are treated as the new methods for biodegradation and an effective technology for reducing the toxicity at a very low level. The modern biotechnology applies technique for the genetic changes of the microorganisms and modifies microbial enzymes to break down the pollutants quickly. Microbial remediation and phytoremediation, phytoextraction, etc. are the future paths of bioremediation.

2 Analysis of Metals in Wastewater

A number of metal ions such as Cr, Pb, Cu, Co, Hg, Cd, As, Fe, Se, V, Ni, and Zn are found in wastewater, and they have hazardous effects on living beings. The metals are mainly discharged from various industries like mining, fertilizer, pesticides, and other chemical manufacturing units. They are polluting the surface and ground water. The elimination of these pollutants at source is required to restrict their mixing with surface water. Arsenic pollution in ground water has become an alarming issue in different parts of the world due to its carcinogenic and other health issues. Common people, especially villagers, are mostly getting affected by the arsenic toxicity, and it is the need of the day to come up with a cheap and simple solution for its removal from drinking water. Here also, bioaccumulation and bioremediation using plant wastes and a variety of microbes may play an important role in a cost-effective way. Newer adsorbents are synthesized from the bio-waste materials by chemical, thermal, and enzymatic treatment to have better adsorption capacity and reusability.

2.1 Toxic Behavior of Heavy Metals

Chromium has two oxidation states in aqueous medium, and Cr(VI) is considered more harmful than Cr(III) by way of producing lung tumors, allergic dermatitis, and even mutation of genes. Cd (II) causes lung fibrosis, liver and kidney diseases, weight loss, hypertension, and many more. Copper is used in household application, but its long exposure even may be harmful. As(III) brings out skin and gastrointestinal effects, liver tumors, and cancer. Lead is very toxic and suspected carcinogen and is the root cause of various diseases like, kidney disease, joint pain, loss of appetite and sterility, etc. Nickel troubles chronic bronchitis and lung and nasal functions. Mercury harms nerve cells, kidney, pulmonary functions, etc. (Ahmad et al. 2005; Saner 1980; Bojes and Pope 2007; Ponder et al. 2000; Uzu et al. 2011; Baum 1982; Kadirvelu 1998; Krishnan and Anirudhan 2002).

3 Removal Techniques

Several methods are available for sequestration of the heavy metals. Among them physical, chemical, and biological treatment processes are mostly in use. The physicochemical treatment methods include reduction and precipitation, coagulation, floatation, electrochemical treatment, ion exchange, cementation, etc., and the physical methods are mainly reverse osmosis, filtration, membrane technology, and adsorption.

3.1 Biological Treatment

Effective trimming down of toxic metal ions by microbial degradation, bioremediation, biosorption, biotransformation, etc. is done using living or dead biomaterials. In this process, inexpensive native materials of biological source, i.e., bacteria, fungi, yeast, algae, and a variety of plant materials, are used as adsorbents, and this can bring down the concentration of toxic elements at a very low level compared to other techniques (Basso et al. 2002). The degrading performance of the microorganisms depends on their growth and metabolism conditions in the toxic effluent and pH; temperature and level of toxicity are the main controlling factors. In bioremediation process, pH plays a major role by changing the solubility of the pollutants. The suitable pH also helps the microorganisms to deliver their optimum performances. The technology is supported by redox reactions which modify the behavior of the toxic pollutants to less harmful and less mobile compound (Tandon and Singh 2016). The mobility and toxicity of Cr(VI), Pb(II), As(III), Ni(II), Cd(II), and some other metal ions depend upon their oxidation states and managed by redox methods. The industrial biomass is also helpful for the degradation process, and seaweed processing plant can be easily cultivated. Simple fermentation process helps the growth of biomass. However, there are certain limitations of biological methods for its commercial use, and researchers are trying to improve it and solve the issues by using hybrid technology (Wang and Chen 2009).

4 Mechanism of Bioremediation Process

The high potential of this process toward metal ion elimination has attracted the researchers to investigate the mechanism of this process so that it becomes a true alternative technology for metal removal from water and wastewater. It is a complex process which combines chemisorptions, complexation, ion exchange, adsorption on surface and within the pores, microprecipitation and condensation, etc. The biomass materials contain mostly lignin, cellulose, and some other constituents like, hemicellulose, lipid, hydrocarbon, sugar, starch, protein, etc. The lingo-cellulosic materials have affinity toward water and thus water infuses through the cell wall and comes in contact with large surface area of the biomass. The amorphous structure of the cellulose further helps in transport across the cell (Demirbas

2008). A large number of different functional groups are present on the cell wall of biomass which helps in metal binding process. Carbonyl, alcoholic, phenolic, amino, amido, and carboxyl group esters are found in the biomass, and their natural attraction for metal ions results metal complexation (Gupta and Ali 2000). To comprehend the method, these functional groups are to be ascertained in each case. Spectroscopic techniques are generally used for identification of the functional groups and to find out their involvement in the metal binding process. Some of the materials are non-selective regarding the metals and can attach a variety of metal ions, whereas some others are found choosy about the metals. The metal adsorption performance may be improved by modification of the functional groups by grafting or exchange (Sud et al. 2008).

4.1 Use of Plant Materials

Materials obtained from various agricultural sources, natural wastes, and residues after processing of crops and fruit wastes are considered as adsorbents due to their natural metal adsorption capacity. These materials can be pre-treated to enhance their adsorption capacity, known as green adsorbents. Though they have less adsorption capacity compared to the commercial adsorbents, their easy availability in large quantities and cost analysis makes them competitive (Kyzas and Kostoglou 2014). In recent years, researchers have tried use of various plant materials to eliminate heavy metals from wastewater. Leaves of different plants were used by Nag et al. for effective remediation of Cr(VI) and Cd(II) (Nag et al. 2016, 2017, 2018a, b). Other agricultural materials such as sugarcane bagasse, rice husk, sawdust, coconut husk, rice mill wastes, neem leaves, fruit seed, and shell and plant roots were studied by researchers (Mohan and Singh 2002; Ajmal et al. 2003; Kadirvelu et al. 2003; Tan et al. 1993; Singha and Das 2012). Sud and her group examined the application agriculture-based unused substances for the treatment of metals from contaminated solutions and reported the potential of the agricultural biomasses (Sud et al. 2008). Rice bran could remove above 80% was of copper, zinc, lead, and cadmium ions (Montanher et al. 2005). Rubber wood sawdust was tried by someothers for Cr (VI) and recorded 60–70% removal (Karthikeyan et al. 2005). Compared to other methods, bioremediation using plant materials produces minimum biological or chemical sludge and has the potential of metal recovery. The metal loaded adsorbents when burnt at high temperature, metal oxides are formed, and these oxides are non-toxic, and metals don't leach out. So, they are safe for disposal in the environment and may be used as fuel or for land filling (Nag et al. 2017).

Biosorption capacities of natural materials are generally low, but the chemical or thermal treatment increases its sorption capacity. When it is treated with acids, such as nitric, hydrochloric, sulfuric, or formic acid, the chemical changes take place, and the functional groups, structure, and pores of the biomass are altered. Different base solutions and metal salts are also used by researchers for activation of adsorbent, and these modifications yield better adsorption capacity. Hydrochloric acid-treated oak sawdust was utilized for the removal of Cu(II), Cr(VI), and Ni(II) by Argun and his

Table 1 Adsorption capacities of different plant-based adsorbents

Adsorbent	Cr(VI)	Zn(II)	Cu(II)	Cd(II)	Pb(II)	Source
Sugar cane bagasse	13.4					Sharma and Forster (1994)
Activated carbon from sugarcane bagasse		31.11		38.03		Mohan and Singh (2002)
Bagasse fly ash modified with hydrogen peroxide	4.35				2.50	Gupta and Ali (2004)
Modified sugarcane Bagasse			114	196	189	Karnitz et al. (2007)
Neem leaf	7.43					Venkateswarlu et al. (2007)
Custard apple peel powder	7.87					Krishna and Sree (2013)
Mango leaf	35.70			16.35		Nag et al. (2017, 2018a)
EDTA treated bagasse (EB)			66.7			Junior et al. (2009)
EDTA treated mercerized cellulose (EMC)			66.7	112.0	232.0	
EDTA treated twice-mercerized bagasse (EMMB)			92.6	149.0	333.0	
Castor leaf powder				38.08		Martins et al. (2013)
Coffee residue					63.00	Boudrahem et al. (2009)

team (Argun et al. 2007). Peanut husk was modified by sulfuric acid and was used for copper, chromium, and lead removal (Li et al. 2007). Some researchers used nitric acid for modification of banana pith (Low et al. 1995). The metal removal capacities of some of the bio-waste materials are shown in Table 1.

4.2 Use of Green Plants

In this case, the green plants are utilized for cleanup, and the process is called phytoremediation. This method is mostly used to attract heavy metals present in the contaminated soil. To get rid of toxic elements from soil near industrial belts, special species of plants which can live on contaminated soils are chosen. In case of phytoextraction, metals are accumulated in the biomass of these plants, and the plant roots remove the pollutants from aqueous medium by the process of rhizofiltration. The phytoremediation encourages the use of some genetically modified plants in case of heavily polluted soils. It is even possible to recover metal ions from the plant tissues (Meagher 2000). It is a cheap method as it requires only the cost of plantation. *Brassica* spp., *Silene vulgaris*, *Arabidopsis halleri*, *Thlaspi caerulescens*, *Alyssum lesbiacum*, etc. are some of the metal tolerant plant species which can survive and accumulate high levels of heavy metals. They are capable to digest copper, aluminum, iron, zinc, cadmium, lead, mercury, etc. (Clemens et al. 2002). Chelating compounds have special roles in metal binding, and the contact of root cells to the heavy metals promotes their production (Cobbett 2000). It is a very slow process and requires several years to eradicate the heavy metals. Researchers are working on its improvement to enhance the natural metal removal capacity (McGrath and Zhao 2003). However, their uses are becoming restricted in several countries due to some other related environmental factors.

4.3 Use of Microbes

The beauty of bioremediation process is that most of the remediates are found in nature with natural metal reduction capacity. In cases, where use of waste is not feasible, microbes are helpful, and the microbial culture seems to be economical (Wang and Chen 2009). The selected natural species, like various kinds of algae, fungi, and bacteria have important metal eradication properties. They can be used to mitigate heavy metals from soil, aquatic, or marine environments. The selected microbes find their importance in case of contaminated industrial wastewater for treatment of radioactive wastes. Microbes pull off the metals by the general bioremediation pathways, i.e., by following one or more of the activities of biosorption, complexation, and bio-assimilation. The surfaces of the microbes are mostly negative due the presence of functional groups and supports in metal binding. The enzymatic transformation has shown its superiority in case of toxic elements. Development of lab-scale bioreactors and their successful capture of metals have shown the possibility of commercial application of the microbes in near future by upscaling the bioreactors. "Oilzapper," a patented consortium of bacteria, was successful to clear out the oils in Mumbai oil spill in 2010 (Sukumaran et al. 2014).

4.4 Use of Bacteria

Present-day researchers are trying a number of living and nonliving bacteria to combat the pollution. *Bacillus cereus*, *Streptomyces*, *Citrobactos*, *Escherichia coli*, etc. are some of them which are preferred for the treatment of metal-tainted effluents (Pan et al. 2007; Jacob et al. 2018). They are mainly collected from the wastes of fermentation process and have a great affinity toward heavy metals. The self-protection mechanism of bacteria helps to change the surface and agglomeration of the cells to capture the pollutants. This was explained and shown by microscopic images by *Bacillus cereus* for Cr(VI) capture. Carboxyl, sulfonate, hydroxyl, amide, etc. are some of the functional groups present in bacteria that take part in metal binding. In case of low level of contaminants, the bacterial biomass works better. Table 2 displays the metal biosorption capacity of various bacterial species.

Table 2 Performance of some bacterial biomass-based adsorbents

Adsorbent	Adsorption capacity in mg/g of adsorbent					Source
	Cr(VI)	Zn(II)	Cu(II)	Cd(II)	Pb(II)	
Bacillus firmus		418.2	381.03		467.0	Salehizadeh and Shojaosadati (2003)
Kocuria *rhizophila*	14.4			9.07		Haq et al. (2016)
Bacillus coagulans	39.9					Srinath et al. (2002)
Alcaligenes sp.					56.8	Jin et al. (2017)
Lactobacillus sp.			0.046			Schut et al. (2011)
Enterobacter sp. J1				46.2		Lu et al. (2006)
Desulfovibrio desulfuricans	99.8		98.2			Kim et al. (2015)
Micrococcus luteus			408.0		1965.0	Puyen et al. (2012)
Bacillus licheniformis	62.0			142.7		Zouboulis et al. (2004)
Enterobacter cloacae					2.3	Kang et al. (2015)

4.5 Use of Fungi

Use of fungal biomass is considered economic with higher metal-binding efficiencies for the treatment of hazardous metals. Literature studies showing the commendable performances of variety of fungi are abridged in Table 3. *Penicillium chrysogenum*, *Rhizopus oryzae*, *Penicillium* spp., *Aspergillus niger* (live), *Penicillium chrysogenum*, and *Aspergillus terreus* are some of the fungi tried by scientists. Existence of metals dictates the metabolic activities of fungi.

4.6 Use of Algae

Algae can prepare food by the process of photosynthesis and have faster growth rate compared to the other members of this group. Its huge availability and potency for detoxification of metal ions encouraged researchers to investigate further. A number of metabolic processes are carried by algae for the bioaccumulation of metals, and modification of cell wall is one of them. This modification helps to increase the metal uptake capacity. Brown and red algae are the two highly capable variety employed for metal attachment (Flores-Chaparro et al. 2017). Deng and his team studied removal of Pb(II) by green algae (Deng et al. 2007). Others tried dried marine green algae for the trimming down copper and zinc (Ajjabi and Chouba 2009). Another group studied cyanobacteria prepared from rice fields of Vietnam for the elimination of Cu(II), Cd(II), and Pb(II) from water and reported the reusability of the bacteria (Tran et al. 2016). Algae find dual use in the field of waste treatment and biofuel production. Its performance may be enhanced by genetically alternations and

Table 3 Bioremediation performance of fungi

Adsorbent	Metal uptake (mg/g) of adsorbent					Source
	Cr(VI)	Cu (II)	Cd (II)	Pb(II)	Others	
Aspergillus niger	11.80					Mondal et al. (2017)
Rhizopus oryzae		34.05				Fu et al. (2012)
Aspergillus niger, *Aspergillus flavus*	Not mentioned			Not mentioned		Dwivedi et al. (2012)
Aspergillus niger				34.4		Dursun et al. (2003)
Lepiota hystrix		8.50		3.89		Kariuki et al. (2017)
Pleurotus platypus					46.7 [Ag (II)]	Das et al. (2010)
Pleurotus sapidus			127.0		287.0 [Hg (II)]	Yalcinkaya et al. (2002)
Aspergillus niger	6.6					Dursun et al. (2003)
Mucor rouxii					0.36 [Ni(II)]	Magyarosy et al. (2002)
Saccharomyces cerevisiae					22.5 [Mn (II)]	Fadel et al. (2017)

other treatments (Favara and Gamlin 2017). The importance of algae is depicted in Table 4 in the field of bioremediation.

5 Conclusion

Environmental contamination is a worldwide problem, and researchers are trying all sorts of cheap and easy to operate technology to come out of the situation. A number of conventional methods were adapted since long, and out of them, biosorption is considered as an innovative and economic technique for the elimination of heavy metals. Some of the waste biomass and microbial agents are considered as good adsorbent because of their effortless accessibility and cost-effectiveness. Wood sawdust, leaves, bacteria, algae, and fungi were utilized for treatment of metal tainted effluent. Microbes are treated as superior members due to their robust growth and multiple chemisorptions sites with potential for redox reactions. Very often, acid, alkali, thermal, or other types of treatment are done to increase the adsorption capacity of these naturally occurring materials. Chemical treatment enhances the surface area and also the efficiency of the active pores. In addition, treatment

Table 4 Adsorption performance of algae-based adsorbents

Adsorbent	Adsorption capacity in mg/g of adsorbent					Source
	Cr(VI)	Cu(II)	Cd(II)	Pb(II)	Others	
Aphanothece halophytica					133 [Zn(II)]	Incharoensakdi and Kitjaharn (2002)
Cladophora sp.				45.4		Lee and Chang (2011)
Chara aculeolata			23.0	105.3	15.2[Zn(II)]	Sooksawat et al. (2016)
Ulva lactuca	60.9	64.5	62.5	68.9		Ibrahim et al. (2016)
Nitella opaca			20.5	104.2	13.4 [Zn (II)]	Sooksawat et al. (2016)
Raw and acid-treated green alga Oedogonium *hatei*	Raw: 31.0 Acid treated: 35.2					Gupta and Rastogi (2009)
Cystoseira barbata			37.6	196.7	78.7 [Ni(II)]	Yalcın et al. (2012)
Cladophora hutchinsiae					74.9 [Se]	Tuzen and Sari (2010)
Eucheuma denticulatum		66.23			81.97 [Pd] 43.48[Zn (II)]	Rahman and Sathasivam (2016)
Raw and acid-treated green alga *Oedogonium hatei*					Untreated alga: 40.9 Acid treated alga 44.9 [Ni(II)]	Gupta et al. (2010)

enhances the strength of the material and its reusability. Enzymatic treatment has also proved its usefulness in many occasions. Literature study reveals enormous uses of plants and microbes for the metal remediation, and the study is still on for the future researchers for their safe and commercial applications.

References

Ahmad R, Rao RAK, Masood MM (2005) Removal and recovery of Cr (VI) from synthetic and industrial wastewater using bark of Pinus roxburghii as an adsorbent. Water Qual Res J Canada 40:462–468

Ajjabi LC, Chouba L (2009) Biosorption of Cu2+ and Zn2+ from aqueous solutions by dried marine green macroalga Chaetomorpha linum. J Environ Manage 90(11):3485–3489

Ajmal M, Rao RAK, Anwar S, Ahmad J, Ahmad R (2003) Adsorption studies on rice husk: removal and recovery of Cd (II) from wastewater. Bioresour Technol 86:147–149

Argun ME, Dursun S, Ozdemir C, Karatas M (2007) Heavy metal adsorption by modified oak sawdust: thermodynamics and kinetics. J Hazard Mater 141(1):77–85

Basso MC, Cerrella EG, Cukierman AL (2002) Lignocellulosic materials as potential biosorbents of trace toxic metals from wastewater. Ind Eng Chem Res 41:3580–3585
Baum SJ (1982) Introduction to organic and biological chemistry. Collier Macmillan, London
Bojes HK, Pope PG (2007) Characterization of EPA's 16 priority pollutant polycyclic aromatic hydrocarbons (PAHs) in tank bottom solids and associated contaminated soils at oil exploration and production sites in Texas. Regul Toxicol Pharmacol 47:288–295
Boudrahem F, Aissani-Benissad F, Aït-Amar H (2009) Batch sorption dynamics and equilibrium for the removal of lead ions from aqueous phase using activated carbon developed from coffee residue activated with zinc chloride. J Environ Manage 90(10):3031–3039
Clemens S, Palmgren MG, Kramer U (2002) A long way ahead: understanding and engineering plant metal accumulation. Trends Plant Sci 7(7):309–315
Cobbett CS (2000) Phytochelatins and their roles in heavy metal detoxification. Plant Physiol 123 (3):825–832
Das D, Das N, Mathew L (2010) Kinetics, equilibrium and thermodynamic studies on biosorption of Ag(I) from aqueous solution by macrofungus Pleurotus platypus. J Hazard Mater 184:765–774
Demirbas A (2008) Heavy metal adsorption onto agro-based waste materials: a review. J Hazard Mater 157:220–229
Deng L, Su Y, Su H, Wang X, Zhu X (2007) Sorption and desorption of lead (II) from wastewater by green algae Cladophora fascicularis. J Hazard Mater 143(1-2):220–225
Dursun AY, Uslu G, Cuci Y, Aksu Z (2003) Bioaccumulation of copper(II), lead(II) and chromium (VI) by growing Aspergillus niger. Process Biochem 38:1647–1651
Dwivedi S, Mishra A, Saini D (2012) Removal of heavy metals in liquid media through fungi isolated from waste water. Int J Sci Res 1(3):181–185
Fadel M, Hassanein NM, Elshafei MM, Mostafa AH, Ahmed MA, Khater HM (2017) Biosorption of manganese from groundwater by biomass of Saccharomyces cerevisiae. HBRC J 13:106–113
Favara P, Gamlin J (2017) Utilization of waste materials, non-refined materials, and renewable energy in in situ remediation and their sustainability benefits. J Environ Manage 204:730–737
Flores-Chaparro CE, Ruiz LFC, de la Torre MCA, Huerta-Diaz MA, Rangel-Mendez JR (2017) Biosorption removal of benzene and toluene by three dried macroalgae at different ionic strength and temperatures: algae biochemical composition and kinetics. J Environ Manage 193:126–135
Fu YQ, Li S, Zhu HY, Jiang R, Yin LF (2012) Biosorption of copper (II) from aqueous solution by mycelial pellets of Rhizopus oryzae. Afr J Biotechnol 11:1403–1411
GLAAS (2012) Global analysis and assessment of sanitation and drinking-water (GLAAS). A report by the challenge of extending and sustaining services. WHO, Geneva, pp 1–112
Gupta VK, Ali I (2000) Utilisation of bagasse fly ash (a sugar industry waste) for the removal of copper and zinc from wastewater. Sep Purif Technol 18(2):131–140
Gupta VK, Ali I (2004) Removal of lead and chromium from wastewater using bagasse fly ash—a sugar industry waste. J Colloid Interface Sci 271(2):321–328
Gupta VK, Rastogi A (2009) Biosorption of hexavalent chromium by raw and acid-treated green alga Oedogonium hatei from aqueous solutions. J Hazard Mater 163(1):396–402
Gupta VK, Rastogi A, Nayak A (2010) Biosorption of nickel onto treated alga (Oedogonium hatei): application of isotherm and kinetic models. J Colloid Interface Sci 342(2):533–539
Haq F, Butt M, Ali H, Chaudhary HJ (2016) Biosorption of cadmium and chromium from water by endophytic Kocuria rhizophila: equilibrium and kinetic studies. Desalin Water Treat 57:19946–19958
Ibrahim WM, Hassan AF, Azab YA (2016) Biosorption of toxic heavy metals from aqueous solution by Ulva lactuca activated carbon. Egypt J Basic Appl Sci 3(3):241–249
Incharoensakdi A, Kitjaharn P (2002) Zinc biosorption from aqueous solution by a halotolerant cyanobacterium Aphanothece halophytica. Curr Microbiol 45(4):261–264

Jacob JM, Karthik C, Saratale RG, Kumar SS, Prabakar D, Kadirvelu K, Pugazhendhi A (2018) Biological approaches to tackle heavy metal pollution: a survey of literature. J Environ Manage 217:56–70

Jin Y, Yu S, Teng C, Song T, Dong L, Liang J, Bai X, Xu X, Qu J (2017) Biosorption characteristic of Alcaligenes sp. BAPb. 1 for removal of lead (II) from aqueous solution. 3 Biotech 7(2):123

Junior OK, Gurgel LVA, de Freitas RP, Gil LF (2009) Adsorption of Cu (II), Cd (II), and Pb (II) from aqueous single metal solutions by mercerized cellulose and mercerized sugarcane bagasse chemically modified with EDTA dianhydride (EDTAD). Carbohydr Polym 77 (3):643–650

Kadirvelu K (1998) Preparation and characterisation of activated carbon from coirpith and its utilization in treatment of metal bearing wastewater. PhD. thesis, Bharathiar University, Coimbatore, Tamil, Nadu, India

Kadirvelu K, Kavipriya M, Karthika C, Radhika M, Vennilamani N, Pattabhi S (2003) Utilization of various agricultural wastes for activated carbon preparation and application for the removal of dyes and metal ions from aqueous solutions. Bioresour Technol 87:129–132

Kang C-H, Oh SJ, Shin Y, Han S-H, Nam I-H, So J-S (2015) Bioremediation of lead by ureolytic bacteria isolated from soil at abandoned metal mines in South Korea. Ecol Eng 74:402–407

Kariuki Z, Kiptoo J, Onyancha D (2017) Biosorption studies of lead and copper using rogers mushroom biomass: Lepiota hystrix. South Afr J Chem Eng 23:62–70

Karnitz O Jr, Gurgel LVA, De Melo JCP, Botaro VR, Melo TMS, de Freitas Gil RP, Gil LF (2007) Adsorption of heavy metal ion from aqueous single metal solution by chemically modified sugarcane bagasse. Bioresour Technol 98(6):1291–1297

Karthikeyan T, Rajgopal S, Miranda LR (2005) Cr (VI) adsorption from aqueous solution by Hevea brasilinesis saw dust activated carbon. J Hazard Mater 124:192–199

Kaur R, Wani SP, Singh AK, Lal K (2012) Wastewater production, treatment and use in India National Report presented at the 2nd regional workshop on safe use of wastewater in agriculture

Kim IH, Choi J-H, Joo JO, Kim Y-K, Choi J-W, Oh B-K (2015) Development of a microbe-zeolite Carrier for the effective elimination of heavy metals from seawater. J Microbiol Biotechnol Food Sci 25:1542–1546

Krishna D, Sree R (2013) Response surface modeling and optimization of chromium (VI) removal from waste water using custard apple peel powder. Walailak J Sci Technol 11:489–496

Krishnan KA, Anirudhan TS (2002) Removal of mercury (II) from aqueous solutions and chlor-alkali industry effluent by steam activated and sulphurised activated carbons prepared from bagasse pith: kinetics and equilibrium studies. J Hazard Mater 92:161–183

Kyzas GZ, Kostoglou M (2014) Green adsorbents for wastewaters: a critical review. Materials 7:333–364

Lee Y-C, Chang S-P (2011) The biosorption of heavy metals from aqueous solution by Spirogyra and Cladophora filamentous macroalgae. Bioresour Technol 102(9):5297–5304

Li Q, Zhai J, Zhang W, Wang M, Zhou J (2007) Kinetic studies of adsorption of Pb (II), Cr (III) and Cu (II) from aqueous solution by sawdust and modified peanut husk. J Hazard Mater B 141:163–167

Low KS, Lee CK, Leo AC (1995) Removal of metals from electroplating wastes using banana pith. Bioresour Technol 51:227–231

Lu W-B, Shi J-J, Wang C-H, Chang J-S (2006) Biosorption of lead, copper and cadmium by an indigenous isolate Enterobacter sp. J1 possessing high heavy-metal resistance. J Hazard Mater 134(1-3):80–86

Magyarosy A, Laidlaw RD, Kilaas R, Echer C, Clark DS, Keasling JD (2002) Nickel accumulation and nickel oxalate precipitation by Aspergillus niger. Appl Microbiol Biotechnol 59:382–388

Managing water under uncertainty and risk (2012) Facts and Figures from the United Nations World Water Development Report 4 (WWDR4)

Martins AE, Pereira MS, Jorgetto AO, Martines MAU, Silva RIV, Saeki MJ, Castro GR (2013) The reactive surface of Castor leaf [Ricinus communis L.] powder as a green adsorbent for the removal of heavy metals from natural river water. Appl Surf Sci 276:24–30

McGrath SP, Zhao F-J (2003) Phytoextraction of metals and metalloids from contaminated soils. Curr Opin Biotechnol 14(3):277–282
Meagher RB (2000) Phytoremediation of toxic elemental and organic pollutants. Curr Opin Plant Biol 3(2):153–162
Mehmet EA, Sukru D, Celalettin O, Mustafa K (2007) Heavy metal adsorption by modified oak sawdust: thermodynamics and kinetics. J Hazard Mater 141:77–85
Mohan D, Singh KP (2002) Single- and multi-component adsorption of cadmium and zinc using activated carbon derived from bagasse—an agricultural waste. Water Res 36:2304–2318
Mondal NK, Samanta A, Dutta S, Chattoraj S (2017) Optimization of Cr(VI) biosorption onto Aspergillus niger using 3-level Box-Behnken design: equilibrium, kinetic, thermodynamic and regeneration studies. J Genet Eng Biotechnol 15:151–160
Montanher SF, Oliveira EA, Rollemberg MC (2005) Removal of metal ions from aqueous solutions by sorption onto rice bran. J Hazard Mater B 117:207–211
Nag S, Mondal A, Mishra U, Bar N, Das SK (2016) Removal of chromium (VI) from aqueous solutions using rubber leaf powder: batch and column studies. Desalin Water Treat 57 (36):16927–16942
Nag S, Mondal A, Bar N, Das SK (2017) Biosorption of chromium (VI) from aqueous solutions and ANN modelling. Environ Sci Pollut Res 24(23):18817–18835
Nag S, Bar N, Das SK (2018a) Sustainable bioremediation of Cd (II) in fixed bed column using green adsorbents: application of kinetic models and GA-ANN technique. Environ Technol Innov 11:83–104
Nag S, Mondal A, Roy DN, Bar N, Das SK (2018b) Sustainable bioremediation of Cd (II) from aqueous solution using natural waste materials: kinetics, equilibrium, thermodynamics, toxicity studies and GA-ANN hybrid modelling. Environ Technol Innov 11:83–104
Pan JH, Liu RX, Tang HX (2007) Surface reaction of Bacillus cereus biomass and its biosorption for lead and copper ions. J Environ Sci 19:403–408
Ponder SM, Darab TG, Mallouk TE (2000) Remediation of Cr(VI) and Pb(II) aqueous solutions using supported, nanoscale zero-valent iron. Environ Sci Technol 34:2564–2569
Puyen ZM, Villagrasa E, Maldonado J, Diestra E, Esteve I, Solé A (2012) Biosorption of lead and copper by heavy-metal tolerant Micrococcus luteus DE2008. Bioresour Technol 126:233–237
Rahman MS, Sathasivam KV (2016) Heavy metal biosorption potential of a Malaysian Rhodophyte (Eucheuma denticulatum) from aqueous solutions. Int J Environ Sci Technol 13(8):1973–1988
Richardson SD, Ternes TA (2014) Water analysis: emerging contaminants and current issues. Anal Chem 86(6):2813–2848
Salehizadeh H, Shojaosadati SA (2003) Removal of metal ions from aqueous solution by polysaccharide produced from Bacillus firmus. Water Res 37:4231–4235
Saner G (1980) Chromium in nutrition and disease. Alan R Liss Inc., New York
Schut S, Zauner S, Hampel G, König H, Claus H (2011) Biosorption of copper by wine-relevant lactobacilli. Int J Food Microbiol 145:126–131
Sharma DC, Forster CF (1994) A preliminary examination into the adsorption of hexavalent chromium using low-cost adsorbents. Bioresour Technol 47:257–264
Singha B, Das SK (2012) Removal of Pb (II) ions from aqueous solution and industrial effluent using natural biosorbents. Environ Sci Pollut Res 19:2212–2226
Sooksawat N, Meetam M, Kruatrachue M, Pokethitiyook P, Inthorn D (2016) Equilibrium and kinetic studies on biosorption potential of charophyte biomass to remove heavy metals from synthetic metal solution and municipal wastewater. Biorem J 20(3):240–251
Srinath T, Verma T, Ramteke PW, Garg SK (2002) Chromium (VI) biosorption and bioaccumulation by chromate resistant bacteria. Chemosphere 48:427–435
Sud D, Mahajan G, Kaur MP (2008) Agricultural waste material as potential adsorbent for sequestering heavy metal ions from aqueous solution: a review. Bioresour Technol 99 (14):6017–6027

Sukumaran S, Mulik J, Rokade MA, Kamble A (2014) Impact of 'Chitra' oil spill on tidal pool macrobenthic communities of a tropical rocky shore (Mumbai, India). Estuaries Coast 37 (6):1415–1431

Tan WT, Ooi ST, Lee CK (1993) Removal of chromium (VI) from solution by coconut husk and palm pressed fibre. Environ Technol 14:277–282

Tandon PK, Singh SB (2016) Redox processes in water remediation. Environ Chem Lett 2016 (14):15–25

Tran HT, Vu ND, Matsukawa M, Okajima M, Kaneko T, Ohki K, Yoshikawa S (2016) Heavy metal biosorption from aqueous solutions by algae inhabiting rice paddies in Vietnam. J Environ Chem Eng 4(2):2529–2535

Tuzen M, Sari A (2010) Biosorption of selenium from aqueous solution by green algae (Cladophora hutchinsiae) biomass: equilibrium, thermodynamic and kinetic studies. Chem Eng J 158 (2):200–206

Uzu G, Sobanska S, Sarret G, Sauvain J-J, Pradere P, Dumat C (2011) Characterization of lead-recycling facility emissions at various workplaces: major insights for sanitary risks assessment. J Hazard Mater 186:1018–1027

Venkateswarlu P, Ratnam MV, Rao DS, Rao MV (2007) Removal of chromium from an aqueous solution using Azadirachta indica (neem) leaf powder as an adsorbent. Int J Phys Sci 2:188–195

Vidali M (2001) Bioremediation. an overview. Pure Appl Chem 73(7):1163–1172

Wang J, Chen C (2009) Biosorbents for heavy metals removal and their future. Biotechnol Adv 27:195–226

Yalcin S, Sezer S, Apak R (2012) Characterization and lead (II), cadmium (II), nickel (II) biosorption of dried marine brown macro algae Cystoseira barbata. Environ Sci Pollut Res 19(8):3118–3125

Yalcinkaya Y, Arica MY, Soysal L, Denzili A, Genc O, Bektas S (2002) Cadmium and mercury uptake by immobilized Pleurotus sapidus. Turk J Chem 26:441–452

Zouboulis AI, Loukidou MX, Matis KA (2004) Biosorption of toxic metals from aqueous solutions by bacteria strains isolated from metal-polluted soils. Process biochem 39(8):909–916

Enzymatic Degradation of Industrial Wastewater Pollutants

Sama A. Al-Mutwalli, Seyda Korkut, Muhammet Samet Kilic, and Derya Y. Imer

Abstract

Enzymes are biocatalysts which intend to accelerate any biochemical reaction. The mutual principle of different enzymes is to form an enzyme-substrate complex. Some enzymes need activators with the intention to become active, while others do not. A decreasing or termination of enzymes activities can be occurred by competitive, noncompetitive, or uncompetitive inhibitions. In the last two decades, enzymes gained various advantages over microorganisms with the purpose of treating different forms of pollutants in different types of wastewater. In addition, the recovery of effective and beneficial compounds from industrial wastewater particularly is another advantage offered by enzymes. The development of competent and suitable solutions is a significant research area for degrading specific compounds in industrial wastewaters. Enzymatic degradation, which is a process of decomposing natural and synthetic polymers, is one of the approaches in order to treat or recover different types of pollutants. Different configurations of enzymatic reactors can be operated with the purpose of conducting the enzymatic degradation process to achieve a promising treatment of industrial wastewater.

S. A. Al-Mutwalli (✉) · D. Y. Imer (✉)
Department of Environmental Engineering, Istanbul Technical University, Istanbul, Turkey
e-mail: imerd@itu.edu.tr

S. Korkut
Department of Environmental Engineering, Zonguldak Bulent Ecevit University, Zonguldak, Turkey
e-mail: s.korkut@beun.edu.tr

M. S. Kilic
Department of Biomedical Engineering, Zonguldak Bulent Ecevit University, Zonguldak, Turkey
e-mail: msametk@beun.edu.tr

M. P. Shah (ed.), *Removal of Emerging Contaminants Through Microbial Processes*,
https://doi.org/10.1007/978-981-15-5901-3_18

Keywords

Enzymatic degradation · Industrial wastewater pollutants · Enzymatic reactors · Enzyme activation · Enzyme inhibition

1 Introduction

Enzymes are biocatalysts which intend to accelerate any biochemical reaction. The mutual principle of different enzymes is to form an enzyme-substrate complex. In the last two decades, enzymes gained various advantages over microorganisms with the purpose of treating different forms of pollutants in wastewater. Enzymes as biocatalysts are stereospecific, nontoxic, environment-friendly, energy-conserving alternatives and the most proficient catalysts, offering much more competitive processes compared to chemical catalysts (Deng et al. 2019). Enzymes have good resistance properties for a wide range of operational conditions (e.g., temperature, ionic strength, and pH) which they can still be activated (Eibes et al. 2007; Pandey et al. 2017). Furthermore, the enzymes differ from microorganisms when microbial growth is not necessary which leads to shortening the required operational time. The reusability of enzymes is another advantage. The enzymatic system never produces the waste sludge (Eibes et al. 2007). Enzymes are highly selective and efficient biocatalysts as they can treat particular pollutants with no side effects on the other compounds by precipitating them or converting them to new products (Pandey et al. 2017).

Enzymatic degradation is a decomposing process of natural and synthetic compounds and is an alternative way in order to treat or remove the different types of pollutants. The specific enzymatic reactors can be designed with the purpose of sustainable enzymatic degradation to achieve a promising treatment of industrial wastewater. By using different types of enzymatic reactors, enzymes are hard to be inhibited when a high concentration of pollutants is subjected. Selecting the most appropriate reactor depends on the characteristics of the enzyme-specific reaction kinetics, operation mode, and flow characteristics. The main factors associated with the choice of the type of reactor to be used are summarized in Fig. 1.

This chapter has some important theoretical and technical aspects involving the enzyme working mechanism and reactor design in the enzymatic treatment of industrial pollutants, showing the present status and the future prospects of their applications. In the theoretical part, the interactions between enzyme and substrate (as pollutants) will be given with the functionality of enzymes and analysis of main parameters influencing enzyme activity. For the technology part, the feasibility and applicability of enzymes in industrial sectors will be discussed with different reactor types and operation modes.

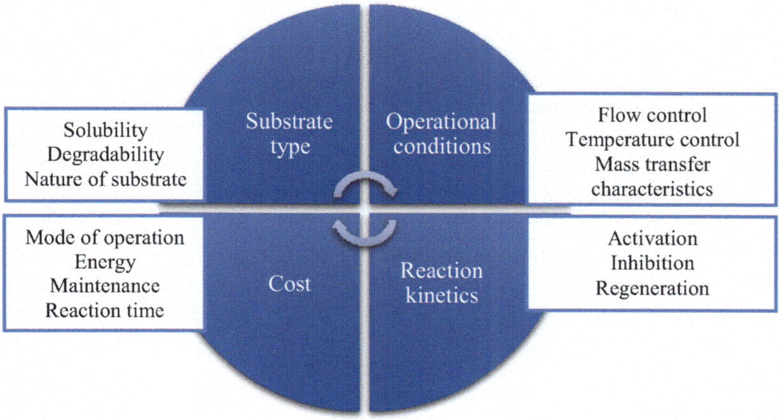

Fig. 1 Reactor selection factors for enzymatic treatment

2 How Do Enzymes Work?

The most powerful catalysts known are enzymes (Voet and Voet 1996). Enzymes are high- efficiency catalysts that operate within the mild conditions of room temperature, mostly at neutral pH, and carry out reactions at a high rate. The catalytic powers of enzymes have made them useful structures in many fields such as applied science, medicine, and industry. Every biological conversion is catalyzed by an enzyme. Enzymes are specific, widely used, and highly effective biocatalysts and react much faster than chemical catalysts. There are more than two thousand known enzymes according to an investigation in the 1980s (Straathof and Adlercreutz 2000). Enzymes are called by adding the suffix "-ase" at the end of the substrate or functions in the reaction they catalyze such as alcohol dehydrogenase, urease, etc. Enzymes can have simply clustered proteins or complex proteins containing more than one subunit. Some enzymes contain nonprotein groups such as NAD, FAD, CoA, Mg, Zn, and Fe. These groups are called cofactors, the protein part is apoenzyme, and the entire molecule is called holoenzyme which is comprised of apoenzyme and cofactor. Some enzymes are in different molecular forms but catalyze the same reaction, and these enzymes are called isoenzymes. Enzymes that can be divided into different groups according to their functions are summarized in Table 1 (Dugas 1996).

2.1 Enzyme-Catalyzed Reactions

Generally, the enzymes bind to the substrate and reduce the activation energy of the reaction and catalyze and increase the reaction rate. It does not affect the change of free energy or the equilibrium constant of the reaction. The studies with X-rays and

Table 1 The classifications of enzymes according to their functions

Oxidoreductases	They are effective in –CH–OH, –C=O, –CH–NH_2, –CH–NH–, NADH groups
Transferases	They provide the transfer of groups containing aldehyde/ketone, acyl, glucosyl, phosphate, and S-
Hydrolases	They are effective in ester, glucoside, peptide, C–N bonds
Lyases	They are often forming a new double bond or a new ring structure such as –C=C–, –C=O, –C=N–
Isomerases	They are enzymes that provide structural rearrangement of isomers
Ligases	An enzyme that combines them by creating a chemical bond between two large molecules. They form the –C–O–, –C–S–, –C–N–, and –C–C– bonds

Raman spectroscopy have shown that the enzyme-substrate (ES) complex is formed. The bonds in the ES complex are weak van der Waals forces and hydrogen bonds, and the ES complex is unstable (Yang et al. 2017). The substrate binds to a specific unit of the enzyme called the active site. This interaction can be explained by the key-lock model as shown in the reaction below (Fischer 1894):

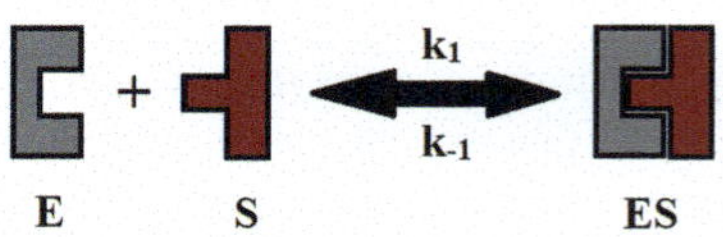

Some enzymes can bind more than one substrate (allosteric regulation of enzymes). Binding of the substrate can change the three-dimensional structure of the enzyme. Binding of other molecules to the enzyme can change the active part properties and decrease or stop the enzymatic reaction rate.

Kinetics of enzymatic reactions are known as Michaelis-Menten or saturation kinetics (Michaelis and Menten 1913). The substrate binds to the enzyme molecule to form the enzyme-substrate (ES) complex. This complex decomposes to yield products (*P*) and the enzyme (*E*) which turns into its native state at the end of the reaction. The general reaction of the enzyme and substrate is given in Eq. (1):

$$E + S \underset{k_{-1}}{\overset{k_1}{\rightleftharpoons}} \mathrm{ES} \overset{k_p}{\rightarrow} E + P \tag{1}$$

There are a certain number of substrate binding sites (active sites) on the enzyme. In the low substrate concentrations, the active sites of the enzyme are empty, and the reaction rate increases with the increasing substrate concentration. At high substrate concentrations, the active sites of the enzyme are fully filled, and the reaction rate is constant (saturation state). The transformation reaction of ES to product is considered unidirectional, and this aspect is only valid for initial rates of reaction where the product concentration is low. The product formation rate in enzymatic reactions is given in the following expression (Eq. 2):

$$V = \frac{dP}{dt} = -\frac{dS}{dt} = k_p \tag{2}$$

V (mole/L s) is the rate of product formation, and it is equal to ($k_p \times$ ES).

The rate of change of the ES complex and conservation of the enzyme are given by Eqs. (3) and (4), respectively:

$$\frac{d(ES)}{dt} = k_1(E)(S) - k_{-1}(ES) - k_P(ES) \tag{3}$$

$$E_0 = E + ES \tag{4}$$

The ES complex is an unstable molecule and is quickly converted to $P + E$. In particular, d(ES)/dt = 0 when enzyme concentration is quite smaller than substrate concentration ((E_0) $\ll$ (S_0)). This is called a quasi-steady-state approach.

For ES, if d(ES)/dt = 0 is placed in Eq. (3), the expression is converted to Eq. (5):

$$ES = \frac{k_1 \times (E_0 - ES) \times S}{k_{-1} + k_p} \tag{5}$$

$$K_m = \frac{k_{-1} + k_p}{k_1} \tag{6}$$

$$ES = \frac{(E_0 - ES) \times S}{K_m} \tag{7}$$

$$ES \times K_m = (E_0 \times S) - (ES \times S) \tag{8}$$

$$ES \times (K_m + S) = E_0 \times S \tag{9}$$

If both sides of this equation are multiplied by k_p,

$$k_p \times ES \times (K_m + S) = k_p \times E_0 \times S \tag{10}$$

$$V = \frac{V_m \times S}{K_m + S} \tag{11}$$

Equation (11) is known as Michaelis-Menten equation. Here, V_m is the maximum reaction rate ($V_m = k_pE_0$). The variation of the reaction rate versus the substrate concentration is shown below. At low substrate concentrations ($S \ll K_m$), the reaction is of the first order and turns to $V = (V_m/K_m)S$. In high substrate concentrations ($S \gg K_m$), the reaction is of the zero-order ($V = V_m$). K_m is the Michaelis constant defined as the decomposition constant of ES complex. Higher K_m values indicate the easily decomposed ES complex, while lower K_m indicates the more stable ES complex. Since the Michaelis constant is equal to the substrate concentration ($S = K_m$) when $V = 1/2V_m$, K_m is also known as the half-saturation constant. Although the curve shown below can be used to calculate the rate constants

(V_m and K_m), the precise determination of those is made from the linear graph constructed by using plots of $1/V$ versus $1/S$:

$$\frac{1}{V} = \frac{K_m}{V_m} \times \frac{1}{S} + \frac{1}{V_m} \tag{12}$$

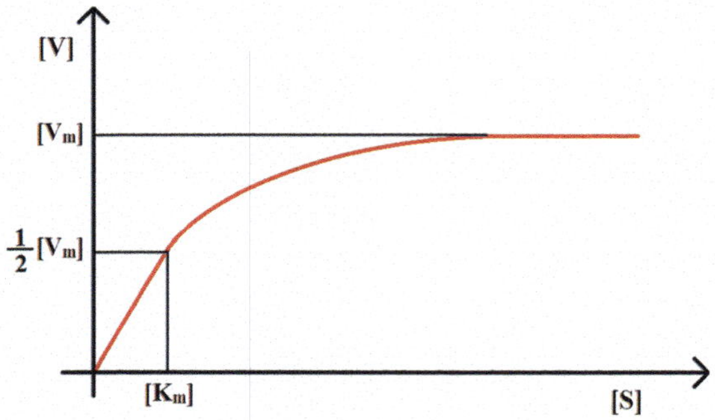

Another approach of the enzymatic reaction kinetics is Briggs-Haldane (Briggs and Haldane 1925). In this approach, E, S, and ES will not reach equilibrium or accumulate to an equilibrium value if the disintegration rate of ES into the product is greater than the decomposition rate of the ES complex to $E + S$. However, if the enzyme is used in a catalytic quantity ($S \gg E_{total}$), a continuous state will be reached in which the ES concentration remains constant very shortly after mixing the enzyme and the substrate. The same kind of rate equation is achieved in both approaches (Eq. 11).

2.2 Activation of Enzyme Reactions

Chemical compounds that increase the enzymatic reaction rate are called enzyme activators, and their effects are opposite to the action of enzyme inhibitors. Examples of activators are small organic molecules, ions, proteins, peptides, and lipids (Lopina 2017). Small inorganic molecules effect on many enzymes and directly activate them. For example, Ca^{2+} cation is a small inorganic molecule that activates enzymes such as protein kinases C, adenylyl cyclases, phospholipases II, etc. In general, there is a special site for the binding of Ca^{2+} in enzymes. The binding of Ca^{2+} leads to a conformational change in enzyme, and then catalytic activity increases (Villarroel et al. 2014). Cations can also bind to the substrate and increase the substrate affinity to the enzyme for enzyme activation. Some enzyme activation processes also occur through the elimination of enzyme inhibitors. A variety of enzymes are inhibited by some of the cations such as heavy metal. Small organic compounds such as chelating agents ethylene glycol-bis(β-aminoethyl ether)-*N,N,N′,N′*-tetraacetic acid (EGTA) and ethylenediaminetetraacetic acid (EDTA) bind these inhibitory cations and, thus, eliminate their inhibitory effects (Lopina 2017).

The process of regulating the enzyme by binding effectors (activators or inhibitors) to another location from the active site of the enzyme is called allosteric modulators. The site where effectors bind is named the allosteric site. Allosteric sites allow the effectors to bind to the protein and often lead to conformational change involving protein dynamics. Effectors that increase and decrease the activity of the protein are called allosteric activators and allosteric inhibitors, respectively (Nelson et al. 2005).

2.3 Inhibition of Enzyme-Catalyzed Reactions

All types of materials which decrease an enzyme-catalyzed reaction rate are called an inhibitor. Inhibition studies give insight into the specificity of the enzyme, the physical/chemical architecture of its active site, and the kinetic mechanisms of the reaction. Enzyme inhibitors are divided into two classes, reversible and irreversible. A reversible inhibitor binds to enzyme reversibly, and its inhibition effects are eliminated by dialysis and simple dilution, to regenerate the enzyme activity, whereas an irreversible inhibitor cannot be removed from the enzyme by dialysis. Reversible inhibitors constitute a balance system with enzyme quickly to show a significant degree of inhibition based on enzyme, substrate, and inhibitor concentrations. This system remains steady for as long as initial speed studies are normally performed, whereas the inhibition degree of the irreversible inhibitors may increase during this time (Segel 1975).

2.3.1 Competitive Inhibition

A competitive inhibitor interacts with the free enzyme to prevent substrate binding. In brief, it competes with the substrate to bind the active center of the enzyme. It can be an analogue of the real substrate which cannot be metabolized, an alternative substrate, and a reaction product. Its other inhibition effect is the substrate binding to the enzyme because of inhibitor's steric hindrance effect which creates a stearic effect on enzyme's active center and competing with the real substrate. In the competitive inhibition, if the substrate concentration is low at a definite inhibitor and enzyme concentration, the competition with substrate realizes easily and inhibition degree will be high. On the contrary, when the substrate concentration is high, the competition is less successful and the inhibition degree will be lower. In this scenario, the inhibitory effect is neglected in excessive substrate concentrations. Therefore, while the V_m value does not change, the K_m value will increase (Segel 1968). Apparent velocity of the enzymatic reaction in the competitive inhibition is expressed as

$$V_{\mathrm{i}} = \frac{V_{\mathrm{m}} \times S}{K_{\mathrm{m}} \times \left(1 + \frac{I}{K_{\mathrm{i}}}\right) + S} \tag{13}$$

In the reaction catalyzed by an enzyme with a single substrate and a single binding site, the reaction scheme in the presence of a competitive inhibitor is presented below:

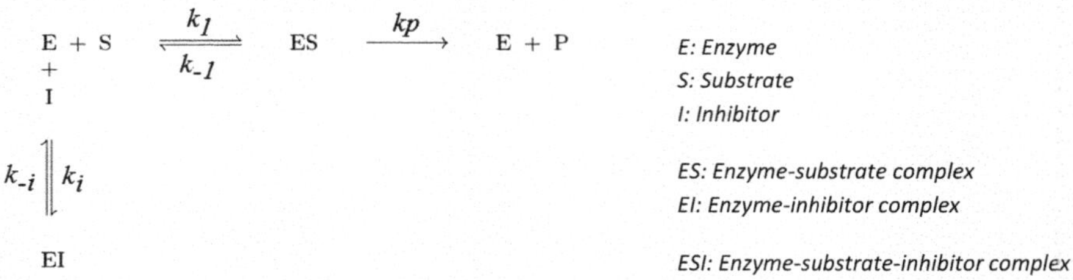

2.3.2 Noncompetitive Inhibition

A noncompetitive inhibitor does not affect the binding of the substrate to the active site of the enzyme. Likewise, the substrate has no effect on inhibitor binding. In this type of inhibition, the inhibitor and the substrate are reversible, incidentally and independently bound to the enzyme in different regions. That is, the inhibitor can be bound to *E* and ES, and *S* can be bound to *E* and EI. In contrast, the resulting ESI complex is inactive, that is, it does not turn into a product. The reaction scheme in the presence of a noncompetitive inhibitor is presented below:

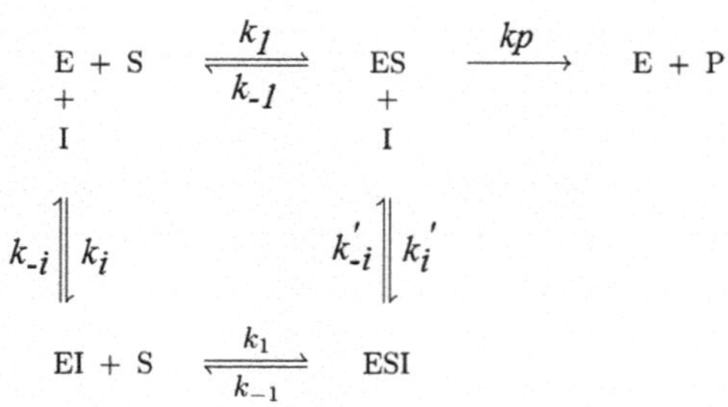

It is clear that even a high substrate concentration at any inhibitor concentration cannot convert the enzyme into a productive ES complex. It will remain the nonproductive ESI complex of the enzyme at any inhibitor concentration. Therefore, in the presence of a noncompetitive inhibitor, the V_{m} value ($V_{\mathrm{m\ app}}$) will be smaller than the V_{m} value observed in the absence of the inhibitor. K_{m} value does not change since *E* and EI complex can combine with *S* with the same affinity. The effect of the noncompetitive inhibitor is to reduce the amount of enzyme that causes product formation (Segel 1968, 1975). Apparent velocity of the enzymatic reaction in the noncompetitive inhibition is expressed as

$$V_{\mathrm{i}} = \frac{V_{\mathrm{m}} \times S}{K_{\mathrm{m}} \times \left(1 + \frac{I}{K_{\mathrm{i}}}\right) + \left(1 + \frac{S}{K_{\mathrm{i}}}\right)} \tag{14}$$

2.3.3 Uncompetitive Inhibition

An uncompetitive inhibitor binds to ES complex reversibly to produce ESI complex. The inhibitor cannot bind to the native enzyme. Since both the ESI complex formed by the effect of the inhibitor and some of the ES remain unproductive, both the V_{m} and K_{m} values of the enzyme decrease in uncompetitive inhibition. The inhibitor and the substrate do not compete with each other to bind to the same site on the enzyme. Therefore, increasing the substrate concentration does not prevent inhibition. The reaction scheme in the presence of an uncompetitive inhibitor and apparent velocity of the enzymatic reaction is shown below:

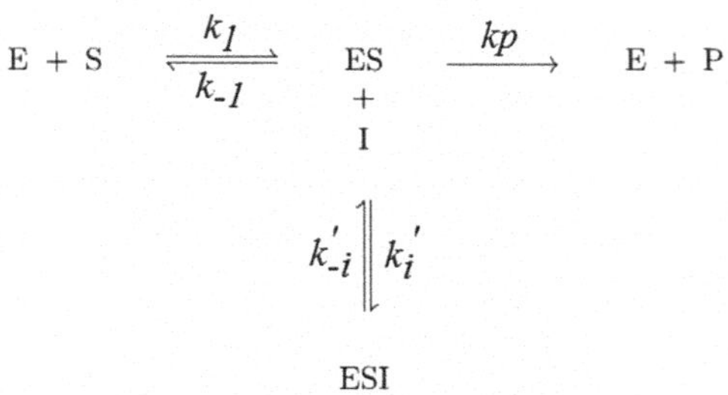

$$V_{\mathrm{i}} = \frac{\frac{V_{\mathrm{m}}}{\left(1+\frac{I}{K_{\mathrm{i}}}\right)} \times S}{\frac{K_{\mathrm{m}}}{\left(1+\frac{I}{K_{\mathrm{i}}}\right)} + S} \tag{15}$$

Identification of the classical types of inhibition from their effects on the plots of velocity versus substrate concentration or 1/velocity versus 1/substrate concentration is given in Fig. 2 (Bhagavan and Ha 2011).

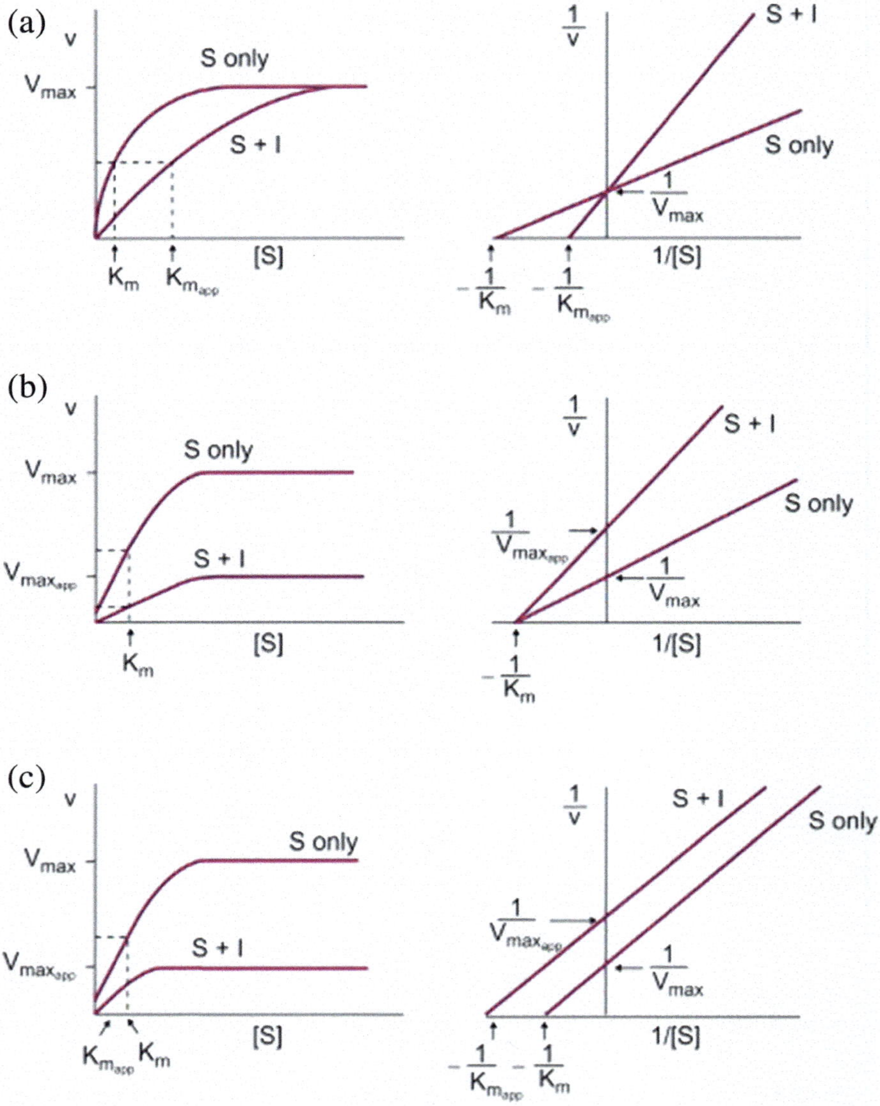

Fig. 2 Substrate concentration versus velocity and Lineweaver-Burk diagrams of competitive (**a**), noncompetitive (**b**), and uncompetitive (**c**) inhibition

3 Factors Influencing Enzyme Activity

3.1 pH Dependence

The pH of a solution plays a vital role in the enzyme. As the pH changes, the charge state of the ionizable groups in an enzyme or the charge of the substrate molecule changes. This can lead to denaturation in the enzyme, non-denaturing structural changes, degradation of oligomers, and/or changes in activity (Grahame et al. 2015). The most intracellular enzymes exhibit optimal activity in a pH range of 5–9. This is due to the fact that proteins function in an environment that reflects this pH. The reason for this narrow pH range is variety of amino acid residues, and the carboxyl and amide termini of proteins have a pK_a range in the range of intracellular pH. As a result, a change in pH can protonate or deprotonate a side group, thereby changing its chemical features. This conformation can cause a decrease in substrate affinity. A dramatic change of pH can change the protein folding, thereby completely deactivating the enzyme or causing irreversible proteolysis (Berg et al. 2007).

3.2 Temperature Dependence

The rate of most chemical reactions increases with increasing temperature. As the increase in temperature provides more kinetic energy to the molecules, more productive collisions are observed per unit time. The enzyme-catalyzed reactions behave similarly. In maintaining the tertiary structure of an enzyme, primarily non-covalent interactions are effective. In practical terms, the enzyme molecule is very fragile. If the enzyme absorbs a lot of energy, the tertiary structure is disrupted and undergoes denaturation; as a result of this, the enzyme loses its catalytic activity. Consequently, the increase in the reaction rate as a result of the increase in the number of collisions of enzyme and substrate molecules with the increase in temperature will not be at the desired level due to the increase in denaturation rate (Segel 1968, 1975). Therefore, the curve obtained by marking the reaction rate versus temperature shows a peak defined as the optimum temperature.

3.3 Substrate Concentration

In an enzyme-catalyzed reaction, if the amount of the enzyme is kept constant and the initial substrate concentration is then gradually increased, the reaction velocity increases until reaching a maximum value (V_m). After this point, increasing substrate concentration does not affect the reaction rate. It is theorized that when this maximum velocity had been reached, all of the available enzymes have been converted to ES. K_m is defined as the substrate concentration at which the reaction rate is half of V_m. A K_m value is an indicator for several things about a particular enzyme: low K_m indicates that the enzyme requires only a low substrate concentration to become saturated. Hence, the V_m is reached at relatively low substrate concentrations, while high K_m indicates that the reaching to V_m is achieved by high substrate concentrations. The lowest K_m value is known to belong to the natural substrate of the enzyme, though this is not true for all enzymes.

The initial rate may sometimes be lower than V_m at excess substrate concentrations. This condition is called substrate inhibition. In substrate inhibition, one substrate molecule binds to one region of the enzyme, while the other substrate molecule binds to another region of the enzyme to form a dead-end complex. This suggests uncompetitive inhibition. The extra substrate molecule binds to the ES complex like an uncompetitive inhibitor.

3.4 Enzyme Concentration

Each enzyme needs time to produce one unit of product. More enzymes are required to generate more products in the same reaction time. If more enzyme molecules are available in the reaction, the substrate can be converted to the product more quickly by assuming that all other factors are constant. Two enzyme molecules produce two units of product in the same reaction time. Thus, increasing enzyme concentration leads to a linear increase in reaction rate in a definite time period. However, if the longer reaction time is selected, it is not possible to maintain linearity at initial rates. Therefore, the first thing to do in the kinetic examination of an enzyme should be to determine the limits of this linearity. In this way, the time required to measure the initial rates that will be used in kinetic studies is determined.

4 Enzymatic Reactors

The decision-making on configurations, designs, and operating strategies of the enzymatic reactors follows various considerations such as enzyme characteristics, restriction of enzyme purifications, proper kinetics, immobilization appropriateness, support materials for immobilization, and operation and materials cost (Eibes et al. 2007; Maria 2012; Arca-Ramos et al. 2017). Furthermore, different factors should be controlled during the operation of enzymatic reactors including enzyme selection, catalytic cycle, operational conditions (e.g., loading rate, hydraulic retention time (HRT), pH, temperature), modeling of enzymatic reactors, and enzyme retaining in the reactor (Eibes et al. 2007).

Two main operational types of reactors are used for different enzymatic applications which are batch and continuous enzymatic reactors (Arca-Ramos et al. 2017; Illanes and Altamirano 2008). In addition, for each operational type of enzymatic reactors, the enzyme can be present as a free enzyme (i.e., soluble enzyme) or being immobilized on a specific support such as alumina, carboxymethyl cellulose, ceramic, clay, ion exchange resins, polymers, silica, starch, and titanium (Maria 2012; Mohamad et al. 2015). Some considerations should be followed in order to find the most appropriate support material for different enzymes and various applications. From those considerations, the support materials should have a high tendency for protein binding, the ability to react with a functional group, regeneration possibilities, biodegradability, nontoxicity, mechanical stability, and rigidity properties.

Enzyme immobilization on specific support can be achieved by four main techniques which are adsorption, entrapment, covalent, and cross-linking (Mohamad et al. 2015). For industrial applications, the covalent method for immobilization purposes is favorable from other methods. Using a covalent technique can provide a strong bond in order to link the enzyme to its support properly and reduce the probability of losing the enzyme by providing a strong bond. This immobilization efficiency increases the possibility of different types and a high amount of pollutants to be degraded (De Cazes et al. 2014).

4.1 Batch Reactors

Enzymatic batch reactors or commonly used as stirred tank reactors (STR) contain the biocatalyst (i.e., enzyme) and the substrates (i.e., pollutants) at the same tank when they are usually added at the beginning of the process. The influent and effluent streams are absent in the batch reactor which is leading to the stabilization of concentration inside the reactor (Woodley 2015; Arca-Ramos et al. 2017). Using the batch reactors allows the researchers to evaluate mechanisms of different reactions in addition to the determination of various kinetic parameters. Furthermore, batch reactors help to investigate the suitability of different technologies for wastewater treatment (Gasser et al. 2014; Arca-Ramos et al. 2017). In contrast, high operating costs can be considered as a drawback for batch reactors compared to continuous reactors (Arca-Ramos et al. 2017).

4.1.1 Free Enzyme Reactors

Free enzymes (i.e., soluble enzymes) are commonly used in batch reactors or also called batch stirred tank reactors (BSTR) which can be used for various applications. BSTR is a simple reactor and mostly used for the substrates with higher viscosity (Moo-Young 2019). However, many drawbacks can arise as the enzymes are not stable in such reactors as well as some difficulties can be faced during the enzymes recovering from the reactors (Illanes and Altamirano 2008). Furthermore, the free enzymes at the end of each process should be separated from the treated wastewater as well from the products obtained from the enzymatic reaction which can be considered as a costly process (De Cazes et al. 2014; Robinson 2015). Various processes can be used in order to eliminate the enzyme such as ammonium sulfate precipitation, ultrafiltration, and thermal denaturation when the products are thermostable (Robinson 2015).

Numerous researches had been done in order to investigate the removal of recalcitrant contaminants by using a free enzyme batch reactor. In Nguyen et al.'s (2014) study, laccase enzyme purified from *Aspergillus oryzae* was used to remove four trace organic contaminants with an initial activity of 90 μM/min. The removal efficiencies of carbamazepine, diclofenac, sulfamethoxazole, and atrazine were achieved as 10%, 21%, 9%, and 14%, respectively, after 24 h by using a batch enzymatic reactor with the free enzyme.

In Al-Kassim et al.'s (1994) study, two different types of peroxidase enzyme (*Coprinus macrorhizus* peroxidase (CMP) and horseradish peroxidase (HRP)) were used in a batch reactor with different time intervals to remove six types of aromatic compounds. Using the HRP enzyme, high removal efficiencies were reached for phenol (99.7%), 2-chlorophenol (96.6%), 3-chlorophenol (98%), and 4-chlorophenol (98.6%), while the removal efficiency was 83% for both 2,4-dichlorophenol and 4-methylphenol. On the other hand, by using CMP enzyme, the removal efficiencies were obtained for phenol as 98.2%, 2-chlorophenol as 94.3%, 3-chlorophenol as 90.9%, 4-chlorophenol as 98%, and 4-methylphenol as 88.3%, while the removal efficiency was 58.4% for 2,4-dichlorophenol which is considerably low (Al-Kassim et al. 1994).

4.1.2 Immobilized Enzyme Reactors

Enzyme immobilization on different supports has different advantages compared to the free enzymes. Using the immobilized enzymes in the reactors helps to enhance the enzyme stability, recover and reuse the enzymes, reduce the possibility of contaminating the products, and reduce the operational costs (Illanes and Altamirano 2008; Pandey et al. 2017). However, one of the disadvantages of using immobilized enzymes is their complexity. On the other hand, the membrane technologies play an effective role with an immobilized enzyme that can be used for recovering the enzymes as well as membranes with different materials that can be used as a support for enzymes to be immobilized (Arca-Ramos et al. 2017).

Batch Immobilized Enzyme Reactors

Immobilized enzyme technique in a batch reactor allows the enzymes to be recovered easily with the conventional processes as the immobilized enzymes have a high stability property (Illanes et al. 2013). After recovering the enzyme, it can be used several times in the other batch reactors (Illanes et al. 2013; Robinson 2015); however, the enzyme should be replaced when a considerable decrease in its activity had been observed. For those reasons, immobilized enzymes are commercially preferred compared to the free enzymes. In addition, immobilization techniques have an ability to improve the stability of different biocatalysts which can lead to enhance catalysis rate, extend the catalysis period, and improve the stability of enzymes to harsh operating conditions such as temperature and pH (Robinson 2015).

Ansari and Husain (2012) investigated two scenarios of *Aspergillus oryzae* β-galactosidase immobilization for lactose hydrolysis in milk and whey. At first scenario, β-galactosidase enzyme was adsorbed on concanavalin A-celite 545 support (AβG) and, secondly, the adsorbed β-galactosidase on concanavalin A-celite 545 support was cross-linked by using glutaraldehyde (CβG). After 10 h in a batch reactor, lactose hydrolysis in the milk was observed as 67% and 72% by using AβG and CβG, respectively. On the other hand, Ansari and Husain (2012) found that the lactose hydrolysis in whey was higher than in milk with the hydrolysis of 75% and 89% by using AβG and CβG, respectively.

Another study had been done by Warburton et al. (1973) in order to hydrolyze the benzylpenicillin by using free and immobilized penicillin amidase obtained from *E. coli*. The experiment was conducted in a batch reactor with a pH of 8.0 and at 37°C temperature for both free and immobilized enzymes. They found that by using the immobilized enzyme, 50 mM benzylpenicillin can be converted with a percentage of 90%. In addition, that conversion percentage was reached with less time compared to the free enzyme by using the same enzyme activity.

Recirculation Batch Reactors (RBRs)

The main application of recirculation batch reactors (RBRs) is to be used when the compounds (i.e., substrates) pass in the enzymatic reactor where the pH should be adjusted (Illanes et al. 2013). The use of RBR allows good temperature control, precise analysis, and ease of determination of reaction rates (Park et al. 1982). The RBR can be used in the form of a packed bed reactor where the biocatalysts have been packed then the solution will be circulated to achieve the required conversion (Illanes and Altamirano 2008). Furthermore, the recirculation rate should be optimized for different compounds as it has a direct effect on hydrolysis efficiency (Sudhakaran et al. 1992).

Penicillin G acylase (PGA) from *E. coli* had been used in recirculated packed bed batch reactor in Sudhakaran et al.'s (1992) study, and glycidyl copolymer was used as support for enzyme immobilization. They investigated the effect of recirculation rate on the hydrolysis of *penicillin G*. The maximum hydrolysis was obtained (more than 95%) when a maximum recirculation rate was applied (3.0 L/min). Using this configuration, extend the activity of the immobilized *penicillin G acylase* by preventing it from alkali contact. Furthermore, the reaction rate in the recirculated packed bed batch reactor had been enhanced as a higher concentration of enzymes can be used. Last but not the least, losing the enzyme and possibility of contamination had been reduced as the enzyme is far from the reaction system (Sudhakaran et al. 1992).

Another application can be used with RBR which is to investigate the effect of external mass transfer of penicillin amidase as a change in pH value is widely happening, and it needs to be controlled. In Park et al. (1982), RBRs with immobilized penicillin amidase from *E. coli* were used to investigate the external mass transfer resistance effect on the reaction rate. They observed that the mass transfer coefficient was higher with the higher flow rate whatever the substrate concentration is. They achieved the maximum mass transfer coefficient with a 16 mL/min flow rate.

4.2 Continuous Reactors

The enzymatic continuous reactors are common configuration to be applied for large-scale processes (Illanes and Altamirano 2008). There are three commonly used continuous reactors which are continuous stirred tank reactors (CSTR), continuous packed bed reactors (CPBR), and continuous fluidized bed reactors (CFBR) could be considered in between (Arca-Ramos et al. 2017). Generally, continuous reactors have lower operational costs compared to their counterparts of the batch reactors (Robinson 2015). The continuous reactor should be applied with high performance in order to maintain the enzymes inside the reactors so as to avoid the high cost that can happen with the frequent addition of enzymes to the reactor as most of the enzymes are expensive (Bódalo et al. 1995; Arca-Ramos et al. 2017). In addition, one of the successful parameters for the continuous reactors is using enzymes with high stability that means the immobilized enzymes are more favorable in those types of reactors (Bódalo et al. 1995; Illanes and Altamirano 2008).

The types of enzymes play an important role in selecting a proper reactor configuration. For instance, CPBR is not suitable for the enzymes with a low resistance to compression as the enzymes would be packed in the reactor. On the other hand, enzymes with poor resistance to shear forces are not appropriate to be used in CSTR configuration. In addition, the characteristics of the substrate solution have a direct effect on the choosing of reactor configuration. The solution with low suspended solids is preferable to be used in CPBR. The productivity might be declined as the operation can be negatively affected by the presence of high suspended materials in the feed solution which will make the regular cleaning crucial (Illanes and Altamirano 2008). Reusing the enzyme for several times is one of the advantages of continuous reactors. Reusing period can be reached to 4 months before it needs to be substituted in case of immobilized glucose isomerase that has been used in syrup production with a high level of fructose (Robinson 2015).

4.2.1 Continuous Stirred Tank Reactors (CSTR)

Enzymes are present in a suspended state inside a flustered tank called a reactor. The operation of CSTR is conducted by flowing the feeding solution having the substrate continuously with a constant flow rate to the reactor which contains the immobilized enzymes in order to generate the targeted products (Illanes and Altamirano 2008; Robinson 2015). The solution inside the reactor should be well mixed in order to allow the enzymes to be linked to the same molar concentration of the substrate (Illanes and Altamirano 2008). Using this type of reactor enables the production costs to be reduced by well controlling the reactor. Furthermore, the improvement of the stirred tank reactor to the continuous-flow stirred tank reactor can be easily accomplished (Robinson 2015).

In Gómez et al.'s (2008) study, the removal of phenol was investigated by using a continuous tank reactor with immobilized soybean peroxidase. The enzyme activity was 108 IU/mg, and the enzyme had been immobilized on a glass support. They optimized different operational parameters such as flow rate, enzyme concentration, and substrate concentration. The removal efficiency of the phenol was achieved as 80% under the optimum operational conditions (Gómez et al. 2008).

4.2.2 Continuous Packed Bed Reactors (CPBR)

The operation of CPBR is conducted by flowing the feeding solution having the substrate continuously with a constant flow rate to the reactor where the immobilized enzymes are packed in order to generate a submerged bed (Illanes and Altamirano

2008). In addition, the stirring process is not necessary in case of using a packed bed reactor as the immobilized enzymes are packed in the reactor (Robinson 2015). One of the challenges with using CPBR is that the operational conditions such as temperature and pH are difficult to be controlled (Illanes and Altamirano 2008).

The flow of the feeding solution can be applied upward or downward inside the reactor. However, the upward direction of the feeding solution is more applicable in lab-scale experiments in order to control the level of the solution beyond the enzyme bed inside the reactor as well as to prevent the bed to be pressed. On the other hand, the downward direction of the feeding solution is more appropriate in pilot- or real-scale applications as the energy used for pumping is less which leads to reduce costs. In addition, the flow can be returned regularly in order to wash the enzymes, and the bed compaction can be managed (Illanes and Altamirano 2008).

In Lloret et al.'s (2012b) study, removal of estrone (E1), 17β-estradiol (E2), and 17α-ethinylestradiol (EE2) by using a continuous packed bed reactor was investigated. The laccase enzyme from *Myceliophthora thermophila* with the activity of 9.7 U/g was immobilized on Eupergit. The experiment had been conducted for 8 h. within room temperature. The removal efficiency of E1 was achieved as 65% while it was around 80% for both E2 and EE2 when the feed concentration for each compound was 5 mg/L.

4.2.3 Continuous Fluidized Bed Reactors (CFBR)

The operation of CFBR is conducted by flowing the feeding solution (with substrate) from the bottom side of the reactor. Flowing the feeding solution in an upward direction provides a hydrodynamic equilibrium state between drag force and gravity which allows the enzymes to be held inside the reactor (Illanes and Altamirano 2008). Furthermore, the flow rate should be controlled properly in order to provide a suitable bed expansion of the enzymes inside the reactor as the high flow rate might cause enzymes escaping out of the reactor (Allen et al. 1979). Similar to the CPBR, the stirring process is not a must in case of using the fluidized bed reactor (Bódalo et al. 1995).

There are various advantages of fluidized bed reactors such as their appropriateness to the substrates having particulate or viscous properties. In addition, the fluidized bed reactors provide a harmonized spreading of the flow inside the reactor as well as decrease the opportunity of pressure loss to occur. Some obstacles can be faced when the CFBR is used in industrial sectors as some factors need to be controlled such as inhibition, diffusion problems, and the ability of reactions to be reversed (Bódalo et al. 1995).

Lloret et al. (2012a) studied the removal of estrogens (estrone (E1), estradiol (E2) and ethinylestradiol (EE2)) by using continuous fluidized bed reactor with immobilized laccase enzyme on Eupergit. The activity of the used laccase was 25 U/g, and the temperature during the experiment was 26 °C. The feed solution containing the different estrogens was entering the reactor from its bottom side. They found that the removal efficiencies of estrogens were between 92% and 100% when the addition rate was high (6 mg/L h for E1 and EE2 and 6.6 mg/L h for E2) at a

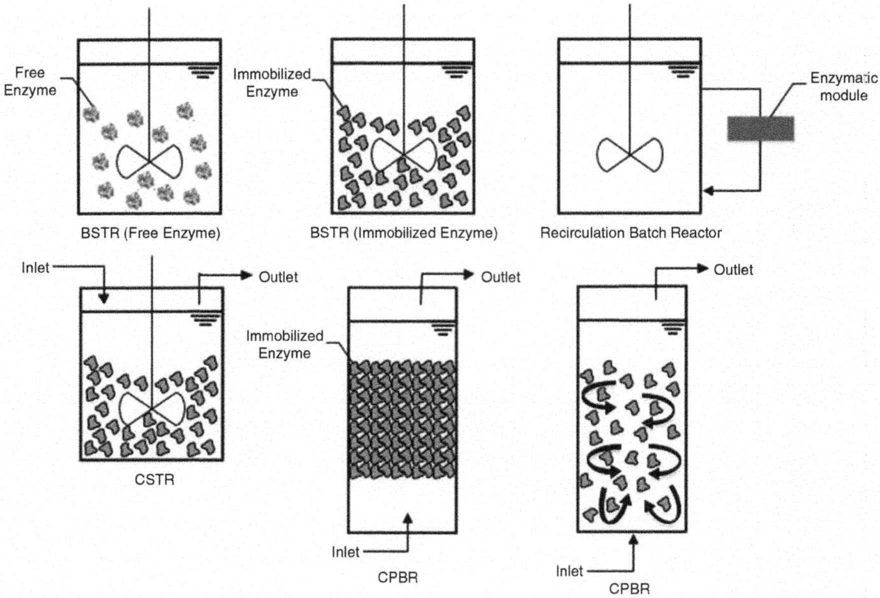

Fig. 3 Enzymatic reactor configurations for batch and continuous reactors

hydraulic retention time of 50 min (Lloret et al. 2012a). As a summary of batch and continuous reactors, Fig. 3 showed different configurations of enzymatic reactors.

4.3 Innovative Enzymatic Reactors

In order to enhance the removal efficiency of different pollutants or compounds, from different wastewaters, some enzymatic reactors should be modified. Different processes or technologies can be incorporated internally or externally to the enzymatic reactors. Some types of adsorptive materials such as granular activated carbon (GAC) can be added, or different types of membranes might be attached to the reactor with the intention of improving the performance of the enzymatic reactor. In the study of Nguyen et al. (2014), they investigated the effect of adding GAC in a batch reactor containing laccase enzyme from *Aspergillus oryzae* in order to remove different types of trace organic contaminants. They observed that the removal efficiencies of trace organic contaminants were higher in GAC + enzyme reactor compared to a reactor with enzyme or GAC alone. The removal efficiencies for carbamazepine, diclofenac, sulfamethoxazole, and atrazine were achieved as 49%, 60%, 53%, and 57%, respectively, when only the GAC was used in a batch reactor. However, the removal efficiencies of carbamazepine, diclofenac, sulfamethoxazole, and atrazine were increased to 52%, 63%, 58%, and 75%, respectively, when the reactor had both GAC and enzyme (Nguyen et al. 2014).

A term of the hybrid system had arisen for the treatment of emerging compounds from different types of wastewater. Ba et al. (2018) applied a hybrid system consisting of two series of treatments starting with continuous stirred tank reactor (CSTR) and followed by a hollow fiber microfiltration (MF) membrane made of polysulfone. The feeding solution mixture contains the wastewater having the pharmaceutical compounds and combined cross-linked tyrosinase and laccase aggregates. The feeding solution flows to the CSTR with 1 h as a hydraulic retention time (HRT). They investigate the performance of the hybrid system to remove three groups of pharmaceutical products which are nonsteroidal anti-inflammatory drugs (NSAIDs), antibiotics, and other pharmaceutical products. The effluent from CSTR is considered as influent of MF membrane while the permeate stream of MF is collected to evaluate the performance of the hybrid system. They achieved almost 100% removal efficiency for the three groups of pharmaceuticals after 120 h by using the hybrid bioreactor (Ba et al. 2018).

Different configurations of membrane bioreactors have been developed with various enzymes in order to improve the performance of wastewater treatment such as extractive membrane bioreactor (EMBR), immobilized enzyme membrane reactor (IEMR), and direct contact membrane reactor (DCMR) (Pandey et al. 2017). The EMBR is a promising configuration for removing toxic compounds from harsh conditions wastewater. Referring to the EMBR term, the membrane used in this reactor is used as an extractor to remove a target compound from the wastewater (Wenten et al. 2020). On the other hand, IEMR is used when the membrane is considered as a support for enzyme immobilization.

5 Feasibility and Applicability of Enzymes in Industrial Sectors

The efficient development of new enzymes provides important tools for recent advances in biotechnology, especially in protein engineering. Since the biocatalyst was first released many centuries ago, a number of enzyme-based processes were commercialized to produce several valuable products. Despite the advantages of enzymes, industrial applications could not reach the desired level mainly due to undesirable properties in terms of stability, catalytic efficiency, and specificity (Deng et al. 2019). In addition, the economic feasibility should be analyzed as the production costs of the enzyme are considerably high (Eibes et al. 2007).

In the industrial sector, there are a variety of enzymes that can be used for treatment and/or recovery of different compounds available in the industrial wastewater as illustrated in Table 2.

Table 2 Types of enzymes used in different industrial applications

Group	Enzyme	Source	Applications	Reference
Oxidoreductases	Laccase	*Trametes versicolor*	Degradation of diclofenac, trimethoprim, carbamazepine, and sulfamethoxazole	Alharbi et al. (2019)
	Laccase	*Trametes versicolor*	Degradation of phenol compounds	Alver and Metin (2017)
	Manganese peroxidase	*Anthracophyllum discolor*	Degradation of polycyclic aromatic hydrocarbons	Acevedo et al. (2010
Transferases	Cyclodextrin glycosyltransferase	*Komagataella phaffii*	Production of high purity α-, β-, or γ-cyclodextrin for different industries (e.g., chemical, pharmaceutical, food, etc.)	Zhang et al. 2019
Hydrolases	α-amylase	*Bacillus licheniformis*	Degradation of starch of bread to oligosaccharides	Demirci et al. (2017)
	Glucoamylase	*Aspergillus niger*	Saccharification process for glucose production from dextrins	Riaukaite et al. (2019)
	Lipases	*Candida cylindracea*	Hydrolysis of olive oil	Hoq et al. (1985)
	Keratinases	–	Hydrolysis of hair keratin and epidermis and break down of the disulfide bonds in the leather industry	de Souza and Gutteres (2012)
Lyases	Acetolactate d~~D~~ecarboxylase	*Bacillus subtilis*	α-A~~a~~cetolactate conversion to acetoin~~.~~	Choi et al. (2015)
			Decreasing the content of diacetyl in beer	
Isomerases	Glucose isomerase	*Streptomyces murinus*	Isomerization of glucose to fructose	Gaily et al. (2010)

6 Enzymes for Recovery Applications in Industry

Enzymes play an important role not for only treatment but for recovery of the valuable compounds from wastewater. Vasileva et al. (2016) studied the possibility of producing glucose-galactose syrup from waste whey by hydrolyzing the lactose with β-galactosidase enzyme from *Escherichia coli*. First, they investigate the effect of operational conditions (e.g., enzyme activity, pH, and temperature) on the performance of lactose hydrolysis by using free and immobilized β-galactosidase enzyme on the polypropylene membrane in a batch reactor. Under different operational conditions, they realized that the immobilized enzyme performance was more advantageous than free enzyme in terms of lactose hydrolysis. Afterward, they conducted another experiment under the optimum conditions (pH 6.8, temperature 40 °C, and flow rate 1.0 mL/min) by using a continuous spirally wound membrane reactor with the immobilized β-galactosidase enzyme. The hydrolysis of lactose was achieved as 91% after 10 h of the experiment. Furthermore, the possibility of reusing the enzyme had been studied as well. They were able to use β-galactosidase enzyme 20 times in 20 days with an operational period of 10 h in each day. It was possible to achieve 69.7% of yield bioreactor at the end of 20 days (Vasileva et al. 2016).

Another study was done by Ghosh et al. (2017) when they were trying to apply another application of enzymes rather than recovery. They investigate protein and lactose hydrolysis in series with the purpose of enhancing the nutritional value in whey to use it for healthy beverages. For protein hydrolysis, they used different concentrations of Flavourzyme enzyme from *Aspergillus oryzae*. The whey containing the enzyme was kept for 30 min in a water bath with 50 °C, and then the pH was adjusted at pH 7 followed by heating for 6–7 min at 85 °C so as to inactivate the enzyme. The higher Flavourzyme enzyme concentration showed maximum protein hydrolysis with an average of 15.11%. Afterward, the lactose in whey with hydrolyzed protein had been hydrolyzed by using different percentages of β-galactosidase enzyme obtained from *Aspergillus oryzae* at 30 °C and pH 5.5 in a batch reactor. In order to inactivate the β-galactosidase enzyme, the hydrolyzed whey had been heated for 5–6 min at 85 °C. The lactose was hydrolyzed with an average of 60.88% at the maximum concentration of β-galactosidase enzyme (Ghosh et al. 2017).

7 Enzymes for Treatment Applications in Industry

Enzymes gain a prestigious reputation in recent years for their applicability to treat or degrade different pollutants in industrial wastewater particularly the recalcitrant compounds. Numerous studies have been conducted in terms of that purpose. Arca-Ramos et al. (2015) studied the applicability of enzymes to remove bisphenol A (BPA) from synthetic wastewater and real treated wastewater with biological treatment. For evaluating pH and enzyme activity, first, they used a batch reactor with laccase enzyme from *Trametes versicolor* and *M. thermophila*. BPA oxidation had been achieved with both types of laccase enzyme. However, they found that the use of laccase from *Trametes versicolor* performed better in terms of BPA conversion with a percentage of 96% after 8 h, while it was only 64% when laccase from *M. thermophila* was used. In addition, the optimum operational condition was achieved with pH 6 and enzyme activity of 1000 U/L. Subsequently, a continuous enzymatic membrane reactor (EMR) was examined, consisting of a stirred tank reactor having a free laccase followed by a ceramic titanium membrane having a high hydrophilic properties, to be able to separate and recycle the laccase enzyme to the stirred tank reactor. For both configurations, the removal efficiency of BPA was achieved as more than 94.5% for the synthetic wastewater. However, the removal efficiency of BPA was considerable with real Wastewater; the enzyme activity had been affected by the availability of particles and ions (Arca-Ramos et al. 2015).

In Shaheen et al.'s (2017) study, the decolorization efficiency was investigated with the free and immobilized enzyme. Free lignin peroxidase from *Ganoderma lucidum* was used, and it had been immobilized on Ca-alginate beads to compare their efficiencies for decolorization of Sandal-fix reactive textile dyes. The batch system was applied by using inoculated flasks, and they were kept at 30°C for 3 days on an orbital shaker with 120 rpm. Decolorization efficiency was achieved with a range of 40–66% by using a free enzyme. On the other hand, by using an immobilized enzyme, the decolorization efficiency had been increased to the range of 70–93% (Shaheen et al. 2017).

Another study had been done in order to evaluate heterocyclic aromatics which is a type of recalcitrant compound. Mashhadi et al. (2019) conducted an experiment by using soybean peroxidase enzyme to remove 3-hydroxyquinoline and 3-aminoquinoline that had been detected in water bodies. A batch reactor had been used in order to conduct this experiment. They achieved a removal efficiency of more than 94% for 3-hydroxy and 3-aminoquinoline at an optimum pH of 8.6 and 5.6, respectively. In the leather industry, proteases hydrolyze the protein fraction of dermatan sulfate, making the collagen more accessible to water and reducing the attachment of the basal layer. Also, they remove globular proteins. Lipases hydrolyze fats, oils, and greases present in the hypoderm. Keratinases hydrolyze the keratin of hair and epidermis and break down the disulfide bonds of this molecule (de Souza and Gutteres 2012).

8 Conclusion

An overview of the enzymes' role in this book chapter led to various conclusions. Enzymes are the most powerful catalysts that have been used in industrial compound applications to treat, recover, or generate new compounds. The effectiveness of enzymes was widely explored in different industrial sectors such as the food industry, pharmaceutical industry, leather industry, etc. Furthermore, the reusability of the enzyme several times makes it an interesting biocatalyst in terms of sustainability and cost minimization. Although a promising performance of enzyme applications was achieved, different parameters could inhibit the enzyme activity.

From different kinds of literature, it was observed that reactor configurations had a great impact on achieving the required enzyme performance. For that reason, proper design of reactors by choosing suitable parameters should be done according to the type of compounds to be treated or recovered as well as the type of enzymes to be used. Additionally, the state of the enzyme during the degradation processes whether it was free (soluble) enzyme or immobilized enzyme had a significant effect during the treatment or recovery. Generally, it was found from different studies that the immobilized enzymes showed a higher efficiency compared to free enzymes in terms of treatment or recovery application in many industrial sectors. Moreover, choosing the support material for enzyme immobilization purposes is another issue to be controlled. In terms of enzyme applications, a wide range of pollutants or compounds in different industrial sectors can be treated, recovered, or enhanced with various types of enzymes from different sources. A considerable elimination had been achieved of different emerging compounds with the aid of enzyme processes. Due to enzyme selectivity, it helps to target a particular compound to be treated.

Last but not the least, a variety of studies might be conducted with the intention of developing and improving the enzymatic applications for treating or recovering different compounds. Different configurations of enzymatic reactors can be improved by integrating new processes or modules together with the enzymatic reactor. Furthermore, applying new materials (i.e., green materials) for enzyme immobilization purposes is another area to be researched. There are limited studies of enzyme applications in advanced degradation of plastics particularly nanoplastics and microplastics that need more investigation.

References

Acevedo F, Pizzul L, González ME, Cea M, Gianfreda L, Diez MC (2010) Degradation of polycyclic aromatic hydrocarbons by free and nanoclay-immobilized manganese peroxidase from Anthracophyllum discolor. Chemosphere 80(3):271–278

Alharbi SK, Nghiem LD, Van De Merwe JP, Leusch FD, Asif MB, Hai FI, Price WE (2019) Degradation of diclofenac, trimethoprim, carbamazepine, and sulfamethoxazole by laccase from Trametes versicolor: transformation products and toxicity of treated effluent. Biocatal Biotransformation 37(6):399–408

Al-Kassim L, Taylor KE, Nicell JA, Bewtra JK, Biswas N (1994) Enzymatic removal of selected aromatic contaminants from wastewater by a fungal peroxidase from Coprinus macrorhizus in batch reactors. J Chem Technol Biotechnol 61(2):179–182

Allen BR, Coughlin RN, Charles M (1979) Fluidized-bed enzyme reactor. Ann N Y Acad Sci 326:105–117

Alver E, Metin AÜ (2017) Chitosan based metal-chelated copolymer nanoparticles: laccase immobilization and phenol degradation studies. Int Biodeter Biodegr 125:235–242

Ansari SA, Husain Q (2012) Lactose hydrolysis from milk/whey in batch and continuous processes by concanavalin A-Celite 545 immobilized Aspergillus oryzae β galactosidase. Food Bioprod Process 90(2):351–359

Arca-Ramos A, Eibes G, Feijoo G, Lema JM, Moreira MT (2015) Potentiality of a ceramic membrane reactor for the laccase-catalyzed removal of bisphenol A from secondary effluents. Appl Microbiol Biotechnol 99(21):9299–9308

Arca-Ramos A, Eibes G, Feijoo G, Lema JM, Moreira MT (2017) Enzymatic reactors for the removal of recalcitrant compounds in wastewater. Biocatal Biotransformation 36(3):195–215

Ba S, Haroune L, Soumano L, Bellenger JP, Jones JP, Cabana H (2018) A hybrid bioreactor based on insolubilized tyrosinase and laccase catalysis and microfiltration membrane remove pharmaceuticals from wastewater. Chemosphere 201:749–755

Berg J, Tymoczko J, Stryer L (2007) Biochemistry, 6th edn. W.H. Freeman & Co Ltd., Basingstoke

Bhagavan NV, Ha CE (2011) Enzymes and enzyme regulation. Essentials of medical biochemistry. Academic Press, New York, pp 47–58

Bódalo A, Gómez JL, Gómez E, Bastida J, Máximo MF (1995) Fluidized bed reactors operating with immobilized enzyme systems: design model and its experimental verification. Enzyme Microb Technol 17(10):915–922

Briggs GE, Haldane JBS (1925) A note on the kinetics of enzyme action. Biochem J 19(2):338

Choi EJ, Ahn HW, Kim WJ (2015) Effect of α-acetolactate decarboxylase on diacetyl content of beer. Food Sci Biotechnol 24(4):1373–1380

De Cazes M, Abejón R, Belleville MP, Sanchez-Marcano J (2014) Membrane bioprocesses for pharmaceutical micropollutant removal from waters. Membranes 4(4):692–729

de Souza FR, Gutteres M (2012) Application of enzymes in leather processing: a comparison between chemical and coenzymatic processes. Braz J Chem Eng 29:473–481

Demirci AS, Palabıyık I, Gümüs T, Özalp Ş (2017) Waste bread as a biomass source: optimization of enzymatic hydrolysis and relation between rheological behavior and glucose yield. Waste Biomass Valorization 8(3):775–782

Deng C, Huang T, Jiang Z, Lv X, Liu L, Chen J, Du G (2019) Enzyme engineering and industrial bioprocess. In: Current developments in biotechnology and bioengineering. Elsevier, Amsterdam, pp 165–188

Dugas H (1996) A chemical approach to enzyme action. Biochem Mol Biol Educ 25:173–188

Eibes G, López C, Moreira MT, Feijoo G, Lema JM (2007) Strategies for the design and operation of enzymatic reactors for the degradation of highly and poorly soluble recalcitrant compounds. Biocatal Biotransformation 25(2-4):260–268

Fischer E (1894) Einfluss der Configuration auf die Wirkung der Enzyme. Ber Dtsch Chem Ges 27:2985–2993. Trans. Holmstedt, B. and Liljestrand, G., Eds. Readings in Pharmacology (1963), 251

Gaily MH, Elhassan BM, Abasaeed AE, Al-Shrhan M (2010) Isomerization and kinetics of glucose into fructose. Int J Eng Technol 10(3):1–6

Gasser CA, Ammann EM, Shahgaldian P, Corvini PFX (2014) Laccases to take on the challenge of emerging organic contaminants in wastewater. Appl Microbiol Biotechnol 98(24):9931–9952

Ghosh BC, Prasad LN, Saha NP (2017) Enzymatic hydrolysis of whey and its analysis. J Food Sci Technol 54(6):1476–1483

Gómez JL, Bódalo A, Gómez E, Hidalgo AM, Gómez M, Murcia MD (2008) A transient design model of a continuous tank reactor for removing phenol with immobilized soybean peroxidase and hydrogen peroxide. Chem Eng J 145(1):142–148

Grahame DAS, Bryksa BC, Yada RY (2015) Factors affecting enzyme activity. In: Improving and tailoring enzymes for food quality and functionality. Woodhead Publishing, Cambridge, pp 11–55

Hoq MM, Yamane T, Shimizu S, Funada T, Ishida S (1985) Continuous hydrolysis of olive oil by lipase in microporous hydrophobic membrane bioreactor. J Am Oil Chem Soc 62(6):1016–1021

Illanes A, Altamirano C (2008) Enzyme reactors. In: Illanes A (ed) Enzyme biocatalysis. Springer, Dordrecht

Illanes A, Wilson L, Vera C (2013) Problem solving in enzyme biocatalysis. Wiley, New York

Lloret L, Eibes G, Feijoo G, Moreira MT, Lema JM (2012a) Continuous operation of a fluidized bed reactor for the removal of estrogens by immobilized laccase on Eupergit supports. J Biotechnol 162(4):404–406

Lloret L, Hollmann F, Eibes G, Feijoo G, Moreira MT, Lema JM (2012b) Immobilisation of laccase on Eupergit supports and its application for the removal of endocrine disrupting chemicals in a packed-bed reactor. Biodegradation 23(3):373–386

Lopina OD (2017) Enzyme inhibitors and activators. IntechOpen, London

Maria G (2012) Enzymatic reactor selection and derivation of the optimal operation policy, by using a model-based modular simulation platform. Comput Chem Eng 36:325–341

Mashhadi N, Taylor KE, Biswas N, Meister P, Gauld JW (2019) Oligomerization of 3-substituted quinolines by catalytic activity of soybean peroxidase as a wastewater treatment. Product formation and computational studies. Chem Eng J 364:340–348

Michaelis L, Menten ML (1913) Kinetics of invertase action. Biochem Z 49:333–369

Mohamad NR, Marzuki NHC, Buang NA, Huyop F, Wahab RA (2015) An overview of technologies for immobilization of enzymes and surface analysis techniques for immobilized enzymes. Biotechnol Biotechnol Equip 29(2):205–220

Moo-Young M (2019) Comprehensive biotechnology. Elsevier, Amsterdam

Nelson, D., Cox, M., & Boyle, J. (2005). Book review: Lehninger principles of biochemistry. Biochemistry and molecular biology education, Freeman and Company New York 33(1), 74-74.

Nguyen LN, Hai FI, Price WE, Leusch FD, Roddick F, Ngo HH et al (2014) The effects of mediator and granular activated carbon addition on degradation of trace organic contaminants by an enzymatic membrane reactor. Bioresour Technol 167:169–177

Pandey K, Singh B, Pandey AK, Badruddin IJ, Pandey S, Mishra VK, Jain PA (2017) Application of microbial enzymes in industrial waste water treatment. Int J Curr Microbiol App Sci 6 (8):1243–1254

Park JM, Choi CY, Seong BL, Han MH (1982) Effect of mass transfer in a recirculation batch reactor system for immobilized penicillin amidase. Biotechnol Bioeng 24(10):2215–2226

Riaukaite J, Basinskiene L, Syrpas M (2019) Bioconversion of waste bread to glucose fructose syrup as a value-added product. In: Conference proceedings. FOODBALT 2019. 13th Baltic conference on food science and technology "FOOD, NUTRITION, WELL-BEING", Jelgava, Latvia, 2-3 May 2019 (pp. 120–124). Latvia University of Life Sciences and Technologies

Robinson PK (2015) Enzymes: principles and biotechnological applications. Essays Biochem 59:1–41

Segel IH (1968) Biochemical calculations: how to solve mathematical problems in general biochemistry. Wiley, New-York

Segel IH (1975) Enzyme kinetics: behavior and analysis of rapid equilibrium and steady state enzyme systems. Wiley, New-York

Shaheen R, Asgher M, Hussain F, Bhatti HN (2017) Immobilized lignin peroxidase from Ganoderma lucidum IBL-05 with improved dye decolorization and cytotoxicity reduction properties. Int J Biol Macromol 103:57–64

Straathof AJ, Adlercreutz P (2000) Applied biocatalysis. CRC Press, Boca Raton

Sudhakaran VK, Deshpande BS, Shewale JG, Naik SR (1992) Production of 6-APA using a recirculated packed bed batch reactor. Biotechnol Lett 14(10):913–918

Vasileva N, Ivanov Y, Damyanova S, Kostova I, Godjevargova T (2016) Hydrolysis of whey lactose by immobilized β-galactosidase in a bioreactor with a spirally wound membrane. Int J Biol Macromol 82:339–346

Villarroel A, Taglialatela M, Bernardo-Seisdedos G, Alaimo A, Agirre J, Alberdi A et al (2014) The ever changing moods of calmodulin: how structural plasticity entails transductional adaptability. J Mol Biol 426(15):2717–2735

Voet D, Voet JG (1996) Biochemistry. Wiley, New York

Warburton D, Dunnill P, Lilly MD (1973) Conversion of benzylpenicillin to 6- aminopenicillanic acid in a batch reactor and continuous feed stirred tank reactor using immobilized penicillin amidase. Biotechnol Bioeng 15(1):13–25

Wenten IG, Friatnasary DL, Khoiruddin K, Setiadi T, Boopathy R (2020) Extractive membrane bioreactor (embr): recent advances and applications. Bioresour Technol 297:122424

Woodley JM (2015) Scale-up and development of enzyme-based processes for large-scale synthesis applications. In: Faber K, Fessner WD, Turner NJ (eds) Science of synthesis: biocatalysis in organic synthesis. Wiley, New York, pp 515–546
Yang J, Dong C, Kirk ML (2017) Xanthine oxidase–product complexes probe the importance of substrate/product orientation along the reaction coordinate. Dalton Trans 46(39):13242–13250
Zhang J, Li M, Mao H (2019) Cross-linked enzyme aggregates of recombinant cyclodextrin glycosyltransferase for high-purity β-cyclodextrin production. J Chem Technol Biotechnol 94(5):1528–1533

Exploring the Use of Biomixtures for the Removal of Structurally Different Pesticides

J. M. Saez, A. L. Bigliardo, E. E. Raimondo, G. E. Briceño, M. A. Polti, and C. S. Benimeli

Abstract

Pesticides are among the most employed compounds worldwide and play an important role in modern agriculture. However, their inadequate management can lead to contamination of the environment. When pesticides enter soils and water bodies, they produce negative impacts on ecosystems through diffuse or point source contamination. The last comprises processes like improper handling or leaks of spraying liquid, remnants, and washes of the spraying equipment, and they can be controlled by agricultural good practices.

In the last decades, with the objectives of reducing pesticides' point source pollution, biobeds or biopurification systems (BPS) were developed as promising and low-cost technologies. They consist of three main components: a clay layer at

J. M. Saez · M. A. Polti
Planta Piloto de Procesos Industriales Microbiológicos (PROIMI-CONICET), Tucumán, Argentina

Facultad de Ciencias Naturales e Instituto Miguel Lillo, Universidad Nacional de Tucumán (UNT), Tucumán, Argentina

A. L. Bigliardo
Planta Piloto de Procesos Industriales Microbiológicos (PROIMI-CONICET), Tucumán, Argentina

E. E. Raimondo
Planta Piloto de Procesos Industriales Microbiológicos (PROIMI-CONICET), Tucumán, Argentina

Facultad de Bioquímica, Química y Farmacia, Universidad Nacional de Tucumán, Tucumán, Argentina

G. E. Briceño
Centro de Investigación Biotecnológica Aplicada al Medio Ambiente (CIBAMA-BIOREN), Universidad de La Frontera, Temuco, Chile

C. S. Benimeli (✉)
Planta Piloto de Procesos Industriales Microbiológicos (PROIMI-CONICET), Tucumán, Argentina

Facultad de Ciencias Exactas y Naturales, Universidad Nacional de Catamarca, Catamarca, Argentina

M. P. Shah (ed.), *Removal of Emerging Contaminants Through Microbial Processes*,
https://doi.org/10.1007/978-981-15-5901-3_19

the bottom; a biomixture; and a grass layer at the surface. The biomixture is the biologically active core of a BPS, and it is composed of three elements which play important roles in the adsorption and degradation of pesticides: a lignocellulosic substrate, a humic rich component, and soil. Since half of the biomixture consists of lignocellulosic substrates, several studies have evaluated the bioaugmentation of biomixtures with ligninolytic fungi obtaining satisfactory results on pesticide removal. Also, the bioaugmentation of BPS with pesticide-degrading bacteria represents a promising approach in the design and optimization of these systems.

This chapter is a short compilation regarding the application of BPS as biotechnological tools for the removal of pesticides. A special emphasis is put on the bioaugmentation of biomixtures with ligninolytic fungi and/or pesticide-degrading bacteria to obtain optimal biodegradation of structurally diverse pesticides.

Keywords

Pesticides · Biopurification systems · Biomixtures · Bioaugmentation

1 Introduction

Pesticides are a class of biocide, responsible for killing, repelling, or controlling any form of plant or animal life that is considered to be a pest. They cover a wide range of products including insecticides, fungicides, and weed killers for agricultural (plant-protection products) and non-agricultural applications (Harrison 2014).

Pesticides play an important role in the success of modern agriculture and food production. However, their inadequate management can cause serious problems of environmental contamination (Diez et al. 2013).

Pollution of the environment can be classified according to its extension and source. Point source pollution appears in areas where there is a direct relationship with the source of the contamination. It is generally associated with localized situations and is produced as a result of inappropriate procedures during filling and cleaning of spray equipment, accidental spillages at farm level, or improper disposal of residues from pesticide applications (Chin-Pampillo et al. 2016). These practices can pose a risk to both soil and water (Saez et al. 2018). On the other hand, diffuse pollution is usually caused by the transport of polluting substances, both soluble and particulate, over large areas often far from the source of origin, through atmospheric deposition, runoff, and leaching, among other forms. Moreover, in certain areas, pollution can be high (in urban areas and industrial complexes), caused both by diffuse and point sources (Harrison 2014).

Diffuse pollution is more difficult to control and eliminate because the sources are not located in space. Several studies have shown that point sources contribute predominantly to the contamination of basic resources such as water and soil (Aparicio et al. 2018; Fuentes et al. 2010). Therefore, it would be more efficient to avoid diffuse contamination by treating the contamination at its point of origin.

Historically, organochlorine pesticides (OPs) have been the most used insecticides; however, since most of them were banned in the 1970s, the use of other types of pesticides, such as organophosphates, triazines, and pyrethroids, has increased dramatically worldwide. These pesticides are relatively more water-soluble, less persistent, and less bioaccumulative than most OPs; however, many of them have demonstrated large-scale atmospheric transport potential under certain conditions (Saez et al. 2018).

In recent years, agricultural producers had to adapt to the standards of international markets, which include requirements related to the quality of products and international agreements signed within the framework of sustainable agriculture and care for the environment. This has led to the important challenge of globalizing markets in the primary food production sector, certifying standards of Good Agricultural Practices (GAP), so that the requirements demanded by the consumer are answered by producers around the world, with a responsible approach regarding the welfare of the worker and the care of the environment.

In this context, in compliance with GAP, point source contamination by pesticides has been rigorously addressed in the last decade, through the evaluation and implementation of biopurification systems or biobeds (Diez et al. 2013; Saez et al. 2018; Urrutia et al. 2013). This technology responds to the need for simple and effective methods to minimize environmental pollution due to the use of pesticides. Biopurification systems were designed to retain and degrade pesticides, during the early stages of preparation and application, avoiding point pollution, and, therefore, it is considered a bioprophylaxis method.

A biopurification system consists of three main components: a layer of clay in the lower part, a biomixture, and a layer of grass covering the surface. Biomixture is the main component of the system, and, therefore, the adequate selection of organic materials for its preparation is a crucial factor for the removal and degradation of pesticides (Diez et al. 2013). It is composed of a lignocellulosic substrate, a humic rich component, and soil. Each component of the biomixture plays an important role in the removal of pesticides, in stimulating the growth of ligninolytic microorganisms, in providing high water retention and adsorption capacity of pesticides, and as an important source of microbiota that can synergistically degrade them (Saez et al. 2018).

Recently, bioaugmentation of biopurification systems with pesticide-degrading microorganisms has been shown to increase the efficiency of these systems (Murillo-Zamora et al. 2017; Saez et al. 2018). Besides, the ability of different microorganisms to degrade or mineralize various pesticides has been demonstrated (Benimeli et al. 2003; Briceño et al. 2012; Fuentes et al. 2013a, b; Karn et al. 2011; Polti et al. 2014; Saez et al. 2012).

In this context, the main objective of this chapter is to compile the state of the art regarding the use of biopurification systems and their bioaugmentation with specific-degrading microorganisms to obtain optimum results about degradation of pesticides of diverse chemical structures, either pure or as mixtures.

2 Pesticides

The constant technological advances have caused the development of an industrialized society. Thus, during the twentieth century, the improvement of chemical industries allowed the production of a large variety of compounds that led to the modernization of the quality of life. However, the growth of the world population and global industrialization have exerted an important pressure on the environment, putting its sustainability at risk, due to the great amount of chemical pollutants that were introduced into the environment, especially in soils and sediments (Iwamoto and Nasu 2001).

The agricultural sector represents an important part of the world economy and provides not only food resources but also commodities necessary to achieve sustainable development. To improve its production, this sector uses diverse resources such as chemical substances, machinery, and specific equipment. Among the chemical compounds widely used are pesticides, so these compounds are ubiquitous in environmental matrices (Kumari et al. 2007).

The International Code of Conduct on the Distribution and Use of Pesticides defines pesticide in Article No 2 as:

> any substance or mixture of substances intended to prevent, destroy, or control any pest, including human or animal disease vectors, undesirable plant or animal species that cause harm or interfere with production, processing, storage, transport or marketing of food, agricultural products, wood and wood products or animal feed, or that can be administered to animals to fight insects, arachnids or other pests in or on their bodies. The term includes substances intended to be used as regulators of plant growth, defoliants, desiccants, agents to reduce the density of fruits, or agents to prevent premature fruit drop, and substances applied to crops before or after harvest to protect the product against deterioration during storage and transport. (FAO 2006)

Since 1940, the pesticide systematic use has undoubtedly produced a great improvement in the quality of people's life by allowing a greater and better production of food and raw materials. Despite the spectacular results obtained by using these compounds in a massive way, their application led to adverse effects on human, animal, and environmental health (Mokarizadeh et al. 2015).

2.1 Classification of Pesticides

Pesticide is a common term that includes a wide range of chemical compounds, which differ in their physical and chemical properties from one class to other. Therefore, it is necessary to classify them based on their properties and characteristics and to study them under their respective groups. There are different criteria for pesticide classification; some of the most popular methods comprise the following: (1) classification based on the chemical composition of the pesticide; (2) classification of pesticides by hazard; (3) classification based on the mode of action; (4) classification based on their target, i.e., the pest they kill; (5) classification based on their method of application; and (6) classification based on their persistence in environment, among others (Yadav and Devi 2017).

2.1.1 Classification Based on Chemical Composition of Pesticides

Classification of pesticides based on their chemical composition and nature of its active ingredients is one of the most useful methods for grouping them. This kind of classification gives information about the efficacy and physical and chemical properties of the different pesticides and brings data such as its application modes, rates, and precautions that need to be taken during its application (Yadav and Devi 2017).

Based on chemical composition, pesticides are classified into two main groups, namely, inorganics and organics. Inorganic pesticides are those that do not contain carbon in their chemical structure. In general, they are salts or mineral elements that are mined from the earth and ground into a fine powder. Inorganic pesticides are the simplest compounds. They have a crystalline, salt-like appearance, are environmentally stable, and usually dissolve readily in water. The earliest chemical pesticides used by man were inorganics. One of the most important inorganic pesticides is sulfur that can be used finely milled (micronized) or as a colloidal solution, to kill plant-feeding mites. Among fungicides, there are various compounds of copper, mainly copper sulfate, which are widely used alone or mixed with organic compounds to control plant diseases. Other inorganics include compounds as arsenic, copper, lead and tin salts, borates, and diatomaceous earths (Eldridge 2008).

Organic pesticides are chemicals having carbon as the basis of their molecular structure, with C–C and C–H bonds. The chemical structure of organic pesticides is more complex than those of inorganic ones and usually does not dissolve easily in water. The vast majority of modern pesticides contain an organic chemical. There are hundreds of pesticides that have been developed based on organic chemicals, often with oxygen, phosphorus, sulfur, or halogens in their molecules, in addition to their basic carbon structure (Eldridge 2008).

Organic pesticides can be subdivided into two additional groups: the natural organics and the synthetic organics. The natural organic pesticides (sometimes just called "organics") are derived from naturally occurring sources such as plants. Rotenone and pyrethrum are examples of natural organic pesticides. Synthetic organic pesticides (usually just called "synthetics") are produced artificially by chemical synthesis. This group comprises the most "modern" pesticides (i.e., discovered or used as insecticides post-World War II) and includes DDT, permethrin, malathion, 2,4-D, glyphosphate, and many others (Eldridge 2008).

3 Biopurification Systems

An important environmental issue is the deterioration of different environmental compartments quality as a consequence of pesticide pollution through diffuse or point-source (Sniegowski et al. 2011). Recently, several studies reported that point-source contamination is the main cause of pesticide accumulation in surface water, groundwater, and large soil volumes (Castillo et al. 2008).

The application of pesticides to crops usually involves three critical steps: (1) dumping and dilution of pesticide concentrates into the spray tank, (2) pesticide spraying on the fields, and (3) handling of pesticide residues that may be retained in the interior and exterior walls of the spray tank. When pesticides are applied in the recommended doses through modern techniques in accordance with good agricultural practices, the risks of environmental contamination from field spraying is small (step 2). In contrast, the risk is higher in steps 1 and 3, which can lead to point-source pollution. Small spills with high pesticide concentrations may occur during step 1 because at this stage it is common to work with concentrated formulas. Meanwhile, step 3 provides higher volumes but with lower pesticide concentrations than step 1, due to the washing and subsequent dilution of the pesticides remaining in the tank (Castillo et al. 2008).

Several physicochemical or biological treatments have been developed for reducing the pesticide contamination effects; however, most of them are expensive or require the implementation of complex technology, hindering their use in farms. In this context, and with the aim of mitigating the point-source contamination produced by improper pesticide handling before or after their field application, biopurification systems (BPS) were developed as a biological technology to collect and treat pesticide polluted wastewater produced during agricultural activities (Chin-Pampillo et al. 2016). BPS are low-cost, simple design, easy to operate, and practical systems, employed by farmers for the accelerated degradation of pesticides (Sniegowski et al. 2011). In addition, they can be adapted to local climatic conditions and legislation (Karanasios et al. 2012). High sorption capacity, active microbial communities to degrade pesticides, and organic carbon sources to support microbial activity are the three main characteristics that BPS should have to work properly.

The first BPS, initially called biobed, was implemented in the 1990s in Sweden by Torstensson and Castillo (1997), and then it spread to some countries in Europe and America. The BPS original design consisted of a pit in the ground (60 cm deep) containing an impermeable clay layer (10 cm) at the base. A solid matrix, called biomixture, was used to fill the remaining volume to provide sorption capacity for pesticide immobilization and microbial activity for degrading these chemical compounds. A grass layer was used to cover the biobed and to regulate the moisture contents of the system (Castillo and Torstensson 2007). Furthermore, the system contained a ramp needed to drive and park the sprayer over the BPS (Fig. 1).

Over the years, different modifications to the original design were made with the intention of meeting the regional requirements (Castillo et al. 2008). Approximately, more than 1500 BPS are being developed in Sweden and similar or modified BPS were formulated in more than 25 countries, with the form of excavations or containers filled with the right components (Karanasios et al. 2012)

Indistinctly its composition, the biomixture represents the biologically active core of the BPS, where pesticides are removed from the wastewater by biodegradation and/or sorption (Castro-Gutiérrez et al. 2018). This filter matrix receives wastewater

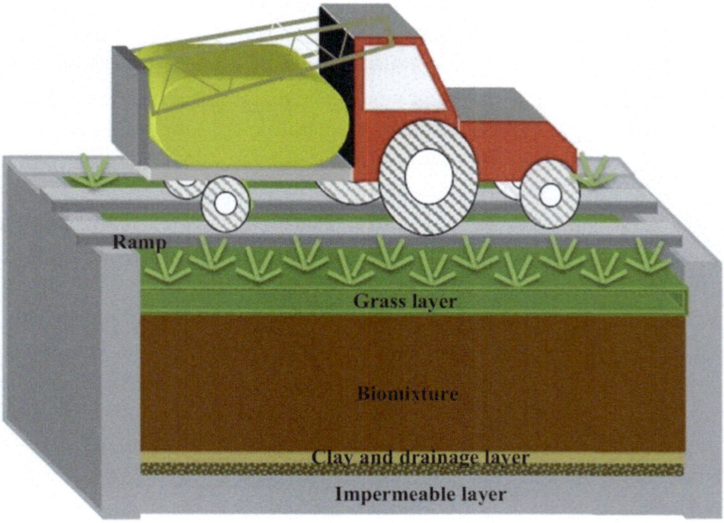

Fig. 1 The BPS original design (Adapted from Castillo et al. 2008)

containing a mixture of several pesticides at high concentrations. Due to the fact that these concentrated residues are collected in a confined space, it is expected that the pesticide concentration be much higher in BPS (in the order of mg kg^{-1}) than in agricultural soil after pesticide application (Castro-Gutiérrez et al. 2018; Chin-Pampillo et al. 2016). For this reason, BPS can act as an optimal microcosm for the development of microorganisms that can metabolize pesticide residues from wastewater (Dunon et al. 2013).

3.1 Components of Biopurification Systems

3.1.1 The Clay Layer

The clay layer, located at the BPS bottom part, acts as an impermeable base to reduce water drainage and increase the pesticide retention time in the matrix, since clay particles present high sorption capacity and low permeability. It is essential that the clay material is moist and swollen, thus avoiding the formation of cracks and preferential flow processes (Castillo et al. 2008).

3.1.2 The Biomixture

The biomixture is the fundamental component of a BPS since the retention and accelerated removal of pesticides occurs in this matrix, so it must have a good capacity of adsorption and high microbial activity (Castro-Gutiérrez et al. 2018). These properties are affected by different factors, such as composition, homogeneity, age, humidity, and temperature (Castillo et al. 2008); however, its composition is the most important factor in BPS performance (Karanasios et al. 2012).

The original Sweden biomixture consisted of straw (lignocellulosic substrate), peat (humic component), and soil, in proportions of 50:25:25 vol%, respectively (Gao et al. 2015). Currently, this composition has been modified to optimize the efficiency of the system. Therefore, in general terms, biomixtures are constituted by three components: a lignocellulosic substrate, a humic component, and soil, each one playing a key role in the capacity of pesticide retention and degradation of the system (Gao et al. 2015; Ruiz-Hidalgo et al. 2016).

The soil, which can be pre-exposed to the pesticides to be treated, provides the sorption capacity and the active edaphic microbiota for pesticide degradation. The soil must be rich in humus and have an adequate clay content to promote microbial activity, without reducing the contaminant bioavailability or hampering the system homogeneity (Castillo et al. 2008). Possibly, the physicochemical characteristics of different soil types would not have a significant effect in a BPS efficiency, due to the destructive mixing process and the inclusion of other components during the biomixture formulation (Saez et al. 2018). In this sense, Fogg et al. (2004) evaluated the use of three soil types (sandy loam, clayey, and clay silty) in BPS performance at laboratory scale and concluded that there were not significant differences between the systems formulated with the different soils. However, clayey soils are not recommended for BPS formulation because the high clay content presents practical management problems and decreases the pesticides availability (Castillo et al. 2008). Considering that the soil type appears to have no significant effect on the removal efficiency of the contaminants in BPS, the modifications in the biomixture formulation mainly focus on the replacement of the lignocellulosic and humic substrates to adapt the system to local agricultural and environmental conditions (Karanasios et al. 2012)

The lignocellulosic substrate aims to promote microbial activity and establishment in the biomixture, especially the colonization and activity of lignin-degrading fungi (like white-rot fungi), which produce phenoloxidases, such as peroxidases and laccases, involved in pesticide degradation (Gao et al. 2015). Although the original lignocellulosic substrate was straw, the existence of other inexpensive lignocellulosic materials in different regions led to its total or partial replacement. The type of substrate used in the biomixture formulation varies according to regional availability; in this sense, by-products such as wood chips, olive leaves, grape stalks, barley husk, coconut fibers, corn cobs, and sunflower residues were used for this purpose, among others (Chin-Pampillo et al. 2015a, 2016; Karanasios et al. 2010b; Urrutia et al. 2013). In Tucumán, Argentina, by-products from the sugar production industry like bagasse are produced in large amounts, so they are considered as adequate substitutes for straw, due to their high availability. This was confirmed by Saez et al. (2018) who observed that laboratory-scale biomixtures containing sugarcane bagasse significantly promoted lindane removal.

Lignocellulosic materials present several advantages such as high availability in farms, high persistence, and low rate of degradation. Thus, they continuously supply carbon, energy, and nutrients for the microbial development (Castillo et al. 2008).

The peat was the original humic component used in the biomixture formulation, and it plays numerous functions: increases sorption capacity, enhances system porosity, controls the moisture content as a consequence of its water retention ability, regulates the pH, and contributes to abiotic degradation of pesticides (Gao et al. 2015). Furthermore, it was shown that high contents of this component confer a low pH that is favorable for the development and activity of fungi. However, too high peat levels (50 vol% or more) reduce microbial activity, possibly due to a significant decrease in pH and water availability. Because of this, 25 vol% of peat is adequate for the formulation of the matrix, because it provides a final pH of 5.9, approximately, suitable to stimulate the development of lignin-degrading fungi (Castillo et al. 2008).

Although peat is very beneficial for BPS, in some regions like Southern Europe, it was necessary to replace this component with alternative materials, due to its high cost and low availability, and also it is not considered adequate for a sustainable agricultural activity (Karanasios et al. 2012). In this context, compost represents a promising alternative to replace the peat, as different compost types are widely available in different geographical regions, in addition to the fact that compost-based biomixtures presented comparable or even better performance than traditional ones based on peat (Chin-Pampillo et al. 2015a; Karanasios et al. 2010b). Composts and peat present different physicochemical characteristics, nutrient contents, and biological activity. These differences may reflect variability in the overall performance of the BPS (Karanasios et al. 2012). Composts provide macronutrients (K, P, N) and extra carbon sources, which can accelerate the pesticide degradation (Ren et al. 2018). Unlike peat, which generally has acidic pH, compost has neutral to basic pH (Karanasios et al. 2012). In addition, composts are known to host several microorganisms with pesticide catabolic activity and to have a good sorption ability for different pesticides, with a retention capacity of over 90% (Monaci et al. 2009). Olive leaves, cotton seed, and cotton crop residue composts are examples of different compost types that significantly promoted pesticide degradation compared to peat-based biomixtures (Karanasios et al. 2010a).

3.1.3 The Grass Layer

In the top part, a grass layer is used to cover the BPS. This layer increases the BPS efficiency, especially that of the upper parts where most of the pesticides are retained and degraded since it regulates moisture content through upward water transport by plant roots, sustains adequate oxygen levels for microorganisms, and produces root exudates that contribute with the co-metabolic processes (Castillo et al. 2008; Coppola et al. 2012). The absence of the grass layer decreases the evapotranspiration, leading to a hydrophobic surface, reducing the microbial activity, and increasing the pesticide lixiviation by promoting the water drainage to the bottom of the BPS (Fogg et al. 2004; Spanoghe et al. 2004). The survival of the grass cover is often used also as an indicator of pesticide spills (Coppola et al. 2012).

4 Application of Biomixtures for the Removal of Different Pesticides

The BPS are being widely studied for their versatile degrading ability to remove different active ingredients (Huete-Soto et al. 2017). Diverse parameters have been studied and optimized for specific uses. The introduction of an SBP in a new country implies intensive research to adapt it to the conditions, practices, and needs of a particular place (Castillo et al. 2008). In this sense, laboratory, pilot, and field studies have been carried out to evaluate the efficiency of different BPS by modifying not only the components of the BPS such as soil type (Fogg et al. 2004; Saez et al. 2018), the lignocellulosic substrate (Chin-Pampillo et al. 2015b, 2016), and the peat (Diez et al. 2013) but also operational parameters such as the use of different hydraulic flows (Diez et al. 2015), temperatures, and doses of pesticides (Karanasios et al. 2012), among others. For instance, moisture is a parameter of great importance in a BPS since it affects the availability of oxygen and microbial activity. However, the saturation of the BPS can lead to the leaching of the pesticide. Therefore, the system must not be located in a low part of the farm to prevent runoff from a large area around the SBP (Castillo et al. 2008). Levels of 60% WHC have been shown to be optimum for the dissipation of most pesticides and are suitable for lignin-degrading fungi (Fernández-Alberti et al. 2012).

Besides, it has been demonstrated that the aging of the biomixture affected the degradation of the pesticides in a BPS. In this sense, the periodic substitution of the biomixture or regular provision of easily degradable organic substrates in BPS should be considered to maintain an adequate depuration capacity on this system (Castro-Gutiérrez et al. 2017).

4.1 Simultaneous Removal of Different Pesticides Using Biomixtures

Taking into account that pesticide application programs include the use of several compounds through the crop production cycle, biobeds may receive high concentrations of relatively complex mixtures of pesticides. So, the optimization of biomixtures should be performed to minimize the residual toxicity of the process and to guarantee the simultaneous elimination of specific pesticide combinations (Huete-Soto et al. 2017). Recent studies evaluated the efficiency of biobed biomixtures for the removal and/or degradation of pesticide mixtures, even those containing structurally diverse compounds. Diez et al. (2017) reported the simultaneous dissipation of atrazine, chlorpyrifos, and iprodione (5 mg kg^{-1}) in a biopurification system, reaching removal yields around 74% and 89%.

Also, some studies have shown evidence that the persistence of pesticides may change when they are used in combination with others (Urrutia et al. 2013). The effects of the concentration and pesticide interaction on degradation rate were investigated by Fogg et al. (2003), demonstrating that at concentrations ranging from half to 20 times the maximum recommended application rate for isoproturon and chlorothalonil, the degradation rate decreased with increasing concentration. Also, they showed that interactions between pesticides are possible. In soil, the time required to degrade 50% of isoproturon increased from 18.5 to 71.5 days in the presence of chlorothalonil, whereas in a biomixture the degradation of each pesticide was unaffected by the presence of the other. These authors suggested that biobeds appear capable of treating high concentrations of more than one pesticide.

Murillo-Zamora et al. (2017) evaluated the simultaneous removal of fungicides belonging to different chemical groups. They found that a biomixture containing coconut fiber, compost, and soil at a volumetric composition of 45:13:42 failed to remove triazole fungicides under the assayed conditions, but it was efficient to eliminate other fungicides such as carbendazim, edifenphos, and metalaxyl. This suggests that the use of different biomixtures for the treatment of different pesticides should be assayed in each case. Similarly, Huete-Soto et al. (2017) assayed the simultaneous removal of structurally diverse pesticides in a biomixture and concluded that the system was highly efficient in the elimination of the herbicides atrazine, linuron, and ametryn (>97% total removal) and some fungicides (metalaxyl and carbendazim) but failed to remove triazole fungicides or neonicotinoid insecticides. In this sense, Ruiz-Hidalgo et al. (2016) remarked the necessity to evaluate new biomixture formulations according to specific pesticide application schemes. Also, it is expected that the use of soil previously exposed to pesticides or successive applications of specific pesticides in the biomixture during the continuous use of the matrices could enhance the removal of such compounds (Lizano-Fallas et al. 2017).

4.2 Bioaugmentation of Biomixtures

The main objective of these inexpensive and simple design systems consists in the sorption and complete mineralization of the toxic compounds. However, incomplete mineralization is common in some cases, which can lead to the formation of even more toxic or recalcitrant metabolites than the parent compound (Coppola et al. 2010). In addition, pesticide degradation is deficient during the initial stages of the process because appropriate microorganisms have to adapt and proliferate in the biomixture before reaching high biodegradation rates (Fogg et al. 2004). In this context, strategies such as biostimulation and bioaugmentation have been explored to enhance pesticides degradation processes in the BPS. For instance, Tortella et al. (2010) proposed that short and fast biostimulation of the microbial population present in a biomixture with an inorganic fertilizer had a significant effect on chlorpyrifos degradation. Later, they demonstrated that the addition of natural wastes rich in terpenes in a peat-based biomixture of a BPS enhanced the biodegradation of atrazine (Tortella et al. 2013).

Meanwhile, a rapid and complete pesticide degradation could be get through bioaugmentation of the biomixture with microorganisms that contain metabolic pathways necessary to complete the pesticide mineralization. Furthermore, the bioaugmentation for enhancing pesticide bioremediation would avoid the toxic metabolite production (Castro-Gutiérrez et al. 2018). The microorganisms inoculated in a biomixture play a key role in the degradation of the pesticides. Microbial communities possess increased metabolic capabilities, allowing members of communities the division of labor and survival to perturbations (Hays et al. 2015).

Genetically modified strains might constitute an efficient strategy for this purpose (Nikel et al. 2014). However, legal restrictions make this solution unfeasible over the short term. In contrast, native organisms pre-exposed to high pesticide concentration are more adequate because they possibly acquired the capability of degrading these toxic substances (Sniegowski et al. 2011). The use of mixed microbial consortia has shown to be more advantageous for improving the degradation capacity of pesticides in BPS compared to pure cultures (Sniegowski and Springael 2015).

Most of the studies have evaluated the bioaugmentation of BPS biomixtures with ligninolytic fungi, considering that half of the biomixture content consists of lignocellulosic substrates. Thus, the biotransformation of lignocellulosic wastes can be attributed to these microorganisms, especially white-rot fungi, due to their extracellular ligninolytic enzymes, such as laccases, lignin peroxidases, and manganese peroxidases, able to attack and transform lignin and thus provide additional sources of nutrients and energy. Besides, these fungi are known to degrade a wide range of organic complex molecules as pollutants, including pesticides (Camacho-Morales et al. 2017; Purnomo et al. 2017). The inoculation of a biomixture with the white-rot fungus *Anthracophyllum discolor* immobilized on an uncoated pelletized support improved atrazine degradation capability of a BPS in 20% and thus could be used to reduce the pesticide migration to surface and groundwater (Elgueta et al. 2016). The use of the ligninolytic white-rot fungus *Trametes versicolor* has been more deeply explored for the bioaugmentation of biomixtures, both as single cultures and co-cultures, for the removal of structurally different pesticides, such as the insecticides carbofuran and lindane, the herbicides atrazine and diuron, and diverse fungicides such as tebuconazole, carbendazim, edifenphos, and metalaxyl (Bending et al. 2002; Madrigal-Zúñiga et al. 2016; Murillo-Zamora et al. 2017; Saez et al. 2018).

Ruiz-Hidalgo et al. (2015) proposed the use of this fungus grown in rice husk as an option of lignocellulosic bioaugmented substrate for its potential application in biomixtures. The lignocellulosic material supported the growth and laccase activity of the fungus. Later, Madrigal-Zúñiga et al. (2016) evaluated the use of this pre-colonized substrate for the removal of carbofuran in two biomixtures of different compositions. They demonstrated that the bioaugmentation improved the performance of the peat-based biomixture rather than the compost-based biomixture, thus highlighting the importance of careful design and evaluation of biomixtures before their large-scale use. In contrast, Murillo-Zamora et al. (2017) found that the bioaugmentation of the biomixture containing coconut fiber, compost, and soil at a volumetric composition of 45:13:42 with *Trametes versicolor* failed to enhance the removal capacity of the matrix, even after biomass re-inoculation. The fungus did not achieve long-term colonization of the biomixture. Similarly, fungal bioaugmentation did not improve the removal of carbofuran in a biomixture composed of rice husk, compost, and soil pre-exposed to this pesticide; however, it favored a lower production of transformation products and hence allowed a decrease in the toxicity of the matrix (Rodríguez-Rodríguez et al. 2017). Lizano-Fallas et al. (2017) hypothesized that the autochthonous microbiota or physicochemical factors of the matrices could have inhibited the proper colonization of the matrices by the fungus, thus minimizing its degrading capacity. Also, the addition of the fungal biomass and/or the production of antagonist compounds could have altered the microbial degrading populations. In this connection, the addition of pesticide-degrading microorganisms can modify the microbial community in contaminated biomixtures to the point that exogenous microorganisms become the main populations in some cases, while in others, it generates undetectable variations in the microbial communities (Castro-Gutiérrez et al. 2018). Verhagen et al. (2013) reported that the bioaugmentation of biomixtures with a mixed culture increased chloropropham removal; however, they found a $65 \pm 5\%$ similarity when comparing the DGGE profiles between the inoculated and the non-inoculated biomixtures, which indicated important changes in the microbial populations. Likewise, Castillo Diaz et al. (2016), through DGGE analysis, found that bands corresponding to the inoculated strains were the main components of the microbial community after bioaugmentation of vermicompost-based biomixtures to accelerate different pesticide removal. Castro-Gutiérrez et al. (2018) informed that bioaugmentation with a carbofuran-degrading bacterial consortium did not affect the pesticide removal performance or microbial community structure, according the DGGE analysis. Instead, biomixture aging was due to the microbial population shifts.

Besides fungi, recent studies have also evaluated the bioaugmentation of biomixtures with bacterial cultures, either single strains or consortia (Briceño et al. 2017; Campos et al. 2017; Karas et al. 2016) or even fungal-bacterial co-cultures (Saez et al. 2018). Verhagen et al. (2013) demonstrated that bioaugmentation of a BPS with a pesticide-degrading bacterial consortium showed an improvement in the BPS functionality for the removal of chlorinated compounds such as chloropropham and 3-chloroaniline, compared to the systems inoculated with a single pesticide-degrading strain or non-inoculated BPS. The efficiency of bioaugmentation to maximize their depuration performance of biobeds may depend on the persistence of the compound to be removed. Generally, the indigenous microbial community of biobeds presents a nonspecific degrading capacity for recalcitrant compounds (Campos et al. 2017). Karas et al. (2016) suggested that the indigenous microbial community of biobeds has, or develops rapidly, the ability to degrade biodegradable compounds such as *ortho*-phenylphenol and diphenylamine, whereas it showed a lower capacity to transform the persistent fungicide thiabendazole (TBZ). Thus, the bioaugmentation of biobeds with TBZ-degrading bacterial inocula resulted in ameliorating the depuration performance against TBZ.

Among bacteria, actinobacteria exhibit a wide ability to grow and degrade several chemical families of pesticides, including organochlorine, triazines, carbamates, organophosphorus, pyrethroids, ureas, and chloroacetanilides (Alvarez et al. 2017; Benimeli et al. 2008; Briceño et al. 2016; Fuentes et al. 2013b). Moreover, Briceño et al. (2017) demonstrated that a *Streptomyces* mixed culture was able to colonize efficiently a biobed biomixture contaminated with diazinon and chlorpyrifos, exhibiting an increase in population, and also increased the dissipation of the pesticides, mainly diazinon, reducing its half-life to 11 days. Based on the capacity of the actinobacteria *Arthrobacter* sp. strain C1 to metabolize iprodione, Campos et al. (2017) evaluated the use of two combined strategies, bioaugmentation with specific degrading bacteria and rhizosphere-assisted biodegradation, to maximize the dissipation capacity of biobeds. The dissipation of the fungicide was accelerated by the bioaugmentation and was further enhanced in the bioaugmented rhizosphere-assisted treatment compared to the non-bioaugmented and non-rhizosphere-assisted controls.

Recently, Saez et al. (2018) demonstrated that the cooperative use of a lindane-degrading actinobacteria mixed culture and the white-rot fungus *Trametes versicolor* SGNG1 improved the pesticide dissipation in different biomixtures, achieving 81–87% of removal at 66 days of incubation and decreased lindane half-life to an average of 24 days, i.e., sixfold less than $t_{1/2}$ of lindane in soils.

5 Concluding Remarks

Biomixtures are the biological core of biopurification systems, initially called biobeds, created for the treatment of pesticide-containing wastewater, to reduce or avoid point-source contamination. The original biomixture of the biobeds consisted of straw, peat, and soil, in proportions of 50:25:25 vol%, respectively. However, several studies and modifications have been performed to optimize the efficiency of the system and to adapt this non-expensive and easy technology to diverse regions, according to the necessities and availabilities of each country.

Different biomixtures have been explored for the treatment of different pesticides, and successful results have been reported, even with complex mixtures containing different groups of pesticides with diverse chemical structures. However, it is necessary to evaluate new biomixture formulations according to specific pesticide uses.

Besides, bioaugmentation of the biomixtures has been proposed as a complementary strategy for improving the efficiency of the BPS on the removal of the pesticides. The bioaugmentation consists of the inoculation of microorganisms, either single cultures or consortia, with specific abilities to degrade pesticides. Based on the fact that the main component of the biomixture is the lignocellulosic substrate, bioaugmentation with ligninolytic fungi has been more deeply studied, mostly with white-rot fungi such as *Anthracophyllum discolor* and *Trametes versicolor*. Moreover, recent studies performed with bacteria, specifically actinobacteria such as *Arthrobacter* and *Streptomyces*, also reported very promising results to maximize the pesticide removal capacity of BPS.

References

Alvarez A, Saez JM, Davila Costa JS, Colin VL, Fuentes MS, Cuozzo SA, Benimeli CS, Polti MA, Amoroso MJ (2017) Actinobacteria: current research and perspectives for bioremediation of pesticides and heavy metals. Chemosphere 166:41–62. https://doi.org/10.1016/j.chemosphere.2016.09.070

Aparicio JD, Raimondo EE, Gil RA, Benimeli CS, Polti MA (2018) Actinobacteria consortium as an efficient biotechnological tool for mixed polluted soil reclamation: experimental factorial design for bioremediation process optimization. J Hazard Mater 342:408–417. https://doi.org/10.1016/J.JHAZMAT.2017.08.041

Bending GD, Friloux M, Walker A (2002) Degradation of contrasting pesticides by white rot fungi and its relationship with ligninolytic potential. FEMS Microbiol Lett 212:59–63. https://doi.org/10.1111/j.1574-6968.2002.tb11245.x

Benimeli CS, Amoroso MJ, Chaile AP, Castro GR (2003) Isolation of four aquatic streptomycetes strains capable of growth on organochlorine pesticides. Bioresour Technol 89:133–138. https://doi.org/10.1016/S0960-8524(03)00061-0

Benimeli CS, Fuentes MS, Abate CM, Amoroso MJ (2008) Bioremediation of lindane-contaminated soil by *Streptomyces* sp. M7 and its effects on *Zea mays* growth. Int Biodeter Biodegr 61:233–239. https://doi.org/10.1016/j.ibiod.2007.09.001

Briceño G, Fuentes MS, Palma G, Jorquera MA, Amoroso MJ, Diez MC (2012) Chlorpyrifos biodegradation and 3,5,6-trichloro-2-pyridinol production by actinobacteria isolated from soil. Int Biodeter Biodegr 73:1–7. https://doi.org/10.1016/j.ibiod.2012.06.002

Briceño G, Schalchli H, Mutis A, Benimeli CS, Palma G, Tortella GR, Diez MC (2016) Use of pure and mixed culture of diazinon-degrading *Streptomyces* to remove other organophosphorus pesticides. Int Biodeter Biodegr 114:193–201. https://doi.org/10.1016/j.ibiod.2016.06.018

Briceño G, Vergara K, Schalchli H, Palma G, Tortella G, Fuentes MS, Diez MC (2017) Organophosphorus pesticide mixture removal from environmental matrices by a soil *Streptomyces* mixed culture. Environ Sci Pollut Res 25(22):21296–22130. https://doi.org/10.1007/s11356-017-9790-y

Camacho-Morales RL, Gerardo-Gerardo JL, Guillén Navarro K, Sánchez JE (2017) Production of ligninolytic enzymes during the degradation of paraquat herbicide by white-rot fungi (in Spanish). Rev Argent Microbiol 49:189–196. https://doi.org/10.1016/j.ram.2016.11.004

Campos M, Perruchon C, Karas PA, Karavasilis D, Diez MC, Karpouzas DG (2017) Bioaugmentation and rhizosphere-assisted biodegradation as strategies for optimization of the dissipation capacity of biobeds. J Environ Manage 187:103–110. https://doi.org/10.1016/j.jenvman.2016.11.031

Castillo MDP, Torstensson L (2007) Effect of biobed composition, moisture, and temperature on the degradation of pesticides. J Agric Food Chem 55:5725–5733. https://doi.org/10.1021/jf0707637

Castillo Diaz JM, Delgado-Moreno L, Núñez R, Nogales R, Romero E (2016) Enhancing pesticide degradation using indigenous microorganisms isolated under high pesticide load in bioremediation systems with vermicomposts. Bioresour Technol 214:234–241. https://doi.org/10.1016/j.biortech.2016.04.105

Castillo Mdel P, Torstensson L, Stenström J (2008) Biobeds for environmental protection from pesticide use: a review. J Agric Food Chem 56(15):6206–6219

Castro-Gutiérrez V, Masís-Mora M, Diez MC, Tortella GR, Rodríguez-Rodríguez CE (2017) Aging of biomixtures: effects on carbofuran removal and microbial community structure. Chemosphere 168:418–425. https://doi.org/10.1016/j.chemosphere.2016.10.065

Castro-Gutiérrez V, Masís-Mora M, Carazo-Rojas E, Mora-López M, Rodríguez-Rodríguez CE (2018) Impact of oxytetracycline and bacterial bioaugmentation on the efficiency and microbial community structure of a pesticide-degrading biomixture. Environ Sci Pollut Res 25:11787–11799. https://doi.org/10.1007/s11356-018-1436-1

Chin-Pampillo JS, Ruiz-Hidalgo K, Masís-Mora M, Carazo-Rojas E, Rodríguez-Rodríguez CE (2015a) Design of an optimized biomixture for the degradation of carbofuran based on pesticide removal and toxicity reduction of the matrix. Environ Sci Pollut Res 22:19184–19193. https://doi.org/10.1007/s11356-015-5093-3

Chin-Pampillo JS, Ruiz-Hidalgo K, Masís-Mora M, Carazo-Rojas E, Rodríguez-Rodríguez CE (2015b) Adaptation of biomixtures for carbofuran degradation in on-farm biopurification systems in tropical regions. Environ Sci Pollut Res 22:9839–9848. https://doi.org/10.1007/s11356-015-4130-6

Chin-Pampillo JS, Masís-Mora M, Ruiz-Hidalgo K, Carazo-Rojas E, Rodríguez-Rodríguez CE (2016) Removal of carbofuran is not affected by co-application of chlorpyrifos in a coconut fiber/compost based biomixture after aging or pre-exposure. J Environ Sci (China) 46:182–189. https://doi.org/10.1016/j.jes.2015.12.026

Coppola L, del Pilar Castillo M, Vischetti C (2010) Degradation of isoproturon and bentazone in peat- and compost-based biomixtures. Pest Manag Sci 67:107–113. https://doi.org/10.1002/ps.2040

Coppola L, Puglisi E, Vischetti C, Trevisan M (2012) Bioremediation and mitigation of organic contaminants in the era of climate changes. In: Environmental adaptations and stress tolerance of plants in the era of climate change. Springer, New York, pp 467–485. https://doi.org/10.1007/978-1-4614-0815-4_22

Diez C, Tortella G, Briceño G, Castillo Mdel P, Diaz J, Palma G, Altamirano C, Calderón C, Rubilar O (2013) Influence of novel lignocellulosic residues in a biobed biopurification system on the degradation of pesticides applied in repeatedly high doses. Electron J Biotechnol 16:11–11. https://doi.org/10.2225/vol16-issue6-fulltext-17

Diez MC, Schalchli H, Elgueta S, Salgado E, Millahueque N, Rubilar O, Tortella GR, Briceño G (2015) Rhizosphere effect on pesticide degradation in biobeds under different hydraulic loads. J Soil Sci Plant Nutr 15(2):410–421. https://doi.org/10.4067/S0718-95162015005000037

Diez MC, Elgueta S, Rubilar O, Tortella GR, Schalchli H, Bornhardt C, Gallardo F (2017) Pesticide dissipation and microbial community changes in a biopurification system: influence of the rhizosphere. Biodegradation 28(5-6):395–412. https://doi.org/10.1007/s10532-017-9804-y

Dunon V, Sniegowski K, Bers K, Lavigne R, Smalla K, Springael D (2013) High prevalence of IncP-1 plasmids and IS1071 insertion sequences in on-farm biopurification systems and other pesticide-polluted environments. FEMS Microbiol Ecol 86:415–431. https://doi.org/10.1111/1574-6941.12173

Eldridge BF (2008) Pesticide application & safety training for applicators of public health pesticides. California Department of Public Health Vector-Borne Disease Section, Sacramento

Elgueta S, Santos C, Lima N, Diez MC (2016) Immobilization of the white-rot fungus *Anthracophyllum disco*lor to degrade the herbicide atrazine. AMB Express 6:104. https://doi.org/10.1186/s13568-016-0275-z

Fernández-Alberti S, Rubilar O, Tortella GR, Diez MC (2012) Chlorpyrifos degradation in a biomix: effect of pre-incubation and water holding capacity. J Soil Sci Plant Nutr 12:785–799. https://doi.org/10.4067/S0718-95162012005000032

Fogg P, Boxall ABA, Walker A (2003) Degradation of pesticides in biobeds: the effect of concentration and pesticide mixtures. J Agric Food Chem 51:5344–5349. https://doi.org/10.1021/jf030060z

Fogg P, Boxall ABA, Walker A, Jukes A (2004) Effect of different soil textures on leaching potential and degradation of pesticides in biobeds. J Agric Food Chem 52:5643–5652. https://doi.org/10.1021/jf040023n

Food and Agriculture Organization of the United Nations (FAO) (2006) International code of conduct on the distribution and use of pesticides. Guidelines on efficacy evaluation for the registration of plant protection products. FAO, Rome

Fuentes MS, Benimeli CS, Cuozzo SA, Amoroso MJ (2010) Isolation of pesticide-degrading actinomycetes from a contaminated site: bacterial growth, removal and dechlorination of organochlorine pesticides. Int Biodeter Biodegr 64:434–441. https://doi.org/10.1016/j.ibiod.2010.05.001

Fuentes MS, Alvarez A, Saez JM, Benimeli CS, Amoroso MJ (2013a) Methoxychlor bioremediation by defined consortium of environmental *Streptomyces* strains. Int J Environ Sci Technol 11:1147–1156. https://doi.org/10.1007/s13762-013-0314-0

Fuentes MS, Briceño GE, Saez JM, Benimeli CS, Diez MC, Amoroso MJ (2013b) Enhanced removal of a pesticides mixture by single cultures and consortia of free and immobilized *Streptomyces* strains. Biomed Res Int 2013:Article ID 392573. https://doi.org/10.1155/2013/392573

Gao W, Liang J, Pizzul L, Feng XM, Zhang K, Castillo Mdel P (2015) Evaluation of spent mushroom substrate as substitute of peat in Chinese biobeds. Int Biodeter Biodegr 98:107–112. https://doi.org/10.1016/j.ibiod.2014.12.008

Harrison RM (2014) Pollution: causes, effects and control, 5th edn. Royal Society of Chemistry, Cambridge

Hays SG, Patrick WG, Ziesack M, Oxman N, Silver PA (2015) Better together: engineering and application of microbial symbioses. Curr Opin Biotechnol 36:40–49. https://doi.org/10.1016/j.copbio.2015.08.008

Huete-Soto A, Masís-Mora M, Lizano-Fallas V, Chin-Pampillo JS, Carazo-Rojas E, Rodríguez-Rodríguez CE (2017) Simultaneous removal of structurally different pesticides in a biomixture: detoxification and effect of oxytetracycline. Chemosphere 169:558–567. https://doi.org/10.1016/j.chemosphere.2016.11.106

Iwamoto T, Nasu M (2001) Current bioremediation practice and perspective. J Biosci Bioeng 92(1):1–8. https://doi.org/10.1263/jbb.92.1

Karanasios E, Tsiropoulos NG, Karpouzas DG, Ehaliotis C (2010a) Degradation and adsorption of pesticides in compost-based biomixtures as potential substrates for biobeds in southern Europe. J Agric Food Chem 58:9147–9156. https://doi.org/10.1021/jf1011853

Karanasios E, Tsiropoulos NG, Karpouzas DG, Menkissoglu-Spiroudi U (2010b) Novel biomixtures based on local Mediterranean lignocellulosic materials: evaluation for use in biobed systems. Chemosphere 80:914–921. https://doi.org/10.1016/j.chemosphere.2010.06.003

Karanasios E, Tsiropoulos NG, Karpouzas DG (2012) On-farm biopurification systems for the depuration of pesticide wastewaters: recent biotechnological advances and future perspectives. Biodegradation 23:787–802. https://doi.org/10.1007/s10532-012-9571-8

Karas PA, Perruchon C, Karanasios E, Papadopoulou ES, Manthou E, Sitra S, Ehaliotis C, Karpouzas DG (2016) Integrated biodepuration of pesticide-contaminated wastewaters from the fruit-packaging industry using biobeds: bioaugmentation, risk assessment and optimized management. J Hazard Mater 320:635–644. https://doi.org/10.1016/j.jhazmat.2016.07.071

Karn SK, Chakrabarti SK, Reddy MS (2011) Degradation of pentachlorophenol by *Kocuria* sp. CL2 isolated from secondary sludge of pulp and paper mill. Biodegradation 22:63–69. https://doi.org/10.1007/s10532-010-9376-6

Kumari B, Madan VK, Kathpal TS (2007) Status of insecticide contamination of soil and water in Haryana, India. Environ Monit Assess 136:239–244. https://doi.org/10.1007/s10661-007-9679-1

Lizano-Fallas V, Masís-Mora M, Espinoza-Villalobos D, Lizano-Brenes M, Rodríguez-Rodríguez CE (2017) Removal of pesticides and ecotoxicological changes during the simultaneous treatment of triazines and chlorpyrifos in biomixtures. Chemosphere 182:106–113. https://doi.org/10.1016/j.chemosphere.2017.04.147

Madrigal-Zúñiga K, Ruiz-Hidalgo K, Chin-Pampillo JS, Masís-Mora M, Castro-Gutiérrez V, Rodríguez-Rodríguez CE (2016) Fungal bioaugmentation of two rice husk-based biomixtures for the removal of carbofuran in on-farm biopurification systems. Biol Fertil Soils 52:243–250. https://doi.org/10.1007/s00374-015-1071-7

Mokarizadeh A, Faryabi MR, Rezvanfar MA, Abdollahi M (2015) A comprehensive review of pesticides and the immune dysregulation: mechanisms, evidence and consequences. Toxicol Mech Methods 25:258–278. https://doi.org/10.3109/15376516.2015.1020182

Monaci E, Coppola L, Casucci C, Perucci P, Vischetti C (2009) Retention capacity of an organic bio-mixture against different mixtures of fungicides used in vineyards. J Environ Sci Health B 44:724–729. https://doi.org/10.1080/03601230903163889

Murillo-Zamora S, Castro-Guti V, Masís-Mora M, Rodríguez-Rodríguez CE (2017) Elimination of fungicides in biopurification systems: effect of fungal bioaugmentation on removal performance and microbial community structure. Chemosphere 186:625–634

Nikel PI, Martínez-García E, De Lorenzo V (2014) Biotechnological domestication of pseudomonads using synthetic biology. Nat Rev Microbiol 12(5):368–379. https://doi.org/10.1038/nrmicro3253

Polti MA, Aparicio JD, Benimeli CS, Amoroso MJ (2014) Simultaneous bioremediation of Cr (VI) and lindane in soil by actinobacteria. Int Biodeter Biodegr 88:48–55. https://doi.org/10.1016/j.ibiod.2013.12.004

Purnomo AS, Nawfa R, Martak F, Shimizu K, Kamei I (2017) Biodegradation of aldrin and dieldrin by the white-rot fungus *Pleurotus ostreatus*. Curr Microbiol 74:320–324. https://doi.org/10.1007/s00284-016-1184-8

Ren X, Zeng G, Tang L, Wang J, Wan J, Wang J, Deng Y, Liu Y, Peng B (2018) The potential impact on the biodegradation of organic pollutants from composting technology for soil remediation. Waste Manag 72:138–149. https://doi.org/10.1016/j.wasman.2017.11.032

Rodríguez-Rodríguez CE, Madrigal-León K, Masís-Mora M, Pérez-Villanueva M, Chin-Pampillo JS (2017) Removal of carbamates and detoxification potential in a biomixture: fungal bioaugmentation versus traditional use. Ecotoxicol Environ Saf 135:252–258

Ruiz-Hidalgo K, Chin-Pampillo JS, Masís-Mora M, Carazo RE, Rodríguez-Rodríguez CE (2015) Degradation of carbofuran by *Trametes versicolor* in rice husk as a potential lignocellulosic substrate for biomixtures: from mineralization to toxicity reduction. Process Biochem 49:2266–2271. https://doi.org/10.1016/j.procbio.2014.10.006

Ruiz-Hidalgo K, Chin-Pampillo JS, Masís-Mora M, Carazo-Rojas E, Rodríguez-Rodríguez CE (2016) Optimization of a fungally bioaugmented biomixture for carbofuran removal in on-farm biopurification systems. Water Air Soil Pollut 227(1):3. https://doi.org/10.1007/s11270-015-2681-2

Saez JM, Benimeli CS, Amoroso MJ (2012) Lindane removal by pure and mixed cultures of immobilized actinobacteria. Chemosphere 89(8):982–987. https://doi.org/10.1016/j.chemosphere.2012.06.057

Saez JM, Bigliardo AL, Raimondo EE, Briceño GE, Polti MA, Benimeli CS (2018) Lindane dissipation in a biomixture: effect of soil properties and bioaugmentation. Ecotoxicol Environ Saf 156:97–105. https://doi.org/10.1016/j.ecoenv.2018.03.011

Sniegowski K, Springael D (2015) Establishment of multiple pesticide biodegradation capacities from pesticide-primed materials in on-farm biopurification system microcosms treating complex pesticide-contaminated wastewater. Pest Manag Sci 71:986–995. https://doi.org/10.1002/ps.3876

Sniegowski K, Bers K, Van Goetem K, Ryckeboer J, Jaeken P, Spanoghe P, Springael D (2011) Improvement of pesticide mineralization in on-farm biopurification systems by bioaugmentation with pesticide-primed soil. FEMS Microbiol Ecol 76:64–73. https://doi.org/10.1111/j.1574-6941.2010.01031.x

Spanoghe P, Maes A, Steurbaut W (2004) Limitation of point source pesticide pollution: results of bioremediation system. Commun Agric Appl Biol Sci 69(4):719–732

Torstensson L, Castillo MDP (Swedish U. of A.S.D. of M.B. 7025, S.-750 07 U. (Sweden)) (1997) Use of biobeds in Sweden to minimize environmental spillages from agricultural spraying equipment. Pestic. Outlook (United Kingdom)

Tortella GR, Rubilar O, Cea M, Wulff C, Martínez O, Diez MC (2010) Biostimulation of agricultural biobeds with NPK fertilizer on chlorpyrifos degradation to avoid soil and water contamination. J Soil Sci Plant Nutr 10:464–475. https://doi.org/10.4067/S0718-95162010000200007

Tortella GR, Rubilar O, Cea M, Briceño G, Quiroz A, Diez MC, Parra L (2013) Natural wastes rich in terpenes and their relevance in the matrix of an on-farm biopurification system for the biodegradation of atrazine. Int Biodeter Biodegr 85:8–15. https://doi.org/10.1016/j.ibiod.2013.06.003

Urrutia C, Rubilar O, Tortella GR, Diez MC (2013) Degradation of pesticide mixture on modified matrix of a biopurification system with alternatives lignocellulosic wastes. Chemosphere 92:1361–1366. https://doi.org/10.1016/j.chemosphere.2013.04.088

Verhagen P, De Gelder L, Boon N (2013) Inoculation with a mixed degrading culture improves the pesticide removal of an on-farm biopurification system. Curr Microbiol 67:466–471. https://doi.org/10.1007/s00284-013-0389-3

Yadav IC, Devi NL (2017) Pesticides classification and its impact on human and environment. In: Environmental science and engineering: Vol. 6: Toxicology. Studium Press LLC, Houston, pp 140–158

Novel Approach for Bioremediation of Perfluorochemicals (PFCs) Using Megasonic Sonochemical Reactor

Vitthal L. Gole, Jyoti, and Rajesh Kumar Yadav

Abstract

With changing lifestyle and facets of life demand of personal care products, drugs, household items, pesticides, surfactants, and intermediate products are increasing on multi-fold basis. Presence of perfluorochemicals (PFC) in products is major challenge for bioremediation. PFCs remain in environment for many centuries and cannot be easily biodegradable results as plastic debris in environment. PFCs are broadly classified base on chain length, long-chain compounds such as perfluorooctanesulfonic acid (PFOS) and perfluorooctanoic acid (PFOA) and traces of fluorotelomers. The long-chain compounds remain in environment for longer duration and cannot be treated with conventional biological routes. Advanced oxidation processes are required for breaking of long fluoro-compounds to smaller-chain compounds. We have addressed the issues of persistence nature of long PFCs by breaking long-chain fluorinated compounds using megasonic sonochemical reactor. It has been observed that megasonic frequency (1 MHz) breaks long-chained PFOS to smaller-chain molecules. Bioremediation of fluorinated carbon is related with half-life, smaller chain compounds have very short half life in mintues to few hours compare to years/centuries of long-chain fluorinated carbon. Novel approach presented in work has given the useful insights on enhancing bioremediation aspects of environmentally persistent long-chain fluorinated carbon.

V. L. Gole (✉) · Jyoti
Department of Chemical Engineering, Madan Mohan Malaviya University of Technology, Gorakhpur, Uttar Pradesh, India

R. K. Yadav
Department of Chemistry and Environmental Sciences, Madan Mohan Malaviya University of Technology, Gorakhpur, Uttar Pradesh, India

M. P. Shah (ed.), *Removal of Emerging Contaminants Through Microbial Processes*,
https://doi.org/10.1007/978-981-15-5901-3_20

Keywords

Perfluorochemicals · Bioremediation · Cavitation · Megasonic frequency · PFCs

1 Introduction

Perfluorochemicals (PFCs) are thermally and chemically stable and have wide application in polymer, lubricants, adhesives, repellents, paper coatings, as well as personal care products and pharmaceuticals (Hughes et al. 2011; Houde et al. 2006; Conder et al. 2008). Electrochemical fluorination and telomerization are main routes for synthesis of PFCs. In electrochemical method, electric current replaced hydrogen by fluoride in –C–H– chain. This method is useful for production of mixture of branched and isomers products of PFCs. Telomerization involves reaction between pentafluoroiodoethane and tetrafluoroethylene to yield mixture of perfluoroalkyl iodides; mostly fluorotelomers are produced via this route (Houde et al. 2006). Bond strength of fluorinated carbon is much higher than –C–H, –C–C–, and other bonds. Breaking the bond required huge quantum of energy, and conventional biological or chemical route is not effective for treatment of PFCs. Due to very strength of fluorinated carbon, it remains in environment for centuries (Conder et al. 2008). There are many studies deals with bioaccumulation of PFCs in water, fish, birds, mammals, and humans. These studies reported that PFOS and PFOA are highly persistent in nature (Li et al. 2019; Jian et al. 2017; Kwon et al. 2014; Kim et al. 2015).

Understanding the environmental fate and effects assessment of PFCs and perfluorinated acids such as PFCAs and PFSAs are mostly commonly studied (Kwon et al. 2014; Kim et al. 2015). These organic acids have fluorinated carbon with carboxylate or sulfonate functional group and remain in anionic form in environment due to low acid dissociation (pK_a) values in range of 2–3, fluorinated carbon length in range of 4–13 with number of structures of linear configuration, and branched isomers (Kwon et al. 2014; Kim et al. 2015). Linear branched configuration observed in biota has very slower degradation rate (Lowe et al. 1998). J. M. Conder et al. (2008) reviewed the detail understanding of bioaccumulation of PFCs based on bond strength, chain length, ionic behavior, pK_a value, hydrophilic, and hydrophobic properties (Conder et al. 2008; Rodriguez-Freire et al. 2015). This analysis gives useful insights on understanding of sink and source of these compounds and potential of bioaccumulation and biomagnification of PFCAs and PFSAs.

Based on the analysis and understanding of literature, the mechanism of bioaccumulation of PFCs and persistency of lipophilic compound has direct relation with fluorinated carbon chain length and hydrophobicity of the compounds (Rodriguez-Freire et al. 2015). Fluorinated chain lengths less than seven are considered as bioaccumulative. Another important parameter that contributes on the nature of bioaccumulation is tale compound of fluorinated carbon. Compounds with

Table 1 Environmental phersistence of PFC (measured in terms of half-life)

Perfluorochemicals		Number of F ion	Number of S ion	Half-life	
				Unit	Value
PFHxA	Perfluorohexanoic acid	11	0	h	0.38
PFHpA	Perfluoroheptanoic acid	13	0	H	0.05–0.1
PFOA	Perfluorooctanoic acid	15	0	year	2.5–3
PFNA	Perfluorononanoic acid	17	0	year	4–5
PFDA	Perfluorohexanoic acid	15	0	year	>4
PFUnDA	Perfluoroundecanoic acid	21	0	year	7–8
PFDoDA	Perfluorodecanoic acid	23	0	year	3–4
PFBS	Perfluorobutanesulfonic acid	9	1	month	>1
PFPeS	Perfluoropentane sulfonate	11	1	year	4–5
PFHxS	Perfluorohexane sulfonate	13	1	year	4–5
PFHpS	Perfluoroheptane sulfonate	15	1	year	2–3
PFOS	Perfluorooctane sulfonate	17	1	year	2–3
6:2FTS	6:2 Fluorotelomer sulfonate	13	2	days	30–45
8:2FTS	6:2 Fluorotelomer sulfonate	17	2	days	8–10

sulfonate tale are highly environmentally persistent in nature compared to compounds with carboxyl tale (Conder et al. 2008). Biomagnification of PFCs study analysis shows that PFCAs did not pose the accumulation behavior of PFCAs and PFSAs for shorter chain compounds of length less than seven (Li et al. 2019; Jian et al. 2017). Half-life of different PFCs shown in Table 1 clearly indicates that technology is needed for breaking long-chain to small-chain compound. Conventional biological routes are not suitable to address the issues of PFCs due to longer processing time, and specific microbes are required for conversion of PFCs. There are good reviews available on biological treatment of PFCs, and most studies have been reported to milligram or gram scale (Houde et al. 2006; Conder et al. 2008; Li et al. 2019). Deposition of PFCs on the solid additives has been explored, but the major shortfall associated with this technique generation of secondary waste and disposal of solid waste itself is a great challenge for the environment (Kim et al. 2015; Lowe et al. 1998; Rodriguez-Freire et al. 2015, 2016; Campbell and Hoffmann 2015). Advanced oxidation process such as sonochemical reactor can be one of the alternatives for treatment of PFCs.

Sonochemical reactors have wide applications for treatment of environmentally persistence and bioaccumulative compounds like (per)fluorochemicals (PFCs) due to its spectacular effects of bubble formation during acoustic cycles of compression and rarefactions. Collapsing acoustic bubble conditions of higher temperature and pressure produced effects of "hot spot." These conditions are highly useful for degradation of pollutants and chemical processing applications (Rodriguez-Freire et al. 2015, 2016). Fluoride ions are liberated from sonochemical treatment of PFCs. Surface cavitation enhances the rate of liberation fluoride ion and further increases with saturation concentration of carbon-fluoride ring on the surface or near vicinity of bubble and life span of the cavitation bubble. It depends on stable cavitational activity in sonochemical reactor. The formation of hydroxyl radicals after the transient collapse of bubble useful for degradation of hydrophilic hydrocarbon and sulfate compounds present in bulk of liquid. Complete mineralization of PFCs depends on the number of stable cavitational bubble events produced per unit volume in sonochemical reactor, uniform pressure field, and microstreaming activities in sonochemical reactor (Rodriguez-Freire et al. 2015, 2016; Campbell and Hoffmann 2015; Fernandez et al. 2016). Frequency of operation, power density, physicochemical properties of liquid medium, and saturation concentration of gas inside bubble are the important parameters for enhancing cavitational efficiency of sonochemical reactor/cavitational activity (Balachandran et al. 2015, 2016). Stable cavitational activity and microstreaming are increased with frequency of operation and power density. It is reported that these activities increase with frequency of operation, maximum effects obtained at megasonic frequency (Keswani et al. 2013, 2014). Monoatomic gases such as argon (saturated aqueous liquid) produced stable cavitation compared to air or oxygen due to higher polytrophic index of argon. Higher power density increased the stable cavitational event due to better nucleation, and growth of bubbles (rectified diffusion or Bjerknes forces), maximum bubble size before collapse, may increase in bubble wall velocity and higher oscillation of amplitude of stable bubble activities (Zhao et al. 2015). Most of studies dealing with sonochemical method are limited to understanding the liberation of fluoride or sulfate ions; bioremediation aspect is lacking.

Present study investigates effect of different sonochemical frequencies on bioremediation of PFCs. The progress of degradation of PFCs was monitored in terms of released concentration of fluoride and sulfate ion and removal of total organic carbon (TOC). Effect of frequency of 1 MHz, 500 kHz, and dual frequency of 1 MHz and 500 kHz was investigated. Bioremediation aspects of PFCs were monitored in terms of breaking of long-chain compound to small-chain compound by monitoring the concentration of different range of PFCs.

2 Materials and Methods

2.1 Materials

Potassium salt of (per)-fluorooctanesulfonic acid (CAS #2795-39-3) purchased from S D Fine Chemicals, India. Distilled water was used preparation of aqueous solution. Hydrochloric acid, sodium hydroxide, total ionic strength adjustment buffer (TISAB-II), sodium fluoride, sodium sulfate, and potassium hydrogen phthalate are required for pH adjustments and analysis, and these chemicals were procured from Merck India Ltd. Nitrogen (ultrapure) required for continuous cooling of transducers and electronic circuits of sonochemical reactor and compressed argon (polytropic index of 1.67) was supplied by local supplier.

2.2 Sonochemical Reactor

Customized dual-frequency sonochemical reactor is used in present work made by Dakshin, Mumbai. It is consisting of two transducers (each having dimensions of 11.8 cm × 9.3 cm) with 1 MHz and 500 kHz operating frequencies placed on opposite side walls at of the reactor. Power supplied was 1.1 and 0.7 kW for 1 MHz and 500 kHz frequency transducer, respectively. Maximum distance between transducers is 26 cm. All the internal parts of sonochemical reactor are made of stainless steel (SS 316). Degradation studies were performed by keeping the pollutant sample in glass tube at center location of transducer at distance of 1.5 cm from surface of transducer. Glass tube was supported by stand and clamp arrangement and covered with cap (gas sealed) with opening on top for bubbling argon. Liquid level which corresponds to the height of transducer (10.4 cm) was maintained in all glass tubes. Side walls of reactor were provided with cooling coils through which cooling water was circulated to maintain constant temperature in reactor. Wave interference due to the samples tubes was neglected. Samples at various intervals of time were taken from all sample tubes.

2.3 Experimental Methodology

All experiments were performed using 2.6 mM concentration of PFOS solutions. Initial solution pH of 4 was maintained for all experiments (Gole et al. 2018b). Aqueous solutions of PFOS were prepared using distilled water. The pH of the solutions was adjusted using 1 M HCl. Initially, the tubes were filled with aqueous solutions of PFOS followed by bubbling the argon gas (polytropic index of 1.67) for 20 min prior to the experiment and maintaining a blanket of this gas during the entire course of experiment. Samples were taken at 30, 60, 150, and 210 min to monitor degradation. To evaluate progress of degradation, the concentration of fluoride ion released (F^-), concentration of sulfate ion (SO_4^{2-}) released, and total organic carbon (TOC) removal were measured as a function of time. Temperature of reactor was maintained at 20 ± 1 °C by circulating cooling water through cooling coils.

2.4 Analysis

Progress of degradation was monitored by measuring concentration of fluoride ion released (F^-) measured by ion-selective electrode (Model 9609BNWP, Thermo Scientific Inc.), concentration of sulfate ion released measured ion chromatography with a Dionex IC-3000 system fitted with a Dionex IonPac AS18 analytical column (4 × 250 mm) and AG18 guard column (4 × 50 mm), and TOC removal measured using a Shimadzu total carbon analyzer VCSH (Columbia, MD, USA). Procedure for measurement of fluoride ion, sulfate ion, and TOC is described by Lucia and coworkers (Rodriguez-Freire et al. 2015, 2016). Optimized result samples were analyzed using ultra-high-performance liquid chromatography to determine the concentration of individual PFCs. Detailed procedure for analysis has been reported in our earlier work (Gole et al. 2018a).

3 Results and Discussion

PFCs act as surfactants due to presence of long-chain carbon-fluoride ring. During the sonochemical degradation of PFCs, cleavage of sulfate group occurred; it occupies the bulk of solution. Due to hydrophobic nature of carbon-fluoride ring, it occupies on gas-liquid interface or close vicinity of bubble. When bubble collapsed high-temperature and high-pressure conditions generated, pollutants near the vicinity of bubble quickly evaporated or break into the smaller chain (Rodriguez-Freire et al. 2015, 2016; Campbell and Hoffmann 2015). Efficiency of sonochemical degradation of PFCs mainly depends on the saturation concentration of pollutants near the vicinity of bubble, life span of bubble (stable cavitational), and cavitational events occurring per unit volume (number of active bubble formation). Cavitational events are classified as stable or transient, and high-temperature elevation can be attained in both cases. Intensity of cavitational event depends on the amplitude and frequency of acoustic wave. The frequency of operation relates with life span of cavity and maximum size of expansion of bubble. Lower frequency of operation is useful for reaction dominated with physical limitation due its nature of longer expansion period, higher cavitational shock waves, and proneness to more violent transient cavitation, whereas higher frequencies of operation produce smaller cavitational events which consider as stable cavitation. During stable cavitation activity, bubble does not collapse but grows and shrinks periodically (Keswani et al. 2014; Zhao et al. 2015; Gole et al. 2018a, b; Sutkar and Gogate 2010; Sutkar et al. 2010). Cavitational effects generated during stable cavitational activity are useful for degradation of hydrophobic nature of PFCs. Significant advantages of cavitational activity at higher frequencies are less bubble coalescence and longer span of life of cavity, and activity exists distant from the surface of transducer. We have investigated effect of two frequencies of 1 MHz and 500 kHz and combination of 500 kHz and 1 MHz on degradation of PFCs (Balachandran et al. 2016; Keswani et al. 2013).

3.1 Effect of 1 MHz Frequency

Results of the effect of 1 MHz frequency on concentration of fluoride ion released during degradation of PFOS is shown in Fig. 1. Fluoride ion released concentration was linearly increasing with time and maximum fluoride ion released was observed at 210 min with further increase in time shown marginal increase in released of fluoride ion. At 1 MHz frequency, maximum concentration of fluoride ion released after 210 min treatment was 506.3 ± 21.1 μM. Similar trends were observed for release of corrected concentration of sulfate ion and moles of TOC removal. Maximum concentration of TOC removal for degradation of PFOS was

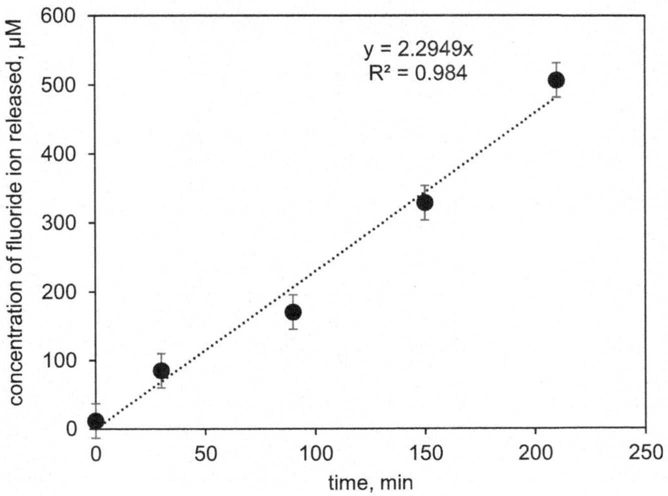

Fig. 1 Effect of 1 MHz frequency on concentration of fluoride ion released from degradation of PFOS at a distance 1:1.5 cm away from transducer surface; pH of solution, 4; initial concentration of PFOS, 2.6 mM

25.5 ± 2.8 μM, and corrected concentration of sulfate ion released for degradation of PFOS was 2.5 ± 0.05 μM. Fluoride or other ion concentration released/removed is dependent on the cavitational activities. This activity depends on the serval parameter, and some activities may have null effect on degradation due to the interference between incident wave and reflected waves (Sutkar and Gogate 2010; Sutkar et al. 2010). Collapsing condition of bubble or maximum temperature inside the bubble is useful for breaking carbon-fluoride chain; these conditions play the significant role for breaking of bond between hydrogen and hydroxyl radicals. Results are closely related with experimental findings; rate formation of hydroxyl radicals (cavitational activity) increased from 0.14 to 0.63 μM min^{-1} with increase in frequency from 37 kHz to 1 MHz, and they also reported higher degradation of *p*-cresol achieved for 1 MHz frequency sonication (85%) compared to 37 kHz frequency (10%) (Fernandez et al. 2016). Megasonic frequency (1.6 MHz, 145 W L^{-1}) is found to be more energy-efficient for degradation of non-polar pollutant DDT dispersed in water compared to lower frequency of 20 kHz (375 W L^{-1}) (Thangavadivel et al. 2009). Cavitational activity decreased with increase in distance away from the transducer surface; the work on theoretical measurement in terms of pressure and liberation of iodine reported that (694 kHz) the liberation of iodine was 7.8, 3.0, 2.9, 3.4, and 2.7 mg/L and pressure field was 0.78, 0.86, 0.61, and 0.62 atm for distances 4, 8, 16, and 20 cm away from transducer surface, respectively. The interference of incident and reflected wave results in net loss of cavitational activites and thus decrease in liberation of iodine concentration (Sutkar and Gogate 2010). Pressure amplitude of ultrasonic horn decrease from 4×10^5 to 0.5×10^5 Pa with increase in distance from very near to irradiating surface to 0.02 m away from surface (Sutkar et al. 2010).

3.2 Effect of 500 kHz Frequency

Concentration of fluoride ions was released during degradation of PFOS as a function of time at frequency of 500 kHz for sample placed at 1.5 cm away from transducer surface. It has been observed that concentration of fluoride released, concentration of TOC removal, and corrected concentration of sulfate released increase linearly with time. Results have been shown in Fig. 2. Maximum concentration of fluoride released, concentration of TOC removal, and corrected concentration of sulfate released after 210 min were 504.8, 38.6, and 0.4 μM, respectively.

Decrease in removal efficiency (cavitational intensity) observed at 500 kHz was compared to 1 MHz. Difference in cavitational activity may be due to the higher power supplied to transducer surface or may be due to the spatial and temporal difference in cavitational activity at different frequencies. Spatial difference correlated with frequency of operation; at higher frequency the acoustic wave pressure field is more uniform with distance away from irradiating source with marginal loss of removal efficiency due smaller bubble size (Entezari and Kruus 1996). Findings of present work are closely related with the reported work, degradation of atrazine at 500 kHz is 7.8 times greater than at 20 kHz, and similar trends were observed for degradation of pentachlorophenol which increased 4.8 times for increase in frequency from 20 to 500 kHz (Petrier et al. 1996). Another work

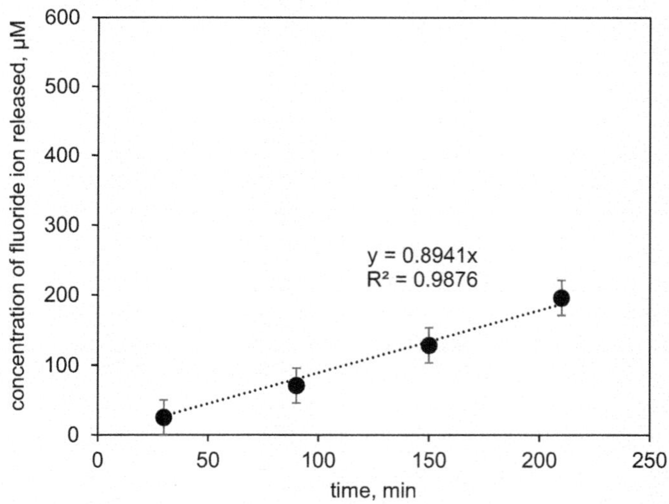

Fig. 2 Effect of 500 kHz frequency on concentration of fluoride ion released from degradation of PFOS at a distance 1:1.5 cm away from transducer surface; pH of solution, 4; initial concentration of PFOS, 2.6 mM

reported that complete mineralization of carbon tetrachloride was observed in 1 h for 500 kHz and 2 h for 20 kHz frequency (Francony and Petrier 1996). Sonochemical efficiency measured in terms of liberated of I_3^- at 500 kHz was 1.3 ± 0.1 times lower with increase height liquid, while the acoustic yield increased 1.2 ± 0.1 with increase in distance away from transducer surface increase from 29 to 252 mm (de La Rochebrochard et al. 2012). The study of sonochemical degradation of potassium iodine at 291 kHz frequency reported that as the liquid height/volume was increased, calorimetric power and energy conversion efficiency increase with power input up to 400 W/L (Lim et al. 2014).

3.3 Dual-Frequency Operation of 1 MHz and 500 kHz

Combination of frequencies enhanced interaction between bubble sound field and bubble-bubble interactions, and these interactions result in more bubble fragmentation due to the waves induced on bubble by surrounding bubbles. Lower- and higher-frequency combination can produce new nuclei or increase the number of nuclei production due to combined effects of two frequencies. Dual-frequency sound field formed the wider range of cavitation bubble radii compared to single-frequency operation. Significant effects of dual-frequency operation give more efficient mass transfer effects than individual irradiation which results in enhanced cavitational yield or hydroxyl radical formation. Synergic of frequencies may result in possible increased cavitational yield and saving in terms of energy and time (Wang et al. 2006; Tiong et al. 2017; Rahimi et al. 2014). Results for released of fluoride ion concentration for dual-frequency operations of 1 MHz and 500 kHz are shown in Fig. 3. It has been observed that the released concentration of fluoride ion after 210 min for degradation was 254.7 ± 18.2 μM. Similar trend observed moles of removal of TOC and corrected concentration of sulfate released. Moles of TOC removal were 36.8 ± 2.3 μM, and corrected concentration of sulfate released was 44.0 ± 1.3 μM. Synergic combination of frequency of operation showed higher released of concentration of fluoride ion than single effect of 1 MHz and 500 kHz of operation.

Degradation studies have shown that 1 MHz frequency operation shows the higher released/removal rate compared to 500 kHz operation; it will be interesting to check the results of synergic effects of combination of two frequencies. Results show that synergic combination did not show additive effect on the rate of removal of pollutants. Study on degradation of pentachlorophenol reported that pseudo-first-order kinetic rate constants were 1.74, 12.69, and 32.93 min^{-1} for frequency of 20, 500 kHz, and combination of 20 and 500 kHz frequencies, respectively and synergistic coefficient was 2.282 (Wang et al. 2006). A study of combined effects multiple frequencies (20, 40 and 70 kHz) arranged in hexagonal geometeries observed that uniform temperature distribution/increment for single, and multiple frequencies. Another study on degradation of rhodamine B observed that sonochemiluminescence effects was not uniform in reactor, certain area in reactor has hot spot. Combination of dual and triple frequencies reduced the production of

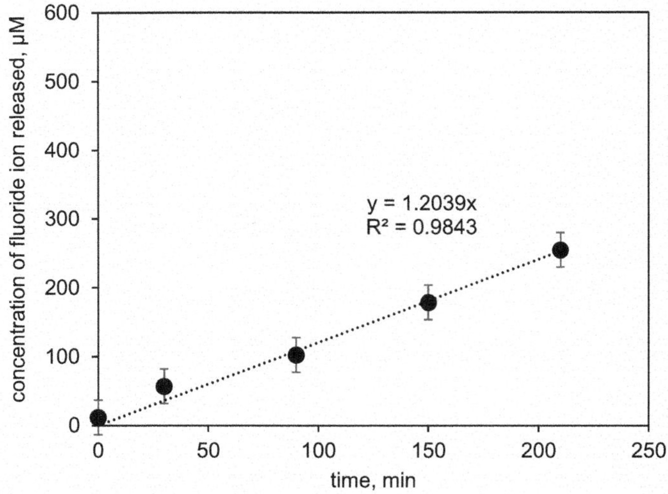

Fig. 3 Effect of dual frequency of 1 MHz and 500 kHz frequency on concentration of fluoride ion released from degradation of PFOS at a distance 1:1.5 cm away from transducer surface; pH of solution, 4; initial concentration of PFOS, 2.6 mM

Table 2 Summary of results

	Frequency of operation		
Parameter	500 kHz	1 MHz	Dual frequency
Fluoride, μM	254.2 ± 3.6	506.3 ± 21.1	254.7 ± 18.2
Sulfate, μM	7.6 ± 0.5	8.1 ± 0.9	4.3 ± 0.02
% TOC removal	19.5 ± 0.2	24.1 ± 0.9	18.6 ± 0.4
TOC removal, μM	38.6 ± 0.9	48.0 ± 3.4	36.8 ± 2.3

hydroxyl radicals due to the wave-cancelling effect (Tiong et al. 2017). Study on degradation of perfluorooctane sulfonic acid (PFOS) and perfluorooctanoic acid (PFOA) reported that degradation kinetic increased with power density and combination of dual frequencies of 20 and 202 kHz (power density 250 W L^{-1}); 12% and 23% enhancement in degradation rate was observed for PFOS and PFOA, respectively (Rahimi et al. 2014). Removal efficiency of ammonia from aqueous solution reduced by 10% for dual frequency of 1.7 MHz and 24 kHz compared to individual effects of frequency (Rahimi et al. 2014).

Synergic combination of 1 MHz and 500 kHz shows the lower efficiency for removal of pollutants compared to individual effect of frequency 1 MHz frequency and shows the better efficiency of removal of pollutant compared to 500 kHz (Table 2). Dual-frequency, 1 MHz and 500 kHz, operation operates on different power inputs, so results should be analyzed based on cavitational yield (moles of pollutant removed/amount of energy utilized). Results of cavitational yield (sum of total moles of fluoride ion released, corrected sulfate ion release, and moles of TOC removed) observed that cavitational yield for 500 kHz was maximum, least for dual-frequency operation and 1 MHz frequency of operation. Cavitational yield was $30.8 \pm 0.9 \times 10^{-3}$, $17.3 \pm 1.6 \times 10^{-3}$, $11.2 \pm 0.08 \times 10^{-3}$ μM/kJ for 500 kHz, 1 MHz, and dual frequencies of 500 kHz and 1 MHz frequencies. Higher cavitational yield confirms that 500 kHz frequency operation may be useful for treatment of PFOS. Dual frequency of operation may not be found suitable for treatment of PFOS in terms of liberation of fluoride ions and cavitational yield (Gogate et al. 2002; Kobayashi et al. 2011). This may be due to the wave cancellation effects of dual-frequency operation (Asakura et al. 2008; Kumar et al. 2007). Combination of different frequencies results in lower cavitational activities.

3.4 Bioremediation Aspects of PFOS

Bioremediation of perfluorochemicals including perfluorinated acids such perfluorinated carboxyl acid (PFCAs) and perfluorinated sulfonate (PFASs) is of great concern, and these are highly persistent compounds (Li et al. 2019). Many global studies confirm the persistency of these compounds for several decades in earth crust which contaminated the groundwater. Understanding bioaccumulation pathways of long-chain PFCs of carboxyl or sulfonate chains is difficult due to differences between partition behaviors of PFCs (Jian et al. 2017; Kwon et al. 2014). In order to understand the partition behavior of these compounds, environmental protection agencies (EPA) and several environmental protection agencies suggested the following guidelines (Conder et al. 2008):

(a) Bioconcentration and bioremediation of PFCs depend on chain length of carbon-fluorine or fluorinated carbon chain.
(b) Carboxyl acid tale length compounds are more bioaccumulative compared to sulfonate tale length compounds.
(c) Carboxyl acid tale length compounds with fluorinated carbon chain length less than 7 can be considered as less bioaccumulative or more potential of bioremediation.

Based on regulatory criteria, it can be considered as fluorinated compounds with chain length less than 7 are bioaccumulative. We did the ultra-high liquid chromatography analysis of sample treated at 1 MHz frequency for 210 min duration (at this condition high liberation of fluoride ion was observed) and detected the concentration of 16 fluorinated compounds such as PFHxA, PFHpA, PFOA, PFNA, PFDA, PFUnDA, PFDoDA, PFBS, PFPeS, PFHxS, PFHpS, PFOS, PFNS, PFDS, 6:2 FTS, and 8:2 FTS from degradation of PFOS, and results have been shown in Table 3. Concentration of PFSA is grouped into the following five categories based on the chained length of compounds (base on regulatory guidelines discussed above):

(a) PFCA ≤ 6: fluorinated compounds with carbon-fluoride chain length less than or equal to 6 and PFHxA and PFHpA compounds with carboxylic tail were detected.

Table 3 Concentration of PFC from sonochemical degradation PFOS at MHz frequency for 210-min treatment

Group	PFC	Concentration (%)
PFCA $\leq$ 6	PFHxA and PFHpA	77.5
PFCA > 6	PFOA, PFNA, PFDA, PFUnDA, PFDoDA	6.6
PFSA $\leq$ 6	PFBS, PFPeS, PFHxS, and PFHpS	10.2
PFSA > 6	PFOS, PFNS, and PFDS	5.2
FTS	6:2FTS, and 8:2FTS1 M	0.5

(b) PFCA > 6: fluorinated compounds with carbon-fluoride chain length greater than 6 and PFOA, PFNA, PFDA, PFUnDA, and PFDoDA compounds with carboxylic tail were detected.
(c) PFSA $\leq$ 6: fluorinated compounds with carbon-fluoride chain length less than or equal to 6 and PFBS, PFPeS PFHxS, and PFHpS compounds with sulfonated tail were detected.
(d) PFSA > 6: fluorinated compounds with carbon-fluoride chain length greater than 6 and PFOS, PFNS, and PFDS compounds with sulfonated tail were detected.
(e) FTS: 6:2 marginal concentration of fluorotelomer thioether amino sulfonate and 8:2 fluorotelomer thioether amino sulfonate were detected.

The results confirm that maximum concentration of PFCA $\leq$ 6, followed by the concentration of PFSA $\leq$ 6. Initial concentration of PFCA > 6 and PFSA > 6 was converted into the smaller-chain compounds of PFCA $\leq$ 6 and PFSA $\leq$ 6. Half-life of smaller-chain compounds is mintues to few hours compare to long-chain compounds in year or centuries. Thus, it confirms that bioremediation of PFOS increased after sonochemical treatment. Temperature and pressure effects of cavitational activities are useful to break down the long-chain compound to smaller-chain compounds. Bond energy of –C–F– and C–C– are –C–H– 485, 346, and 413 kJ/mol (Rodriguez-Freire et al. 2015, 2016). More amount of energy is required for breaking –C–F bond compared to other bond. Most of –C–F compounds are hydrophobic nature and easily saturated on cavitating bubble. At collapsing stage temperature inside bubble is 5000–6000 K, whereas at bubble-liquid interface 1200–1500 K (Rahimi et al. 2014; Gogate et al. 2002; Kobayashi et al. 2011; Asakura et al. 2008; Kumar et al. 2007). Temperature at bubble is sufficient for breaking the –C–F bond and useful for converting the longer chain C–F to smaller-chain compounds.

4 Concluding Remarks

Present investigation confirms that the degradation efficiency of PFCs using sonochemical reactor is highly dependent on the frequency of operation. Removal efficiency of PFCs increased with increasing the frequency of operation. Dual frequency has shown increasing removal efficiency, but synergic effects were much lesser than the algebraic sum of effects of individual frequency. Megasonic frequencies have shown more uniformity of cavitational activity, due to uniform acoustic wave distribution at megasonic frequencies. Present work has given useful insights on degradation of PFCs at megasonic scale which will be highly useful for development of megasonic scale of operation. Work has given the useful insights on the bioremediation of PFCs; sonochemical-treated aqueous solution of PFOS was broken down to smaller-chain compound which has higher biodegradability compared to long-chain PFCs such as PFOS. Time required for bioremediation is reduced from years to hours.

Nomenclatures

6:2 FTS 6:2 Fluorotelomer thioether amino sulfonate
8:2 FTS 8:2 Fluorotelomer thioether amino sulfonate
AFFF Aqueous film forming foams
PFBS Perfluorobutanesulfonic acid
PFCA Perfluorocarboxylic acid
PFCs Perfluorochemicals
PFDA Perfluorooctanoic acid
PFDoDA Perfluorodecanoic acid
PFDS Perfluorodecane sulfonate
PFHpA Perfluoroheptanoic acid
PFHpS Perfluoroheptane sulfonate
PFHxA Perfluorohexanoic acid
PFHxS Perfluorohexane sulfonate
PFNA Perfluorononanoic acid
PFNS Perfluorononane sulfonate
PFOA Perfluorooctanoic acid
PFOS Perfluorooctane sulfonate
PFPeS Perfluoropentane sulfonate
PFSA Perfluorosulfonic acid
PFUnDA Perfluoroundecanoic acid

References

Asakura Y, Nishida T, Matsuoka T, Koda S (2008) Effects of ultrasonic frequency and liquid height on sonochemical efficiency of large-scale sonochemical reactors. Ultrason Sonochem 15:244–250

Balachandran R, Zhao M, Yam P, Zanelli C, Keswani M (2015) Characterization of stable and transient cavitation in megasonically irradiated aqueous solutions. Microelectron Eng 133:45–50

Balachandran R, Patterson Z, Deymier P, Snyder SA, Keswani M (2016) Understanding acoustic cavitation for sonolytic degradation of p-cresol as a model contaminant. Chemosphere 147:52–59

Campbell T, Hoffmann MR (2015) Sonochemical degradation of perfluorinated surfactants: power and multiple frequency effects. Sep Purif Technol 156:1019–1027

Conder JM, Hoke RA, De Wolf W, Russell MH, Buck RC (2008) Are PFCAs bioaccumulative? A critical review and comparison with regulatory criteria and persistent lipophilic compounds. Environ Sci Technol 42:995–1003

de La Rochebrochard S, Suptil J, Blais J-F, Naffrechoux E (2012) Sonochemical efficiency dependence on liquid height and frequency in an improved sonochemical reactor. Ultrason Sonochem 19:280–285

Entezari MH, Kruus P (1996) Effect of frequency on sonochemical reactions II. Temperature and intensity effects. Ultrason Sonochem 3:19–24

Fernandez NA, Rodriguez-Freire L, Keswani M, Sierra-Alvarez R (2016) Effect of chemical structure on the sonochemical degradation of perfluoroalkyl and polyfluoroalkyl substances (PFASs). Environ Sci Water Res Technol 2:975–983

Francony A, Petrier C (1996) Sonochemical degradation of carbon tetrachloride in aqueous solution at two frequencies: 20 kHz and 500 kHz. Ultrason Sonochem 3:S77–S82

Gogate PR, Tatake PA, Kanthale PM, Pandit AB (2002) Mapping of sonochemical reactors: review, analysis, and experimental verification. AIChE J 48:1542–1560

Gole VL, Fishgold A, Sierra R, Deymier P, Keswani M (2018a) Treatment of (per) fluorooctanesulfonic acid (PFOS) using a large-scale sonochemical reactor. Sep Purif Technol 194:104–110

Gole VL, Peng H, Sierra R, Giesy JP, Deymier P, Keswani M (2018b) Sonochemical treatment of (per)fluorinated compounds in aqueous film forming foams using a large-scale acoustic reactor. Ultrason Sonochem 45:213–222

Houde M, Martin JW, Letcher RJ, Solomon KR, Muir DG (2006) Biological monitoring of polyfluoroalkyl substances: a review. Environ Sci Technol 40:3463–3473

Hughes D, Clark BR, Murphy CD (2011) Biodegradation of polyfluorinated biphenyl in bacteria. Biodegradation 22:741–749

Jian J-M, Guo Y, Zeng L, Liu L-Y, Lu X, Wang F, Zeng EY (2017) Global distribution of perfluorochemicals (PFCs) in potential human exposure source—a review. Environ Int 108:51–62

Keswani M, Raghavan S, Deymier P (2013) Characterization of transient cavitation in gas sparged solutions exposed to megasonic field using cyclic voltammetry. Microelectron Eng 102:91–97

Keswani M, Raghavan S, Deymier P (2014) Electrochemical investigations of stable cavitation from bubbles generated during reduction of water. Ultrason Sonochem 21:1893–1899

Kim JH, Ok YS, Choi G-H, Park B-J (2015) Residual perfluorochemicals in the biochar from sewage sludge. Chemosphere 134:435–437

Kobayashi D, Sano K, Takeuchi Y, Terasaka K (2011) Effect of irradiation distance on degradation of phenol using indirect ultrasonic irradiation method. Ultrason Sonochem 18:1205–1210

Kumar A, Gogate PR, Pandit AB (2007) Mapping the efficacy of new designs for large scale sonochemical reactors. Ultrason Sonochem 14:538–544

Kwon BG, Lim H-J, Na S-H, Choi B-I, Shin D-S, Chung S-Y (2014) Biodegradation of perfluorooctanesulfonate (PFOS) as an emerging contaminant. Chemosphere 109:221–225

Li P, Zhi D, Zhang X, Zhu H, Li Z, Peng Y, He Y, Luo L, Rong X, Zhou Y (2019) Research progress on the removal of hazardous perfluorochemicals: a review. J Environ Manage 250:109488

Lim M, Ashokkumar M, Son Y (2014) The effects of liquid height/volume, initial concentration of reactant and acoustic power on sonochemical oxidation. Ultrason Sonochem 21:1988–1993

Lowe KC, Davey MR, Power JB (1998) Perfluorochemicals: their applications and benefits to cell culture. Trends Biotechnol 16:272–277

Petrier C, David B, Laguian S (1996) Ultrasonic degradation at 20 kHz and 500 kHz of atrazine and pentachlorophenol in aqueous solution: preliminary results. Chemosphere 32:1709–1718

Rahimi M, Safari S, Faryadi M, Moradi N (2014) Experimental investigation on proper use of dual high-low frequency ultrasound waves-advantages and disadvantages. Chem Eng Process 78:17–26

Rodriguez-Freire L, Balachandran R, Sierra-Alvarez R, Keswani M (2015) Effect of sound frequency and initial concentration on the sonochemical degradation of perfluorooctane sulfonate (PFOS). J Hazard Mater 300:662–669

Rodriguez-Freire L, Abad-Fernández N, Sierra-Alvarez R, Hoppe-Jones C, Peng H, Giesy JP, Snyder S, Keswani M (2016) Sonochemical degradation of perfluorinated chemicals in aqueous film-forming foams. J Hazard Mater 317:275–283

Sutkar VS, Gogate PR (2010) Mapping of cavitational activity in high frequency sonochemical reactor. Chem Eng J 158:296–304

Sutkar VS, Gogate PR, Csoka L (2010) Theoretical prediction of cavitational activity distribution in sonochemical reactors. Chem Eng J 158:290–295

Thangavadivel K, Megharaj M, Smart RSC, Lesniewski PJ, Naidu R (2009) Application of high frequency ultrasound in the destruction of DDT in contaminated sand and water. J Hazard Mater 168:1380–1386

Tiong TJ, Liew DKL, Gondipon RC, Wong RW, Loo YL, Lok MST, Manickam S (2017) Identification of active sonochemical zones in a triple frequency ultrasonic reactor via physical and chemical characterization techniques. Ultrason Sonochem 35:569–576

Wang S, Huang B, Wang Y, Liao L (2006) Comparison of enhancement of pentachlorophenol sonolysis at 20 kHz by dual-frequency sonication. Ultrason Sonochem 13:506–510

Zhao M, Balachandran R, Madigappu PR, Yam P, Zanelli C, Sierra R, Keswani M (2015) Characterization of cavitation in ultrasonic or megasonic irradiated gas saturated solution using hydrophone. Solid State Phenomena 219:165–169

Macrophyte Importance in Contaminant Treatment and Biomonitoring

H. R. Hadad, M. A. Maine, M. M. Mufarrege, G. A. Di Luca,
G. C. Sanchez, and E. Nocetti

Abstract

Constructed wetlands are systems based on nature, in which macrophytes are the main biological community. According to the diversity of wastewaters that these systems can treat, its use has expanded all over the world. Treatment wetlands are successfully used in the removal of emerging contaminants. However, the knowledge of the removal mechanisms is still scarce. The use of macrophytes as biomonitors should be considered as a useful tool for the management of aquatic systems contaminated by different sources. In macrophyte ecotoxicological studies, not only the measurement of antioxidant enzymes but also other basic biological parameters, such as changes in different biomass compartments and the internal morphology, should be considered. This chapter discusses the importance of macrophytes in wetlands constructed for the treatment of different effluents, their use as biomonitors in natural wetlands that receive urban pollutants, and some ecotoxicological aspects applied to phytoremediation.

Keywords

Aquatic plants · Wetland systems · Industrial wastewaters · Emerging contaminants · Toxic effects

H. R. Hadad (✉)
Química Analítica Ambiental, Instituto de Química Aplicada del Litoral (IQAL, CONICET-UNL), Facultad de Ingeniería Química, Universidad Nacional del Litoral (UNL)-Consejo Nacional de Investigaciones Científicas y Técnicas (CONICET), Santa Fe, Argentina

Departamento de Ciencias Naturales, Facultad de Humanidades y Ciencias, UNL, Santa Fe, Argentina

M. A. Maine · M. M. Mufarrege · G. A. Di Luca · G. C. Sanchez · E. Nocetti
Química Analítica Ambiental, Instituto de Química Aplicada del Litoral (IQAL, CONICET-UNL), Facultad de Ingeniería Química, Universidad Nacional del Litoral (UNL)-Consejo Nacional de Investigaciones Científicas y Técnicas (CONICET), Santa Fe, Argentina

M. P. Shah (ed.), *Removal of Emerging Contaminants Through Microbial Processes*,
https://doi.org/10.1007/978-981-15-5901-3_21

1 Introduction

Macrophytes present a high diversity explained by different life forms and morphological and physiological characteristics. They have evolutionary adaptations that allow them to grow in strictly aquatic environments, in aquatic/terrestrial transition environments, and even in terrestrial environments in periods of low waters (Hadad and Mufarrege 2017).

As a key biological community in natural wetlands, macrophytes produce the highest biomass influencing the system dynamics (Esteves 1988). Similarly, in wetlands constructed for wastewater treatment, the macrophytes are essential components because they accumulate contaminants in their tissues, influence the sediment biogeochemistry, and enhance chemical and biochemical reactions in the root zone improving the contaminant removal (Jenssen et al. 1993). In general, they show fast growth and high tolerance when exposed to different pollutants. The treatment of sewage, agricultural and urban runoff wastewaters using constructed wetlands, has the advantage that nutrients are the main contaminants being assimilated by plants in high proportions. In the case of the treatment of metallurgical effluents, metals affect the biological plant responses to the environmental conditions of the system (Hadad et al. 2010; Maine et al. 2013).

Having into account the diversity of wastewaters that constructed wetlands can treat, its use has expanded all over the world (Avila et al. 2017; Kadlec and Wallace 2009; Maine et al. 2017, 2019; Nivala et al. 2019; Zhang et al. 2020). Because treatment wetlands are subjected to plant biological cycles and environmental changes, they are dynamic systems that should be monitored for a long time during the treatment. Besides, plant role should be studied deeper. More studies focused on the plant tolerance to different contaminants and new emerging contaminants and the search for species that are not commonly used in treatment wetlands should be carried out.

The analysis of the plant functions and their responses in wetland systems are necessary to improve phytoremediation techniques (Fig. 1). This chapter discusses the importance of macrophytes in wetlands constructed for the treatment of different effluents, their use as biomonitors in natural wetlands that receive pollutants, and some ecotoxicological aspects applied to phytoremediation.

2 Constructed Wetlands for Effluent Treatment

Constructed wetlands are wastewater treatment systems based on nature, in which macrophytes are the main biological community. These systems try to reproduce what happens in natural ecosystems, but under controlled conditions to optimize contaminant removal processes. Microorganisms and substrate are involved, in addition to the macrophytes. These interactions are key mechanisms to understand wetland functioning (Ahmed et al. 2017). The selection of the wetland type, the macrophytes, and substrates to be used for the treatment of a specific effluent should be the result of previous studies conducted carefully.

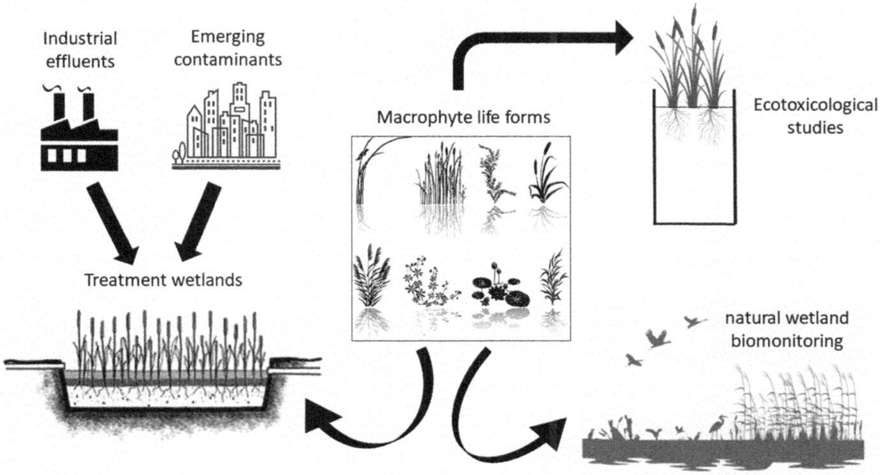

Fig. 1 Macrophyte importance in phytoremediation techniques

The different life forms of aquatic plants acquire great importance with respect to their functions in treatment wetlands. In a natural wetland, a sequence of life forms can be found along a gradient that extends from the coast to the deepest zones. In treatment wetlands, this gradient is attempted to be represented by planting the different types of macrophytes in the areas corresponding to their form of growth. For example, at the edges of a free-surface flow wetland, it would be convenient to plant emergent species, while in the deeper central areas, free-floating species would develop. In order to achieve a higher cover of emergent plants, a strategy to be applied could be the decrease of water depth by the construction of mounds of land (Maine et al. 2009).

On the other hand, since in subsurface flow wetlands emergent species are always used, plants should be capable to grow in the substrate used. Examples of emergent macrophytes commonly used in treatment wetlands are *Typha* spp., *Arundo donax*, and *Phragmites australis* (Kadlec and Wallace 2009). The different macrophytes that inhabit a region can be used in constructed wetlands since they have the advantage of being adapted to the climate and the prevailing water and edaphic conditions. However, not all species have the same efficiency in the removal of contaminants. The plant species used in the treatment will imply differences in removal efficiency (Brisson 2013; Gersberg et al. 1986). It is necessary to find local plant species that have the ability to survive the potentially toxic effects of the effluent to be treated and its variability. During a sanitary effluent treatment where the contaminants to be treated are nutrients, plant growth is assured. However, during the treatment of industrial effluents, the plants not only should achieve a good efficiency in the removal of pollutants but also survive the physical and chemical matrix of the effluent (Hadad et al. 2006; Maine et al. 2009, 2013).

Besides, the basis of wastewater treatment using constructed wetlands is the cooperative growth of plants and microorganisms. It has been thought that most of the responsibility for biodegradation is carried out by microorganisms that live on the plant roots. Once these microorganisms are established in the roots, they generate a symbiotic relationship with the macrophytes. This relationship has a synergistic effect that results in an increase in the rate of degradation and removal of pollutants in the root zone.

For the treatment of effluents containing metals, it is important to highlight that constructed wetlands are final treatments (secondary or tertiary treatments). Therefore, the metal concentrations in the effluents are relatively low. Commonly, free-water surface wetlands are used to treat effluents containing metals. In Argentina, our research group has gained experience during the last decades studying free-water surface wetlands for the treatment of effluents from different metallurgical industries (Hadad et al. 2006, 2018; Maine et al. 2006, 2007, 2009, 2013, 2017, 2019). All the studied parameters in the effluent decreased their variabilities and mean values after the treatment. Metal removal efficiencies were high, and therefore, metal water concentrations were found under the Argentinian law limits for discharge. Due to the chemical composition of the effluent (high pH, alkalinity, Fe, Ca, and ionic concentrations), metals were mainly accumulated in the sediment of the inlet zone and in the roots of the macrophytes (Hadad et al. 2006; Maine et al. 2009, 2017). Di Luca et al. (2011) proposed that if the effluent chemical composition and the environmental conditions of the system are maintained, metals will remain bound to sediment and will not release into the water column. During the long-term study in one of the treatment wetlands, changes in vegetation dominance were observed (Maine et al. 2009). Several macrophytes were transplanted to this wetland initially. Brisson (2013) proposed that a high diversity of plants has advantages, such as a higher efficiency in contaminant removal, disturbance resilience, better habitat, etc. *Eicchornia crassipes* was the dominant species for 2 years. But then, *Typha domingensis* became the dominant species for the last 15 years (Maine et al. 2017). High pH and salinity but not metal concentrations were the cause of disappearance of *E. crassipes* (Hadad et al. 2006). We could see that plant diversity is not easy to maintain for a long time due to the fact that one plant species becomes dominant. pH and salinity are key factors for plant tolerance.

The changes in the dominant plant species determined different contaminant retention mechanisms. Table 1 shows a mass balance of the studied contaminants. During the dominance of *E. crassipes*, metals and P were mainly accumulated in the

Table 1 Contaminant accumulation (%) in plants and sediment during three different stages of macrophyte dominance in a constructed wetland for the treatment of a metallurgical effluent (extracted from Maine et al. 2017)

Dominant species	Cr		Ni		Zn		TP	
	Sediment	Tissues	Sediment	Tissues	Sediment	Tissues	Sediment	Tissues
E. crassipes	11	89	7	93	5	95	2	98
T. domingensis	73	27	87	13	71	29	62	38

plant tissues. Therefore, the plant harvest could remove the metals and P from the system. Contrarily, during the dominance period of the emergent *T. domingensis*, metals were accumulated and phytostabilized in the sediment within the wetland (Hadad et al. 2018).

During the long-term studies, a remarkable fact was that the free-water flow wetlands suffered accidental metal dumps or animal depredation. Since the macrophyte *T. domingensis* showed to be resilient after these events, the systems were capable of recovering their performance, demonstrating its robustness (Maine et al. 2013, 2017).

Among recent works, Xu et al. (2019) who studied a free-water surface wetland which is constructed to remove Cu and Zn from storm runoff water in the Savannah River Site (USA) can be mentioned. Metal removals were 64 and 71% for Cu and Zn, respectively, being their concentrations under the legal limits. Zhang et al. (2020) compared the efficiency of three different substrates in hybrid treatment wetlands planted with *Canna indica*. These wetlands were constructed for the treatment of mixed domestic-industrial effluents discharged into peri-urban water bodies. The wetland systems were satisfactorily efficient in the removal of Pb, Cd, Cu, and Zn, reaching values higher than 85% during the first 24 h of the treatment. The most efficient substrates were gravel and lava rock.

Constructed wetlands can be used for the treatment of different wastewaters, such as municipal, sewage, and domestic effluents. In these cases, macrophyte growth is favored because the wastewaters to be treated contain nutrients in large proportions. García-Ávila et al. (2019) compared the efficiency of two vertical flow wetlands planted with different macrophytes. The wetlands receive municipal wastewater with primary treatment and were planted with *Cyperus papyrus* and *Phragmites australis*. The wetlands planted with the first one showed significantly higher removals of biochemical oxygen demand (81%), chemical oxygen demand (70%), ammoniacal nitrogen (70%), total phosphorus (50%), total coliforms (98%), and fecal coliforms (96%). Contrarily, the wetland planted with *P. australis* showed significantly higher retention of solids. Nivala et al. (2019) compared the removal efficiency of 15 subsurface flow pilot-scale wetlands with different designs. The wetlands were used to treat domestic wastewaters and were planted with *P. australis*. The macrophyte importance on the removal efficiencies was also assessed by using wetlands without plants. There were observed differences among the different wetland designs for the removal efficiencies of the parameter measured. Therefore, it was concluded that design complexity enhanced treatment efficiency. The presence of macrophytes enhanced the contaminant removals in the wetlands.

Landfill leachate treatment is one of the most difficult environmental problems due to its great variability in its composition and in the expected volume that depends on the age of the landfill, the climatic conditions, and the degradation rate of solid wastes. High concentrations of ammonia, recalcitrant COD, and the presence of metals in low concentrations are the main characteristics of leachate (Wojciechowska et al. 2016). Ammonium concentration found in landfill leachate is 0.01–1900 mg L^{-1} NH_4^+-N (Kadlec and Wallace 2009). Clarke and Baldwin (2002) proposed that the toxic ammonium concentration for plants is above

200 mg L^{-1} N. Natural wetlands were often the landfill leachate fate because they were located on land adjacent to them. It was observed that the chemical quality of the leachate improved after crossing them (Higgins and Lugowski 1996; La Forge 1997). Based on these results, the use of constructed wetlands for the treatment of this type of wastewater began to develop (Kadlec and Wallace 2009). Vertical flow wetlands have proven effective and are the most commonly used for the treatment of leachate (Kadlec and Zmarthie 2010; Stefanakis et al. 2014). Yalcuk and Ugurlu (2009) compared vertical flow and horizontal flow wetlands for the treatment of landfill leachate and obtained better ammonium removals in the first one. A et al. (2017) studied at laboratory scale a vertical flow wetland for the treatment of synthetic leachate and found ammonium removals of 44–73% in systems planted with *Juncus effusus* and 46–76% in systems planted with *P. australis*. Lavrova (2016) studied the treatment efficiency of leachate through two laboratory-scale vertical flow wetlands with and without additional carbon source. The significant removal efficiency of chemical and biochemical oxygen demand was achieved: 95% and 96%, respectively. Camaño Silvestrini et al. (2019) compared the system performance of wetlands constructed for the treatment of landfill leachate at mesocosms scale using the species *T. domingensis* and *C. indica*. There were no observed differences for COD, ammonia, nitrate, and TN removals between the systems planted with the two macrophytes. Although both species were tolerant to the wastewater studied, *T. domingensis* presented a significantly higher growth rate in comparison with *C. indica*.

3 Constructed Wetlands for the Treatment of Emerging Contaminants

Emerging contaminants are a current concern related to the quality of water bodies. Among others, these contaminants can be organic compounds such as agrochemicals, medicines, antibiotics, hormones, drugs, and personal care products (Galindo-Miranda et al. 2019). Emerging contaminants are commonly found at trace concentrations in waters, ranging in a scale magnitude from ng L^{-1} to μg L^{-1} (Luo et al. 2014). From an ecological point of view, wildlife and flora are in potential risk due to emerging contaminants that could negatively affect their metabolism and behavior. Treatment wetlands have demonstrated high efficiency in reducing not only common contaminants, such as nitrogen, phosphorus, and metals, but also emerging pollutants (Avila et al. 2017; Carvalho et al. 2014; Chaves-Barquero et al. 2018; Chen et al. 2016, 2019; Pei et al. 2019; Vymazal 2005; Yi et al. 2017). However, the knowledge of the removal mechanisms of emerging contaminants in treatment wetlands is still scarce. As it occurs in different types of treatment wetlands during the treatment of different wastewaters, operation and design factors, such as the types of wetlands, substrates, plant species, hydraulic retention time, hydraulic loading, solar irradiation, and temperature, could affect the performance of the treatment. Further studies should be carried out in order to gain knowledge to optimize the efficiency and operation of the treatment wetlands.

During the last years, the study of the treatment of emerging contaminants has increased significantly. Avila et al. (2015) studied a hybrid wetland system constructed for the treatment of wastewater and reuse in small communities. The hybrid system was composed of a vertical flow, a horizontal flow, and a free-water surface flow wetland. These authors reported a removal of above 80% of endocrine-disrupting compounds and pharmaceuticals and personal care products, which was explained by photodegradation, biodegradation, sorption, volatilization, and hydrolysis. The treatment wetland operation should be taken into account to obtain high removal efficiencies. Regarding hydraulic and carbon loading rates, Sharif et al. (2014) concluded that to obtain a high contaminant removal, the carbon loading rates should be changed without increasing the cumulative hydraulic loading rates. Matamoros et al. (2008) studied a free-water flow wetland for the removal of pharmaceuticals and personal care products, obtaining a removal of 90% when increasing the hydraulic retention time. A hybrid constructed wetland system was studied by Avila et al. (2014) to compare the removals of emerging contaminants using different hydraulic loading rates. These authors concluded that the full-scale system was adequately efficient for the removal of the emerging contaminants from the wastewater at high hydraulic loading rates.

Arginine is widely used as a dietary supplement. Its removal in treatment wetlands is affected by retention time and hydraulic load (Miller et al. 2015). Besides, Decamp and Warren (2000) reported that, in comparison with free-water surface wetlands, subsurface flow wetlands are more efficient in the inactivation and elimination of arginines. Chen et al. (2016) compared six different flow wetlands and obtained a significantly higher arginine removal (64–84%) in wetlands with subsurface flow in comparison with free-water flow wetlands. Previous works reported that zeolite is an efficient substrate to be used in vertical flow wetlands for the treatment of wastewaters containing arginines (Liu et al. 2013). Treatment wetlands have shown removal efficiencies near to 100% of the analgesics ibuprofen, diclofenac, and naproxen. For these compounds, free-water surface wetlands have demonstrated to be the most effective (Matamoros et al. 2008). Li et al. (2014) reported that studies focused on the use of constructed wetlands for the treatment of emerging contaminants are carried out only at microcosms or mesocosms-scales. Also, these authors suggested that different types of treatments, such as tidal flow and dewatered alum-sludge-based wetlands, could be suitable for the treatment of pharmaceuticals. Further studies are necessary because new emerging pollutants are being determined in different wastewaters.

Plants used in constructed wetlands for the treatment of wastewaters containing emerging contaminants affect different environmental variables, favoring indirectly the contaminant removal. For example, macrophytes provide substrate areas for the development of microorganisms (Fang et al. 2017) and oxygenate the microorganisms that are present in the system, which will carry out the arginine removal (Huang et al. 2015). In comparison with other macrophytes, *Phragmites* spp. are widely used in treatment wetlands due to enhanced reduction of arginines and 16sRNA (Yi et al. 2017). Chen et al. (2016) proposed that macrophytes such as *Pontederia cordata*, *Myriophyllum verticillatum*, and *Cyperus alternifolius* are capable to enhance the arginine removal from wastewaters. Rühmland et al. (2015) studied a subsurface flow wetland and a pond with floating macrophytes for the treatment of 18 different pharmaceuticals and 11 different human metabolites. Despite the diversity of the treated compounds, these authors reported good performance of both studied systems. Christofilopoulos et al. (2019) studied the removal of bisphenol A and the antibiotics ciprofloxacin and sulfamethoxazole in a treatment wetland planted with the macrophyte *Juncus acutus*. At the end of the study, the system showed a removal efficiency of 76 and 94% for bisphenol A and ciprofloxacin, respectively, while sulfamethoxazole was not efficiently removed from the wastewater. De la Paz et al. (2019) compared systems planted with *P. australis* with systems without plants constructed for the removal of benzotriazole, sulfamethoxazole, carbamazepine, bisphenol A, and diclofenac. In comparison with the unvegetated systems, the vegetated wetlands showed removals 200% higher. These authors concluded that the higher efficiency of systems with plants was due to root exudates which are released that enhances the metabolism of microorganisms.

4 Contaminant Biomonitoring Using Macrophytes in Natural Wetlands

Natural wetlands receive different sources of pollution, such as industrial effluents, sewage, urban, and road runoff waters (Rodak et al. 2019), among others. The wastewaters that reach the wetlands may contain contaminants of different types, such as metals, organic matter, nutrients, and emerging contaminants. Wetland biomonitoring is an effective tool for contamination assessment (Zhou et al. 2008). Macrophytes are efficient contaminant accumulators being tolerant to the different contaminants; whereby, they are used as efficient contaminant biomonitors (Bonanno et al. 2017; Cardwell et al. 2002). The contaminant accumulation in plant tissues of different macrophytes was studied in wetlands of different parts of the world (Alonso et al. 2018; Batzias and Siontorou 2008; Costa et al. 2018; Bertrand et al. 2019; Farias et al. 2018; Mechora et al. 2014; Wach et al. 2019; Wang et al. 2014; Zhou et al. 2008). The knowledge acquired is key to be applied in the management of contaminated wetlands and in phytoremediation techniques (Bonanno et al. 2017; Hadad et al. 2018).

Alonso et al. (2018) studied peri-urban wetlands located in the Middle Paraná River floodplain (Argentina). In that work, nutrients and metals in tissues of *T. domingensis*, *E. crassipes*, *Alternanthera philoxeroides*, and *Pistia stratiotes* were measured. The roots of *A. philoxeroides* showed the highest phosphorus concentrations in the tissues (Table 2). Also, the tissues of this macrophyte showed higher metal concentrations than that of measured in sediment. Therefore, these authors proposed this scarcely studied macrophyte as an efficient metal hyperaccumulator. These authors also studied root anatomical parameters, concluding that of *T. domingensis* and *E. crassipes* roots which were sensitive to nitrate and ammonium concentrations in water. Having into account the contaminant plant

Table 2 TP and metal concentrations (mg g^{-1} d.w.) in tissues of plants monitored at peri-urban wetlands and a control wetland

	Leaves	Roots	Leaves	Roots	Leaves	Roots
Plant species	TP		Cr		Cu	
T. domingensis (Rincón)	2.05–2.78	1.15–4.36	ND (0.002)	ND (0.002)–0.008	ND (0.002)	0.036–0.072
E. crassipes (Arroyo Leyes)	1.14–1.79	3.15–3.93	ND (0.002)	ND (0.002)–0.003	ND (0.002)	ND (0.002)–0.046
A. philoxeroides (Cayastá)	4.41–5.45	3.09–3.94	ND (0.002)	ND (0.002)	ND (0.002)	0.05
P. stratiotes (RECU)	1.01–1.31	0.923–2.27	ND (0.002)	ND (0.002)–0.005	ND (0.002)	0.028–0.031
	Ni		Pb		Zn	
T. domingensis (Rincón)	ND (0.002)	ND (0.002)–0.008	ND (0.002)	0.029–0.089	ND (0.003)	0.056–0.209
E. crassipes (Arroyo Leyes)	ND (0.002)	ND (0.002)–0.004	ND (0.002)	0.029–0.107	ND (0.003)	0.029–0.059
A. philoxeroides (Cayastá)	ND (0.002)	ND (0.002)	ND (0.002)	0.05	ND (0.003)	0.133
P. stratiotes (RECU)	ND (0.002)	ND (0.002)–0.001	ND (0.002)	0.031–0.032	ND (0.003)	0.043–0.094

ND = not detected, values in parentheses are the detection limits of the method (extracted from Alonso et al. 2018)

tolerance, Alonso et al. (2018) proposed that macrophytes from the Middle Paraná River floodplain are efficient biomonitors.

Baldantoni et al. (2018) carried out the biomonitoring of the Sarno River (Italy) using the macrophytes *Apium nodiflorum* and *Potamogeton pectinatus*. As a result of the measured contaminants in tissues, these authors concluded that this river has a severe pollution degree caused by Cd, Cr, Pb, and V, followed by Cu, Fe, Mn, Na, Ni, and Zn originated from agricultural and urban activities. Some plant taxonomical groups are scarcely studied. This is the case of aquatic bryophytes. Favas et al. (2018) studied the accumulation of 46 elements in tissues of *Fontinalis squamosa*, *Brachythecium rivulare*, *Platyhypnidium riparioides*, and *Thamnobryum alopecurum* in Góis mine region (Central Portugal). These mosses were efficient biomonitors due to high contaminant concentrations in their tissues. However, due to their low biomass, they are not efficient bioaccumulators to be considered in phytoremediation techniques. Farias et al. (2018) measured the concentrations of As, Cd, Cu, Pb, Se, and Zn in 12 macrophyte species from the Derwent estuary (Australia). Due to the different species that showed different metal concentrations in tissues, these authors concluded that the biomonitoring of multiple macrophyte species is needed to understand the metal pollution of the wide estuary. Similar work was carried out by Bonanno et al. (2017) in different wetlands from Sicily (Italy). These authors measured the concentrations of As, Cd, Cr, Cu, Hg, Mn, Ni, Pb, and Zn in the seagrasses *Posidonia oceanica* and *Cymodocea nodosa* and in the wetland plants *P. australis*, *Arundo donax*, *T. domingensis*, *A. nodiflorum*, and

Nasturtium officinale. Their results demonstrated that metal biomonitoring was species-specific, and it was not dependent on the type of environment and the plant life forms. Therefore, each plant species responses are different faced with metal exposure in the different water bodies.

Based on the above, the use of macrophytes as biomonitors should be considered as a useful tool for the management of aquatic systems contaminated by different sources and in the design and operation of treatment wetlands.

5 Ecotoxicology Applied to Phytoremediation

Commonly, the changes in different antioxidant enzymes involved in the protection against metal stress are evaluated to assess the macrophyte tolerance to metal exposure. Among the works focused on free-floating macrophytes, it can be mentioned the study of the enzymatic detoxification strategies of *E. crassipes* when it was exposed to different water concentrations of Ni (González et al. 2015) and Zn (González et al. 2018). During the first hours of the Ni exposure, a significant increase of the photosynthetic pigment concentrations was observed indicating the continuity of the metabolic functions. Malondialdehyde (MDA) concentration in leaves increased significantly with a simultaneous increase of the antioxidant enzyme activities demonstrating a compensatory protective mechanism. In the Zn exposure, Chl-b and carotenoid concentrations increase at the end of the study. MDA in plant tissues did not change during the study; however, a significant increase in the enzymatic activity of the antioxidant system was observed. The results reported by Gonzalez et al. (2015, 2018) indicate that *E. crassipes* is tolerant to Ni and Zn exposure increasing the activity of its antioxidant system against metal stress. The physiological effects of Cu on *P. pectinatus* were studied by Brandão Costa et al. (2018). After an exposure of 96 h, chlorophyll and carotenoid concentrations decreased, affecting the photosystem II and causing photosynthesis inhibition. According to the sensitivity to Cu, these authors propose that this macrophyte can be a suitable species for monitoring water bodies contaminated with this metal. Cosio and Renault (2020) compared the physiological effects of methyl-Hg, inorganic Hg, and Cd on the submerged species *Elodea nuttallii*. The exposure to these contaminants did not affect photosynthetic mechanisms nor antioxidant enzymes. These authors reported that the exposure to methyl-Hg affects aminoacyl-tRNA biosynthesis, glycine, serine and threonine metabolism, nitrogen metabolism, arginine and proline metabolism, and cyanoamino acid metabolism. The Cd exposure produced changes in aminoacyl-tRNA biosynthesis and branched-chain amino acid pathways. Contrarily, inorganic Hg did not show metabolic effects. Cosio and Renault (2020) stated that their work was the first one that applied metabolomics in *E. nuttallii* concluding that these data complement the knowledge gained by transcriptomics and proteomics.

Vidal Ribeiro et al. (2019) studied the toxicity of hexazinone, an herbicide commonly applied in sugarcane production, on the free-floating macrophytes *P. stratiotes* and *E. crassipes*. These authors concluded that concentrations of

111 and 333 μg L^{-1} were toxic to the studied species, having into account the changes in biomass and foliar morphological parameters (adaxial and abaxial epidermis, palisade parenchyma, aerenchyma, and leaf blade). The toxic effects of toluene, ethylbenzene, and xylene on the submerged macrophyte *Hydrilla verticillata* were studied by Yan and Zhou (2011). After the exposure at different concentrations of these organic compounds, chlorophyll concentration, lipid peroxidation, and antioxidant enzymes showed changes. Using linear regression analyses, the changes observed indicated that the contaminant concentrations expected to protect the studied macrophyte were 7.30, 1.15, and 2.36 mg L^{-1}, for toluene, ethylbenzene, and xylene, respectively.

Parameters such as foliar injury, chlorophyll concentrations, and biomass changes can indicate the toxic effects of metal exposure on macrophytes. Besides, root morphological plasticity is an important mechanism to regulate the contaminant uptake and to enhance the plant tolerance (Hadad et al. 2007; Kapitonova 2002; Kolotov et al. 2003; Mufarrege et al. 2018; Zhou et al. 2008). The Cr, Ni, and Zn toxic effects on the productivity and root morphology of the free-floating macrophyte *E. crassipes* were studied by Hadad et al. (2011). This species was exposed to 1 mg L^{-1} of each contaminant. Biomass did not decrease in any treatment; however, the three studied metals produced a decrease in the leaf chlorophyll concentrations. Figure 2 shows that the cross-sectional area of roots (CSA) and the number of metaxylematic vessels increased, while the root length decreased during the Zn exposure.

Without a doubt, *Typha* spp. are one of the most studied wetland plants, being also one of the most used macrophytes in treatment wetlands all over the world. Hadad et al. (2010) studied the tolerance and the root morphology of *T. domingensis* in a constructed wetland for the treatment of a metallurgical effluent. In the inlet zone of the constructed wetland, plant height, dry biomass, the cross-sectional area of the roots and stele, and the number of metaxylematic vessels were significantly higher than the measured in the outlet zone and in a natural control wetland. However, the chlorophyll concentration measured in the plants from the inlet zone was significantly lower than the obtained in the plants from the outlet and control wetland (Fig. 3). The morphological root modifications observed in the plants sampled from the inlet zone enhanced the contaminant tissue accumulation demonstrating the adaptability of the studied species to the environmental wetland conditions.

In wetlands studied at mesocosms scale, Mufarrege et al. (2014) exposed *T. domingensis* plants to a metal-combined solution constituted by 100 mg Cr L^{-1} + 100 mg Ni L^{-1} + 100 mg Zn L^{-1}. The studied metals caused a decrease in chlorophyll concentration and growth inhibition. Regarding the root anatomical parameters, they were significantly lower in comparison with the obtained in the control wetlands. These authors concluded that *T. domingensis* could accumulate metals efficiently in its tissues and tolerate the conditions imposed by the presence of contaminants in treatment wetlands, despite the sublethal effects observed. The toxic effects of the chemical species Cr(VI) on *T. domingensis* in treatments with the addition of organic matter were studied by Mufarrege et al. (2018). Plants were exposed to Cr(VI) solutions of 15, 30, and 100 mg L^{-1}. Despite that the organic

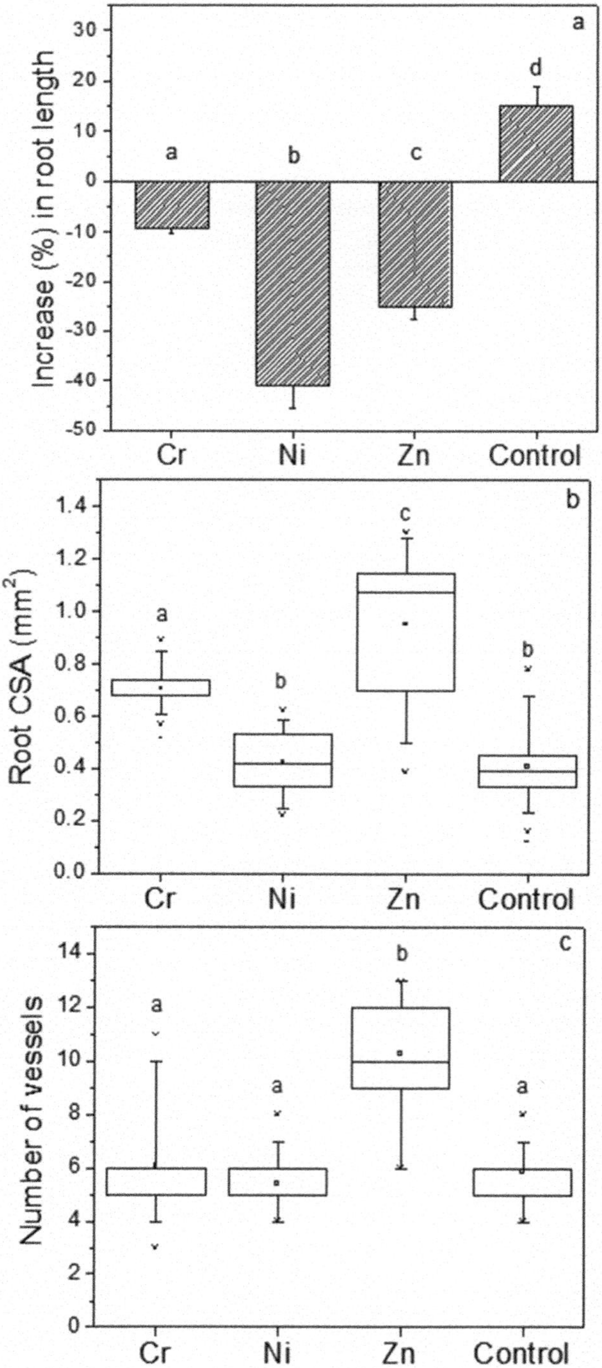

Fig. 2 Percent increase of the root length (a), box and whisker plots of root cross-sectional areas, and number of vessels in roots (c) of *E. crassipes* exposed to Cr, Ni, and Zn. Different letters represent statistically significant differences among treatment (extracted from Hadad et al. 2011)

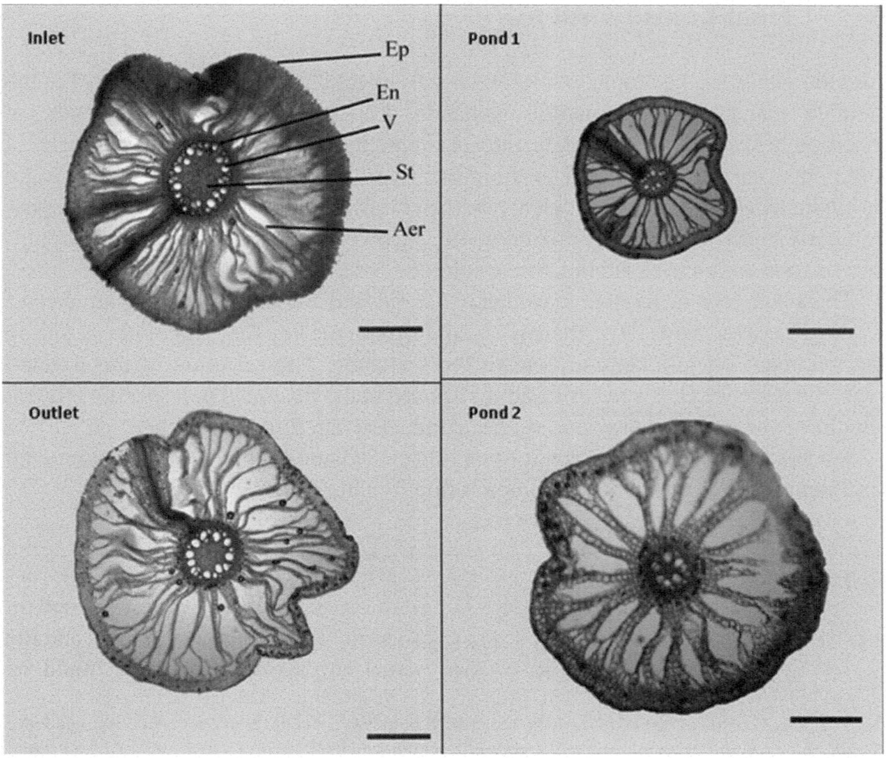

Fig. 3 Optical microscopy image of cross-sectional *T. domingensis* roots from the inlet and outlet areas of a treatment wetland constructed for the treatment of a metallurgical effluent and from natural ponds. *Ep* epidermis, *En* endodermis, *V* metaxylem vessels, *St* stele, *Aer* aerenchyma. Bar = 650 μm (extracted from Hadad et al. 2010)

matter addition increased the plant biomass and the chlorophyll concentrations, the exposure to Cr(VI) affected the root morphology. The ecotoxicological studies focused on the plant root morphology indicated that macrophytes have a considerable root morphological plasticity, which enhances the contaminant uptake and tolerance in wetland systems such as constructed wetlands.

To assess the effects of toxic compounds, different living organisms are used at a laboratory-scale, from bacteria and algae to organisms of higher biological complexity level (Guilizzoni 1991). Generally, studies using submerged and low biomass species are carried out, being focused on the enzymatic activity faced to the exposure to specific contaminants. However, it should be necessary to carry out studies focused on wetland plants that are commonly used in phytoremediation techniques such as constructed wetlands. Besides, these studies should consider other biological variables, such as changes in biomass or growth rate, in order to obtain basic information to be applied not only in wastewater treatments by phytoremediation but also in biomonitoring studies at field scale.

6 Final Considerations

Aquatic plants are key tools to monitor environmental pollution and to carry out the suitable management of natural wetlands. For the ecotoxicological study of macrophytes, it should be taken into account that not only the measurement of antioxidant enzymes involved in the protection against contaminant stress should be evaluated. Other basic biological parameters such as changes in different biomass compartments (roots, leaves, rhizomes, etc.), growth rates, and the internal morphology of roots and leaves should be considered.

The plant role in treatment wetlands has been discussed in a large number of works. However, nowadays, there is no doubt about the key function of plants during the treatment of pollutants in constructed wetlands. The selection of the wetland type, the macrophytes, and substrates to be used in the treatment of a specific effluent should be the result of previous studies conducted carefully.

Studies focused on the search for plant species tolerant for emerging contaminant treatment in CW and for their biomonitoring in natural wetlands are needed.

References

A D, Oka M, Fujii Y, Soda S, Ishigaki T, Machimura T, Ike M (2017) Removal of heavy metals from synthetic landfill leachate in lab-scale vertical flow constructed wetlands. Sci Total Environ 584–585:742–750

Ahmed MB, Zhou JL, Ngo HH, Guo W, Thomaidis NS, Xu J (2017) Progress in the biological and chemical treatment technologies for emerging contaminant removal from wastewater: a critical review. J Hazard Mater 323:274–298

Alonso X, Hadad HR, Córdoba C, Polla W, Reyes MS, Fernández V, Granados I, Marino L, Villalba A (2018) Macrophytes as potential biomonitors in peri-urban wetlands of the Middle Parana River (Argentina). Environ Sci Pollut Res 25(1):312–323

Avila C, Matamoros V, Reyes-Contreras C, Pina B, Casado M, Mita L, Rivetti C, Barata C, García J, Bayona JM (2014) Attenuation of emerging organic contaminants in a hybrid constructed wetland system under different hydraulic loading rates and their associated toxicological effects in wastewater. Sci Total Environ 470–471:1272–1280

Avila C, Bayona JM, Martin I, Salas JJ, Garcia J (2015) Emerging organic contaminant removal in a full-scale hybrid constructed wetland system for wastewater treatment and reuse. Ecol Eng 80:108–116

Avila C, Pelissari C, Sezerino PH, Sgroi M, Roccaro P, García J (2017) Enhancement of total nitrogen removal through effluent recirculation and fate of PPCPs in a hybrid constructed wetland system treating urban wastewater. Sci Total Environ 584–585:414–425

Baldantoni D, Bellino A, Lofrano G, Libralato G, Pucci L, Carotenuto M (2018) Biomonitoring of nutrient and toxic element concentrations in the Sarno River through aquatic plants. Ecotoxicol Environ Saf 148:520–527

Batzias AF, Siontorou CG (2008) A new scheme for biomonitoring heavy metal concentrations in semi-natural wetlands. J Hazard Mater 158(2–3):340–358

Bertrand L, Monferrán MV, Valdés ME, Amé MV (2019) Usefulness of a freshwater macrophyte (*Potamogeton pusillus*) for an environmental risk assessment in a multi-source contaminated basin. Chemosphere 222:1003–1016

Bonanno G, Borg JA, Di Martino V (2017) Levels of heavy metals in wetland and marine vascular plants and their biomonitoring potential: a comparative assessment. Sci Total Environ 576:796–806

Brandão Costa M, Valêncio Tavares F, Bueno Martinez C, Gonçalves Colares I, de Martinez Gaspar Martins C (2018) Accumulation and effects of copper on aquatic macrophytes *Potamogeton pectinatus* L.: potential application to environmental monitoring and phytoremediation. Ecotoxicol Environ Saf 155:117–124

Brisson J (2013) Ecosystem services of wetlands: does plant diversity really matter? In: Chazarenc F, Gagnon V, Méchineau M (eds) Proceedings of 5th international symposium on wetland pollutant dynamics and control, WETPOL 2013, Ecole des Mines de Nantes-GEPEA Nantes (Francia), 13-17 October, 2013, pp 10–11

Camaño Silvestrini NE, Maine MA, Hadad HR, Nocetti E, Campagnoli MA (2019) Effect of feeding strategy on the performance of a pilot-scale vertical flow wetland for the treatment of landfill leachate. Sci Total Environ 648:542–549

Cardwell AJ, Hawker DW, Greenway M (2002) Metal accumulation in aquatic macrophytes from southeast Queensland, Australia. Chemosphere 48:653–663

Carvalho PN, Basto MCP, Almeida CMR, Brix H (2014) A review of plant–pharmaceutical interactions: from uptake and effects in crop plants to phytoremediation in constructed wetlands. Environ Sci Pollut Res Int 21(20):11729–11763

Chaves-Barquero LG, Luong KH, Rudy MD, Frank RA, Hanson ML, Wong CS (2018) Attenuation of pharmaceuticals, nutrients, and toxicity in a rural sewage lagoon system integrated with a subsurface filtration technology. Chemosphere 209:767–775

Chen J, Ying GG, Wei XD, Liu YS, Liu SS, Hu LX, He LY, Chen ZF, Chen FR, Yang YQ (2016) Removal of antibiotics and antibiotic resistance genes from domestic sewage by constructed wetlands: effect of flow configuration and plant species. Sci Total Environ 571:974–982

Chen J, Liu YS, Deng WJ, Ying GG (2019) Removal of steroid hormones and biocides from rural wastewater by an integrated constructed wetland. Sci Total Environ 660:358–365

Christofilopoulos S, Kaliakatsos A, Triantafyllou K, Gounaki I, Venieri D, Kalogerakis N (2019) Evaluation of a constructed wetland for wastewater treatment: addressing emerging organic contaminants and antibiotic resistant bacteria. N Biotechnol 52:94–103

Clarke E, Baldwin AH (2002) Responses of wetland plants to ammonia and water level. Ecol Eng 18:257–264

Cosio C, Renault D (2020) Effects of cadmium, inorganic mercury, and methyl-mercury on the physiology and metabolomic profiles of shoots of the macrophyte *Elodea nuttallii*. Environ Pollut 257:113557

Costa MB, Tavares FV, Martinez CB, Colares IG, Martins CMG (2018) Accumulation and effects of copper on aquatic macrophytes *Potamogeton pectinatus* L.: potential application to environmental monitoring and phytoremediation. Ecotoxicol Environ Saf 155:117–124

De la Paz A, Salinas N, Matamoros V (2019) Unraveling the role of vegetation in the attenuation of contaminants of emerging concern from wetland systems: preliminary results from column studies. Water Res 166:115031

Decamp O, Warren A (2000) Investigation of *Escherichia coli* removal in various designs of subsurface flow wetlands used for wastewater treatment. Ecol Eng 14:293–299

Di Luca GA, Maine MA, Mufarrege MM, Hadad HR, Sánchez GC, Bonetto CA (2011) Metal retention and distribution in the sediment of a constructed wetland for industrial wastewater treatment. Ecol Eng 37(9):1267–1275

Esteves FA (1988) Fundamentos de limnología. Interciencia, FIMEP, Río de Janeiro

Fang H, Zhang Q, Nie X, Chen B, Xiao Y, Zhou Q, Liao W, Liang X (2017) Occurrence and elimination of antibiotic resistance genes in a long-term operation integrated surface flow constructed wetland. Chemosphere 173:99–106

Farias DR, Hurd CL, Eriksen RS, Macleod CK (2018) Macrophytes as bioindicators of heavy metal pollution in estuarine and coastal environments. Mar Pollut Bull 128:175–184

Favas PJC, Pratas J, Rodrigues N, D'Souza R, Varun M, Paul MS (2018) Metal(loid) accumulation in aquatic plants of a mining area: potential for water quality biomonitoring and biogeochemical prospecting. Chemosphere 194:158–170

Galindo-Miranda JM, Guízar-González C, Becerril-Bravo EJ, Moeller-Chávez G, León-Becerril E, Vallejo-Rodríguez R (2019) Occurrence of emerging contaminants in environmental surface waters and their analytical methodology—a review. Water Sci Technol Water Suppl 19 (7):1871–1884

García-Ávila F, Patiño-Chávez J, Zhinín-Chimbo F, Donoso-Moscoso S, Flores del Pino L, Avilés-Añazco A (2019) Performance of *Phragmites australis* and *Cyperus papyrus* in the treatment of municipal wastewater by vertical flow subsurface constructed wetlands. Int Soil Water Cons Res 7(3):286–296

Gersberg RM, Elkins BV, Lyon SR, Goldman CR (1986) Role of aquatic plants in wastewater treatment by artificial wetlands. Water Res 20:363–368

González CI, Maine MA, Cazenave J, Hadad HR, Benavides MP (2015) Ni accumulation and its effects on physiological and biochemical parameters of *Eichhornia crassipes*. Environ Exp Bot 117:20–27

González CI, Maine MA, Hadad HR, Sanchez GC, Benavides MP, Campagnoli MA (2018) Effects on *Eichhornia crassipes* under Zn stress. Environ Sci Pollut Res 25(27):1–8

Guilizzoni P (1991) The role of heavy metals and toxic materials in the physiological ecology of submersed macrophytes. Aquat Bot 41:87–109

Hadad HR, Mufarrege MM (2017) Response of macrophytes and their role in constructed wetlands (Ch. 417). In: Singhal JC, Sharma UC, Bhola R, Gurjar J, Govil N (eds) Environment science and engineering. Vol. 4: Water pollution & waste water treatment. Studium Press LLC, Houston, pp 381–406

Hadad HR, Maine MA, Bonetto CA (2006) Macrophyte growth in a pilot-scale constructed wetland for industrial wastewater treatment. Chemosphere 63(10):1744–1753

Hadad HR, Maine MA, Natale GS, Bonetto C (2007) The effect of nutrient addition on metal tolerance in *Salvinia herzogii*. Ecol Eng 31(2):122–131

Hadad HR, Mufarrege MM, Pinciroli M, Di Luca GA, Maine MA (2010) Morphological response of *Typha domingensis* to an industrial effluent containing heavy metals in a constructed wetland. Arch Environ Contam Toxicol 58(3):666–675

Hadad HR, Maine MA, Mufarrege MM, Del Sastre MV, Di Luca GA (2011) Bioaccumulation kinetics and toxic effects of Cr, Ni and Zn on *Eichhornia crassipes*. J Hazard Mater 190 (1–3):1016–1022

Hadad HR, Mufarrege MM, Di Luca GA, Maine MA (2018) Long-term study of Cr, Ni, Zn, and P distribution in *Typha domingensis* growing in a constructed wetland. Environ Sci Pollut Res 25 (18):1–8

Higgins J, Lugowski A (1996) The use of a natural forested wetland for landfill leachate polishing in a cold climate. Presented at the constructed wetlands in cold climates: design, operation, performance symposium; The Friends of St. George: Niagaraon-the-lake, Ontario, Canada

Huang X, Liu C, Li K, Su J, Zhu G, Liu L (2015) Performance of vertical up-flow constructed wetlands on swine wastewater containing tetra-cyclines and tet genes. Water Res 70:109–117

Jenssen P, Maehlum T, Krogstad T (1993) Potential use of constructed wetlands for wastewater treatment in northern environments. Water Sci Technol 28:149–157

Kadlec RH, Wallace SD (2009) Treatment wetlands, 2nd edn. CRC Press, Boca Raton

Kadlec RH, Zmarthie LA (2010) Wetland treatment of leachate from a closed landfill. Ecol Eng 36:946–957

Kapitonova OA (2002) Specific anatomical features of vegetative organs in some macrophyte species under conditions of industrial pollution. Russ J Ecol 33(1):59–61

Kolotov BA, Demidov VV, Volkov SN (2003) Chlorophyll content as a primary indicator of the environment degradation due to contamination with heavy metals. Dokl Biol Sci 393:550–552

La Forge MJ (1997) Attenuation of landfill leachate by a natural marshland system. In: Proceedings of a leachate wetlands conference; Romulus, Michigan

Lavrova S (2016) Treatment of landfill leachate in two stage vertical-flow wetland system with/without addition of carbon source. J Chem Technol Metall 51(2):223–228

Li Y, Zhu G, Ng WJ, Tan SK (2014) A review on removing pharmaceutical contaminants from wastewater by constructed wetlands: design, performance and mechanism. Sci Total Environ 468–469:908–932

Liu L, Liu C, Zheng J, Huang X, Wang Z, Liu Y, Zhu G (2013) Elimination of veterinary antibiotics and antibiotic resistance genes from swine wastewater in the vertical flow constructed wetlands. Chemosphere 91:1088

Luo Y, Guo W, Ngo H, Nghiem L, Hai F, Zhang J, Liang S, Wang X (2014) A review on the occurrence of micropollutants in the aquatic environment and their fate and removal during wastewater treatment. Sci Total Environ 473–474:619–641

Maine MA, Suñe N, Hadad H, Sánchez G, Bonetto C (2006) Nutrient and metal removal in a constructed wetland for wastewater treatment from a metallurgic industry. Ecol Eng 26 (4):341–347

Maine MA, Suñe N, Hadad HR, Sánchez G, Bonetto C (2007) Removal efficiency of a constructed wetland for wastewater treatment according to vegetation dominance. Chemosphere 68 (6):1105–1113

Maine MA, Suñe N, Hadad HR, Sánchez G, Bonetto C (2009) Influence of vegetation on the removal of heavy metals and nutrients in a constructed wetland. J Environ Manage 90 (1):355–363

Maine MA, Hadad HR, Sánchez GC, Mufarrege MM, Di Luca GA, Caffaratti SE, Pedro MC (2013) Sustainability of a constructed wetland faced with a depredation event. J Environ Manage 128:1–6

Maine MA, Hadad HR, Sánchez GC, Di Luca GA, Mufarrege MM, Caffaratti SE, Pedro MC (2017) Long-term performance of two free-water surface wetlands for metallurgical effluent treatment. Ecol Eng 98:372–377

Maine MA, Hadad HR, Sánchez GC, Mufarrege MM, Di Luca GA (2019) 6.11-Case study 10-Bahco treatment wetland for effluent final polishing (Argentina). In: Langergraber G, Dotro G, Nivala J, Rizo A, Stein O (eds) Wetland technology. Practical information on the design and application of treatment wetlands. Scientific and technical report no.27. IWA Publishing, London, pp 147–148

Matamoros V, Garcia J, Bayona JM (2008) Organic micropollutant removal in a full-scale surface flow constructed wetland fed with secondary effluent. Water Res 42:653–660

Mechora Š, Germ M, Stibilj V (2014) Monitoring of selenium in macrophytes—the case of Slovenia. Chemosphere 111:464–470

Miller JH, Novak JT, Knocke WR, Pruden A (2015) Elevation of antibiotic resistance genes at cold temperatures: implications for winter storage of sludge and biosolids. Lett Appl Microbiol 59:587–593

Mufarrege MM, Hadad HR, Di Luca GA, Maine MA (2014) Metal dynamics and tolerance of *Typha domingensis* exposed to high concentrations of Cr, Ni and Zn. Ecotoxicol Environ Saf 105(1):90–96

Mufarrege MM, Hadad HR, Di Luca GA (2018) Organic matter effects on the Cr (VI) removal efficiency and tolerance of *Typha domingensis*. Water Air Soil Pollut 229:1–12

Nivala J, Boog J, Headley T, Aubron T, Wallace S, Brix H, Sibylle M, van Afferden M, Müller RA (2019) Side-by-side comparison of 15 pilot-scale conventional and intensified subsurface flow wetlands for treatment of domestic wastewater. Sci Total Environ 658:1500–1513

Pei M, Zhang B, He Y, Su J, Gin K, Lev O, Shen G, Hu S (2019) State of the art of tertiary treatment technologies for controlling antibiotic resistance in wastewater treatment plants. Environ Int 131:105026

Rodak CM, Moore TL, David R, Jayakaran AD, Vogel JR (2019) Urban stormwater characterization, control, and treatment. Water Environ Res 91(10):1034–1060

Rühmland S, Wick A, Ternes TA, Barjenbruch M (2015) Fate of pharmaceuticals in a subsurface flow constructed wetland and two ponds. Ecol Eng 80:1–15

Sharif F, Westerhoff P, Herckes P (2014) Impact of hydraulic and carbon loading rates of constructed wetlands on contaminants of emerging concern (CECs) removal. Environ Pollut 185:107–115

Stefanakis A, Akratos CS, Tsihrintzis VA (2014) Vertical flow constructed wetlands. Eco-engineering systems for wastewater and sludge treatment. Elsevier, Amsterdam

Vidal Ribeiro VH, Barbalho Alencar BT, Correa dos Santos NM, Martins da Costa VA, Barbosa dos Santos J, Teodoro Francino DM, de Freitas M, de Souza Valadão Silva D (2019) Sensitivity of the macrophytes *Pistia stratiotes* and *Eichhornia crassipes* to hexazinone and dissipation of this pesticide in aquatic ecosystems. Ecotoxicol Environ Saf 168:177–183

Vymazal J (2005) Constructed wetlands for wastewater treatment: five decades of experience. CRC Crit Rev Environ Control 25:475–477

Wach M, Guéguen J, Chauvin C, Delmas F, Dagens N, Feret T, Loriot S, Tison-Rosebery J (2019) Probability of misclassifying river ecological status: a large-scale approach to assign uncertainty in macrophyte and diatom-based biomonitoring. Ecol Indic 101:285–295

Wang Z, Yao L, Liu G, Liu W (2014) Heavy metals in water, sediments and submerged macrophytes in ponds around the Dianchi Lake, China. Ecotoxicol Environ Saf 107:200–206

Wojciechowska E, Gajewska M, Ostojski A (2016) Reliability of nitrogen removal processes in multistage treatment wetlands receiving high-strength wastewater. Ecol Eng 98:365–371

Xu X, Mills G, Lindell A, Peck E, Korotasz A, Burgess E (2019) The performance of a free surface and metal-removing constructed wetland: how a young wetland becomes mature. Ecol Eng 133:32–38

Yalcuk A, Ugurlu A (2009) Comparison of horizontal and vertical constructed wetland systems for landfill leachate treatment. Bioresour Technol 100(9):2521–2526

Yan S, Zhou Q (2011) Toxic effects of *Hydrilla verticillata* exposed to toluene, ethylbenzene and xylene and safety assessment for protecting aquatic macrophytes. Chemosphere 85 (6):1088–1094

Yi X, Tran NH, Yin T, He Y, Gin KY (2017) Removal of selected PPCPs, EDCs, and antibiotic resistance genes in landfill leachate by a full-scale constructed wetlands system. Water Res 121:46

Zhang X, Wang T, Xu Z, Zhang L, Dai Y, Tang X, Tao R, Lia R, Yang Y, Tai Y (2020) Effect of heavy metals in mixed domestic-industrial wastewater on performance of recirculating standing hybrid constructed wetlands (RSHCWs) and their removal. Chem Eng J 379:122363

Zhou Q, Zhang J, Fu J, Shi J, Jiang G (2008) Biomonitoring: an appealing tool for assessment of metal pollution in the aquatic ecosystem. Anal Chim Acta 606(2):135–150

Decontamination of Aqueous Heavy Metal Ions by Valence Regulation Strategy

Dongli Guo, Fuqiang Liu, Fang Li, and Yanbiao Liu

Abstract

Recently, the widespread heavy metal ions pollution of water bodies requires urgent solutions. The properties of heavy meatal ions are strongly dependent on their physicochemical properties. For example, the toxicity of heavy metal ions may vary vastly at different valence states. Therefore, to regulate the valence state of heavy metal ions may provide a promising and feasible solution for their decontamination from water. This chapter summarizes the recent advances on the state-of-the-art detoxification strategies of heavy metal ions based on the valence regulation mechanism. Their applications in detoxification and sequestration of toxic heavy metal ions are systematically reviewed. We then discuss the existing limitations and perspectives for future research.

Keywords

Heavy metal ions · Toxicity · Valence regulation · Detoxification · Sequestration

D. Guo · F. Liu
Textile Pollution Controlling Engineering Center of Ministry of Environmental Protection, College of Environmental Science and Engineering, Donghua University, Shanghai, China

F. Li · Y. Liu (✉)
Textile Pollution Controlling Engineering Center of Ministry of Environmental Protection, College of Environmental Science and Engineering, Donghua University, Shanghai, China

Shanghai Institute of Pollution Control and Ecological Security, Shanghai, China
e-mail: yanbiaoliu@dhu.edu.cn

M. P. Shah (ed.), *Removal of Emerging Contaminants Through Microbial Processes*,
https://doi.org/10.1007/978-981-15-5901-3_22

1 Introduction

Heavy metal pollution has become a social and environmental issue that requires urgent solutions (Bolisetty et al. 2019; Rosenberg 2015; Fu and Wang 2011). Heavy metal ions in nature tend to form various complexes or minerals with different valence states, which demonstrated varying physicochemical properties as well. For example, Cr(VI) is soluble and has acute toxic and carcinogenic effects, while its Cr(III) counterpart is less toxic and less soluble. Similarly, in comparison with Sb (V), Sb(III) is ten times more toxic, and its removal efficacy by sorption is generally low due to a predominant charge-neutral state across a broad pH range (Filella et al. 2012). Therefore, to regulate the valance state of heavy metal ions provides a promising route for the decontamination of toxic heavy metal ions.

Conventional technologies for heavy metal ions removal include precipitation/coagulation, ion exchange, and physical adsorption. The precipitation/coagulation technology has been considered as a low-cost, simple, and effective process and been widely used for the removal heavy metal ions from water (Matlock et al. 2002; Ghurye et al. 2004). However, their limited selectivity, formation of toxic by-products, and additional sludge disposal are often regarded as the main drawbacks for heavy metal ions decontamination. For ion exchange technology, despite its wide applicability (De Dardel and Arden 2000), the compulsory pretreatment rendered increased operational costs. The performance of ion exchange technology was highly specific to target ions, which make them unsuitable for simultaneous ion removal from complex water streams. Among these existing technologies viable for the removal of heavy metal ions from water bodies, adsorption holds considerable promise due to its cost-effectiveness and simplicity (Ungureanu et al. 2015). Till now, several sorbents (such as nanoscale metal oxides, activated carbons, metal organic frameworks, clays, and zeolites) with affinity to heavy metal ions have been developed (Bolisetty et al. 2019; Ungureanu et al. 2015; Efome et al. 2018). To further improve the metal uptake capacity, complex functional groups may also be introduced into the sorbents surface. Despite recent advances on the boost of sorption capacity, the existing sorbents only demonstrated limited regulation capability towards heavy metal's valence state. Since for some highly toxic heavy metal ions, the decontamination not only requires adsorptive removal but also needs further detoxification step to meet the rising environmental standards. It is, therefore, highly desirable to develop advanced technology that enables regulation of the valence state of heavy metal ions and further removal of the products as well.

It is of note that the efficacy of different approaches is also depended on the physicochemical properties of the target heavy metal ions (Ungureanu et al. 2015). Hence, to regulate the valence state of heavy metal ions could be an effective solution. Among the currently available methods, electrochemical processes have great potential in the detoxification of heavy metal ions. However, to meet the strict criteria of increasing standard of wastewater discharge, combining electrochemistry with other processes may be more feasible. For example, the electroactive membrane technology, which combines electrochemistry with membrane separation process, has attracted increasing attention from the scientific community. This technology provides a promising strategy for the simultaneous detoxification and sequestration of heavy metal ions. Besides, the photocatalytic oxidation and photocatalytic reduction were also applied for the efficiently detoxification of heavy metal ions. The rapid development of the photocatalytic technologies for heavy metal ions removal can be reflected by the increase of published paper over the years. However, specific focus on the detoxification of heavy metal ions via the valence regulation mechanism is yet to be reviewed. This chapter will summarize the recent advances on the decontamination of heavy metal ions through valence regulation strategy. The valence regulation mechanisms based on oxidative and reductive detoxification routes were discussed in detail. Furthermore, existing challenges and perspectives on this technology were discussed as well.

2 Oxidative Detoxification of Heavy Metal Ions

The physicochemical properties (e.g., solubility, mobility, and toxicity) of heavy metal ions in water bodies are strongly governed by their valence state. For common heavy metal ions, like As and Sb, their lower valence state possesses much higher toxicity than their higher valence state counterparts. Therefore, researchers proposed an oxidative pretreatment for these oxidizable heavy metal ions (e.g., Sb(III) and As (III)) that carry elevated toxicity to higher valence state to achieve detoxification. Then, other operational unit was used to deal with those less toxic counterparts. Among those available technologies, electrochemical and photocatalytic processes have gained much attention to serve as viable oxidation step.

2.1 Electrochemical Oxidation

The electrochemical process is a simple and environmentally friendly process with tunable oxidative capabilities. However, the high cost of conventional electrochemical processes has greatly hampered their wide applications. The combination of electrochemical technology with other processes provides a promising solution to alleviation this limitation (Chaplin 2019). Among them, electrochemical oxidation coupled with membrane separation, which known as electrofiltration (Zhu and Jassby 2019), has been regarded as a promising approach by integrating the advantages of both membrane filtration with electrochemistry (Liu et al. 2020). In 2010, Vecitis et al. proposed a flow-through design based on highly conductive carbon nanotubes (CNT) (Vecitis et al. 2011). In this design, the CNT filter was utilized as an anode to adsorb and oxidize the target compounds; a perforated stainless steel was used as cathode to support reduction reaction. Besides degradation of organic compounds, this technology also demonstrated to be effective for the decontamination of several heavy metal ions. Although CNT itself doesn't have affinity to heavy metal ions, after modification with certain metal ion-specific nanoscale metal oxides, a hybrid filter may serve as high-performance functional

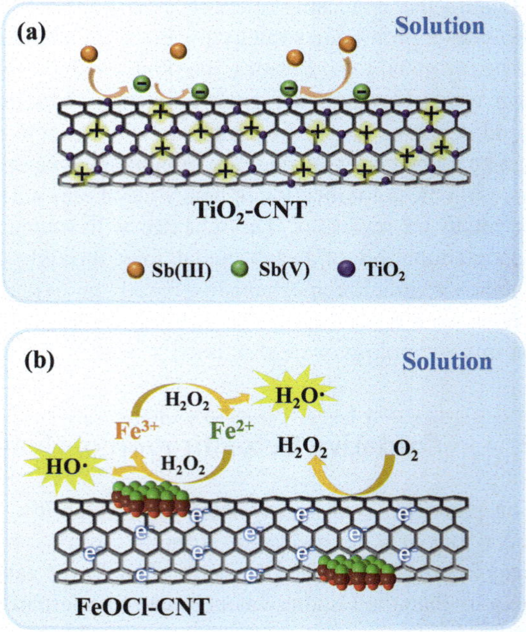

Fig. 1 The mechanism of oxidation and adsorption simultaneously for highly toxic Sb(III) using electroactive membranes TiO_2-CNT (**a**) and FeOCl-CNT (**b**). Reproduced with permission. (**a**) Copyright 2019, American Chemical Society. (**b**) Copyright 2020, Elsevier

membranes. Such supported design provides viable solution towards practical environmental applications, since the post-separation of nanoparticles was avoided. For example, Liu et al. recently developed a electroactive CNT filter functionalized with TiO_2 nanoparticles and applied for the decontamination of highly toxic Sb(III) (Liu et al. 2019a). The as-prepared hybrid filter not only maintains excellent electrical conductivity of CNT networks but also provides sufficient sorption sites (owing to these binder-free TiO_2). Upon application of an external electrical field (e.g., 2 V), in situ transformation of Sb(III) to less toxic Sb(V) can be achieved, which are further sequestered by TiO_2 (Fig. 1a). The Sb(III) removal efficiency is as high as >95% over 8 h continuous filtration, and Sb(V) become the dominant Sb specie rather than the highly toxic Sb(III). This indicates that the electro-induced oxidation reaction is active during the filtration process. Besides, Sb(III) removal kinetics and capacity increased with applied voltage and flow rate. This can be explained for the synergistic effects of the filter's flow-through design, electrochemical reactivity, small pore size, and increased number of exposed sorption sites. Similar designs can also be extended to other nanoscale Sb-specific sorbents such as titanate nanowires and iron oxides; comparable or even higher performance can still be achieved in those cases (Liu et al. 2019b, c).

In another design, an electroactive CNT filter functionalized with iron oxychloride (FeOCl) nanoparticles has been developed (Li et al. 2020). The FeOCl are promising and novel Fe-based catalysts enabling the decomposition of H_2O_2 to $HO^{\bullet}$, i.e., Fenton reaction (Sun et al. 2018). In this design, the FeOCl-modified CNT filter served as a functional cathode as well as a filter. Dissolved O_2 in the influent can be in situ reduced to H_2O_2 upon application of a proper cathode potential. These nanoscale FeOCl catalyst not only catalyze the $HO^•$ production but also facilitate the cycling of Fe^{3+}/Fe^{2+} when assisted by electric field (Fig. 1b) (Liu et al. 2019d). At −0.4 V vs. Ag/AgCl, an ultra-rapid conversion of Sb(III) to less toxic Sb(V) can be achieved (>99%) just by a single pass through the filter system ($\tau < 2$ s). Compared with the conventional batch reactor, such continuous flow system demonstrated a much enhanced Sb(III) detoxification kinetics due to the convection-enhanced mass transport. Moreover, the proposed flow-through E-Fenton system works effectively across a wide pH range (e.g., Fu and Wang 2011; Filella et al. 2012; Matlock et al. 2002; Ghurye et al. 2004; De Dardel and Arden 2000; Ungureanu et al. 2015;Efome et al. 2018). The rapid oxidation kinetics and superior performance of CNT filters have shown that this technology could potentially serve as a potentially simple, effective, and affordable option to remove heavy metal ions from water.

2.2 Photocatalytic Oxidation

As one of the most popular advanced oxidation processes (AOPs), photocatalytic technology has been widely studied (Sillanpää et al. 2011.) Upon irradiation, photogenerated electrons are excited from the valence band of the photocatalyst to conduction band, generating electron-hole pairs. These holes and electrons both can be used for the detoxification of heavy metal ions via oxidation or reduction reaction because the holes and electrons are powerful oxidants and reductants, respectively. Up to now, various nanoscale photocatalysts (e.g., titanium dioxide, TiO_2; tungsten trioxide, WO_3; zinc oxide, ZnO; ferric oxide, Fe_2O_3; cadmium sulfide, CdS) have been applied for detoxification purposes under irradiation (Sillanpää et al. 2011; Schneider et al. 2014; Zhang and Itoh 2006; Herrmann 1999). Among them, TiO_2 demonstrated robust and desirable photocatalytic performances compared with other counterparts under UV irradiation (Schneider et al. 2014). However, despite these achievements, the practical application of TiO_2 is still restricted by its wide bandgap and high-charge recombination rate. To overcome these drawbacks, ion doping has been extensively studied to expand the light response range of TiO_2. Zhang et al. fabricated nitrogen-doped TiO_2 fibers for photocatalytic oxidation of arsenic under visible light illumination (Fig. 2) (Zhang et al. 2015). These TiO_2 fibers exhibited superior photooxidation efficiency of As(III) in both the UV and visible light regions. Besides, combination with narrow bandgap semiconductors of TiO_2 is another feasible route (Sannino et al. 2011). Vaiano et al. adopted a wet impregnation method to prepare MoO_x/TiO_2 composites and applied for the detoxification of As(III) to less toxic As(V) (Vaiano et al. 2014). Results showed that, compared to TiO_2 alone, a complete conversion of As(III) to As(V) can be achieved over 60 min reaction in the presence of MoO_x/TiO_2 catalyst. Furthermore, the overall As (V) generated on the MoO_x/TiO_2 catalyst were completely released into solution and replenished the surface active sites of MoO_x/TiO_2 catalyst for further reuse.

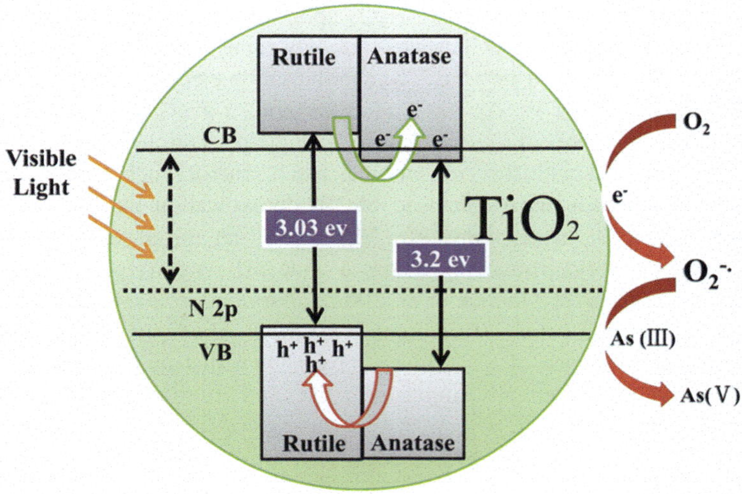

Fig. 2 Mechanism for the photooxidation of As(III) catalyzed by N-doped TiO_2 fibers under visible light irradiation, Reproduced with permission, copyright 2015, American Chemical Society

Photo-induced transformation of heavy metal ions on minerals has also been regarded as an effective and affordable remediation strategy. A previous study revealed that significant amounts of Sb(III) on a goethite suspension could be converted to Sb(V) under light irradiation (Fan et al. 2014). The underlying mechanism for Sb(III) oxidation is likely due to $HO^•$ produced by goethite under illumination. Moreover, Kong et al. reported that the specie of Fe(III) dictate the Sb(III) photooxidation process (Kong et al. 2016). For example, $FeCl^{2+}$ and $FeOH^{2+}$ are the main species of Fe(III) under acidic condition (pH 1–3), so that dichloro radicals ($^•Cl_2^-$) and $HO^•$ generated by the photocatalysis of $FeCl^{2+}$ and $FeOH^{2+}$ were the main oxidants for Sb(III) oxidation. While under elevated pH conditions (Filella et al. 2012; Matlock et al. 2002; Ghurye et al. 2004; De Dardel and Arden 2000; Ungureanu et al. 2015; Efome et al. 2018; Chaplin 2019), photooxidation of Sb(III) occurred through electron transfer from Sb(III) to Fe(III) along with the reduction of Fe(III) to Fe(II) through a ligand-to-metal charge-transfer (LMCT) procedure. Hence, the photocatalytic oxidation is a promising approach for the decontamination of heavy metal ions.

3 Reductive Detoxification of Heavy Metal Ions

Contrary to these heavy metal ions stated above, some other heavy metal ions such as chromium (Cr) possess increased toxicity at their higher valence state. Hence, the transformation of the highly toxic Cr(VI) to less toxic Cr(III) is the most preferred methods to achieve effective remediation of Cr(VI)-contained wastewater. In this section, we briefly discussed the reductive detoxification strategies of heavy metal ions.

3.1 Electrochemical Reduction

It is well documented that the valence states of chromium (Cr) range from −2 to +6. Among them, Cr(VI) and Cr(III) are environmentally stable, and Cr(VI) is 500–1000 times more toxic and far more mobile than Cr(III) (Costa 2003; Elwakeel 2010). Accordingly, there is strict regulations on the discharge of Cr(VI) in water bodies, and reducing Cr(VI) to Cr(III) is a commonly adopted strategy.

Based on our previous studies on the detoxification of Sb(III) by the "oxidation-adsorption" strategy (Liu et al. 2019a, c, d; Li et al. 2019), we also applied the hybrid filter to serve as a functional filter for the "reduction-sorption" purposes. In a recent work, we developed a dual-functional electroactive filters that enables Cr (VI) detoxification and sequestration (Fig. 3) (Liu et al. 2019e). An electroactive filter consisting of nanoscale polyaniline-modified carbon nanotubes (PANI-CNT) was rationally designed to reduce Cr(VI) to Cr(III). These amine and imine moieties of PANI can chelate and further reduce Cr(VI) at strong acidic pH conditions. To further expand the working pH range, the authors proposed an electric field-assisted flow-through system by employing the PANI-CNT hybrid membrane as functional

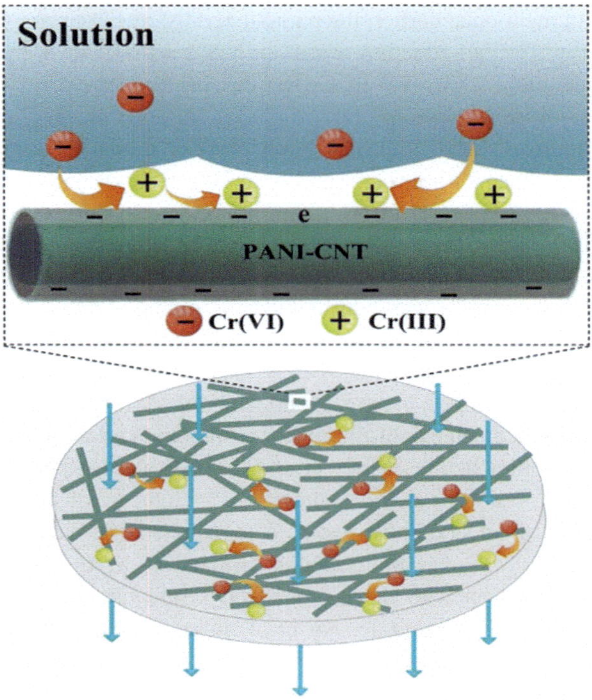

Fig. 3 The mechanism of reduction highly toxic Cr (VI) using electroactive PANI-CNT membranes. Copyright 2019, Elsevier

cathode. The reduction kinetics for the Cr(VI) was enhanced with flow rate and voltage increased. A complete Cr(VI) reduction can be observed at 2.5 V and pH 7, much higher than that of 78.0% at 0 V under similar conditions. Moreover, the removal efficiency of the total chromium (or Cr_{total}) can be further improved by applying an electric field. At pH 7, Cr_{total} removal efficiency increased from 29.3% at 0 V to 70.2% at 2.5 V. What's more, the energy consumption of this system was only 0.032 kWh/m^3, much lower than other state-of-the-art treatment technologies. Similarly, a polyvinyl alcohol (PVA)-CNT composite ultrafiltration (UF) membrane was also employed by Duan et al. for the electrochemical reduction and removal of highly toxic Cr(VI) (Duan et al. 2017). The removal mechanism was highly dependent on the solution conductivity, with higher conductivity leads to increased Cr (VI) reduction and Cr(III) precipitation performance on the membrane surface. The highest Cr(VI) removal efficiency of 95% can be achieved upon the application of an external potential of 7 V when using the membrane as cathode. Meanwhile, the energy needed to remove 1 ppm of Cr(VI) (>99%) from tap water was determined to be 1.48 kWh/m^3, less than other alternative processes (e.g., ion exchange and reduction/coagulation processes). Thus, the electroactive membrane technology can be considered as a promising and cost-effective strategy for the detoxification of heavy metal ions.

3.2 Photocatalytic Reduction

As mentioned earlier, the photogenerated electrons are strong reductants that can be used for the reduction of certain heavy metal ions. For example, the photocatalytic reduction of Cr(VI) to Cr(III) has been extensively studied using TiO_2 or other modified composite photocatalysts. Ku and Jung reported that Cr(VI) can be efficiently reduced in an UV/TiO_2 process at acidic conditions with 0.5% ethanol as scavenger for holes (Ku and Jung 2001). However, as stated previously, the limited visible light absorption of TiO_2 significantly restricts the practical applications of TiO_2. The light response of TiO_2 can be extended to visible light range by modification with quantum dots or dyes. Among these reported photosensitizers, silver sulfide (Ag_2S) was considered as a potential candidate for the photosensitization of TiO_2 because of its low bandgap and desired band positions relative to TiO_2, making it active in the visible range of the solar spectrum. For instance, Ghafoor et al. developed the Ag_2S/TiO_2 nanofibers for the photoreduction of Cr(VI) under simulated solar light illumination and obtained a 90% Cr(VI) photocatalytic reduction efficiency (Ghafoor et al. 2018). In comparison, hollow spherical Ag-Ag_2S/TiO_2 composite particles under visible light irradiation were capable to reduce only 36% of Cr(VI). These results demonstrated that the significantly enhanced capacity of the TiO_2 nanofibers photosensitized with Ag_2S to reduce Cr(VI). In addition, combination with narrow bandgap semiconductors of TiO_2 was another feasible strategy. For example, Ku et al. fabricated the NiO/TiO_2-coupled photocatalysts using a sol-gel method, which exhibited excellent photocatalytic activity for Cr (VI) reduction (Ku et al. 2011). Results revealed that a >95% Cr

(VI) photoreduction efficiency can be observed within the first 20 min after adding ethanol as sacrificial agent. Furthermore, it is noteworthy that although TiO_2 and TiO_2-based materials exhibit favorable photocatalytic performance in the detoxification of heavy metals, these nanoscale sorbents have rarely been applied in practical applications due to additional efforts that are necessary for post-separation. Such issues might be partially addressed by attaching these sorbents onto support materials or blend into membranes for photocatalytic detoxification of heavy metals in future (Leong et al. 2014).

Recently, as a newly emerged type of functional inorganic-organic hybrid materials, metal organic frameworks (MOFs) have attracted wide attention for photocatalytic applications due to their merits of desirable topologies, high surface areas along with efficient light harvest (Li et al. 2018; Kaur et al. 2018). Several MOFs have been reported to be effective to act as high-performance photocatalyst under UV light irradiation. To further expand the light response range of MOFs, modifying the inorganic unit or organic linker (like functionalization) through synthetic screening has been proposed as effective approaches (Wang et al. 2016, 2019; Du et al. 2019; Chen et al. 2019). For instance, Wang et al. compared the photocatalytic Cr(VI) reduction performance of Ti-benzenedicarboxylate (MIL-125 (Ti)) and amino-functionalized Ti-benzenedicarboxylate (NH_2-MIL-125(Ti)) (Wang et al. 2015). They found that the NH_2-MIL-125(Ti) exhibited more higher Cr (VI) reduction efficiency under visible light irradiation compared with the MIL-125(Ti). At acidic pH conditions, the highest Cr(VI) reduction efficiency can be further boosted from 76% to 91% just by adding ethanol as hole scavenger for the NH_2-MIL-125(Ti). Similarly, apart from Ti source, Shi and co-workers fabricated a novel amine-functionalized NH_2-MIL-88B (Fe) via a microwave-assisted solvothermal synthesis method (Shi et al. 2015). The prepared amine-functionalized NH_2-MIL-88B (Fe) showed improved stability and efficiency for the photocatalytic Cr(VI) reduction. Under visible light illumination, both 2-amino-benzenedicarboxylic acid (NH_2-BDC) and Fe_3-μ_3-oxo clusters in NH_2-MIL-88B (Fe) were excited. The electrons from the excited Fe_3-μ_3-oxo clusters as well as the photogenerated electrons were transferred from NH_2-BDC to Fe_3-μ_3-oxo clusters for the reduction of Cr(VI). Besides, similar catalytic mechanisms were also reported by combining MIL-53(Fe) with g-C_3N_4. The introduction of g-C_3N_4 on the surface of MIL-53(Fe) would improve the separation and migration rate of photo-induced charges, consequently resulting in promoting the photocatalytic efficiency (Fig. 4) (Huang et al. 2017). Moreover, Chen et al. fabricated the Z-scheme MIL-100(Fe)/PANI composite photocatalysts from MIL-100(Fe) and polyaniline (PANI) by ball-milling (Chen et al. 2019). In this design, the PANI not only served as an electron donor but also act as excellent hole acceptor. Therefore, the as-obtained MIL-100 (Fe)/PANI composite exhibited complete Cr(VI) reduction, much higher than that in the absence of PANI. Although those MOFs exhibited excellent photocatalytic performance, it was still a big challenge to post-separate the nanoscale materials for recycle (Leong et al. 2014). Therefore, Zhao et al. proposed a membrane design by depositing UiO-66-NH_2(Zr/Hf) onto an α-Al_2O_3 support by a reactive seeding method (Du et al. 2019). The membranes demonstrated excellent photocatalytic

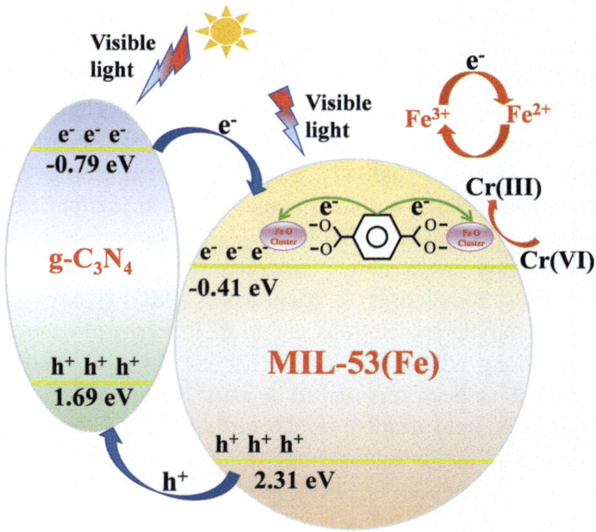

Fig. 4 A schematic illustration of the reduction of Cr(VI) over CMFe-3 nanocomposite under visible light irradiation. Reproduced with permission. Copyright 2017, Elsevier

performance towards Cr(VI) reduction under both simulated and real sunlight irradiation. The overall results suggest that the MOF-based membrane photocatalysts may provide a feasible approach for the efficient decontamination of heavy metal ions from water.

4 Challenges and Future Perspectives

This chapter discussed the recent advances on the decontamination of heavy metal ions by valence regulation. Though several promising technologies have been developed recently, extensive and large-scale applications of these technologies on pollution control in practical engineering remain limited. For instance, for the electroactive filter technology, novel porous conductive electrode materials that are affordable and robust during operation are still highly desirable. For example, although the price of CNTs has decreased significantly compared with that of a decade ago owing to advanced mass production technologies, their price is still much higher than activated carbon and other widely applied carbon counterparts. Besides, the electrochemical oxidation process may not be able to directly mineralize some refractory compounds into CO_2 and potentially may lead to secondary pollution of water due to the generation of by-products or intermediates (Du et al. 2017). Therefore, it is critical to find affordable methods to solve these problems associated with electrochemical filter technology. For the MOFs-based photocatalytic technology, besides the complicated fabrication process, the long-term stability of MOFs materials under light irradiation should be reconsidered. Since incidental pyrolyzation may cause damage to the framework of MOFs due to the thermal effect of photons. Further efforts should be directed towards designing and preparing

MOFs with high thermal stability towards practical applications. What's more, the serious photocarriers recombination issue was another obstacle to be overcome in practice.

Furthermore, it is of note that most of these technologies were conducted only in laboratory conditions or at bench scale. To facilitate the technology transfer process from a technical hype into a real applications, extensive efforts are needed. For example, to expand the applicability of these technologies on real wastewater and compare the performances of emerging and conventional technologies under identical conditions; to evaluate the benefits and risks of these technologies and explore solutions for controlling the formation of undesirable by-products or intermediates.

Acknowledgments This research was funded by the Natural Science Foundation of Shanghai, China (No. 18ZR1401000), and the Shanghai Pujiang Program (No. 18PJ1400400).

References

Bolisetty S, Peydayesh M, Mezzenga R (2019) Sustainable technologies for water purification from heavy metals: review and analysis. Chem Soc Rev 48(2):463–487

Chaplin BP (2019) The prospect of electrochemical technologies advancing worldwide water treatment. Acc Chem Res 52(3):596–604

Chen DD, Yi XH, Zhao C, Fu H, Wang P, Wang CC (2019) Polyaniline modified MIL-100(Fe) for enhanced photocatalytic Cr(VI) reduction and tetracycline degradation under white light. Chemosphere 245:125659

Costa M (2003) Potential hazards of hexavalent chromate in our drinking water. Toxicol Appl Pharmacol 188(1):1–5

De Dardel F, Arden TV (2000) Ion exchangers. In: Ullmann's encyclopedia of industrial chemistry. Wiley-VCH Verlag GmbH, Weinheim

Du Y, Lv XT, Wu QY, Zhang DY, Zhou YT, Peng L, Hu HY (2017) Formation and control of disinfection byproducts and toxicity during reclaimed water chlorination: a review. J Environ Sci 58:51–63

Du XD, Yi XH, Wang P, Zheng WW, Deng JG, Wang CC (2019) Robust photocatalytic reduction of Cr(VI) on UiO-66-NH_2(Zr/Hf) metal-organic framework membrane under sunlight irradiation. Chem Eng J 356:393–399

Duan WY, Chen GD, Chen CX, Sanghvi R, Iddya A, Walker S, Liu HZ, Ronen A, Jassby D (2017) Electrochemical removal of hexavalent chromium using electrically conducting carbon nanotube/polymer composite ultrafiltration membranes. J Membr Sci 531:160–171

Efome JE, Rana D, Matsuura T, Lan CQ (2018) Metal–organic frameworks supported on nanofibers to remove heavy metals. J Mater Chem A 6(10):4550–4555

Elwakeel KZ (2010) Removal of Cr(VI) from alkaline aqueous solutions using chemically modified magnetic chitosan resins. Desalination 250(1):105–112

Fan JX, Wang YJ, Fan TT, Cui XD, Zhou DM (2014) Photo-induced oxidation of Sb(III) on goethite. Chemosphere 95:295–300

Filella M, Belzile N, Chen YW (2012) Human exposure to antimony. II. Contents in some human tissues often used in biomonitoring (hair, nails, teeth). Crit Rev Environ Sci Technol 42 (10):1058–1115

Fu FL, Wang Q (2011) Removal of heavy metal ions from wastewaters: a review. J Environ Manage 92(3):407–418

Ghafoor S, Hussain SZ, Waseem S, Arshad SN (2018) Photo-reduction of heavy metal ions and photo-disinfection of pathogenic bacteria under simulated solar light using photosensitized TiO_2 nanofibers. RSC Adv 8(36):20354–20362

Ghurye G, Clifford D, Tripp A (2004) Iron coagulation and direct microfiltration to remove arsenic from groundwater. J AWWA 96(4):143–152

Herrmann HJ (1999) Heterogeneous photocatalysis: fundamentals and applications to the removal of various types of aqueous pollutants. Catal Today 53(1):115–129

Huang WY, Liu N, Zhang XD, Wu MH, Tang L (2017) Metal organic framework g-C_3N_4/MIL-53 (Fe) heterojunctions with enhanced photocatalytic activity for Cr (VI) reduction under visible light. Appl Surf Sci 425:107–116

Kaur H, Venkateswarulu M, Kumar S, Krishnan V, Koner RR (2018) A metal-organic framework based multifunctional catalytic platform for organic transformation and environmental remediation. Dalton Trans 47(5):1488–1497

Kong LH, He MC, Hu XY (2016) Rapid photooxidation of Sb(III) in the presence of different Fe (III) species. Geochim Cosmochim Acta 180:214–226

Ku Y, Jung IL (2001) Photocatalytic reduction of Cr (VI) in aqueous solutions by UV irradiation with the presence of titanium dioxide. Water Res 35(1):135–142

Ku Y, Lin CN, Hou WM (2011) Characterization of coupled NiO/TiO_2 photocatalyst for the photocatalytic reduction of Cr(VI) in aqueous solution. J Mol Catal A: Chem 349(1-2):20–27

Leong S, Razmjou A, Wang K, Hapgood K, Zhang XW, Wang HT (2014) TiO_2 based photocatalytic membranes: a review. J Membr Sci 472:167–184

Li H, Yao S, Wu HL, Qu JY, Zhang ZM, Lu TB, Lin WB, Wang EB (2018) Charge-regulated sequential adsorption of anionic catalysts and cationic photosensitizers into metal-organic frameworks enhances photocatalytic proton reduction. Appl Catal Environ 224:46–52

Li MH, Liu YB, Shen CS, Li F, Wang CC, Huang MH, Yang B, Wang ZW, Yang JM, Sand W (2019) One-step Sb(III) decontamination using a bifunctional photoelectrochemical filter. J Hazard Mater 389:121840

Li ZZ, Shen CS, Liu YB, Ma CY, Li F, Yang B, Huang MH, Wang ZW, Dong LM, Sand W (2020) Carbon nanotube filter functionalized with iron oxychloride for flow-through electro-Fenton. Appl Catal Environ 260:118204

Liu YB, Wu P, Liu FQ, Li F, An XQ, Liu JS, Wang ZW, Shen CS, Sand W (2019a) Electroactive modified carbon nanotube filter for simultaneous detoxification and sequestration of Sb (III). Environ Sci Technol 53(3):1527–1535

Liu YB, Yao JY, Liu FQ, Shen CS, Li F, Yang B, Huang MH, Sand W (2019b) Nanoscale iron (oxyhydr)oxide-modified carbon nanotube filter for rapid and effective Sb(III) removal. RSC Adv 9(32):18196–18204

Liu YB, Liu FQ, Qi ZL, Shen CS, Li F, Ma CY, Huang MH, Wang ZW, Li JJ (2019c) Simultaneous oxidation and sorption of highly toxic Sb(III) using a dual-functional electroactive filter. Environ Pollut 251:72–80

Liu YB, Zhang J, Liu FQ, Shen CS, Li F, Huang MH, Yang B, Wang ZW, Sand W (2019d) Ultrarapid detoxification of Sb(III) using a flow-through electro-fenton system. Chemosphere 245:125604

Liu YB, Liu FQ, Ding N, Shen CS, Li F, Dong LM, Huang MH, Yang B, Wang ZW, Sand W (2019e) Boosting Cr(VI) detoxification and sequestration efficiency with carbon nanotube electrochemical filter functionalized with nanoscale polyaniline: performance and mechanism. Sci Total Environ 695:133926

Liu FQ, Liu YB, Yao QF, Wang YX, Fang XF, Shen CS, Li F, Huang MH, Wang ZW, Sand W, Xie JP (2020) Supported atomically-precise gold nanoclusters for enhanced flow-through electro-Fenton. Environ Sci Technol. 54(9): 5913–5921

Matlock MM, Howerton BS, Atwood DA (2002) Chemical precipitation of heavy metals from acid mine drainage. Water Res 36(19):4757–4764

Rosenberg E (2015) Heavy metals in water: presence, removal and safety. Johnson Matthey Technol Rev 59(4):293–297

Sannino D, Vaiano V, Ciambelli P, Eloy P, Gaigneaux EM (2011) Avoiding the deactivation of sulphated MoO_x/TiO_2 catalysts in the photocatalytic cyclohexane oxidative dehydrogenation by a fluidized bed photoreactor. Appl Catal A Gen 394(1-2):71–78

Schneider J, Matsuoka M, Takeuchi M, Zhang JL, Horiuchi Y, Anpo M, Bahnemann DW (2014) Understanding TiO_2 photocatalysis: mechanisms and materials. Chem Rev 114(19):9919–9986

Shi L, Wang T, Zhang HB, Chang K, Meng XG, Liu HM, Ye JH (2015) An amine-functionalized iron (III) metal–organic framework as efficient visible-light photocatalyst for Cr (VI) reduction. Adv Sci 2(3):1500006

Sillanpää ME, Kurniawan TA, Lo WH (2011) Degradation of chelating agents in aqueous solution using advanced oxidation process (AOP). Chemosphere 83(11):1443–1460

Sun M, Zucker I, Davenport DM, Zhou XC, Qu J, Elimelech M (2018) Reactive, self-cleaning ultrafiltration membrane functionalized with iron oxychloride nanocatalysts. Environ Sci Technol 52(15):8674–8683

Ungureanu G, Santos S, Boaventura R, Botelho C (2015) Arsenic and antimony in water and wastewater: overview of removal techniques with special reference to latest advances in adsorption. J Environ Manage 151:326–342

Vaiano V, Iervolino G, Sannino D, Rizzo L, Sarno G, Farina A (2014) Enhanced photocatalytic oxidation of arsenite to arsenate in water solutions by a new catalyst based on MoO_x supported on TiO_2. Appl Catal Environ 160-161:247–253

Vecitis CD, Schnoor MH, Rahaman MS, Schiffman JD, Elimelech M (2011) Electrochemical multiwalled carbon nanotube filter for viral and bacterial removal and inactivation. Environ Sci Technol 45(8):3672–3679

Wang H, Yuan XZ, Wu Y, Zeng G, Chen X, Leng LJ, Wu ZB, Jiang LB, Li H (2015) Facile synthesis of amino-functionalized titanium metal-organic frameworks and their superior visible-light photocatalytic activity for Cr(VI) reduction. J Hazard Mater 286:187–194

Wang CC, Du XD, Li J, Guo XX, Wang P, Zhang J (2016) Photocatalytic Cr(VI) reduction in metal-organic frameworks: a mini-review. Appl Catal Environ 193:198–216

Wang X, Liu W, Fu H, Yi XH, Wang P, Zhao C, Wang CC, Zheng W (2019) Simultaneous Cr (VI) reduction and Cr(III) removal of bifunctional MOF/titanate nanotube composites. Environ Pollut 249:502–511

Zhang FS, Itoh H (2006) Photocatalytic oxidation and removal of arsenite from water using slag-iron oxide-TiO_2 adsorbent. Chemosphere 65(1):125–131

Zhang G, Sun M, Liu Y, Lang XF, Liu LM, Liu HJ, Qu JH, Li JH (2015) Visible-light induced photocatalytic activity of electrospun-TiO_2 in arsenic(III) oxidation. ACS Appl Mater Interfaces 7(1):511–518

Zhu XB, Jassby D (2019) Electroactive membranes for water treatment: enhanced treatment functionalities, energy considerations, and future challenges. Acc Chem Res 52(5):1177–1186

Mechanistic Insight to Bioremediation of Hazardous Metals and Pesticides from Water Bodies by Microbes

Priyadarshini Dey, Deepak Gola, Nitin Chauhan, Randhir Kumar Bharti, and Anushree Malik

Abstract

In the last few decades, there has been a surge in the hazardous chemicals contamination in the water bodies in various parts of India. Consequently, bioremediation of these hazardous chemicals by microorganisms has emerged as a lucrative option. Furthermore, advancement in the various physicochemical and robust molecular techniques such as 1D-SDS-PAGE followed by LC-MS/MS analysis as well as metabolite identification by ^{1}H-NMR has shed light on the various cellular strategies adopted by a microorganism for remediation of a cocktail of metals and pesticide that were either poorly highlighted or unexplained.

With this background the present chapter aims to understand the complete process of a metal or cocktail of multimetal and pesticide mixture remediation by

Priyadarshini Dey, Deepak Gola, Nitin Chauhan and Randhir Kumar Bharti contributed equally with all other contributors.

P. Dey (✉)
Department of Biotechnology, Ramaiah Institute of Technology, Bengaluru, India

D. Gola
Department of Biotechnology, Noida Institute of Engineering and Technology, Greater Noida, India

N. Chauhan
Department of Microbiology, Shaheed Rajguru College of Applied Sciences for Women, University of Delhi, New Delhi, India

R. K. Bharti
University School of Environmental Management, Guru Gobind Singh Indraprastha University, New Delhi, India

A. Malik
Applied Microbiology Laboratory, Center for Rural Development and Technology, Indian Institute of Technology Delhi, Delhi, India

M. P. Shah (ed.), *Removal of Emerging Contaminants Through Microbial Processes*,
https://doi.org/10.1007/978-981-15-5901-3_23

microorganisms such as bacteria, fungus or algae at molecular level by proteomic and metabolomic techniques. Further the scale up of this bioremediation process by bioreactors would be detailed.

Keywords

Proteomic · Metabolomic · Mechanism · Multimetal · Pesticide

1 Introduction

The organizations like Central Pollution Control Board (CPCB), India and Food and Agriculture Organization (FAO), monitor the level of toxic pollutants such as cadmium, total chromium, copper, nickel, lead and zinc in environment. Various industries viz. preservatives, fertilizer, pesticide, electroplating, batteries and textile etc., have been identified that discharge these pollutants into the environment. These heavy metals released anthropogenically have imposed limits on their discharge to the water bodies (lake, river, stream, etc.) which is utilized for irrigational purpose. Despite strict effluent treatment guidelines, laid down by pollution control board agencies, many small and medium scale industries do not comply appropriately with the effluent discharge limits (as per guidelines). Further, due to the lack of expensive and effective discharge facilities, these industries discharge heavy metal effluents, that seep into the major rivers like Yamuna and other water bodies (Rawat et al. 2003; Bhattacharya et al. 2015; Sheng et al. 2016) consequently exposing every consumer to unknown quantities of hazardous heavy metals in the water they consume (Central Pollution Control Board 2011; Delhi Pollution Control Committee 2011).

The residues of organophosphorus pesticides (Malathion and Chlorpyrifos), organochlorine pesticides {gamma-hexachlorocyclohexane (lindane)} and DDT persist in the environment. The Central Insecticides Board and Registration Committee (CIBRC), India, has banned their use. However, as these chemicals are cheaper, farmers are still using them excessively in agriculture, in order to control weeds, pests and diseases in crops. Further, due to agriculture runoffs, these chemicals accumulate in water bodies. Moreover, the crops growing in these fields become the carrier of heavy metals leading to their entry in food chain, posing major health problems such as cancer, neurological diseases, etc. as enlisted by the Agency for Toxic Substances and Disease Registry (ATSDR) and US Environmental Protection Agency (EPA) (Kim et al. 2017; Mostafalou and Abdollahi 2017; Carvalho 2017; Honeycutt and Weber 2019; Pellicer-Castell et al. 2020). Therefore, it becomes important to remove these harmful pollutants from these natural ecosystems ex situ.

Microorganisms such as algae, bacteria and fungi have already been identified as key players for remediation of heavy metals and pesticide-contaminated soil and water. Therefore, a large amount of research findings are available on the use of biological systems (microorganisms) as a bioremediating agent (Mishra and Malik 2013; Gola et al. 2016b, 2018; Mathur et al. 2018). Due to various advantages associated with microorganisms, i.e. efficient action, economical and eco-friendliness, it is now considered as a beneficial substitute to physico-chemical methods for remediation of heavy metal and pesticide contamination. There is a great scope of research and development in the area of microbial assisted bioremediation. As most of the studies available rely on single metal/pesticide remediation and very few of the studies have employed growing microbial cells, harbouring multiple mechanisms of metal/pesticide sequestration through production of enzymes, which can be used to uptake large amounts of metal/pesticide (Malik 2004; Mishra and Malik 2013). Based on the review of past studies in the last few years, importance has now shifted on using microbial remediation of environments co-contaminated with heavy metals and pesticides. Further, in comparison to traditional physical and chemical treatment methods such as filtration, evaporation, oxidation and reduction, ion exchange, reverse osmosis and chemical precipitation, microbial assisted processes of bioremediation carry great potential with additional benefits of environment safety. The occurrences/incidences of the heavy metals and pesticide in the Indian water bodies are compiled here. The developments made in the field of bioremediation of contaminants, heavy metals and pesticides in the recent past have also been summarised.

2 Incidences of Heavy Metals Contamination in India

Many researchers (Begum et al. 2009; Kaushik et al. 2009; Subramanian 2010; Chabukdhara and Nema 2012; Lokhande et al. 2011; Kumar et al. 2013; Rosin et al. 2013; Shekhar and Sarkar 2013; Dwivedi and Vankar 2014) and government agencies (CPCB 2008; River Data Directorate Planning & Development Organisation 2014) have investigated the occurrence of heavy metals in the Indian water bodies which are utilized for irrigational purpose. These studies evaluated the existing metal pollution levels which pose potential ecological risk.

A research survey conducted by Centre for Environmental Studies and the Energy Resources and Institute, New Delhi (Sehgal et al. 2012) on the river Yamuna in Delhi, India revealed increased levels of nickel, manganese, lead and mercury in agricultural soil along the river. Small-scale industries such as electroplating, metal alloys and pigments for paints are responsible for addition of heavy metal concentrations in the wastewater. The mixing of waste effluent with the river leads to the contamination of the river water and also the flood plains. It enters the food chain through the vegetables that are grown on the river banks and affects the people habiting the catchment area. The vegetables such as spinach, cauliflower and radish grown in such kind of flood plain soils were analysed for the contaminants. The amount of heavy metals was found to be the higher in spinach, followed by cauliflower and the least in radish. This is in agreement with the research findings where leafy vegetables are high accumulators of heavy metals (Singh et al. 2010;

Table 1 Sources of heavy metal contamination and their permissible limits (Gola et al. 2016b)

Heavy metals	Major sources	Maximum permissible limit in irrigation water (mg/L) according to FAO
Chromium	Glass making, ceramics, chrome electroplating, paints, cement	0.1
Copper	Pesticides, fungicides, electrical appliance, electroplating	0.2
Lead	Ceramic glazer, smoking, jet engines emission, mining, burning of coal	5.0
Arsenic	Glass, textiles, pulp and paper, electronics metal	0.1
Mercury	Electrical appliances, switches, fungicides, antiseptics, thermometers, barometers, preservatives and pharma	– –
Cobalt	Alloy, petroleum, chemical, porcelain, pottery, glass and jewellery	0.05
Lithium	Ceramics, glass, electronic items, medicines, lubricants and battery	2.5

Ghosh et al. 2012; Bhatti et al. 2016). Table 1 depicts the compilation of the sources of the selected heavy metals and their permissible limits in irrigational water according to FAO.

Figure 1a–e depicts the percentage of heavy metals released from various industrial units' effluent. In this study (Singare and Dhabarde 2014), Cr is released maximum from oil surfactant (31%) and fine chemicals (30%) effluents. Cu is discharged maximum from textile (22%) and dyeing industries (22%). Ni is majorly released from oil surfactant (17%) and dyeing (15%). Pb is contributed maximum by dyeing (65%) while Zn from dyeing (26%) and fine chemicals (27%). It was further observed that the majority of the average concentration of the metals except Ni was above the maximum permissible limit set for the discharge of industrial effluents into the inland surface water.

3 Occurrence of Pesticide in the Water Bodies

Many studies indicated the occurrence of organochlorine pesticides in water bodies (Dikshith et al. 1990; Jani et al. 1991; Mukherjee and Gopal 2002; Bakore et al. 2004; Sankararamakrishnan et al. 2005; Shukla et al. 2006; Kaushik et al. 2010; Mutiyar et al. 2011).

Table 2 depicts the compilation of occurrence of the pesticide lindane (HCH) in different water bodies.

In a study undertaken by a research group from School of Environmental Sciences, Jawaharlal Nehru University, New Delhi, pesticides such as endosulfan, DDT, lindane and methoxychlor were detected in consequential concentrations at all the sites of the river Yamuna in Delhi (Pandey et al. 2011), in all the seasons,

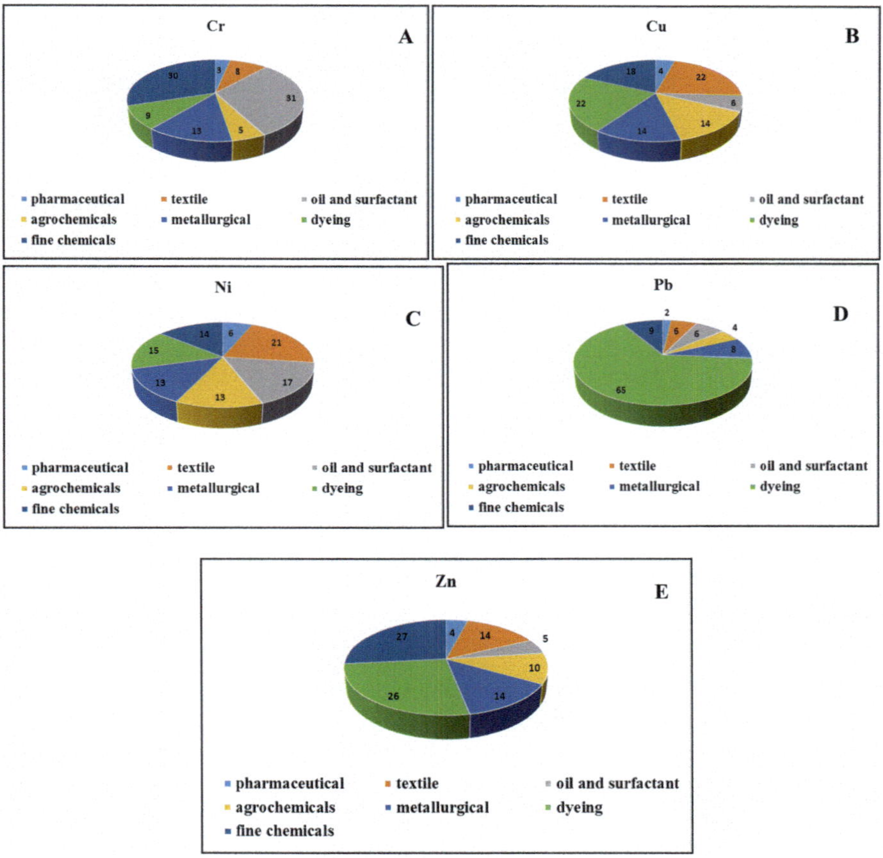

Fig. 1 (**a–e**) Percentage of heavy metals present in the pharmaceutical, textile, oil and surfactant, agrochemicals, metallurgical, dyeing and fine chemicals effluents Source Singare and Dhabarde (2014)

indicating their wide use. Farmers try mixture of pesticides, not being aware of their residual effect in soil and food. Moreover, use of pesticides enables farmers to fetch higher price of vegetable crops owing to the clean and fresh look of produce after the spray. The vegetables such as cauliflower, brinjal, okra, cabbage, etc. grown in such kind of flood plain soils were analysed for the contaminants. The concentration of many pesticides were found in these vegetables (Mukherjee 2003; Singh and Singh 2006; Mandal and Singh 2010). However, the pesticide lindane was found to be below the maximum residue limits.

4 Remediation of Contaminants

Heavy metals such as Cd, Cr, Cu, Ni, Pb and Zn are a key group of toxic pollutants due to their recalcitrance nature. Heavy metals have high atomic weights and density greater than 5.0 g/cm^3. Heavy metals cannot be biodegraded and tend to accumulate

Table 2 Global occurrence of lindane (ng/L) in different water bodies

Location	Type of water body	HCH	References
India (Devprayag)	Ganges River	7.24	Mutiyar and Mittal (2012)
India (Ambala)	Groundwater	147.1	Kaushik et al. (2012)
India (Nagaon)	River, ponds	4911	Mishra and Sharma (2011)
India (Dibrugarh)	River, ponds	4403	Mishra and Sharma (2011)
India (Pilbhit)	Gomti River	46.69	Malik et al. (2009)
India (Karnataka)	Cauvery River	2430	Begum et al. (2009)
India (Kolkata)	Groundwater	100	Ghosh et al. (2009)
Northwest China	Weihe River	2.41–178.18	Wang et al. (2016)
Mexico	Culiacan River	0.0045	Morales et al. (2017)
China	Shaying River	14.3	Bai et al. (2018)
India (Odisha)	Chilika Lake	0.025–23.4	Nag et al. (2020)

in the ecosystem. Carcinogenic heavy metals are particularly zinc, copper, nickel, mercury, cadmium, lead and chromium.

Pesticides such as hexachlorocyclohexane (HCH)/lindane are a cyclic, saturated and chlorinated hydrocarbon. It has a molecular weight of 290.8, molecular formula $C_6H_6Cl_6$ and density of 1.85 g/cm^3. The solubility of lindane in water is 0.7–5.59 mg/L. Lindane has been banned in India since 1997. However, lindane is still produced as a by-product in many chemical processes, such as the manufacture of chlorinated aromatics, chlorinated solvents and pesticides. In addition, by incomplete combustion it is released from the old dump sites (Tremolada et al. 2008). Lindane formulations are to curb vector-borne diseases and for pest control in few crops (Gupta 2004). The volatilization, diffusion, decomposition, retention of lindane in the soil and its subsequent transportation by water from soil to groundwater are the main concerns as lindane is a potential carcinogen (Yadav et al. 2015).

In contrast to the pesticides which undergo degradation, although extremely slowly, heavy metals cannot be eliminated from water bodies and therefore persist in the sediments and are released into the water slowly. Thus, the bioaccumulation of the heavy metals and pesticides in the aquatic environment such as river bodies due to agricultural runoff and effluent discharge and their biomagnification in food chain is a major cause of concern as they affect reproductive system, supress immunity and are carcinogen.

Table 3 Advantages and limitations of various chemical and physical methods of heavy metal removal

Physicochemical method	Advantages	Disadvantages
Chemical precipitation	Low investment and simple to operate	Generation of sludge
Ion exchange	No sludge formation	High cost, all ion exchange resin not suitable for metal removal
Adsorption	Low investment, easy to operate, functional in a wide pH range, high metal-binding capacity	Low selectivity, generation of waste products
Membrane filtration	Higher selectivity for separation	Incur of high cost due to membrane fouling
Coagulation-flocculation	Less time for settling of suspended solids	Sludge production
Flotation	Low cost, shorter hydraulic retention time	Further treatments required to improve removal efficiency

Source: Fu and Wang (2011)

4.1 Chemical and Physical Means of Remediation

As heavy metals and pesticides are the environmental priority pollutants posing threat to the environment, many methods are used to remediate heavy metal ions such as adsorption, membrane filtration, chemical precipitation, ion exchange, electrochemical treatment technologies, etc. Table 3 shows few techniques for the removal of heavy metal ions from wastewater. Their advantages and disadvantages in application are also mentioned.

4.2 Microbial Means of Remediation

To overcome the limitations of the physicochemical treatments of heavy metals/ pesticide, a large section of studies have investigated microbial agents like algae, bacteria and fungus as bio-remediating agents (Say et al. 2001; Sheng et al. 2004; Muñoz et al. 2006; Preetha and Viruthagiri 2007; Gupta and Rastogi 2008; Choudhary and Sar 2009; Subbaiah et al. 2011). Few studies employed growing microbial cells which constitute the mechanisms of metal remediation including production of enzymes have greater potential for metal uptake capacities. Active growing cells are self-renewing through the metabolic uptake of the metal contaminants (Malik 2004). Also, bioremediation of metals by growing cells can get rid of the cultivation of the biomass, harvesting it, drying it, processing it and storing it prior to use (Chojnacka 2010).

4.2.1 Metal Bioremoval

Fungi are superior than the single-celled microorganisms. Fungi are pervasive and multifaceted and have better ability to adjust to environmental limitations like extremes of temperature, pH, nutrient availability and high metal concentrations due to their morphological diversity (Anand et al. 2006; Kaushik and Malik 2010). Moreover fungi have a tendency to produce a suite of extracellular enzymes, exopolysaccharides, organic acids, proteins and other metabolites that help in metal bioremediation (Gadd 2007; Guria et al. 2014; Gola et al. 2016a).

Hence studies related to removal of heavy metals by fungus are growing steadily. The bioaccumulation capacity of total chromium and Cu (II) ions by *A. terreus* AML02 was studied (Mishra and Malik 2014). They reported substantial removal of metals (Cr, 95.57%; Cu, 65.77 %) at the incubation period of 5 days at 100 mg/L initial individual metal concentrations by the strain *A. terreus* AML02. Further, mannan, chitin, glucan and proteins are the constituents of cell walls of fungus. Additionally, it also contains other lipids, polysaccharides and pigments (melanin) which enable binding of many metal ions. Carboxyl, phenolic, hydroxyl, carbonyl and methoxyl groups are significant as they bind to the oxygen binding sites which are present in the phenolic polymers and melanins of the fungal cell wall (Gadd 2004). Hence metal remediation by fungal biomass is receiving considerable attention.

4.2.2 Pesticide Bioremoval

Another advantage of utilizing fungus as a bio-remediating agent is that the fungal mycelia growth facilitates both the mechanical and enzymatic contact with the pollutant due to increased cell-to-surface ratio (Sagar and Singh 2011). Karas et al. (2011) demonstrated three white-rot fungal strains: *Phanerochaete chrysosporium*, *Trametes versicolor* and *Pleurotus ostreatus*, and *Aspergillus niger* could break down the pesticides thiabendazole, imazalil, thiophanate methyl, *ortho*-phenylphenol, diphenylamine and chlorpyrifos in straw extract medium and soil extract medium. They determined that peroxidase (LiP, MnP) and laccase (Lac) activities were involved in pesticide degradation. Pinto et al. (2012) showed *Penicillium brevicompactum* PP0021, *Lecanicillium saksenae* PP0011, *Fusarium oxysporum* CCMI866, *Aspergillus oryzae* CCMI 125 and *Lentinula edodes* EL1 fungal strains degraded the pesticides terbuthylazine, difenoconazole and pendimethalin from liquid culture medium. They concluded the presence of desethyl-terbuthylazine in the samples of *L. edodes* liquid cultures which indicated that terbuthylazine dealkylation is one of the plausible metabolic pathways for degradation of terbuthylazine. Sagar and Singh (2011) showed that *Fusarium* sp. was able to efficiently degrade lindane by about 60% removal in reasonable time. Also, the utilization of non-white-rot fungi for their enzymatic biodegradation potential of the pesticides was highlighted (Marco-Urrea et al. 2015).

4.2.3 Metal Pesticide Mixture

River bodies are often contaminated with a mixture of organic and inorganic chemicals, such as heavy metals and pesticides. Thus the ability to extract more than one pollutant is a remarkable process advantage. Nevertheless, microbial potential varies when co-pollutants exist together in the environment in comparison to the occurrence of a single toxic compound. The effect of alachlor and zinc on the growth of the filamentous fungus *Paecilomyces marquandii* and its potential to degrade alachlor and accumulate zinc was studied (Słaba et al. 2009). The herbicide removal was not inhibited by zinc up to 1.0 mM of the metal content in the culture medium and 5.0–7.5 mM of the metal limited alachlor consumption by 30–50%, whereas a higher zinc concentration ceased this process. Furthermore, no study is present which investigated the simultaneous uptake of multimetals in the presence of pesticide by fungus.

5 Insight into the Mechanism of Individual Metals and Multimetal Pesticide Mixture Uptake

The occurrence of the fungus in the metal-contaminated sites could be due to the metal resistance and tolerance mechanisms inherent in them (Baker 1987; Gadd and White 1993).

The fungi can broadly counteract heavy metals by two ways. First, by preventing the metal uptake and its transportation inside the fungal cell, e.g. by reducing the uptake of metals or by increasing the efflux of metals, or by biosorbing to the impermeable cell walls by metal binding peptides, polysaccharides or by forming complexes extracellularly and by release of organic acid. Extracellular mechanisms aim to avoid the entry of metal. Second, the microorganism can survive in the presence of high metal concentrations by tolerance after its detoxification process, e.g. chelation of metals intracellularly through the synthesis of ligands such as phytochelatins, metallothioneins, polyphosphates or compartmentalization within the cellular organelles such as vacuoles. Intracellular mechanisms are mainly to reduce the metal burden in the cytosol (Sandau et al. 1996; Gadd 2000, 2007, 2010).

In addition to these mechanisms, detoxification by antioxidation production allow the microorganisms to tackle the reactive oxygen species which accumulate in the cell in response to the metals. The implication of these processes may differ as a function of the metal involved, metal's location and its concentration in the cell.

5.1 Extracellular Metal Interactions

5.1.1 Organic Acid Chelation of Metals

Different extracellular fungal products of organic molecules such as oxalic acid, citric acid and malic acid which are excreted into the cell are effective metal chelators. Oxalic acids chelate metals and forms insoluble oxalate crystals which tend to deposit around the cell walls of the fungi (Gadd and White 1993). The production of H_2S also aids in precipitation of metals with the formation of insoluble sulphides.

5.1.2 Metal Binding by Fungal Cell Wall Components and Exopolymers

Eukaryotic fungi have been effectively able to bind metals to cell walls which act as the main barrier to the entry of the toxic metals inside the cell. All filamentous fungi cell is composed of chitin with (l-3)-α-glucan as skeletal components. The cell wall is mainly composed of polysaccharides, some maybe associated with protein. Other components like lipids and melanins may also be present in the cell wall (Gadd and White 1993). Functional groups like free hydroxyl, phosphate, carboxyl and amine, sulphydryl groups present in the cell wall and mercapto groups in the cell wall polymers and proteins are potential sites that are involved in metal sequestration. The primary interactions however involve binding to phosphate and carboxyl groups which are raised by the electrostatic attraction to the other negatively charged functional groups (Gadd and White 1993). Cell wall binding is called biosorption (Gadd and White 1993), which is cell metabolic independent. Heavy metals biosorption may lead to reduction in the intracellular metal accumulation and decrease the toxicity of metals in the cytoplasm (Brown and Wilkins 1985). Blaudez et al. (2000) investigated the substantial accumulation of Cd in the cell wall of *Paxillus involutus*, a mycorrhizal fungus which exhibited higher tolerance to Cd. In another study glomalin, a glycoprotein synthesized and excreted by the arbuscular mycorrhizal fungal hyphae, was able to bind metal ions Cd, Cu and Pb found at the metal-contaminated soils (Wright and Upadhyaya 1998). Gonzalez-Chavez et al. (2002) exhibited that mycorrhizal fungal specie *Glomus*, isolated from the copper mine, biosorbed copper onto their hyphae. *Glomus species* have a high tendency to bind Zn and Cd. Furthermore, Bhanoori and Venkateswerlu (2000) showed the complex formation between cell wall chitin of the specie *Neurospora crassa* and the metal ion Cd based on the analysis by X-ray diffraction, infrared spectroscopy and ^{13}C-NMR spectroscopy.

5.1.3 Transport Phenomena of Metal Ions

The proteins that are involved in the transportation of metals operate in two ways. One is by forcing the toxic metal ions out of the cell from the cytosol. Second is by allowing the metal ion sequestration in the cellular compartments (Hall 2002).

There are two kinds of transporter proteins embedded in the plasma membrane of a fungal cell.

Carrier Proteins

In this metal transport mechanism, changes in the conformation in the transport protein results in the exposure of its binding site on the membrane. Carrier proteins mainly comprises of metabolically coupled as well as H^+-gradient-driven transport systems.

Channel Proteins

Channel proteins are present as pores in the plasma membrane that permit the flow of ions across a chemical or electrical gradient. These ion selective proteins contains pores which allow solute to pass at high-flux rates when the channel gates open. The main features of channel proteins are selectivity of solute, permit faster rate of solute and regulate solute permeation through gating. Few important channel proteins include Ca^{2+} channel protein and Na^{+} channel proteins.

5.2 Intracellular Metal Complexation

Despite the resistance mechanisms of extracellular metal chelation and cell wall binding of metals, high volume of metals make way into the cells. Metal chelation in the cytoplasm by ligands which have high affinity is an important process of detoxification of metals and metal tolerance. The ligands comprises of organic acids, amino acids and the peptides metallothioneins and phytochelatins (Gadd and White 1993). Also the antioxidants and molecular chaperones play important role in metal detoxification process.

5.2.1 Metallothioneins

Metallothioneins (MTs) are cysteine rich with low-molecular-weight (3500–14,000 Da) proteins. MTs have the ability to chelate the metals which are of physiological importance, viz. Cu, Se and Zn and the xenobiotic of Ag, Cd and Hg. The MTs bind the heavy metals through the thiol group present in the cysteine residues. MTs are hypothesized to fix the essential trace elements, viz. copper and zinc intracellularly. They control the concentrations of the free ions and aid regulating the flow of these ions to their respective cellular destinations. MTs also neutralize the toxic effects of elements such as cadmium and mercury which protect the cell from many stressful conditions (Gadd and White 1993).

MTs in fungi have been mostly characterized in yeasts which are activated when exposed to heavy metals. Specific genes such *CUP1* and *CRS5* loci have been identified in baker's *Saccharomyces cerevisiae* that code for MTs and help in detoxification pathway of metals (Butt et al. 1984).

5.2.2 Phytochelatins

Phytochelatins (PCs) are peptide molecules that are rich in thiol. These are produced on induction by metals, viz. As, Cd, Cu and Zn (Grill and Zenk 1987). PCs are synthesized from the reduction of glutathione (GSH) by transpeptidation of γ-glutamyl-cysteinyl dipeptides by the action of enzyme PC synthase (Vatamaniuk et al. 2001). PCs are mainly produced in plants, algae, yeasts and many filamentous fungi. PCs form complexes with Ag, Hg and Pb in vitro (Rauser 1999). PCs carrying metal ions are transported to the vacuoles where the metal ions are stored away from the proteins in the cytoplasm.

5.2.3 Antioxidant Production to Counter Metal Toxicity

Heavy metals can initiate the formation of free radical species which cause damage to the cellular components in a cell. These free radicals can be formed directly or by indirect method (Bellion et al. 2006). Heavy metals are generally categorized into redox-active (Fe, Cr, Cu, V) and redox-inactive metals (Cd, Hg, Ni, Zn). The redox-active metals directly produce reactive oxygen species (ROS), viz. superoxide ($O_2^{\bullet -}$) substrates and reactive hydroxyl ($HO^{\bullet}$) radical from H_2O_2. These ROS are generated through Haber-Weiss and Fenton reactions and are responsible for the peroxidation of the lipid which leads to damage and death of the cell (Halliwell and Gutteridget 1986). On the other hand, non-redox metals bind to the thiol groups occurring in the active sites of glutathione and antioxidants enzymes that inactivates the antioxidant defence mechanism system of an organism that results in accumulation of ROS and redox imbalance. ROS species lead to oxidation of membrane lipids and proteins and cause injury to DNA. This oxidation process is tackled by antioxidative systems which are composed of enzymes, viz. superoxide dismutase, peroxidases, catalases that scavenge the activated oxygen (Zhang et al. 2014) or metabolites, viz. glutathione, tocopherol and ascorbate.

5.2.4 Molecular Chaperones

Heat shock proteins are a class of proteins that ensures survival of the microorganism under any stressful conditions such as heavy metal stress that would otherwise lead to cell death. Heat shock proteins have role in synthesis of proteins and their transport and folding and are known as molecular chaperones. In eukaryotes, heat shock response is mediated by transcriptions factors known as heat shock factors (HSF). In an unstressed condition, HSF are present as monomers in nucleus as well as in cytosol with no DNA-binding activity. In a stressed situation, the HSF assembles to form a trimer and collects in the nucleus where it binds to the specific heat shock gene promoter sequences and thus phosphorylates that lead to enhanced levels of Hsp70 (Morimoto 1993).

6 Tools to Study the Metal Uptake in Microbe

With the aid of sophisticated techniques such as scanning electron microscope (SEM), transmission electron microscope (TEM), energy-dispersive X-ray spectroscopy (EDAXS), X-ray and electron diffraction analysis (XRD) and Fourier transform-infrared spectroscopy (FTIR), the mechanism of localization of metals in the cells and also speciation in X-ray photoelectron spectroscopy (XPS) can be understood from Table 4.

In addition, molecular tools like proteomic and metabolomic have been utilised to study the effect of toxicity of heavy metals on microbial communities (Azarbad et al. 2015; Dhindwal et al. 2016). With help of proteomic techniques, reductases, laccases and many more such novel proteins could be identified (Table 4). The extracellular enzymes such as laccase produced by fungi enable them to remediate xenobiotic compounds (Mougin et al. 2000). Furthermore, no study is currently available that investigated the simultaneous uptake of multimetals in the presence of pesticide by fungal strain at physiological, proteomic and metabolomics level.

Table 4 Mechanism of metal accumulation and pesticide degradation

Organism	Metals/pesticides	Localization	Technique employed	Mechanism	Reference
Aspergillus niger	Ni	Cell wall and intracellular	EDXS, XRD	Metabolism dependent uptake	Magyarosy et al. (2002)
Rhodobacter sphaeroides	Co, Ni	Extracellular	ICP-AES, ATR-FTIR	R-COO-play important role in metal immobilization	Italiano et al. (2009)
Pseudomonas aeruginosa	Ni	Restricted to periplasm and cell membrane	TEM, EDX, XRD	Potential role of phosphoryl and carboxyl/carbonyl group of cell wall/membrane	Sar et al. (2001)
Bradyrhizobium japonicum	Cd, Cu, Pb, Zn	Lipopolysaccharide (LPS)	SDS-PAGE of isolated LPS	LPS has metal-binding property	Oh et al. (2002)
Rhizoctonia solani	*p*-Anisidine, Cu(II), Mn(II), Ca ionophore A23187, $LiCl_2$, $CaCl_2$, cyclic AMP (cAMP), caffeine, amphotericin B, paraquat, ethanol and isopropanol	–	Spectrophotometrically	Production of the inducible form of laccases	Meburn Ellis (2001)
Phanerochaete chrysosporium and Trametes versicolor	Herbicide isoxaflutole	–	Spectrophotometrically	Enhancement of laccase activity	Mougin et al. (2002)
Bacillus megaterium TKW3	Cr(VI) in concomitance with Se^{4+}, Se^{6+} and As^{5+}	–	(2-DE) followed by MALDI-TOF-MS	Isolation of a novel Cr^{6+} reductases	Cheung and Gu (2007)
Pseudomonas aeruginosa	Chromium(VI)	Production of exopolysaccharides	MALDI–TOF MS, LC–MS/MS	Cr chelation by EPS	Kiliç et al. (2010)
Chlorella vulgaris	Cd, Cu, Pb	Production of metabolite	^{1}H-NMR	Production of antioxidants	Zhang et al. (2014)

(continued)

Table 4 (continued)

Organism	Metals/pesticides	Localization	Technique employed	Mechanism	Reference
Paecilomyces marquandii	Alachlor	–	MALDI–TOF MS and LC–MS/MS	Upregulation of proteins related to energy, sugar metabolism and ROS production	Szewczyk and Sobo (2015)
Aspergillus sp. Adapted from Mishra and Malik (2013)*WU 243*	Co	–	^{1}H-NMR and ^{13}C-NMR	Production of metabolite of polyketide-terpenoid, aspergstressin	Ding et al. (2016)

Adapted from Mishra and Malik (2013)

7 Proteomic Study

1D-SDS-PAGE and LC-MS/MS are a better approach to study the functional aspects of the differentially expressed proteins by the fungal strain under various metal/pesticide exposures. Proteomics studies the proteins by biochemical methods. A mass spectrometry (MS)-based approach is highly versatile that generates valuable information such as mass, fragmentation pattern of the analysed molecule as well as the isotopic signature of the compound with high sensitivity. Tandem MS (MS/MS) analyses involve the collision of ions in a collision cell which are fragmented with argon gas before it is analysed by the detector. These fragmented patterns are unique for every compound that leads to the identification of the compound. Szewczyk and Sobo (2015) studied the degradation pathway of the herbicide alachlor by the analysis of the proteome of *Paecilomyces marquandii* by LC-MS/MS. The large-scale analysis of gene products of the *A. fumigatus* under various metal/pesticide exposures is largely understudied.

8 Metabolomic Study

^{1}H-NMR is a robust analytical tool to analyse a large number of metabolites produced by the fungal strain. This process involves the study of fingerprint of water-soluble compounds and to verify the pattern among the metabolites produced by the fungal strain. NMR spectroscopy depends on the magnetic properties of an atomic nucleus in a molecule. Every compound generates a unique spectrum that tells about the chemical nature of the surrounding atoms. This leads to elucidation of complete structure. Ding et al. (2016) studied the structure of a unique metabolite polyketide-terpenoid, aspergstressin expressed in the presence of the metal cobalt by *Aspergillus* sp. WU 243 by ^{1}H-NMR and ^{13}C-NMR. ^{1}H-NMR metabolomics fingerprinting of *A. fumigatus* under various metal/pesticide exposures has never been studied before.

9 Conclusion

Many studies have been conducted for the removal of individual metal, individual pesticide or in few cases binary metal/pesticide mixture or single metal and single pesticide mixture by a microorganism. However, the study of mechanism of removal of multiple metals in the presence of a pesticide is scare. Furthermore, many studies dealt with the mechanism of metal removal by biosorption method only, viz. binding of metals onto the microorganism passively. Also, few researchers attempted to study the metal uptake by a microorganism at molecular level only. Thus, this review was undertaken to have a complete and comprehensive picture of the metal uptake and pesticide degradation by microbes by the active process of bioaccumulation along with the metal removal by the passive process of biosorption.

Acknowledgments The authors are grateful to the Department of Biotechnology, Ramaiah Institute of Technology, Bengaluru; Department of Biotechnology, Noida Institute of Engineering and Technology, Greater Noida; University School of Environmental Management, GGSIPU, New Delhi and Department of Microbiology, Shaheed Rajguru College of Applied Sciences for Women, University of Delhi, New Delhi; and Applied Microbiology Laboratory, Center for Rural Development and Technology Indian Institute of Technology Delhi, New Delhi. One of the authors (Randhir K. Bharti) thanks D.S Kothari Post-Doctoral Fellowship (BL/18-19/0164), UGC, Govt. of India.

References

Anand P, Isar J, Saran S, Saxena RK (2006) Bioaccumulation of copper by Trichoderma viride. Bioresour Technol 97:1018–1025. https://doi.org/10.1016/j.biortech.2005.04.046

Azarbad H, Niklińska M, Laskowski R et al (2015) Microbial community composition and functions are resilient to metal pollution along two forest soil gradients. FEMS Microbiol Ecol 91:1–11. https://doi.org/10.1093/femsec/fiu003

Bai Y, Ruan X, van der Hoek JP (2018) Residues of organochlorine pesticides (OCPs) in aquatic environment and risk assessment along Shaying River, China. Environ Geochem Health 40 (6):2525–2538

Baker AJM (1987) Metal tolerance. New Phytol 106:93–111

Bakore N, John PJ, Bhatnagar P (2004) Organochlorine pesticide residues in wheat and drinking water samples from Jaipur, Rajasthan, India. Environ Monit Assess 98:381–389. https://doi.org/10.1023/B:EMAS.0000038197.76047.83

Begum A, Ramaiah M, Harikrishna et al (2009) Heavy metal pollution and chemical profile of cauvery river water. E-Journal Chem 6:47–52. https://doi.org/10.1155/2009/154610

Bellion M, Courbot M, Jacob C et al (2006) Extracellular and cellular mechanisms sustaining metal tolerance in ectomycorrhizal fungi. FEMS Microbiol Lett 254:173–181. https://doi.org/10.1111/j.1574-6968.2005.00044.x

Bhanoori M, Venkateswerlu G (2000) In vivo chitin-cadmium complexation in cell wall of Neurospora crassa. Biochim Biophys Acta 1523:21–28

Bhattacharya A, Dey P, Gola D et al (2015) Assessment of Yamuna and associated drains used for irrigation in rural and peri-urban settings of Delhi NCR. Environ Monit Assess 187:4146. https://doi.org/10.1007/s10661-014-4146-2

Bhatti SS, Sambyal V, Nagpal AK (2016) Heavy metals bioaccumulation in Berseem (Trifolium alexandrinum) cultivated in areas under intensive agriculture, Punjab, India. Springerplus 5:173. https://doi.org/10.1186/s40064-016-1777-5

Blaudez D, Botton B, Chalot M (2000) Cadmium uptake and subcellular compartmentation in the ectomycorrhizal fungus Paxillus involutus. Microbiology 146:1109–1117

Brown MT, Wilkins DA (1985) Zinc tolerance of Amanita and Paxillus. Trans Br Mycol Soc 84:367–369. https://doi.org/10.1016/S0007-1536(85)80096-6

Butt TR, Sternberg EJ, Gormant JA et al (1984) Copper metallothionein of yeast, structure of the gene, and regulation of expression. Biochemistry 81:3332–3336

Carvalho FP (2017) Pesticides, environment, and food safety. Food Energy Secur 6:48–60

Central Pollution Control Board (2011) Polluted river stretches in india criteria and status

Chabukdhara M, Nema AK (2012) Heavy metals in water, sediments, and aquatic macrophytes: river Hindon, India. J Hazardous Toxic Radioact Waste 16:273–281. https://doi.org/10.1061/(ASCE)HZ.2153-5515.0000127

Cheung KH, Gu J-D (2007) Mechanism of hexavalent chromium detoxification by microorganisms and bioremediation application potential: a review. Int Biodeter Biodegr 59:8–15. https://doi.org/10.1016/j.ibiod.2006.05.002

Chojnacka K (2010) Biosorption and bioaccumulation—the prospects for practical applications. Environ Int 36:299–307. https://doi.org/10.1016/j.envint.2009.12.001

Choudhary S, Sar P (2009) Characterization of a metal resistant Pseudomonas sp. isolated from uranium mine for its potential in heavy metal (Ni2+, Co2+, Cu2+, and Cd2+) sequestration. Bioresour Technol 100:2482–2492. https://doi.org/10.1016/j.biortech.2008.12.015

CPCB (2008) Status of water treatment plants in India. Minist Environ For Cent Pollut Control Board, pp 1–107

Dhindwal S, Gomez-Gil L, Neau DB et al (2016) Structural basis of the enhanced pollutant-degrading capabilities of an engineered biphenyl dioxygenase. J Bacteriol 198:1499–1512. https://doi.org/10.1128/JB.00952-15

Dikshith TSS, Raizada RB, Kumar SN et al (1990) Residues of DDT and HCH in major sources of drinking water in Bhopal, India. Bull Environ Contam Toxicol 45:389–393. https://doi.org/10.1007/BF01701162

Ding C, Wu X, Auckloo BN et al (2016) An unusual stress metabolite from a hydrothermal vent fungus Aspergillus sp. WU 243 induced by cobalt. Molecules 21(1):105. https://doi.org/10.3390/molecules21010105

DPCC (2011) Action plan: abatement of pollution in critically polluted area of Najafgarh drain basin including Okhla, Naraina, Anand Parbat and Wazirpur Industrial areas

Dwivedi AK, Vankar PS (2014) Source identification study of heavy metal contamination in the industrial hub of Unnao, India. Environ Monit Assess 186:3531–3539. https://doi.org/10.1007/s10661-014-3636-6

Fu F, Wang Q (2011) Removal of heavy metal ions from wastewaters: a review. J Environ Manage 92:407–418. https://doi.org/10.1016/j.jenvman.2010.11.011

Gadd GM (2000) Bioremedial potential of microbial mechanisms of metal mobilization and immobilization. Curr Opin Biotechnol 11:271–279. https://doi.org/10.1016/S0958-1669(00)00095-1

Gadd GM (2004) Microbial influence on metal mobility and application for bioremediation. Geoderma 122:109–119. https://doi.org/10.1016/j.geoderma.2004.01.002

Gadd GM (2007) Geomycology: biogeochemical transformations of rocks, minerals, metals and radionuclides by fungi, bioweathering and bioremediation. Mycol Res 111:3–49. https://doi.org/10.1016/j.mycres.2006.12.001

Gadd GM (2010) Metals, minerals and microbes: geomicrobiology and bioremediation. Microbiology 156:609–643. https://doi.org/10.1099/mic.0.037143-0

Gadd GM, White C (1993) Microbial treatment of metal pollution—a working biotechnology? Trends Biotechnol 11:353–359. https://doi.org/10.1016/0167-7799(93)90158-6

Ghose NC, Saha D, Gupta A (2009) Synthetic detergents (surfactants) and organochlorine pesticide signatures in surface water and groundwater of greater Kolkata, India. J Water Resour Protect 4:290–298. C GN Protection SD-J of WR and, 2009 undefined

Ghosh AK, Bhatt MA, Agrawal HP (2012) Effect of long-term application of treated sewage water on heavy metal accumulation in vegetables grown in Northern India. Environ Monit Assess 184:1025–1036. https://doi.org/10.1007/s10661-011-2018-6

Gola D, Dey P, Bhattacharya A et al (2016a) Multiple heavy metal removal using an entomopathogenic fungi Beauveria bassiana. Bioresour Technol 218:388–396. https://doi.org/10.1016/j.biortech.2016.06.096

Gola D, Malik A, Shaikh ZA, Sreekrishnan TR (2016b) Impact of heavy metal containing wastewater on agricultural soil and produce: relevance of biological treatment. Environ Process 3:1063–1080. https://doi.org/10.1007/s40710-016-0176-9

Gola D, Malik A, Namburath M, Ahammad SZ (2018) Removal of industrial dyes and heavy metals by Beauveria bassiana: FTIR, SEM, TEM and AFM investigations with Pb(II). Environ Sci Pollut Res 25:20486–20496. https://doi.org/10.1007/s11356-017-0246-1

Gonzalez-Chavez C, Haen JD, Vangronsveld J, Dodd JC (2002) Copper sorption and accumulation by the extraradical mycelium of different Glomus spp. (arbuscular mycorrhizal fungi) isolated from the same polluted soil. Plant Soil 240:287–297

Grill E, Zenk MH (1987) Phytochelatins, a class of heavy-metal-binding peptides from plants, are functionally analogous to metallothioneins. Proc Natl Acad Sci U S A 84:439–443

Gupta PK (2004) Pesticide exposure--Indian scene. Toxicology 198(1-3):83–90
Gupta VK, Rastogi A (2008) Biosorption of lead from aqueous solutions by green algae Spirogyra species: kinetics and equilibrium studies. J Hazard Mater 152:407–414. https://doi.org/10.1016/j.jhazmat.2007.07.028
Guria MK, Guha AK, Bhattacharyya M (2014) A green chemical approach for biotransformation of Cr(VI) to Cr(III), utilizing Fusarium sp. MMT1 and consequent structural alteration of cell morphology. J Environ Chem Eng 2:424–433. https://doi.org/10.1016/j.jece.2014.01.016
Hall JL (2002) Cellular mechanisms for heavy metal detoxification and tolerance. J Exp Bot 53:1–11
Halliwell B, Gutteridget JMC (1986) Invited paper. Arch Biochem Biophys 246:501–514
Honeycutt R, Weber JB (2019) Properties and behavior of pesticides in soil. In: Mechanisms of pesticide movement into ground water. CRC Press, Boca Raton, pp 15–42
Italiano F, Buccolieri A, Giotta L et al (2009) International biodeterioration & biodegradation response of the carotenoidless mutant Rhodobacter sphaeroides growing cells to cobalt and nickel exposure. Int Biodeter Biodegr 63:948–957. https://doi.org/10.1016/j.ibiod.2009.05.001
Jani JP, Raiyani CV, Mistry JS et al (1991) Residues of organochlorine pesticides and polycyclic aromatic hydrocarbons in drinking water of Ahmedabad City, India. Bull Environ Contam Toxicol 47:381–385
Karas PA, Perruchon C, Exarhou K et al (2011) Potential for bioremediation of agro-industrial effluents with high loads of pesticides by selected fungi. Biodegradation 22:215–228. https://doi.org/10.1007/s10532-010-9389-1
Kaushik P, Malik A (2010) Alkali, thermo and halo tolerant fungal isolate for the removal of textile dyes. Colloids Surf B Biointerfaces 81:321–328. https://doi.org/10.1016/j.colsurfb.2010.07.034
Kaushik A, Kansal A, Santosh et al (2009) Heavy metal contamination of river Yamuna, Haryana, India: assessment by metal enrichment factor of the sediments. J Hazard Mater 164:265–270. https://doi.org/10.1016/j.jhazmat.2008.08.031
Kaushik A, Sharma HR, Jain S et al (2010) Pesticide pollution of River Ghaggar in Haryana, India. Environ Monit Assess 160:61–69. https://doi.org/10.1007/s10661-008-0657-z
Kaushik CP, Sharma HR, Kaushik A (2012) Organochlorine pesticide residues in drinking water in the rural areas of Haryana, India. Environ Monit Assess 184:103–112. https://doi.org/10.1007/s10661-011-1950-9
Kiliç NK, Stensballe A, Otzen DE, Dönmez G (2010) Proteomic changes in response to chromium (VI) toxicity in Pseudomonas aeruginosa. Bioresour Technol 101:2134–2140. https://doi.org/10.1016/j.biortech.2009.11.008
Kim KH, Kabir E, Jahan SA (2017) Exposure to pesticides and the associated human health effects. Sci Total Environ 575:525–535
Kumar RN, Solanki R, Kumar JIN (2013) Seasonal variation in heavy metal contamination in water and sediments of river Sabarmati and Kharicut canal at Ahmedabad, Gujarat. Environ Monit Assess 185:359–368. https://doi.org/10.1007/s10661-012-2558-4
Lokhande RS, Singare PU, Pimple DS (2011) Pollution in water of Kasardi River flowing along Taloja industrial area of Mumbai, India. World Environ 1(1):6–13
Magyarosy A, Laidlaw R, Kilaas R et al (2002) Nickel accumulation and nickel oxalate precipitation by Aspergillus niger. Appl Microbiol Biotechnol 59:382–388. https://doi.org/10.1007/s00253-002-1020-x
Malik A (2004) Metal bioremediation through growing cells. Environ Int 30:261–278. https://doi.org/10.1016/j.envint.2003.08.001
Malik A, Ojha P, Singh KP (2009) Levels and distribution of persistent organochlorine pesticide residues in water and sediments of Gomti River (India)—a tributary of the Ganges River. Environ Monit Assess 148:421–435. https://doi.org/10.1007/s10661-008-0172-2
Mandal K, Singh B (2010) Magnitude and frequency of pesticide residues in farmgate samples of cauliflower in Punjab, India. Bull Environ Contam Toxicol 85:423–426. https://doi.org/10.1007/s00128-010-0107-9

Marco-Urrea E, García-Romera I, Aranda E (2015) Potential of non-ligninolytic fungi in bioremediation of chlorinated and polycyclic aromatic hydrocarbons. N Biotechnol 32:620–628. https://doi.org/10.1016/j.nbt.2015.01.005
Mathur M, Gola D, Panja R et al (2018) Performance evaluation of two Aspergillus spp. for the decolourization of reactive dyes by bioaccumulation and biosorption. Environ Sci Pollut Res 25 (1):345–352. https://doi.org/10.1007/s11356-017-0417-0
Meburn Ellis (2001) Inventorship and ownership. Mewburn Ellis 67:2088–2094. https://doi.org/10.1128/AEM.67.5.2088
Mishra A, Malik A (2013) Recent advances in microbial metal bioaccumulation. Crit Rev Environ Sci Technol 43:1162–1222. https://doi.org/10.1080/10934529.2011.627044
Mishra A, Malik A (2014) Novel fungal consortium for bioremediation of metals and dyes from mixed waste stream. Bioresour Technol 171:217–226. https://doi.org/10.1016/j.biortech.2014.08.047
Mishra K, Sharma RC (2011) Contamination of aquatic system by chlorinated pesticides and their spatial distribution over North-East India. Toxicol Environ Health Sci 3:144–155. https://doi.org/10.1007/s13530-011-0092-3
Morales ML, Belisario J, Torres V, Benigno J, Bastidas B, De Jesús P et al (2017) Universidad de Guanajuato. Acta Univ [Internet] 27(1):45–54. [Cited 2020 Jun 20]
Morimoto RI (1993) Cells in stress: transcription activation of heat shock genes. JSTOR 259:1409–1410
Mostafalou S, Abdollahi M (2017) Pesticides: an update of human exposure and toxicity. Arch Toxicol 91:549–599
Mougin C, Boyer FD, Caminade E, Rama R (2000) Cleavage of the diketonitrile derivative of the herbicide isoxaflutole by extracellular fungal oxidases. J Agric Food Chem 48:4529–4534. https://doi.org/10.1021/jf000397q
Mougin C, Kollmann A, Jolivalt C (2002) Enhanced production of laccase in the fungus Trametes versicolor by the addition of xenobiotics. Biotechnol Lett 24:139–142
Mukherjee I (2003) Pesticides residues in vegetables in and around Delhi. Environ Monit Assess 86:265–271. https://doi.org/10.1023/A:1024057420937
Mukherjee I, Gopal M (2002) Organochlorine insecticide residues in drinking and ground water in and around Delhi. Environ Monit Assess 76:185–193. https://doi.org/10.1023/A:1015527929438
Muñoz R, Alvarez MT, Muñoz A et al (2006) Sequential removal of heavy metals ions and organic pollutants using an algal-bacterial consortium. Chemosphere 63:903–911. https://doi.org/10.1016/j.chemosphere.2005.09.062
Mutiyar P, Mittal AK (2012) Status of organochlorine pesticides in Ganga river basin: anthropogenic or glacial? Drink Water Eng Sci 6:69–80
Mutiyar PK, Mittal AK, Pekdeger A (2011) Status of organochlorine pesticides in the drinking water well-field located in the Delhi region of the flood plains of river Yamuna. Drink Water Eng Sci 4:51–60. https://doi.org/10.5194/dwes-4-51-2011
Nag SK, Saha K, Bandopadhyay S, Ghosh A, Mukherjee M, Raut A, Raman RK, Suresh VR, Mohanty SK (2020) Status of pesticide residues in water, sediment, and fishes of Chilika Lake, India. Environ Monit Assess 192(2):1–10
Oh ET, Yun HS, Heo TR, Koh SC, Oh KH, So JS (2002) Involvement of lipopolysaccharide of Bradyrhizobium japonicum in metal binding. J Microbiol Biotechnol 12:296–300
Pandey P, Khillare PS, Kumar K (2011) Assessment of organochlorine pesticide residues in the surface sediments of river Yamuna in Delhi, India. J Environ Prot (Irvine, Calif) 2:511–524. https://doi.org/10.4236/jep.2011.25059
Pellicer-Castell E, Belenguer-Sapiña C, Amorós P et al (2020) Comparison of silica-based materials for organophosphorus pesticides sampling and occupational risk assessment. Anal Chim Acta 1110:26–34. https://doi.org/10.1016/j.aca.2020.03.008

Pinto AP, Serrano C, Pires T et al (2012) Degradation of terbuthylazine, difenoconazole and pendimethalin pesticides by selected fungi cultures. Sci Total Environ 435–436:402–410. https://doi.org/10.1016/j.scitotenv.2012.07.027

Preetha B, Viruthagiri T (2007) Bioaccumulation of chromium(VI), copper(II) and nickel(II) ions by growing Rhizopus arrhizus. Biochem Eng J 34:131–135. https://doi.org/10.1016/j.bej.2006.11.022

Rauser W (1999) Structure and function of metal chelators produced by plants. Cell Biochem Biophys 31:19–48

Rawat M, Moturi MCZ, Subramanian V (2003) Inventory compilation and distribution of heavy metals in wastewater from small-scale industrial areas of Delhi, India. J Environ Monit 5:906–912. https://doi.org/10.1039/b306628b

Rawat M, Singh UK, Subramanian V (2010) Movement of toxic metals from small-scale industrial areas: a case study from Delhi, India. Int J Environ Waste Manag 5:224–236

River Data Directorate Planning & Development Organisation (2014) Status of trace and toxic metals in Indian rivers. Central Water Commission, p 207

Rosin KG, Kaur R, Singh SD et al (2013) Groundwater vulnerability to contaminated irrigation waters—a case of peri-urban agricultural lands around an industrial district of Haryana, India. Procedia Environ Sci 18:200–210. https://doi.org/10.1016/j.proenv.2013.04.026

Sagar V, Singh DP (2011) Biodegradation of lindane pesticide by non white-rots soil fungus Fusarium sp. World J Microbiol Biotechnol 27:1747–1754. https://doi.org/10.1007/s11274-010-0628-8

Sandau E, Sandau P, Pulz O (1996) Heavy metal sorption by microalgae. Acta Biotechnol 16:227–235. https://doi.org/10.1002/abio.370160402

Sankararamakrishnan N, Kumar Sharma A, Sanghi R (2005) Organochlorine and organophosphorus pesticide residues in ground water and surface waters of Kanpur, Uttar Pradesh, India. Environ Int 31:113–120. https://doi.org/10.1016/j.envint.2004.08.001

Sar P, Kazy SK, Singh SP (2001) Intracellular nickel accumulation by *Pseudomonas aeruginosa* and its chemical nature. Lett Appl Microbiol 32:257–261. https://doi.org/10.1046/j.1472-765X.2001.00878.x

Say R, Denizli A, Arica MY (2001) Biosorption of cadmium(II), lead (II) and copper(II) with the filamentous fungus Phanerochaete chrysosporium. Bioresour Technol 76:67–70. https://doi.org/10.1016/S0960-8524(00)00071-7

Sehgal M, Garg A, Suresh R, Dagar P (2012) Heavy metal contamination in the Delhi segment of Yamuna basin. Environ Monit Assess 184:1181–1196. https://doi.org/10.1007/s10661-011-2031-9

Shekhar S, Sarkar A (2013) Hydrogeological characterization and assessment of groundwater quality in shallow aquifers in vicinity of Najafgarh drain of NCT Delhi. J Earth Syst Sci 122:43–54. https://doi.org/10.1007/s12040-012-0256-9

Sheng PX, Ting YP, Chen JP, Hong L (2004) Sorption of lead, copper, cadmium, zinc, and nickel by marine algal biomass: characterization of biosorptive capacity and investigation of mechanisms. J Colloid Interface Sci 275:131–141. https://doi.org/10.1016/j.jcis.2004.01.036

Sheng J, Qiu W, Xu B et al (2016) Monitoring of heavy metal levels in the major rivers and in residents' blood in Zhenjiang City, China, and assessment of heavy metal elimination via urine and sweat in humans. Environ Sci Pollut Res 23(11):11034–11045. https://doi.org/10.1007/s11356-016-6287-z

Shukla G, Kumar A, Bhanti M et al (2006) Organochlorine pesticide contamination of ground water in the city of Hyderabad. Environ Int 32(2):244–247

Singare PU, Dhabarde SS (2014) Toxic metals pollution due to industrial effluents released along Dombivali Industrial Belt of Mumbai, India. Eur J Environ Saf Sci Eur Sci Res Inst 2:5–11

Singh SP, Singh NK (2006) Pesticide residues in farm gate vegetable samples in Bihar. Pest Manag Hortic Ecosyst 12:152–155

Singh A, Sharma RK, Agrawal M, Marshall FM (2010) Health risk assessment of heavy metals via dietary intake of foodstuffs from the wastewater irrigated site of a dry tropical area of India. Food Chem Toxicol 48:611–619. https://doi.org/10.1016/j.fct.2009.11.041

Słaba M, Szewczyk R, Bernat P, Długoński J (2009) Simultaneous toxic action of zinc and alachlor resulted in enhancement of zinc uptake by the filamentous fungus Paecilomyces marquandii. Sci Total Environ 407:4127–4133. https://doi.org/10.1016/j.scitotenv.2009.03.023

Subbaiah MV, Yuvaraja G, Vijaya Y, Krishnaiah A (2011) Equilibrium, kinetic and thermodynamic studies on biosorption of Pb(II) and Cd(II) from aqueous solution by fungus (Trametes versicolor) biomass. J Taiwan Inst Chem Eng 42:965–971. https://doi.org/10.1016/j.jtice.2011.04.007

Szewczyk R, Sobo A (2015) Mechanism study of alachlor biodegradation by Paecilomyces marquandii with proteomic and metabolomic methods. J Hazard Mater 291:52–64. https://doi.org/10.1016/j.jhazmat.2015.02.063

Tremolada P, Villa S, Bazzarin P et al (2008) POPs in mountain soils from the Alps and Andes: suggestions for a "precipitation effect" on altitudinal gradients. Water Air Soil Pollut 188:93–109. https://doi.org/10.1007/s11270-007-9527-5

Vatamaniuk OK, Bucher EA, Ward JT, Rea PA (2001) A new pathway for heavy metal. J Biol Chem 276:20817–20821. https://doi.org/10.1074/jbc.C100152200

Wang D, Yang S, Wang G, Gao L, Wang Y, Jiang Q, Chen Y (2016) Residues and distributions of Organochlorine pesticides in China's Weihe River. Pol J Environ Stud 25(3):1285–1292

Wright SF, Upadhyaya A (1998) A survey of soils for aggregate stability and glomalin, a glycoprotein produced by hyphae of arbuscular mycorrhizal fungi. Plant Soil 198:97–107

Yadav IC, Devi NL, Syed JH et al (2015) Current status of persistent organic pesticides residues in air, water, and soil, and their possible effect on neighboring countries: a comprehensive review of India. Sci Total Environ 511:123–137. https://doi.org/10.1016/j.scitotenv.2014.12.041

Zhang W, Tan GJ, Li SFY (2014) NMR-based metabolomics and LC-MS/MS quantification reveal metal-specific tolerance and redox homeostasis in Chlorella vulgaris. Mol Biosyst 10:149–160. https://doi.org/10.1039/c3mb70425d

Reuse of Water: An Integral Approach for Survival

Komal Agrawal and Pradeep Verma

Abstract

Reutilization of water has become an integral part of the current decade globally. Various methodologies and techniques have been developed for the treatment of wastewater followed by its reutilization in various segments. Polluted water has tremendous hazards associated with it such as the loss of flora and fauna and health-related issues in the areas where humans come in direct contact with the pollutant, e.g., near water bodies where the majority of industrial, domestic, and agricultural waste are discharged without its prior treatment. Thus, the treatment of water prior to its release should be mandatory which will eventually result in its reutilization. The reuse of water will help treat the polluted water, thereby reducing pollution and simultaneously the load on environment as well. The main problem which is faced for the reuse of treated water is the public perception which poses a major challenge for its effective reutilization. Thus, the present chapter would focus on various hazardous and biological pollutants, technologies involved in its treatment, areas where the water is reused followed by the limitation which are prevalent in the technical sectors of treatment, as well as its future prospect.

Keywords

Wastewater · Reuse · Hazard · Treatment · Global scenario

K. Agrawal · P. Verma (✉)
Bioprocess and Bioenergy Laboratory, Department of Microbiology, Central University of Rajasthan, Ajmer, Rajasthan, India
e-mail: pradeepverma@curaj.ac.in

M. P. Shah (ed.), *Removal of Emerging Contaminants Through Microbial Processes*,
https://doi.org/10.1007/978-981-15-5901-3_24

1 Introduction

Water is widely used in various types of economic activity but mostly is used for consumption purposes from various sources among which surface water is utilized after treatment that ensure the reduction of pollutants (Pawełczyk 2012). Despite of so many sources of water, there is still shortage of soft water in the world due to population explosion and excessive utility in industries. The physicochemical aspect of any form of water body plays a very essential role in maintaining the fragile ecosystem which eventually maintains several life forms globally. Various sources of water are polluted due to multiple factors which include contamination by pathogenic agents, chemicals, heavy metals, pesticides, and water disinfectants. In addition, industrial and agricultural activities, leaching from soil and rocks, atmospheric deposition, and other human activities have become a potent problem, contaminating all the basic element of nature necessary for the survival of life entities.

More than one billion people all over the world do not have ready accessibility to enough and safe water supply. According to the World Health Report (WHR-1998), water supply differs widely in terms of regional aspect. In the 1970s, of the approximately 2.5 billion populations in the developing world, only 38% had safe drinking water. At the beginning of the 1980s, water supply was 75% in urban areas and 46% in rural areas (Kumar and Puri 2012). In developing countries, only 75% of the population has access to water supply as a result of which there is always a risk of deadly diseases. In modern era there is a large decline in freshwater due to population increase, urbanization, industrialization, and concentrated agricultural sector. The limited availability of surface water makes people dependent on groundwater resources to complete their needs. A number of towns and cities fulfill their demand from groundwater and surface water through municipality and also from several concealed boreholes. Regular water quality monitoring of the water resources is very necessary to calculate the quality of water for ecosystem health and hygiene, industrial use, agricultural use, and domestic use (Kumar and Puri 2012; Pawełczyk 2012). The social, economic, and environmental aspects of historic water resource development practices and the inevitable prospects of water scarcity are taking the shift to a new paradigm in water resource management. The new approach should be applied based on the principles of sustainability, ethics of the environment, and public participation. Sustainable water resource management tries to make integrated and adaptable systems, increasing efficiency of water use and making regular efforts toward protecting ecosystems (Baron et al. 2002). Thus, the present ster deals with various hazardous pollutants, microbial pollutant, and its treatment for the effective reutilization of water.

2 Hazardous Pollutants

The various chemicals which contaminate in the water bodies and have imposed threat to the flora and fauna both on land and in water are as follows (Fig. 1).

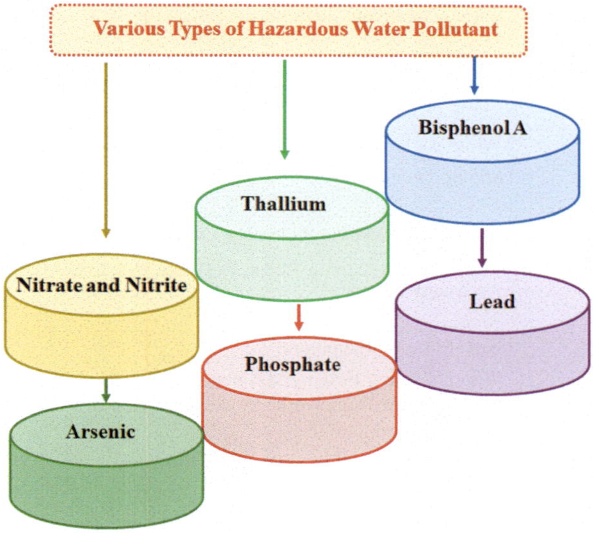

Fig. 1 Various types of hazardous pollutant responsible for water contamination

2.1 Bisphenol A (BPA)

Bisphenol A (BPA) is the common name for 2,2-(4,4′-dihydroxydiphenyl) propane, 4,4′-isopropyllidenediphenol, or 2,2′-bis(4-hydroxyphenyl) propane, an organic compound with two phenol moieties. Its important properties comprise of low vapor pressure, mild water solubility, and low volatility and assume solid state at room temperature (Tsai 2006).

Wastewater from an industrial park situated in southern Taiwan contains high level (16 μg/L) of BPA, which was 16 times higher than the global BPA level in environmental waters. In China it has been proved that BPA can cause oxidative stress on human embryo liver L-02 cells by damaging DNA, although vitamin C (an antioxidant) can decrease DNA damage. Many animal studies showed that BPA exposure reduces sperm production (Zhang et al. 2005), enhances the stillbirth rate (Wang et al. 2005), affects embryonic development (Pei et al. 2003), and diminishes the viability of mesencephalic neuronal cell (Lin et al. 2006) (Table 1).

2.2 Thallium

Thallium, while being a highly toxic element, has been studied to a comparatively less than other toxic elements such as lead, cadmium, or mercury. One of main reason could be that the classical analytical methods have poor sensitivity to thallium, and secondly as it has already been stated as being highly toxic, further studies were not performed (Peter and Viraraghavan 2005). It is suggested that thallium is mobilized in waters in the mining districts usually as sulfates, arsenates, and chloride complexes. Its distribution in the world oceans therefore looks similar

Table 1 Various treatment methods employed for the removal of bisphenol A

S. no.	Treatment process	Reference
1.	Thermally activated persulfate oxidation process	Olmez-Hanci et al. (2013)
2.	White-rot basidiomycetes, *Pleurotus ostreatus*	Hirano et al. (2000)
3.	*Sphingomonas bisphenolicum* AO1	Matsumura et al. (2015)
4.	Immobilized marine microalga *Pavlova* sp.	Shimoda and Hamada (2009)
5.	Activated carbon	Bautista-Toledo et al. (2005)
6.	Nanofiltration membrane	Zhang et al. (2006)
7.	Removal using $FexCo_3$-xO_4 nanocages derived from nanoscale metal-organic frameworks by the activation of peroxymonosulfate	Li et al. (2016)
8.	Oxidative removal by manganese dioxide	Lin et al. (2009)
9.	Membrane bioreactor	Chen et al. (2008)
10.	UV-C/peroxymonosulfate (PMS)	Sharma et al. (2015)

Table 2 Various treatment methods employed for the removal of thallium

S. no.	Treatment process	Reference
1.	*Aspergillus niger* biomass	Peter and Viraraghavan (2008)
2.	Modified sugar beet pulp	Zolgharnein et al. (2011)
3.	Fe-Mn binary oxides	Li et al. (2017)
4.	Iron metal and hydrogen peroxide	Kikuchi et al. (1990)
5.	Combination of persulfate oxidation and iron coagulation	Li et al. (2018)
6.	Eucalyptus leaves powders	Khavidaki et al. (2013)

to that of alkali metals, which are its principal biogeochemical entity (Zhang et al. 1998). Concentrations of thallium in river water and groundwater of 0.04 and 800 Ag/L, respectively, have been reported (Delvalls et al. 1999). Thallium was detected in natural waters (including sea water) by first pre-concentrating it by adsorption from oxidizing medium onto a strongly basic anion exchanger as the tetrachlorothallate (3+) ion (Riley and Siddiqui 1986). The other method used for the detection of thallium includes selective separation and purification which enables detection even at low concentrations. Thus, techniques have to be developed for the detection of thallium for its effective treatment (Table 2).

2.3 Nitrate and Nitrite

Nitrate and nitrite are naturally occurring ions that are components of nitrogen cycle; however, when the nitrate level in drinking water exceeds 50 mg/L, drinking water becomes the main source of concern. High level of nitrate in drinking water can be caused due to excessive use of agricultural fertilizers, decayed vegetable water, domestic effluent, sewage disposal, and industrial discharges, leachable from refuse dumps and atmospheric and atmospheric precipitation (Makhijani and Manoharan 1999). Nitrites on the other hand cause serious condition in fish called "brown blood disease." Nitrites also combine directly with hemoglobin in human blood and other warm-blooded animals to make methemoglobin. Methemoglobin restricts the ability of red blood cell to transport oxygen. This condition is especially significant in babies under 3 months of age. It causes a disease known as methemoglobinemia or "blue baby disease" (Kumar and Puri 2012).

2.4 Lead

Lead compounds has its use in plumbing fittings and as solder in water distribution systems. Lead is found in tap water to some extent as a result of its dissolution from natural sources but primarily from household plumbing systems, e.g., pipe, solder, and fittings. PVC pipes also have lead compounds and result in high concentration of lead in drinking water. The general forms of Pb, i.e., ionic lead, Pb(II), lead oxides and hydroxides, and lead metal oxyanion complexes, are released in the land and water systems of which two forms, i.e., Pb(II) and lead-hydroxy complexes, are stable forms. Lead (II) forms mono- and polynuclear oxides and hydroxides which are the most reactive forms of Pb. According to India's Bureau of Indian Standards (1991) for standard drinking water specification, the highest desirable amount of lead in drinking water was 0.05 ppm (parts per million), and no relaxation for maximum permissible limit for the element (Kumar and Puri 2012). Lead exposure has enormous biological implications being directly proportional to the concentration and duration of exposure (Table 3).

Table 3 Various treatment methods employed for the removal of lead

S. no.	Treatment process	Reference
1.	Biochars prepared from hydrothermal liquefaction of biomass	Liu and Zhang (2009)
2.	Granulated blast furnace slag	Dimitrova and Mehandgiev (1998)
3.	Granular slag columns	Dimitrova (2002)
4.	Carbon nanotubes	Li et al. (2006a, b)
5.	Fixed-bed columns by zeolite and sepiolite	Turan et al. (2005)

2.5 Phosphate

Phosphates in soil are contributed from the wash of farm soils which enters the nearby water bodies. In required concentration phosphate is necessary for the survival of plankton and aquatic plants which supply food for fish. However, large quantities of phosphate in water bodies will enable the growth of algal and aquatic plants randomly, thereby clogging the water way and use up large amounts of oxygen. Digestive problems could be there from extremely high level of phosphate (Kumar and Puri 2012) (Table 4).

2.6 Arsenic

Arsenic contamination in drinking water has been reported in different parts of the world and prominently in China (WHO 1996). In India, it had been found to be abundant in different region of the West Bengal due to dissolution of arsenic containing bed rocks. The WHO has prescribed a provisional guideline value of 10 μg/L in potable water, and according to Indian standard drinking water specification (1991), the maximum considerable limit of 50 μg/L has no relaxation for maximum permissible limit. Chronic exposure due to arsenic-contaminated drinking water identifies dermal lesions, peripheral neuropathy, skin cancer, and peripheral vascular disease (Saha et al. 1999) (Table 5).

Table 4 Various treatment methods employed for the removal of phosphate

S. no.	Treatment process	Reference
1.	Red mud	Huang et al. (2008)
2.	Electrocoagulation	Bektaş et al. (2004)
3.	Fe-Mn binary oxide adsorbent	Zhang et al. 2009
4.	Blast furnace slags and opoka-mechanisms	Johansson and Gustafsson (2000)
5.	Raw and activated red mud and fly ash	Li et al. (2006b)

Table 5 Various treatment methods employed for the removal of arsenic

S. no.	Treatment process	Reference
1.	Electrocoagulation	Kumar et al. (2004)
2.	Reverse osmosis	Ning (2002)
3.	Nanocrystalline magnetite	Mayo et al. (2007)
4.	Iron-modified activated carbon	Chen et al. (2007)
5.	Polymeric/inorganic hybrid sorbent	DeMarco et al. (2003)

3 Microbiological Hazard

Fecal coliform bacteria are a collection of comparatively harmless microorganisms that live in large quantity in the intestines of the warm- and cold-blooded animals and support in the digestion of food. There is a specific subgroup of this collection: the fecal coliform bacteria, the most common and known member being *Escherichia coli*. The presence of fecal coliform bacteria in aquatic environmental shows that the water has been polluted with the fecal material of man or other animals. The source water may have been contaminated by pathogens or disease-producing bacteria or viruses which are also there in fecal material. Some waterborne pathogenic diseases are typhoid fever, viral and bacterial gastroenteritis, and hepatitis A. The presence of fecal contamination is an indicator that a potential health danger is there for individuals exposed to water (Kumar and Puri 2012). In developing countries, untreated or incompletely treated municipal wastewater remains a risk for public health. The potential spread of infectious disease by pathogenic organisms is the most common issue in water reclamation and reuse. The main infectious agents that may be found in untreated municipal wastewater can be grouped into three: bacteria, protozoa, and viruses (Fig. 2) (Metcalf and Eddy, Inc, et al. 2007).

3.1 Bacteria

The various bacteria which are known to hazardous biological pollutants and common are as follows.

3.1.1 *Shigella*

Shigella infects humans and primates and causes shigellosis bacillary dysentery. *S. sonnei* causes a number of waterborne infections, although all four subgroups (*S. dysenteriae*, *S. flexneri*, *S. boydii*, and *S. sonnei*) have been procured during various disease outbreaks (Moyer 1999; WHO 2005).

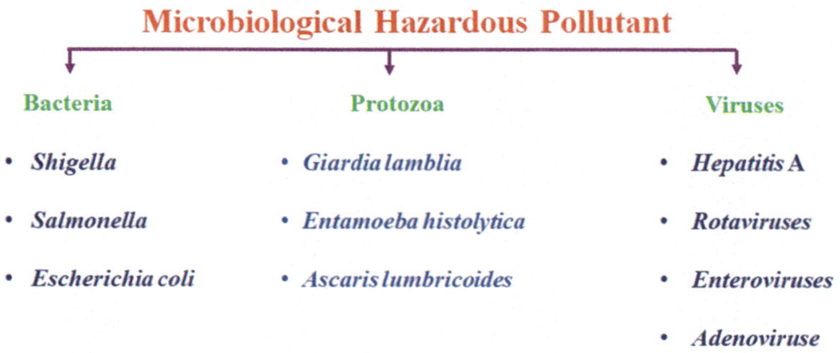

Fig. 2 The various microbiological hazardous pollutants responsible for water contamination

3.1.2 *Salmonella*

Over 2200 known serotypes of *Salmonella* exist in the environment, all of which are pathogenic to humans. They generally cause gastrointestinal illness; however, a few can cause other forms of disease, such as typhoid (*S. typhi*) and paratyphoid (*S. paratyphi*) fevers. The latter two species infect humans only, while the others are carried by both humans and animals (Metcalf and Eddy, Inc, et al. 2007; Chiu et al. 2004).

3.1.3 *Escherichia coli*

E. coli is a member of the fecal coliform group of bacteria usually found in the intestinal tracts of humans and warm-blooded animals and is generally harmless. This organism in water indicates fecal contamination. Some strains of *E. coli* are pathogenic and cause gastroenteritis. A particular strain, *E. coli* O157:H7 (Perna et al. 2001), causes acute bloody diarrhea and abdominal cramps (enterohemorrhagic) and in some situations have resulted in hemolytic uremic syndrome (HUS), in which red blood cells are destructed and the kidneys fail. One of the highest death rates of all waterborne diseases is due to HUS (Hunter 2003; Noris and Remuzzi 2009).

3.2 Protozoa

Protozoa are generally present in fresh and marine water, and some can grow in soil and other locations. The enteric protozoan parasites generate cysts or oocysts that help in their survival in wastewater and under unfavorable conditions in the aquatic environment. Common pathogenic protozoa include *Giardia lamblia*, *Cryptosporidium parvum*, and *Entamoeba histolytica* (Khalifa et al. 2001; Adam 2001; Abrahamsen et al. 2004; Loftus et al. 2005).

3.2.1 *Giardia lamblia*

Waterborne giardiasis, caused by the protozoan *G. lamblia*, is identified as the most common protozoan infection in the United States and exists a major public health concern. The reported evidence of waterborne giardiasis, a gastrointestinal disease established by diarrhea, fatigue, and cramps, has increased in the United States since 1971 (Adam 2001).

3.2.2 *Entamoeba histolytica*

E. histolytica can cause amoebic dysentery, with symptoms ranging from acute bloody diarrhea and fever to mild gastrointestinal illness. The organism can cause ulcers sometimes and then invade the bloodstream, causing more serious effects (Haque et al. 1995).

3.2.3 *Ascaris lumbricoides*

The infectious disease established by *A. lumbricoides* (an intestinal roundworm) is known as ascariasis. In its mild form ascariasis is characterized by digestive and nutritional problems, abdominal pain, vomiting, and the presence of live worms in stools or vomit. Serious cases involving the liver can cause death also (O'lorcain and Holland 2000).

3.3 Viruses

Viruses are composed of a nucleic acid core and are covered by an outer coat of protein. Enteric viruses obligate human pathogens, which mean they multiply only in the human host. Their simple structure, a protein coat surrounding a core of genetic material (DNA or RNA), provides them a better survival in the environment. Some common enteric viruses can cause potential waterborne diseases which are discussed below (Cohn et al. 1999).

3.3.1 Hepatitis A

Although all enteric viruses are potentially spread by drinking water, evidence of this route of infection is prominent for hepatitis A virus (HAV). The HAV causes infectious hepatitis, an illness signified by inflammation and necrosis of the liver. Symptoms are fever, weakness, nausea, vomiting, diarrhea, and occasionally jaundice (Nainan et al. 2006).

3.3.2 Rotaviruses

Rotaviruses cause acute gastroenteritis, majorly in children. Almost all children have been infected at least once by the age of 5 years. In developing countries, rotavirus infections are a significant cause of infant mortality. Rotaviruses are transmitted by fecal-oral transmission and have been identified in municipal wastewater, lakes, rivers, and even tap water (Odimayo 2005).

3.3.3 Enteroviruses

The enteroviruses include groups like polioviruses, coxsackieviruses, and echoviruses. Enteroviruses are present in wastewater and surface water and in drinking water occasionally. Vaccination with oral poliovirus vaccine (OPV) was not continued in the United States since 2000. In 2005, however, four unvaccinated children in Minnesota were found infected by poliovirus, raising concerns regarding the spread of poliovirus to other communities with low levels of vaccination comparatively and the chance for an outbreak in the United States (CDC 2005).

3.3.4 Adenoviruses

Various types of adenoviruses are reasonable for upper respiratory illness, including the common cold. However, all types may be present in feces and may be spread by the fecal-oral route. Although adenoviruses have been found in wastewater, surface water, and drinking water, data on their presence in water are limited. Adenoviruses are relatively resistant to disinfectants and may not be inactivated or killed by traditional treatment strategies (Ginsberg 2013).

4 Treatment of Polluted Water

A number of methods are utilized for the removal of various pollutants from contaminated water which includes variations of the activated sludge process for denitrification and submerged attached growth processes (Pedros et al. 2008; Gerardi 2003). On the other hand, certain phosphorus accumulating bacteria can be utilized to release phosphorus under specific redox conditions as component of the activated sludge process. After the phosphorus uptake has occurred, these organisms must be removed from the process to accomplish phosphorus removal. Phosphorus removal can be done by the addition of salts of the multivalent metal ions that produce precipitates of readily soluble phosphates. The general processes by which phosphorus can be removed may be classified as precipitation, coprecipitation, and postprecipitation (Metcalf and Eddy, Inc, et al. 2007). Thus, it has been observed that various treatment methods are used for the removal of various pollutants and are described as follows.

4.1 Depth Filtration

Depth filtration was created originally for the treatment of surface water for potable uses and adapted for wastewater treatment applications later. Depth filtration is utilized in reuse applications to provide supplemental removal of suspended solids (including particulate BOD) from wastewater effluents for more effective disinfection, as a pretreatment procedure for subsequent treatment steps such as carbon adsorption, membrane filtration, or advanced oxidation and to remove chemically precipitated phosphorus (Metcalf and Eddy, Inc, et al. 2007).

4.2 Surface Filtration

Surface filtration is used to release the residual suspended solids from secondary effluents and stabilization of pond effluents and is being used as an option to depth filtration as pretreatment for membrane filtration. Surface filtration, a new technology, includes a sieving action similar to a kitchen colander (Metcalf and Eddy, Inc, et al. 2007).

4.3 Membrane Filtration

Membrane filtration with (membrane filter) MF and (ultrafiltration) UF membranes is being used for water and wastewater applications. Microfiltration and UF membrane filters are also surface filtration devices but are differentiated on the criteria of the sizes of the pores in the filter medium; the pore size can differ from 0.005 to 2.0 mm. In water reuse applications, MF and UF usually assume biological treatment and are used to filter particulate matter (including pathogens), organic matter, and some nutrients not filtered by secondary clarification. Product water from MF and UF may be used (Metcalf and Eddy, Inc, et al. 2007).

4.4 Nanofiltration

It used to filter particles in the 300–1000 molecular weight (MW) range, removing selected salts and most organics and microorganisms, operating at greater recovery rates and at lower pressures than RO systems. Even though most inorganic and organic components and microorganisms are removed, disinfection is needed to ensure system reliability in the event of a defect in the membrane. Nanofiltration is used where the high salt rejection of RO is not important although NF has an ability of removing hardness elements in water such as calcium and magnesium. Nanofiltration membranes are sometimes used for removing salts to meet TDS requirements for groundwater recharge and for softening of water (Metcalf and Eddy, Inc, et al. 2007).

4.5 Dissolved Air Flotation (DAF)

Flotation is a gravity separation process in which gas bubbles cling to solid particles to cause the density of the bubble-solid agglomerates to be lighter compared to water. In DAF, bubbles are generated by the reduction of pressure in a water stream saturated with air, similar to the bubble production in a carbonated beverage when the top is opened. Dissolved air flotation has been utilized in water treatment as an option for the sedimentation and treatment of nutrient-rich water bodies having heavy algae blooms and for low alkalinity (AWWA 1999).

4.6 Reverse Osmosis

Reverse osmosis has numerous applications, e.g., desalination of sea and brackish water and industrial process water for removing the dissolved constituents. It effectively removes 95–99.5% of the TDS and 95–97% of dissolved organic matter. It is also used to remove dissolved materials, commonly salts, under pressures having a range from 1200 to 1800 kPa for low TDS water (1000–2500 mg/L) to 5500–8500 kPa for seawater and at flux rates ranging from about 12 to 200 L/m^2 per hour. Generally, the ranges and mean geometric standard deviations for the RO process are higher than those for the other membrane processes. Although several different statistical methods are present for dealing with undetected values, it is common practice, for the purpose of data analysis, to assign a value equal to one-half of the detection limit for the test to a non-detect value (Metcalf and Eddy, Inc, et al. 2007).

4.7 Adsorption

It is used for the separation of substances that are in solution by adsorbing them on a solid phase. Adsorption is assumed to be a mass transfer operation as a component is transferred from a liquid phase to a solid phase. The *adsorbate* is the substance that is being removed from the liquid or gas phase at the interface. The *adsorbent* is the solid, liquid, or gas phase onto which the adsorbate sticks. Although adsorption is used at the air-liquid interface in the flotation method, only the situation of adsorption at the liquid-solid interface is assumed in this section. Activated carbon is the primary adsorbent used in adsorption processes. Adsorption treatment of reclaimed water is generally considered as a polishing process for water that has already received usual biological treatment. Adsorption has been used for the removal of refractory organic components, residual inorganic constituents such as nitrogen, sulfides, and heavy metals. Under optimum conditions, adsorption can be used to remove the effluent COD to less than 10 mg/L (Metcalf and Eddy, Inc, et al. 2007)

4.8 Ion Exchange

This involves exchange of an ion in the aqueous phase for an ion in a solid phase. The solid phase ion replacement material is insoluble and can be natural such as kaolinite and montmorillonite minerals or a manmade material such as a polymeric resin. The exchange materials have fixed charged functional entities located on their external and/or internal surface, and aligned with these groups are ions of opposite charge called "counter ions." The mobile counter ions are attached by electrostatic attraction to each of the charged functional groups to complete the criterion that electroneutrality is maintained at all times within the replacement material as well as in the bulk aqueous solution. Based on the charge of the functional group on exchanger, the counter ion can either be a cation if the functional entity is negative or an anion if the functional entity is positive and can exchange with another counter ion in the aqueous phase (Metcalf and Eddy, Inc, et al. 2007).

4.9 Chemical Oxidation

It is used for several applications in water reclamation, involving the degradation of anthropogenic or synthetic toxic organics. Process in which oxidizing chemicals are put into water that directly reacts with the constituents in water is called conventional oxidation processes. Process in which oxidizing chemicals are added to water to aid the formation of hydroxyl radicals which then react with the components in water is known as advanced oxidation processes (AOPs). A dot is put after the radical species to show that there is a single unpaired electron in the outer orbital. The principal applications of conventional chemical oxidation in water reutilization are for (1) odor control, (2) hydrogen sulfide control, (3) color removal, (4) iron and manganese removal, (5) disinfection, (6) control of biofilm formation and biofouling in treatment operations and distribution system components, and (7) oxidation of certain trace organic constituents. Oxidants that are usually used in water reclamation are (1) hydroxyl radical, (2) chlorine, (3) ozone, (4) chlorine dioxide, (5) permanganate, and (6) hydrogen peroxide (Metcalf and Eddy, Inc, et al. 2007).

4.10 Photolysis

It is a process by which components are broken down by exposure and absorption of photons/light from a light source. In nature, sunlight is the source of light for photolysis reactions; however, in engineered systems, UV lamps are used to generate the photonic energy. The photons that are absorbed cause the electrons in the outer orbital of some compounds to become unstable and separate or become reactive. The effectiveness of the photolysis process is based on the characteristics of the reclaimed water, structure of the components, structure of the photolysis reactor, and amount and wavelength of the applied light. The photolysis rate can be calculated from the rate at which the compound absorbs light and the photonic capability of the reaction (quantum yield) (Metcalf and Eddy, Inc, et al. 2007).

4.11 Chemical Agents

Chlorine and its derivative and ozone are the principal chemical compounds used for the disinfection of water under treatment. Other chemical agents that have been utilized as disinfectants in various applications include (1) bromine, (2) iodine, (3) phenol and phenolic compounds, (4) alcohols, and (5) heavy metals. Disinfection with chemical agents is achieved by mixing completely the diluted disinfecting agent with the liquid to be disinfected and allowing enough time for the disinfectant to react with the microorganisms that may be there in the liquid (Metcalf and Eddy, Inc, et al. 2007)

4.12 Radiation

The prominent types of radiation are electromagnetic, acoustic, and particle. For example, the degradation of microorganisms is there in oxidation ponds to their exposure to the ultraviolet (UV) light component of sunlight of the electromagnetic spectrum. Special lamps developed to emit UV light have been applied succzessfully to disinfect treated water. Disinfection with UV light is done by exposing the microbes in the liquid to UV light (Metcalf and Eddy, Inc, et al. 2007).

4.13 Membrane Modules

Membrane modules have varied from being utilized solely in tertiary wastewater treatment to being implemented into secondary wastewater treatment. These systems are generally referred as membrane bioreactors (MBRs). There are a number of advantages associated with the MBR which make it a good alternative over other treatment techniques. The potential for using the MBR at very high solid retention times without having the problem of settling allows high biomass concentrations in the bioreactor. MBR systems were primarily used for municipal wastewater treatment, initially in the area of water reuse and recycling. Compactness, generation of reusable water, and smooth operation made the MBR an ideal process for recycling municipal wastewater space limited environments. High organic matter load and difficult to treat compounds are two major characteristics of industrial waste streams that require the use of alternative treatment techniques such as the MBR desirable. Since traditionally wastewater with high COD content was purified under anaerobic conditions, initial cases of MBR applications for industrial wastewater were in the field of anaerobic treatment (Cicek 2003).

4.14 Fenton's Reagent

The Fenton's chemistry started at the end of the nineteenth century, when Fenton published, in a pioneering work, a detailed study on the application of a mixture of H_2O_2 and Fe^{2+} (later called the Fenton's reagent) for the oxidation and degradation of tartaric acid (Fenton 1894). Basically, the Fenton process contains several important advantages for water/wastewater treatment (Bautista et al. 2008): (1) a simple and flexible process permitting easy implementation in existing plants and (2) easy and relatively inexpensive chemicals. Nevertheless, the following drawbacks have been also noted (Tarr 2003): (1) rather high cost and risks due to the storage and transportation of hydrogen peroxide and (2) need of large quantity of chemicals for acidifying effluents at pH 2–4 before decontamination and/or for neutralizing treated solutions before discarding (Gadipelly et al. 2014).

4.15 Biosorption

It is rather hard to define because many mechanisms may aid to the overall process depending on the substance to be sorbed, the biosorbent used, environmental factors, and the presence or absence of metabolic processes in the case of living organisms. Biosorption of the base metal cations is depending on pH and usually takes place in the range pH 3–7. Biosorptive processes may be a part of varying importance in any form of primary or secondary biological treatment operations for aqueous waters and process streams including domestic, municipal, and industrial wastes and, in some cases, solid wastes. In biological treatment processes, a shear advantage is that metabolic activities aid in the degradation of organic materials. Some examples have been mentioned where organic and inorganic transformations are closely linked to biosorptive separation, e.g., rotating biological contactors to treat dilute metal-containing mine waste streams (Gadd 2009).

5 Reutilization of Water

Given the huge number of treatment technologies applied in the preceding sections, it can be said that decentralized systems can be used to virtualize any water reuse design. The most common decentralized water reutilization projects include systems for landscape irrigation with drip emitters, groundwater recharge, non-potable indoor reutilization, and habitat development. Most decentralized wastewater treatment systems make use of the local soil for spread of effluent. These soil infiltration systems are not produced usually to make use of plant uptake of water or nutrients found in wastewater. Therefore, the arrival of wastewater to groundwater and plant uptake of wastewater in soil infiltration systems may be assumed unplanned water reuse. The application of planned decentralized water reuse systems for the irrigation of landscapes is used in many locations because it decreases water demands and is considered to be safe, and the point of reuse is sometimes near to the point of reclaimed water generation. Decentralized wastewater systems designed for water reuse by landscape irrigation are utilized to apply reclaimed water and nutrients at rates enough for plant uptake and typically make use of shallow effluent distribution methods, such as drip irrigation. Generally, wastewater from an apartment building, commercial facility, or a cluster of homes remains in a large watertight septic tank or other solids separation unit, and all or a part of the flow may be used for tree and median strip watering by applying subsurface drip irrigation. Because the nutrients in the wastewater are useful, the effluent from the septic tank is used after it has passed an effluent filter to remove coarse solid particles larger than about 2–3 mm (Metcalf and Eddy, Inc, et al. 2007). The wastewater produced from bathing and washing facilities that does not contain concentrated human waste or food waste is commonly known as grey water. In some cases, grey water is applied for subsurface irrigation without treatment, however, for some reuse applications. Grey water may need to be treated before use. The treatment systems utilized for grey water include biological treatment, particle removal, and disinfection. If untreated grey water is stored in a holding tank for any length of time, anaerobic conditions may develop, similar to what happens in a septic tank. In some cases, grey water can be supplied without treatment for tree irrigation using subsurface dispersal systems (Ludwig 1997). The concentration of pathogens, organic matter, trace constituents, and nutrients are assumed to be significantly lower than those there in combined wastewater; however, the amount of sodium and other minerals may be increased depending on the type and nature of detergents and cleaning agents used in the home, as well as the presence of a salt-based water softener (Metcalf and Eddy, Inc, et al. 2007). Wastewater treatment processes that utilize plants and other ecological components are commonly referred as natural treatment systems. Natural treatment systems may enhance the growth of beneficial plants and trees that are used as sanctuary for several organisms. In addition, reduced accessibility to wastewater treatment and reuse areas decreases disturbance to wildlife in these parts. Because natural processes typically depend on solar energy and gravity flow, the cost of running the operation and maintenance may be significantly less than a high rate process, but the treatment processes runs at much lower rates and are more sensitive to environmental and seasonal changes.

5.1 Reutilization for Agricultural Applications

Agricultural irrigation evolved from the initial practice of sewage farming when untreated municipal wastewater (sewage) was directly added to crops. In some parts of the world that practice is still applied, in spite of unfavorable health and environmental impacts. Agricultural irrigation is the largest sector of reclaimed water utilized in the United States and in most of the world. Depending on the application of the crop for food and other purposes and how the irrigation water is to be applied, i.e., by ridge and furrows, sprinklers, or surface/ subsurface drip irrigation, the degree of the treatment required will differ. In almost all irrigation applications in the United States, a threshold of secondary treatment is required. For spray application to food crops, if needed, higher levels of treatment including disinfection are necessary. Other assumptions in agricultural irrigation include the type of crop; topography and soil properties; effect of water quality, particularly dissolved solids, on soils, crops, and the groundwater; runoff and drainage water management; and infrastructure requirements. It is after the application of these techniques that the water is reused, for agricultural purposes.

5.2 Reutilization for Landscape Irrigation

Landscape irrigation is the second largest consumer of reclaimed water in the United States and is being applied increasingly for several locations involving golf courses, parks, residential areas, roadway medians and roadside plantings, and cemeteries. Because public contact with the used water is considered as a potential health hazard, reclaimed water has to have higher water quality levels for suspended solids and microbial contamination, as compared to some agricultural applications. Many of the other physical and chemical properties of the reclaimed water in landscape applications are similar to those for agricultural process. In irrigation system processes, considerations include reducing the formation and spread of aerosols, managing application rates to stop ponding and runoff and controlling chlorine residuals to maintain disinfection (Niu and Cabrera 2010).

5.3 Reutilization for Industrial Uses

The major industrial users of treated water are power plants, oil refineries, and manufacturing facilities where water is needed principally for cooling purposes. Water quality, particularly total dissolved solids, chlorides, and dissolved oxygen, is an important issue because of potential scaling or corrosion in piping systems and heat exchangers. Residual organic matter may also aid to biological growths in heat exchangers and cooling towers. Additional treatment may be important at the point of use based on the water quality requirements for a particular industrial process. Other important consideration involves matching demand with supply, system reliability, and disposal of cooling tower blow down (Metcalf and Eddy, Inc, et al. 2007).

5.4 Reutilization for Urban Non-irrigation

Urban non-irrigation uses cover a large variety of applications involving air conditioning cooling water, fire protection, toilet and urinal flushing, ornamental water features, and maintenance. Commercial utilization of reclaimed water such as a car washing and commercial laundries is practiced usually in urban areas, and they are assumed as part of urban non-irrigation. Urban use is limited commonly to high-density development such as office buildings and apartments where there is economic advantage for installing a dual distribution system, one system for potable water, and one for treated water. Water required for most urban non-irrigation water reuse applications is small, and usually, multiple water reuse applications are involved including landscape irrigation. The principal issues with dual distribution systems are the large cost of infrastructure and the limitation of cross-connection between the two water supplies. High-quality and well-disinfected treated water must also be used to ensure public health protection (Eslamian et al. 2010).

6 Overview of Wastewater Reuse

Reusing wastewater is executed to take advantage of water resource in other sectors than those from which it was obtained but by previously subjecting it to a process of purification. In Latin America, about 400 $m^3\ s^{-1}$ of wastewater are released into surface water sources; for example, in Colombia, 1,230,193 ha are irrigated with wastewater, of which 27% is reutilized treated wastewater and 73% untreated but usually added with surface water (Silva et al. 2008). However, there is no proper and reliable information about wastewater reutilization in Latin America, and only 8% of the total wastewater that is produced daily is treated. Thus, irrigation of agricultural areas is done usually with untreated wastewater. Mexico produces more than half of the wastewater of all Latin America (Post 2006). The irrigation infrastructure in Mexico is one of the largest irrigation infrastructures in the world, covering 6.46 million ha, but water availability is decreasing under due to increasing population.

The distribution of water relies on the demand from the industrial, agricultural, and municipal sectors, since there is the need to generate food and provide basic services to raising human population (Elizondo et al. 2004). Conagua (2009) reported that, in 2007, Mexico produced 243 $m^3 s^{-1}$ of urban wastewater, of which 32.6% was treated. However, only 0.53 km^3 were treated completely, and at least 6.8 km of untreated wastewater were released into the environment. In 2009, the 2029 water treatment plants used in the country treated 42% of wastewater, which is equivalent to 88.1 $m^3 s^{-1}$ of the total of 209.1 $m^3 s^{-1}$ of the wastewater operated in sewer systems (Mejía et al. 2012). In 2011, 36% of all municipal wastewater was processed, removing 1.34 million tons of MO, which mentions that the production of polluting matter will always be greater than the capacity to separate it, even if there was a notable rise in the number of the water treatment plants, which could be 2709 by the year 2030, but the volume of wastewater treated would usually be well below the total volume produced (Sina 2012). The agricultural sector needs the largest volume of water, ranging from 76.7% to 80% of the total water available; however, it only requires 46%, meaning that something less than half of the water is utilized for crop production. In contrast, the allotment of public water distribution is about 14% (Mejía et al. 2012). The reutilization of nutritional components and water from domestic treated wastewater to urban agriculture is often assumed as a potential food, energy, water, health (FEW-health) nexus strategy that may utilize energy and rise in GHG emissions while advancing water cycling and nutrient reutilization and also providing fresh/healthy foods (Hanjra et al. 2015; Makoni et al. 2016).

7 Limitations and Future Prospect

There are a number of advantages and disadvantages of irrigating urban farms with untreated or incompletely treated wastewater. The merits, particularly in developing countries, involve water conservation, nutrient recycling (Van der Hoek et al. 2002; Qadir et al. 2007), restricting fertilizer usage (Pitterle and Ramaswami 2009; Corominas et al. 2013), land treatment of wastewater (Raschid-Sally and Jayakody 2009), spatial and temporary access of water for irrigation, reduced need for expensive refrigerated mobilization or storage facilities enhanced nutrition for urban residents, and better livelihoods (Qadir et al. 2007; Kilelu 2004; Raschid-Sally and Jayakody 2009). The demerits include increased health risks and decreased environmental aspects as water, soil, and crops become increasingly contaminated with pathogens, metals, etc.; pathogens, e.g., bacteria, viruses, and protozoa, can induce acute health problems (Qadir et al. 2007, Ensink et al. 2008). Engineered physical and biochemical treatment operations can help to reduce pathogens, thereby mitigating some of the health dangers of urban wastewater agriculture. The construction and operation of these centralized wastewater treatment plants (WWTPs) apply energy and release associated greenhouse gases (GHGs), which are assumed as an investment that can mitigate the health problems associated with resource reuse when wastewater is processed and reused in agriculture. Indeed, this is often showed as a key chance when exploring the FEW-health nexus from a city perspective.

However, the interaction, trade-offs, and co-benefits of WWTP energy application and urban wastewater agriculture risks have not been calculated in the field to ensure these expectations, primarily because wastewater agriculture is being utilized in developing world cities where WWTPs are only now being established at an important scale. As more and more cities in the developing world apply WWTP infrastructure to reutilize sewage-polluted water, the trade-offs between water, energy, and health in the aspect of food-borne disease must be calculated at the intersection of urban agriculture and urban water systems (including groundwater, wastewater, and surface flows) (Miller-Robbie et al. 2017).

Process related to the urban FEW-health nexus draws upon several strands of literature, emerging at three major fronts. First, regarding life cycle assessment (LCA) analysis of WWTP technologies, many studies involves embodied energy and GHGs along with restricted fertilizer (Corominas et al. 2013), and many have been analyzed in developing cities (Friedrich et al. 2009; Zhang et al. 2010; Fine and Hadas 2012; Cornejo et al. 2014; Li et al. 2013). However, only a few involves the straight link with urban farming uses within the cities (Verbyla et al. 2013; Symonds et al. 2014). In particular, the bio-solid pathway does not assume the case in developing cities where untreated or treated wastewater is directly being applied in urban agriculture. There are no LCA studies of WWTPs in developing countries that involve a consequential study of the wastewater effluent above the WWTP fence line in the context of direct utilization of nutrient-rich effluent in urban agriculture. Beyond a traditional WWTP LCA, methane (CH_4) and nitrous oxide (N_2O) are released from wastewater effluent in streams and from farming, and the latter has not been modeled.

Second, concerning GHG leaching from agriculture, national research has been done; however, vegetable crops that are usual in urban agriculture are not involved in these studies. The DAYCENT model, applied in the study, calculates GHG emissions from cropped lands (EPA A 2011) but has not been used in urban agriculture crops (herbs, vegetables) grown with nutritional wastewater.

Third, there have been several studies in developed countries establishing general water reclamation standards, involving urban agriculture, and the general instructions in the United States are to generate levels similar to drinking water standards; in developing countries, this is an area where the studies are few and guidance is sought to involve the complex (Miller-Robbie et al. 2017).

Furthermore, there are few field studies analyzing pathogens on crops treated with irrigation water from various sources. For example, only one study has calculated bacteria content of the crop (peppers) following irrigation with treated water (Dagianta et al. 2014); however, the entire system was not analyzed, including agriculture with untreated wastewater or the system effects wherein the energy utilization and emissions from WWTP with agricultural impacts are related to the crop quality. Integrated field studies are required that connect systems study around WWTPs with actual field measurements and related health advantages of the resulting urban agriculture system. Recently, a few studies have investigated WWTP effluent water properties and its suitability for reutilization in agriculture (Norton-Brandão et al. 2013; Becerra-Castro et al. 2015; Bunani et al. 2015; Kihila et al. 2014; Abdul Mojid and Wyseure 2014; Myszograj et al. 2014; Quist-Jensen et al. 2015; Woltersdorf et al. 2015). However, implications from WWTP effluent reuse in agriculture were not fully studied in the aspect of GHG emissions from such agriculture or crop pathogen quality (Miller-Robbie et al. 2017).

References

Abdul Mojid M, Wyseure GC (2014) Fertility response of potato to municipal wastewater and inorganic fertilizers. J Plant Nutr 37(12):1997–2016

Abrahamsen MS, Templeton TJ, Enomoto S, Abrahante JE, Zhu G, Lancto CA, Deng M, Liu C, Widmer G, Tzipori S, Buck GA (2004) Complete genome sequence of the apicomplexan, Cryptosporidium parvum. Science 304(5669):441–445

Adam RD (2001) Biology of Giardia lamblia. Clin Microbiol Rev 14(3):447–475

AWWA (1999) Waterborne pathogens, AWWA M48. American Water Works Association, Denver

Baron JS, Poff NL, Angermeier PL, Dahm CN, Gleick PH, Hairston NG Jr, Jackson RB, Johnston CA, Richter BD, Steinman AD (2002) Meeting ecological and societal needs for freshwater. Ecol Appl 12(5):1247–1260

Bautista P, Mohedano AF, Casas JA, Zazo JA, Rodriguez JJ (2008) An overview of the application of Fenton oxidation to industrial wastewaters treatment. J Chem Technol Biotechnol 83 (10):1323–1338

Bautista-Toledo I, Ferro-Garcia MA, Rivera-Utrilla J, Moreno-Castilla C, Vegas Fernández FJ (2005) Bisphenol A removal from water by activated carbon. Effects of carbon characteristics and solution chemistry. Environ Sci Technol 39(16):6246–6250

Becerra-Castro C, Lopes AR, Vaz-Moreira I, Silva EF, Manaia CM, Nunes OC (2015) Wastewater reuse in irrigation: a microbiological perspective on implications in soil fertility and human and environmental health. Environ Int 75:117–135

Bektaş N, Akbulut H, Inan H, Dimoglo A (2004) Removal of phosphate from aqueous solutions by electro-coagulation. J Hazard Mater 106(2-3):101–105

Bunani S, Yörükoğlu E, Yüksel Ü, Kabay N, Yüksel M, Sert G (2015) Application of reverse osmosis for reuse of secondary treated urban wastewater in agricultural irrigation. Desalination 364:68–74

Centers for Disease Control and Prevention (CDC) (2005) Poliovirus infections in four unvaccinated children-Minnesota, August-October 2005. MMWR Morb Mortal Wkly Rep 54(41):1053

Chen W, Parette R, Zou J, Cannon FS, Dempsey BA (2007) Arsenic removal by iron-modified activated carbon. Water Res 41(9):1851–1858

Chen J, Huang X, Lee D (2008) Bisphenol A removal by a membrane bioreactor. Process Biochem 43(4):451–456

Chiu CH, Su LH, Chu C (2004) *Salmonella enterica* serotype Choleraesuis: epidemiology, pathogenesis, clinical disease, and treatment. Clin Microbiol Rev 17(2):311–322

Cicek N (2003) A review of membrane bioreactors and their potential application in the treatment of agricultural wastewater. Can Biosyst Eng 45:6–37

Cohn PD, Cox M, Berger PS (1999) Health and aesthetic aspects of water quality, water quality and treatment: a handbook of community water supplies, 2–7 and 2–8. Association, American Water Works

CONAGUA (2009) Situación del subsector agua potable, alcantarillado y saneamiento. – Comisión Nacional del Agua. Edición 2009, SEMARNAT, México, 167

Cornejo PK, Santana MV, Hokanson DR, Mihelcic JR, Zhang Q (2014) Carbon footprint of water reuse and desalination: a review of greenhouse gas emissions and estimation tools. J Water Reuse Desalin 4(4):238–252

Corominas L, Foley J, Guest JS, Hospido A, Larsen HF, Morera S, Shaw A (2013) Life cycle assessment applied to wastewater treatment: state of the art. Water Res 47(15):5480–5492

Dagianta E, Goumas D, Manios T, Tzortzakis N (2014) The use of treated wastewater and fertigation in greenhouse pepper crop as affecting growth and fruit quality. J Water Reuse Desalin 4(2):92–99

Delvalls TA, Saenz V, Arias AM, Blasco J (1999) Thallium in the marine environment: first ecotoxicological assessments in the guadalquivir estuary and its potential adverse effect on the doñana European natural reserve after the.... Cienc Mar 25(2):161–175

DeMarco MJ, SenGupta AK, Greenleaf JE (2003) Arsenic removal using a polymeric/inorganic hybrid sorbent. Water Res 37(1):164–176

Dimitrova SV (2002) Use of granular slag columns for lead removal. Water Res 36(16):4001–4008

Dimitrova SV, Mehandgiev DR (1998) Lead removal from aqueous solutions by granulated blast-furnace slag. Water Res 32(11):3289–3292

Elizondo AJL, Esquivel GG, Delgado WM, Cruz QR, Sánchez GG (2004) Prospectiva de la demanda de agua en México 2000-2030, Fundación Gonzalo Río Arronte, Fundación Javier Barros Sierra, 105

Ensink JH, Blumenthal UJ, Brooker S (2008) Wastewater quality and the risk of intestinal nematode infection in sewage farming families in Hyderabad, India. Am J Trop Med Hyg 79 (4):561–567

EPA A (2011) Inventory of US greenhouse gas emissions and sinks: 1990-2009. In: US Environmental Protection Agency. Washington, DC

Eslamian S, Tarkesh-Isfahani S, Malekpour I (2010) Investigating heavy metals concentration of a wastewater treatment plant for agricultural and landscape reuses. In: Dryland hydrology: global challenges local solutions, September, 1-4

Fenton HJH (1894) LXXIII.–Oxidation of tartaric acid in presence of iron. J Chem Soc Trans 65 (0):899–910

Fine P, Hadas E (2012) Options to reduce greenhouse gas emissions during wastewater treatment for agricultural use. Sci Total Environ 416:289–299

Friedrich E, Pillay S, Buckley CA (2009) Carbon footprint analysis for increasing water supply and sanitation in South Africa: a case study. J Clean Prod 17(1):1–12

Gadd GM (2009) Biosorption: critical review of scientific rationale, environmental importance and significance for pollution treatment. J Chem Technol Biotechnol 84(1):13–28

Gadipelly C, Pérez-González A, Yadav GD, Ortiz I, Ibáñez R, Rathod VK, Marathe KV (2014) Pharmaceutical Industry wastewater: review of the technologies for water treatment and reuse. Ind Eng Chem Res 53(29):11571–11592

Gerardi MH (2003) Nitrification and denitrification in the activated sludge process. Wiley, New York

Ginsberg HS (ed) (2013) The adenoviruses. Springer Science & Business Media, Berlin

Hanjra MA, Drechsel P, Wichelns D, Qadir M (2015) Transforming urban wastewater into an economic asset: opportunities and challenges. In: Wastewater. Springer, Dordrecht, pp 271–278

Haque R, Neville LM, Hahn P, Petri WA (1995) Rapid diagnosis of Entamoeba infection by using *Entamoeba* and *Entamoeba histolytica* stool antigen detection kits. J Clin Microbiol 33 (10):2558–2561

Hirano T, Honda Y, Watanabe T, Kuwahara M (2000) Degradation of bisphenol A by the lignin-degrading enzyme, manganese peroxidase, produced by the white-rot basidiomycete, *Pleurotus ostreatus*. Biosci Biotechnol Biochem 64(9):1958–1962

Huang W, Wang S, Zhu Z, Li L, Yao X, Rudolph V, Haghseresht F (2008) Phosphate removal from wastewater using red mud. J Hazard Mater 158(1):35–42

Hunter PR (2003) Drinking water and diarrhoeal disease due to Escherichia coli. J Water Health 1 (2):65–72

Johansson L, Gustafsson JP (2000) Phosphate removal using blast furnace slags and opoka-mechanisms. Water Res 34(1):259–265

Khalifa AM, El MT, Abou IEN (2001) Effect of ozone on the viability of some protozoa in drinking water. J Egypt Soc Parasitol 31(2):603–616

Khavidaki HD, Aghaie M, Shishehbore MR, Aghaie H (2013) Adsorptive removal of thallium (III) ions from aqueous solutions using eucalyptus leaves powders. Indian J Chem Technol 20 (6):380–384

Kihila J, Mtei KM, Njau KN (2014) Wastewater treatment for reuse in urban agriculture; the case of Moshi Municipality, Tanzania. Phys Chem Earth Parts A/B/C 72:104–110

Kikuchi E, Itoh K, Fujishima A, Yonezawa T, Kimura T (1990) Removal of thallium from waste water by using the iron metal and hydrogen peroxide. Chem Lett 19(2):253–254

Kilelu CW (2004) Wastewater irrigation, farmers' perceptions of health risks and institutional perspectives: a case study in Maili Saba, Nairobi. Cities feeding people series; rept. 38

Kumar M, Puri A (2012) A review of permissible limits of drinking water. Indian J Occup Environ Med 16(1):40

Kumar PR, Chaudhari S, Khilar KC, Mahajan SP (2004) Removal of arsenic from water by electrocoagulation. Chemosphere 55(9):1245–1252

Li Y, Liu C, Luan Z, Peng X, Zhu C, Chen Z, Zhang Z, Fan J, Jia Z (2006a) Phosphate removal from aqueous solutions using raw and activated red mud and fly ash. J Hazard Mater 137 (1):374–383

Li YH, Zhu Y, Zhao Y, Wu D, Luan Z (2006b) Different morphologies of carbon nanotubes effect on the lead removal from aqueous solution. Diamond Relat Mater 15(1):90–94

Li Y, Luo X, Huang X, Wang D, Zhang W (2013) Life Cycle Assessment of a municipal wastewater treatment plant: a case study in Suzhou, China. J Clean Prod 57:221–227

Li X, Wang Z, Zhang B, Rykov AI, Ahmed MA, Wang J (2016) FexCo3− xO4 nanocages derived from nanoscale metal–organic frameworks for removal of bisphenol A by activation of peroxymonosulfate. Appl Catal Environ 181:788–799

Li H, Chen Y, Long J, Li X, Jiang D, Zhang P, Qi J, Huang X, Liu J, Xu R, Gong J (2017) Removal of thallium from aqueous solutions using Fe-Mn binary oxides. J Hazard Mater 338:296–305

Li K, Li H, Xiao T, Zhang G, Long J, Luo D, Zhang H, Xiong J, Wang Q (2018) Removal of thallium from wastewater by a combination of persulfate oxidation and iron coagulation. Process Saf Environ Protect 119:340–349

Lin Y, Zeng XG, Wu DS, Wang X (2006) Study on bisphenol A induced primary cultured mesencephalic neuronal cell injury by oxidative stress. Wei sheng yan jiu J Hyg Res 35 (4):419–422

Lin K, Liu W, Gan J (2009) Oxidative removal of bisphenol A by manganese dioxide: efficacy, products, and pathways. Environ Sci Technol 43(10):3860–3864

Liu Z, Zhang FS (2009) Removal of lead from water using biochars prepared from hydrothermal liquefaction of biomass. J Hazard Mater 167(1-3):933–939

Loftus B, Anderson I, Davies R, Alsmark UCM, Samuelson J, Amedeo P, Roncaglia P, Berriman M, Hirt RP, Mann BJ, Nozaki T (2005) The genome of the protist parasite *Entamoeba histolytica*. Nature 433(7028):865–868

Ludwig A (1997) Create an oasis with greywater. Distributed by Oasis Design, Santa Barbara, CA

Makhijani SD, Manoharan A (1999) Nitrate pollution problem in drinking water sources: monitoring and surveillance. In: Workshop water quality field test kits for arsenic, fluoride and nitrate held from, 8-9 September 1999

Makoni FS, Thekisoe OM, Mbati PA (2016) Urban wastewater for sustainable urban agriculture and water management in developing countries. In: Sustainable water management in urban environments. Springer, Cham, pp 265–293

Matsumura Y, Akahira-Moriya A, Sasaki-Mori M (2015) Bioremediation of bisphenol-A polluted soil by *Sphingomonas bisphenolicum* AO1 and the microbial community existing in the soil. Biocontrol Sci 20(1):35–42

Mayo JT, Yavuz C, Yean S, Cong L, Shipley H, Yu W, Falkner J, Kan A, Tomson M, Colvin VL (2007) The effect of nanocrystalline magnetite size on arsenic removal. Sci Technol Adv Mater 8(1-2):71

Mejía ME, Siebe C, Paillés CA (2012) Proyecto de desarrollo de capacidades para el uso seguro de aguas servidas en agricultura. – (FAO, WHO, UNEP, UNUINWEH, UNW-DPC, IWMI e ICID), México

Metcalf and Eddy, Inc, Asano T, Burton FL, Leverenz H, Tsuchihashi R, Tchobanoglous G (2007) Water reuse. McGraw-Hill Professional Publishing, New York

Miller-Robbie L, Ramaswami A, Amerasinghe P (2017) Wastewater treatment and reuse in urban agriculture: exploring the food, energy, water, and health nexus in Hyderabad, India. Environ Res Lett 12(7):075005

Moyer NP (1999) Shigella. Waterborne pathogens, vol 48. American Water Works Association, AWWA M, Denver, pp 115–117

Myszograj S, Qteishat O, Sadecka Z, Jędrczak A, Suchowska-Kisielewicz M (2014) Possibilities of reuse of treated wastewater for irrigation purposes in the Northern Jordan Valley. Environ Prot Eng 40(2):17–31

Nainan OV, Xia G, Vaughan G, Margolis HS (2006) Diagnosis of hepatitis A virus infection: a molecular approach. Clin Microbiol Rev 19(1):63–79

Ning RY (2002) Arsenic removal by reverse osmosis. Desalination 143(3):237–241

Niu G, Cabrera RI (2010) Growth and physiological responses of landscape plants to saline water irrigation: a review. HortScience 45(11):1605–1609

Noris M, Remuzzi G (2009) Atypical hemolytic–uremic syndrome. N Engl J Med 361 (17):1676–1687

Norton-Brandão D, Scherrenberg SM, van Lier JB (2013) Reclamation of used urban waters for irrigation purposes——a review of treatment technologies. J Environ Manage 122:85–98

O'lorcain P, Holland CV (2000) The public health importance of Ascaris lumbricoides. Parasitology 121(S1):S51–S71

Odimayo MS (2005) Rotavirus induced diarrhoea among children under 5 years in Ilorin. Nigeria, Faculty of Pathology

Olmez-Hanci T, Arslan-Alaton I, Genc B (2013) Bisphenol A treatment by the hot persulfate process: oxidation products and acute toxicity. J Hazard Mater 263:283–290

Pawełczyk A (2012) Assessment of health hazard associated with nitrogen compounds in water. Water Sci Technol 66(3):666–672

Pedros PB, Onnis-Hayden A, Tyler C (2008) Investigation of nitrification and nitrogen removal from centrate in a submerged attached-growth bioreactor. Water Environ Res 80(3):222–228

Pei XR, Li Y, Long DX, Chen X, Gao LF, Chen XG (2003) In vitro study of bisphenol-A toxicity on the early embryo development of mice. Chin J Reprod Health 14(1):34–37

Perna NT, Plunkett G, Burland V, Mau B, Glasner JD, Rose DJ, Mayhew GF, Evans PS, Gregor J, Kirkpatrick HA, Pósfai G (2001) Genome sequence of enterohaemorrhagic *Escherichia coli* O157: H7. Nature 409(6819):529–533

Peter AJ, Viraraghavan T (2005) Thallium: a review of public health and environmental concerns. Environ Int 31(4):493–501

Peter AJ, Viraraghavan T (2008) Removal of thallium from aqueous solutions by modified *Aspergillus niger* biomass. Bioresour Technol 99(3):618–625

Pitterle M, Ramaswami A (2009) h. University of Colorado Denver, Denver

Post J (2006) Wastewater treatment and reuse in the eastern Mediterranean region. Water 21:36–41

Qadir M, Sharma BR, Bruggeman A, Choukr-Allah R, Karajeh F (2007) Non-conventional water resources and opportunities for water augmentation to achieve food security in water scarce countries. Agric Water Manag 87(1):2–22

Quist-Jensen CA, Macedonio F, Drioli E (2015) Membrane technology for water production in agriculture: desalination and wastewater reuse. Desalination 364:17–32

Raschid-Sally L, Jayakody P (2009) Drivers and characteristics of wastewater agriculture in developing countries: Results from a global assessment, vol 127. IWMI, Colombo

Riley JP, Siddiqui SA (1986) The determination of thallium in sediments and natural waters. Anal Chim Acta 181:117–123

Saha JC, Dikshit AK, Bandyopadhyay M, Saha KC (1999) A review of arsenic poisoning and its effects on human health. Crit Rev Environ Sci Technol 29(3):281–313

Sharma J, Mishra IM, Dionysiou DD, Kumar V (2015) Oxidative removal of bisphenol A by UV-C/peroxymonosulfate (PMS): kinetics, influence of co-existing chemicals and degradation pathway. Chem Eng J 276:193–204

Shimoda K, Hamada H (2009) Bioremediation of bisphenol A and benzophenone by glycosylation with immobilized marine microalga *Pavlova* sp. Environ Health Insights 3:89–94

Silva J, Torres P, Madera C (2008) Reuso de aguas residuales domésticas en agricultura. Una revisión Agronomía Colombiana 26(2):347–359

SINA (2012) Atlas Digital del Agua México, Sistema Nacional de Información del Agua. – CONAGUA, Subdirección General de Programación, México, D.F

Symonds EM, Verbyla ME, Lukasik JO, Kafle RC, Breitbart M, Mihelcic JR (2014) A case study of enteric virus removal and insights into the associated risk of water reuse for two wastewater treatment pond systems in Bolivia. Water Res 65:257–270

Tarr MA (ed) (2003) Chemical degradation methods for wastes and pollutants: environmental and industrial applications. CRC Press, Boca Raton

Tsai WT (2006) Human health risk on environmental exposure to Bisphenol-A: a review. J Environ Sci Health C 24(2):225–255

Turan M, Mart U, Yüksel B, Celik MS (2005) Lead removal in fixed-bed columns by zeolite and sepiolite. Chemosphere 60(10):1487–1492

Van der Hoek W, Hassan MU, Ensink JH, Feenstra S, Raschid-Sally L, Munir S, Aslam R, Ali N, Hussain R, Matsuno Y (2002) Urban wastewater: a valuable resource for agriculture: a case study from Haroonabad, Pakistan, vol 63. IWMI, Colombo

Verbyla ME, Oakley SM, Mihelcic JR (2013) Wastewater infrastructure for small cities in an urbanizing world: integrating protection of human health and the environment with resource recovery and food security. Environ Sci Technol 47(8):3598–3605

Wang L, He XB, Li FH, Liu C (2005) A serological survey of Bisphenol A in Shenzhen City. Pract Prev Med 12:241–242

Woltersdorf L, Liehr S, Scheidegger R, Döll P (2015) Small-scale water reuse for urban agriculture in Namibia: modeling water flows and productivity. Urban Water J 12(5):414–429

WHO (1996) Guildlines for Drinking Water Quality, vol 2, Second edn. Health Criteria and Other Supporting Information WHO, Geneva

World Health Organization (2005) Guidelines for the control of shigellosis, including epidemics due to Shigella dysenteriae type 1. World Health Organization, Geneva

Zhang Z, Zhang B, Long J, Zhang X, Chen G (1998) Thallium pollution associated with mining of thallium deposits. Sci China Ser D Earth Sci 41(1):75–81

Zhang JH, Shi D, Jiang M, Hao QL, Zhou YK, Lu B (2005) Study on the DNA damage and repair effect of Bisphenol A in human embryo liver L-02 cell line. J Environ Occup Med 22:197–199

Zhang Y, Causserand C, Aimar P, Cravedi JP (2006) Removal of bisphenol A by a nanofiltration membrane in view of drinking water production. Water Res 40(20):3793–3799

Zhang G, Liu H, Liu R, Qu J (2009) Removal of phosphate from water by a Fe–Mn binary oxide adsorbent. J Colloid Interface Sci 335(2):168–174

Zhang QH, Wang XC, Xiong JQ, Chen R, Cao B (2010) Application of life cycle assessment for an evaluation of wastewater treatment and reuse project—case study of Xi'an, China. Bioresour Technol 101(5):1421–1425

Zolgharnein J, Asanjarani N, Shariatmanesh T (2011) Removal of thallium (I) from aqueous solution using modified sugar beet pulp. Toxicol Environ Chem 93(2):207–214

Toxicity of Hexavalent Chromium and Its Microbial Detoxification Through Bioremediation

Sarika Chaturvedi, Aradhana Khare, and S. M. Paul Khurana

Abstract

Chromium exists in different oxidation states ranging from –IV to +VI, in relation to human exposure. Chromium contaminated ecosystems through ore refining, chemical and refractory processing, cement plants, automobile brake lining and catalytic converters for automobiles, leather tanneries and chrome pigments. It's having carcinogenic and mutagenic effects upon both acute and chronic exposures because of its high water solubility, etc. Hexavalent chromium (Cr^{6+}) is known to have 100-fold more toxicity than its other forms. Chromium reduction has been evaluated in both aerobic and anaerobic conditions by a group of microorganisms. Bioremediation is a viable, environment-friendly technology for cleaning up the chromium-contaminated sites. Bacterial strains such as *Klebsiella pneumoniae*, *Bacillus firmus*, *Pseudomonas maltophilia* and *Mycobacterium* sp. are capable of absorbing Cr^{6+} efficiently into their biomass. The fungal strains, *Aspergillus flavus*, *Aspergillus* sp. and *A. niger*, are also capable of transforming Cr^{6+} to Cr^{3+} relative to cell-wall-binding properties by their soluble enzymes and membrane-associated reductases of electron transfer systems. Infrared spectral analysis showed that bacterial isolates having the binding groups –OH, $-NH_2$ and C=O conjugated with –NH were responsible for Cr^{6+} adsorption within the cell. Cr^{6+} serves as electron acceptor involving cytochromes b and c. Genetic and protein engineering may further enhance the Cr^{6+} reductase efficiency. The direct application of Cr^{6+} reductases may be a promising approach both in situ and ex situ bioremediation in a wide range of environments.

S. Chaturvedi · S. M. P. Khurana (✉)
Amity Institute of Biotechnology, Amity University Haryana, Gurgaon, India

A. Khare
Amity School of Applied Sciences, Amity University, Mumbai, India

M. P. Shah (ed.), *Removal of Emerging Contaminants Through Microbial Processes*,
https://doi.org/10.1007/978-981-15-5901-3_25

Keywords

Cr^{6+} · Cr^{3+} · Bioremediation · Biotransformation · Bioreduction · Bioabsorption · Cytochromes

1 Introduction

Chromium is increasing in the environment by different industries as a result of metallurgical refractory, chemical industries, tannery, agriculture and many more anthropogenic activities. It is one of the most toxic metals and abundant pollutants in both aquatic and terrestrial ecosystem (Costa et al. 2003). It exists in environment in different oxidation states; it is found that hexavalent Cr(VI) and trivalent Cr(III) are more stable in nature and enter in bacterial and mammalian cells via transport system (Ackerley et al. 2004). Cr(III) is a less toxic, trace element essential for functioning of the environment. It is readily converted into Cr(VI) under natural conditions, and it reacts with nucleic acid and other cell components and causes mutagenic, carcinogenic effects on biological system (Wang et al. 2009). Cr bioremediation mechanisms include biosorption, bioaccumulation, complexation, electrostatic attraction and ion exchange. Conversion of Cr (VI) to Cr (III) by means of reduction, lowers the concentration of Cr(VI), decreases its toxicity, and makes it environmentally benign.

Heavy metals are essential for plant growth at low concentration but at higher concentration are essential for plant growth but at higher are toxic, carcinogenic, mutagenic, malfunctioning and teratogenic and cause heavy metal pollution which is rising due to industries. It mostly affects the neurological, liver and kidney functions. Therefore, heavy metal remediation is a matter of serious concern for healthy environment.

Various techniques are used for removal of metals from environment, and among them bioremediation is found to be the most eco-friendly, low-cost and energy-saving technique.

Bioremediation is a process using microorganisms to convert organic contaminants such as heavy metals into nontoxic products such as water and carbon dioxide (USEPA 2001; Sharma and Reddy 2004) in presence of oxygen (aerobic condition) or without oxygen (anaerobic condition), respectively, or enzymes to clean up polluted environment, as shown in the Eq. (1) (Okoduwa et al. 2017).

$$\text{Organic contamimnants} + O_2^{-} \rightarrow H_2O + CO_2 + \text{Cell material} + \text{energy} \quad (1)$$

Simplified chemical equation of aerobic bioremediation (Sharma and Reddy 2004).

Bioremediation can also be defined as any process that uses microorganisms or their enzymes to remove and/or neutralize contaminants within the environment (i.e. within soil and water) to their original condition. The steps involved in the mechanism of bioremediation of metals from contaminated soil are shown below in (Fig. 1). Figure 2 shows the basic concept of bioremediation and role of organic contaminant in the process.

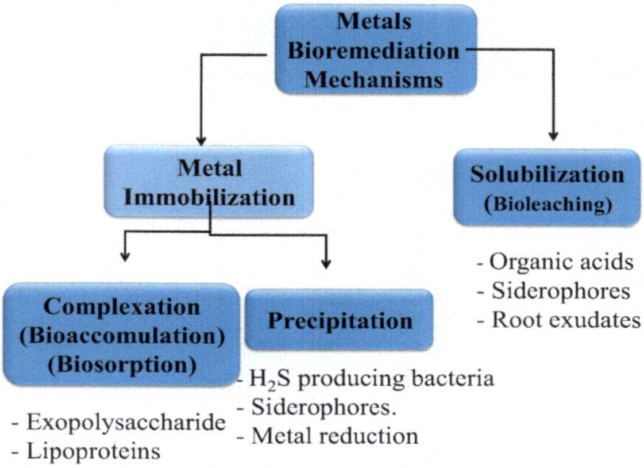

Fig. 1 Mechanism of bioremediation for metal-contaminated soil

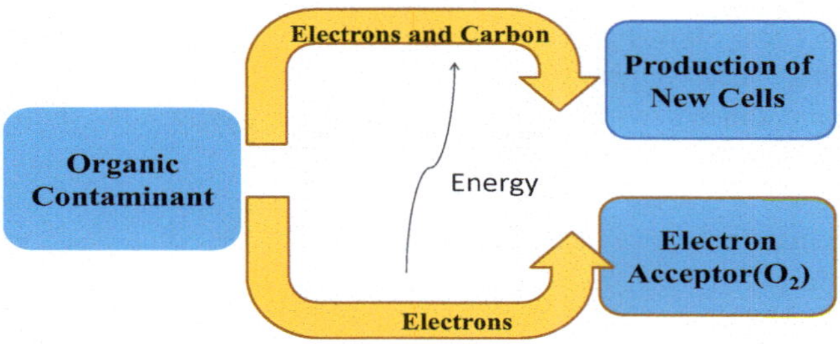

Fig. 2 Basic concept of bioremediation (USEPA 2001)

In the process organic contaminants serve two purposes:

- They act as a source of carbon that can be used as basic building blocks of new cell constituent.
- They supply electrons that are used by organisms to obtain energy by breaking chemical bonds. In this process microbes gain energy that helps to grow and reproduce into more cells.

The optimum conditions considered for effective bioremediation process are as follows (Sharma and Reddy 2004):

- Temperature range—15 to 45 °C.
- pH range—5.5 to 8.5.
- Moisture levels—40–80% of field capacity.
- Oxygen concentration—>2 mg/L (aerobic).
- Nutrients including carbon, hydrogen, oxygen, nitrogen and phosphorous should be present in abundance.

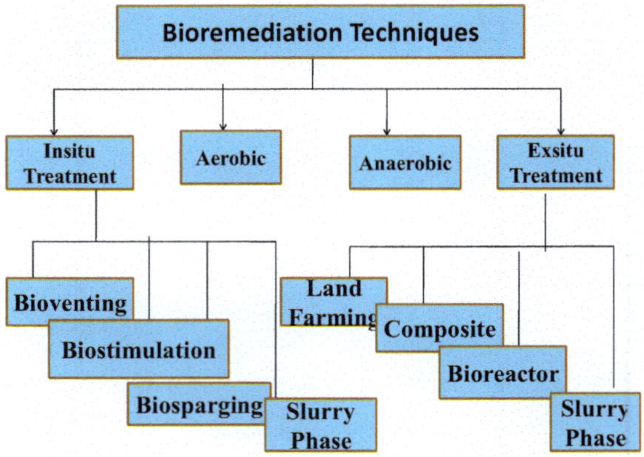

Fig. 3 Different types of bioremediation techniques

- Use of indigenous microorganisms (native population) and in some cases exogenous (foreign) microorganisms.
- Use of microbes showing limited efficiency of metals, including poor competitiveness and excessive heavy metal concentration.
- Several amendments with inorganic nutrients, biosurfactants, bulking agents, compost and biochar (Truu et al. 2015).

There are different types of bioremediation techniques (Fig. 3) depending upon the conditions.

Aerobic treatment This process is normally used for groundwater and amendments (oxygen-releasing compounds ORC and H_2O_2) introduced either by injection or water recirculating systems. This process transforms organic contaminants into carbon dioxide and water at a very fast speed.

Anaerobic treatment Carbon sources such as molasses or hydrogen-releasing compounds (HRC) as amendments are introduced by injection or water recirculating systems. In this process contaminants are converted into methane, carbon dioxide and traces of hydrogen.

In situ treatment

1. **Bioventing** In this process microbial activity is increased by introducing oxygen into the soil using forced air. Nitrogen and phosphorus as nutrients are also added into soil to increase the growth rate of microorganisms.
2. **Slurry-phase lagoon aeration** Air and soil are combined in a lagoon to support fast microbial activity.
3. **Biostimulation/injection or pumping of oxygen-releasing compounds (ORCs)** It is used in groundwater contamination as ORCs are present in liquid form only. ORCs are introduced into the soil to encourage microbial activity. It can be achieved by adding *amendments* hydrogen peroxide (H_2O_2) or

oxygen-releasing compounds (ORCs) to the soil, and the process is known as *biostimulation*.

4. **Biosparging** For this air is injected into the groundwater to enhance the number of microorganisms to increase biodegradation.

Ex situ soil treatment

1. **Biopiles** In this process soil is excavated and mixed with necessary amendments, after which the loads are placed in an enclosed building and aerated using blowers and/or pumps.
2. **Compositing** Contaminated soil is excavated and combined with bulking agents and organic amendments such as hay, wood chips, vegetative waste and manure that increase microorganism activity. It is a controlled biological process either aerobic or anaerobic under thermophilic conditions (54–65 °C) to convert organic contaminants into mild stabilized byproducts. There are three most effective methods of composting:
   (a) Aerated static pile composting (compost is formed into piles and aerated with blowers or vacuum pumps)
   (b) Mechanically agitated in-vessel composting (compost is placed in a reactor vessel where it is mixed and aerated)
   (c) Windrow composting (compost is placed in long piles known as windrows and periodically mixed with mobile equipment): (a) open windrow system in which the compost is stacked in elongated piles and aerated by tearing down and rebuilding piles and (b) static windrow system in which the compost piles are aerated by a forced air system

   Among these windrow composting is usually considered to be the most cost-effective composting alternative.
3. **Land farming** Soil is excavated and spread in lined treatment beds, and amendments are not necessarily added, but the soil is tilled to aerate to increase the microbial activity. This treatment is enhanced from photooxidation by the sun.

Slurry-phase treatment This involves the controlled treatment of excavated soil. In this a mixture of soil and water is mixed with amendments to a predetermined concentration dependent upon the concentration of the contaminants, the rate of biodegradation and the physical nature of the soils. Clean sand may then be discharged, and normally, a slurry contains from 10 to 30% solids by weight. This mixture is stirred continuously so that solids get suspended and maximize area between microorganisms and soil particles. Slurry-phase bioreactors are used to treat PCBs, halogenated SVOCs, pesticides and ordnance compounds found in excavated soils.

Contaminants of different types are remediated by this method such as petroleum hydrocarbon (gasoline and diesel fuel), nonchlorinated solvents (acetone and other ketones), wood-preserving wastes, creosote and pentachlorophenol (PCP), some chlorinated aromatic compounds (chlorobenzenes, azo dyes and biphenyls) having

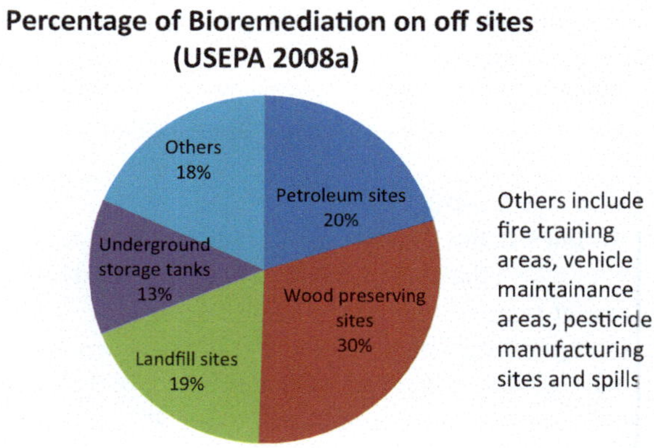

Fig. 4 Bioremediation based on off-sites

less than five chlorine atoms per molecule, underground storage tanks and some chlorinated aliphatic compounds like trichloroethene (TCE) (Sarika 2019), and their percentage is shown in Fig. 4. The process of pollutant removal depends primarily on the nature of the pollutant. There are some criteria on which bioremediation techniques are based as nature of pollutants, depth, and degree of pollution, type of environment, location, cost, and environmental policies (Smith et al. 2015).

2 Characteristics of Chromium

Chromium (Cr) is a silvery, shiny, hard, brittle and seventh most abundant critical metal on Earth and 21st in the Earth's crust with an average concentration of 100 ppm. It occurs rarely in free or native state (Kohl 1967). Surface elements react strongly with atmospheric oxygen and convert into strong, dense, passive and nonporous thick layer of Cr(III) oxide, which stops it from further corrosion and imparts a metallic sheen. Chromium occurs in oxidation states from −2 to +6; however the most stable oxidation state is +3 and +6. The oxidation number of Cr depends upon the pH and redox potential (Eh) of the medium (Cotton and Wilkinson 1980).

At room temperature, Cr exists as solid and reacts readily with halogen gases to form bright-coloured compounds. Readily it gets dissolve in nonoxidizing mineral acids like HCl and sulphuric acid but not with nitric acid due to its oxidation into chromic oxide which is passive in nature. It is resistant to normal corrosive agents hence extensively used in electroplating as protective coating on other reactive metals. At high temperatures it combines directly with nonmetals like halogens, sulphur, nitrogen, boron, oxygen, etc. Chromium is one of the chief ingredients of coloured gem stones, mineral and metals. Coloured gemstones having Cr are emeralds, ruby, alexandrite, chrome garnet and some sapphires (Jacques et al. 2004).

3 Speciation of Chromium

In nature chromium exists in various forms in different oxidation states varying from −2 to +6, but the most stable is +6. The different oxidation states of Cr (stable/unstable) are found in environment as Cr compounds mentioned in Table 1.

The most common and stable forms are trivalent Cr(III) and hexavalent Cr (VI) species (Fendorf 1995). Cr(III) is most stable because of low kinetic energy potential and reacts slowly to form complexes required for functioning of living organisms, and Cr(VI) is a strong oxidizing agent and is not as stable as Cr(III). It has toxic effect in biological systems (Fendorf 1995). Hexavalent chromium is readily soluble in water in the full pH range, while trivalent chromium tends to get at any pH to be adsorbed on soil surface or forms precipitate as chromium hydroxide in slightly acidic and alkaline environment. As a result it has the high potential for environmental contamination of aquifers and surface water. Environmental Protection Agency (EPA) has declared chromium as a Group 'A' human carcinogen and is one of the main pollutants (Dhal et al. 2013).

The most prevalent forms of Cr(VI) species are CrO_4^{2-}, $HCrO_4^-$ and $Cr_2O_7^{2-}$ depending upon the pH of the medium (Table 2) and total concentration of Cr(VI) as shown in predominance diagram in Fig. 5 (Greenwood and Earnshaw 1997; Cotton and Wilkinson 1980).

Much equilibrium exists between hexavalent chromium, proton and water in aqueous medium, but the most important are as follows:

$$H_2CrO_4 \rightarrow H^+ + HCrO_4^-, \quad K_1 = 10^{-0.75}$$

Table 1 Different oxidation states of chromium species (Thacker et al. 2007)

Oxidation state	Compound	Nature
−2	$Na_2[Cr(CO)_5]$	Unstable
−1	$Na_2[Cr_2(CO)_{10}]$	Unstable
0	[Cr(CO)6]	Unstable
+1	$[Cr(bipy)_3]^+$	Unstable
+2	$[Cr(CH_3COO)_4$, CrO, $CrSO_4]$	Unstable (chromous)
+3	$[CrCl_3$, Cr_2O_3, $Cr_2(SO_4)_3]$	Stable (chromic)
+4	CrO_2	Unstable
+5	$[CrF_5]$	Unstable
+6	$[K_2Cr_2O_7$, $K_2Cr_2O_4$, $CrO_3]$	Stable

Table 2 Predominance of Cr species vs. pH (Cotton and Wilkinson 1980)

pH <1(strongly acidic)	H_2CrO_4
pH >1 or pH = 2–6 (acidic)	$HCrO_4^- \leftrightarrow Cr_2O_7^{2-}$ (in equilibrium)
pH >6 (alkaline)	CrO_4^{2-}

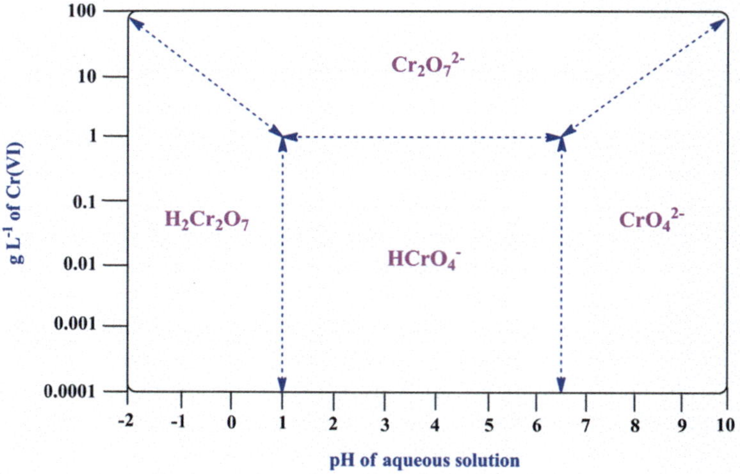

Fig. 5 Predominance diagram of chromate ion (Dinker and Kulkarni 2015)

$$HCrO_4^- \rightarrow H^+ + CrO_4^{2-}, \quad K_2 = 10^{-5.9}$$

$$2HCrO_4^- \rightarrow Cr_2O_7^{2-} + H_2O, \quad K_3 = 10^{-2.2}$$

It is also shown in the predominance diagram below (www.thermofisher.com/chromatography).

The predominance diagram shows that the chromate ion is the most predominant species in alkaline solutions, while dichromate ions can become predominant in acidic solutions (Dinker and Kulkarni 2015).

4 Sources of Hexavalent Chromium

Naturally occurring Cr(VI) is present in groundwater and surface waters. Cr(III) as particular chromite is present in rock strata rich in bearing minerals. They are universally found in the areas that occur near convergent plate margins (Table 3).

Table 3 Difference between natural and anthropogenic sources of chromium

Natural	1. Occurrence of natural sources with high concentration of Cr is rare. However, mafic and ultramafic rocks generally show higher concentrations of chromium than other rocks (Chrysochoou and Johnston 2012)
	2. One of the important and natural occurrences of chromium is in fly ash, bituminous and subbituminous coal (Fig. 6) (Saha et al. 2011; Groves et al. 2007)
Anthropogenic	Mining, sewage, sludge, manure, phosphate fertilizers as impurities, metallurgy, chemical dying, tanning, electroplating, pigments, photography, anticorrosive, oil well drilling
	Emissions from chemical plants and incineration facilities, cement dust, contaminated landfill; effluents from chemical plants, asbestos lining erosion; road dust from catalytic converter erosion and asbestos brakes, tobacco and topsoil and rocks

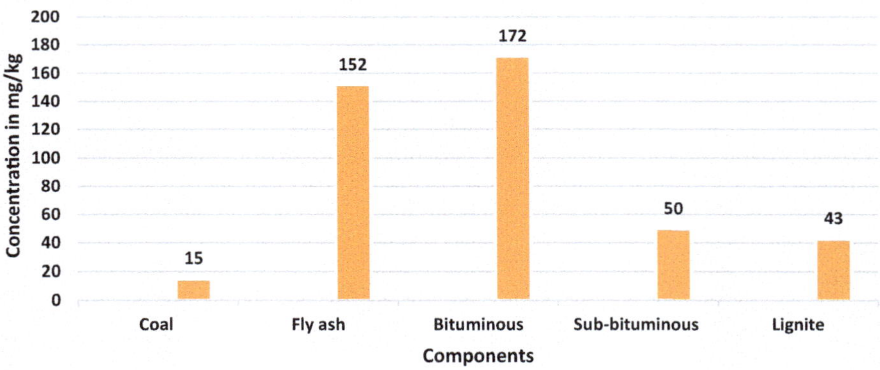

Fig. 6 Concentration (mg/kg) of Cr in coal and fly ash (Saha et al. 2011)

5 Distribution of Hexavalent Chromium

Chromium is known as a redox-active metal that is distributed in all the components of environment, viz. air, soil, surface water and groundwater, through the above-mentioned sources, as Cr(III) and Cr(VI). Both the forms have different mobility and toxicity. Cr(III) is immobile and benign, whereas Cr(VI) is highly toxic and can be readily transported (Tarrah 2004). Cr mainly occurs in the form of chromite ore. Its production in last 10 years is mentioned in Fig. 7, whereas the presence of Cr(VI) in different environmental media is shown in Table 4.

In Air The presence of Cr(VI) in the air is observed as suspended particle which is contributed mainly by combustion of fuels, iron and steel industry and chemical manufacturing industries. Industrial exhaust emission from various industries like cement production and consumable products is another source of Cr(VI) in the air.

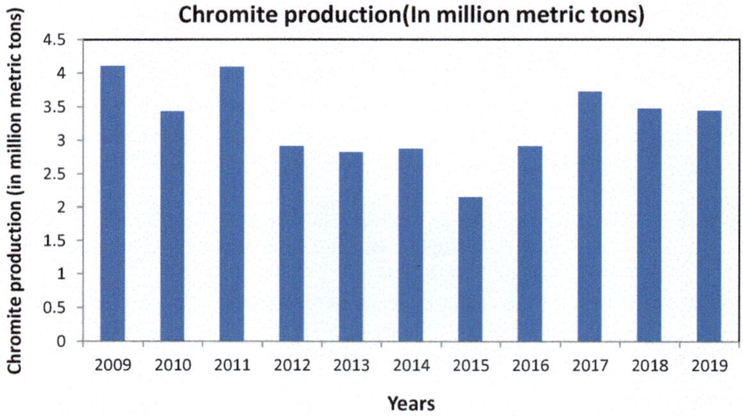

Fig. 7 Chromite production (million metric/ton) over a decade (FY 2009 to FY 2019)

Table 4 Distribution of Cr(VI) in different environmental media

	Cr(VI) μg L^{-1}	References
Drinking water	0.057 and 7.51	Eaton et al. (2018)
Rainwater	0.2 to <1	WRc (2015)
Sea water	0.04–0.5	WRc (2015)
Surface water	0.5–2	WRc (2015)
Natural stream waters (India)	<2	EFSA (2014)
Groundwater	2–10	ATSDR (2012)
Food	0.0005–1.3	ATSDR (2012), EFSA (2014)
Seafood, fish	>0.97 to 1.70	Copat et al. (2013)
Alcohol	135–400	USEPA (1984)
Ambient air	0.003	USEPA (1984)
Nonindustralized area	0.0001	NAS (1980)
Indoor air	0.1–0.6	Li et al. (2002)
Outdoor air	2.97	Li et al. (2002)

During chromium electroplating process, bubbles of gas are emitted from the surface of the bath and get mixed into the atmosphere (Saha et al. 2011).

In Soil Direct percolation of leachates from landfill disposal of municipal solid waste for composting, sewage sludge and industry effluents can contaminate soil. Leaching from topsoil and rocks to the waterbodies is the most important natural source of chromium entry. The industries mainly responsible for Cr (VI) contamination of soil are metal plating especially chrome plating, ink, dye, tanning, textile, corrosion inhibitors, etc. (Steritt et al. 1990). Disposal of trapped and bottom fly ash on soil is also one of the ways for higher concentration of Cr(VI) in soil (Stowe 1987). Cr(VI) in the road side soil comes due to wear and tear of chromium containing asbestos brake linings in vehicles as well as from the aerosols containing chromium catalyst as used to treat exhaust system.

In Waterbodies Chromium does not occur freely in water, whereas their compound, mainly chromite, occurs in water in traces. The main sources of Cr in surface water are from industries handling as leather, tanning, textile, electroplating and chemical as well as leaching from rocks, top soil, etc. Improper disposal of solid waste from Cr processing industries can also be the source of groundwater contamination (www.atsdr.cdc.gov›csem›chromium›docs›chromium; https://www.lenntech.com/periodic/water/chromium/chromium-and-water.htm#ixzz6DpJDf6My).

Cr concentration in surface water is estimated to be 1 ppb and in sea water approximately 0.6 ppb. Under reducing conditions, when pH is less than 6, Cr exists as relatively immobile form Cr(III), as the pH is more than 6, it oxidizes to highly mobile and toxic form Cr(VI) (Tarrah et al. 2004; Saha et al. 2011).

6 Transformation of Chromium

Due to widespread anthropogenic activities, chromium enters into the atmosphere and contaminates air, soil and waterbodies. Once it enters the natural water or soil system, it exists primarily in two states, less mobile Cr(III) and more soluble Cr(VI). The less soluble form retains Cr in solid state as precipitate or colloid, but the highly soluble form Cr(VI) gets transported into various systems before being reduced back to Cr(III). This transformation of Cr(III) to Cr(VI) and vice versa is affected by pH/Eh and carried out by chemicals mainly like Fe and Mn and various microorganisms as well.

6.1 Chemical Transformation of Cr in Soil and Water

Chemical transformation of Cr(VI) to Cr(III) and vice versa occurs simultaneously in presence of Fe(II), Mn(IV) and some organic matter. The extent of oxidation of soluble Cr(III) to Cr(VI) depends upon the factors like form of Cr(III) present in the soil whether soluble or insoluble, inorganic or organically complexed form, concentration of oxides and hydroxides of Mn(III, IV) present in the soil. Reduction of Cr (VI) to Cr(III) may also take place simultaneously in presence of Fe(II), Mn(III, IV) and some organic matter. Both the processes are regulated by pH variation of the soil as shown in Fig. 6 where interconversion of Cr(III) to Cr(VI) is coupled with redox cycle of Mn, Fe and some organic matter like alkyl-substituted diphenols (James 1996) (Figs. 8 and 9).

Microbial Transformation of Cr Besides the chemical transformation, microorganisms can also play an important role in interconversion of Cr(VI) and Cr(III). There are many microorganisms which have been discovered to convert Cr

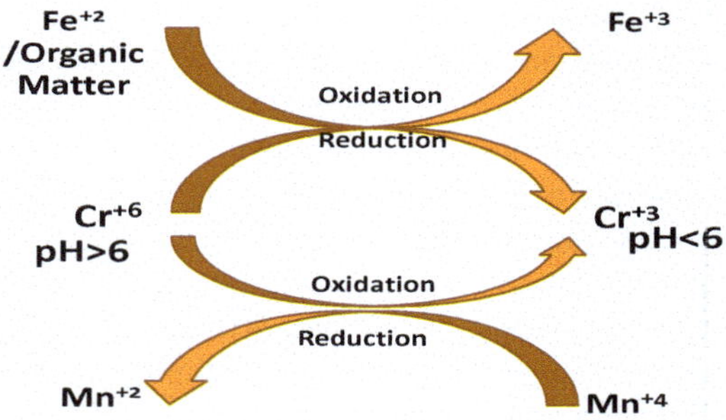

Fig. 8 Chemical transformation cycle of Cr in soil (https://www.statista.com/statistics/667460/india-chromite-production-volume, Vaibhav Asher 2019)

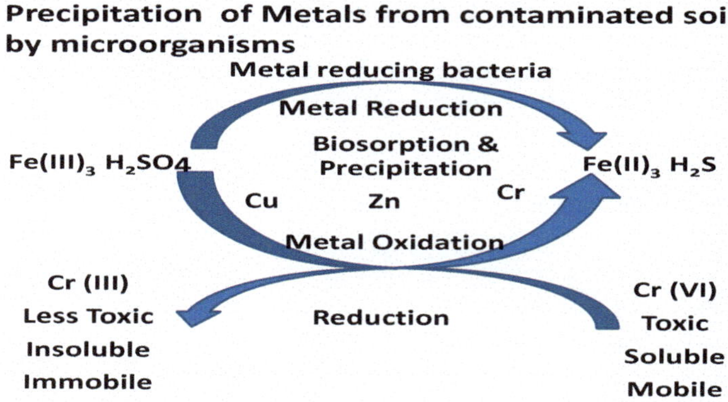

Fig. 9 Microbial transformation of Cr in the soil (Jim and Chen Oliver 1998)

(VI) to Cr(III) within a range of Cr(VI) concentration through plasmid or membrane-associated phenomenon. The biomass concentration, initial Cr(VI) level, carbon source, pH and temperature, oxidation-reduction potential, oxyanions and metal cations are the factors affecting reduction of Cr(VI). Biological transformation of Cr provides an effective pathway for bioremediation of Cr(VI) from the environment (Jim and Chen Oliver 1998).

7 Toxicity of Metals and Its Microbial Detoxification Through Bioremediation

The discharge of wastes from different industries containing biotoxic substances of heavy metals is one of the most important environmental and health challenges in our environment. Hence, there is a growing need for the development of novel, efficient, eco-friendly and cost-effective approaches for bioremediation of the inorganic metals (Cr, Hg, Cd and Pb) released into the situation and to prevent the degradation of ecosystem. Cr is one of the major heavy metals which is generated mostly through anthropogenic activities (Stern 1982) as well as the effluents from various industries that enter the environment causing serious pollution in sediments, soil, water and air (Saha et al. 2011). Maximum quantity of Cr is present in wastewater which causes toxicological risk to human, animals and plants (Barceloux and Barceloux 1999). Since Cr(VI) has greater solubility, motility and bioavailability, hence it easily gets dispersed in sediments and surface water. Trivalent Cr(III) naturally abundant predominates in the environment (Languard 1990).

Microorganisms as bacteria produce metal chelating substances as siderophores, which help to increase the mobility and reduce bioavailability of metals resulting in pollutant exclusion from soil. *Desulfovibrio desulfuricans*, *Actinobacteria* and sulphate-reducing bacteria convert sulphate to hydrogen sulphate reacting with heavy metals such as Cd and Zn and form insoluble sulphides (Chibuike and Obiora 2014; Sarika 2019). Biosorption is one of the techniques used for removal of heavy metals, compounds and particulates using dead mass or natural materials with higher degradative ability (Srivastava and Dwivedi 2015). It could be dependent on or independent of cell's metabolism or area of metal removal, so it may be extracellular accumulation/precipitation, cell surface sorption/precipitation and intracellular accumulation (Beiyuan et al. 2017).

7.1 Factors Affecting Microbial Remediation of Heavy Metals

The factors which affect the efficiency of bioremediation are temperature, pH, redox potential, nutritional status, moisture and chemical composition of heavy metals (Shukla et al. 2013). The effectiveness of microbes can be achieved by several amendments with inorganic nutrients, fertilizer and organic substrates, biosurfactants, bulking agents, compost and biochar (Wiszniewska et al. 2016). Bioaugmentation is achieved by addition of either compound-degrading microbes or organic amendments containing active microorganisms, e.g. activated sludge or compost (Kastner and Miltner 2016). Compost/farmyard manure addition and composting with various organic supplements have been found to be very effective and resourceful for soil bioremediation (Tables 4 and 5).

7.2 Microorganisms Used in Heavy Metal Remediation

Microbial bioremediation is an active and growing technology (Hennebel et al. 2015). Microbe-metal interaction exists in both natural and anthropogenic environment for removal of metals as remediation and depollution. Microbial-electrochemical technologies (METs) by using solid state electrodes as electron donors or acceptors are a new technology (Thrash and Coates 2008). Metallic nanoparticles (NPs) synthesis is a green approach by the application of microorganisms (Klaus-Joerger et al. 2001). Genetically modified microorganisms, genetic engineering, dialysis sac cultures and chemical modification have also been used as a remediation technique (Paliwal et al. 2012). These may alter the

Table 5 Factors that influence bioremediation of heavy metal (Boopathy 2000)

Factors	Activities
Microbial concentration	Production of toxic metabolites, enzyme reduction, mutation and horizontal gene transfer
Substrate	Chemical structure of contaminants, less concentration of contaminants, toxicity of contaminants, solubility of contaminants
Mass transfer limitation	Oxygen diffusion and solubility, solubility/miscibility with water, diffusion of nutrients
Environment	Inhibitory environmental conditions, depletion of substrate, lack of nutrients
Cometabolism	Microbial interaction, concentration, alternate carbon source present
Biological aerobic vs. anaerobic process	Microbial population present in the site, oxidation/reduction potential, availability of electron acceptors

components of cell surface and can efficiently improve the adsorption capacity and selectivity to target metal species (Table 6).

Dialysis sac culture is a method for removal of toxic hexavalent chromium (chromate: CrO_4^{2-}) by using chromate-reducing strain of *Enterobacter cloacae* (HO1). In this method *E cloacae* strain, HO1 was put into dialysis sac that is submerged into water containing CrO_4^{2-}, chromate ion gets diffused into the culture, and reduced anaerobically to insoluble chromium hydroxides in the dialysis sac, hence remove toxic chromium from the surrounding water. This described method shows about 90% of the total chromium could be removed from water if the initial concentration of CrO_4^{2-} was less than 208 ppm as chromium (Komori et al. 1990a, b) (Table 7).

7.3 Bacterial Remediation Capacity of Heavy Metals

Different bacteria as *Escherichia coli* (Shen and Wang 1993), *Pseudomonas putida* (Ishibashi and Cervantes 1990), *Desulfovibrio* sp. (Amanda and Lynne 2001), *Bacillus* sp. (Yun et al. 2006), *Shewanella* sp. (Myers et al. 2000), *Arthrobacter* sp. (Nino et al. 2004), *Streptomyces* sp. Mc1 (Marta et al. 2010) and *Mycobacterium* sp. CR07 (Zhaoming et al. 2012) are used for bioremediation of heavy metals. Generally, heavy metals are the elements of high molecular mass, exist naturally in the soil, and have a density almost five times than water such as Fe (atomic weight higher than 55.8 g mol^{-1} and density 5.0 g cm^{-3}), Cr, As and Se. Heavy metals may serve as micronutrients, but exposure to higher levels can be detrimental to human health.

Table 6 Inhibitory effects of heavy metals on the growth and development of microbes

Heavy metals	Effects on microbes	Reference
Arsenic	Deactivation of enzymes	Sankarammal et al. (2014)
Aluminium	DNA damage, configurational changes in enzymes	Booth et al. (2015)
Cadmium	Denatures protein, destroys nucleic acid, hinders cell division and transcription	Fashola et al. (2016)
Chromium	Growth inhibition, elongation of lag phase, inhibition of oxygen uptake	Cervantes and Devars (2001)
Copper	Disrupts cellular function, inhibits enzyme activities	Fashola et al. (2016)
Selenium	Inhibits growth rate	Dixit et al. (2005)
Lead	Destroyed nucleic acid and protein, inhibits enzyme actions and transcription	Fashola et al. (2016)
Mercury	Denatures protein, inhibits enzyme function, disrupts cell membrane	Fashola et al. (2016)
Nickel	Upsets cell membrane, hinders enzyme activities and oxidative stress	Malik (2004)
Silver	Cell lysis, inhibits cell transduction and growth	Prabhu and Poulose (2012)
Zinc	Decrease in biomass, inhibits growth death	Malik (2004)

8 Bioremediation of Cr(VI)

Cr reduction by microorganisms is a bioredox process, i.e. complex, multiple and poor process. *Pseudomonas*, *Aeromonas*, *Enterobacter* spp., *Streptomyces* sp. MC1 and sulphate-reducing bacteria have reduced Cr(VI) enzymatically.

Bacterial degradation Kafilzadeh et al. isolated 80 different strains that are used for biodegradation in which ten genera were the best degrading, viz. *Bacillus*, *Corynebacterium*, *Staphylococcus*, *Streptococcus*, *Shigella*, *Alcaligenes*, *Acinetobacter*, *Escherichia*, *Klebsiella*, *Enterobacter*, hydrogen-degrading bacteria (Kafilzadeh et al. 2011). *Pseudomonas*, *Mycobacterium*, *Corynebacterium*, *Aeromonas*, *Rhodococcus* and *Bacillus* are used for aromatic hydrocarbon degradation (Mrozik et al. 2003). Mixed microbial communities having different genetic information are used for complex mixtures of organic compounds in contaminated areas (Fritsche and Hofrichter 2005). Reduction of metal as Cr(VI) to Cr(III) occurs in both aerobic *Enterococcus gallinarum* (Sayel et al. 2012) and anaerobic *Achromobacter* sp. strain Ch1 (Zhu et al. 2008). Bacterial consortium is used to remove heavy metals as Cd, Cr, Ni and Pb from synthetic waste water effluent (Adarsh et al. 2007). Indigenous bacteria and plant growth-promoting rhizobacteria enriched from chromium-contaminated biotopes remove Cr(VI) in multi-contaminated solution and remove heavy metals from contaminated soils (Nezha 2011).

Table 7 Various microorganisms used for bioremediation of different metals

Heavy metal	Microorganisms	References
Bacteria		
As	*Sporosarcina ginsengisoli*	Achal et al. (2012)
Cr(VI)	*Bacillus cereus strain* XMCr-6	Dong et al. (2013)
	Bacillus cereus	Kanmani et al. (2012)
	Bacillus subtilis	Balamurugan et al. (2014)
	Enterobacter cloacae B2-DHA	Rahman et al. (2015)
	Pseudomonas putida	Balamurugan et al. (2014)
	Escherichia coli VITSUKMW3	Samuel et al. (2013)
	Enterobacter cloacae (HO1)	Komori et al. (1990a, b)
	Streptococcus thermocarboxydus NBRC 16323	Terahara et al. (2015)
Cu	*Kocuria flava*	Coelho et al. (2015)
Cd, Zn, Cu	*Pseudomonas veronii*	Vullo et al. (2008)
Cr(VI)	*Pseudomonas aeruginosa* Rb-1 and *Ochrobactrum intermedium* Rb-2	Batool et al. (2014)
Pb	*Enterobacter cloacae* KJ-46 and KJ-47	Chang et al. (2015)
Ni, Cr, Cd	*Pseudomonas aeruginosa* KP717554, *Alcaligenes feacalis* KP717561 and *Bacillus subtilis* KP717559	
Fungi		
Cr(VI)	*Gloeophyllum sepiarium*	Achal et al. (2011)
	Rhizopus oryzae (MPRO)	Sukumar (2010)
Ni, Cu	*Aspergillus versicolor*	Taştan et al. (2010)
Pb	*Aspergillus fumigates*	Kumar et al. (2011)
Yeast		
Pb, Cd	*Saccharomyces cerevisiae*	Farhan and Khadom (2015)
Algae		
As	*Hydrodictyon*, *Oedogonium* and *Rhizoclonium* spp.	Srivastava and Dwivedi (2015)
Cr, Cu, Fe, Mn, Zn	*Spirogyra* spp. and *Spirulina* spp.	Mane and Bhosle (2012)
Pb(II), Cu (II)	*Spirogyra* spp. and *Cladophora* spp.	Lee and Chang (2011)

Lactobacillus sps. (P1) and *Bacillus* sps. (P2, P3 and P4) isolated from the pulp and paper processing industrial effluents showed very high degree of resistance to the chromium. These strains are exposed to high level of the chromium metal concentration in the environment which have adapted to this stress. The rate of Cr (VI) reduction increased with decreasing concentration of Cr(VI) and increasing time interval for all the bacteria isolates. All the bacterial isolates showed the best chromate reduction for different concentration of Cr(VI) (Biradar et al. 2012).

Degradation by Fungi Fungi can grow in low moisture condition and low pH, having multienzyme complexes used in breaking down natural polymeric compounds. They form mycorrhizal intracellular association, arbuscular mycorrhizal fungi (AMF) or extracellular ectomycorrhizal fungi that are used in bioremediation. This process of phytoremediation using fungi is named as mycorrhizoremediation (Khan 2006) used to eliminate the hazardous wastes from the environment (Fritsche and Hofrichter 2005). Yeasts are better accumulators than bacteria and capable of accumulating heavy metals (Wang and Chen 2006). *P. anomala* is capable of removing Cr(VI) (Bahafid et al. 2011). Several other yeast species, viz. *Cyberlindnera fabianii*, *S. cerevisiae*, *Hansenula polymorpha*, *P. guilliermondii*, *Rhodotorula pilimanae*, *Wickerhamomyces anomalus*, *Yarrowia lipolytica* and *C. tropicalis*, are used for biosorption of Cr(VI) and to reduce Cr (VI) to Cr(III) (Bahafid et al. 2012 Ksheminska et al. 2006). Immobilized cells of *Schizosaccharomyces pombe* are used for removal of metals (Sai Subhashini et al. 2011). Fungi facilitate in bioremediation by all the three strategies:

1. The target compound is used as a carbon source.
2. The target compound is enzymatically attacked but is not used as a carbon source (cometabolism).
3. The target compound is not metabolized at all yet taken up and get concentrated within the organism (bioaccumulation; Abruscia et al. 2007).

Cladophialophora, *Exophiala*, *Leptodontium* and *Pseudeurotium zonatum* use toluene as sole carbon and energy source (Francesc et al. 2001). Filamentous fungi *Aspergillus*, *Cladosporium*, *Cunninghamella*, *Penicillium*, *Fusarium* and *Aspergillus* can take part in aromatic hydrocarbon decomposition participating in aliphatic hydrocarbon biodegradation (Steliga 2012). *Aspergillus niger* AB10 and *Rhizopus arrhizus* M1 act as cadmium and lead metal resistant, respectively, and ensure the better accumulation of these metals from the aqueous culture media (Pal et al. 2010).

Chlorella, *Anabaena inaequalis*, *Westiellop sis prolifica*, *Stigeoclonium lenue*, *Synechococcus* sp. and *marine algae* which tolerate heavy metals are used for removal of heavy metals (Dwivedi 2012) through adsorption. *Chlorella sorokiniana* (LSIBCS) was investigated as a new biosorbent for the removal of Cr(III) from aqueous solution (Akhtar et al. 2008). The microalga *S. incrassatulus* and green algae *C. sorokiniana* were used to remove Cr(VI) in continuous cultures (Pena-Castro et al. 2004) for bioremediation of heavy metals.

8.1 Sequestration

Sequestration is the process of binding metal ions and microbial colonies as insoluble compounds. It is of two types: intracellular and extracellular. In intracellular sequestration metal ions interact with various compounds in the cell cytoplasm resulting slow transport and increase in concentration of metals within the microbial cells. This technique helps to accumulate metals intracellularly in bacterial cells, and it is applicable in treatment of effluent. Extracellular sequestration is the accretion of metal ions by cellular components in the periplasm or complexation of metal ions as insoluble compounds.

Pseudomonas putida strain has ability of intracellular sequestration of copper, cadmium and zinc ions with the help of cysteine-rich low molecular weight proteins (Higham et al. 1986). Microbes as fungi, its cell wall, spores, and mycelia accumulating metals in the exterior of cell walls work like ligand to eliminate inorganic metals and help to decontaminate metal ions by energy uptake, extracellular and intracellular precipitation and valency conversion (Xie et al. 2016). Different components of cell wall such as peptidoglycan, polysaccharide and lipid are rich in metal-binding ligands (e.g. –OH, –COOH, HPO_4^{2-}, SO_4^{2-}, $-RCOO^-$, $-NH_2$ and –SH) via electrostatic interaction and cationic metal species through surface complexation.

Pseudomonas syringae strains produce copper-inducible proteins CopA, CopB (periplasmic proteins) and CopC (outer membrane protein) which bind copper ions and microbial colonies (Chal and Cooksey 1991). Other bacterial cells can eject metal ions from the cytoplasm to sequester the metal within the periplasm. *Synechocystis* PCC 6803 strain ejects zinc ions from the cytoplasm by efflux system in the periplasm (Igiri et al. 2018). Another technique of extracellular sequestration is precipitation of metal. *Geobacter* spp. and sulphur-reducing bacterium like *Desulfuromonas* spp. are capable of reducing harmful metals to less or nontoxic metals. *Klebsiella planticola* reduces cadmium ions as insoluble sulphide under anaerobic conditions and generating hydrogen sulphide (Sharma et al. 2000). *P. aeruginosa* precipitates cadmium under aerobic conditions (Wang et al. 2002).

Chromium is non-biodegradable and environmentally persistent. Many yeasts, bacteria and fungal species have been assessed for their suitability to reduce or remove Cr(VI) contamination that consists of active efflux of Cr compounds, metabolic decline of Cr(VI) to Cr(Ill) and either intercellular or extracellular precipitation. It involves three stages:

1. Binding of chromium to the cell surface
2. Translocation of chromium into the cell
3. Reduction of Cr(VI) to Cr(Ill)

Cr(VI) reduction by microorganisms may continue either on the surface or 440 outside the cell/intracellularly. Cr(VI) reduction (intracellular /extracellular) has been processed either direct via chromate reductase enzymes or indirect via metabolite reduction. Bioremediation of chromium involves different strategies as biosorption, bioaccumulation, bioreduction and immobilization of biomaterial(s). Biosorption, bioaccumulation & bioreduction rely on free cells for bioremediation and cause cell damage. The uptake of chromium ions is a multiphasic process. First step is biosorption, a nondirected physiochemical interaction and metabolic energy-dependent (living biomass is employed) as well as independent process involving dead cell biomass. It occurs in between metal species and the cellular components of biological species. Independent process involving dead cell biomass is much more effective than dependent on biosorption of heavy metals, including chromium. Bioaccumulation is a slow process and dependent on cell metabolic activity in living organism's involving adsorption, intracellular accumulation, and bioprecipitation mechanisms. In bioreduction processes, microorganisms alter the oxidation/reduction state of toxic metals through direct or indirect biological and chemical process. Bioreduction of Cr(VI) to Cr(III) decreases the chromium toxicity to living organisms and helps to precipitate chromium at a neutral pH for further physical removal. Bioremediation using immobilization of microbial cell is a cost-effective more viable method, having the ability of easier separation of biosorbents, to complete multiple biosorption cycles and desorption (elution) of metal(s) from matrices for reuse. Thus, microbial bioremediation can be a cost-competitive and beneficial bioresource approach for removing many hazardous contaminants from industrial wastes (Garg et al. 2012).

8.2 Engineered Microorganisms

In this technique, microorganisms' genetic material gets changed by natural or artificial genetic exchange between microorganisms. This is known as Recombinant DNA Technology. It involves the insertion of the appropriate gene for the production of a particular enzyme to degrade various pollutants, whereas the indigenous microbes cannot degrade the pollutants. It helps in the removal of pollutants such as hydrocarbons and petrol discharges like xylene, toluene, octane, naphthalene, and salicylate which are coded on bacterial plasmids for successful degradation of the environment. A variety of molecular tools such as molecular cloning, horizontal transfer of DNA in bacteria, electroporation, protoplast transformation, biolistic transformation, conjugation and transformation of competent cells are available for the successful construction of GMOs (Menn et al. 2008).There are four main approaches for developing genetically engineered microorganisms for bioremediation as:

1. Modification of enzyme specificity and affinity
2. Pathway construction and regulation
3. Bioprocess development, monitoring and control
4. Bioaffinity bioreporter sensor application for chemical sensing, toxicity reduction and end point analysis

8.3 Natural Attenuation

Natural attenuation is also known as intrinsic bioremediation and is an in situ bioremediation technique. Natural attenuation is likely to occur at chromium-contaminated sites involving passive remediation, without individual intrusion, and is less expensive. It involves both aerobic and anaerobic processes to biodegrade polluting molecules and recalcitrant. Monitored natural attenuation (MNA) is a more holistic approach to intrinsic bioremediation process to monitor to establish that bioremediation is ongoing and sustainable.

But there are some circumstances for natural attenuation:

(a) These are natural reductants present within the aquifer that can transform the more toxic hexavalent form of chromium to the less toxic trivalent form.
(b) The amount of Cr(VI) and other reactive constituents does not exceed the capacity of the aquifer to reduce them
(c) The time scale required to achieve the reduction of Cr(VI) to the target concentration is less than the time scale for the transport of the aqueous Cr(VI) from source area to the point of compliance.
(d) The Cr(III) will remain immobile, and Cr(III) precipitates as a fairly insoluble hydroxide, thereby immobilizing it within the soil.
(e) There is no net oxidation of Cr(III) to Cr(VI). It is not easy to know the time scales for reduction and oxidation of chromium in the soil (Palmer and Puls 1994).

According to the US National Research Council (US NRC), there are three criteria that include:

1. Expression of loss of contaminants from the contaminated sites.
2. Exhibit the microorganisms isolated from polluted sites have the usual potentials to biodegrade or transform contaminants present at contaminated site and facts of biodegradation potentials in the field (Philp and Atlas 2005), e.g. hydrocarbon-degrading bacteria isolated from refinery oil-contaminated soil and the biodegradation potentials of the isolates by using mineral salt medium with saturated and unsaturated hydrocarbon substrates as sole carbon sources and also by their capacities to reduce hydrocarbon concentrations (Mrassi et al. 2015).
3. In chronically polluted marine coastal environment, monitoring of intrinsic bioremediation showed higher total bacterial diversity, abundance and culturable hydrocarbon degraders and contributed to natural attenuation of such site. It shows that bacterial communities could be used as **sensitive indicators** of contamination in marine sediment (Catania et al. 2015).

Chlorinated compounds of groundwater contaminated with trichloroethene (TCE) are bioremediated by three treatments (biostimulation, biostimulation-bioaugmentation and monitored natural attenuation) (Adetutu et al. (2015). Among them MNA is the best mechanism used in most European countries except extreme cold climate condition of pollutant removal (Declercq et al. 2012).

One of the major limitations of intrinsic bioremediation is that it might take much longer time to achieve the target level of pollutant concentration, given that no external force is incorporated to expedite the remediation. It thus follows that prior to application of intrinsic bioremediation, risk assessment needs to be carried out to ensure that remediation time would be less than the time stipulated for pollutant to reach exposure point relative to the closest human and animal populations. Further, it was reported that intrinsic bioremediation does not result in adequate polyaromatic hydrocarbon (PAH) removal and corresponding reduction in ecotoxicity of polluted soil (García-Delgado et al. 2015).

8.4 Direct Cr^{6+} Reduction

Toxic forms of hexavalent chromium such as chromate and dichromate are released from electroplating facilities and certain types of chemical plants. This hexavalent Cr form must be reduced before discharged into environment. This reduction process is achieved in two steps:

1. Hexavalent chromium (Cr^6) is reduced to trivalent chromium (Cr^3).
2. Trivalent chromium (Cr^3) is precipitated as chromium hydroxide.

A new oxidative **polymerization method** using nanocomposite and polyaniline/manganese dioxide/titanium dioxide ($PANI/MnO_2/TiO_2$) is used to convert Cr^{6+} to benign Cr^{3+}. The rate of reaction is dependent on the substrate concentration, oxidant (HCOOH), catalyst, pH, temperature and foreign ions.

Rotating disc electrode (RDE) and rotating cylinder electrode (RCE) cell is used to electro-reduce toxic hexavalent chromium to trivalent state. Using galvanostatic and linear sweep voltametric techniques, the effect of chromium, sulphuric acid concentration, current density, time and speed of cathode rotation on current efficiency, cell voltage and energy consumption have been investigated.

8.5 Cr^{6+} Reductase Enzyme

Bioremediation with the enzyme was observed in noncompatible cell reproduction systems, conditions frequently found in contaminated environments. Enzymatic reduction is a less costly and more eco-friendly process for biological transformation of Cr. *Pseudomonas maltophilia* O-2 and *Bacillus megaterium* TKW3 utilize membrane-associated reductases. Cr reductases are generally isolated and purified from aerobic bacteria but process involved in reduction, is completed under anaerobic conditions. Most specific protein used is flavoprotein reductase NAD(P)H dependent. Chromate reductase enzyme isolated from Halomonas sp. TA-04 from polluted marine sediments and Chr B, act as a protein substitute such as chromate sensitivity regulator (Rita et al. 2005). The best mechanism is to include the efflux of chromate ion from the cytoplasm to reduce Cr(III) to Cr(VI). This has been identified in *Pseudomonas aeruginosa* and *Cupriavidus metallidurans* by a membrane potential, so reduction of chromate is completed by chromate reductase isolated from bacterial species.

Chromate reductase enzyme is mainly associated with biomass and it is isolated from Streptomyces sp. MC1 as a constitutive enzyme, requiring NAD(P)H as an electron donor for remediation of pollutants. It was active over a broad temperature (19–39 °C) and pH (5–8) range, and optimum conditions were 30 °C and pH 7. The enzyme was present in supernatant, pellet and cell-free extract.

Both aerobic and anaerobic microorganisms are able to reduce Cr(VI) to Cr(III). In general aerobic condition microbial metabolism is used for reduction of Cr(VI) (Losi and Frankenberger Jr. 1994) by enzyme, but in some cases P*seudomonas maltophilia* O-2 and *Bacillus megaterium* TKW3 use membrane-associated reductases. In anaerobic condition both soluble and membrane-associated enzymes are used in which Cr(VI) functions as electron acceptor and the process involving carbohydrates, proteins, fats, hydrogen, NAD(P)H and endogenous electron reserves as electron donor (Wang 2000). Therefore Cr use of reductases could be an important approach for bioremediation of Cr(VI) (Cheung and Ji-Dong 2007).

Chromate reductase of *Streptomyces* sp. MC1 is a constitutive enzyme, able to reduce Cr(VI) present in a synthetic medium, soil extract and soil samples (Marta et al. 2010). It was mainly associated with biomass and required NAD(P)H as an electron donor, found active over a broad temperature (19–39 °C) and pH (5–8) range, while optimum conditions were 30 °C and pH 7.

8.6 Biosorption Mechanism and Biotransformation

Several different strategies are adopted by microorganisms for removal of Cr (VI) mostly involving both biosorption and biotransformation. This review article focuses on chromium pollution problem, its chemistry, sources, effects, remediation strategies by biological agents and detailed chromium detoxification mechanism in microbial cell. A summary of chromium bioremediation technologies applied in situ and ex situ is also listed. This can be helpful for developing technologies to be more efficient for Cr(VI) removal, thereby bridging the gap between laboratory findings and industrial application for chromium remediation (Fig. 10).

9 Case Study

1. In this case study, Cr(VI) was reduced by Actinobacteria, isolated from polluted sediments near a stainless steel plant in the lower Hackensack River, New Jersey. Whereas the total Cr and Cr(VI) were obtained from the sediment porewater in the lower Hackensack River to study the relationship between sediment geochemistry and chromium speciation, that controls the mobility, bioavailability,

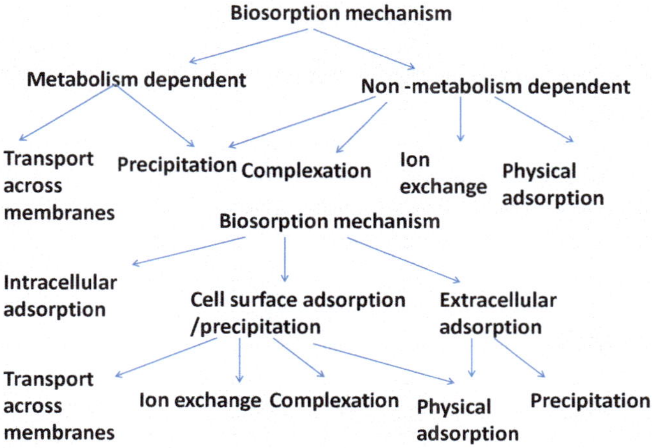

Fig. 10 Mechanism of biosorption (Gupta et al. 2016)

and toxicity of chromium. Approximately 100 surface samples were tested for total chromium and other chemical. These were collected by centrifugation or in situ samples where total chromium and Cr(VI) concentrations ranged from 5 to 9190 mg/kg dry weight (dw) and from <0.47 to 31 mg/kg dw, respectively. Sediment pore water concentrations ranged from <10 to 83 μg/L for total chromium. *Nereis virens* and *Macoma nasuta* exposed for 28 days to sediment containing 135 and 1780 mg/kg dw total chromium in whole cells after 24-h depuration ranged between 1.2 and 14.8 mg/kg wet weight (ww; median 1.6 mg/kg ww) total chromium. In whole tissues of indigenous microorganisms, total chromium concentrations were 1.0 and 37.5 mg/kg ww. The consequences of this study are constant with sediment studies conducted to another place indicating low chromium bioavailability in sediment under reducing conditions (Martello et al. 2007).

2. Wastes from stainless steel plants generated wastes mainly with higher concentrations of Cr(VI) that contaminated sediments, soils and waterbodies near industrialized areas (Martello et al. 2007). A halophilic Cr(VI)-resistant bacterial strain *Holomonas sp* TA 04 was isolated from stainless steel plant in A halophilic Cr(VI)-resistant bacterial strain Holomonassp TA 04 was isolated from stainless steel plant in Southern Italy, near industrialized areas. Cr(VI) was removed with a minimum inhibition concentration 200 μg/mL and reduction in the presence of 80 g/L of NaCl by immobilized cells with maximum activity at pH6.5 and 28 °C. This isolate is characterized by phylogenetic and physiological features. Therefore this result shows that isolated strain is best used in bioremediation processes, in saline-polluted environments (Silvia et al. 2012). These are the highlights of this isolate:
   (a) It is an efficient hexavalent chromium reduction by the whole cells and also cell-free extract of *Halomonas aquamarina*.

   (b) Immobilized *Halomonas aquamarina* cells have chromate removal capacity of 91.2% from sediments leachate.
   (c) Reduction of hexavalent chromium at moderate halophilic conditions by the whole cells.
3. New isolates *Actinobacteria*, *Microbacterium* and *Cellulomonas* are isolated from sediments contaminated by Cr(VI) . These strains showed resistance to Cr (VI) at different temperature and help in reducing Cr(VI) to Cr(III) for possible bioremediation of contaminated sites.

10 Conclusion

Many challenges exist for bioremediation—sometimes degradation depends upon the concentration of other nutrients that are in limited supply. In bioremediation of xenobiotics, sometimes microbes do not evolve with the biochemical pathway to degrade the compound, while all require a consortium of microbes. Cr(VI) is generated from industrial processes and causes toxicity in environment affecting plant, animal and human. Microbial bioremediation is a promising method due to their diversity, versatility and adaptability. Microorganisms used both enzymatic as aerobic and anaerobic and non-enzymatic methods as membrane-bound reductases (flavin reductases, cytochromes, hydrogenases) of chromate reduction.

Acknowledgement Authors are thankful to the Amity University, Haryana, authorities for the facilities and constant encouragement for research and innovation.

Conflict of Interest: We declare that we have no conflict of interest.

References

Abruscia C, Marquinaa D, Del Amob A, Catalina F (2007) Biodegradation of cinematographic gelatin emulsion by bacteria and filamentous fungi using indirect impedance technique. Int Biodeter Biodegr 60:137–114

Achal V, Kumari D, Pan X (2011) Bioremediation of chromium contaminated soil by a brown-rot fungus, *Gloeophyllum sepiarium*. Res J Microbiol 6:166–172

Achal V, Pan X, Fu Q, Zhang D (2012) Biomineralization based remediation of As (III) contaminated soil by *Sporosarcina ginsengisoli*. J Hazard Mater 201:178–184

Ackerley DF, Gonzalez CF, Park CH, Blake R, Keyhan M, Matin A (2004) Chromate-reducing properties of soluble flavoproteins from *Pseudomonas putida* and *Escherichia coli*. Appl Environ Microbiol 70:873–882

Adarsh VK, Mishra M, Chowdhury S, Sudarshan M (2007) Studies on metal microbe interaction of three bacterial isolates from East Calcutta Wetland. OnLine J Biol Sci 7(2):80–88

Adetutu EM, Gundry TD, Patil SS, Golneshin A, Adigun J, Bhaskarla V, Aleer S, Shahsavari E, Ross E, Ball AS (2015) Exploiting the intrinsic microbial degradative potential for field-based in situ dechlorination of trichloroethene contaminated groundwater. J Hazard Mater 300:48–57

Agency for Toxic Substance and Disease Registry (2012) Toxicological profile for chromium. Agency for Toxic Substances and Disease Registry, U.S. Department of Health and Human Services, Atlanta, Georgia. CAS#: 7440-47-3 Available at: www.atsdr.cdc.gov/toxprofiles/tp.asp?id=62&tid=17

Akhtar N, Iqbal M, Iqbal ZS, Iqbal J (2008) Biosorption characteristics of unicellular green alga *Chlorella sorokiniana* immobilized in loofa sponge for removal of Cr (III). J Environ Sci 20:231–239

Amanda NM, Lynne EM (2001) A novel isolate of *Desulfovibrio sp.* with enhanced ability to reduce Cr (VI). Biotechnol Lett 23(9):683–687

Bahafid W, Sayel H, Tahri-Joutey N, EL Ghachtouli N (2011) Removal mechanism of hexavalent chromium by a novel strain of *Pichia anomala* isolated from industrial effluents of Fez (Morocco). J Environ Sci Eng 5:980–991

Bahafid W, Tahri-Joutey N, Sayel H, Iraqui-Houssaini M, El Ghachtouli N (2012) Chromium adsorption by three yeast strains isolated from sediments in Morocco. Geomicrobiol J 30 (5):422–429

Balamurugan D, Udayasooriyan C, Kamaladevi B (2014) Chromium (VI) reduction by *Pseudomonas putida* and *Bacillus subtilis* isolated from contaminated soils. Int J Environ Sci 5:522–534

Barceloux DG, Barceloux D (1999) Chromium. Clin Toxicol 37:173–194

Batool R, Yrjala K, Hasnain S (2014) Impact of environmental stress on biochemical parameters of bacteria reducing chromium. Braz J Microbiol 45(2):573–583

Beiyuan J, Awad YM, Beckers F, Tsang DC, Ok YS, Rinklebe J (2017) Mobility and phytoavailability of As and Pb in a contaminated soil using pine sawdust biochar under systematic change of redox conditions. Chemosphere 178:110–118

Biradar NV, Sindagi AS, Reddy J, Ingalhalli SS (2012) Bioremediation of chromium in pulp and paper processing industrial effluents by indigenous microbes. J Bioremed Biodegrad 3(12):1–7

Boopathy R (2000) Factors limiting bioremediation technologies. Bioresour Technol 74(1):63–67

Booth SC, Weljie AM, Turner RJ (2015) Metabolomics reveals diferences of metal toxicity in cultures of Pseudomonas pseudoalcaligenes KF707 grown on different carbon sources. Front Microbiol 6:827–841

Catania V, Santisi S, Signa G, Vizzini S, Mazzola A, Cappello S, Yakimov MM, Quatrini P (2015) Intrinsic bioremediation potential of a chronically polluted marine coastal area. Mar Pollut Bull 99(1-2):138–149

Cervantes C, Campos-García J, Devars S (2001) Interactions of chromium with microorganisms and plants. FEMS Microbiol Rev 25(3):335–347

Chal JS, Cooksey DA (1991) Copper resistance in *Pseudomonas syringae* mediated by periplasmic and outer membrane proteins. Proc Natl Acad Sci U S A 88(20):8915–8919

Chang HK, Soo JO, Yu JS, Sang HH, In HN, Jae SS (2015) Bioremediation of lead by ureolytic bacteria isolated from soil at abandoned metal mines in South Korea. Ecol Eng 74:402–407

Cheung KH, Ji-Dong G (2007) Mechanism of hexavalent chromium detoxification by microorganisms and bioremediation application potential: a review. Int Biodeter Biodegr 59 (1):8–15

Chibuike G, Obiora S (2014) Heavy metal polluted soils: effect on plants and bioremediation methods. Appl Environ Soil Sci 2014:1–24

Chrysochoou M, Johnston CP (2012) Reduction of chromium(VI) in saturated zone sediments by calcium polysulfide and nano scale zero valent iron derived from green tea extract. In: Hryciw RD, Athanasopoulos-Zekkos A, Yesiller N (eds) Geo congress 2012: state of the art and practice in geotechnical engineering 2012, p 3959

Coelho LM, Rezende HC, Coelho LM, de Sousa PA, Melo DF, Coelho NM (2015) Bioremediation of polluted waters using microorganisms. In: Shiomi N (ed) Advances in bioremediation of wastewater and polluted soil. InTech, Shanghai

Copat C, Arena G, Fiore M, Ledda C, Fallico R, Sciacca S, Ferrante M (2013) Heavy metals concentrations in fish and shellfish from eastern Mediterranean Sea: consumption advisories. Food Chem Toxicol 53:33–47

Costa HM, Rodrigues RC, Mattos MG, Ribeiro RF (2003) Evaluation of the adaptation interface of one piece implant supported superstructures obtains in Ni-Cr-Ti and Pd-Ag alloys. Braz Dent J 14:197–202

Cotton FA, Wilkinson G (1980) Chromium. In: Advanced inorganic chemistry, a comprehensive text, 4th edn. John Wiley, New York, pp 719–736

Declercq I, Cappuyns V, Duclos Y (2012) Monitored natural attenuation (MNA) of contaminated soils: state of the art in Europe—a critical evaluation. Sci Total Environ 426:393–405

Determination of Cr(VI) in water, wastewater, and solid waste extracts (n.d.). www.thermofisher.com/chromatography

Dhal B, Thatoi HN, Das NN, Pandey BD (2013) Chemical and microbial remediation of hexavalent chromium from contaminated soil and mining/metallurgical solid waste: a review. J Hazard Mater 250–251:272–291

Dinker MK, Kulkarni PS (2015) Recent advances in silica-based materials for the removal of hexavalent chromium: a review. J Chem Eng Data 60(9):2521–2540

Dixit R, Wasiullah, Malaviya D (2005) Bioremediation of heavy metals from soil and aquatic environment: an overview of principles and criteria of fundamental processes. Sustainability 7 (2):2189–2212

Dong G, Wang Y, Gong L, Wang M, Wang H, He N, Zheng Y, Li Q (2013) Formation of soluble Cr (III) end-products and nanoparticles during Cr (VI) reduction by *Bacillus cereus* strain XMCr-6. Biochem Eng J 70:166–172

Dwivedi S (2012) Bioremediation of heavy metal by algae: current and future perspective. J Adv Lab Res Biol 3(3):195–199

Eaton A, Bartrand T, Rosen S (2018) Detailed analysis of the UCMR 3 database: implications for future groundwater monitoring. J Am Water Works Assoc 110(4):13–25

European Food Safety Authority (2014) Panel on Contaminants in the Food Chain (CONTAM). Scientific opinion on the risks to public health related to the presence of chromium in food and drinking water. EFSA J 12(3):3595

Farhan SN, Khadom AA (2015) Biosorption of heavy metals from aqueous solutions by *Saccharomyces cerevisiae*. Int J Ind Chem 6:119–130

Fashola MO, Ngole-Jeme VM, Babalola OO (2016) Heavy metal pollution from gold mines: environmental effects and bacterial strategies for resistance. Int J Environ Res Public Health 13(11):1047–1063

Fendorf SE (1995) Surface reactions of chromium in soils and waters. Geoderma 67(1–2):55–71

Francesc X, Prenafeta B, Andrea K, Dion MAML, Heidrun A, van Groenestijn JW, de Bont JAM (2001) Isolation and characterization of fungi growing on volatile aromatic hydrocarbons as their sole carbon and energy source. Mycol Res 105(4):477–484

Fritsche W, Hofrichter M (2005) Aerobic degradation of recalcitrant organic compounds by microorganisms. In: Jordening H-J, Winter J (eds) Environmental biotechnology: concepts and applications AND environmental processes. Wiley-VCH Verlag GmbH, Weinheim, pp 145–167

García-Delgado C, Alfaro-Barta I, Eymar E (2015) Combination of biochar amendment and mycoremediation for polycyclic aromatic hydrocarbons immobilization and biodegradation in creosote-contaminated soil. J Hazard Mater 285:259–266

Garg SK, Tripathi M, Srinath T (2012) Strategies for chromium bioremediation of tannery effluent. Rev Environ Contam Toxicol 217:75–140

Greenwood, NN, Earnshaw A (1997) Chemistry of the elements 2nd ed Butterworth-Heinemann Oxford 4(4): 637-643

Groves WA, Kecojevic VJ, Komljenovic D (2007) Analysis of fatalities and injuries involving mining equipment. J Safety Res 38:461–470

Gupta A, Joia J, Sood A, Sood R, Sidhu C, Kaur G (2016) Microbes as potential tool for remediation of heavy metals: a review. J Microb Biochem Technol 8:364–372

Hennebel T, Boon N, Maes S, Lenz M (2015) Biotechnologies for critical raw material recovery from primary and secondary sources: R&D priorities and future perspectives. N Biotechnol 32 (1):121–127

Higham DP, Sadler PJ, Scawen MD (1986) Cadmium-binding proteins in *Pseudomonas putida:* pseudothioneins. Environ Health Perspect 65:5–11

Igiri BE, Okoduwa IR, Idoko GO, Akabuogu EP, Adeyi AO, Ejiogu IK (2018) Toxicity and bioremediation of heavy metals contaminated ecosystem from tannery wastewater: a review. J Toxicol 2018:1–16

Ishibashi Y, Cervantes CS (1990) Silver chromium reduction in *Pseudomonas putida*. Appl Environ Microbiol 56(7):2268–2270

Jacques G, James AJ, Cynthia SP (2004) Avakian chromium hand book, written by Independent Environmental Technical Evaluation Group. CRC Press, Boca Raton, pp 272–289

James BR (1996) Environ. The challenge of remediating chromium-contaminated soil. Environ Sci Technol *30*(6):248–251

Jim M, Chen Oliver JH (1998) Microbial chromium (VI) reduction. Crit Rev Environ Sci Technol 28(3):219–251

Kafilzadeh F, Sahragard P, Jamali H, Tahery Y (2011) Isolation and identification of hydrocarbons degrading bacteria in soil around Shiraz Refinery. Afr J Microbiol Res 4(19):3084–3089

Kanmani P, Aravind J, Preston D (2012) Remediation of chromium contaminants using bacteria. Int J Environ Sci Technol 9:183–193

Kastner M, Miltner A (2016) Application of compost for effective bioremediation of organic contaminants and pollutants in soil. Appl Microbiol Biotechnol 100(8):3433–3449

Khan AG (2006) Mycorrhiza-remediation an enhanced form of phytoremediation. J Zhejiang Univ Sci 7(7):503–514

Klaus-Joerger T, Joerger R, Olsson E, Granqvist C (2001) Bacteria as workers in the living factory: metal-accumulating bacteria and their potential for materials science. Trends Biotechnol 19 (1):15–20

Kohl WH (1967) The electron microscope, Handbook of materials & technology for vacuum device. Reinhold Publishing, New York, pp 161–201

Komori K, Rivas A, Toda K, Ohtake H (1990a) A method for removal of toxic chromium using dialysis-sac cultures of a chromate-reducing strain of Enterobacter cloacae. Appl Microbiol Biotechnol 33:117–119

Komori K, Rivas A, Toda K, Ohtake H (1990b) Biological removal of toxic chromium using an *Enterobacter cloacae* strain that reduces chromate. Microb Ecol 7:245–252

Ksheminska HP, Taras MH, Galyna ZG, Mykhailo VG (2006) Extra-cellular chromate-reducing activity of the yeast cultures. Cent Eur Sci J 1(1):137–149

Kumar RR, Congeevaram S, Thamaraiselvi K (2011) Evaluation of isolated fungal strain from e-waste recycling facility for effective sorption of toxic heavy metal Pb (II) ions and fungal protein molecular characterization—a mycoremediation approach. Asian J Exp Biol Sci 2:342–347

Languard S (1990) One hundred years of chromum and cancer a review of epidemoliogical evidence and selected case reports. Am J Ind Med 17:189–214

Lee YC, Chang SP (2011) The biosorption of heavy metals from aqueous solution by Spirogyra and Cladophora filamentous macroalgae. Bioresour Technol 102:5297–5304

lenntech (n.d.). https://www.lenntech.com/periodic/water/chromium/chromium-andwater.htm#ixzz6DpJDf6My

Li Y, Pradhan NK, Foley R, Low GKC (2002) Selective determination of airborne hexavalent chromium using inductively coupled plasma mass spectrometry. Talanta 57(6):1143–1153

Losi ME, Frankenberger WT Jr (1994) Chromium-resistant microorganisms isolated from evaporation ponds of a metal processing plant. Water Air Soil Pollut 74(3–4):405–413

Malik (2004) Metal bioremediation through growing cells. Environ Int 30(2):261–278

Mane P, Bhosle A (2012) Bioremoval of some metals by living algae *Spirogyra* sp. and *Spirulina* sp. from aqueous solution. Int J Environ Res 6:571–576

Marta AP, María JA, Carlos MA (2010) Chromate reductase activity in *Streptomyces* sp. MC1. J Gen Appl Microbiol 56(1):11–18

Martello L, Fuchsman P, Sorensen M, Magar V, Wenning RJ (2007) Chromium geochemistry and bioaccumulation in sediments from the Lower Hackensack River. New Arch Environ Contam Toxicol 53(3):337–350

Menn FM, Easter JP, Sayler GS (2008) Genetically engineered microorganisms and bio-remediation. In: Rehm H-J, Reed G (eds) Biotechnology: environmental processes II; 11b, 2nd edn. Wiley-VCH Verlag GmbH, Weinheim, pp 443–457

Mrassi AG, Bensalah F, Gury J, Duran R (2015) Isolation and characterization of different bacterial strains for bioremediation of n-alkanes and polycyclic aromatic hydrocarbons. Environ Sci Pollut Res Int 22:15332–15346

Mrozik A, Piotrowska Seget Z, Labuzek S (2003) Bacterial degradation and bioremediation of polycyclic aromatic hydrocarbons. Polish J Environ Stud 12(1):15–25

Myers CR, Carstens BP, Antholine WE, Myers JM (2000) Chromium(VI) reductase activity is associated with the cytoplasmic membrane of anaerobically grown *Shewanella putrefaciens* MR-1. J Appl Microbiol 88:98–106

National Academy of Sciences (1980) Problems in risk estimation. Drinking water and health, vol 3. National Academy Press, Washington, DC, pp 25–65

Nezha TJ, Wifak B, Hanane S, Soumya EA, Naïma EG (2011) Remediation of hexavalent chromium by consortia of indigenous bacteria from tannery waste-contaminated biotopes in Fez, Morocco. Int J Environ Stud 68(6):901–912

Nino VA, Marina KA, Tamar MK, Bakradze NG, Nelly AS, Nelly Ya T, Leila VT, Lia VL, Lali LA, Hoi-Ying H (2004) Effect of chromium(VI) action on *Arthrobacter oxydans*. Curr Microbiol 49(5):321–326

Okoduwa SIR, Igiri B, Udeh CB, Edenta C, Gauje B (2017) Tannery effluent treatment by yeast species isolates from watermelon. Toxics 5(1):1–6

Pal TK, Bhattacharyya S, Basumajumdar A (2010) Cellular distribution of bioaccumulated toxic heavy metals in *Aspergillus niger* and *Rhizopus arrhizus*. Int J Pharma Bio Sci 1(2):1–6

Paliwal V, Puranik S, Purohit HJ (2012) Integrated perspective for effective bioremediation. Appl Biochem Biotechnol 166(4):903–924

Palmer CD, Puls RW (1994) Natural attenuation of hexavalent chromium in groundwater and soils EPA ground water issue, October EPA 540-5-94-505, 12 pp

Pena-Castro JM, Martínez-Jerónimo F, Esparza-García F, Canizares-Villanueva RO (2004) Heavy metals removal by the microalga *Scenedesmus incrassatulus* in continuous cultures. Bioresour Technol 94:219–222

Philp JC, Atlas RM (2005) Bioremediation of contaminated soils and aquifers. In: Atlas RM, Philp JC (eds) Bioremediation: applied microbial solutions for real-world environmental cleanup. American Society for Microbiology (ASM) Press, Washington, DC, pp 139–236

Pierzynski GM, Sims JT, Vance GF (2000) Soils and environmental quality, 2nd edn. CRC Press, Boca Raton

Prabhu S, Poulose EK (2012) Silver nanoparticles: mechanism of antimicrobial action, synthesis, medical applications, and toxicity effects. International Nano Letters 2(1):1–10

Rahman A, Nahar N, Nawani NN, Jass J, Hossain K, Saud ZA, Saha AK, Ghosh S, Olsson B, Mandal A (2015) Bioremediation of hexavalent chromium (VI) by a soil-borne bacterium *Enterobacter cloacae* b2-dha. J Environ Sci Health Part A 50:1136–1147

Rita B, Ana-Paula C, Antonio V, Paula VM (2005) Impact of chromium-contaminated wastewaters on the microbial community of a river. FEMS Microbiol Ecol 54(1):5–46

Saha R, Nandi R, Saha B (2011) Sources and toxicity of hexavalent chromium. J Coord Chem 1 (64):1782–1806

Sai Subhashini S, Kaliappan S, Velan M (2011) Removal of heavy metal from aqueous solution using *Schizosaccharomyces pombe* in free and alginate immobilized cells. In: 2nd international conference on environmental science and technology 6107, pp 111

Samuel J, Paul ML, Ravishankar H, Mathur A, Saha DP, Natarajan C, Mukherje A (2013) The differential stress response of adapted chromite mine isolates *Bacillus subtilis* and *Escherichia coli* and its impact on bioremediation potential. Biodegradation 24(6):829–842

Sankarammal M, Thatheyus A, Ramya D (2014) Bioremoval of cadmium using *Pseudomonas fluorescens*. Open J Water Pollut Treat 4(2):92–100

Sarika C (2019) Azo dyes decolorization using white rot fungi. Int J Appl Nat Sci 8(4):33–46

Sayel H, Bahafid W, Joutey NT, Derraz K (2012) Cr(VI) reduction by *Enterococcus gallinarum* isolated from tannery waste-contaminated soil. Ann Microbiol 62(3):1269–1277
Sharma HD, Reddy KR (2004) Geo-environmental engineering: site remediation waste containment and emerging waste management technologies geo-environmental engineering, vol 35. Wiley.. Environmental International, Hoboken, pp 50–55
Sharma PK, Balkwill DL, Frenkel A, Vairavamurthy MA (2000) A new *Klebsiella planticola* strain (Cd-1) grows anaerobically at high cadmium concentrations and precipitates cadmium sulfide. Appl Environ Microbiol 66(7):3083–3087
Shen H, Wang Y (1993) Characterization of enzymatic reduction of hexavalent chromium by Escherchia coli ATCC 33456. Appl Environ Microbiol 59:3171–3777
Shukla KP, Sharma S, Singh KN, Singh V, Bisht S, Kumar V (2013) Rhizoremediation: a promising rhizosphere technology. In: Patil YB, Rao P (eds) Applied bioremediation: active and passive approaches. Intech, Rijeka, pp 331–347
Silvia F, Milva P, Giacomo L, Simone G, Marcella R, Pamela Di B, Silvano EF (2012) Hexavalent chromium reduction by whole cells and cell free extract of the moderate halophilic bacterial strain *Halomonas* sp. TA-04. Int Biodeter Biodegr 66(1):63–70
Smith E, Thavamani P, Ramadass K, Naidu R, Srivastava P, Megharaj M (2015) Remediation trials for hydrocarbon-contaminated soils in arid environments: evaluation of bioslurry and biopiling techniques. Int Biodeter Biodegr 101:56–65
Srivastava S, Dwivedi AK (2015) Biological wastes the tool for biosorption of arsenic. J Bioremed Biodegr 7:323–326
Steliga T (2012) Role of fungi in biodegradation of petroleum hydrocarbons in drill waste. Polish J Environ Stud 21(2):471–479
Steritt RM, Brown MJ, Lester JN (1990) Environ Pollut 24:313
Stern RM (1982) Chromium compounds: production and occupational exposure. In: Langard S (ed) Biological and environmental aspects of chromium. Elsevier Biomedical Press, New York, pp 5–47
Stowe CW (ed) (1987) Evolution of chromium ore fields. Van Nostrand Reinhold, New York, pp 1–22
Sukumar M (2010) Reduction of hexavalent chromium by *Rhizopus oryzae*. Afr J Environ Sci Technol 4:412–418
Tarrah DH, Martin GS, Chet A, Simion V, Weber JV (2004) In: Guertin J, Jacobs JA, Avakian CP (eds) Chromium(VI) Handbook. CRC Press, New York, pp 94–116
Taştan BE, Ertugrul S, Donmez G (2010) Effective bioremoval of reactive dye and heavy metals by *Aspergillus versicolor*. Bioresour Technol 101:870–876
Terahara T, Xu X, Kobayashi T, Imada C (2015) Isolation and characterization of Cr(VI)-reducing actinomycetes from estuarine sediments. Appl Biochem Biotechnol 175(7):3297–3309
Thacker U, Parikh R, Shouche Y, Madamwar D (2007) Reduction of chromate by cell-free extract of *Brucella sp*. isolated from Cr(VI) contaminated sites. Bioresour Technol 98:1541–1547
Thrash JC, Coates JD (2008) Review: direct and indirect electrical stimulation of microbial metabolism. Environ Sci Technol 42(11):3921–3931
Truu J, Truu M, Espenberg M, Nolvak H, Juhanson J (2015) Phytoremediation and plant-assisted bioremediation in soil and treatment wetlands: a review. Open Biotechnol J 9(1):85–92
United States Environmental Protection Agency (USEPA) (1984) Environmental Monitoring Systems Laboratory. Frequency distributions by site/year for chromium, the results of samples collected at National Air Surveillance Network sites. Research Triangle Park, NC, US Environmental Protection Agency
United States Environmental Protection Agency (USEPA) (2001) A citizen's guide to bioremediation EPA 524-F-01-001
Vullo DL, Ceretti HM, Daniel MA, Ramírez SA, Zalts A (2008) Cadmium, zinc and copper biosorption mediated by Pseudomonas veronii 2e. Bioresour Technol 99:5574–5581
Wang YT (2000) Microbial reduction of chromate. In: Lovely DR (ed) Environmental microbe-metal interactions. ASM, Washington, DC, pp 225–254

Wang J, Chen C (2006) Biosorption of heavy metals by *Saccharomyces cerevisiae*: a review. Biotechnol Adv 24:427–451

Wang CL, Ozuna SC, Clark DS, Keasling JD (2002) A deep-sea hydrothermal vent isolate, *Pseudomonas aeruginosa* CW961, requires thiosulfate for Cd^{2+} tolerance and precipitation. Biotechnol Lett 24(8):637–641

Wang Y, Xu W, Luo Y, Ma L, Li Y, Yang S (2009) Bio effects of chromium (III) on the growth of *Spirulina platensis* and its biotransformation. J Sci Food Agric 89:947–952

Wiszniewska A, Hanus-Fajerska E, MuszyNska E, Ciarkowska K (2016) Natural organic amendments for improved phytoremediation of polluted soils: a review of recent progress. Pedosphere 26(1):1–12

WRc (2015) Understanding the significance of chromium in drinking water. Defra-893004, Swindon, Wiltshire

Xie Y, Fan J, Zhu W (2016) Effect of heavy metals pollution on soil microbial diversity and bermudagrass genetic variation. Front Plant Sci 7:775–791

Yun G, Liu WH, Xu GM, Zeng XL, Hui G (2006) Cr(VI) reduction by *Bacillus* sp. isolated from chromium landfill. Process Biochem 41(9):1981–1986

Zhaoming L, Yan W, Chengfeng L, Pengming L, Meiying G (2012) Chromate reduction by a chromate-resistant bacterium, *Microbacterium* sp. World J Microbiol Biotechnol 28 (4):1585–1592

Zhu W, Chai L, Ma Z, Wang Y, Xiao H, Zhao K (2008) Anaerobic reduction of hexavalent chromium by bacterial cells of *Achromobacter* sp. strain Ch1. Microbiol Res 163:616–623

Correction to: New Trends in Removing Heavy Metals from Industrial Wastewater Through Microbes

Mohammad Haris, Adnan Shakeel, Touseef Hussain, Gufran Ahmad, Moh. Sajid Ansari, and Abrar Ahmad Khan

Correction to:
Chapter 9 in: M. P. Shah (ed.), *Removal of Emerging Contaminants Through Microbial Processes*,
https://doi.org/10.1007/978-981-15-5901-3_9

Owing to an oversight, this chapter was initially published with incorrect authorship. One of the author's names, Moh. Sajid Ansari, was inadvertently removed during the production process. The chapter has now been amended with the correct authorship.

The updated online version of this chapter can be found at:
https://doi.org/10.1007/978-981-15-5901-3_9

M. P. Shah (ed.), *Removal of Emerging Contaminants Through Microbial Processes*,
https://doi.org/10.1007/978-981-15-5901-3_26

The manufacturer's authorised representative in the EU is Springer Nature Customer Service Centre GmbH, Europaplatz 3, 69115 Heidelberg, Germany. If you have any concerns regarding our products, please contact ProductSafety@springernature.com

Printed and bound by CPI Group (UK) Ltd, Croydon, CR0 4YY
30/09/2026
02225247-0019